Springer Series in
OPTICAL SCIENCES 108

Founded by H.K.V. Lotsch

Springer Series in

OPTICAL SCIENCES

The Springer Series in Optical Sciences, under the leadership of Editor-in-Chief *William T. Rhodes*, Georgia Institute of Technology, USA, provides an expanding selection of research monographs in all major areas of optics: lasers and quantum optics, ultrafast phenomena, optical spectroscopy techniques, optoelectronics, quantum information, information optics, applied laser technology, industrial applications, and other topics of contemporary interest.
With this broad coverage of topics, the series is of use to all research scientists and engineers who need up-to-date reference books.

The editors encourage prospective authors to correspond with them in advance of submitting a manuscript. Submission of manuscripts should be made to the Editor-in-Chief or one of the Editors. See also http://springeronline.com/series/624

Norman Hodgson   Horst Weber

Laser Resonators and Beam Propagation

Fundamentals, Advanced Concepts and Applications

Second Edition

With 587 Figures

Springer

Dr. rer. nat. Norman Hodgson
Coherent, Inc.
5100 Patrick Henry Drive
Santa Clara, CA 95054
USA
norman.hodgson@coherent.com

Prof. Dr. Ing. Horst Weber
Optisches Institut
Technische Universität Berlin
Strasse des 17. Juni 135
10623 Berlin
Germany
weber@physik.tu-berlin.de

Library of Congress Control Number: 2005923403

ISBN-10: 0-387-40078-8 e-ISBN 0-387-25110-3 Printed on acid-free paper.
ISBN-13: 978-0387-40078-5

Printed in the United States of America. (MVY)

9 8 7 6 5 4 3 2 1

springeronline.com

Preface

Since its first demonstration in 1960, the laser has found widespread application in diverse areas including medicine, materials processing, optical communications and information technology. The number of engineers and scientists working on lasers or in laser related fields is continuously increasing as new applications for this exciting technology are being discovered. This also means that more and more people need to gain a detailed knowledge of lasers and their characteristics.

The basic understanding of the properties of lasers and their radiation requires knowledge of the physics of optical resonators. The laser beam characteristics as well as efficiency and sensitivity against misalignment are determined mainly by the resonator. Despite this important role optical resonators play in laser engineering, most publications treat them either on a too basic and incomplete level or in the form of a theoretical presentation that is only useful to academics. The result is that very often an engineer or physicist confronted with a laser resonator problem will have difficulties finding information in scientific publications unless he is able to derive his own equations or successfully link the publication's results to his own unique problem.

It is for this reason that we decided to write this overview on laser resonators which covers basics as well as the latest research results. Although the emphasis was put on application and laser engineering problems, the book should also satisfy readers seeking a more thorough background in the field. The first part, entitled "The Electromagnetic Field", provides the reader with the theoretical background necessary for the mathematical description of resonators. We tried to keep the mathematical level as low as possible, e.g. the Kirchhoff Integral is derived in an empirical way instead of using the common approach of applying Green's theorem to the wave equation. Ray transfer matrices in geometrical optics as well as basic and advanced concepts in diffraction theory, beam propagation and coherence are presented here. However, it is not necessary to work oneself through this part to make use of the rest of the book. All succeeding parts can be used without having read the theoretical part. But the reader who seeks a better understanding of the derivation and applicability of the presented equations will get help here. Anyone new to the field of lasers and laser resonators should certainly go through the theory part to get familiar with the general mathematical concepts of optics.

The outline of the book was chosen such that the subject matter becomes more specialized with proceeding chapters. We will start in Part II with the Fabry-Perot Interferometer to discuss the basic resonator properties such as loss, gain, threshold, and line width. The following part will deal with passive (no active medium) resonators. Here we deal with linear stable and unstable resonators which represent probably 95% of all resonators currently used in lasers. Leaving the active medium out of the treatment is the classical approach to the subject since the gain generally only perturbs the physical properties of the resonator rather than completely changing them. The influence of the medium on the resonator properties will be discussed in Part IV. This part also reviews the physics of laser emission and presents output power calculation models, the effects of gain on the mode structure, as well as pulsed operation of laser resonators via Q-switching or mode-locking..

A collection of special resonator concepts is presented in Part V. These concepts are either only used in a limited number of applications or laser designs, or might play an important role in the near future. Resonator schemes such as prism resonators, Fourier transform resonators, hybrid resonators, resonators for annular gain media, and micro-resonators fall into this category. We also included the ring resonator into this part although some readers might argue that it deserves its own part since it is a widely used scheme and probably more important than any other resonator presented here.

A collection of major measurement techniques is given in Part VI. This will help the practicing engineer to make a detailed analysis of his laser system. A mong others, techniques for measuring gain, losses, and beam quality are invaluable for anyone designing and working with laser systems.

A detailed reference list will help the reader to get more information on a preferred subject. We have included the titles of the publications as a help, and publications which give a good review or are a must to read are referred to in the text. We certainly do not claim completeness but to the best of our knowledge we have covered as many publications as possible. The references are listed in their chronological order to give the reader a feeling for the historical development in the specific area.

We hope that this monograph will help you to get more insight into laser resonators and assist you in analyzing and solving the problems you are facing as a laser engineer or physicist. We also hope that, after having worked with this book, you will love laser resonators and laser beams as much as we do!!

This book represents a revised version of "Optical Resonators", published by Springer-Verlag in 1996. Obsolete material has been deleted and new information has been added. In particular, the following areas have been expanded or added:

- In Part I, "The electromagnetic field", we have added an overview of temporal and spatial coherence and their mathematical description incorporating advanced mathematical concepts such as the global degree of coherence and the Wigner function. In addition, angular momentum of beams and the beam propagation in anisotropic crystals are treated as well.

- In Part IV," Open Resonators with Gain", a chapter on Q-switched laser resonators has been added. The chapter on nonlinear optical elements has been expanded to include third harmonic generation.

- In Part V," Special Resonator Concepts", a paragraph on micro-optical resonators has been included.

- In Part VI, " Measurement Techniques", the chapter on beam quality measurement has been rewritten to reflect the ISO standard ISO 11146:1999. We have also added a paragraph on the experimental determination of the phase.

In addition to these expansions, we have updated the reference lists for all chapters and incorporated new material in many chapters. This book would not have been possible without the many contributions that have appeared in the scientific literature. No reference list can ever be complete, and at this point we apologize to any of our colleagues whose work has not been acknowledged or adequately represented. We also would like to thank those readers that have pointed out errors in the first version of this book.

Our special thanks are due to Kathleen M. Millar for proof-reading the original manuscript and to Dr. Keith Murdoch, Dr. Lukas E. Hunziker, Dr. Edward J. Reed, all of Coherent, Inc. and Dr. Hanna J. Hoffman of Spectra-Physics for proof-reading selected chapters of the final manuscript. We also thank Dr. Hans Koelsch, Margaret Mitchell and Virginia Lipscy of Springer-Verlag New York, Inc. for their support and assistance in preparing this book.

July 2004

Belmont, CA — Dr. rer. nat. Norman Hodgson
Berlin, Germany — Prof. Dr. Ing. Horst Weber

Contents

List of Symbols and Abbreviations

Abbreviations

ADP	ammonium dihydrogen phosphate
AD*P	deuterated ammonium dihydrogen phosphate
AIP	American Institute of Physics
Al	aluminum
Al_2O_3	alumina, aluminum oxide
Ar	argon
Au	gold
Banana	barium sodium niobate
BBO	beta barium borate
BeO	beryllia, beryllium oxide
BPP	beam parameter product
$C_2Cl_3F_3$	freon 113
C_2F_6	hexafluorethane
cc	complex conjugate
CDA	cesium dihydrogen arsenate
CD*A	deuterated cesium dihydrogen arsenate
CCl_4	tetra carbon chloride
CBO	cesium borate
CLBO	cesium lithium borate
CO_2	carbon dioxide
Cr	chromium
CS_2	carbon disulfide
Cu	copper
cw	continuous wave, clockwise
ccw	counterclockwise
FPI	Fabry Perot Interferometer
FR	Faraday rotator
FWHM	full width half maximum
$GdVO_4$	gadolinium vanadium oxide
GGG	gadolinium gallium garnet
GRM	graded reflectivity mirror
GSGG	gadolinium scandium gallium garnet
H_2O	water
HeNe	helium neon
HfO_2	hafnium dioxide
HSURIA	half symmetric unstable resonator with intra-cavity axicon
ICSHG	intracavity second harmonic generation
IEEE	The Institute of Electrical and Electronics Engineers, Inc.

ISO	International Standardization Organization
KDA	pottasium dihydrogen arsenate
KDP	pottasium dihydrogen phosphate
KD*P	deuterated potassium dihydrogen phosphate
KrF	krypton fluoride
KTP	potassium titanyl phosphate
LAP	L-arginine phosphate
LBO	lithium triborate
LH	left handed
$LiNbO_3$	lithium niobate
LiSAF	lithium scandium fluoride
LSB	lanthanum scandium borate
MgF_2	magnesium fluoride
Na_3AlF_6	cryolite
NBUR	negative branch unstable resonator
NCUR	near concentric unstable resonator
Nd	neodymium
OC	output coupler
OSA	Optical Society of America
PBUR	positive branch unstable resonator
POM	3-methyl-4-nitropyridine-1-oxide
QR	quartz rotator
RH	right handed
RIUR	rod imaging unstable resonator
SBS	stimulated Brillouin scattering
SRS	stimulated Raman scattering
SF_6	sulfur hexafluoride
SFUR	self-filtering unstable resonator
SiO_2	silicon dioxide
SPIE	Society of Photo-Optical Instrumentation Engineers
Ta_2O_5	tantalum pentoxide
TiO_2	titanium dioxide
TFR	tightly folded resonator
VRM	variable reflectivity mirror
XeCl	xenon chloride
XeF	xenon fluoride
YAG	yttrium aluminum garnet
YAP	yttrium aluminum perovskite
Yb	Ytterbium
YLF	yttrium lithium fluoride
YVO_4	yttrium vanadium oxide (vanadate)
ZnS	zinc sulphide
ZrO_2	zirconium dioxide

Symbols

a, a_1, a_2	aperture width, aperture radius
a	lateral shift of probe beam
a	inner radius of annular gain medium
a	real constant
a	length of micro-cavity
A	cross sectional area
A	ray transfer matrix element
A	2x2 ray transfer submatrix
A_b	cross sectional area of laser beam
A_b^*	cross sectional area of beam in active medium
A_n	electric field
A_{nm}	normalization constant, waveguide modes
b	aperture height
b	radius of laser rod
b	outer radius of annular gain medium
b	proportionality factor, real constant
b_i	normalized coefficients
B	ray transfer matrix element
B	2x2 ray transfer submatrix
B_{nm}	normalization constant, waveguide modes
c	speed of light in homogeneous medium
c_0	speed of light in vacuum ($=3\times10^8$ m/s)
c_{pn}	mode expansion coefficient
C	proportionality constant
C	ray transfer matrix element
C	2x2 ray transfer submatrix
$\mathbf{C}_p$	mode expansion vector, waveguide
$d, d_0, d_1, d_2, d_3, d_4, d_L$	beam diameter
d	wall thickness of annular medium
d_0	center thickness of birefringent element
d, d_1, d_2	distances
d_{pm}	mode expansion coefficient
D	ray transfer matrix element
D	refractive power
D_r	refractive power for radially polarized light
D_Φ	refractive power for azimuthally polarized light
D	2x2 ray transfer submatrix
e	propagation direction unit vector
e	charge of electron (1.6021×10^{-19} As)
E, E_0, $E(x,y)$, $E(r,\Phi)$	electric field (scalar)
E_F	far field
E_p	pulse energy

$\mathbf{E}$, $\mathbf{E}(x,y)$	electric field (vector)
f, f_1, f_2	focal length
f	repetition rate
$f(\nu)$	line shape function
$f(t)$	temporal shape function
F	finesse of FPI
g	gravitational acceleration (9.81 m/s^2)
g	g-parameter of resonator mirror
g_1,g_2	g-parameters of resonator mirrors 1,2
g_1^*,g_2^*	g-parameters with internal lenses
g_0, $g_0(\nu)$	small-signal gain coefficient
$g_0\ell$	small-signal gain
$g_0\ell_{th}$	small-signal gain at the laser threshold
g_B	gain coefficient for SBS
G	gain factor
G	equivalent g-parameter
$G(\omega)$	power spectrum
G_0	small-signal gain factor
h	Planck constant (=6.626×10^{-34} Js)
$h(r,\theta)$	Wigner-function
h,h_1,h_2	distance from lens surface to principal plane
H	magnetic field (amplitude)
H_m	Hermite polynomial of order m
$\mathbf{H}$	magnetic field (vector)
I,I^+,I^-,I_0	intensity of electromagnetic field
$I(z)$	intensity of electromagnetic field at coordinate z
I_S	saturation intensity
I_{SE}	intensity of spontaneous emission
$\mathbf{I}$	unity matrix
$\mathbf{J}$	angular momentum flux
J_ℓ	Bessel function of order ℓ
k	real number
k	Boltzmann constant (=1.381×10^{-23} J/K)
k, k_0	wave number
$\mathbf{k}$	wave vector
$\mathbf{K}$	diffraction integral operator
K	global degree of coherence
l,ℓ	length
l_t	temporal coherence length
ℓ	azimuthal index for circularly symmetric modes
ℓ_{zz}	geometrical length of optical axis in zig-zag slab
ℓ_{glob}	global or transverse coherence length
ℓ_{loc}	local coherence length
$\mathbf{L}$	angular momentum

L	length
L_{eff}	effective resonator length
L_i	distance
L_0	geometrical resonator length
L_{01}, L_{02}	distance from mirror to beam waist
L_{opt}	optical resonator length
L^*	effective resonator length with internal lenses
L'	effective waveguide length
$L_{p\ell}$	Laguerre polynomial of order p,ℓ
m	mode number (integer)
m	mass
m	slope in Findlay-Clay diagram
M, M_0, M_1, M_2	magnification
M^2, M^2_x, M^2_y	beam propagation factor
$\mathbf{M}$	ray transfer matrix
$\mathbf{M}^P$	Jones matrix
n	mode number (integer)
n, n_0, n_1, n_2	index of refraction
n_1, n_2	number of atoms
n_A	index of refraction of air
n_i	index of refraction of coating *i*
n_S	index of refraction of substrate
$n_r(r)$	index of refraction for radial polarization
$n_\theta(r)$	index of refraction for azimuthal polarization
N	number of folds in folded resonators
N_0	density of atoms/ions/molecules
N_{eff}	effective Fresnel number
N_{eq}	equivalent Fresnel number
p	integer
p	radial index for circularly symmetric modes
p	gas pressure
P, P_1, P_2	power
P	degree of polarization
P_N	normalization constant, proportional to power
$\mathbf{P}$	second order moment matrix, variance matrix
$\mathbf{P}_m$	momentum
$\mathbf{P}_j$	propagation matrix for coating *j*
P_{th}	threshold power for SBS phaseconjugation
P_{electr}	electrical pump power
P_{out}	output power of laser resonators
$P_{out,max}$	maximum output power of laser resonators
q	mode number (integer)
q	number of photons
q, q_1, q_2	beam parameter of Gaussian beam

Q	cavity quality
$\mathbf{Q}^{-1}$	complex beam matrix
r, r_1, r_2	radial coordinate
$\mathbf{r}$	vector
r_0	radius of mirror vertex
r, r_x, r_y	amplitude reflectance
R	intensity reflectance
R_{max}	maximum reflectance of FPI
R_{opt}	optimum reflectance for maximum laser power
R_1, R_2	reflectance of mirror 1,2
$R, R_1, R_2, R_t, R_1^*, R_2^*$	radius of curvature of wavefront
R_0	center reflectance of variable reflectivity mirror
R	radius of curvature (electric field)
$R_{xx}, R_{xy}, R_{yx}, R_{yy}$	elements of 2x2 curvature matrix (electric field)
$\mathbf{R}$	2x2 curvature matrix (electric field)
s	length
$\mathbf{s}$	vector
S	shift sensitivity
$\mathbf{S}$	stack matrix for optical coatings
S, S_T, S_z	Poynting vector
t	amplitude transmission (electric field)
t	time
t_D	delay time
T, T(ν)	intensity transmission
T	temperature
$\mathbf{T}_{ij}$	transition matrix from coating *i* to coating *j*
T_{max}	maximum transmission
U	voltage
$U_{\lambda/4}$	quarter wave voltage (Pockels cell)
$\mathbf{v}$	ray vector
v	electron velocity
V	Verdet constant
V	visibility or contrast
V	volume of cavity
V, V_1, V_2, V_3, V_4	loss factor (=1-loss)
V_S	loss factor due to scattering
V_D	loss factor due to diffraction
V_{00}	mode volume of TEM_{00} mode
V_{ol}	mode volume in medium with beam overlap
V_{tot}	total mode volume in medium
w	beam radius, profile radius
w_g	radius of gain profile
w_0	waist radius
W	pump rate

x, x_1, x_2	cartesian coordinate
x, x_1, x_2	distance in the x direction
x	factor characterizing line broadening
y, y_1, y_2	cartesian coordinate
z, z'	cartesian coordinate
z_0	Rayleigh range
z_M, z_M'	multimode Rayleigh range
Z_A	ambiguity function
α_0	loss coefficient of active medium
$\alpha_0 \ell$	loss of active medium
$\alpha, \alpha_1, \alpha_2$	ray slope in x-direction
α	thermal lensing coefficient
α	angle of rotation
α_{mn}	amplitude absorption coefficient
β, β_1, β_2	ray slope in y-direction
β	rotation angle of Faraday rotator
β	beam magnification in phase cojugator
β_{mn}	waveguide mode propagation constant
γ	eigenvalue of diffraction integral
γ	fill factor
γ	relative threshold difference (ring-linear)
γ	correction factor (knife edge method)
$\gamma_1, \gamma_2, \gamma_3$	correction factors for power of folded resonator
γ_{11}	complex degree of coherence
γ_{12}	normalized cross-correlation function
$\Gamma, \Gamma_1, \Gamma_2$	power coupling loss of EH_{11} mode
Γ, Γ_C	decay rate
Γ_{11}	autocorrelation function
Γ_{12}	cross-correlation function
$\Delta, \Delta_x, \Delta_y$	shift
$\Delta\alpha$	angular width of light source
$\Delta\phi$	phase shift
$\Delta\nu$	line width, free spectral range
$\Delta\omega$	bandwidth
ΔD	difference in refractive power
ΔL	distance
Δn	difference in refractive index
ΔN	population inversion density
ΔP_{electr}	range of electrical pump power
ΔP_{out}	range of output power
Δt	pulse duration
Δt_p	pulse width
Δt_{coh}	coherence time

ΔV	loss
Δx	spatial width of light source
δ	phase
δ	complex number
δ	gain profile shape factor
$\delta\nu$	frequency shift, bandwidth
ϵ	dielectric constant of homogeneous medium
ϵ_0	permittivity of vacuum (=8.854×10^{-12} As/(Vm))
η	efficiency
η_{excit}	excitation efficiency
η_{extr}	extraction efficiency
$\eta_{extr,max}$	maximum extraction efficiency
η_{tot}	total efficiency
θ	angle of rotation
$\theta_1, \theta_2, \theta_3, \theta_4$	ray inclination in tilted optics
θ	divergence angle (half angle)
θ_0	divergence angle of Gaussian beam (half angle)
λ	wavelength
λ_0	center wavelength
λ_q	wavelength of axial mode of order q
Λ_p	eigenvalue, p-th waveguide resonator mode
μ	matrix eigenvalue
μ_0	permeability of vacuum (=1.257×10^{-10} Vs/(Am))
μ_{12}	dipole moment
ν	light frequency
ν_0	center frequency of atomic transition
ν_{hom}	homogeneous line width
ν_{inhom}	inhomogeneous line width
ν_q	frequency of axial resonator mode of order q
ν_q^*	axial mode frequency in resonators with gain
π	= 3.141592...
ρ, ρ_1, ρ_2	radius of curvature of surfaces
σ	cross section of stimulated emission
σ_{eff}	effective cross section of stimulated emission
σ_0	cross section of stimulated emission at line center
σ^*	collision cross section
τ	lifetime, duration of light burst
τ^*	decay time due to atomic collisions
τ_B	phonon lifetime (Brillouin scattering)
Φ	spherical coordinate (angle)
Φ	full angle of divergence (86.5% power content)
Φ	phase
Φ_n	field of waveguide eigenmode of order n

ϕ, ϕ_1, ϕ_2	phase induced by retarder
χ	atomic susceptibility
χ_1	real part of atomic susceptibility
χ_2	imaginary part of atomic susceptibility
ψ	phase
Ψ_m	field of free space eigenmode of order m
Ω	angular beat frequency
ω	angular frequency

Introduction

Laser radiation exhibits several properties which are not provided by conventional light sources such as incandescent or fluorescent lamps:

- Narrow frequency bandwidth (as low as 10^{-3} Hz) and high temporal coherence
- Low beam divergence (on the order of milliradians) and high spatial coherence
- High intensity in focal spots (up to 10^{15} W/cm^2)

Due to these beam characteristics the laser has found application in technologies where material has to be removed or transformed within defined, often small, areas (e.g. laser surgery, material processing) or where a narrow spectral frequency band is required (e.g. remote sensing and spectroscopy). In addition, the unique properties of laser diodes (small size, high efficiency, high modulation rates) have been the enabling force in the burgeoning fields of optical communications and optical processing.

Without the optical resonator, the radiation emitted by the laser medium (solid state material, gas or liquid dye) could hardly be used for any application. Since the active atoms generate photons mainly through the process of spontaneous emission, the radiation (superluminescence) does not differ from the light emitted by thermal light sources as far as spatial coherence and focusability are concerned. This is due to the fact that the photons are emitted independently over a range of light frequencies that are characteristic for the laser material. By feeding back the emitted photons into the active medium, this lack of synchronization between emitting atoms can be overcome to generate the exciting properties of laser beams. One could say that the feedback forces the output to stabilize on one preferred channel.

In all laser systems this feedback is accomplished by the optical resonator. Beam properties such as spatial intensity structure, output power, and minimum focal spot size attainable with a focusing lens, are mainly determined by the optical resonator. This crucial role played by the resonator in laser systems is the reason that even after forty years of research in this field, scientists are still working on optical resonators to further understand and improve laser beam characteristics.

The principle of feedback to stabilize the output of a physical system to a desired level is not only limited to optics but has been applied to different engineering areas for eighty years. The best known example is the Meissner circuit, invented in 1913 by the physicist Alexander Meissner, which generates undamped electromagnetic oscillations (Fig.1).

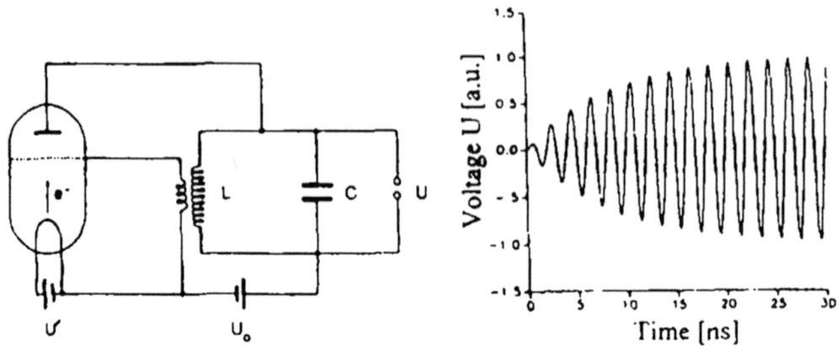

Fig.1 Meissner Circuit for generating undamped electrical oscillation as an example for the feedback principle. The right hand graph shows the normalized output voltage U as a function of time for the case that the initial gain factor overcompensates the loss in the LC circuit.

The attenuation losses of a LC-oscillation circuit are compensated by feeding back the right amount of electrons by means of a triode. Without feedback, the output voltage U would oscillate sinusoidally with an exponential amplitude decay. The feedback is accomplished by coupling the voltage at the inductance L to the acceleration grid of the triode. As soon as the grid voltage turns positive, electrons are forced to enter the LC-circuit to compensate for resistance losses. This process maintains a temporally constant amplitude of the oscillating output voltage U. By using this circuit, electric oscillations with frequencies of up to 10^9 Hz can be generated. For higher frequencies the time of flight of the electrons in the triode becomes a problem.

In steady state the gain factor G ($G>1$) by which the voltage is increased per period by the triode exactly compensates the attenuation losses by the loss factor V ($V<1$). The steady state oscillation condition therefore reads:

$$G\ V = 1$$

The gain factor G is a function of the acceleration voltage at the triode grid and thus defined by the characteristics of the triode and the coupling between the inductors. In general, the gain factor increases as the voltage increases but it reaches a saturation value at higher voltages. Only if the gain factor G is greater than the loss factor V, which means $GV>1$, the oscillation will build up as shown in Fig.1. As the output voltage increases, the gain factor saturates. The steady state output voltage amplitude is reached as soon as the condition $GV=1$ is realized. Since the gain factor G can be expressed as a function of the output voltage U (the grid acceleration voltage and the output voltage are proportional!) the steady state output voltage U_S is given by:

$$G(U_S) = \frac{1}{V}$$

This equation can be used to calculate the steady state output voltage amplitude since the expressions for both the gain and loss as a function of output voltage U can easily be derived.

The laser can be considered as the extension of the feedback principle into the frequency range of 10^{15} Hz, the frequency range of UV, visible, and infrared light. There is a strong similarity between the laser and the Meissner circuit since the principle of operation is basically the same. The triode is now represented by the active medium and the oscillation of electrons in the LC circuit is now replaced by the oscillation of light between the resonator mirrors (Fig.2). The light first generated by spontaneous emission in the active medium bounces back and forth inside the resonator. The light intensity is decreased by a loss factor of R_l (mirror reflectivity) when it bounces off the output coupling mirror 1 and increased by a gain factor G when it travels through the active medium. Similar to the triode the gain factor is a function of intensity and it decreases as the intensity increases. The laser is forced to attain a steady state intensity I_0 inside the resonator at which loss and gain per round trip are balanced. Since the light travels twice through the active medium before experiencing a loss, the steady state condition reads:

$$G(I_0)\ G(I_0)\ R_1 = 1 \quad \Rightarrow \quad G(I_0) = \frac{1}{\sqrt{R_1}}$$

The gain factor G is generated by an external pumping process by which energy is transferred into the active medium to generate a population inversion between atomic or molecular energy levels. In solid state lasers, this is accomplished by using high power flashlamps or diodes; gas lasers are generally pumped by means of a high voltage gas discharge, and a diode laser by an electric current.

The laser starts oscillating as soon as the gain factor is greater than $1/(V\sqrt{R_1})$ with all other losses caused by diffraction, scattering, and absorption being included in the loss factor per transit V. The light intensity experiences an amplification with every round trip inside the resonator. With increasing intensity, however, the gain factor G starts to decrease and the intensity will reach the steady state solution I_0 for which the above shown steady state condition holds. A higher intensity can only be achieved if the pumping power is increased by increasing the flashlamp power, the pump diode current, or the discharge voltage, respectively.

The reader should always keep in mind that our definition of an optical resonator includes the active medium. This means that we are basically dealing with a laser. However, the laser is only a part of the whole field since optical resonators can also be used without an active medium. Examples in this area are interferometers, multi-pass optical delay lines, and optical coatings. The calculation of the properties of an optical resonator thus far seems to be as straightforward as a design of a Meissner circuit. With our current knowledge we can determine the steady state intensity I_0 inside the resonator which means that we also know the output power emitted by the laser. Unfortunately, this model is far too simplified to be useful for the description of optical resonators.

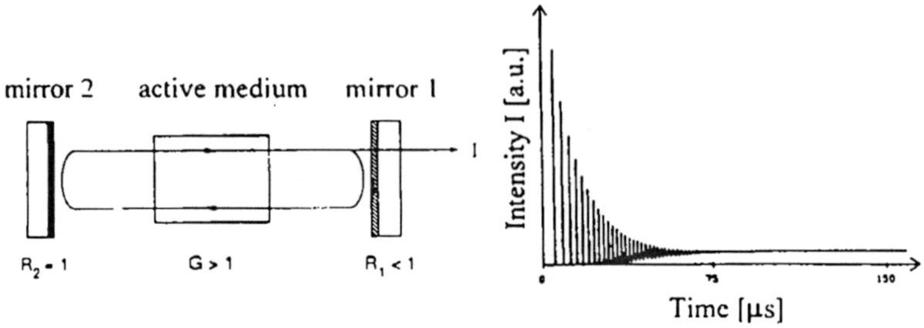

Fig.2 Principle of an optical resonator. In order to attain laser oscillation, the gain factor G of the active medium must be greater than the loss factor caused by the output coupling at mirror 1. After an initial intensity oscillation, a steady state intensity is reached for which the gain exactly compensates the output coupling losses (right hand graph). The laser then emits light in continuous wave (cw) operation.

In contrast to the Meissner circuit where we only have to deal with one parameter, the voltage amplitude, electromagnetic radiation exhibits more parameters than only the amplitude of the electric field or the intensity. Steady state solutions also exist for the spatial field structure, the polarization, and the phase distribution of the electric field. Furthermore, the shape of the resonator mirrors, the size of internal apertures as well as the properties of polarizing optics inside the resonator have a considerable influence on the beam properties. The nonlinear interaction of the oscillating electric field with the atoms in the active medium makes the treatment even more challenging.

This complexity of optical resonators makes it necessary to first discuss the physical properties of light and to derive methods for their calculation. The equations for the calculation of beam propagation, transverse beam structure, polarization, coherence and their application to optical resonators are derived in Part I; "The Electromagnetic Field". The mathematical content of this chapter was kept as low as possible by translating exact mathematical derivations into a more intuitive physical language. All equations printed in this book can be derived by using the mathematical tools presented in Part I. Parts II-V, which deal with optical resonators and their design, contain only short descriptions of how to obtain the result rather than going through the whole derivation.

Part II starts with the most basic optical resonator, the Fabry-Perot-Interferometer, and will help readers new to the field to become familiar with the properties of optical resonators. After discussing linear resonators without the influence of the active medium (Part III), the properties of the laser medium and its effect on the laser characteristics will be presented in Part IV. Part V is a collection of resonator concepts that exhibit specific advantages, like low misalignment sensitivity, narrow bandwidth, or excellent beam quality. These resonators have either a limited applicability or represent new concepts which may become more important in the near future. Readers interested in designing and optimizing laser resonators may find the measurement techniques presented in Part VI helpful. A detailed reference list will help readers to find more information on their chosen subject.

Part I

The Electromagnetic Field

Chapter 1

Geometrical Optics

1.1 General Aspects

Light is an electromagnetic wave which differs from the waves known in radiofrequency technology only by a higher frequency ν and a shorter wavelength λ [1.1-1.5]. In both cases, the electromagnetic field is characterized by (Fig.1.1):

the electric field $\boldsymbol{E}$	...	(V/m)
the magnetic field $\boldsymbol{H}$	....	(A/m)
the wave vector $\boldsymbol{k}=2\pi/\lambda\ \boldsymbol{e}$	.	(1/m)

In homogeneous, isotropic, and unconfined media, all three vectors are perpendicular to one another and the wave vector $\boldsymbol{k}$ points into the direction of propagation $\boldsymbol{e}$ of the wave. The energy flow is characterized by the Poynting vector $\boldsymbol{S}=\boldsymbol{E}x\boldsymbol{H}$. The physical property that humans detect, with the eye or with a light-sensitive detector, is always the time averaged intensity:

$$I = \langle S \rangle \tag{1.1}$$

The power content of the electromagnetic wave is obtained by integrating the intensity over the area perpendicular to the propagation direction:

$$P = \int_A \langle S \rangle \, dA = \int_A \langle \boldsymbol{E} \times \boldsymbol{H} \rangle \, dA \tag{1.2}$$

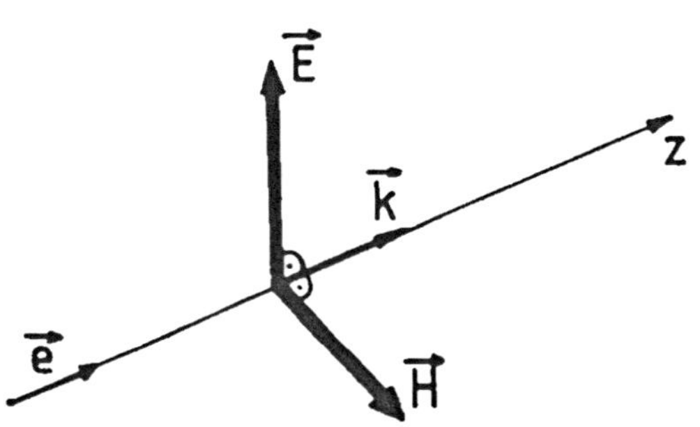

Fig. 1.1 The electromagnetic wave is completely determined by the field vectors ***E***, ***H***, and the wave vector ***k*** pointing into the direction of propagation.

Note that this expression contains both the electric and magnetic field. Within the scope of this book we are only interested in non-magnetic media with low absorption losses. In this case the magnetic field $\boldsymbol{H}$ is linked to the electric field vector $\boldsymbol{E}$ via:

$$\boldsymbol{H} = n\, c_0\, \epsilon_0\, [\boldsymbol{e} \times \boldsymbol{E}]$$

The intensity then reads:

$$I = \frac{1}{2}\, n\, c_0\, \epsilon_0\, |E|^2 \tag{1.3}$$

with:		
	$c_o = 3\mathrm{x}10^8$ m/s	speed of light in vacuum
	$\epsilon_o = 8.85\mathrm{x}10^{-12}$ As/(Vm) ...	permittivity in vacuum
	n : dimensionless number =	refractive index of the medium
	$E = 2E_o/(1+n)$	field inside medium, where E_o is the field in vacuum

Thus it is sufficient in most cases to use only the electric field to describe the properties of the electromagnetic wave. It can easily be shown that the magnetic and the electric field both contribute the same amount to the total power of (1.2).

In the following we consider only the electric field $\boldsymbol{E}$ which depends on the spatial coordinates x,y,z and is assumed to show a purely periodic oscillation with a frequency ν. It is common to chose the z-axis as the main direction of propagation. The general electromagnetic wave can then be expressed as:

$$\boldsymbol{E} = \boldsymbol{E}_0(x,y,z)\, \cos(\omega t - kz) \tag{1.4}$$

with $\omega = 2\pi\nu$. Since we consider only linear media, the frequency ν remains constant as the electromagnetic wave propagates through areas of different indices of refraction n. Linearity means that the index of refraction does not depend on the intensity I. This is in contrast to the propagation of high intensity beams through nonlinear materials such as KTP or BBO, which exhibit a change of the refractive index with the intensity, resulting in frequency conversion. The wavelength λ and the speed of light c, however, do change in linear media:

$$\lambda = \lambda_0/n \qquad c = c_0/n \qquad \nu = c/\lambda = c_0/\lambda_0 \tag{1.5}$$

with λ_0, c_0 being the wavelength and the speed of light both in vacuum.

Examples:

1) On a bright summer day in California the intensity of the sun light is about 500 W/m^2 for normal incidence. By using (1.1) we get an electric field of 614 V/m.

2) For a 1kW CO_2 laser with a beam diameter of 10mm the intensity is 12.7 MW/m^2. The corresponding electric field is 98,935 V/m.

1.2 Ray Transfer Matrices

In the following we will discuss the propagation of light geometrically by analyzing the propagation of rays [1.10]. In this geometrical optics approximation the spatial structure of the electromagnetic wave as well as diffraction effects caused by apertures or edges encountered by the wave are not taken into account. This approximate description of light propagation can be applied as long as one deals with a light beam whose characteristic parameter N, called the Fresnel number, is much greater than one:

$$N = \frac{a^2}{\lambda L} \gg 1 \tag{1.6}$$

with a: beam radius, λ: wavelength L: distance in propagation direction

The meaning of the Fresnel number will be discussed in later sections (see, for instance, Sec. 2.2). Light can only be described in terms of geometrical optics if we have a beam that is not too long or too thin. If the Fresnel number is close to one or lower, the beam propagation has to be calculated using diffraction theory. Unfortunately, there is no well defined Fresnel number which separates geometrical optics from diffraction theory. For Fresnel numbers N greater than twenty, however, geometrical optics can definitely be applied.

Example:
A light beam in the green spectral range (λ=500nm) with a diameter of $2a$=5mm can be described geometrically over a distance of about 1m (N=12.5).

1.2.1 One-Dimensional Optical Systems

In the geometrical approximation, light rays propagate along straight lines (in free space) and experience a declination if they pass through an optical element like a lens. If the ray starts in a plane perpendicular to the z-axis, it is completely determined by its starting points x,y and the inclination angles α in x-direction and β in y-direction. The z-axis is generally chosen such that it coincides with the optical axis. The optical axis is defined by the center of symmetry of the first optical element and is perpendicular to this elements' front surface. Thus, a ray can be expressed mathematically as a vector:

$$V = \begin{pmatrix} x \\ y \\ \alpha \\ \beta \end{pmatrix} \tag{1.7}$$

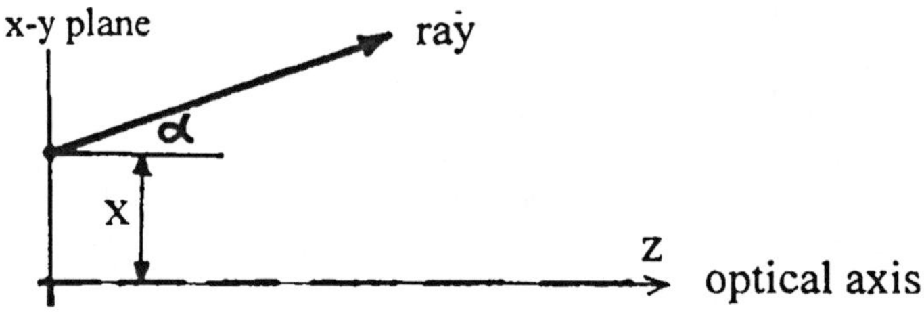

Fig. 1.2 In geometrical optics light is described by rays propagating along straight lines in free space.

If all optical elements in the propagation direction exhibit rotational symmetry (for example a spherical lens) the cartesian coordinate system can always be chosen such that the ray is completely determined by its starting point x and its inclination angle in x-direction α. This is due to the fact that the ray propagates only in the x-z-plane. In the following we will restrict our discussion to these one-dimensional optical systems. Ray propagation in two-dimensional optical systems such as cylinder lenses will be treated in Sec. 1.2.4.

The problem we have to solve is: How does this vector change as the ray propagates through an optical system? As far as the propagation in a medium with index of refraction n is concerned, the calculation of this problem is quite straightforward (Fig.1.3a). For the propagation over a distance L a ray starting at point x_1 under a small angle α_1 will end up at point x_2 with an inclination angle α_2, with:

$$x_2 = x_1 + L\cdot\alpha_1 \tag{1.8}$$

$$\alpha_2 = \alpha_1 \tag{1.9}$$

This can be written in form of a matrix equation:

$$\begin{pmatrix} x_2 \\ \alpha_2 \end{pmatrix} = M_{FS} \begin{pmatrix} x_1 \\ \alpha_1 \end{pmatrix} \qquad M_{FS} = \begin{pmatrix} 1 & L \\ 0 & 1 \end{pmatrix} \tag{1.10}$$

For every linear optical element such as a lens or a mirror one can find a matrix M that mathematically describes the transformation of a ray while going through this element [1.6,1.10]. This matrix M is called the ray transfer matrix. In order to find such a constant matrix - constant means that the matrix elements do not depend on the ray parameters - the approximation of small angles α has to be made (as we did in (1.8)). The ray transfer matrix theory is only applicable to the analysis of an optical system if the following relation holds for the maximum angle:

$$\sin\alpha \approx \tan\alpha \approx \alpha \tag{1.11}$$

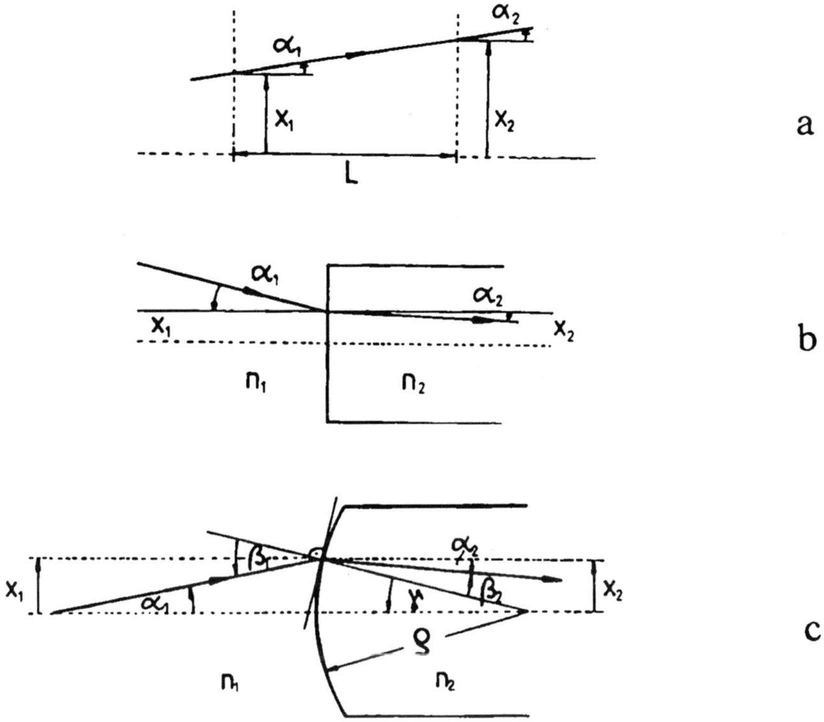

Fig. 1.3 Propagation of light rays. **a)** free space propagation over a distance L, **b)** refraction at the planar boundary surface between two media with different index of refraction, **c)** refraction at a spherical surface.

As a rule of thumb the maximum angle should not exceed 15° (α=0.262rad, sinα=0.259, tanα=0.268). This is not a serious limitation as far as optical resonators are concerned since the angle under which the light propagates inside a resonator is rarely greater than a couple of degrees. This limitation to small angles is called the paraxial approximation. It generates a linear transformation between ray vectors. If a light beam passes from one medium with index of refraction n_1 to another with index of refraction n_2, the rays are refracted at the interface (Fig.1.3b). In the paraxial approximation the refraction law, also known as Snell's law, reads:

$$\frac{\alpha_1}{\alpha_2} \simeq \frac{\sin\alpha_1}{\sin\alpha_2} = \frac{n_2}{n_1} \tag{1.12}$$

Thus, the ray transfer matrix M_R for refraction at a planar interface is fully determined since the position x of the ray at the surface remains unchanged:

$$M_R = \begin{pmatrix} 1 & 0 \\ 0 & \dfrac{n_1}{n_2} \end{pmatrix} \tag{1.13}$$

A third important optical transition is the refraction at a spherical surface as depicted in Fig.1.3c. The surface geometry is characterized by the radius of curvature ρ. For a convex surface which means that the center of curvature is to the right of the surface and the ray is arriving from the left (as shown), the radius of curvature is positive. A negative radius of curvature defines a concave surface. Again the ray position at the surface remains fixed and Snell's law (1.12) now holds for the angles β_1 and β_2. The angle α_2 can be calculated by using the geometrical relations $\gamma = x_1/\rho$, $\alpha_1 = \beta_1 - \gamma$, and $\alpha_2 = \beta_2 - \gamma$. The final result reads:

$$x_2 = x_1 \tag{1.14}$$

$$\alpha_2 = \frac{n_1 - n_2}{n_2 \rho} x_1 + \frac{n_1}{n_2} \alpha_1 \tag{1.15}$$

The ray transfer matrix for refraction at a spherical interface is given by:

$$M_{RS}(\rho) = \begin{pmatrix} 1 & 0 \\ \dfrac{n_1 - n_2}{n_2 \rho} & \dfrac{n_1}{n_2} \end{pmatrix} \tag{1.16}$$

As to be expected, this ray transfer matrix is transformed into matrix (1.13) if we set the radius of curvature ρ of the surface to infinity.

So far we have found two fundamental ray transfer matrices to describe ray propagation in free space $M_{FS}(L)$ and refraction at a spherical dielectric interface $M_{RS}(\rho)$. The knowledge of these two matrices is sufficient to describe arbitrary optical systems since ray transfer matrices for all optical elements can be derived by using these two fundamental matrices. Before we discuss this in more detail the reader should memorize the following rules to avoid confusion in later sections:

1) light rays always propagate from left to right.
2) convex dielectric surfaces have a positive radius of curvature, concave dielectric surfaces a negative radius of curvature. However, this is reversed for mirrors!
3) Angle orientations are defined mathematically. This means that in Fig.1.3c α_1 is positive and α_2 is negative.

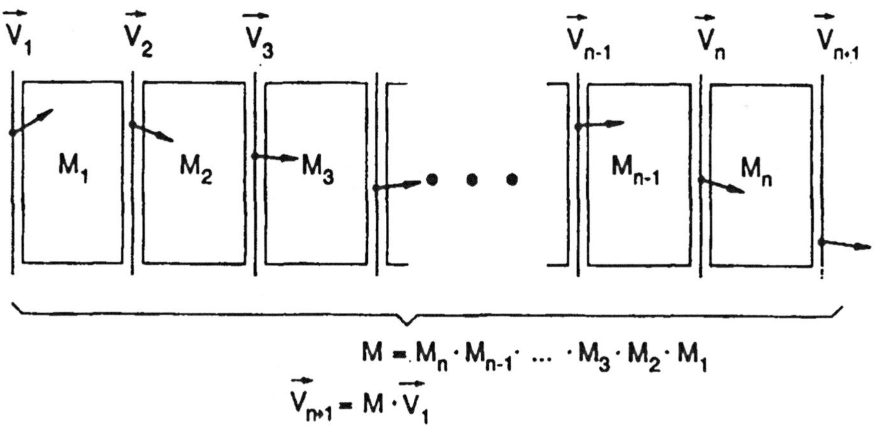

Fig. 1.4 Propagation of a light ray through a series of optical elements. The individual ray matrices can be combined to a resulting matrix $\boldsymbol{M}$.

If the ray propagates through several optical elements their ray transfer matrices may be combined into a single one. This procedure is visualized in Fig. 1.4. The ray starts on the left plane having a ray vector $\mathbf{v}_1$. This ray vector is transformed into the ray vector $\mathbf{v}_2$ by the first optical element, the second element generates the ray vector $\mathbf{v}_3$ on the third plane and so forth. If we have n optical elements we get n equations:

$$\begin{aligned} V_2 &= M_1\, V_1 \\ V_3 &= M_2\, V_2 \\ &\vdots \\ V_{n+1} &= M_n\, V_n \end{aligned}$$

and therefore:

$$V_{n+1} = M_n M_{n-1} M_{n-2} \dots M_2 M_1\, V_1 = M\, V_1 \quad . \tag{1.17}$$

Thus, the resulting ray transfer matrix $\boldsymbol{M}$ is obtained by multiplying all individual matrices in the opposite order (i.e. right to left) of the ray propagation. In other words, the first ray matrix 'seen' by the ray is on the right side of the matrix product, the last one on the left side. Now we have a powerful tool in hand to determine the ray transfer matrices of more complicated optical systems.

Examples (see also Fig.1.5)

1) Thin Lens

In general a spherical lens is determined by its index of refraction n_2, the thickness L and the curvatures of the front and the back surface. In the thin lens approximation, any change in ray position or angle inside the medium is neglected which means that we do not propagate the ray between the two surfaces. Thus the ray transfer matrix M_{TL} of a thin lens is the product of two transfer matrices for refraction at a spherical interface (1.16). If ρ_1, ρ_2 denote the radii of curvature of the front and back surface and the lens is surrounded by a medium with index of refraction n_1, the resulting ray transfer matrix reads:

$$M_{TL} = M_{RS}(\rho_2)\, M_{RS}(\rho_1)$$

$$= \begin{pmatrix} 1 & 0 \\ \dfrac{n_2 - n_1}{n_1 \rho_2} & \dfrac{n_2}{n_1} \end{pmatrix} \begin{pmatrix} 1 & 0 \\ \dfrac{n_1 - n_2}{n_2 \rho_1} & \dfrac{n_1}{n_2} \end{pmatrix}$$

$$= \begin{pmatrix} 1 & 0 \\ -\dfrac{1}{f} & 1 \end{pmatrix} \tag{1.18}$$

with a focal length f given by:

$$\frac{1}{f} = \frac{n_2 - n_1}{n_1} \left(\frac{1}{\rho_1} - \frac{1}{\rho_2} \right) \tag{1.19}$$

Note that the curvature ρ is positive for convex surfaces (center of curvature to the right of the interface) and negative for concave surfaces.

2) Plane Dielectric Slab

We consider a slab of length L with parallel surfaces and an index of refraction n_2. The ray transfer matrix is found by combining the two matrices M_R for refraction at the surfaces with the free space propagation matrix M_{FS}:

$$M_P = \begin{pmatrix} 1 & 0 \\ 0 & n_2/n_1 \end{pmatrix} \begin{pmatrix} 1 & L \\ 0 & 1 \end{pmatrix} \begin{pmatrix} 1 & 0 \\ 0 & n_1/n_2 \end{pmatrix} = \begin{pmatrix} 1 & n_1 L/n_2 \\ 0 & 1 \end{pmatrix} \tag{1.20}$$

Interestingly, this is exactly the matrix for free space propagation over an effective distance n_1L/n_2. For a glass plate in air, for instance, this effective distance is smaller than the actual physical thickness L of the plate. This means that objects will appear closer to the eye if we look through the plate (this is, of course, only true for near objects which are viewed under an angle).

3) Spherical Mirror

A spherical mirror with radius of curvature ρ provides the same ray transformation properties as a lens except for the fact that the beam propagation direction is reversed. As shall be discussed later in more detail, beam reversal needs not to be incorporated into the ray matrices if the coordinate system is mirrored together with the beam. In other words, we always ride with the ray. Within this approach the ray transfer matrix of a mirror is identical to that of a thin lens (1.18) with focal length $f=\rho/2$:

$$M_M = \begin{pmatrix} 1 & 0 \\ -\frac{2}{\rho} & 1 \end{pmatrix} \tag{1.21}$$

The radius of curvature ρ is positive for a convex mirror surface which means the mirror opens to the left, towards the incoming ray (as shown in Fig. 1.5). It is most important for the reader to realize that this sign convention is contrary to the one used for dielectric interfaces. Keeping this in mind will save you a lot of calculation time.

A collection of commonly used ray transfer matrices is presented in Fig.1.5. It should be noted that from a mathematical point of view, ray transfer matrices can only be defined for optical elements that have parabolic index profiles (like the thermal lens in Fig.1.5) or parabolic surfaces (like an aberration free lens). Fortunately, near its center of symmetry a parabolic surface hardly differs from a spherical surface. Since only paraxial rays are considered we are limited to the central area of the optical elements. Thus, from a practical point of view, the ray transfer matrices can also be applied to optics with spherical, elliptical, or hyperbolic surfaces and index profiles without introducing noticeable errors.

The paraxial approximation also implies that the optical systems considered exhibit perfect imaging properties. For perfectly aligned optics a point will be imaged to a point because aberration is not incorporated into ray transfer matrix theory. For analyzing aberrations in optical systems numerical ray tracing algorithms are required (i.e. the commercially available optics design software like CODE V, ZEMAX, or OSLO). The impetus for computer solutions is that when the paraxial approximation (1.11) is no longer applicable, the time-efficient solution of the trigonometric functions for a great number of rays (on the order of 1,000) requires a numerical approach.

Astigmatism, however, which is a general feature of two-dimensional optics (cylindrical lenses) and is also induced in one-dimensional optical elements by tilt (i.e. a Brewster plate), can be analyzed by using 4x4 ray matrices (see Sec. 1.2.4).

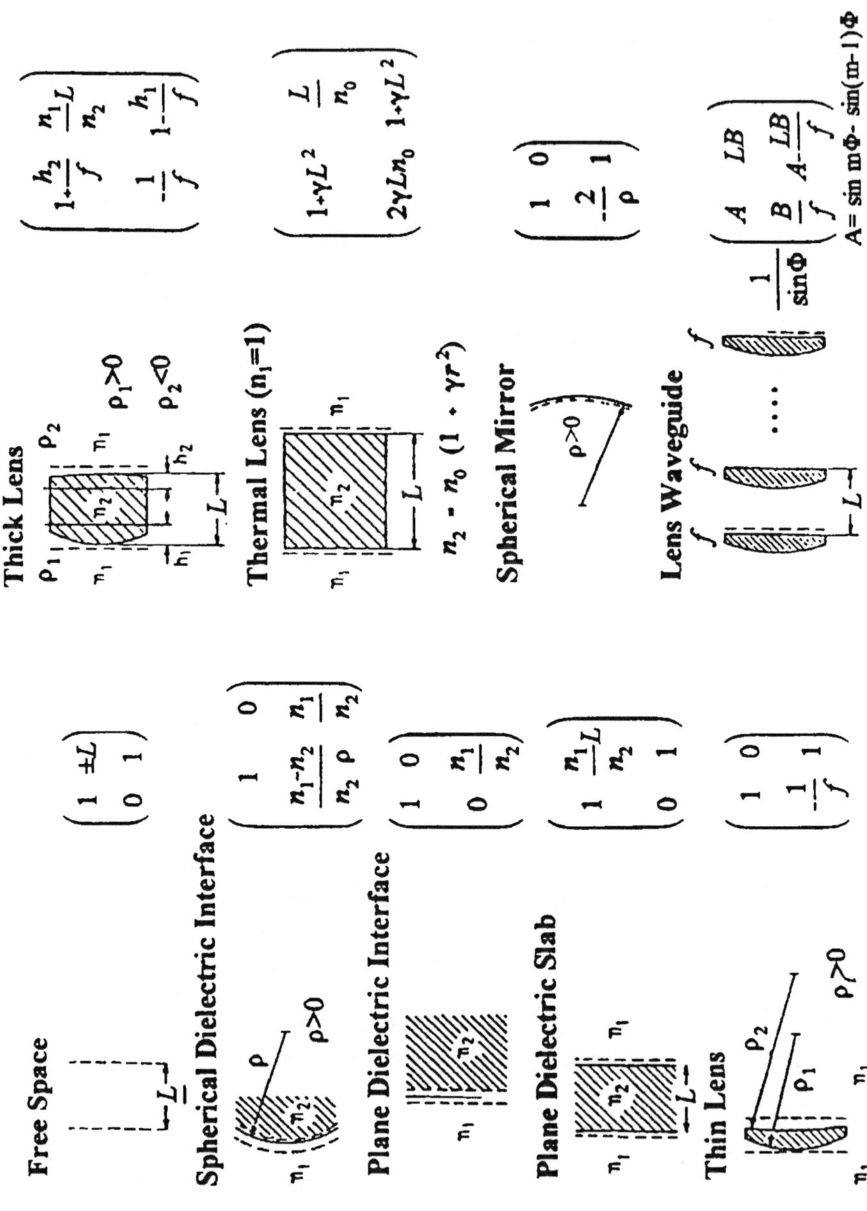

Fig. 1.5 Common optical elements and their ray transfer matrices. See (1.31) for the focal length and the principal planes of the thick lens.

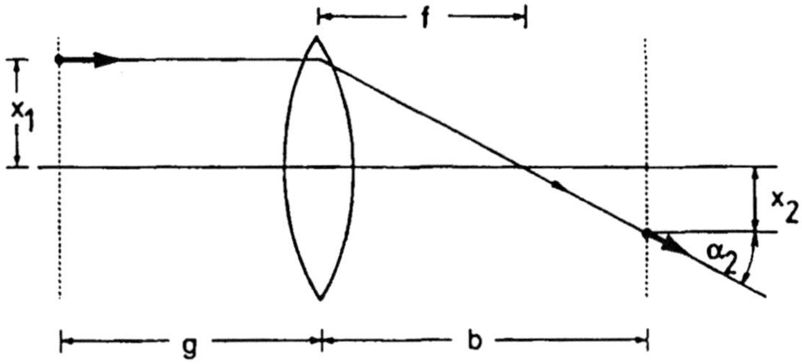

Fig. 1.6 Imaging with a biconvex focusing lens.

In order to get more experience and confidence in the utilization of ray transfer matrices the reader should first go through the two examples presented below. After this we will proceed with a more generalized presentation of ray matrix properties.

Imaging with a focusing lens

We are looking for the imaging condition for a focusing lens. The goal is to find the relation between the object distance g, the image distance b and the focal length f. Using (1.17) we first determine the resulting ray transfer matrix:

$$M_{IM} = \begin{pmatrix} 1 & b \\ 0 & 1 \end{pmatrix} \begin{pmatrix} 1 & 0 \\ -1/f & 1 \end{pmatrix} \begin{pmatrix} 1 & g \\ 0 & 1 \end{pmatrix}$$

$$= \begin{pmatrix} 1-\frac{b}{f} & g+b-\frac{gb}{f} \\ -\frac{1}{f} & 1-\frac{g}{f} \end{pmatrix} \tag{1.22}$$

A ray starting on the left plane with parameters x_1, α_1 will intersect the image plane having the parameters:

$$x_2 = (1-\frac{b}{f})\, x_1 + (g+b-\frac{bg}{f})\, \alpha_1 \tag{1.23}$$

$$\alpha_2 = -\frac{1}{f}\, x_1 + (1-\frac{g}{f})\, \alpha_1 \tag{1.24}$$

This result can be easily verified geometrically by using a ray that propagates parallel to the optical axis ($\alpha_1=0$), as depicted in Fig. 1.6, and a second ray emerging from the same point but going through the center of the lens ($\alpha_1=-x_1/g$). Now we have to find a condition so that every point x_1 is imaged onto one point x_2. This means that all rays starting at x_1 have to end up at x_2 regardless of their angle α_1. A look at (1.23) indicates that this can only be accomplished if

$$g + b - \frac{bg}{f} = 0 \quad \longrightarrow \quad \frac{1}{b} + \frac{1}{g} = \frac{1}{f} \tag{1.25}$$

This is easily recognized as the imaging condition.

Simulation of a Spherical Mirror

While setting up an important optical experiment a student realizes that he does not have the spherical mirror needed. All he can find is a plane mirror. Is there a way to simulate a spherical mirror with a certain radius of curvature ρ just by placing a focusing lens at a distance L in front of the plane mirror (Fig. 1.7) ?

Here we are again confronted with the beam reversal problem and we should use this example to understand how the reflection of rays is dealt with. The basic idea is that we always ride with the ray. We can then mirror the optical system at the mirror plane and propagate through the whole system only in the forward direction. This method is shown in Fig. 1.7. The lens-mirror assembly with spacing L can be replaced by two lenses separated by twice the distance. In both presentations we get the same ray vector at the output plane. The ray transfer matrix for a round trip starting at the lens now reads:

$$M_{LM} = \begin{pmatrix} 1 & 0 \\ -1/f & 1 \end{pmatrix} \begin{pmatrix} 1 & 2L \\ 0 & 1 \end{pmatrix} \begin{pmatrix} 1 & 0 \\ -1/f & 1 \end{pmatrix} = \begin{pmatrix} 1-2L/f & 2L \\ -2/f\,(1-L/f) & 1-2L/f \end{pmatrix}$$

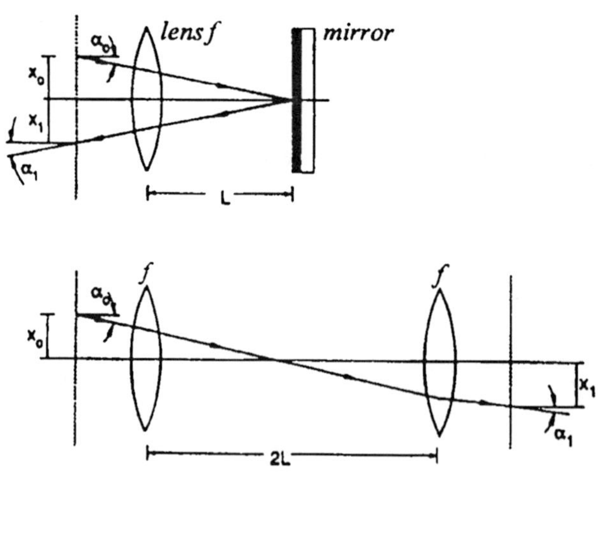

Fig. 1.7 Reflection at mirrors is dealt with by mirror imaging the optical system at the mirror plane. Thus, the lens-mirror combination is equivalent to the dual lens system.

A comparison with the matrix of a spherical mirror (1.21) indicates that we have to glue the lens right onto the plane mirror ($L=0$) to simulate a mirror curvature of $\rho=f$.

1.2.2 Matrix Elements and Liouville's Theorem

In order to get a better understanding of ray transfer matrices, we will now discuss the meaning of the matrix elements as far as ray transformation is concerned. This approach is very helpful for the initial layout of an optical system since each matrix element represents a characteristic property of the beam transformation. The most efficient way to visualize the meaning of the matrix elements is to set them to zero and analyze the propagation changes generated. In order to keep the discussion as general as possible we use the following form of the ray transfer matrix $\boldsymbol{M}$:

$$\boldsymbol{M} = \begin{pmatrix} A & B \\ C & D \end{pmatrix}$$

or

$$\begin{aligned} x_2 &= A\,x_1 + B\,\alpha_1 \\ \alpha_2 &= C\,x_1 + D\,\alpha_1 \end{aligned} \tag{1.26}$$

a) A=0

The relationship between input and output ray parameters then reads:

$$x_2 = B\,\alpha_1 \tag{1.27}$$

$$\alpha_2 = C\,x_1 + D\,\alpha_1 \tag{1.28}$$

The position x_2 of the ray does not depend on the initial position x_1. All rays going through the system under an angle α_1 end up at the same coordinate x_2. This means that a parallel beam will be focused (Fig.1.9a).

Example: At which distance L are all parallel incoming rays focused by a lens with focal length f? By multiplying the ray transfer matrices for a thin lens and for free space propagation, the matrix elements A,B are found to be:

$$A = 1 - \frac{L}{f} \qquad B = L$$

Matrix element A goes to zero if we go into the focal plane of the lens ($L=f$). In this case a parallel beam (see Fig. 1.9a), hitting the lens at an angle α_1 will be **focused** into the point $x_2=f\alpha_1$. This means that we obtain information on the angular ray distribution by looking at the intensity pattern in the focal plane of a focusing lens. We actually see the Fourier transform of the incoming beam (see Sec. 2.4).

b) B=0

The output ray parameters are now given by:

$$x_2 = A\,x_1 \tag{1.29}$$

$$\alpha_2 = C\,x_1 + D\,\alpha_1 \tag{1.30}$$

All rays starting at point x_1 under arbitrary angles will be reunified in point x_2. Thus setting B equal to zero creates an **imaging system** (Fig.1.9b), with the lateral magnification A.

Example: We have already derived the imaging condition for a thin lens. Let us now do the same for a thick lens (Fig. 1.8) since it is an important optical element in resonator physics. All laser media can be described in a first approach as a thick lens due to a refractive power induced by the pumping process. We utilize the ray matrix of a thick lens presented in Fig. 1.5 and first calculate the ray transfer matrix for the propagation from the input to the output plane as shown in Fig. 1.8. The resulting ray transfer matrix reads:

$$M = \begin{pmatrix} 1 & b \\ 0 & 1 \end{pmatrix} \begin{pmatrix} 1+h_2/f & Ln_1/n_2 \\ -1/f & 1-h_1/f \end{pmatrix} \begin{pmatrix} 1 & g \\ 0 & 1 \end{pmatrix}$$

$$= \begin{pmatrix} 1 + \dfrac{h_2 - b}{f} & L\dfrac{n_1}{n_2} + b\left(1 - \dfrac{h_1}{f}\right) + g\left(1 + \dfrac{h_2 - b}{f}\right) \\ -\dfrac{1}{f} & 1 - \dfrac{h_1 + g}{f} \end{pmatrix} . \tag{1.31}$$

with $\quad \dfrac{1}{f} = \dfrac{n_2 - n_1}{n_1}\left[\dfrac{1}{\rho_1} - \dfrac{1}{\rho_2} + \dfrac{n_2 - n_1}{n_2}\dfrac{L}{\rho_1\rho_2}\right]$; $h_i = \dfrac{f\,L}{\rho_j}\dfrac{n_1 - n_2}{n_2}$ $\quad i, j = 1,2;\ i \neq j$

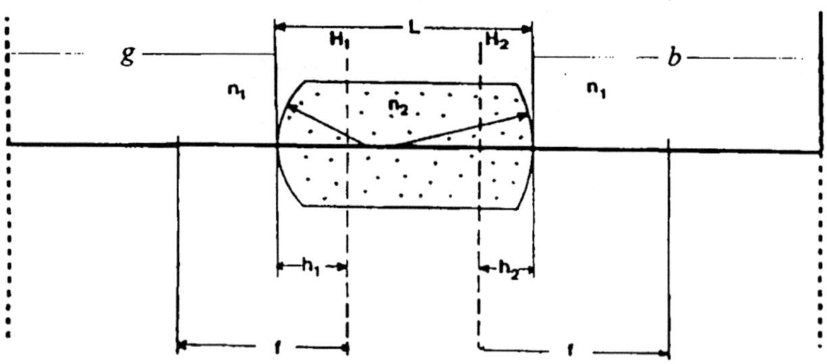

Fig. 1.8 Imaging by means of a thick lens. H_i denote the principal planes. The focal length f is measured from the principal planes. $h_j > 0$, if the principal plane is located on the right hand side of the surface.

Setting matrix element B equal to zero yields:

$$b+g-\frac{gb}{f} = \frac{h_1 b-h_2 g}{f} - L\frac{n_1}{n_2}$$

which is equivalent to:

$$\frac{1}{g'} + \frac{1}{b'} = \frac{1}{f} \qquad \textit{with} \quad g'=g+h_1 \quad b'=b-h_2 \tag{1.32}$$

We get the same imaging condition as compared to a thin lens if we measure the object and image distance from the principal planes H_i. Note that h_2 is negative (ρ_i is positive) !

<u>c) C=0</u>
At the output plane the ray parameters read:

$$x_2 = A\,x_1 + B\,\alpha_1 \tag{1.33}$$

$$\alpha_2 = D\,\alpha_1 \tag{1.34}$$

Parallel rays will still be parallel after passage through the optical system, but their inclination angle has changed by a factor of D. If we look through this optical system at a distant object, the size of the object is magnified by $/D/$. All telescopic systems such as the Galilean or the astronomical telescope have a zero C-component in their ray transfer matrix. D is the angle magnification.

<u>d) D=0</u>
At the output plane the ray parameters read:

$$x_2 = A\,x_1 + B\,\alpha_1 \tag{1.35}$$

$$\alpha_2 = C\,x_1 \tag{1.36}$$

We see that optical systems having $D=0$ generate a collimated beam from a divergent ray pattern. We can take the thin lens of Fig. 1.6 as an example. $D=0$ requires $g=f$. All rays starting in the point x_1 at the front focal plane will be collimated to a parallel beam with an angle α_2 with respect to the z-axis.

Figure 1.9 gives an overview of the four cases discussed. The reader should keep the properties of the matrix elements in mind since this is very helpful in designing an optical system to provide a desired beam transformation. Generally, the optical properties of the components (distances, focal lengths, etc.) need to be varied until the corresponding matrix element goes to zero.

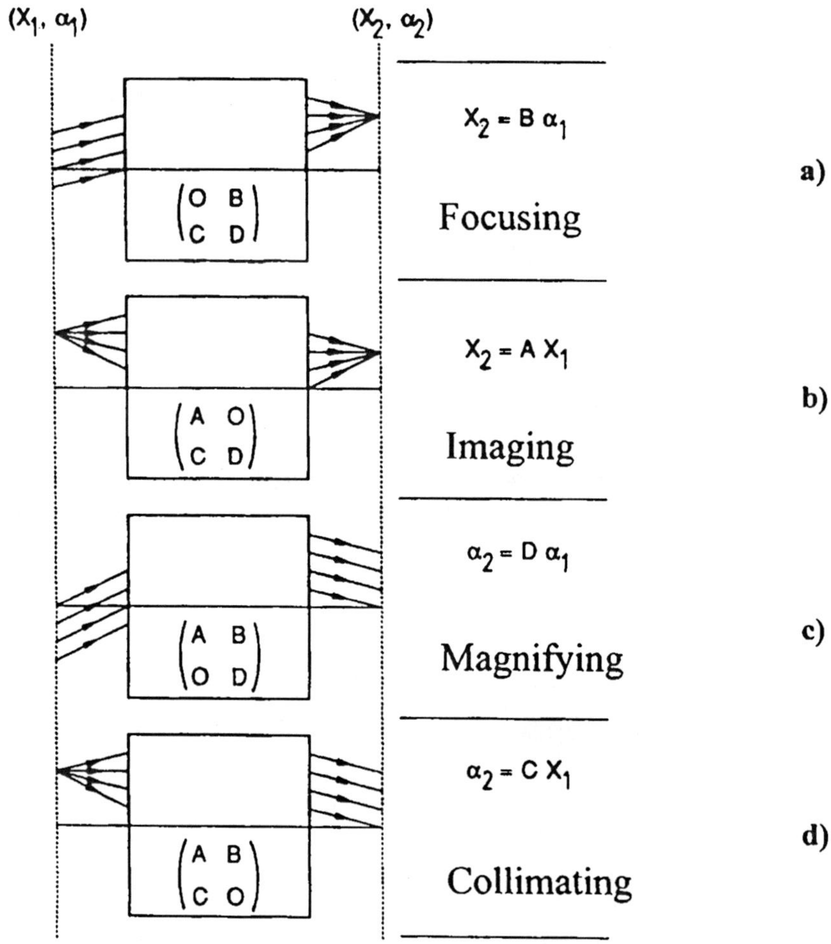

Fig. 1.9 Transformation of rays by optical systems having one vanishing element in their ray transfer matrix. The ray transfer matrix describes the ray propagation from left to right between the planes indicated by vertical lines.

Phase Space and Liouville's Theorem

So far we have discussed the propagation of single rays emerging from a light source through an optical system. In general the light source will emit over a spatial dimension Δx with an angular distribution at each point. An elegant way to visualize the effect of a beam propagation on the spatial and angular distribution of the rays is the presentation in phase space (Fig. 1.10). Each ray with starting point x and angle α can be represented by a point in phase space where the angular coordinate is plotted versus the spatial coordinate. As shown in Fig. 1.10 an extended light source of dimension Δx and angular width $\Delta\alpha$ is described by a rectangular area with height $\Delta\alpha$ and width Δx. Propagation through an optical system to a new plane now means that each point in this area will be imaged onto a new point in phase space. Therefore, the area will move as the light travels down the optical system.

It can be proven mathematically that the volume in a phase space cannot change as the physical system develops (Liouville's Theorem [1.7]). In our application Liouville's Theorem holds for light propagation in free space and through parabolic phase elements (e.g. aberration free lens or mirror). The area A in phase space must remain constant as long as the light propagates within the same medium. Let A_1 and A_2 be two areas in phase space with:

$$A_1 = \iint dx_1 d\alpha_1 \quad , \quad A_2 = \iint dx_2 d\alpha_2$$

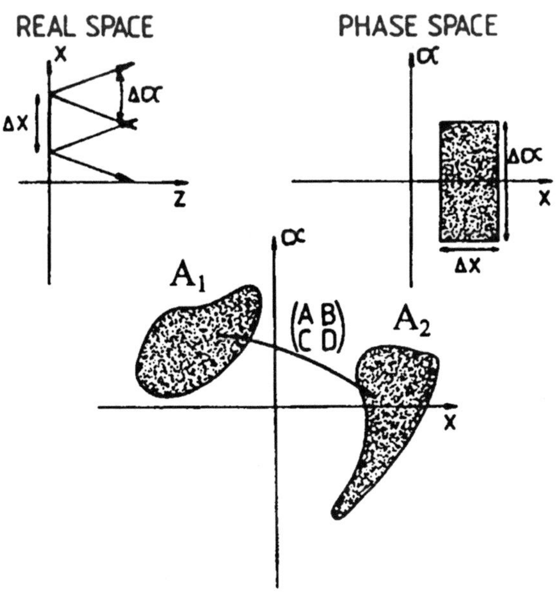

Fig. 1.10 Phase space presentation of a light source (upper graphs). The area moves in phase space as light propagates through optical systems.

The transformation of the differential is given by the determinant of the matrix $\boldsymbol{m}$:

$$\boldsymbol{m} = \begin{pmatrix} \dfrac{\delta x_2}{\delta x_1} & \dfrac{\delta x_2}{\delta \alpha_1} \\ \dfrac{\delta \alpha_2}{\delta x_1} & \dfrac{\delta \alpha_2}{\delta \alpha_1} \end{pmatrix} \tag{1.37}$$

By using the linear transformation (1.26) this results in:

$$dx_2 d\alpha_2 = (AD-BC)\; dx_1 d\alpha_1 \tag{1.38}$$

which means that the determinant of the ray transfer matrix must be equal to 1 to fulfill Liouville's Theorem $A_1 = A_2$ (the refractive index is constant).

An example is shown in Fig. 1.11. Each point of the extended light source emits rays within an angular width $\Delta\alpha_1$. In phase space this source is represented by a rectangular area that has the width Δx_1, and height $\Delta\alpha_1$. The angular width remains constant as the rays propagate in free space, but the spatial extent of the beam increases from Δx_1 to Δx_i at the lens. In contrast to the light source the angular width at the lens is different for each point. The points P_1 and P_2 emit with zero divergence, whereas points in the center emit within the full angular width. In phase space the intermediate field at z_i is thus represented by a parallelogram. If the field is imaged by the lens, all beams starting from point Q_1 of the source are collimated into Q_2. Each point of the image now emits within the angular width $\Delta\alpha_2$. The image is again represented by a rectangular area in phase space with the same area as the source:

$$\Delta x_2 \Delta\alpha_2 = \Delta x_1 \Delta\alpha_1 \tag{1.39}$$

Note that this equation holds only for imaging! In the intermediate region z_i, the area in phase space is the same as that of the source or the image, but it cannot be written as a product of a spatial and an angular width since the area is not rectangular. The more general form of (1.39) is known as Abbe's sine law:

$$\Delta x_1\; n_1\; \sin\Delta\alpha_1 = \Delta x_2\; n_2\; \sin\Delta\alpha_2 \tag{1.40}$$

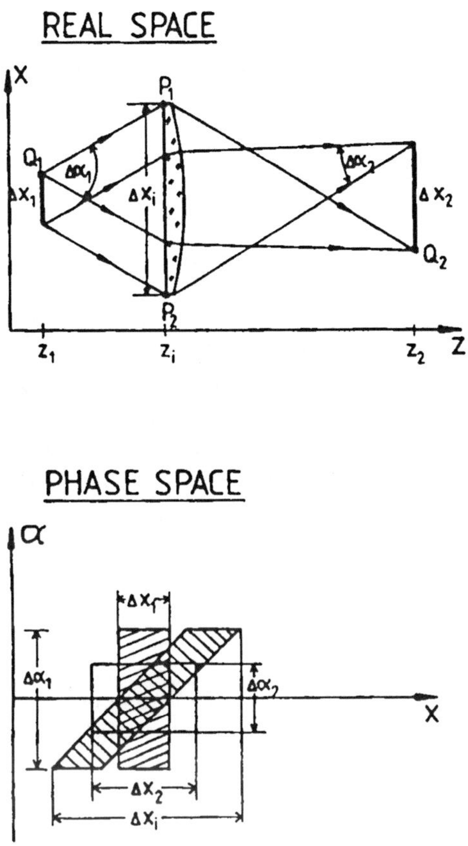

Fig. 1.11 Imaging by means of a focusing lens and the corresponding phase space presentation.

In the general case that light propagates from one plane (index n_1) to another (index n_2) and the ray propagation is described by the ray transfer matrix M, the following relation always holds:

$$\det M = AD - BC = \frac{n_1}{n_2} \qquad (1.41)$$

The condition on the determinant also arises since energy conservative optical systems are reversible. An optical system can only exist if (1.41) holds for its ray transfer matrix. Thus it is not possible to build an optical funnel which transforms a large diameter divergent beam into a collimated thin ray . The product of beam size and beam divergence is a constant in imaging and is called beam parameter product $\Delta x \Delta\alpha/4$ (product of the beam radius and the half angle of divergence). This product cannot be changed by any optical element without decreasing the power content of the beam (e.g. by using an aperture), if the light is completely incoherent. For coherent light, it is possible to convert any field distribution into a fundamental mode without loss of power by inserting suitable phase plates [1.18,1.20]. As will be discussed in detail in Sec.2.6.3, the beam parameter product is a direct measure for the beam quality, since a small focal spot size and a high depth of field can only be attained for small beam parameter products.

Example (Fig. 1.12):
A beam with diameter d and angle of divergence $\Delta\alpha$ is focused by means of a biconvex lens. Since the incoming rays are not all parallel to the optical axis they are not combined in one spot in the focal plane. We can easily calculate the resulting focus diameter by applying the phase space concept. In plane 1, the beam is characterized by a rectangle in phase space with area $d\Delta\alpha$. In the focal plane only rays with an angle α smaller than $d/2f$ can be detected (we assume that the beam diameter on the lens is also d). This means that the rectangular area in phase space has increased its height from $\Delta\alpha$ to d/f. According to (1.41) the area in phase space remains constant. This yields for the spot diameter s in the focal plane:

$$s\,\frac{d}{f} = d\,\Delta\alpha \quad\text{-----}\!\!\rightarrow\quad s = f\,\Delta\alpha \tag{1.42}$$

Let us also look at the depth of field z_0 in the focal plane. We define the depth of field as the distance from the focal plane at which the beam diameter has increased by a factor of 2:

$$s + z_0\frac{d}{f} = 2s \quad\text{-----}\!\!\rightarrow\quad z_0 = \frac{sf}{d} \tag{1.43}$$

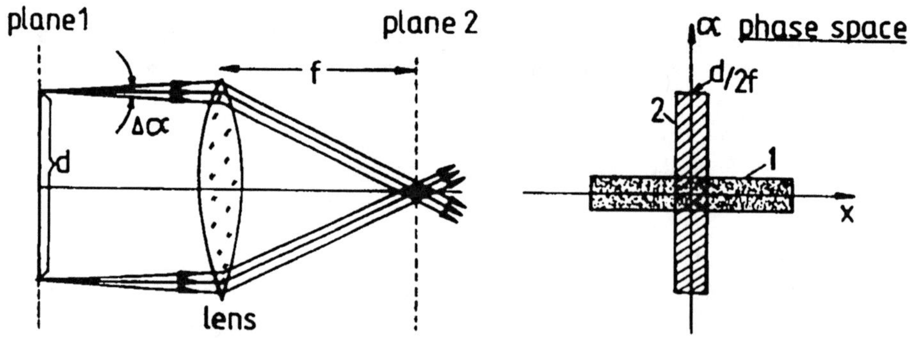

Fig. 1.12 Focusing of a beam with angular divergence Δα by a lens. The right hand graph shows the effect of the propagation from plane 1 to plane 2 in phase space.

The ratio of the focal spot area to the depth of field z_0 is a constant of the beam since it is proportional to the beam parameter product, or the phase space area:

$$\frac{\pi s^2}{4z_0} = \frac{\pi s d}{4f} = \pi\,\frac{d\,\Delta\alpha}{4} \qquad (1.44)$$

Since the beam parameter product $d\Delta\alpha/4$ is constant, the depth of field decreases quadratically with the spot size. This fact clearly indicates that the beam parameter product $\Delta x\Delta\alpha/4$ needs to be as small as possible to attain good focusability. However, as we will see in Chapter 2, there is a general lower bound for the beam parameter product BPP or the area A in phase space, determined by the wavelength λ:

$$BPP = \frac{\Delta x\Delta\alpha}{4} \geq \frac{\lambda}{\pi} \qquad (1.45)$$

In the green spectral range (λ=500nm), the minimum beam parameter product is 0.16mm mrad. The beam parameter products of lasers are close to this diffraction-limit. Conventional light sources, however, exhibit beam parameter products that are several orders of magnitude higher. Table 1.1 presents typical beam parameter products of different light sources.

Table 1.1 Beam parameter products of light sources.

light source	wavelength [nm]	beam parameter product [mm mrad]
200mW Nd:YVO4 laser (quadrupled)	266	0.09
20 W Nd:YVO4 laser (tripled)	355	0.13
20 W argon laser	488	0.16
100 mW laser diode	808	0.26
100 W cw Yb fiber laser	1,030	0.35
10 kW CO_2 laser	10,600	10
1 kW Yb:YAG disk laser	1,030	10
1 kW Nd:YAG rod laser	1,064	15
superbright LED (2 mW)	830-870	30
40 W fiber-coupled diode bar	808	50
conventional LED (2 mW)	630-670	200
flashlight (1 W)	400-800	1,000

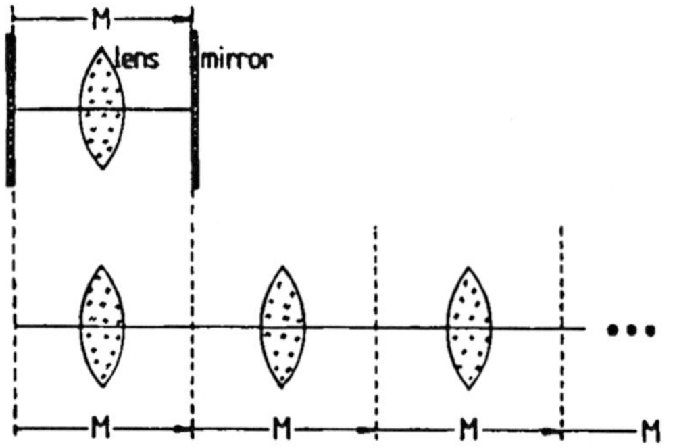

Fig. 1.13 Bouncing between two mirrors can be described as a straightforward propagation in a periodic lens guide. After propagating a distance containing N repetitions of the base unit cell described by the ray matrix M, the resulting matrix is M^N.

Sylvester's Theorem

A ray transfer matrix for periodic optical systems was derived by Tovar amd Casperson [1.17], based on Sylvester's formula [1.12]. This theorem is especially useful for the treatment of optical resonators since the back and forth bouncing of a ray can be equivalently described as a straightforward propagation in a periodic lens guide (Fig. 1.13). For the symmetric resonator shown, the transit from one mirror to the other is the basic optical element, described by the ray matrix M. In order to determine the ray parameters after N subsequent travels, the N-th power of the fundamental ray matrix M has to be calculated. If we again take the general form of the ray transfer matrix M and restrict ourselves to matrices with determinant $AD-BC=1$, the N-th power of M reads:

$$M^N = \frac{1}{\sin\Phi} \begin{pmatrix} A\sin[N\Phi] - \sin[(N-1)\Phi] & B\sin[N\Phi] \\ C\sin[N\Phi] & D\sin[N\Phi] - \sin[(N-1)\Phi] \end{pmatrix} \tag{1.46}$$

with: $\cos\Phi = \dfrac{A+D}{2}$ and $M = \begin{pmatrix} A & B \\ C & D \end{pmatrix}$

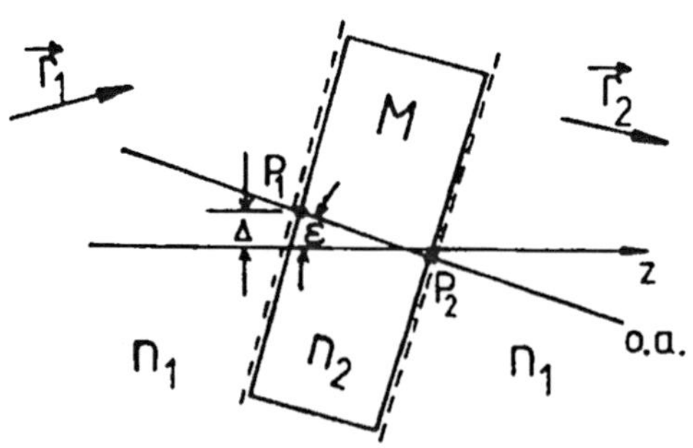

Fig. 1.14 An optical element with the optical axis, o.a., tilted by an angle ϵ about the front nodal vertex point P_1 and shifted by Δ perpendicularly to the z-direction ($\Delta>0$, $\epsilon<0$).

1.2.3 Misaligned Optical Elements

The matrices discussed so far hold for aligned optical elements. This means that the optical axis ***a*** of each element coincides with the reference axis z. The ray **a**, which represents the optical axis and intersects the surfaces of the optical element at the two vertex points P_1,P_2. This special ray is not affected by the element, it is neither shifted nor deflected. For a singlet, the spherical interface (see Fig. 1.5), the optical axis is degenerated. Each ray crossing the center of curvature, is optical axis; for the spherical lens the connecting line of the two centers of curvature is the optical axis. For a bifocal spherical interface the optical axis is defined by the ray, which is perpendicular to both center-lines of curvature. For a sequence of misaligned optical elements the axis depends on the misalignment parameters and will be calculated in this section. The optical axis of a resonator is defined by the ray, which is reproduced after one roundtrip.

Perfect alignment is impossible to achieve in reality, especially for a sequence of elements. In this section the influence of a shift and a tilt on the ray propagation will be briefly discussed. We assume an element which is tilted by an angle ϵ about the front vertex point P_1 and shifted by Δ, as shown in Fig. 1.14. The paraxial approximation holds with $|\epsilon| << 1$. For larger angles, the situation becomes more complicated because trigonometric functions are involved.

The misalignment of a single element is of no interest since the reference axis z can always be chosen along the optical axis of the element, and the ray transfer matrix of the aligned system still applies. However, for a sequence of optical elements with different alignments, the following relations become useful [1.13]. A ray $\boldsymbol{r}_1$ is incident on the optical system with:

$$\boldsymbol{r}_1 = \begin{pmatrix} x_1 \\ \alpha_1 \end{pmatrix} \tag{1.47}$$

This vector transforms in the misaligned system to:

$$\boldsymbol{r}'_1 = \boldsymbol{r}_1 - \boldsymbol{s} \tag{1.48}$$

with $\boldsymbol{s}$ being the misalignment vector:

$$\boldsymbol{s} = \begin{pmatrix} \Delta \\ \epsilon \end{pmatrix} \tag{1.49}$$

The ray transfer matrix of the optical element transforms $\boldsymbol{r}'_1$ into $\boldsymbol{r}'_2$:

$$\boldsymbol{r}'_2 = \boldsymbol{M}\, \boldsymbol{r}'_1 \tag{1.50}$$

This vector can be transformed back into the z-system by taking into account the propagation of vector s over the distance L between the reference planes:

$$r_2 = r'_2 + M_{FS}\, s = M r_1 - (M - M_{FS})\, s \tag{1.51}$$

If the resulting misalignment vector is introduced:

$$s_M = (M - M_{FS})\, s \tag{1.52}$$

the ray vector behind a misaligned optical element reads:

$$r_2 = M\, r_1 - s_M \tag{1.53}$$

The misalignment vector can be calculated for any element by using the matrices shown in Fig. 1.5:

<u>Thick lens with focal length f, thickness L, and principal plane distances h_i:</u>

$$M_{FS} = \begin{pmatrix} 1 & L \\ 0 & 1 \end{pmatrix} \quad , \quad s_M = \begin{pmatrix} \Delta\dfrac{h_2}{f} + \epsilon L(\dfrac{n_1}{n_2} - 1) \\ \dfrac{-\Delta}{f} - \epsilon\dfrac{h_1}{f} \end{pmatrix}$$

<u>Plane slab of length L:</u>

$$M_{FS} = \begin{pmatrix} 1 & L \\ 0 & 1 \end{pmatrix} \quad , \quad s_M = \begin{pmatrix} \epsilon L(\dfrac{n_1}{n_2} - 1) \\ 0 \end{pmatrix}$$

Example 1:
Let us consider a symmetric optical resonator with an internal lens as shown in Fig. 1.15. If all elements are aligned, the optical axis of the resonator, defined by the line going through both centers of curvature P_1, P_2 of the mirrors, coincides with the optical axis of the lens which is defined by the two focal points. Which ray represents the optical axis when the lens is shifted by Δ?

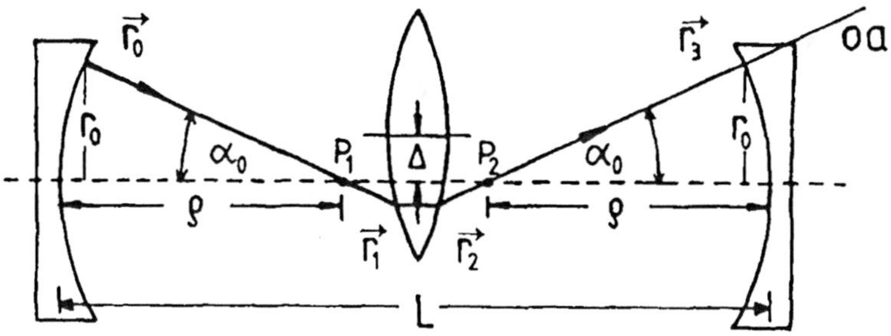

Fig. 1.15 A symmetric resonator with a shifted lens.

The optical axis is defined by the ray which travels back and forth along the same path, which means that the optical axis must be perpendicular to both mirror surfaces. The ray representing the optical axis on mirror 1 thus reads:

$$\boldsymbol{r_0} = \begin{pmatrix} r_0 \\ -\dfrac{r_0}{\rho} \end{pmatrix}$$

The propagation of this ray to mirror 2 is given by:

$$\boldsymbol{r_1} = \boldsymbol{M_{FS}}(L/2)\ \boldsymbol{r_0}$$

$$\boldsymbol{r_2} = \boldsymbol{M_{TL}}\ \boldsymbol{r_1} - \boldsymbol{s_{TL}}$$

$$\boldsymbol{r_3} = \boldsymbol{M_{FS}}(L/2)\ \boldsymbol{r_2}$$

where $\boldsymbol{M_{TL}}$ is the ray transfer matrix for the thick lens. The combination of these three equations together with the fact that $\boldsymbol{r_3}$ must be equal to $\boldsymbol{r_0}$ (with a different sign of the angle α_0) yields the displacement and the tilt of the optical axis:

$$r_0 = \frac{\Delta}{1 - \dfrac{L}{2\rho} + \dfrac{2f}{\rho}} \quad , \qquad \alpha_0 = \frac{\Delta}{\rho - \dfrac{L}{2} + 2f}$$

For plane mirrors with $\rho \to \infty$, r_0 is equal to Δ and α_0 equals zero, as it should be. The above relation indicates that the sensitivity of a resonator to lens shifting depends on the particular configuration.

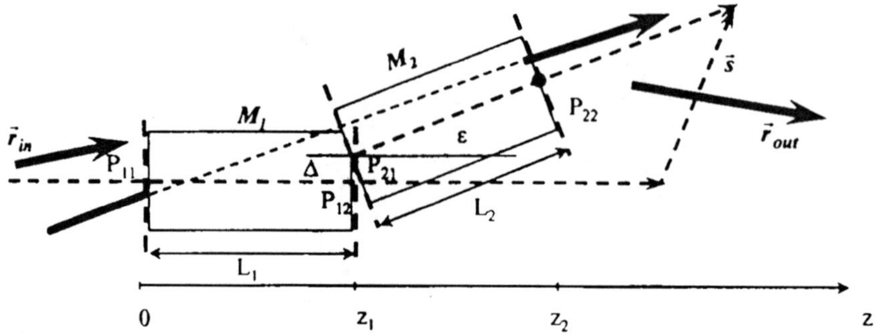

Fig. 1.16 The optical axis ā of a misaligned optical system. P_{ij} are the vertex points of the optical elements. The arrows indicate the direction of the vectors, not their moduli. $\bar{a}_i$ are the axes of the single elements.

The Optical Axis of Misaligned Systems

Two optical elements M_1, M_2 with geometrical lengths L_1, L_2 and optical axes a_1, a_2 are shifted and tilted against each other as shown in Fig. 1.16. The misalignment again is characterized by the vector s of (1.49), which now reads:

$$s = a_2 - a_1 \tag{1.54}$$

The element M_1 is aligned with respect to the z-axis, which determines the vector a_1 to be:

$$a_1 = \begin{pmatrix} 0 \\ 0 \end{pmatrix}$$

and $a_2 = s$. In the a_1-reference system the output ray is related to the input ray according to (1.53):

$$r_{out} = M_2 M_1 \, r_\epsilon - (M_2 - M_{FS2}) \, s \tag{1.55}$$

If r_{in} is the ray, which represents the optical axis a of the complete system, the following relations must hold:

$$r_{in} = a_{z=0} \qquad\qquad r_{out} = a_{z=z2} = M_{FS2} M_{FS1} \, a_{z=0}$$

and (1.55) delivers an equation to determine the optical axis $a_{z=0}$ of the misaligned system at the position z=0:

$$(M_2M_1 - M_{FS2}M_{FS1})\, a_{z=0} = (MM_2 - M_{FS2})\, s \tag{1.56}$$

If the ray vectors r_{in}, r_{out} are transformed into the a-system, according to

$$r'_{in} = r_{in} - a_{z=0} \qquad\qquad r'_{out} = r_{out} - a_{z=z2} = r_{out} - M_{FS2}M_{FS1}\, a_{z=0}$$

the simple matrix law can again be applied for the ray-vectors:

$$r'_{out} = M_2M_1\, r'_{in} \tag{1.57}$$

which can be proved easily from (1.55) and (1.56). This procedure can be continued for more elements. Any sequence of misaligned optical elements has a resulting optical axis. If the ray vectors are transformed into this reference system, the normal matrix laws hold.

Example 2:
As an example the resulting optical axis of a misaligned three-lens system as shown in Fig. 1.17 will be calculated. The matrix M_1 of the first optical element includes the two lenses on the left and the two distances L. The focal lengths of all three lenses are the same and equal to f. Then the two matrices read:

$$M_1 = \begin{pmatrix} (1-L/f)^2 - L/f & L(2-L/f) \\ (-2/f + L/f^2) & 1-L/f \end{pmatrix} \qquad\qquad M_2 = \begin{pmatrix} 1 & 0 \\ -1/f & 1 \end{pmatrix}$$

The free-space matrices and the misalignment vector are

$$M_{FS1} = \begin{pmatrix} 1 & 2L \\ 0 & 1 \end{pmatrix} \qquad M_{FS2} = \begin{pmatrix} 1 & 0 \\ 0 & 1 \end{pmatrix} \qquad s = \begin{pmatrix} \Delta \\ 0 \end{pmatrix}$$

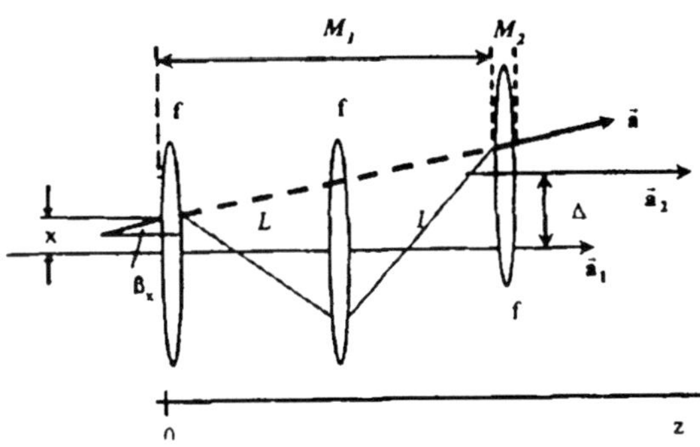

Fig. 1.17 The optical axis ā of a misaligned three-lens system.

With (1.56) the vector $\boldsymbol{a}$ of the optical axis $z = 0$ is obtained:

$$\boldsymbol{a} = \begin{pmatrix} x \\ \beta_x \end{pmatrix} = \begin{pmatrix} \dfrac{\Delta/2}{L/f-3} \\ \Delta/(2L) \end{pmatrix}$$

Misalignment Matrices

The matrix of an optical element can be defined in such a way that the misalignment is included. Then 4x4-matrices are required, which is not very convenient. It becomes more difficult for two-dimensional rays, where 8x8 matrices now appear [1.19].

1.2.4 Two-Dimensional Optical Systems

In previous sections, we have restricted the calculation of ray propagation to one dimension, the x-z plane. This is sufficient as long as the radiation field and the optical elements exhibit rotational symmetry. In general, laser beams can be astigmatic (e.g. laser diodes) and parabolical optical elements can have different radii of curvature in x and y-directions as shown in Fig. 1.18. In this case, it is necessary to track all ray parameters. An arbitrary ray, starting at a x-y plane, is described by a four-dimensional ray vector (Fig. 1.18):

$$\boldsymbol{v}_1 = \begin{pmatrix} x_1 \\ y_1 \\ \alpha_1 \\ \beta_1 \end{pmatrix} = \begin{pmatrix} \boldsymbol{r}_1 \\ \boldsymbol{\gamma}_1 \end{pmatrix} \tag{1.58}$$

with $\boldsymbol{r}_1$, $\boldsymbol{\gamma}_1$ being two-dimensional vectors. Thus, we need a 4x4 matrix to describe the propagation from plane 1 to plane 2:

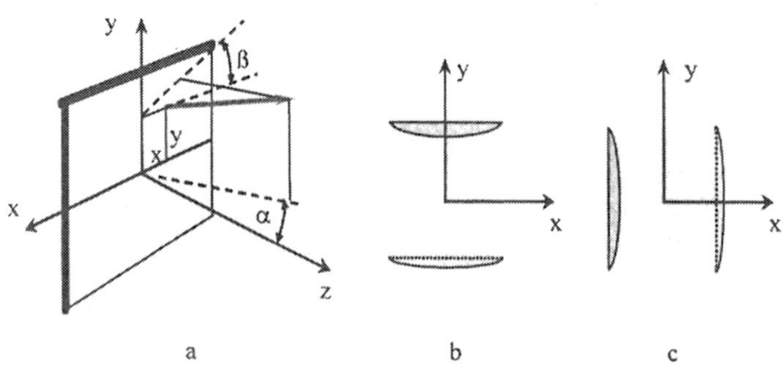

Fig. 1.18 A ray is generally described by four parameters (a). All four parameters have to be tracked when the ray passes through two-dimensional optics such as cylinder lenses (b,c).

$$\begin{pmatrix} x_2 \\ y_2 \\ \alpha_2 \\ \beta_2 \end{pmatrix} = \begin{pmatrix} A_{xx} & A_{xy} & B_{xx} & B_{xy} \\ A_{yx} & A_{yy} & B_{yx} & B_{yy} \\ C_{xx} & C_{xy} & D_{xx} & D_{xy} \\ C_{yx} & C_{yy} & D_{yx} & D_{yy} \end{pmatrix} \begin{pmatrix} x_1 \\ y_1 \\ \alpha_1 \\ \beta_1 \end{pmatrix} \tag{1.59}$$

By defining the sub-matrices $\boldsymbol{A},\boldsymbol{B},\boldsymbol{C},\boldsymbol{D}$ this can be rewritten in an already familiar form:

$$\begin{pmatrix} \boldsymbol{r}_2 \\ \boldsymbol{\gamma}_2 \end{pmatrix} = \begin{pmatrix} \boldsymbol{A} & \boldsymbol{B} \\ \boldsymbol{C} & \boldsymbol{D} \end{pmatrix} \begin{pmatrix} \boldsymbol{r}_1 \\ \boldsymbol{\gamma}_1 \end{pmatrix} \tag{1.60}$$

The multiplication of matrices, whose elements are are again matrices is carried out the same way as for normal matrices. Relation (1.60) is a very convenient way to express ray transformations in two-dimensional optics since all relations that hold in one dimension can be generalized by using submatrices as matrix elements. An optical system is now described by a 4x4 matrix with 16 elements, but there are constraints to this matrix as was shown by Nemes [1.15,1.16]. The first is the determinant relation (1.41), which now reads:

$$\boldsymbol{A}\boldsymbol{D}^T - \boldsymbol{B}\boldsymbol{C}^T = \frac{n_1}{n_2}\boldsymbol{I} \tag{1.61}$$

with $\boldsymbol{I}$ being the identity matrix and $\boldsymbol{D}^T$ is the transposed matrix of $\boldsymbol{D}$ with:

$$\boldsymbol{D} = \begin{pmatrix} D_{xx} & D_{xy} \\ D_{yx} & D_{yy} \end{pmatrix} \qquad \boldsymbol{D}^T = \begin{pmatrix} D_{xx} & D_{yx} \\ D_{xy} & D_{yy} \end{pmatrix}$$

The two other constraints are:

$$\boldsymbol{A}\boldsymbol{B}^T = \boldsymbol{B}\boldsymbol{A}^T \tag{1.62}$$

$$\boldsymbol{C}\boldsymbol{D}^T = \boldsymbol{D}\boldsymbol{C}^T \tag{1.63}$$

These three conditions reduce the number of independent elements to ten. This means that in the most general case of parabolic optical elements ten parameters are required for a complete description. The number of elements is further reduced if the optical system exhibits symmetry with respect to the coordinate system. For rotational symmetry, only four independent elements exist in general. If the refractive indices before and after the element are the same, only three independent elements remain.

For optical elements having mirror symmetry with respect to the x-axis or y-axis, the 4x4 ray transfer matrix can be easily derived by treating the x- and y-directions separately. In this case the sub-matrices $\boldsymbol{A},\boldsymbol{B},\boldsymbol{C},\boldsymbol{D}$ are diagonal and we thus have only six independent elements (for equal refractive indices). In the following we list the matrices for common optics:

a) Propagation in isotropic, homogeneous medium over geometrical distance L

$$M_{FS} = \begin{pmatrix} 1 & 0 & L & 0 \\ 0 & 1 & 0 & L \\ 0 & 0 & 1 & 0 \\ 0 & 0 & 0 & 1 \end{pmatrix} \tag{1.64}$$

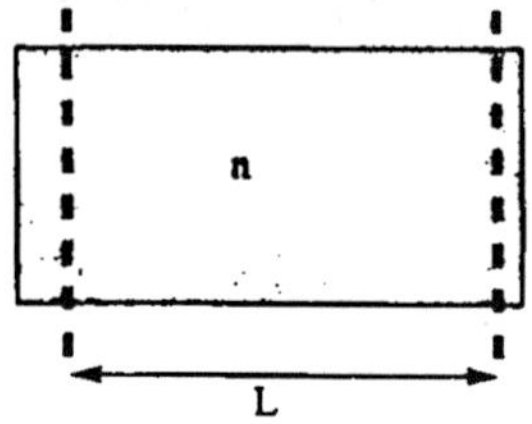

b) Aligned bifocal spherical interface (see Eq. 1.16)

$$M_{BSI} = \begin{pmatrix} 1 & 0 & 0 & 0 \\ 0 & 1 & 0 & 0 \\ \dfrac{n_1-n_2}{n_2\rho_x} & 0 & \dfrac{n_1}{n_2} & 0 \\ 0 & \dfrac{n_1-n_2}{n_2\rho_y} & 0 & \dfrac{n_1}{n_2} \end{pmatrix} \tag{1.65}$$

c) Aligned bifocal thin lens

It has the refractive powers D_x in x-direction and D_y in y-direction. Special cases are the cylinder lenses with D_x or D_y equal to zero.

$$M_{BTL} = \begin{pmatrix} 1 & 0 & 0 & 0 \\ 0 & 1 & 0 & 0 \\ -D_x & 0 & 1 & 0 \\ 0 & -D_y & 0 & 1 \end{pmatrix} \tag{1.66}$$

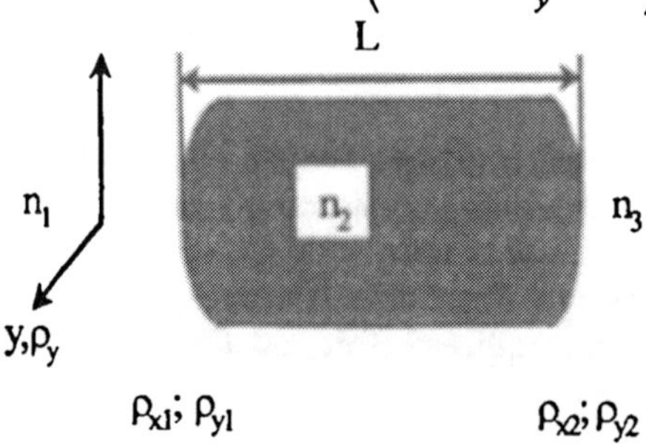

Fig.1.19 Aligned bifocal thick lens.

d) Aligned bifocal thick lens (Fig. 1.19)

The principal axes of both bifocal transitions are parallel to the x,y-axes. In the two-dimensional case, no principal planes exit. The system is characterized by seven independent parameters: n_1/n_2, n_2/n_3, ρ_{x1}, ρ_{y1}, ρ_{x2}, ρ_{y2}, and L. with $a=(n_1/n_2-1)$ and $b=(n_2/n_3-1)$.

$$M_{BFL} = \begin{pmatrix} 1+\frac{aL}{\rho_{x1}} & 0 & \frac{n_1}{n_2}L & 0 \\ 0 & 1+\frac{aL}{\rho_{y1}} & 0 & \frac{n_1}{n_2}L \\ \frac{b}{\rho_{x2}}+\frac{a}{\rho_{x1}}\left(\frac{n_2}{n_3}+\frac{bL}{\rho_{x2}}\right) & 0 & \frac{n_1}{n_3}+\frac{bL}{\rho_{x2}}\frac{n_1}{n_2} & 0 \\ 0 & \frac{b}{\rho_{y2}}+\frac{a}{\rho_{y1}}\left(\frac{n_2}{n_3}+\frac{bL}{\rho_{y2}}\right) & 0 & \frac{n_1}{n_3}+\frac{bL}{\rho_{y2}}\frac{n_1}{n_2} \end{pmatrix} \tag{1.67}$$

e) Free space propagation with different lengths in x and y

This special element consists of four aligned cylinder lenses to generate different propagation distances $L_x=L$ and $L_y=L+8f_x$ in x- and y-direction, respectively.

$$M_{FSA} = \begin{pmatrix} 1 & 0 & L_x & 0 \\ 0 & 1 & 0 & L_y \\ 0 & 0 & 1 & 0 \\ 0 & 0 & 0 & 1 \end{pmatrix} \tag{1.68}$$

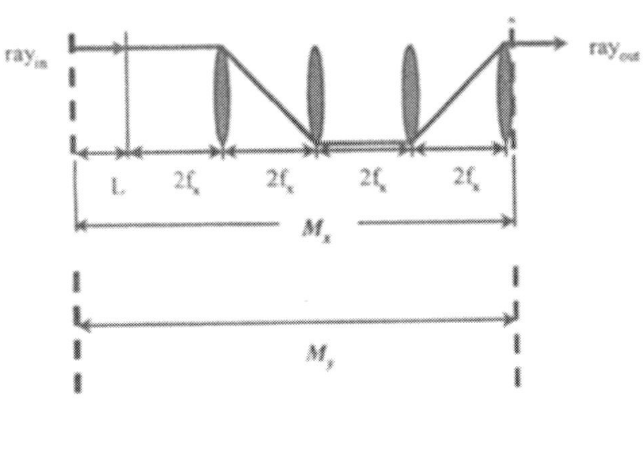

Fig. 1.20 Cylindrical telescope system with different propagation distances in x- and y-direction.

1.2.5 Rotation and Misalignment

If the optical element is rotated by an angle θ as shown for the cylinder lens in Fig. 1.21, the original ray transfer matrix $\boldsymbol{M}$ has to be transformed into an angle dependent transfer matrix $\boldsymbol{M}(\theta)$ to describe the ray propagation through the element. In the following, the transformation rule for ray matrices under rotation is derived.

We know that the optical element can still be described by the initial ray matrix $\boldsymbol{M}$ if we use the rotated coordinate axes as the reference coordinate system (indicated by asterisks). Ray propagation can then be written as:

$$\begin{pmatrix} x_2^* \\ y_2^* \\ \alpha_2^* \\ \beta_2^* \end{pmatrix} = \boldsymbol{M} \begin{pmatrix} x_1^* \\ y_1^* \\ \alpha_1^* \\ \beta_1^* \end{pmatrix} \tag{1.69}$$

The initial coordinate axes are related to the rotated ones (for ccw rotation) via:

$$\begin{pmatrix} x^* \\ y^* \end{pmatrix} = \begin{pmatrix} \cos\theta & \sin\theta \\ -\sin\theta & \cos\theta \end{pmatrix} \begin{pmatrix} x \\ y \end{pmatrix} \tag{1.70}$$

This matrix equation also holds for the angles α,β since they are defined through x,y-coordinates. Thus the transformation rule for the ray vector reads:

$$\begin{pmatrix} x_1^* \\ y_1^* \\ \alpha_1^* \\ \beta_1^* \end{pmatrix} = \begin{pmatrix} \cos\theta & \sin\theta & 0 & 0 \\ -\sin\theta & \cos\theta & 0 & 0 \\ 0 & 0 & \cos\theta & \sin\theta \\ 0 & 0 & -\sin\theta & \cos\theta \end{pmatrix} \begin{pmatrix} x_1 \\ y_1 \\ \alpha_1 \\ \beta_1 \end{pmatrix} := \boldsymbol{R} \begin{pmatrix} x_1 \\ y_1 \\ \alpha_1 \\ \beta_1 \end{pmatrix} \tag{1.71}$$

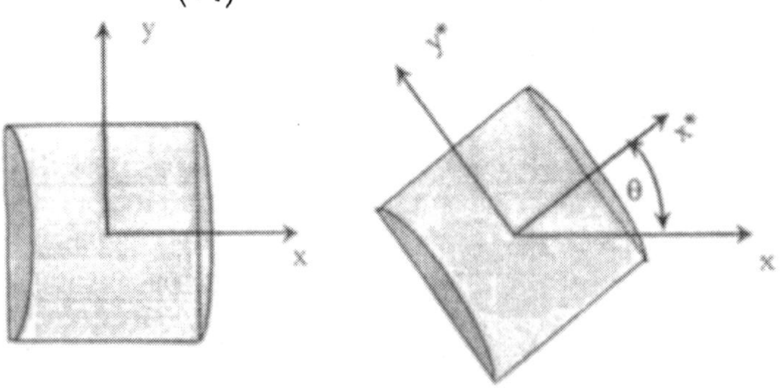

Fig. 1.21 Aligned cylinder lens (left) and rotated cylinder lens (right).

$$\begin{pmatrix} x_1 \\ y_1 \\ \alpha_1 \\ \beta_1 \end{pmatrix} = \begin{pmatrix} \cos\theta & -\sin\theta & 0 & 0 \\ \sin\theta & \cos\theta & 0 & 0 \\ 0 & 0 & \cos\theta & -\sin\theta \\ 0 & 0 & \sin\theta & \cos\theta \end{pmatrix} \begin{pmatrix} x_1^{\bullet} \\ y_1^{\bullet} \\ \alpha_1^{\bullet} \\ \beta_1^{\bullet} \end{pmatrix} := \boldsymbol{R}^{-1} \begin{pmatrix} x_1^{\bullet} \\ y_1^{\bullet} \\ \alpha_1^{\bullet} \\ \beta_1^{\bullet} \end{pmatrix} \tag{1.72}$$

By multiplying (1.69) with $\boldsymbol{R}^{-1}$ from the left and using the relation $\boldsymbol{RR}^{-1}=\boldsymbol{I}$ we obtain:

$$\begin{pmatrix} x_2 \\ y_2 \\ \alpha_2 \\ \beta_2 \end{pmatrix} = \boldsymbol{R}^{-1}\boldsymbol{MR} \begin{pmatrix} x_1 \\ y_1 \\ \alpha_1 \\ \beta_1 \end{pmatrix} \tag{1.73}$$

This is already what we are looking for! The ray transfer matrix $\boldsymbol{M(\theta)}$ for optics rotated in ccw direction thus reads:

$$\boldsymbol{M}(\theta) = \boldsymbol{R}^{-1}\boldsymbol{MR} = \begin{pmatrix} \cos\theta & -\sin\theta & 0 & 0 \\ \sin\theta & \cos\theta & 0 & 0 \\ 0 & 0 & \cos\theta & -\sin\theta \\ 0 & 0 & \sin\theta & \cos\theta \end{pmatrix} \boldsymbol{M} \begin{pmatrix} \cos\theta & \sin\theta & 0 & 0 \\ -\sin\theta & \cos\theta & 0 & 0 \\ 0 & 0 & \cos\theta & \sin\theta \\ 0 & 0 & -\sin\theta & \cos\theta \end{pmatrix} \tag{1.74}$$

Example : Stokes Lens Pair (Fig. 1.22)
The Stokes Lens Pair consists of a negative and a positive cylinder lens with equal refractive power D [1.8]. The lenses can be rotated by an angle θ in opposite directions around the z-axis. We will show in the following that this element is equivalent to a combination of a spherical lens and a cylindrical lens under 45°, both with varying refractive power. Let us first consider the negative lens which is rotated counterclockwise. By using (1.66) and (1.74) we obtain:

$$\boldsymbol{M}_{-CYL} = \begin{pmatrix} \cos\theta & -\sin\theta & 0 & 0 \\ \sin\theta & \cos\theta & 0 & 0 \\ 0 & 0 & \cos\theta & -\sin\theta \\ 0 & 0 & \sin\theta & \cos\theta \end{pmatrix} \begin{pmatrix} 1 & 0 & 0 & 0 \\ 0 & 1 & 0 & 0 \\ D & 0 & 1 & 0 \\ 0 & 0 & 0 & 1 \end{pmatrix} \begin{pmatrix} \cos\theta & \sin\theta & 0 & 0 \\ -\sin\theta & \cos\theta & 0 & 0 \\ 0 & 0 & \cos\theta & \sin\theta \\ 0 & 0 & -\sin\theta & \cos\theta \end{pmatrix}$$

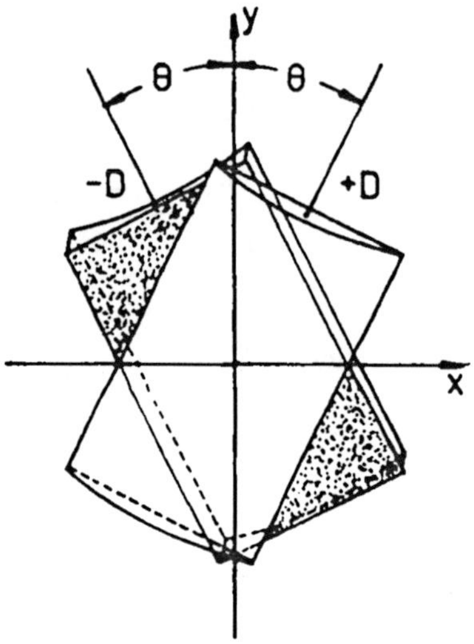

Fig. 1.22 A Stokes lens pair as an example of a rotated optics. This lens pair is used in ophthalmic devices (phoropters, refractors) to measure cylinder errors (astigmatism) of human eyes.

$$= \begin{pmatrix} 1 & 0 & 0 & 0 \\ 0 & 1 & 0 & 0 \\ D\cos^2\theta & D\sin\theta\cos\theta & 1 & 0 \\ D\sin\theta\cos\theta & D\sin^2\theta & 0 & 1 \end{pmatrix} \qquad D > 0 \qquad (1.75)$$

The matrix M_{+CYL} for the positive cylinder lens can be found by replacing D with $-D$ and θ with $-\theta$ in M_{-CYL}. The resulting matrix for the combined lens pair is given by:

$$M_{STOKES} = M_{-CYL}\, M_{+CYL} = \begin{pmatrix} 1 & 0 & 0 & 0 \\ 0 & 1 & 0 & 0 \\ 0 & D\sin2\theta & 1 & 0 \\ D\sin2\theta & 0 & 0 & 1 \end{pmatrix} \qquad D>0 \qquad (1.76)$$

Now we compare the Stokes Lens Pair with a system consisting of a negative cylinder lens rotated by 45° with refractive power $-D^*$ $(D^*>0)$ combined with a positive spherical lens with refractive power $D^*/2$. Matrix (1.71) can be used to determine the ray transfer matrix of the cylinder lens $M_{-CYL}(45°)$. The ray transfer matrix of the equivalent system is:

$$M_{EQU.} = \begin{pmatrix} 1 & 0 & 0 & 0 \\ 0 & 1 & 0 & 0 \\ -\frac{D^*}{2} & 0 & 1 & 0 \\ 0 & -\frac{D^*}{2} & 0 & 1 \end{pmatrix} \begin{pmatrix} 1 & 0 & 0 & 0 \\ 0 & 1 & 0 & 0 \\ \frac{D^*}{2} & \frac{D^*}{2} & 1 & 0 \\ \frac{D^*}{2} & \frac{D^*}{2} & 0 & 1 \end{pmatrix} = \begin{pmatrix} 1 & 0 & 0 & 0 \\ 0 & 1 & 0 & 0 \\ 0 & \frac{D^*}{2} & 1 & 0 \\ \frac{D^*}{2} & 0 & 0 & 1 \end{pmatrix} \qquad (1.77)$$

This is equivalent to (1.76) if D^* equals $2D\sin(2\theta)$. A Stokes lens pair can thus be made equivalent to 45°-cylinder lenses with refractive powers D^* between $-2D$ and $+2D$. By adding a second lens pair rotated at 45° with respect to the first one, both pairs together can be made equivalent to any angle cylinder lens with any power between $\pm 2D$.

Example: The Phase Space Beam Analyzer

The Phase Space Beam Analyzer is a fascinating optical system which enables one to record the phase space presentation of a one-dimensional light source (Fig. 1.23) [1.11,1.14,1.21]. A small slit in plane 1 generates a line source so that only rays with $y=0$ can enter the system. The second slit, which is preferably located close to the spherical lens ($c=0$), selects only rays with zero inclination in y-direction. This means that the whole setup is only sensitive to rays having ray vectors $v=(x, \alpha, 0, 0)$. The basic principle of the device is that on plane 4 the y-coordinate depends only on the angle α whereas in the x-direction we see the image of the first slit. Imaging of the slit is accomplished by the spherical lens and the quadrupole lens (a Stokes lens pair with $\theta=-45°$) translates the angle α into a shift in the y-direction.

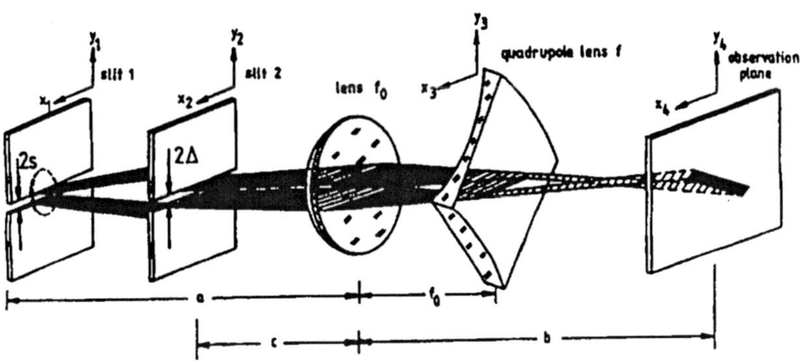

Fig. 1.23 The Phase Space Beam Analyzer.

By multiplying all ray transfer matrices between plane 1 and 4, the resulting matrix reads:

$$M_{PSBA} = \begin{pmatrix} 1 & 0 & b-f_0 & 0 \\ 0 & 1 & 0 & b-f_0 \\ 0 & 0 & 1 & 0 \\ 0 & 0 & 0 & 1 \end{pmatrix} \begin{pmatrix} 1 & 0 & 0 & 0 \\ 0 & 1 & 0 & 0 \\ 0 & -1/f & 1 & 0 \\ -1/f & 0 & 0 & 1 \end{pmatrix} \begin{pmatrix} 1 & 0 & f_0 & 0 \\ 0 & 1 & 0 & f_0 \\ 0 & 0 & 1 & 0 \\ 0 & 0 & 0 & 1 \end{pmatrix} \begin{pmatrix} 1 & 0 & 0 & 0 \\ 0 & 1 & 0 & 0 \\ -1/f_0 & 0 & 1 & 0 \\ 0 & -1/f_0 & 0 & 1 \end{pmatrix} \begin{pmatrix} 1 & 0 & a & 0 \\ 0 & 1 & 0 & a \\ 0 & 0 & 1 & 0 \\ 0 & 0 & 0 & 1 \end{pmatrix}$$

$$= - \begin{pmatrix} P & 0 & 0 & S \\ 0 & P & S & 0 \\ 1/f_0 & 0 & 1/P & f_0/f \\ 0 & 1/f_0 & f_0/f & 1/P \end{pmatrix} \qquad \text{with} \quad P = \frac{b-f_0}{f_0} \quad ; \quad S = \frac{f_0}{f}(b-f_0) \tag{1.78}$$

Since only rays with $\mathbf{v}=(x, 0, \alpha, 0)$ can reach the observation plane, (1.78) yields for the ray coordinates in plane 4:

$$x_4 = -P\,x_1 \quad , \quad y_4 = -S\,\alpha_1 \tag{1.79}$$

In plane 4 we observe the phase space presentation of the horizontal line source in plane 1, scaled by the factor P and S. In order to visualize the phase space presentation for different parts of the light source, the phase space analyzer has to be rotated around the z-axis.

Meridional Rays and Skew Rays

If the ray is propagating in a plane that contains the z-axis, it is called a meridional ray. By rotating the reference frame, the four-dimensional meridional ray can be transformed into a two-dimensional ray. With $y_2=0$ and $\beta_2=0$, the transformation (1.71) then yields:

$$\begin{pmatrix} x_2 \\ 0 \end{pmatrix} = \begin{pmatrix} \cos\theta & \sin\theta \\ -\sin\theta & \cos\theta \end{pmatrix} \begin{pmatrix} x_1 \\ y_1 \end{pmatrix} \qquad \begin{pmatrix} \alpha_2 \\ 0 \end{pmatrix} = \begin{pmatrix} \cos\theta & \sin\theta \\ -\sin\theta & \cos\theta \end{pmatrix} \begin{pmatrix} \alpha_1 \\ \beta_1 \end{pmatrix}$$

This can be accomplished by chosing the angle of rotation θ as:

$$\tan\theta = \frac{y_1}{x_1} = \frac{\alpha_1}{\beta_1}$$

If this condition holds the ray is a meridional ray, if not (i.e. $y_1/x_1 \neq \alpha_1/\beta_1$) it is a skew ray as shown in Fig.1.24.

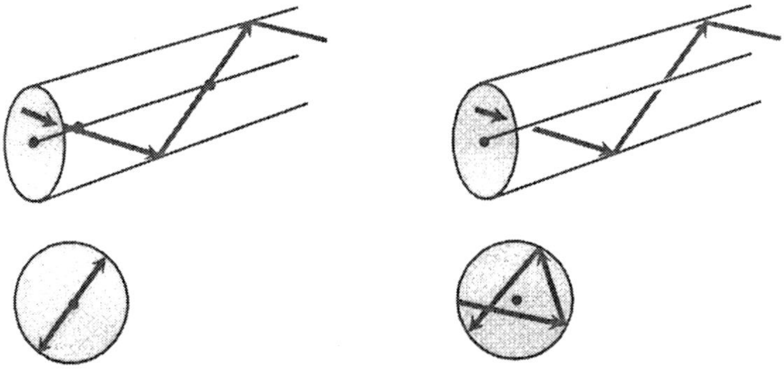

Fig.1.24 Meridional (left) and skew rays (right) in a step-index fiber. The meridional ray propagates in a plane that contains the symmetry axis of the fiber. The skew ray is propagating around the axis.

Misalignment of Two-Dimensional Optical Elements

Misalignment of two-dimensional elements can be described in the same way as for one-dimensional elements in Sec. 1.2.3. Equations (1.50)-(1.53) also hold for four-dimensional vectors and matrices. The misalignment vector now reads:

$$s = \begin{pmatrix} \Delta_x \\ \Delta_y \\ \epsilon_x \\ \epsilon_y \end{pmatrix}$$

The tilt introduced in Sec. 1.2.3 was assumed to be very small. For large tilt angles the trigonometric functions have to be considered. We define a tilt as a rotation by an angle θ around the x-axis as shown for the lens in Fig. 1.25. The projection of the lens onto the x-y plane now is an ellipse with the smaller axis along the y-axis. The plane defined by the small ellipse axis and the z-axis is called the *tangential plane*, the plane defined by the z-axis and the large ellipse axis is the *sagittal plane*. In the following we restrict the discussion to optical elements that exhibit mirror symmetry in x- and y-direction. We are thus dealing with the ray transfer matrix:

$$\begin{pmatrix} A_{xx} & 0 & B_{xx} & 0 \\ 0 & A_{yy} & 0 & B_{yy} \\ C_{xx} & 0 & D_{xx} & 0 \\ 0 & C_{yy} & 0 & D_{yy} \end{pmatrix}$$

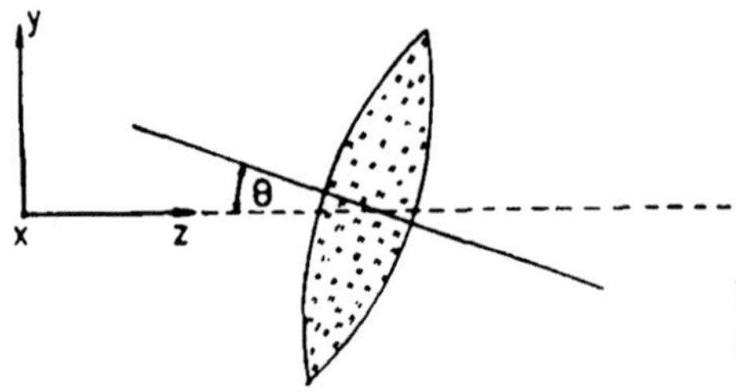

Fig. 1.25 A tilted focusing lens (rotation around x-axis) as seen in the tangential plane.

Since all submatrices are diagonal we can treat ray propagation separately for the two planes. The ray matrices of the most common tilted optics can be derived by using the matrix of a tilted spherical dielectric interface (Fig. 1.26). If θ_1 denotes the angle of incidence of the optical axis with respect to the surface normal (this is also the tilt angle) and θ_2 is the corresponding angle of refraction, the ray transfer matrix for the tangential plane (y-z plane) reads:

$$M^{T}_{SI,TILT} = \begin{pmatrix} A_{yy} & B_{yy} \\ C_{yy} & D_{yy} \end{pmatrix} = \begin{pmatrix} \dfrac{\cos\theta_2}{\cos\theta_1} & 0 \\ \dfrac{n_2\cos\theta_2 - n_1\cos\theta_1}{n_2\rho\,\cos\theta_1\cos\theta_2} & \dfrac{n_1\cos\theta_1}{n_2\cos\theta_2} \end{pmatrix} \tag{1.80}$$

and for the sagittal plane (x-z plane) we get:

$$M^{S}_{SI,TILT} = \begin{pmatrix} A_{xx} & B_{xx} \\ C_{xx} & D_{xx} \end{pmatrix} = \begin{pmatrix} 1 & 0 \\ \dfrac{n_1\cos\theta_1 - n_2\cos\theta_2}{n_2\rho} & \dfrac{n_1}{n_2} \end{pmatrix} \tag{1.81}$$

with $$\frac{\sin\theta_2}{\sin\theta_1} = \frac{n_1}{n_2}$$

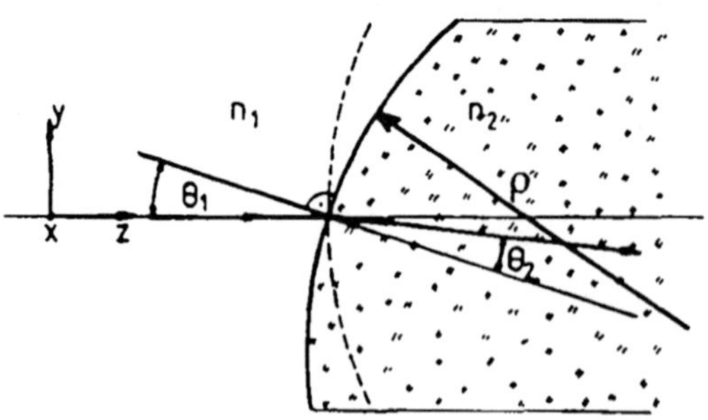

Fig. 1.26 The spherical dielectric interface, tilted by an angle θ_1 around the x-axis.

The 4x4 ray transfer matrix for the spherical interface tilted around the x-axis is given by:

$$M_{SI,TILT} = \begin{pmatrix} 1 & 0 & 0 & 0 \\ 0 & \dfrac{\cos\theta_2}{\cos\theta_1} & 0 & 0 \\ \dfrac{n_1\cos\theta_1 - n_2\cos\theta_2}{n_2\,\rho} & 0 & \dfrac{n_1}{n_2} & 0 \\ 0 & \dfrac{n_2\cos\theta_2 - n_1\cos\theta_1}{n_2\rho\,\cos\theta_1\cos\theta_2} & 0 & \dfrac{n_1\cos\theta_1}{n_2\cos\theta_2} \end{pmatrix} \tag{1.82}$$

The knowledge of this ray matrix enables us to derive the 4x4 ray transfer matrix for two important tilted optical elements used in laser resonators: the tilted thin lens, and the tilted slab (Fig. 1.27).

Tilted Thin Lens

The lens with index of refraction n_2 is surrounded by a medium with index n_1 and tilted by an angle θ_1 around the x-axis as shown in Fig. 1.27. In order to simplify the discussion we assume that both interfaces have the same radii of curvature $|\rho_1|,|\rho_2|$. The ray transfer matrix in the y-direction (tangential plane) is obtained by multiplying two matrices (1.80) for a tilted spherical interface:

$$M^T_{TL,TILT} = \begin{pmatrix} \dfrac{\cos\theta_4}{\cos\theta_3} & 0 \\ \dfrac{n_1\cos\theta_4 - n_2\cos\theta_3}{n_1\rho\,\cos\theta_3\cos\theta_4} & \dfrac{n_2\cos\theta_3}{n_1\cos\theta_4} \end{pmatrix} \begin{pmatrix} \dfrac{\cos\theta_2}{\cos\theta_1} & 0 \\ \dfrac{n_2\cos\theta_2 - n_1\cos\theta_1}{n_2\rho\,\cos\theta_1\cos\theta_2} & \dfrac{n_1\cos\theta_1}{n_2\cos\theta_2} \end{pmatrix} \tag{1.83}$$

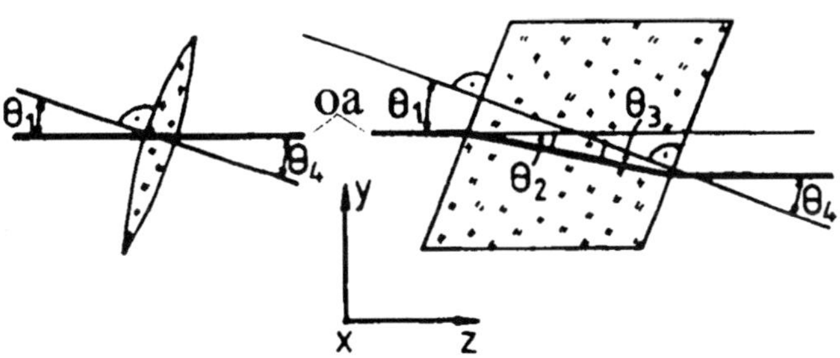

Fig. 1.27 Tilted thin lens and tilted slab (oa: optical axis).

with θ_1, θ_3 being the angles of incidence and θ_2, θ_4 the angles of refraction of the optical axis. Since both curvatures are equal and the lens has no thickness, the relations $\theta_1 = \theta_4$ and $\theta_2 = \theta_3$ hold. Insertion into (1.83) yields:

$$M^T_{TL,TILT} = \begin{pmatrix} 1 & 0 \\ \dfrac{1}{\cos\theta_1}\left[\dfrac{n_2\cos\theta_2}{n_1\cos\theta_1} - 1\right]\left[\dfrac{1}{\rho_1} - \dfrac{1}{\rho_2}\right] & 1 \end{pmatrix} \tag{1.84}$$

The ray transfer matrix for the sagittal plane (x-z plane) can be found in a similar way by using the product of two ray transfer matrices (1.81):

$$M^S_{TL,TILT} = \begin{pmatrix} 1 & 0 \\ \cos\theta_1\left[\dfrac{n_2\cos\theta_2}{n_1\cos\theta_1} - 1\right]\left[\dfrac{1}{\rho_1} - \dfrac{1}{\rho_2}\right] & 1 \end{pmatrix} \tag{1.85}$$

Tilted Slab

We can use the ray matrices for the spherical interface with infinite radius of curvature and combine two of those with the ray transfer matrix for free space propagation. In the tangential and the sagittal plane we get:

$$M^T_{SLAB,TILT} = \begin{pmatrix} 1 & \dfrac{Ln_1}{n_2}\left[\dfrac{\cos\theta_1}{\cos\theta_2}\right]^2 \\ 0 & 1 \end{pmatrix} \tag{1.86}$$

$$M^S_{SLAB,TILT} = \begin{pmatrix} 1 & \dfrac{Ln_1}{n_2} \\ 0 & 1 \end{pmatrix} \tag{1.87}$$

In the sagittal plane, we obtain the ray transfer matrix of the aligned slab, whereas the tilt decreases the effective length of the slab in the tangential plane. For a slab inserted at the Brewster angle, the 4x4 ray transfer matrix is given by:

$$M_{Brewster\ Slab} = \begin{pmatrix} 1 & 0 & Ln_1/n_2 & 0 \\ 0 & 1 & 0 & L(n_1/n_2)^3 \\ 0 & 0 & 1 & 0 \\ 0 & 0 & 0 & 1 \end{pmatrix} \tag{1.88}$$

1.2.6 The ABCD Law for the Radius of Curvature

The knowledge about ray matrices we have acquired enables us to send rays into optical systems and calculate the coordinates and the angles at which they will emerge. In some applications, however, it is more interesting to find a relation among radii of curvature of spherical waves. In geometrical optics a spherical wave is defined by rays having a virtual source at the center of curvature (Fig.1.28). After passage through the optics the wavefront will still be spherical, but with a changed radius of curvature R. This occurs since all optical elements are assumed to have parabolic surfaces or index profiles. The relationship between the two curvatures is called the *ABCD law*. For the derivation we first consider the case of one-dimensional optics and express the radius of curvature R_1 of the incident wave in terms of the ray parameters of the surface normals (for small angles α):

$$R_1 = x_1/\alpha_1 \tag{1.89}$$

Accordingly, at the other side of the optical system with ray transfer matrix $\boldsymbol{M}$ the radius of curvature is $R_2=x_2/\alpha_2$. We find the relationship between the two curvatures by dividing the two equations for the ray parameters:

$$\frac{x_2}{\alpha_2} = \frac{Ax_1 + B\,\alpha_1}{Cx_1 + D\,\alpha_1}$$

which yields the ABCD law:

$$R_2 = \frac{A\,R_1 + B}{C\,R_1 + D} \tag{1.90}$$

Note that the radius of curvature is positive for a divergent wave (center of curvature is to the left of the wavefront) and negative for a convergent wave.

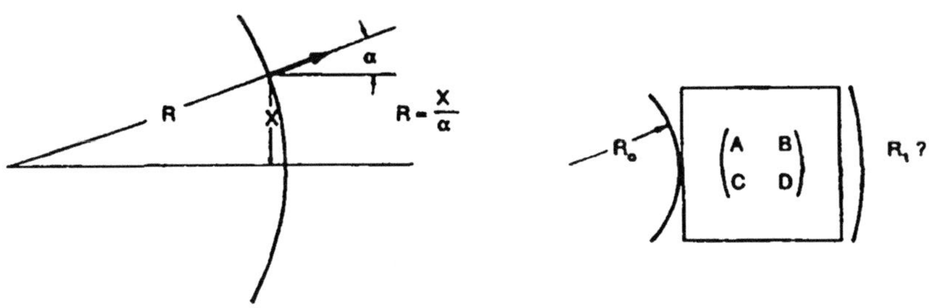

Fig. 1.28 Definition of a spherical wavefront. The radius of curvature is changed after passage through the optical system.

Example: Using the ray matrix for a thin lens (1.18), the ABCD law (1.90) yields

$$\frac{1}{R_1} - \frac{1}{R_2} = \frac{1}{f}$$

This is equivalent to the imaging condition since the object distance is R_1 and the image distance is $-R_2$.

Similar to the determinant relation (1.41), the ABCD law can be extended to the case of arbitrary two-dimensional optical systems. The two-dimensional ABCD law can be written as (the superscript *-1* denotes the inverse matrix):

$$\boldsymbol{R}_2 = (\boldsymbol{A}\ \boldsymbol{R}_1 + \boldsymbol{B})(\boldsymbol{C}\ \boldsymbol{R}_1 + \boldsymbol{D})^{-1} \tag{1.91}$$

with $\boldsymbol{A},\boldsymbol{B},\boldsymbol{C},\boldsymbol{D}$ being the 2x2 submatrices according to (1.60), and the curvature matrix is given by:

$$\boldsymbol{R} = \begin{pmatrix} R_{xx} & R_{xy} \\ R_{yx} & R_{yy} \end{pmatrix} \tag{1.92}$$

If the wavefront exhibits symmetry along the x- and y-direction, the curvature matrix is diagonal and R_{xx}, R_{yy} denote the radii of curvature along the x-axis and y-axis respectively. If (1.92) is not diagonal, the coordinate frame has to be rotated around the z-axis to find the new x,y-frame in which the wavefront becomes symmetric. This procedure is equivalent to finding the eigenvalues and eigenvectors of the curvature matrix (1.92). The eigenvalues give the radii of curvature R^*_{xx} and R^*_{yy} along the x*- and y*-axis of the new symmetric coordinate system with:

$$R^*_{xx} = \frac{R_{xx} + R_{yy}}{2} + \sqrt{\left(\frac{R_{xx} - R_{yy}}{2}\right)^2 + R_{xy}R_{yx}} \tag{1.93}$$

$$R^*_{yy} = \frac{R_{xx} + R_{yy}}{2} - \sqrt{\left(\frac{R_{xx} - R_{yy}}{2}\right)^2 + R_{xy}R_{yx}} \tag{1.94}$$

and the eigenvectors $\boldsymbol{e}^*_x$, $\boldsymbol{e}^*_y$ represent the new coordinate axes expressed in the initial coordinate frame:

$$e_x^* = \begin{pmatrix} 1 \\ \dfrac{R_{xx}^* - R_{xx}}{R_{xy}} \end{pmatrix} \qquad e_y^* = \begin{pmatrix} 1 \\ \dfrac{R_{yy}^* - R_{xx}}{R_{xy}} \end{pmatrix} \tag{1.95, 1.96}$$

Example: We want to calculate the change in spherical wave curvature that is induced by a cylindrical lens rotated clockwise by an angle of 45° with respect to the y-axis. Without rotation the lens focuses in the x-direction with a refractive power D $(D>0)$ (see the positive lens in Fig. 1.21). We assume that the incident spherical wave has the same curvature with respect to x- and y-axis and its curvature matrix is thus given by:

$$R_1 = \begin{pmatrix} R & 0 \\ 0 & R \end{pmatrix}$$

The submatrices for the rotated cylinder lens are:

$$A = D = \begin{pmatrix} 1 & 0 \\ 0 & 1 \end{pmatrix} \quad B = \begin{pmatrix} 0 & 0 \\ 0 & 0 \end{pmatrix} \quad C = \begin{pmatrix} -D\cos^2\theta & \dfrac{D}{2}\sin(2\theta) \\ \dfrac{D}{2}\sin(2\theta) & -D\sin^2\theta \end{pmatrix}$$

Insertion into the ABCD law (1.91) yields for the new curvature matrix (with θ=45°):

$$R_2 = \begin{pmatrix} R & 0 \\ 0 & R \end{pmatrix} \begin{pmatrix} -RD/2 + 1 & RD/2 \\ RD/2 & -RD/2 + 1 \end{pmatrix}^{-1} .$$

$$= \frac{R}{1-RD} \begin{pmatrix} 1-RD/2 & -RD/2 \\ -RD/2 & 1-RD/2 \end{pmatrix} = \begin{pmatrix} R_{xx} & R_{xy} \\ R_{yx} & R_{yy} \end{pmatrix}$$

With (1.93) - (1.96) we get:

$$R_{xx}^* = \frac{R}{1-RD} , \; e_x^* = \begin{pmatrix} 1 \\ -1 \end{pmatrix} \qquad R_{yy}^* = R , \; e_y^* = \begin{pmatrix} 1 \\ 1 \end{pmatrix}$$

As expected the new coordinate frame is rotated by -45° with respect to the old one and we obtain the imaging condition $1/R - 1/R_{xx}^* = D$ in the x*-direction.

1.2.7 Eigensolutions and Eigenvalues

As mentioned in the introductory remarks on optical resonators we are mainly interested in light intensity distributions on the resonator mirrors that reproduce themselves after each round trip in the resonator. In terms of ray propagation we are looking for spherical wavefronts whose radii of curvature are not affected by the resonator round trip.

If $\boldsymbol{M}$ denotes the general ray transfer matrix of an optical system, we can always find self-reproducing spherical waves with radius of curvature R by applying the ABCD law (1.90) with $R_2=R_1=R$:

$$R = \frac{A\,R + B}{C\,R + D} \tag{1.97}$$

There are generally two solutions to this equation and these two self-reproducing wave curvatures R_a,R_b are linked to the two eigenvectors $\boldsymbol{v}_a, \boldsymbol{v}_b$ and eigenvalues μ_a,μ_b of the ray transfer matrix $\boldsymbol{M}$ defined by:

$$\mu_{a,b}\boldsymbol{v}_{a,b} = \boldsymbol{M}\ \boldsymbol{v}_{a,b} \tag{1.98}$$

The effect of the optical system on rays defined by eigenvectors is a multiplication of both the coordinate x and the angle α by a factor of μ. The absolute value of the eigenvalue is therefore called the magnification of the optical system. All rays defining a self-reproducing spherical wavefront according to (1.97) are automatically eigenvectors of the ray transfer matrix, since:

$$R_2 = \frac{x_2}{\alpha_2} = \frac{\mu x_1}{\mu\alpha_1} = \frac{x_1}{\alpha_1} = R_1 \tag{1.99}$$

For every 2x2 ray transfer matrix $\boldsymbol{M}$ *with* $C\neq 0$ the eigenvalues and corresponding reproducing radii of curvature are given by:

$$\mu_a = \frac{A+D}{2} + \sqrt{\left(\frac{A+D}{2}\right)^2 - 1} \qquad R_a = \frac{A-D + \sqrt{(A+D)^2 - 4}}{2C} \qquad \text{(1.100)}\ \text{(1.101)}$$

$$\mu_b = \frac{A+D}{2} - \sqrt{\left(\frac{A+D}{2}\right)^2 - 1} \qquad R_b = \frac{A-D - \sqrt{(A+D)^2 - 4}}{2C} \qquad \text{(1.102)}\ \text{(1.103)}$$

An incident beam with diameter d and radius of curvature $R_{a,b}$ will emerge from the optical system without a change in curvature but with a diameter $|\mu_{a,b}|d$.

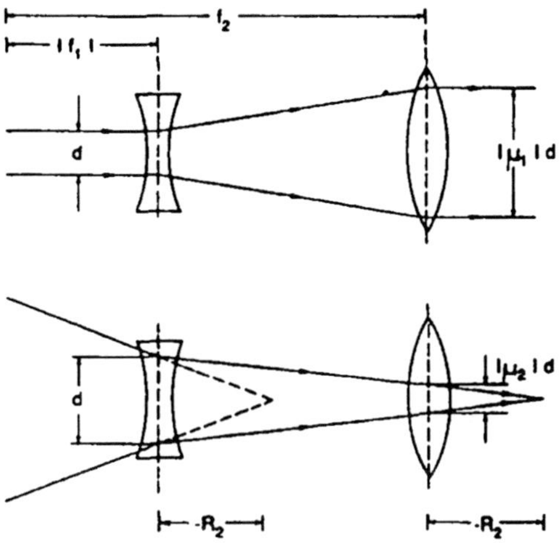

Fig. 1.29 Self-reproducing spherical waves in a Galilean Telescope.

Example: Galilean Telescope

The Galilean Telescope comprises a focusing lens with focal length f_2 and a negative lens with focal length $-f_1$ ($f_1>0$) which are spaced such that they have a common focal point (Fig. 1.29). The ray transfer matrix for the whole system reads:

$$M_{GT} = \begin{pmatrix} \frac{f_2}{f_1} & f_2-f_1 \\ 0 & \frac{f_1}{f_2} \end{pmatrix}$$

Equations (1.100-1.103) yield for the eigensolutions:

$$\mu_a = \frac{f_2}{f_1} \qquad R_a = \infty$$

$$\mu_b = \frac{f_1}{f_2} \qquad R_b = -\frac{(f_1 f_2)^2}{f_2-f_1} < 0$$

The eigensolution given by the first row is probably familiar to all readers: a plane wave will reemerge with a plane wavefront, but the beam diameter has increased by f_2/f_1. There is, however, a second spherical wavefront that reproduces its radius of curvature. This eigensolution of the telescope is often referred to as the convergent wave and its beam diameter has decreased by f_1/f_2.

1.3 Optical Resonators and Ray Transfer Matrices

An optical resonator usually consists of two spherical mirrors with radius of curvatures ρ_1, ρ_2 separated by a distance L (Fig.1.30). It is customary to replace each mirror by two lenses with focal length $f_i = \rho_i$ and locate the reference planes in between. This technique generally simplifies the ray transfer matrix *(A=D)* and allows for an easier memorizing of the matrices. The reference planes are the mirror surfaces, which means a plane wave in this representation is actually a spherical wave having the curvature of the mirror in real space. The ray transfer matrix for the round trip inside an arbitrary optical resonator starting at mirror 1 is given by:

$$M_{RES} = \begin{pmatrix} 2g_1g_2-1 & 2Lg_2 \\ \dfrac{(2g_1g_2-1)^2-1}{2Lg_2} & 2g_1g_2-1 \end{pmatrix} \tag{1.104}$$

with $g_i = 1 - \dfrac{L}{\rho_i}$; $i=1,2$

The parameters g_1 and g_2 are called the g-parameters of the optical resonator. Note that the radius of curvature is positive for a concave mirror (focusing mirror) and negative for a convex mirror! According to (1.104), the imaging properties of two-mirror resonators are fully defined by the g-parameters and the mirror spacing L. We can further simplify the ray transfer matrix M_{RES} by introducing the *equivalent G-Parameter* $G=2g_1g_2-1$:

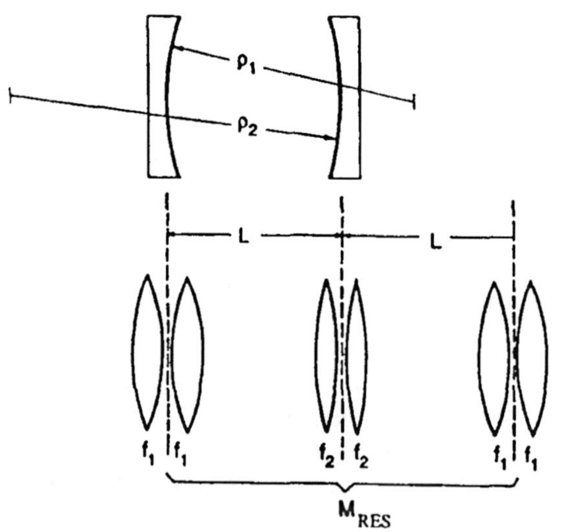

Fig. 1.30 A round trip in an optical resonator can be described as a transit in an equivalent lens waveguide. Each mirror has been replaced by a lens pair with focal length $f_i = \rho_i$.

$$M_{RES} = \begin{pmatrix} G & 2Lg_2 \\ \dfrac{G^2-1}{2Lg_2} & G \end{pmatrix} \tag{1.105}$$

We can now use this matrix to find eigensolutions of the resonator. We are looking for spherical wavefronts starting on mirror 1 which reproduce themselves after one round trip. Application of (1.100)-(1.103) yields for the eigenvalues and the corresponding self-reproducing wave curvatures:

$$\mu_a = G + \sqrt{G^2 - 1} \qquad R_a = \frac{+2Lg_2}{\sqrt{G^2 - 1}} \qquad (1.106)\ (1.107)$$

$$\mu_b = G - \sqrt{G^2 - 1} \qquad R_b = \frac{-2Lg_2}{\sqrt{G^2 - 1}} \qquad (1.108)\ (1.109)$$

Note that R_a and R_b refer to the dotted planes in Fig. 1.30. According to these eigensolutions we can distinguish between three different types of optical resonators:

1) |G|>1, equivalent to $|g_1g_2|>1$
We can find two spherical waves with real values for the radii of curvature $R_{a,b}$ reproducing themselves inside the resonator. If the beam diameter on mirror 1 is d, the spherical wave with radius of curvature R_a increases its diameter by $|\mu_a|$ every round trip (Fig. 1.31a). This eigensolution is called the divergent wave. The second eigensolution, the convergent wave leads to a decrease of the beam diameter by $|\mu_b|$ per round trip. Resonators having these ray propagation properties are called **unstable resonators**.

2) |G|=1, equivalent to $|g_1g_2|=1$
Both radii of curvature R_a and R_b are infinite and both eigenvalues are equal to 1. This means that a plane wavefront is coming back planar after the round trip without change in diameter. These resonators are referred to as the **resonators on the stability boundaries**. The plane-plane resonator in Fig. 1.31b is one example of such a resonator.

3) |G|<1, equivalent to $|g_1g_2|<1$
The radii of curvature and the eigenvalues are all complex numbers. This is a very puzzling result since we do not know what to make of complex curvatures. We can only interpret it in such a way that in geometrical optics no eigensolutions can be found in this type of resonator. We shall see in Section 2 that eigensolutions exist, but they can only be calculated by applying diffraction theory. These resonators are called **stable resonators**.

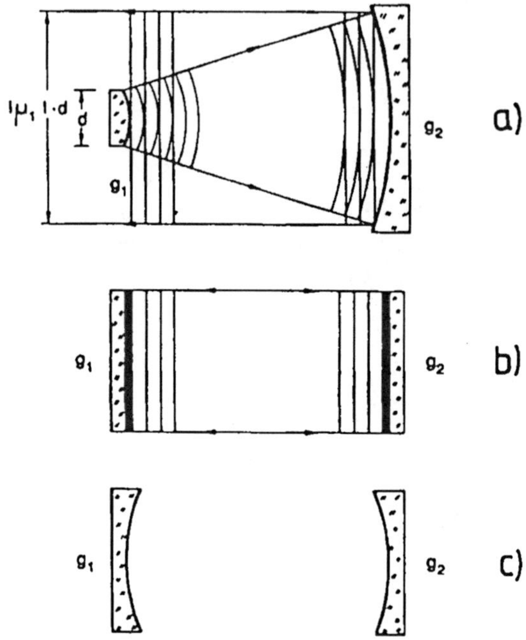

Fig. 1.31 The three types of optical resonators with spherical mirrors. a) unstable resonator, b) plane-plane resonator as an example of a resonator on the stability boundaries, c) stable resonator having no eigensolutions in geometrical optics.

There is a different and more commonly used approach to define the three resonator types. It is the tracking of rays launched into the resonator parallel to the optical axis as a function of the number of round trips. Using Sylvester's Theorem (1.46) the ray coordinate after N round trips reads:

$$x_{N+1} = (G \sin[N\Phi] - \sin[(N-1)\Phi])\, x_1 \tag{1.110}$$

with $\cos\Phi = G$

The ray stays confined within the resonator if $|G| \leq 1$ holds. For $|G| > 1$, $\cos\Phi$ has to be replaced by the hyperbolic function, which means that x_N increases exponentially with the number of round trips. The ray leaves the system. This makes it more understandable why resonators with $|G| < 1$ are called stable.

Optical resonators with two mirrors can be visualized in a diagram where the g-parameters represent the coordinate axes (Fig. 1.32). This diagram is referred to as the g-diagram or, more often, as the stability diagram. A resonator defined by the g-parameters g_1 and g_2 is represented by a point in the stability diagram. Unfortunately, this representation is not unambiguous because the mirror spacing is not included.

A special class of resonators are the confocal resonators with both mirrors having a common focal point. The confocal condition reads:

$$g_1 + g_2 = 2g_1g_2 \tag{1.111}$$

Examples:

ρ_1=1m, ρ_2=1m,	L=1m	⇒	g_1=0.0, g_2=0.0	confocal
ρ_1=2m, ρ_2=∞,	L=0.5m	⇒	g_1=0.5, g_2=1.0	stable
ρ_1=-0.5m, ρ_2=1.5m,	L=0.5m	⇒	g_1=2.0, g_2=0.66	unstable, confocal

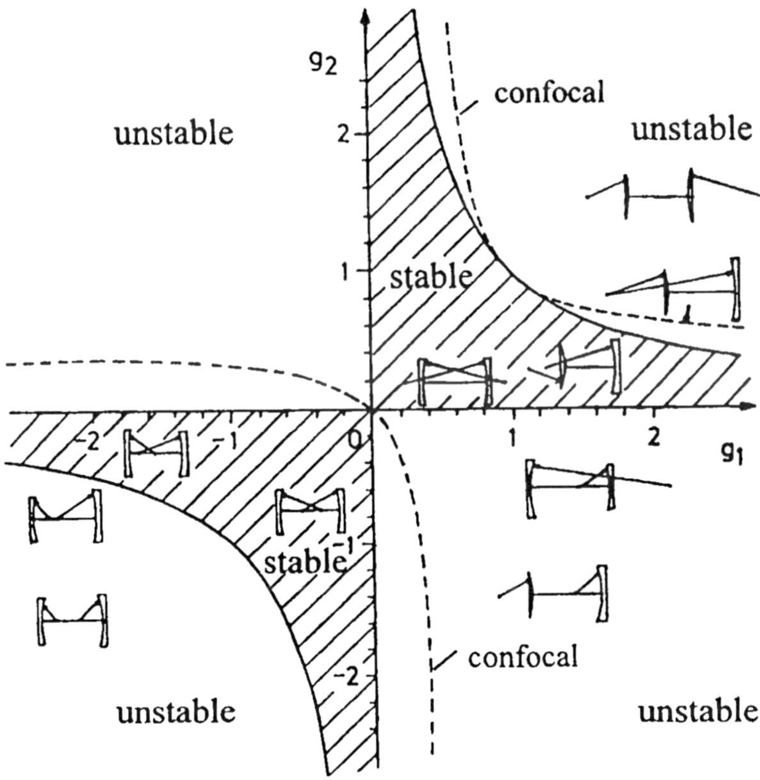

Fig. 1.32 The stability diagram of optical resonators with two spherical mirrors.

Chapter 2

Wave Optics

2.1 Huygen's Principle and Diffraction Integral

The geometrical treatment of the propagation of light presented in the preceding chapter is only an approximate description. It does not take the spatial amplitude distribution of the electromagnetic wave into account. Geometrical optics can be applied as long as the wavelength is small compared to the lateral extent of the wave. This restriction, as already discussed, is equivalent to a large Fresnel number N. The exact description of the propagation of light is obtained by utilizing Maxwell's equations to derive the wave equations for the electric and the magnetic fields. If we neglect the vector properties of the field, the wave equation for the electric field E in homogeneous, isotropic, loss-free, dielectric media reads [1.1,1.3,1.63]:

$$\frac{\partial^2 E}{\partial x^2} + \frac{\partial^2 E}{\partial y^2} + \frac{\partial^2 E}{\partial z^2} - \frac{1}{c^2}\frac{\partial^2 E}{\partial t^2} = 0 \tag{2.1}$$

with c being the speed of light in the medium. This equations holds for each of the three components of the field. In general an infinite number of solutions to this wave equation exists since all fields E for which the relations:

$$E(x,y,z,t) = E(\omega t - k_x x - k_y y - k_z z) \tag{2.2}$$

$$|\boldsymbol{k}|^2 = k_x^2 + k_y^2 + k_z^2 = \frac{\omega^2}{c^2} \tag{2.3}$$

hold are solutions of the wave equation (2.1). One well-known electromagnetic field is the plane wave, which reads in the real presentation:

$$E_r(x,y,z,t) = E_0 \cos(\omega t - k_x x - k_y y - k_z z) \tag{2.4}$$

with:

	$\omega = 2\pi\nu$	angular frequency
	$\nu = c/\lambda$	light frequency
	$\boldsymbol{k} = (k_x, k_y, k_z)$	wave vector with $\lvert\boldsymbol{k}\rvert = 2\pi/\lambda$

The wave vector $\boldsymbol{k}$ points into the direction of propagation of the wave and is perpendicular to the planes of constant phase. Another common solution is the spherical wave:

$$E(r,t) = E_0 \frac{\lambda}{r} \cos(\omega t - kr) \qquad r > \lambda \tag{2.5}$$

with r being the distance from the source. Although both waves are solutions of the wave equation, they cannot be realized since no spatial confinement is allowed. Without lateral confinement the waves are infinite in extent and contain infinite power. Therefore, infinite waves do not make physical sense. However, they can be considered as very good approximations applicable to the description of many optical phenomena.

Before we discuss more realistic electromagnetic waves, we will now introduce the complex notation for the electric field which is very useful to simplify analytical derivations. Since it is much more convenient to replace the cosine by the complex exponentials, the plane wave (2.4) can be rewritten in the form:

$$E_r(x,y,z,t) = E_0 \frac{1}{2} \left[\exp[i(\omega t - k_x x - k_y y - k_z z)] + \exp[-i(\omega t - k_x x - k_y y - k_z z)]\right] = \frac{1}{2} \left[E + E^*\right]$$

where, as usual, the asterisk indicates a complex conjugate. It is sufficient to consider only one of the two complex electric fields on the right hand side. The real, physical field has been replaced by a complex field. In a similar way, one obtains the complex field for the spherical wave (2.5):

$$E = E_0 \frac{\lambda}{r} \exp[i(\omega t - kr)] \tag{2.6}$$

The real physical field E_r is linked to the complex field E through:

$$E_r = \frac{1}{2} \left[E + E^*\right] \tag{2.7}$$

and the time-averaged intensity is given by:

$$I = \frac{1}{2} c \epsilon_0 \, E E^*$$

Note that throughout this book we will always present the electric field in its complex notation. For any optical system only one electric field out of the infinite number of solutions of the wave equation describes the light propagation. This unique wave is determined by the boundary conditions of the optical system. This means that at a certain time the electric field in a certain area is given and the propagation in space and time is calculated by using the wave equation. In the following we will describe this time-space development of electric fields. We will find a mathematical way to determine the electric field at any point in space for an initial field distribution given in a confined area (e.g. a resonator mirror). The propagation of a given confined electric field into space is called *diffraction*. The propagation of electric fields can be evaluated by using Huygen's Principle: *An electric field distribution E(x,y) given on a surface propagates into space in such a way that each point (x,y) can be considered as the source of a spherical wave with amplitude E(x,y). At a point P=(x,y,z) the resulting electric field is the superposition of all of these spherical waves.* As shown in Fig. 2.1 we have to subdivide the starting plane into points $(x_i, y_i, 0)$. Without lack of generality, we assume in the following that the points are equally spaced in both directions. At point P the spherical wave originating from one point on the source plane reads:

$$E_i(P) = C\ E(x_i,y_i)\ \frac{\exp[i(\omega t - kr_i)]}{r_i}\ \cos\theta_i\ \Delta x \Delta y \tag{2.8}$$

with r_i being the distance from point $(x_i, y_i, 0)$ to P and $\Delta x, \Delta y$ denoting the point spacing in the x- and y-direction. The term $\cos\theta_i$ accounts for the radiation pattern of a Lambertian source: the energy flow is maximum in the direction of the surface normal ($\cos\theta = 1$) and is equal to zero tangential to the surface. The constant C is as yet an unknown proportionality factor which ensures that the total energy is conserved. In order to attain the resulting field at point P we have to take the sum over all starting points. For a monochromatic source the summation over N points yields

$$E(P) = \sum_{i=1}^{N} E_i(P) = C\ \exp(i\omega t) \sum_{i=1}^{N} E(x_i,y_i)\ \frac{\exp(-ikr_i)}{r_i}\ \cos\theta_i\ \Delta x \Delta y \tag{2.9}$$

This calculation of the field in point P will, of course, become more and more accurate as the number of points is increased. In the limit $N \rightarrow \infty$, the summation is replaced by the integration over the starting surface. This discussion of diffraction is purely phenomenological. However, compared to the mathematical derivation of the diffraction integral, this empirical approach provides a better feeling for the physical meaning of the diffraction integral.

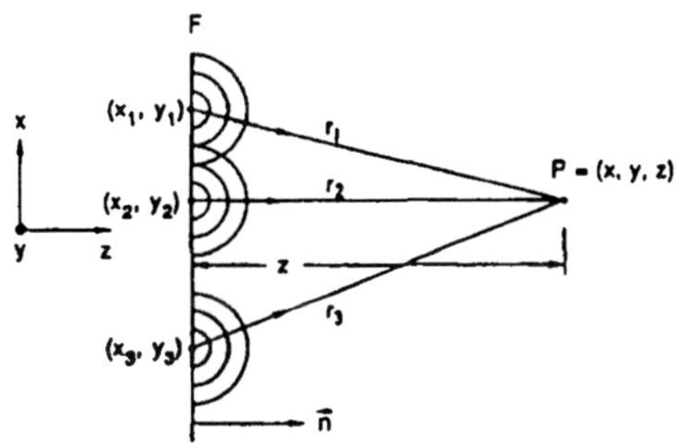

Fig. 2.1 The propagation of an electric field distribution given on a plane at $z=0$ can be dealt with by adding the spherical waves emerging from all points on the plane.

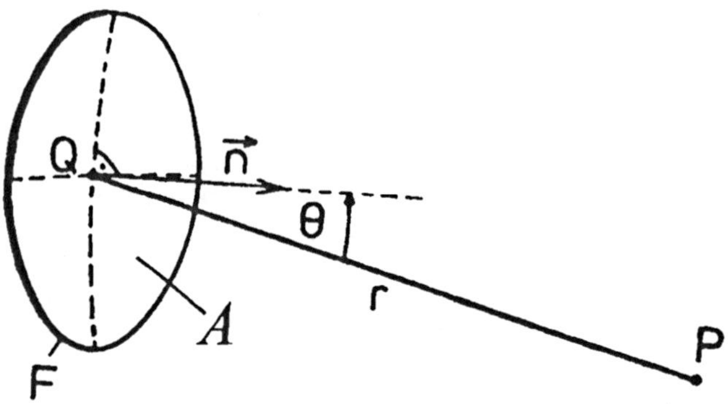

Fig. 2.2 The electric field E at a point P in front of a mirror can be calculated by applying the Kirchhoff integral (2.10) to the field distribution E_l on the mirror. The paraxial approximation ($\cos\theta \approx 1$) must hold and the diameter of the mirror needs to be much greater than the wavelength.

A mathematical derivation*[1.1,1.4,1.23,1.63] yields for the electric field at point P:

$$E(P) = \frac{i}{\lambda} \exp(i\omega t) \int_A E_1(Q) \frac{\exp(-ikr)}{r} dA \tag{2.10}$$

This surface-integral is referred to as the *Kirchhoff integral.*

Comparing the integral with (2.9), we notice that the cosine term has disappeared. This stems from the paraxial approximation ($\cos\theta \approx 1$) used in the mathematical derivation. A second approximation made is that the extent of the surface A, defined as the square root of the surface area, has to be much greater than the wavelength. As far as optical resonators are concerned both restrictions are easily satisfied. Mirror size and spacing are generally large enough to ensure the validity of both approximations. The paraxial approximation also assumes that the distance r does not change considerably during the integration. It is for this reason that the term $1/r$ is commonly placed outside the integral.

*A strict mathematical derivation of the Kirchhoff-integral is not possible since a fundamental theorem of mathematics is violated. In spite of this mathematical inconsistency, the integral is used in the presented form since the experimental results are in agreement with its theoretical predictions. An exact treatment of diffraction was performed by Arnold Sommerfeld [1.4]. This exact diffraction integral differs from the Kirchhoff integral only for small distances from the surface or for large distances from the optical axis [1.27]. With the above-mentioned restrictions that apply to the Kirchhoff integral, both diffraction integrals provide the same results.

2.2 Diffraction

In the following, we apply the Kirchhoff integral to two common diffraction geometries and discuss the properties of the diffracted field at different distances from the aperture. The time-dependent exponential in (2.10) will be dropped for convenience.

2.2.1 Rectangular Aperture

We consider a rectangular aperture with width *2a* and height *2b*, as depicted in Fig. 2.3, and we calculate the field distribution $E_2(x_2,y_2)$ for different distances L from the aperture by applying the Kirchhoff integral (2.10) to the field distribution $E_1(x_1,y_1)$ inside the aperture. The distance r between two points in plane 1 and plane 2 reads:

$$r = L\sqrt{1 + \frac{(x_2-x_1)^2 + (y_2-y_1)^2}{L^2}} \tag{2.11}$$

The field $E_2(x_2,y_2)$ is given by the Kirchhoff integral (2.10), which reads:

$$E_2(x_2,y_2) = \frac{i}{\lambda L}\int_{-b}^{b}\int_{-a}^{a} E_1(x_1,y_1)\,\exp\left[-ikL\sqrt{1 + [(x_2-x_1)^2+(y_2-y_1)^2]/L^2}\right] dx_1 dy_1 \tag{2.12}$$

The $1/r$ term in the integrand of (2.10) was replaced by $1/L$ outside of the integral due to the validity of the paraxial approximation. This integral cannot be solved analytically in its present form. However, if we restrict ourselves to distances large compared to the aperture dimensions $(L \gg x_1, y_1)$ and assume that the field distribution does not spread too fast laterally $(x_2/L, y_2/L \ll 1)$, the square root in the exponential of (2.12) can be expanded into a series:

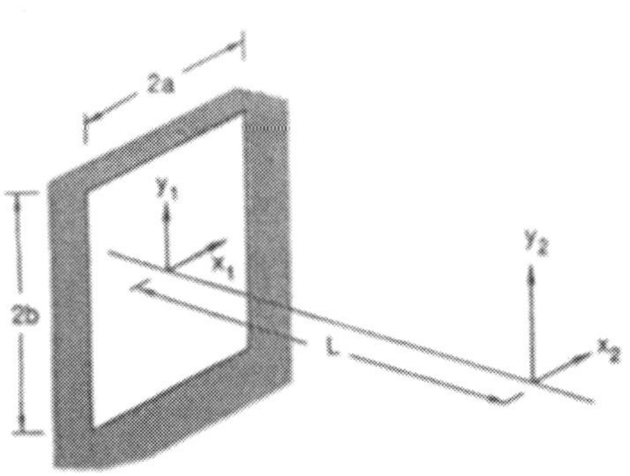

Fig. 2.3 Geometry of a rectangular aperture. The electric field at a plane separated by a distance L from the aperture plane is being calculated.

$$r = L\left[1 + \frac{1}{2}\left(\frac{x_2-x_1}{L}\right)^2 + \frac{1}{2}\left(\frac{y_2-y_1}{L}\right)^2 - \frac{1}{8}\left(\frac{x_2-x_1}{L}\right)^4 - \frac{1}{8}\left(\frac{y_2-y_1}{L}\right)^4 + \ldots\right] \tag{2.13}$$

The following three areas can be distinguished:

a) Fraunhofer-Diffraction

For distances L very large compared to the aperture size, the quadratic terms x_1^2/L, y_1^2/L become negligibly small and the expression (2.13) for the distance r can be reduced to:

$$r = L - \frac{x_1x_2+y_1y_2}{L} + \frac{x_2^2+y_2^2}{2L} \tag{2.14}$$

The quadratic term of (2.14) represents the phase curvature of a spherical wave with radius of curvature L. This term is usually disregarded since it vanishes when the intensity of the field is calculated. By using the angular coordinates $\theta_x=x_2/L$ and $\theta_y=y_2/L$, the normalized coordinates $x_1{}^*=x_1/a$ and $x_2{}^*=x_2/b$, and Eq. (2.14), the Kirchhoff integral (2.12) reads:

$$E_2(\theta_x,\theta_y) = i\frac{ab}{\lambda L}\exp(-i\phi)\int_{-1}^{1}\int_{-1}^{1} E_1(x_1^*,y_1^*)\exp[ik(a\theta_x x_1^* + b\theta_y y_1^*)]\,dx_1^*dy_1^* \tag{2.15}$$

with $$\phi = kL + \frac{x_2^2+y_2^2}{2L}$$

This integral equation can be separated into two equations, one for the x-direction and one for the y-direction, if the field E_1 can be written as a product:

$$E_1(x_1^*,y_1^*) = u_1(x_1^*)\,v_1(y_1^*)$$

The resulting field $E_2(\theta_x,\theta_y) = u_2(\theta_x)\,v_2(\theta_y)$ is given by the two integral equations:

$$u_2(\theta_x) = \exp\left[-i\left(\frac{\phi}{2}-\frac{\pi}{4}\right)\right]\sqrt{N_x}\int_{-1}^{1} u_1(x_1^*)\exp[ikax_1^*\theta_x]\,dx_1^* \tag{2.16}$$

$$v_2(\theta_y) = \exp\left[-i\left(\frac{\phi}{2}-\frac{\pi}{4}\right)\right]\sqrt{N_y}\int_{-1}^{1} v_1(y_1^*)\exp[ikby_1^*\theta_y]\,dy_1^* \tag{2.17}$$

with $N_x=a^2/\lambda L$ and $N_y=b^2/\lambda L$ being the Fresnel numbers in x-direction and y-direction, respectively. The field distributions (2.16) and (2.17) are referred to as the far field of the field distribution $E_l(x,y)$. The angular coordinates θ_x, θ_y denote the angles under which the field diverges into space, also called the divergence angles. If the far field is observed at the back focal plane of a lens, the distance L has to be replaced by the focal length f. Mathematically, these equations correspond to the Fourier transformation.

Example: Homogeneously Illuminated Aperture

For a homogeneous field distribution $E_l(x,y)=const.=E$, the integral equation (2.16) and (2.17) can be solved analytically. This yields for the field intensity in x-direction (excluding a proportionality factor):

$$I_x(\theta_x) = N_x \frac{\sin^2(ka\theta_x)}{(ka\theta_x)^2} \tag{2.18}$$

and the corresponding expression for the y-direction. The intensity distribution at a sufficiently large distance from the aperture is thus given by:

$$I(\theta_x,\theta_y) = I_0 N_x N_y \frac{\sin^2(ka\theta_x)}{(ka\theta_x)^2} \frac{\sin^2(kb\theta_y)}{(kb\theta_y)^2} \tag{2.19}$$

with I_0 being the intensity across the aperture. Figure 2.4 shows the one-dimensional intensity distribution (2.18) in the x-direction. The intensity minima are determined by:

$$\theta_x = n\frac{\lambda}{2a} \qquad \theta_y = m\frac{\lambda}{2b} \tag{2.20}$$

with n,m being integer numbers greater than zero.

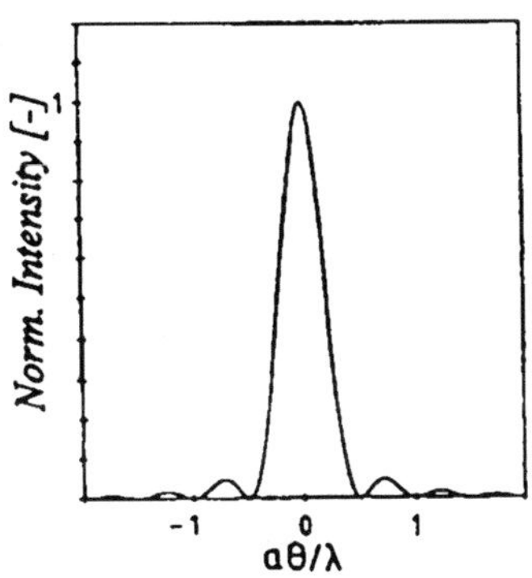

Fig. 2.4 One-dimensional intensity distribution in the far field of a homogeneously illuminated rectangular aperture.

The full width half maximum (FWHM) diameters of the central lobe of the intensity distribution (2.19) are:

$$\Delta\theta_x = 0.44\,\frac{\lambda}{a} \qquad \Delta\theta_y = 0.44\,\frac{\lambda}{b} \tag{2.21}$$

In conclusion, the field distribution at a sufficiently large distance L from the aperture is given by the Kirchhoff integral with the Fraunhofer approximation (2.15) (often referred to as the Fraunhofer integral). This field distribution, which remains constant as the distance L is increased, is called the far field. The far field is usually expressed in terms of the divergence angles θ_x and θ_y. The lateral position of the field minima and maxima can be determined by multiplying the divergence angles by the distance L. The Fraunhofer approximation can be applied if the Fresnel numbers $N_{x,y}$ are much smaller than 1; a "safe" limit is $N_{x,y}<0.2$.

Example: For λ=500nm and an aperture width of $2a$=2mm, the Fraunhofer-approximation can be used if the distance L is greater than 10m. At a distance of L=40m the position of the first intensity minimum with respect to the optical axis is x=10mm. Propagation over a distance of L=4km results in a minimum position of x=1m. At both distances, the shapes of the intensity distributions are the same.

b) Fresnel-Diffraction

If at least one of the Fresnel numbers $N_{x,y}$ is greater than one (which means that we get closer to the aperture), the linear approximation (2.14) of the distance r is not applicable. We have to add the quadratic terms of x_1 and y_1 in order to be able to calculate the field distribution. Thus, we insert the second order series of the distance:

$$r = L - \frac{x_1x_2+y_1y_2}{L} + \frac{x_1^2+y_1^2}{2L} + \frac{x_2^2+y_2^2}{2L} \tag{2.22}$$

into the integral (2.12) and we obtain:

$$E_2(x_2,y_2) = i\,\exp[-ikL]\,\sqrt{N_xN_y}\;\bullet \tag{2.23}$$

$$\int_{-1}^{1}\int_{-1}^{1} E_1(x_1^*,y_1^*)\,\exp[-i\pi N_x(x_1^{*2}+x_2^{*2}-2x_1^*x_2^*)\;-i\pi N_y(y_1^{*2}+y_2^{*2}-2y_1^*y_2^*)]\;dx_1^*dy_1^*$$

with the normalized coordinates $x^*=x/a$ and $y^*=y/b$. The Fresnel approximation can be applied if the Fresnel number N is smaller than 100. Similar to the Fraunhofer integral

(2.15), the Fresnel integral can be separated into two one-dimensional integrals if the electric field can be expressed as a product of one-dimensional fields. Unfortunately, there is no analytical solution to the Fresnel integral and the electric field distribution has to be computed numerically. In the case of a homogeneously illuminated rectangular aperture the values of the Fresnel integral can be found in mathematical handbooks such as [1.28]. Figures 2.5 and 2.6 show numerically calculated and experimentally recorded intensity distributions for different Fresnel numbers for a homogeneously illuminated square aperture. In the experiment a metal aperture of side length 2mm was inserted into a collimated HeNe-laser beam (λ=632.8nm) with a beam diameter expanded to 15mm by means of a telescope. Note the agreement between the calculated and the recorded intensity distributions and how the intensity distributions approach the intensity distribution in the far field as the Fresnel number decreases.

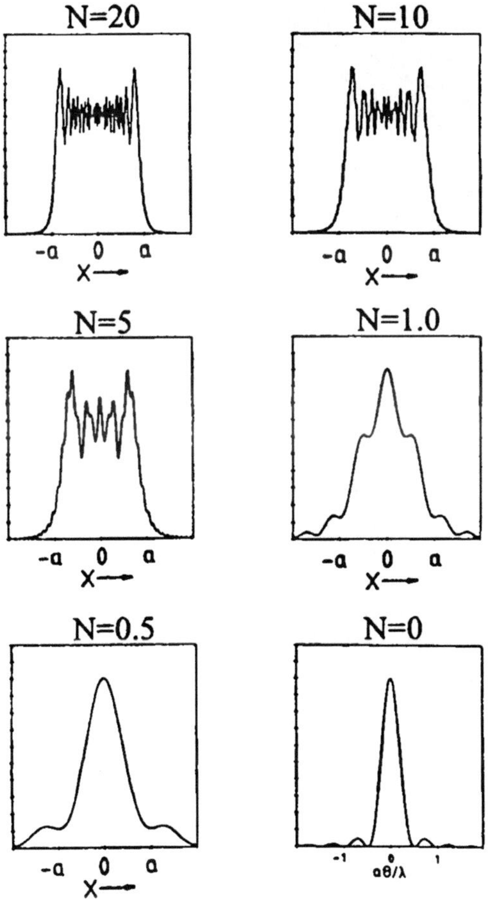

Fig. 2.5 Calculated normalized one-dimensional intensity distributions for a homogeneously illuminated aperture with width *2a* and different Fresnel numbers *N*. The lower right graph shows the far field intensity distribution.

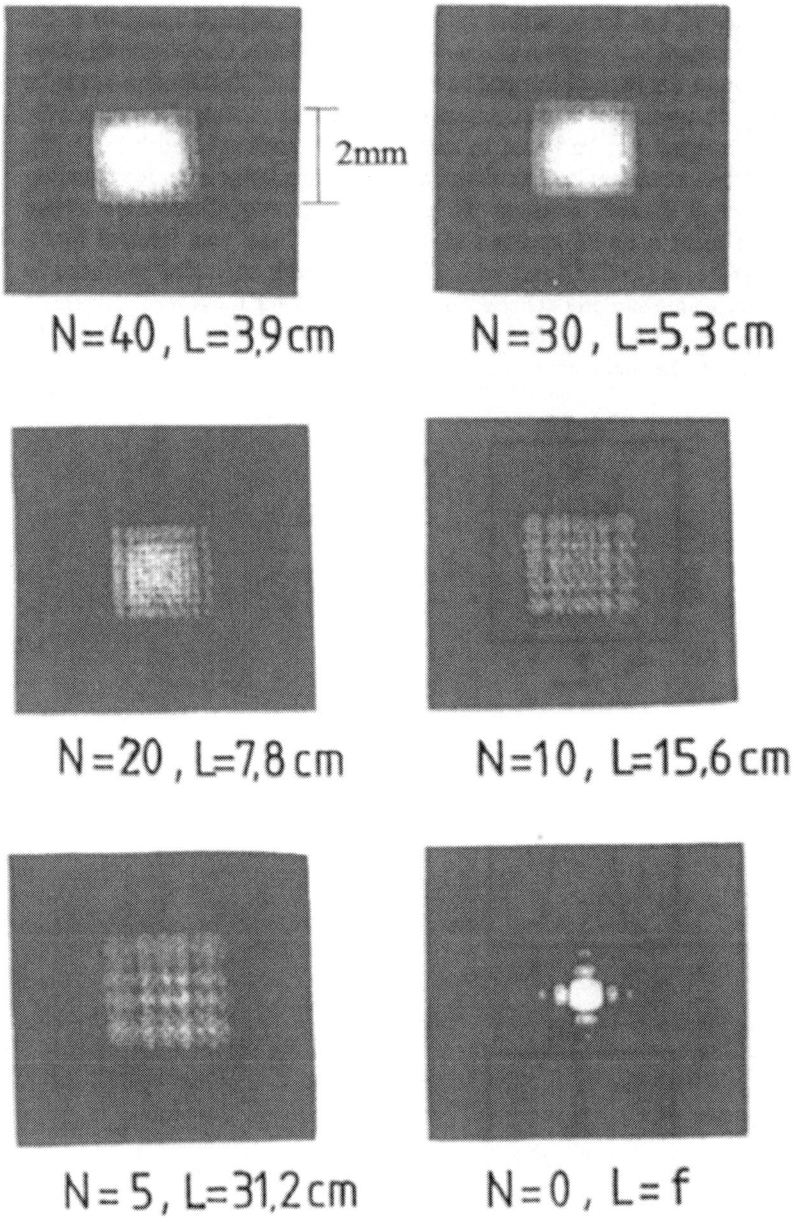

Fig. 2.6 Intensity distributions photographed at different distances L behind a square aperture with side length 2mm. The aperture was homogeneously illuminated with a HeNe-laser beam (λ=632.8nm) and the intensity distributions were recorded by means of a CCD-camera. The lower right photograph shows the far-field, recorded in the focal plane of a focusing lens. Here the CCD camera was overexposed to show the side lobes more clearly.

c) Geometrical Optics

If we approach the aperture even closer, so that the Fresnel numbers N_x, N_y become greater than 100, the quadratic series (2.22) for the distance r is not accurate enough. Unfortunately, simply adding the next term (fourth power) to the expansion, does not solve the problem. The Kirchhoff integral (2.10) is not applicable to such high Fresnel numbers since we are too close to the aperture and the paraxial approximation *(cos* θ*=1)* does not hold anymore. A detailed investigation, however, indicates that the field distributions do not change considerably for large Fresnel numbers. Additionally, the shape of the intensity distribution is almost identical to that of the initial distribution across the aperture (see $N=40$ in Fig.2.6). Field propagation for Fresnel numbers greater than 100 can be calculated by using the laws of geometrical optics as long as one is not interested in fine structures near the aperture edges, or by using the place wave approximation (see Sec. 2.9).

2.2.2 Circular Aperture

If the electric field is diffracted by a circular aperture with radius R (Fig. 2.7), it is convenient to express the Kirchhoff integral in terms of polar coordinates r, Φ with

$$x_i = r_i \cos\Phi_i \,, \qquad y_i = r_i \sin\Phi_i \,, \qquad 0 \le r_i \le R$$

In the Fraunhofer approximation one obtains with $\theta = r_2/L$: (2.24)

$$E_2(\theta,\Phi_2) = i\, N \exp[-ikL] \int_0^{2\pi}\!\!\int_0^1 E_1(r_1^*,\Phi_1) \exp[-ikR\theta r_1^* \cos(\Phi_2-\Phi_1)]\, r_1^* dr_1^* d\Phi_1$$

and the Fresnel integral reads: (2.25)

$$E_2(r_2^*,\Phi_2) = iN \exp[-ikL] \int_0^{2\pi}\!\!\int_0^1 E_1(r_1^*,\Phi_1) \exp[-i\pi N(r_1^{*2} + r_2^{*2} - 2r_1^* r_2^* \cos(\Phi_2-\Phi_1))]\, r_1^* dr_1^* d\Phi_1$$

with $N=R^2/(\lambda L)$: Fresnel number
$r_i^*=r/R$: normalized radial coordinate

For field distributions E_i that can be expressed as products of a purely radial function $u(r)$ and an oscillating angular term:

$$E_i(r_i^*,\Phi_i) = u_i(r_i^*) \exp[-im\Phi_i] \,, \qquad m = 0,1,2,\ldots \,;\; i = 1,2$$

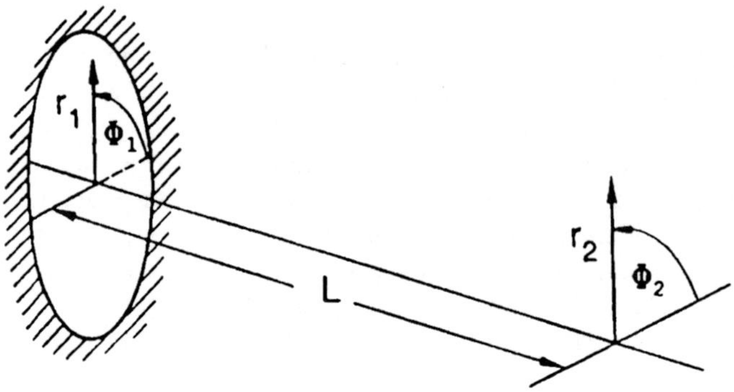

Fig. 2.7 Diffraction at a circular aperture. The field distributions at both planes are expressed in terms of polar coordinates r, Φ.

both integrals can be simplified by performing the angular integration. The final result with the Fraunhofer approximation is given by:

$$u_2(\theta) = -(-i)^{m+1}\ 2\pi N \exp[-ikL] \int_0^1 u_1(r_1^*)\ J_m(\frac{2\pi}{\lambda} R\theta r_1^*)\ r_1^* dr_1^* \qquad (2.26)$$

and the Fresnel integral is given by:

$$u_2(r_2^*) = -(-i)^{m+1}\ 2\pi N \exp[-ikL] \int_0^1 u_1(r_1^*) \exp[-i\pi N(r_1^{*2}+r_2^{*2})]\ J_m(2\pi N r_1^* r_2^*)\ r_1^* dr_1 \qquad (2.27)$$

with $J_m(x)$: Bessel function of order m [1.28].

Example: Far field of a homogeneously illuminated circular aperture with radius R. By using equation (2.26) with $m=0$ and $u(r)=E_0=const.$ one obtains:

$$u_2(\theta) = 2\pi\ N\ E_0 \frac{J_1(2\pi R\theta/\lambda)}{2\pi R\theta/\lambda} \qquad (2.28)$$

The corresponding intensity distribution is shown in Fig. 2.8 together with the photograph of a far field pattern of a HeNe-laser diffracted by a round aperture with radius *R=1mm*. The far field intensity distribution of a circular aperture is called the Airy pattern. It looks very similar to the far field of a rectangular aperture in one dimension. The position of the intensity minima, height of the intensity maxima, and the energy content in the central lobe are, however, quite different as Table 2.1 indicates.

Table 2.1 Comparison of far field properties of rectangular and circular apertures. Positions and diameters are given in terms of divergence. *The height of the central lobe is normalized to 1.0.

	rectangular aperture width 2a	circular aperture diameter 2R
FWHM diameter	$0.44\lambda/a$	$0.52\lambda/R$
Position 1. minimum	$0.50\lambda/a$	$0.61\lambda/R$
Position 2. minimum	$1.00\lambda/a$	$1.12\lambda/R$
Height of 1. maximum*	0.04718	0.01753
Height of 2. maximum*	0.01694	0.00522
Diameter containing 86% of total power	$1.05\ \lambda/a$	$1.61\ \lambda/R$
Power content of central peak	81.5%	84%
Peak intensity of central peak I_{max}/I_0	$(4ab/\lambda f)^2$	$(\pi\rho^2/\lambda f)^2$

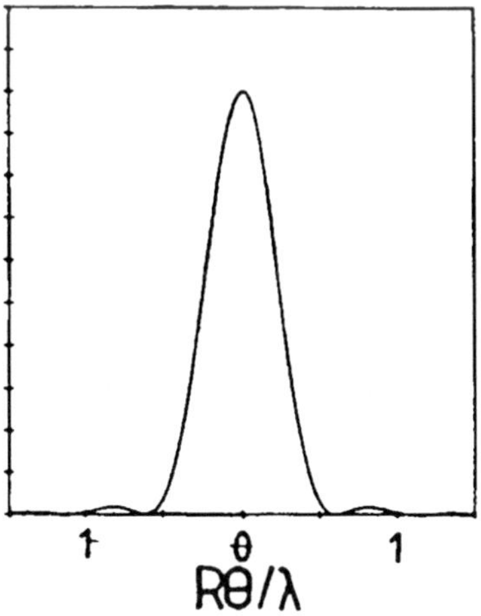

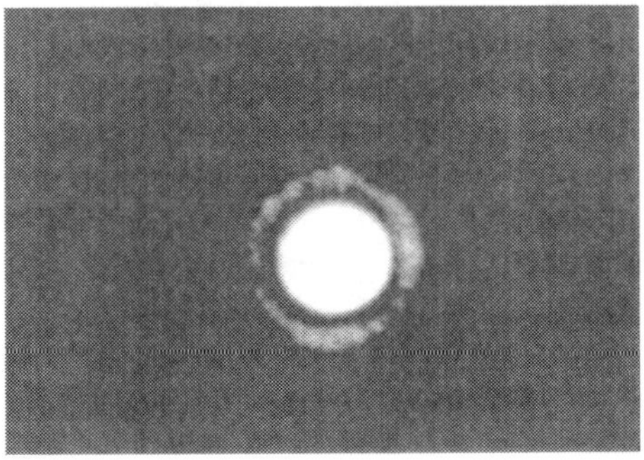

Fig 2.8 Calculated and measured far field intensity distributions for a circular aperture with radius *a*. For the photograph a pinhole with radius 1mm was inserted into an expanded and collimated HeNe- laser beam (λ=632.8nm). The far field was generated in the focal plane of a positive lens.

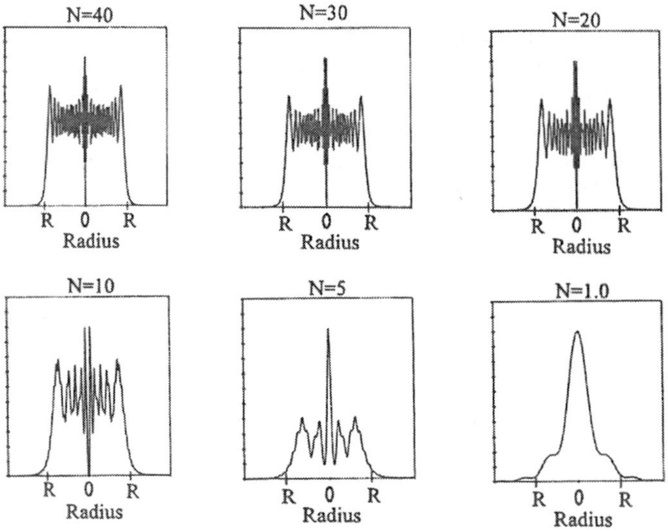

Fig. 2.9 Calculated radial intensity distributions for different Fresnel numbers N of a homogeneously illuminated circular aperture of radius R. With decreasing Fresnel number the distributions become less spiky and for Fresnel numbers of less than 1, the far field intensity distribution is approached.

For Fresnel numbers greater than about 0.2, the Fresnel integral should be used to calculate the field distribution. Similar to the rectangular symmetry, this can only be accomplished by using numerical methods. Figure 2.9 shows the calculated radial intensity distributions for a homogeneously illuminated circular aperture with radius R for different Fresnel numbers N. The diffraction patterns are similar to those of a rectangular aperture (Fig. 2.5), but now the rotational symmetry results in a high intensity peak on the optical axis. In high power lasers having circular output beams this so called "Spot of Arago" can lead to the destruction of optical components. For both rectangular and circular apertures, the far field depends only on the product of the aperture diameter and the far field angle θ (assuming a constant wavelength λ). Decreasing the aperture width thus results in the increase of the beam divergence. Again we observe that the beam parameter product is a constant of the optical system. This is quite a satisfying result since it is in agreement with our discussion of the phase space transformations (see Sec. 1.2.2). Let us define the far field half angle of divergence θ via the 86% power content (see Table 2.1) and the beam radius w in a similar way ($w=0.86a$ for the rectangular aperture and $w=0.93R$ for the circular aperture). We then get for the beam parameter products in the two geometries:

$$\textit{rectangular}:\quad w\theta = 1.418\,\frac{\lambda}{\pi} \qquad\qquad \textit{circular}:\quad w\theta = 2.337\,\frac{\lambda}{\pi}$$

This is already close to the minimum beam parameter product of λ/π.

2.3 Collins Integral

2.3.1 One-Dimensional Optical Systems

We have so far only discussed the free propagation of the electric field without any optical elements like lenses and mirrors being in the path. We have found that the electric field after propagation is linked to the initial field via the Kirchhoff integral. It is possible to incorporate any optical system with parabolic surfaces into the propagation and derive a corresponding diffraction integral. This generalized Kirchhoff integral is referred to as the Collins integral. The derivation of the Collins integral is beyond the scope of this book, the interested reader is referred to the original paper [1.34]. The basic principle of the derivation is to find the shortest distance between points on two separated planes where arbitrary optical elements may be located between those planes. By applying Fermat's principle one can derive an expression for the distance r which depends on the elements of the ray transfer matrix $\boldsymbol{M}$. We will first restrict the discussion to one dimensional optical systems, which means optics that exhibit rotational symmetry and can thus be described by a 2x2 ray transfer matrix. As already shown in the last chapter, the generalization to two dimensions can be performed by replacing the matrix elements with the corresponding submatrices.

If the electric field propagates from a plane 1 to a plane 2 and an optical system with a ray transfer matrix $\boldsymbol{M}$ is located between the planes, the Kirchhoff integral in the Fresnel approximation reads (Fig.2.10):

$$E_2(x_2,y_2) = \frac{i}{\lambda B}\exp[-ikL]\int\int E_1(x_1,y_1)\exp[-i\frac{\pi}{\lambda B}(Ax_1^2+Dx_2^2-2x_1x_2 + Ay_1^2+Dy_2^2-2y_1y_2)]\,dx_1dy_1 \tag{2.29}$$

with A,B,D being the elements of the ray transfer matrix $\boldsymbol{M}$ and L is the optical path along the optical axis.

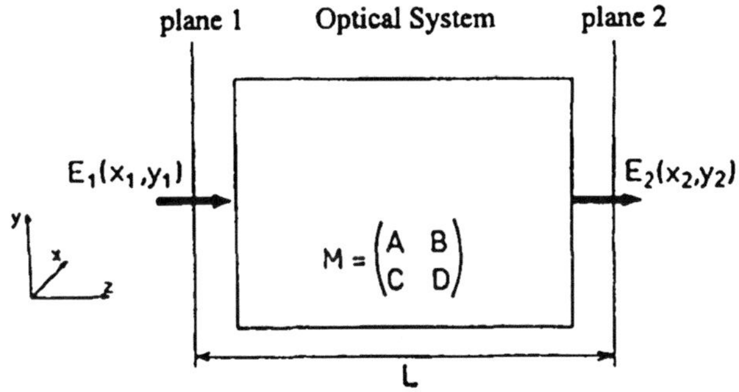

Fig. 2.10 Propagation of an electric field through an optical system. The fields at plane 1 and plane 2 are linked to each other via the Collins integral (2.29).

This integral is referred to as the Collins integral. It does not differ from the Kirchhoff integral with the Fresnel approximation, except for the fact that the shortest distance between two points is already incorporated with the ray transfer matrix elements being system parameters. Similar to the Kirchhoff integral the derivation is based on the applicability of the paraxial approximation. This means that the Collins integral is valid for Fresnel numbers lower than 100.

Examples:

1) Free Space Propagation:
We consider the simplest case of an empty optical system with length L. By inserting the ray matrix elements $A=1$, $B=L$ and $D=1$ into (2.29) we obtain, as to be expected, the Fresnel integral (2.23).

2) Field Distribution at the Focal Plane of a Lens:
A field distribution E_1 is given at a plane in front of focusing lens with focal length f. The distance from the plane to the focusing lens is d. We want to calculate the resulting field distribution at the focal plane of the lens. The ray transfer matrix elements for the propagation to the focal plane of the lens are found to be $A=0$, $B=f$ and $D=1-d/f$. Insertion into (2.29) yields:

$$E_2(\theta_x,\theta_y) = i\,\frac{\exp(-ik\delta)}{\lambda f}\int_{-\infty}^{\infty}\!\!\int_{-\infty}^{\infty} E_1(x_1,y_1)\,\exp[ik(\theta_x x_1 + \theta_y y_1)]\,dx_1 dy_1 \tag{2.30}$$

with $$\delta = d + f + \frac{1}{2}(\theta_x^2 + \theta_y^2)(f-d)$$

and $$\theta_x = \frac{x_2}{f}\ ,\quad \theta_y = \frac{y_2}{f}\ ,\quad k = \frac{2\pi}{\lambda}$$

A comparison with (2.15) indicates that this integral is equivalent to the Kirchhoff integral in the Fraunhofer approximation. This means that at the focal plane of a focusing lens the far field intensity distribution of the electric field E_1 can be measured. This is the common experimental technique used to determine angles of divergence of laser beams. In the special case of $d=f$, the parabolic phase term in front of the integral disappears and the exact Fourier transform of the field is observed in the back focal plane. The integral in (2.30) is referred to as a Fourier transform. Fourier transforms play an important role in diffraction theory not only because of their linkage with the far field, but also because they can provide a better understanding of the properties of electric fields and their propagation. We shall discuss the Fourier transforms and the mathematics involved in more detail in Sec. 2.4.2.

2.3.2 Two-Dimensional Optical Systems

The Collins integral formalism can be generalized for arbitrary optical systems provided they have parabolic interfaces or parabolic index profiles in paraxial approximation. As was discussed in the preceding chapter, these optics are described by 4x4 ray transfer matrices M. Propagation of rays through these optical systems can be written in the form:

$$\begin{pmatrix} r_2 \\ \gamma_2 \end{pmatrix} = \begin{pmatrix} A & B \\ C & D \end{pmatrix} \begin{pmatrix} r_1 \\ \gamma_1 \end{pmatrix} \tag{2.31}$$

with $\quad r_i = (x_i, y_i) \,, \quad \gamma_i = (\alpha_i, \beta_i) \quad , \; i=1,2$

and A,B,C,D being the 2x2 matrices presented in Sec. 1.2.4.

By using the same generalization we have already applied to the matrix determinant (1.61), we can find the generalized Collins integral for two-dimensional optical systems:

$$E_2(x_2,y_2) = \frac{i\,\exp(-ikL)}{\lambda\sqrt{\det B}} \iint E_1(x_1,y_1)\,\exp[-i\frac{\pi}{\lambda}(r_1 B^{-1} A r_1 + r_2 D B^{-1} r_2 - 2 r_1 B^{-1} r_2)]\; dx_1 dy_1 \tag{2.32}$$

with B^{-1} being the inverse of matrix B and $detB$ denoting its determinant.

The beauty of this equation lies in the fact that we can easily memorize the diffraction integral for two-dimensional systems due to its similarity to the one-dimensional equation (2.29). The generalized Collins integral (2.32) enables us to calculate even complex diffraction problems like the propagation through rotated and tilted optical elements as the following examples will show:

Examples:

1) Phase Space Beam Analyzer

We use the Phase Space Beam Analyzer as discussed in Sec. 1.2.4 and calculate the light propagation from the entrance plane 1 to the recording plane 4 (see Fig. 1.23). We neglect the presence of the second slit and determine the phase term $\exp[-i\pi\Phi/\lambda]$ in the integrand of (2.32). By using the resulting ray transfer matrix (1.78) we get (note that the coordinates in the recording plane are denoted as x_4, y_4):

$$\Phi = \begin{pmatrix} x_1 \\ y_1 \end{pmatrix} \begin{pmatrix} P & 0 \\ 0 & P \end{pmatrix} \begin{pmatrix} 0 & 1/S \\ 1/S & 0 \end{pmatrix} \begin{pmatrix} x_1 \\ y_1 \end{pmatrix} + \begin{pmatrix} x_4 \\ y_4 \end{pmatrix} \begin{pmatrix} 1/P & f_0/f \\ f_0/f & 1/P \end{pmatrix} \begin{pmatrix} 0 & 1/S \\ 1/S & 0 \end{pmatrix} \begin{pmatrix} x_4 \\ y_4 \end{pmatrix}$$

$$- 2 \begin{pmatrix} x_1 \\ y_1 \end{pmatrix} \begin{pmatrix} 0 & 1/S \\ 1/S & 0 \end{pmatrix} \begin{pmatrix} x_4 \\ y_4 \end{pmatrix} \tag{2.33}$$

which can be rewritten as:

$$\Phi = \frac{2P}{S}x_1y_1 + \frac{2}{PS}x_4y_4 - \frac{2}{S}(x_1y_4+x_4y_1) + \frac{f_0}{fS}(x_4^2+y_4^2) \tag{2.34}$$

The diffraction integral (2.32) then reads with $detB$=-S^2: (2.35)

$$E_4(x_4,y_4) = \frac{\exp[-ik(a+b)]}{S} \cdot$$

$$\iint_{-\infty-\infty}^{+\infty+\infty} E_1(x_1,y_1)\exp\left[-i\frac{\pi}{\lambda}\left[\frac{2P}{S}x_1y_1 + \frac{2}{PS}x_4y_4 - \frac{2}{S}(x_1y_4+x_4y_1) + \frac{f_0}{fS}(x_4^2+y_4^2)\right]\right] dx_4dy_4$$

This integral can only be solved numerically, except for Gaussian input beams.

2) Far Field Measurement with Tilted Lens

We want to record the far field intensity distribution in the focal plane of a focusing lens with focal length f. The electric field incident on the lens is E_1. Unfortunately, the surface normal of the lens is tilted by an angle α around the x-axis with the z-axis being the optical axis (see Fig. 1.20). By using (1.84) and (1.85), the ray transfer matrix for the passage through the lens and propagation to the focal plane can be approximated by:

$$M = \begin{pmatrix} A & B \\ C & D \end{pmatrix} = \begin{pmatrix} 1-\cos\alpha & 0 & f & 0 \\ 0 & 1-\dfrac{1}{\cos\alpha} & 0 & f \\ \dfrac{-\cos\alpha}{f} & 0 & 1 & 0 \\ 0 & \dfrac{-1}{f\cos\alpha} & 0 & 1 \end{pmatrix} \tag{2.36}$$

With $\sqrt{\det B} = f$ and $B^{-1} = \begin{pmatrix} 1/f & 0 \\ 0 & 1/f \end{pmatrix}$, we get from (2.32):

$$E_2(x_2,y_2) = \frac{i}{\lambda f}\exp[-ik(f+\frac{1}{2}(x_2^2+y_2^2))] \cdot$$

$$\iint E_1(x_1,y_1)\exp[-i\frac{\pi}{\lambda f}((1-\cos\alpha)x_1^2+(1-1/\cos\alpha)y_1^2-2(x_1x_2+y_1y_2))]\, dx_1dy_1 \tag{2.37}$$

By introducing the variables $\theta_x = x_2/f$ and $\theta_y = y_2/f$ we can rewrite this as:

$$E_2(\theta_x,\theta_y) = i\,\frac{\exp(ik\delta)}{\lambda f}\iint E_1^{*}(x_1,y_1)\,\exp[-ik(\theta_x x_1 + \theta_y y_1)]\,dx_1 dy_1 \tag{2.38}$$

with $$E_1^{*}(x_1,y_1) = E_1(x_1,y_1)\,\exp[-i\frac{\pi}{\lambda f}((1-\cos\alpha)x_1^2 + (1-1/\cos\alpha)y_1^2)]$$

and $$\delta = f + \frac{f}{2}(\theta_x^2 + \theta_y^2)$$

A comparison with (2.30) indicates that we measure the far field intensity distribution of the field E_1' which is the product of the original field and the parabolic phase terms induced by the lens tilt. Fortunately, there are transformation rules available that simplify the analytical calculation of Fourier integrals like this. This will be shown in the following section.

2.4 Collins Integral and Vanishing Ray Matrix Elements

2.4.1 Imaging Condition (B=0)

We know from our discussion of ray transfer matrices in the preceding chapter that a vanishing B-component represents an imaging optical system. The matrix component A represents the magnification by which the object is scaled in the image plane. In terms of wave optics, imaging means that the initial field distribution $E_1(x_1,y_1)$ at the input plane 1 is reproduced at plane 2. Following relation then holds for the field $E_2(x_2,y_2)$ at plane 2:

$$E_2(x_2,y_2) = C\,\frac{1}{A}\,E_1(\frac{x_2}{A},\frac{y_2}{A}) \tag{2.39}$$

The constant C can be a complex function of x_2 y_2 with $CC^*=1$, and the factor $1/A$ is necessary to conserve the total power. It is possible to prove that the Collins integral (2.29) yields the imaging condition (2.39) in the limit $B \rightarrow 0$ [1.72]. In order to show this we rewrite Eq. (2.29) as:

$$E_2(x_2,y_2) = \frac{i}{\lambda B}\exp[-ikL]\,\exp[-i\frac{\pi}{\lambda B}(D-\frac{1}{A})(x_2^2+y_2^2)]\;\bullet \tag{2.40}$$

$$\iint E_1(x_1,y_1)\,\exp[-i\frac{\pi}{\lambda B}[A(x_1-\frac{x_2}{A})^2 + A(y_1-\frac{y_2}{A})^2]\,dx_1 dy_1$$

By using one of the definitions of the Dirac delta function:

$$\delta(z) = \lim_{B \to 0} \sqrt{\frac{i}{2\pi B}}\, \exp\left[-i\frac{z^2}{2B}\right] \tag{2.41}$$

$$\int_{-\infty}^{+\infty} f(z)\delta(z-z_o)dz = f(z_o) \tag{2.42}$$

the Collins integral (2.40) in the limit $B \rightarrow 0$ can be written as:

$$E_2(x_2,y_2) = \frac{-2\pi}{\lambda} \exp[-ikL] \exp[-i\frac{\pi}{\lambda B}(D-\frac{1}{A})(x_2^2+y_2^2)] \;\bullet$$

$$\bullet \iint E_1(x_1,y_1)\; \delta(i\sqrt{kA}(x_1-\frac{x_2}{A}))\delta(i\sqrt{kA}(y_1-\frac{y_2}{A}))\; dx_1dy_1$$

Performing the integration and applying the relationship $\delta(az)=\delta(z)/|a|$ yields the final result:

$$E_2(x_2,y_2) = \exp[-ikL] \exp[-i\frac{\pi}{\lambda B}(D-\frac{1}{A})(x_2^2+y_2^2)]\; \frac{1}{A}\; E_1(\frac{x_2}{A},\frac{y_2}{A})$$

Equation (2.39) holds for $B \rightarrow 0$ (imaging and the original plane at $z=0$) if the paraxial approximation is fulfilled.

2.4.2 Fourier Transformation (A=0)

We have already seen that in the focal plane of a focusing lens, the Collins integral becomes proportional to the Fourier transform of the incident field E_1. This means that the intensity distribution in the focal plane equals the intensity distribution in the far field. For an arbitrary optical system, the general condition for propagation to a focal plane is given by $A=0$. In this case the Collins integral (2.29) reads:

$$E_2(\theta_x,\theta_y) = \frac{i}{\lambda B} \exp[-ik(L+\frac{DB}{2}(\theta_x^2+\theta_y^2))] \iint E_1(x_1,y_1) \exp[ik(x_1\theta_x+y_1\theta_y)]\; dx_1dy_1$$

$$= \frac{i}{\lambda B} \exp[-ik(L+\frac{DB}{2}(\theta_x^2+\theta_y^2))]\; \mathcal{F}\big(E_1(x_1,y_1)\big) \tag{2.43}$$

which is equivalent to the Kirchhoff integral in Fraunhofer approximation and thus represents the far field distribution. If the optical system is chosen such that in addition to A, both A and D are equal to 0, the far field amplitude is proportional to the Fourier transform. The useful feature of the Fourier transform is the fact that a light source is completely characterized either by its electric field E or by the Fourier transform $\boldsymbol{F}(\boldsymbol{E})$ of the field. It is for this reason that Fourier transforms and the knowledge of their transformation properties are more important in optics than their basic relationship to the far field. Diffraction calculations can be considerably simplified if the light is represented by its Fourier transform rather than by the electric field itself. This will become more clear when we apply the mathematics behind Fourier transforms to diffraction problems [1.30].

Properties of Fourier Transforms

The Fourier Transform $\boldsymbol{F}$ of a function $f(x)$ is defined by:

$$A(\theta) = \boldsymbol{F}(f(x)) = \frac{k}{2\pi}\int_{-\infty}^{+\infty} f(x)\,\exp[ik\theta x]\,dx\,,\quad k\in\mathbf{R} \tag{2.44}$$

$A(\theta)$ is called the Fourier spectrum or Fourier transform of $f(x)$. For the Fourier transform to exist, the function $f(x)$ has to meet certain requirements:

a) $f(x)$ is continuous and its derivative df/dx exists at all but a finite number of points,

b) $f(x)$ is square-integrable, which means that it has to decrease faster than $1/x^2$.

The inverse Fourier transformation, $\boldsymbol{F}^{-1}$, restores the original function $f(x)$:

$$f(x) = \boldsymbol{F}^{-1}(A(\theta)) = \int_{-\infty}^{+\infty} A(\theta)\,\exp[-ik\theta x]\,d\theta \tag{2.45}$$

If $A(\theta)$, $B(\theta)$ are the Fourier transforms of the complex functions $f(x)$ and $g(x)$, respectively, and a,b are complex numbers, the following rules apply [1.1]:

Function	Fourier Spectrum	Name of Operation
$af(x)+bg(x)$	$aA(\theta)+bB(\theta)$	Linearity
$f(ax)$	$A(\theta/a)\ 1/a$	Reciprocity
$f(x-x_0)$	$A(\theta)\ exp[i\theta x_0]$	Shifting
$f(x)\ exp[-i\theta_0 x]$	$A(\theta-\theta_0)$	Shifting
$f(x)g(x)$	$A(\theta)\odot B(\theta)$	Convolution
$f(x)\odot g(x)$	$A(\theta)B(\theta)$	Convolution
$f(x)\otimes g(x)$	$A(\theta)B^*(\theta)$	Correlation

Convolution and correlation are defined as:

$$\textit{Convolution:} \quad A(\theta)\odot B(\theta) = \int_{-\infty}^{+\infty} A(\phi)\, B(\theta-\phi)\, d\phi \tag{2.46}$$

$$\textit{Correlation:} \quad f(x)\otimes g(x) = \int_{-\infty}^{+\infty} f(x')\, g^{*}(x'-x)\, dx' \tag{2.47}$$

Table 2.2 Common Fourier transforms (*a*, *b*, and *k* are real and *c* is a complex number) [1.1]:

f(x)	range	A(θ)=***F***(f(x))
c	for all x	$c\,\delta(\theta)$
$c\,\delta(x)$		c
c	for $a<x<b$	
0	else	$-ic[\exp(ikb\theta)-\exp(ika\theta)]/(2\pi\theta)$
$\exp[ik\theta_0 x]$	for all x	$\delta(\theta-\theta_0)$
$\cos(k\theta_0 x)$	for all x	$[\delta(\theta-\theta_0)+\delta(\theta+\theta_0)]/2$
$\sin(k\theta_0 x)$	for all x	$[\delta(\theta-\theta_0)-\delta(\theta+\theta_0)]/(2i)$
$\exp(-\pi(x/a)^2)$	for all x	$k\lvert a\rvert/(2\pi)\exp[-\pi(ka\theta/(2\pi))^2]$
$\exp(-x/a)$	for $x>0$	
0	else	$k/(2\pi/(1/a+ik\theta))$

Examples:

1) Far Field of a Sinusoidal Amplitude Grating

We consider a slit of infinite height in the y-direction and width $2a$ in the x-direction which is homogeneously illuminated by a plane wave with wavelength λ. Thus, we have an electric field $E(x)$ which is equal to a constant amplitude E_0 for $|x|\leq a$ and exactly zero for $|x|>a$. The amplitude transmission $t(x)$ inside the slit varies sinusoidally between 0 and 1, according to

$$t(x) = \frac{1}{2}(1+\cos[N2\pi x/a]) = \frac{1}{2}(1+g(x)), \qquad N=0,1,2,...$$

and

$$g(x) = \cos[N2\pi x/a]$$

The transmission always exhibits maxima at the edges and center of the slit while N controls the number of oscillations in between. The far field $E_F(\theta)$ is proportional to the Fourier transform of the product $E(x)t(x)$ (with $k=2\pi/\lambda$ in (2.44)):

$$E_F(\theta) = c\ F[E(x)t(x)] = c\ \frac{1}{2}F[E(x)(1+g(x))]$$

By using the convolution theorem and the linearity theorem we can rewrite this as:

$$E_F(\theta) = c\ \frac{1}{2}(F[E(x)]+F[E(x)]\odot F[g(x)])$$

We can look up both Fourier transforms in the Table 2.2 (with $\theta_0=N\lambda/a$) and together with (2.46) we obtain:

$$E_F(\theta) = \frac{cE_0}{\lambda}\left[\frac{\sin(ka\theta)}{k\theta} + \int_{-\infty}^{+\infty}\frac{\sin(ka\phi)}{k\phi}\ \frac{1}{2}(\delta(\theta-\phi-\theta_0)+\delta(\theta-\phi+\theta_0))\ d\phi\right]$$

$$= \frac{caE_0}{\lambda}\left[\frac{\sin(ka\theta)}{ka\theta} + \frac{1}{2}\frac{\sin(ka(\theta-\theta_0))}{ka(\theta-\theta_0)} + \frac{1}{2}\frac{\sin(ka(\theta+\theta_0))}{ka(\theta+\theta_0)}\right]$$

The sinusoidal transmission generates peaks in the far field at angles $\theta=\pm\theta_0=\pm N\lambda/a$.

2) Collins Integral Written In Terms of Fourier Transforms

We can rewrite the Collins-Integral (2.29) in terms of Fourier transforms and inverse Fourier transforms of the electric field and the parabolic phase factors. This rearrangement is very useful for a fast numerical calculation of diffraction integrals since time-efficient algorithms for the computation of Fourier transforms exist (such as the Fast Fourier Transform Algorithm). The basic idea is to transform the Collins integral into a convolution integral and apply the Fourier transform convolution theorem. A straightforward rearrangement of the Collins integral (2.29) results in:

$$E_2(x_2,y_2) = \frac{i}{\lambda B}\exp[-ikL]\ \exp[-i\frac{\pi}{\lambda B}(D-1)(x_2^2+y_2^2)]\ \bullet$$

$$\iint E_1(x_1,y_1)\ \exp[-i\frac{\pi}{\lambda B}(A-1)(x_1^2+y_1^2)]\ \ \exp[-i\frac{\pi}{\lambda B}[(x_1-x_2)^2+(y_1-y_2)^2]]\ dx_1dy_1$$

|-----------$f(x_1,y_1)$-----------------| |---------$g(x_1-x_2,y_1-y_2)$--------|

which is a two-dimensional convolution integral.

We know that the Fourier transform of a convolution integral equals the product of the individual Fourier transforms $F(f)$ and $F(g)$. The inverse Fourier transform of this product is therefore equivalent to the convolution integral itself:

$$E_2(x_2,y_2) = \frac{i}{\lambda B} \exp[-ikL] \exp[-i\frac{\pi}{\lambda B}(D-1)(x_2^2+y_2^2)] \, F^{-1}\left[F(f(x_1,y_1)) \, F(g(x_1,y_1))\right]$$

The Fourier transform of $g(x_1,y_1)$ can be calculated analytically and the final form reads:

$$E_2(x_2,y_2) = \frac{i}{\lambda B} \exp[-ikL] \exp[-i\frac{\pi}{\lambda B}(D-1)(x_2^2+y_2^2)] \, F^{-1}\left[F(f(x_1,y_1)) \exp[-i\frac{\pi B}{\lambda}(\theta_x^2+\theta_y^2)]\right]$$

2.5 Gaussian Beams

2.5.1 Gaussian Beams in One-Dimensional Optical Systems

The Collins integral,describes the changes in field structure as the electromagnetic field propagates through an optical system. In general the amplitude as well as the shape of the electric field distribution varies during propagation. We can, however, find a special class of fields for which the shape of the distribution remains constant and only the amplitude and lateral extent of the field is changed. These fields are referred to as eigensolutions of the diffraction integral. In the case of an unconfined electric field (which means that the integration is performed from -∞ to +∞), the eigensolutions of the Collins integral can be found analytically. Again, we first deal with one-dimensional optics which can be described by 2x2 ray transfer matrices and which exhibit rotational symmetry. An electric field $E_1(x,y)$ is an eigensolution of the Collins integral (2.29) if the following relation holds:

$$E_2(x_2,y_2) = \gamma \, E_1(\frac{x_2}{\sigma},\frac{y_2}{\sigma}) = \tag{2.48}$$

$$\frac{i}{\lambda B} \exp[-ikL] \int\int E_1(x_1,y_1) \exp[-i\frac{\pi}{\lambda B}(Ax_1^2+Dx_2^2-2x_1x_2 + Ay_1^2+Dy_2^2-2y_1y_2)] \, dx_1 dy_1$$

The form of the integral leads us to the assumption that a Gaussian field distribution

$$E_1(x_1,y_1) = E_0 \exp\left[\frac{-ik}{2q_1}(x_1^2+y_1^2)\right] \tag{2.49}$$

with q_1 being a complex number, might be a solution of the integral equation (2.48).

Insertion of (2.49) into (2.48) yields:

$$E_2(x_2,y_2) = \frac{E_0}{A-B/q_1} \exp\left[\frac{-ik}{2q_2}(x_2^2+y_2^2)\right] \tag{2.50}$$

with
$$q_2 = \frac{Aq_1 + B}{Cq_1 + D} \tag{2.51}$$

This fundamental eigensolution of the diffraction integral is called the *Gaussian beam*. A Gaussian beam stays a Gaussian beam as the field propagates through an optical system with parabolic surfaces or parabolic index profiles. Only the amplitude and the characteristic parameter q, called the beam parameter, are changed. The transformation rule (2.51) is called the ABCD law of Gaussian optics. This ABCD law can be considered as a generalization of the geometrical optics ABCD law (1.90) already discussed in Section 1.2.6. In the limit $\lambda \to 0$, the geometrical optics ABCD law is obtained from (2.51) and all imaging characteristics of Gaussian beams can then be described by geometrical optics, as will be discussed in further detail below.

Free Space Propagation of Gaussian Beams

In the simplest case, the field distribution in plane 1 (at $z=0$) is a real Gaussian distribution with a beam radius w_0 (Fig. 2.11):

$$E_1(x_1,y_1) = E_0 \exp\left[-\frac{x_1^2+y_1^2}{w_0^2}\right] \tag{2.52}$$

A comparison with (2.49) indicates that the Gaussian beam parameter q_1 is given by:

$$q_1 = i\frac{\pi w_0^2}{\lambda} = iz_0 \tag{2.53}$$

Propagation over a distance z will change the beam parameter according to the ABCD law (2.51). By using the ray transfer matrix for free space propagation (1.10), the new beam parameter at a distance z is found to be:

$$q(z) = q_1 + z \tag{2.54}$$

In order to obtain the new Gaussian beam according to (2.50), the term $1/q(z)$ has to be determined. A straightforward calculation yields:

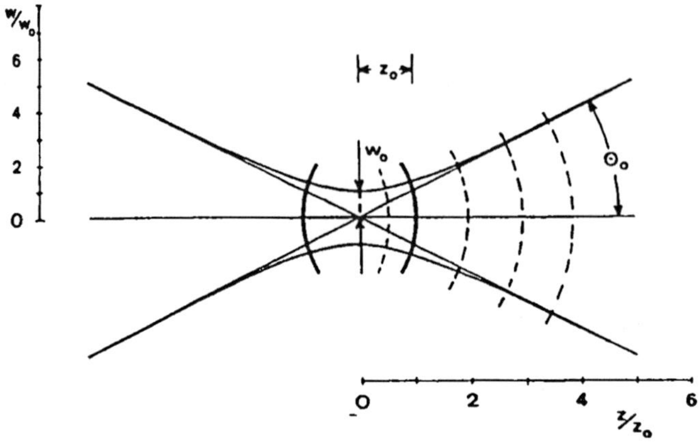

Fig. 2.11 Free space propagation of a Gaussian beam.

$$\frac{1}{q(z)} = \frac{1}{z_0(z/z_0 + z_0/z)} - \frac{i\lambda}{\pi w_0^2[1+(z/z_0)^2]} \tag{2.55}$$

The real and imaginary parts of *1/q(z)* determine the phase and the beam radius of the Gaussian beam, respectively. To show this, we define two new quantities,

$$R(z) = z_0\left[\frac{z}{z_0} + \frac{z_0}{z}\right], \qquad w(z) = w_0\sqrt{1+\left(\frac{z}{z_0}\right)^2} \tag{2.56, 2.57}$$

We can then write the beam parameter $q(z)$ as:

$$\frac{1}{q(z)} = \frac{1}{R(z)} - \frac{i\lambda}{\pi w(z)^2} \tag{2.58}$$

The Gaussian beam at a distance z from plane 1 thus reads:

$$E_2(x_2,y_2) = \frac{E_0}{1-iz/z_0}\exp\left[-\frac{ik}{2}\frac{x_2^2+y_2^2}{R(z)}\right]\exp\left[-\frac{x_2^2+y_2^2}{w(z)^2}\right] \tag{2.59}$$

We see that the $R(z)$ represents the radius of curvature of the phase front and $w(z)$ is the beam radius, defined by a $1/e^2$-decrease of the intensity with respect to the on- axis value. The Gaussian beam diverges as it propagates along the z-axis according to (2.57). A characteristic parameter of the Gaussian beam is the Rayleigh range z_0 with:

$$z_0 = \frac{\pi w_0^2}{\lambda} \tag{2.60}$$

which is also referred to as the depth of field, or sometimes as the confocal parameter. The Rayleigh range denotes the distance from the origin (location of the beam waist w_0) at which the beam radius has increased by a factor of $\sqrt{2}$. At distances z much greater than the Rayleigh range the Gaussian beam exhibits the behavior of a spherical wave. In this case, the imaginary part of $1/q(z)$ becomes negligibly small and the radius of curvature $R(z)$ increases linearly with the distance z. The Gaussian ABCD law (2.51) will then be transformed into the geometrical ABCD law (1.90). The same behavior can be found if the wavelength λ is increased, since this is equivalent to decreasing the Rayleigh range.

For smaller distances z, however, the propagation behavior is completely different than that of spherical waves in geometrical optics. The radius of curvature $R(z)$ of a Gaussian beam shows a minimum value of $2z_0$ at $z=z_0$, whereas a spherical wave increases its phase curvature proportionally to the distance from the origin. This is why Gaussian beams exhibit imaging properties that are quite different from those of geometrical optics.

For large distances $z \gg z_0$ the beam radius $w(z)$ approaches an asymptote at a finite angle θ_0 This angle, known as the divergence angle, reads:

$$\theta_0 = \lim_{z \to \infty} \frac{w(z)}{z} = \frac{w_0}{z_0} = \frac{\lambda}{\pi w_0} \tag{2.61}$$

The product of the divergence angle and the beam waist is called the *beam parameter product*. All Gaussian beams have the same beam parameter product of:

$$w_0\, \theta_0 = \frac{\lambda}{\pi} \tag{2.62}$$

which represents the minimum value possible for any field distribution if a suitable definition for the beam radii is used (see Sec. 2.6). The beam parameter product is a constant of the beam and is not changed by propagation through an optical system, provided that the optical elements can be described by ray transfer matrices.

Example: For waist w_0 of 1mm and a wavelength of λ=500nm we get:

Rayleigh range:	6.283 m
Divergence angle:	0.159 mrad
Distance z with $w(z)$=1m:	6.283 m.

Transformation Rules for Gaussian Beams

In the general case of an arbitrary optical system, the relationships for the beam radius and the radius of curvature become more complicated. If the ray transfer matrix of the optical system is given by:

$$M = \begin{pmatrix} A & B \\ C & D \end{pmatrix}$$

the new beam radius and phase curvature can be calculated by applying the ABCD law (2.51). The Gaussian beam entering the optical system is described by the q-parameter $q=z+iz_0$, and at the exit plane a new Gaussian beam emerges with the beam parameter $q'=-z'+iz_0'$ (Fig. 2.12). The distances z and z' denote the location of the beam waist of the initial and the transformed Gaussian beams, respectively. The ABCD law yields for the new waist location and the new Rayleigh range:

$$z' = \frac{(Az+B)(Cz+D) - ACz_0^2}{C^2 z_0^2 + (Cz+D)^2} \qquad \text{if } C \neq 0 \tag{2.63}$$

$$z' = -\frac{Az+B}{D} \qquad \text{if } C=0 \tag{2.64}$$

$$z_0' = z_0 \frac{Cz'+A}{Cz+D} \qquad \text{with } z' \text{ from (2.63/2.64)} \tag{2.65}$$

Note that the new beam waist is located left of the exit plane if z' is negative. A positive value of z' refers to a beam waist location to the right of the optical system, as it is the case in Fig. 2.12.

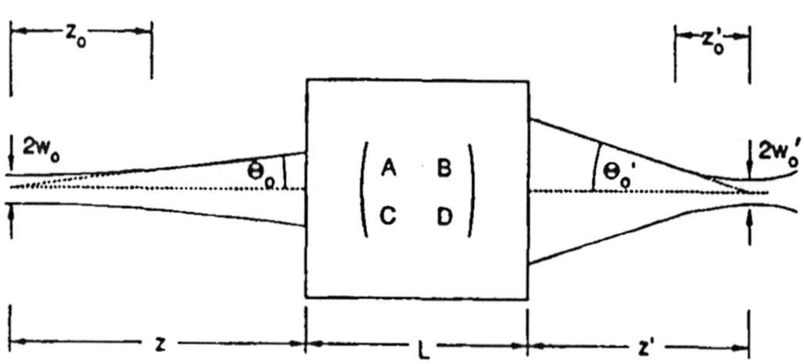

Fig. 2.12 The propagation through an optical system of length L and ray transfer matrix M transforms the Gaussian beam into a new Gaussian beam with beam waist location z', Rayleigh range z_0', and divergence angle θ_0'.

The new angle of divergence θ'_0 and the new beam waist w'_0 can be calculated from:

$$\theta'_0 = \sqrt{\frac{\lambda}{\pi z'_0}} \quad , \qquad w'_0 = \sqrt{\frac{\lambda z'_0}{\pi}}$$

which means that the beam parameter product $\theta'_0 w'_0$ remains constant ($=\lambda/\pi$).

A different approach to calculating the Gaussian beam transformation is to link the beam radii and angles of divergence at the exit plane of the optical system to those at the entrance plane. If w_1, R_1, and θ_{01} denote the beam radius, radius of curvature, and divergence angle at the entrance plane, respectively, then the beam radius w_2, the radius of curvature R_2, and divergence angle θ_{02} at the exit plane are given by:

$$w_2^2 = A^2 w_1^2 + 2\,ABw_1^2/R_1 + B^2\theta_{01}^2 \tag{2.66}$$

$$\theta_{02}^2 = C^2 w_1^2 + 2\,CDw_1^2/R_1 + D^2\theta_{01}^2 \tag{2.67}$$

$$R_2 = \frac{w_2^2}{\sqrt{w_2^2\theta_{02}^2 - (\lambda/\pi)^2}} \tag{2.68}$$

The new beam waist location z', the Rayleigh range z'_0, and the beam waist radius w'_0 can be found by using the relations:

$$z' = \frac{-1}{R_1}\,\frac{1}{\dfrac{1}{R_1^2} + \dfrac{\lambda^2}{\pi^2 w_2^4}} \quad , \qquad z'_0 = \frac{\lambda}{\pi\theta_{02}^2} \quad , \qquad w'_0 = \frac{\lambda}{\pi\theta_{02}}$$

In the following we will present the transformation rules for common optical systems:

a) Plane Dielectric Slab

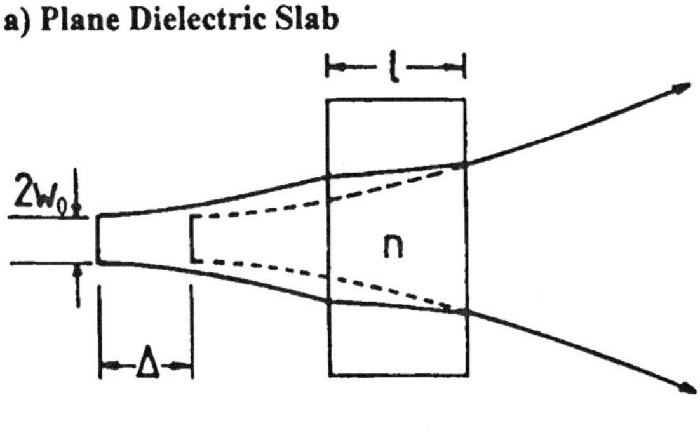

Fig. 2.13 Propagation of a Gaussian beam through a plane dielectric slab with refractive index *n*.

The ray transfer matrix of the plane slab (see Fig. 1.5) yields with (2.63)-(2.65):

$$z'_0 = z_0 \qquad z' = -z - \frac{l}{n}$$

$$\theta'_0 = \theta_0 \qquad w'_0 = w_0$$

If the refractive index of the slab were $n=1$, we would get the new waist location $z' = -z-l$, which is just the location of the original beam waist. For indexes other than unity, the new beam waist is shifted to the right (as seen looking back into the slab) by:

$$\Delta = l\left(\frac{n-1}{n}\right)$$

but the Gaussian beam remains unchanged.

b) Propagation Through a Telescope

Telescopes are widely used in laser systems to magnify the beam and thus decrease the angle of divergence (Fig. 2.14). A lower divergence results in a smaller spot size in the focal plane of a focusing lens. With (2.63)-(2.65) we get:

$$z' = -z\left[\frac{f_2}{f_1}\right]^2 + \frac{f_2}{f_1}(f_1+f_2)$$

$$w'_0 = w_0\frac{f_2}{f_1}\,, \qquad \theta'_0 = \theta_0\frac{f_1}{f_2} \qquad z'_0 = z_0\left[\frac{f_2}{f_1}\right]^2$$

A telescope with magnification $M=f_2/f_1$ decreases the divergence by a factor $1/M$ and increases the beam radius by M. Again, the beam parameter product stays constant.

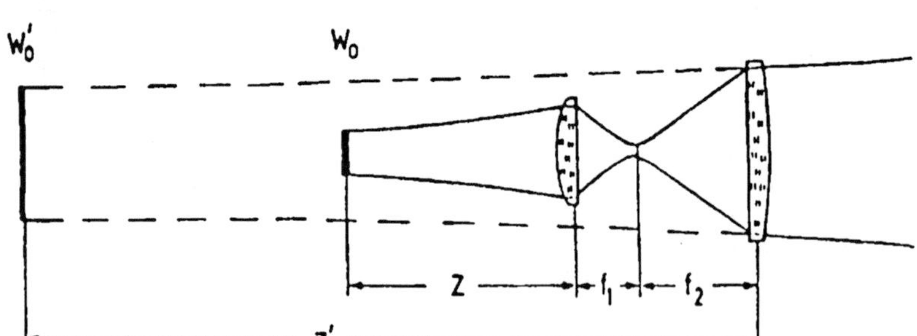

Fig. 2.14 Gaussian Beam propagation inside a telescope (not to scale).

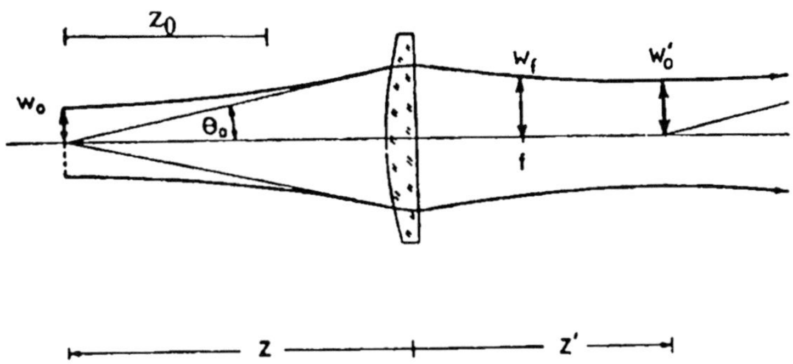

Fig. 2.15 Gaussian beam propagation through a focusing lens. An image of the beam waist is generated at a distance z' behind the lens.

c) Imaging/Focusing with a Lens

We are looking for the location and the size of the new beam waist when the Gaussian beam is transformed by a lens. For a lens with positive focal length f, the Gaussian beam is focused as depicted in Fig. 2.15. Since the focusing of Gaussian beams plays an important role in laser optics, we will discuss this beam propagation problem in more detail.

Note that we use the sign convention of geometrical optics for the distances z and z': both distances are positive if the object (beam waist w_0) and the image (beam waist w'_0) are located to the left and to the right of the lens, respectively. The imaging condition of geometrical optics:

$$\frac{1}{z} + \frac{1}{z'} = \frac{1}{f}$$

does not apply to the waist of Gaussian beams since it is only valid for spherical waves whose radii of curvatures increase proportionally to the distance z. For Gaussian beams the different phase front propagation behavior generates an additional term in the imaging condition. By using the ABCD law (2.51) and setting the radii of curvature to infinity at both planes, the imaging condition can be found to be:

$$\frac{1}{z} + \frac{1}{z'} = \frac{1}{f} + \frac{z_0^2}{z(z^2 + z_0^2 - zf)} \tag{2.69}$$

with z' being the distance of the new beam waist location from the lens. If z' is positive the beam waist is located at the right hand side of the lens.

The following relations hold for the parameters depicted in Fig. 2.15:

$$w'_0 = w_0 \frac{f}{\sqrt{z_0^2 + (z-f)^2}} \tag{2.70}$$

$$z'_0 = z_0 \frac{f^2}{z_0^2 + (z-f)^2} \tag{2.71}$$

$$w_f = f\,\theta_0 \tag{2.72}$$

Figures 2.16 and 2.17 show the graphic presentation of (2.69) and (2.70), respectively. The imaging condition for geometrical optics is reached in the limit $z_0/z=0$ which means that the distance of the object beam waist has to be much greater than the Rayleigh range. This is to be expected because for large distances from the beam waist, the Gaussian beam behaves like a spherical wave as far as propagation is concerned.

The reader should note that there are major differences in the imaging of the waist of Gaussian beams as compared to geometrical optics. One example worth mentioning is the beam waist of the Gaussian beam located in the front focal plane $(z=f)$. According to (2.69), the image waist is found in the back focal plane $(z'=f)$. This is quite surprising since from our experience in geometrical optics we would expect that placing a source in the front focal plane would produce a collimated beam (image at infinity)! This focusing experiment can serve as a quick check to determine whether a laser beam can be described as a Gaussian beam or as a geometrical optics light source (at the exit aperture of a fiber for example).

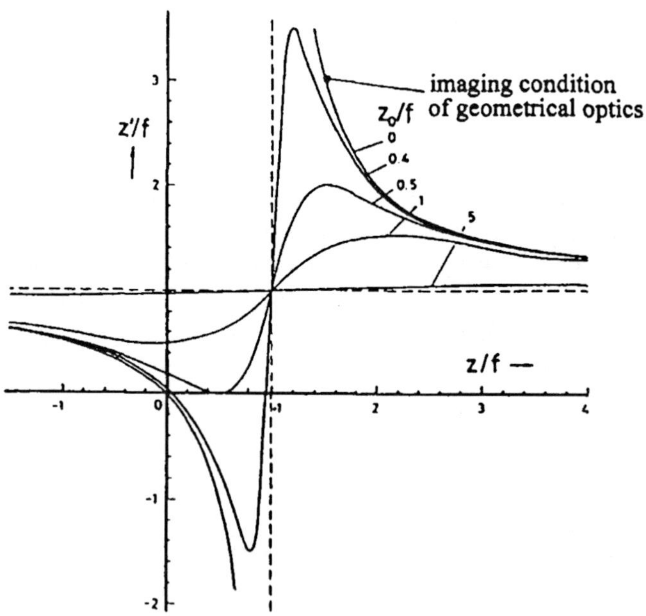

Fig. 2.16 Imaging of Gaussian beams according to (2.69).

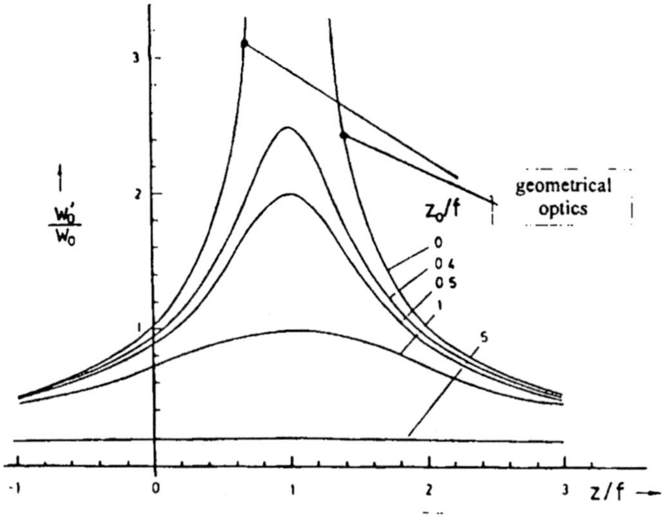

Fig. 2.17 Imaging of Gaussian beams. Shown is the relationship between the magnification and the distance z of the beam waist from the focusing lens according to (2.70).

Equations (2.70) and (2.71) indicate that we cannot simultaneously attain a small focal spot radius w'_0 and a large Rayleigh range z'_0. Division of (2.70) by (2.71) yields:

$$\frac{w'^2_0}{z'_0} = \frac{w_0^2}{z_0} = \frac{\lambda}{\pi} = w_0\theta_0 \tag{2.73}$$

In addition to the beam parameter product $w_0\theta_0$, the ratio of focal spot size to Rayleigh range is a constant of the beam. If we try to decrease the focal spot radius (by choosing a lens with a shorter focal length or by increasing the Rayleigh range z_0 of the object beam with a telescope), we will reduce the Rayleigh range proportional to the area of the focal spot. Equation (2.73) holds only for Gaussian beams. Laser beams usually have beam parameter products that are larger than λ/π (by a factor 3-100, depending on the type of laser and the output power). Equation (2.73) then reads:

$$\frac{w'^2_0}{z'_0} = \frac{w_0^2}{z_0} = M^2\frac{\lambda}{\pi} = M^2\, w_0\theta_0 \tag{2.74}$$

with M^2 being the beam propagation factor. It is for this reason that in laser material processing it is difficult to generate thin, deep cuts into a work piece. The quality of a laser cut is commonly defined by the aspect ratio which is the depth of the cut over the width. In order to get a high aspect ratio it is a prerequisite to have a laser beam with a low beam parameter product. Laser beam quality can thus be characterized by the beam parameter product. High beam quality is equivalent to a low beam parameter product (beam propagation factor, M^2, near unity) or, described in other words, by a large Rayleigh range for a given focal spot size. It is interesting that the beam radius in the focal plane (2.72) depends only on the divergence angle of the Gaussian beam in front of the lens. This is a direct consequence of the fact that the Fourier transform of the beam is generated in the focal plane.

In most cases the beam waist w_0 and the Rayleigh range z_0 of the beam are not known and only the beam radius w_L at the lens can be determined and (2.70) cannot be used to calculate the focal spot radius w'_0. Fortunately, if the object distance z is large compared to the focal length f ($z>5f$ is a good rule of thumb), we can derive an approximate formula for the spot size w'_0. According to (2.57) the beam radius on the lens is given by:

$$w_L = w_0 \sqrt{1+(z/z_0)^2}$$

By using this expression and the assumption $z \gg f$, we can rewrite (2.70) and (2.71):

$$w'_0 = \frac{w_0}{z_0} \frac{f}{\sqrt{1+((z-f)/z_0)^2}} = \frac{\lambda f}{\pi w_0} \frac{1}{\sqrt{1+((z-f)/z_0)^2}} \approx \frac{\lambda f}{\pi w_L} \tag{2.75}$$

$$z' \approx f \tag{2.76}$$

Example: A Nd:YAG laser whose output beam is a Gaussian beam with $w_0=0.582mm$, $z_0=1m$, and $\lambda=1.064\mu m$ is focused by means of a $50mm$ lens. The lens is at $z=1m$ in front of the beam waist. Using (2.69) and (2.70) we get for the position and the size of the focal spot:

$z' = 51.25mm \qquad w'_0 = 0.0211$ mm

The approximated equations (2.75) and (2.76) provide the following solutions:

$z' = 50.00mm \qquad w'_0 = 0.0206$ mm

d) Penetration into a Medium with a Higher Index of Refraction

We have a Gaussian beam hitting a dielectric planar interface with refractive index n. The radius of curvature and beam radius of the beam at the interface are R_1 and w_1, respectively. The ray transfer matrix for the refraction at the interface reads:

$$M = \begin{pmatrix} 1 & 0 \\ 0 & 1/n \end{pmatrix}$$

The ABCD law (2.51) yields for the new q-parameter q_2:

$$\frac{1}{q_2} = \frac{1}{nq_1}$$

Separation into the radius of curvature R and the beam radius w according to (2.58) yields:

$$R_2 = n\,R_1 \;, \qquad w_2^2 = w_1^2$$

The refraction leads to an increase in curvature, but the beam radius at the interface remains unchanged.

2.5.2 Elliptical Gaussian Beams

So far one-dimensional Gaussian beams or Gaussian beams of rotational symmetry have been discussed. They are completely characterized by the complex beam parameter q with:

$$\frac{1}{q} = \frac{1}{R} - \frac{i\lambda}{\pi w^2}$$

By measuring the real part and the imaginary part of q at a position z in the reference frame, the beam is completely determined. Equations (2.56) to (2.58) provide the position z_w of the waist and the Rayleigh range z_0. If it is known that the beam is a TEM_{00} mode with rotational symmetry, the waist radius w_0 is obtained from (2.60), and (2.61) provides the far field divergence angle θ_0. Such a beam is called a *stigmatic fundamental mode*. It exhibits a circular spot and a spherical wave front at every distance z. The curves of constant intensity (isophotes) and the lines of constant phase (isophases) in the x-y plane are always circles.

However, in some lasers, such as diode lasers and slab lasers, the fundamental mode exhibits an elliptical cross section. Let us assume a stigmatic TEM_{00} mode, generated by a HeNe laser, passing through a thin cylinder lens with refractive power D_y and the focal line along the x-axis (see Fig. 1.18b). The spot size directly behind the lens is unchanged and the radius of curvature in the x-direction is unaffected. The radius of curvature in the y-direction now reads:

$$\frac{1}{R_y} = \frac{1}{R} - D_y$$

The isophases are now ellipses. Due to the focusing in y-direction, the beam radius w_y will differ from w_x with increasing distance z from the lens (Fig. 2.18). The isophotes, which are circles right behind the lens, become elliptical. Such a beam is called simple astigmatic. The ellipses of constant intensity and constant phase have the same orientation (along the principal axes of the cylinder lens) as the beam propagates.

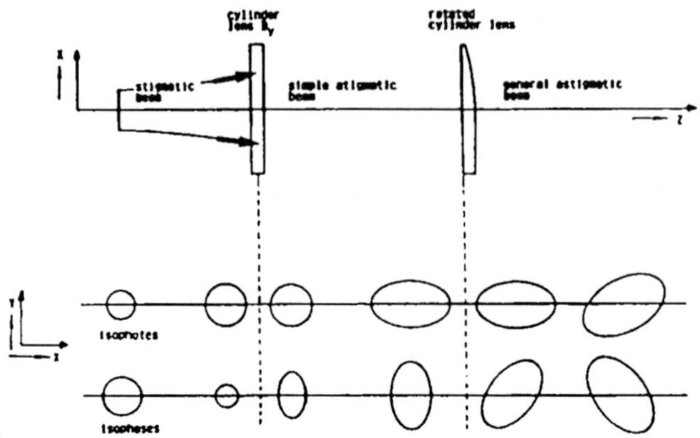

Fig. 2.18 Transformation of a stigmatic beam into a simple astigmatic beam and into a general astigmatic beam. For a general astigmatic beam, the isophotes and the isophases have different orientation and the ellipses rotate as the beam propagates.

The two ellipses are determined by 4 parameters plus one additional parameter for the orientation of the ellipses with respect to the reference frame. The field of a simple astigmatic beam can still be separated in the form:

$$E(x,y) = E(x)\ E(y)$$

The Gaussian beam propagation law (2.51) can be applied separately in the x- and the y-direction for the q- parameters q_x and q_y, respectively. If such a beam is focused with a spherical lens, two line foci will appear at different distances from the lens.

After a distance L the simple astigmatic beam passes a rotated cylinder lens. The ellipse of constant intensity is not affected by this, but the ellipse of constant phase is rotated. The two ellipses now have different orientations. Such a beam is called a general astigmatic beam. The main axes of the two ellipses form an angle to each other and the orientation does not remain constant during the propagation. A general astigmatic beam is characterized by six parameters, three for each ellipse (two for the main axes and one for the orientation).

The concept of the complex beam parameter can still be applied if we introduce a complex beam matrix Q^{1} with [1.35,1.36]:

$$Q^{-1} = \begin{pmatrix} \dfrac{1}{q_{xx}} & \dfrac{1}{q_{xy}} \\ \dfrac{1}{q_{xy}} & \dfrac{1}{q_{yy}} \end{pmatrix} \qquad (2.77)$$

Table 2.3 Properties of Gaussian beams

	constant intensity	constant phase	free parameters
stigmatic	$\frac{x^2}{w^2(z)} + \frac{y^2}{w^2(z)} = C$	$\frac{kx^2}{R(z)} + \frac{ky^2}{R(z)} = C$	2
simple astigmatic	$\frac{x^2}{w_x^2(z)} + \frac{y^2}{w_y^2(z)} = C$	$\frac{kx^2}{R_x(z)} + \frac{ky^2}{R_y(z)} = C$	5
general astigmatic	$\frac{x^2}{w_x^2(z)} + \frac{xy}{w_{xy}^2(z)} + \frac{y^2}{w_y^2(z)} = C$	$\frac{kx^2}{R_x(z)} + \frac{kxy}{R_{xy}(z)} + \frac{ky^2}{R_y(z)} = C$	6

The general astigmatic fundamental mode is given by:

$$E = E_0 \exp\left[\frac{-ik}{2} \boldsymbol{r}\boldsymbol{Q}^{-1}\boldsymbol{r}\right] \quad , \quad \boldsymbol{r} = (x,y) \tag{2.78}$$

with
$$\boldsymbol{r}\boldsymbol{Q}^{-1}\boldsymbol{r} = \frac{x^2}{q_{xx}} + \frac{2xy}{q_{xy}} + \frac{y^2}{q_{yy}} \tag{2.79}$$

The beam is determined by the three different complex parameters of the $\boldsymbol{Q}^{-1}$ matrix. The propagation of such a beam through ABCD-type optical systems can be evaluated by applying the Collins integral (2.32). A straightforward but troublesome calculation yields the generalized ABCD law [1.53]:

$$\boldsymbol{Q}_2^{-1} = (\boldsymbol{C} + \boldsymbol{D}\boldsymbol{Q}_1^{-1})(\boldsymbol{A} + \boldsymbol{B}\boldsymbol{Q}_1^{-1})^{-1} \tag{2.80}$$

with $\boldsymbol{A},\boldsymbol{B},\boldsymbol{C},\boldsymbol{D}$ being the 2x2 submatrices introduced in Sec. 1.2.4.

Examples:

1) Transformation of a general astigmatic Gaussian beam by an astigmatic thin lens.
By using (2.80) with:

$$\boldsymbol{A} = \begin{pmatrix} 1 & 0 \\ 0 & 1 \end{pmatrix} = \boldsymbol{I} \ , \ \boldsymbol{B} = \begin{pmatrix} 0 & 0 \\ 0 & 0 \end{pmatrix} = \boldsymbol{0} \ , \ \boldsymbol{C} = \begin{pmatrix} -1/f_x & 0 \\ 0 & -1/f_y \end{pmatrix} , \ \boldsymbol{D} = \begin{pmatrix} 1 & 0 \\ 0 & 1 \end{pmatrix} = \boldsymbol{I}$$

we obtain: $$Q_2^{-1} = C + Q_1^{-1} \tag{2.81}$$

This relation is equivalent to the three equations:

$$\frac{1}{q_{xx,2}} = -\frac{1}{f_x} + \frac{1}{q_{xx,1}}$$

$$\frac{1}{q_{xy,2}} = \frac{1}{q_{xy,1}}$$

$$\frac{1}{q_{yy,2}} = -\frac{1}{f_y} + \frac{1}{q_{yy,1}}$$

The real parts of $1/q_{xx}$ and $1/q_{yy}$, which represent the radii of curvatures, are changed.

2) Propagation of a general astigmatic Gaussian beam in free space over a distance *z*. The four submatrices are ***A=I, B=zI, C=0, D=I.*** The propagation law (2.80) yields:

$$\frac{1}{q_{xx,2}} = \frac{1}{N}\left[\frac{1}{q_{xx,1}}\left(1+\frac{z}{q_{yy,1}}\right) - \frac{z}{q_{xy,1}^2}\right]$$

$$\frac{1}{q_{xy,2}} = \frac{1}{N}\frac{1}{q_{xy,1}}$$

$$\frac{1}{q_{yy,2}} = \frac{1}{N}\left[\frac{1}{q_{yy,1}}\left(1+\frac{z}{q_{xx,1}}\right) - \frac{z}{q_{xy,1}^2}\right]$$

with $$N = \left(1 + \frac{z}{q_{xx,1}}\right)\left(1 + \frac{z}{q_{yy,1}}\right) - \frac{z^2}{q_{xy}^2}$$

This complicated set of equations reflects the fact that the two ellipses are rotating while the beam propagates in the z-direction. The simple case of the stigmatic beam is immediately obtained for $1/q_{xx,1}=1/q_{yy,1}=1/q_1$, $1/q_{xy,1}=0$ and the upper equation results in (2.51).

A stigmatic beam can be transformed into a general astigmatic beam by a suitable set of lenses as shown in Fig. 2.18. The inverse transformation can also be applied, but only for Gaussian beams, as was shown in [1.36]. First, the matrix Q^{-1} has to be diagonalized by a rotation. The angle of rotation is given by:

$$\tan(2\phi) = \frac{2/q_{xy}}{1/q_{xx} - 1/q_{yy}} \tag{2.82}$$

Since this equation generally yields a complex angle, a simple astigmatic lens cannot transform a general astigmatic beam into a stigmatic one. These transformation properties were first investigated in detail by Arnaud [1.32]. It is interesting to note that the complex angle ϕ is constant for free space propagation. This can be easily proved by inserting the three equations of example 2 into (2.82).

The transformation of a general astigmatic Gaussian beam into a stigmatic one has to be done in two steps. First we generate a simple astigmatic beam by a rotation and in a second step the simple astigmatic beam is transformed into a stigmatic one. The ellipse of constant intensity is rotated with respect to the reference frame by an angle θ_w, given by:

$$\tan(2\theta_w) = \frac{2\ Im(1/q_{xy})}{Im(1/q_{xx} - 1/q_{yy})} = \frac{2/w_{xy}^2}{1/w_{xx}^2 - 1/w_{yy}^2} \tag{2.83}$$

and the ellipse of constant phase is rotated by θ_p, with:

$$\tan(2\theta_p) = \frac{2\ Re(1/q_{xy})}{Re(1/q_{xx} - 1/q_{yy})} = \frac{2/R_{xy}}{1/R_{xx} - 1/R_{yy}} \tag{2.84}$$

If θ_p is equal to θ_w, we already have a simple astigmatic beam, and by rotating the reference frame by this angle we obtain the beam in the main axes presentation. If the two angles are not equal, we have to rotate the axis of the phase ellipse by a suitable bifocal or cylindrical lens. In both cases we obtain a simple astigmatic beam. If this beam is focused, two foci appear at different distances from the focusing lens. By using a set of cylindrical lenses the foci can be imaged into one circular waist [1.57].

Note that all the equations discussed above hold only for Gaussian beams. For other beams such as higher order modes, these relations are not valid! The case of general fields will be discussed in the next section.

2.6 Intensity Moments and Beam Propagation

2.6.1 Stigmatic and Simple Astigmatic Beams

We have seen in the last section how to calculate the beam radius w and the divergence angle θ of Gaussian beams in any plane within an optical system by using the ABCD law for the beam parameter q. If we start at the plane of the beam waist w_0 and propagate through an optical system with ray matrix elements A, B, C, D, the beam radius w at the new plane and the divergence angle θ of the new Gaussian beam read according to (2.66) and (2.67)

$$w^2 = A^2 w_0^2 + B^2 \theta_0^2 \tag{2.85}$$

$$\theta^2 = C^2 w_0^2 + D^2 \theta_0^2 \tag{2.86}$$

assuming the simplest case of a stigmatic Gaussian beam (circular symmetry). For simple astigmatic beams the above relations can be applied separately for the two main axes. Beam radius and divergence angle are both defined by the $1/e^2$-decrease of the intensity. For free space propagation over a distance z with $A=1$, $B=z$, $C=0$, and $D=1$, we obtain the familiar result:

$$w = w_0\sqrt{1 + (z/z_0)^2} \quad , \qquad \theta = \theta_0$$

or

$$w^2 = w_0^2 + z^2\theta_0^2$$

The majority of laser beams, however, cannot be described by Gaussian beams. The application of the above shown propagation rules is restricted to lasers working in fundamental mode operation like HeNe lasers or single emitter diode lasers. Fortunately, the ABCD law can be generalized to arbitrary, partially coherent beams by defining beam sizes via the intensity moments of the beam [1.53,1.58,1.60,1.62,1.64]. In this section we will discuss simple astigmatic beams. These are elliptical beams whose main axes of the phase and the intensity ellipses coincide with the x,y reference frame, the equivalent to the simple astigmatic Gaussian beam of Sec. 2.5.2. The field can then be factorized into a product of functions depending on x or y. The propagation of the x and the y terms can be treated separately. The stigmatic beam with circular symmetry is a special case of the astigmatic beam. In the following only the x terms are considered. By replacing x with y the corresponding formulas for the y-direction are obtained. The general case in which a separation in x and y terms is not possible will be discussed in Sec. 2.6.2.

First Order Moments

The first order moments are defined as:

$$\langle x\rangle = \frac{1}{P_N}\int\limits_{-\infty}^{+\infty}\int\limits_{-\infty}^{+\infty} x|E(x,y)|^2 dx dy \tag{2.87}$$

$$\langle\theta_x\rangle = \frac{1}{P_N}\int\limits_{-\infty}^{+\infty}\int\limits_{-\infty}^{+\infty} \theta_x|E_F(\theta_x,\theta_y)|^2 d\theta_x d\theta_y \tag{2.88}$$

with P_N being a normalization factor, related to the total power in the beam:

$$P_N = \int_{-\infty}^{+\infty}\int_{-\infty}^{+\infty} |E(x,y)|^2 dxdy = \lambda^2 \int_{-\infty}^{+\infty}\int_{-\infty}^{+\infty} |E_F(\theta_x,\theta_y)|^2 d\theta_x d\theta_y \tag{2.89}$$

The Fourier transform in (2.88) can be replaced by the first derivative of the field. Application of (2.44) and (2.45) yields:

$$<\theta_x> = \frac{-i\lambda}{2\pi P_N} \int_{-\infty}^{+\infty} E(x,y) \frac{\delta E^*(x,y)}{\delta x} dxdy \tag{2.90}$$

This equation is sometimes more convenient for analytical calculations. The above definitions require that the field decreases sufficiently fast with x, otherwise (2.87) will diverge. Furthermore, in order to get a finite value in (2.88) the field must be continuous. Note that although the integrals have infinite limits, the paraxial approximation has to hold for the electric field. This means that the far field exists only for small values of θ_x.

The first moments provide the center of gravity of the near field and the propagation direction (far field). These moments can be measured by recording the intensity distributions with a CCD camera and processing the image mathematically using image processing software. It is interesting to look at the propagation of the first moments. By applying the Collins integral, (2.87) and (2.88) yield after some troublesome but straightforward calculations:

$$<x_2> = A<x_1> + B<\theta_{x1}> \tag{2.91}$$

$$<\theta_{x2}> = C<x_1> + D<\theta_{x1}> \tag{2.92}$$

and the corresponding expressions for the y-direction. These equations are identical with those we obtained for the propagation of rays in geometrical optics, as discussed in Sec. 1.2. The center of gravity of an arbitrary field propagates through optical systems in the same way as rays, a very simple and satisfying result.

Second Order Moments

For simple astigmatic beams the following relations can be applied separately for the x- and the y-direction. Beam radius and divergence angle are defined by:

$$<w_x^2> = \frac{4}{P_N} \int_{-\infty}^{+\infty}\int_{-\infty}^{+\infty} x^2 |E(x,y)|^2 dxdy \tag{2.93}$$

$$<\theta_x^2> = \frac{4\lambda^2}{P_N} \int\int_{-\infty}^{+\infty} \theta_x^2 |E_F(\theta_x,\theta_y)|^2 d\theta_x d\theta_y \qquad (2.94)$$

Equation (2.94) can again be transformed by using the Fourier transform of (2.87):

$$<\theta_x^2> = \frac{\lambda^2}{\pi^2 P_N} \int\int_{-\infty}^{+\infty} \left|\frac{\delta E(x,y)}{\delta x}\right|^2 dxdy \qquad (2.95)$$

A third, mixed moment of second order exists which is defined by:

$$<w_x\theta_x> = \frac{-i\lambda}{\pi P_N} \int\int_{-\infty}^{+\infty} x \; [E(x,y)\frac{\delta E^*(x,y)}{\delta x} - E^*(x,y)\frac{\delta E(x,y)}{\delta x}] \; dxdy \qquad (2.96)$$

The factor 4 in the above equations is introduced to adapt the results to the waist radius and the divergence angle of a Gaussian beam. If the electric field of a Gaussian beam, which may be astigmatic but aligned, is inserted one obtains:

$$<w_{x,y}^2> = w_{x,y}^2$$

$$<\theta_{x,y}^2> = \theta_{0\;x,y}^2$$

$$<w_{x,y}\theta_{x,y}> = \frac{w_{x,y}^2}{R_{x,y}}$$

The second intensity moments of an arbitrary field can be identified with the beam radius, the angle of divergence, and the radius of curvature. It is assumed that the origin of the reference frame coincides with the centers of gravity, which means that the first moments of the field vanish. At the beam waist the radius of curvature is infinite. For an arbitrary field the waist can thus be defined by the condition:

$$<w_x\theta_x> = 0$$

If the beam and the optical system both exhibit rotational symmetry (a one-dimensional optical system with a 2x2 ray transfer matrix), the second moment definitions of the beam radii read:

$$<w_r^2> = \frac{4\pi}{P_N} \int_0^{\infty} r^3 |E(r)|^2 dr \qquad (2.97)$$

$$<\theta^2> = \frac{4\pi\lambda^2}{P_N} \int_0^\infty \theta^3 |E_F(\theta)|^2 d\theta \tag{2.98}$$

$$<w_r \theta_r> = \frac{-i\lambda}{P_N} \int_0^\infty r^2 \, [E \frac{\delta E^*}{\delta r} - E^* \frac{\delta E}{\delta r}] \, r \, dr \tag{2.99}$$

Using these definitions of the beam radii and divergence angles the generalized ABCD law for one-dimensional optics can be derived by applying the Collins integral or the Wigner function presented in Sec.2.7. The ABCD law holds in rectangular symmetry as well as for the radial components:

$$<w_2^2> = A^2 <w_1^2> + 2AB <w_1 \theta_1> + B^2 <\theta_1^2> \tag{2.100}$$

$$<w_2 \theta_2> = AC <w_1^2> + (AD+BC) <w_1 \theta_1> + BD <\theta_1^2> \tag{2.101}$$

$$<\theta_2^2> = C^2 <w_1^2> + 2CD <w_1 \theta_1> + D^2 <\theta_1^2> \tag{2.102}$$

It is convenient to refer to the beam waist with $<w_1^2> = <w_{01}^2>$ *and* $<w_1 \theta_1> = 0$. The above equations then reduce to:

$$<w_2^2> = A^2 <w_{01}^2> + B^2 <\theta_1^2> \tag{2.103}$$

$$<w_2 \theta_2> = AC <w_{01}^2> + BD <\theta_1^2> \tag{2.104}$$

$$<\theta_2^2> = C^2 <w_{01}^2> + D^2 <\theta_1^2> \tag{2.105}$$

We can define a Rayleigh range z_0 by generalizing the expression from Gaussian optics:

$$<z_0^2> = <w_0^2> / <\theta_0^2> \tag{2.106}$$

Be aware that the Rayleigh range can be different for the x- and the y-direction if the beam is elliptical. By using this definition, the generalized law for free space propagation of an arbitrary field over a distance *z* has the same form as for Gaussian beam propagation:

$$<w^2> = <w_0^2> \left[1 + \frac{z^2}{<z_0^2>} \right] \tag{2.107}$$

The product of the waist $\sqrt{<w_0^2>}$ and the divergence $\sqrt{<\theta^2>}$ is called the beam parameter product. It characterizes the beam quality. In the best case of a Gaussian beam it results in:

$$\left[<w_0^2><\theta^2>\right]_{TEM_{00}} = \frac{\lambda^2}{\pi^2}$$

For higher order modes (see Sec. 5.2.1) one obtains in rectangular symmetry:

$$\left[<w_0^2><\theta^2>\right]_{TEM_{m}} = (2m+1)^2\,\frac{\lambda^2}{\pi^2}$$

and in circular symmetry:

$$\left[<w_0^2><\theta^2>\right]_{TEM_{p\ell}} = (2p+\ell+1)^2\,\frac{\lambda^2}{\pi^2}$$

Arbitrary fields, especially high power laser beams, are a mixture of modes and the beam parameter product is defined by:

$$\left[<w_0^2><\theta^2>\right]_{general} = M^4\,\frac{\lambda^2}{\pi^2} \tag{2.108}$$

with M^2 being the beam propagation factor. With the exception of the fundamental mode (Gaussian beam, $M^2=1$), the beam propagation factor is always larger than one. M^2 is a constant of the beam and does not change during propagation through ABCD-type optical systems.

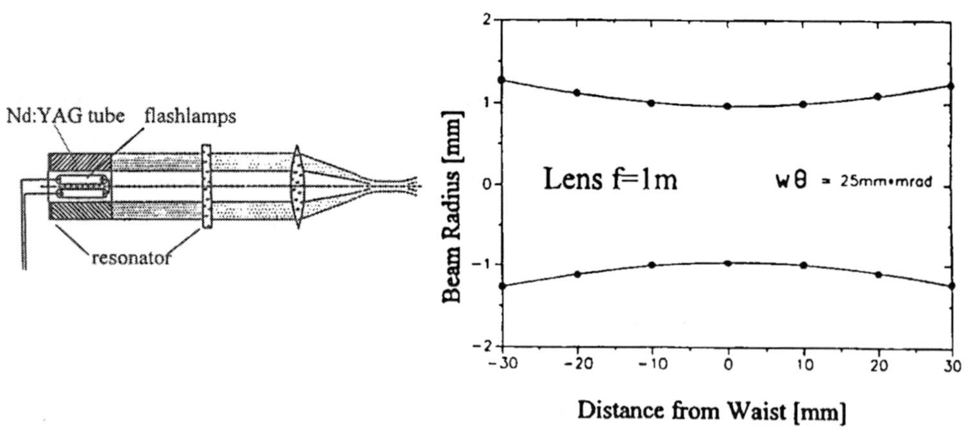

Fig. 2.19 Second moment beam radii in the vicinity of the beam waist of a focused Nd:YAG tube laser beam. A parabolic fit using (2.107) yields a beam parameter product of 25 mm mrad.

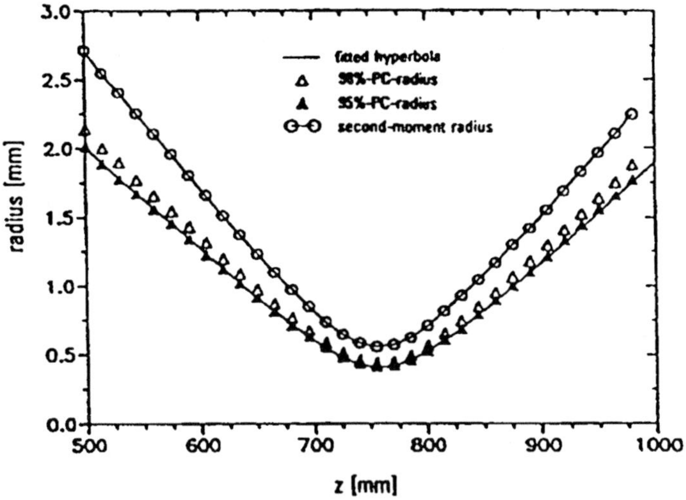

Fig. 2.20 Measured beam radii of a focused Nd:YAG laser beam (stable resonator in multimode operation) using the second moment definition (2.97) and two different power content (PC) definitions of the beam radius [1.70] (© SPIE 1994).

Figure 2.19 shows an experimental application of the generalized propagation law (2.103). The beam from an Nd:YAG tube laser (outer beam diameter at the output coupler: 55mm, inner beam diameter: 35mm) was focused by means of a focusing lens with focal length f=1m. The second moment beam radii were determined in the vicinity of the focal spot according to (2.97) by using a CCD camera and an image processing system. The solid line in the right hand graph represents the parabola, fitted according to (2.107). This parabolic fit enables one to determine the waist radius $\sqrt{<w_0^2>}$ and the Rayleigh range $\sqrt{<z_0^2>}$. The knowledge of these two parameters is sufficient to calculate the divergence angle and the beam parameter product [1.66]. It should be noted that the second moment beam radii do not exist for intensity distributions that decrease slower than $1/x^2$. This is the reason second order moment radius definitions cannot be applied to beams diffracted by hard edge apertures (the second moment diverges in the far field).

In practice, beam radii and divergence angles are defined via the confinement of 86.5% of the total beam power. In most cases beam radii defined by the second order intensity moment are larger than those containing 86.5% of the total power [1.70] (see Fig. 2.20). The generalized ABCD law, therefore, is a powerful tool to understand the characteristics of electromagnetic fields and define the beam quality of laser beams [1.64,1.71,1.73,1.86,1.107], but it does only provide limited information the focusability of laser beams in material processing applications.

The Complex Beam Parameter q

As was shown for the Gaussian beam, the mixed moment $<w\theta>$ is directly related to the radius of curvature. For arbitrary fields a generalized radius of curvature can be defined:

$$R_g = \frac{<w^2>}{<w\theta>} \tag{2.109}$$

By using the generalized q-parameter [1.60]:

$$\frac{1}{q_g} = \frac{1}{R_g} - \frac{i\lambda M^2}{\pi<w^2>}, \tag{2.110}$$

we can rewrite (2.100)-(2.102) into the convenient form of the well-known ABCD law:

$$q_{g,2} = \frac{Aq_{g,1} + B}{Cq_{g,1} + D} \tag{2.111}$$

Unfortunately, this generalized ABCD law cannot be applied to general two-dimensional beams as will be discussed in the next section.

2.6.2 Generalized Astigmatic Beams

In Sec. 2.5.2 the generalized Gaussian beam was discussed, characterized by different ellipses for constant intensity and constant phase. Additionally, these ellipses are rotated with respect to each other and with respect to the reference frame. This most general Gaussian beam is determined by six free parameters. In the case of an arbitrary beam with general astigmatism, we will have more parameters since the divergence is no longer related to the beam waist, but is given by (2.95).

First Order Moments

Two-dimensional beams are characterized by four first order moments which transform like rays as discussed in Sec. 1.2.4. The center of gravity of arbitrary fields propagates through two-dimensional optics on a trajectory that is defined by the ray transformation:

$$\begin{pmatrix} <x_2> \\ <y_2> \\ <\theta_{x2}> \\ <\theta_{y2}> \end{pmatrix} = \begin{pmatrix} \boldsymbol{A} & \boldsymbol{B} \\ \boldsymbol{C} & \boldsymbol{D} \end{pmatrix} \begin{pmatrix} <x_1> \\ <y_1> \\ <\theta_{x1}> \\ <\theta_{y1}> \end{pmatrix} \tag{2.112}$$

With the ray transfer matrix:

$$M = \begin{pmatrix} A_{xx} & A_{xy} & B_{xx} & B_{xy} \\ A_{yx} & A_{yy} & B_{yx} & B_{yy} \\ C_{xx} & C_{xy} & D_{xx} & D_{xy} \\ C_{yx} & C_{yy} & D_{yx} & D_{yy} \end{pmatrix} = \begin{pmatrix} A & B \\ C & D \end{pmatrix}$$

Second Order Moments

They can be defined in the same way as given by (2.93), (2.94), and (2.96). But now ten second order moments exist which are conveniently arranged in a 4x4 matrix $\boldsymbol{P}$, called the second order moment matrix or the variance matrix [1.71]:

$$\boldsymbol{P} = \begin{pmatrix} <w_x^2> & <w_x w_y> & <w_x \theta_x> & <w_x \theta_y> \\ <w_x w_y> & <w_y^2> & <w_y \theta_x> & <w_y \theta_y> \\ <w_x \theta_x> & <w_y \theta_x> & <\theta_x^2> & <\theta_x \theta_y> \\ <w_x \theta_y> & <w_y \theta_y> & <\theta_x \theta_y> & <\theta_y^2> \end{pmatrix} \tag{2.113}$$

with the newly introduced second moments:

$$<w_x w_y> = \frac{4}{P_N} \int_{-\infty}^{+\infty}\int_{-\infty}^{+\infty} xy|E(x,y)|^2 dxdy \tag{2.114}$$

$$<w_x \theta_y> = \frac{-i\lambda}{\pi P_N} \int_{-\infty}^{+\infty}\int_{-\infty}^{+\infty} x \, [E(x,y)\frac{\delta E^*(x,y)}{\delta y} - E^*(x,y)\frac{\delta E(x,y)}{\delta y}] \, dxdy \tag{2.115}$$

$$<\theta_x \theta_y> = \frac{4}{P_N} \int_{-\infty}^{+\infty}\int_{-\infty}^{+\infty} \theta_x \theta_y |E_F(\theta_x,\theta_y)|^2 d\theta_x d\theta_y \tag{2.116}$$

By using the Fourier transform, the last equation can also be written as:

$$<\theta_x \theta_y> = \frac{\lambda^2}{\pi^2 P_N} \int_{-\infty}^{+\infty}\int_{-\infty}^{+\infty} \frac{\delta E(x,y)}{\delta x} \frac{\delta E^*(x,y)}{\delta y} \, dxdy \tag{2.117}$$

The physical meaning of the mixed second order moments are as follows:

$<w_x w_y>$: characterizes the rotation of the near field intensity ellipse with respect to the reference frame

$<\theta_x \theta_y>$: characterizes the rotation of the far field intensity ellipse with respect to the reference frame

$<w_x \theta_y>, <w_y \theta_x>$: orbital angular momentum of the field (twist)

$<w_{x,y} \theta_{x,y}> = <w^2_{x,y}>/R_{x,y}$: characterizes the radius of curvature in x-and y-direction, respectively

In total sixteen second order moments exist which, due to symmetry relations, reduce to a maximum of ten different moments. In the most general case, an arbitrary radiation is thus characterized by ten different parameters. These parameters have to be determined experimentally by intensity measurements. By using the two-dimensional ABCD-propagation law in free space and by measuring $<w_x^2(z)>$, $<w_y^2(z)>$ and $<w_x w_y(z)>$ at a reasonable number of positions z along the focal region of the beam, the following nine parameters can be determined: $<w_x^2(0)>$, $<w_y^2(0)>$, $<w_x w_y(0)>$, $<w_x \theta_x(0)>$, $<w_y \theta_y(0)>$, $<\theta_x^2>$, $<\theta_x \theta_y>$, $<\theta_y^2>$, and the sum $<w_x \theta_y(0)> + <w_y \theta_x(0)>$. One additional measurement is required to separate the sum. This can be done by using a cylindrical lens. Details are discussed in [1.71,1.125].

Special Beams

The different types of beams discussed in Sec. 2.5.2 appear again in the general case. The most simple one is the beam of circular symmetry, the stigmatic beam, with beam radius w and half angle of divergence θ. Its second order moment matrix reads:

$$P_S = \begin{pmatrix} <w^2> & 0 & <w\theta> & 0 \\ 0 & <w^2> & 0 & <w\theta> \\ <w\theta> & 0 & <\theta^2> & 0 \\ 0 & <w\theta> & 0 & <\theta^2> \end{pmatrix} \tag{2.118}$$

The stigmatic beam is characterized by three parameters. This is one more than for the stigmatic Gaussian beam, because the waist radius w_0 and the half angle of divergence are no longer related by (2.62) but by:

$$w_0 \theta = M^2 \frac{\lambda}{\pi}$$

The simple astigmatic beam, which is aligned and can be factorized, is described by the second order matrix:

$$P_{SA} = \begin{pmatrix} \langle w_x^2 \rangle & 0 & \langle w_x\theta_x \rangle & 0 \\ 0 & \langle w_y^2 \rangle & 0 & \langle w_y\theta_y \rangle \\ \langle w_x\theta_x \rangle & 0 & \langle \theta_x^2 \rangle & 0 \\ 0 & \langle w_y\theta_y \rangle & 0 & \langle \theta_y^2 \rangle \end{pmatrix} \tag{2.119}$$

Now six parameters are required to characterize this beam, two more than for a Gaussian beam with simple astigmatism. The beam propagation factor can be different in the x- and the y-direction.

The next, more complicated, beam is the rotated simple astigmatic beam. The corresponding matrix is obtained by applying the rotation matrix $\boldsymbol{R(\alpha)}$ of (1.71) and (1.72). This results in:

$$P_{SA}(\alpha) = R^{-1}(\alpha)\, P_{SA}\, R(\alpha) \tag{2.120}$$

Now all elements of the rotated second order matrix differ from zero, but only seven independent parameters exist. In addition to the six parameters of the simple astigmatic beam, the rotation angle α is a new parameter. In contrast to more complicated beams, the matrix $\boldsymbol{P_{SA}(\alpha)}$ exhibits the following restrictions:

$$\langle w_x\theta_y \rangle = \langle w_y\theta_x \rangle$$

$$\frac{\langle xy \rangle}{\langle w_x^2 \rangle - \langle w_y^2 \rangle} = \frac{\langle w_x\theta_y \rangle}{\langle w_x\theta_x \rangle - \langle w_y\theta_y \rangle} = \frac{\langle \theta_x\theta_y \rangle}{\langle \theta_x^2 \rangle - \langle \theta_y^2 \rangle} = \frac{1}{2}\tan(2\alpha)$$

The beam quality, characterized for simple astigmatic beams by the propagation factors in the x- and the y-direction, now becomes more complicated. As was shown by Nemes [1.71], the generalized beam propagation factor is a constant of propagation:

$$M^4 = (\pi/\lambda)^2 \sqrt{\det \boldsymbol{P}} \geq 1 \tag{2.121}$$

For the simple astigmatic beam, it can be written as the product of the beam propagation factors in the x- and the y-direction:

$$M^4 = M_x^2 M_y^2 \tag{2.122}$$

$$M_x^2 = \frac{k}{2}\sqrt{<w_x^2><\theta_x^2> - <w_x\theta_x>^2} \geq 1 \qquad (2.123)$$

$$M_y^2 = \frac{k}{2}\sqrt{<w_y^2><\theta_y^2> - <w_y\theta_y>^2} \geq 1 \qquad (2.124)$$

A second interesting invariant quantity is the astigmatism factor T given by:

$$T = \frac{\pi^2}{2\lambda^2}\Big([<w_x^2><\theta_x^2> - <w_x\theta_x>^2] + [<w_y^2><\theta_y^2> - <w_y\theta_y>^2] + $$
$$+ 2[<w_x w_y><\theta_x\theta_y> - <x\theta_y><y\theta_x>]\Big) \qquad (2.125)$$

The astigmatism factor characterizes the astigmatism of the beam. For stigmatic beams, T is equal to the generalized beam propagation factor:

$$T_S = M^4$$

and for simple astigmatic beams one gets:

$$T_{SA} = \frac{1}{2}\left(M_x^4 + M_y^4\right)$$

Rotated and general astigmatic beams are characterized by the parameters T and M^4.

Beam Propagation

The beam propagation of a general astigmatic beam cannot be described by the generalized ABCD law (2.111) since the generalized q-parameter q_g does not provide all information on the beam properties. The transformation rule of the second order moment matrix for propagation from plane 1 to plane 2 through an optical system described by a 4x4 ray transfer matrix $\boldsymbol{M}$ according to (2.112) can be derived using the Wigner function (see Sec. 2.7) [1.109,1.110]:

$$\boldsymbol{P}_2 = \boldsymbol{M}\boldsymbol{P}_1\boldsymbol{M}^T \qquad (2.126)$$

with $\boldsymbol{M}^T$ being the transpose of $\boldsymbol{M}$. This equation can be written in a form similar to the linear case given by (2.100)-(2.102). The first two of the ten equations read:

$$<w_{x2}^2> = <(A_{xx}w_{x1} + A_{xy}w_{y1} + B_{xx}\theta_{x1} + B_{xy}\theta_{y1})^2>$$

$$<w_{x2}w_{y2}> = <(A_{xx}w_{x1} + A_{xy}w_{y1} + B_{xx}\theta_{x1} + B_{xy}\theta_{y1}) \cdot (A_{yx}w_{x1} + A_{yy}w_{y1} + B_{yx}\theta_{x1} + B_{yy}\theta_{y1})>$$

In the most general case, each parameter of the beam depends on all ten parameters of the incident beam. The equation arises now as to whether a general astigmatic beam can be transformed into a stigmatic one by using suitable ABCD optics. This is a delicate problem which is beyond the scope of this book [1.67]. Under certain conditions this transformation is possible, but not in general.

2.6.3 Beam Quality

We have seen that the propagation of partially coherent fields can be characterized by the second order intensity moments. If the intensity $I(x,y)$ in the beam waist is known, the beam radii at any distance z from the waist as well as the angle of divergence in the far field can be calculated using the generalized ABCD law (2.103/105). Laser beams can, in general, be described by partially coherent fields. Therefore, we can apply the generalized propagation laws to investigate the focusing properties of laser beams and define a measure for laser beam quality.

As shown in Fig. 2.21, the focusing of a laser beam with beam parameter product $w\theta$ is equivalent to generating a beam waist with a small waist radius w_0. Note that we define beam radii of the laser beam by the squareroot of the second intensity moments, according to (2.93) or (2.97). At a distance z from the plane of this focal spot, the beam radius $w(z)$ is given by the generalized ABCD law:

$$<w^2(z)> = <w_0^2>\left[1 + \frac{z^2}{<z_0^2>}\right] \tag{2.127}$$

At the distance of one Rayleigh range z_0 from the focal spot, the cross sectional area has increased by a factor of 2. The Rayleigh range and the beam waist radius are related to the divergence angle via:

$$<\theta_0^2> = \frac{<w_0^2>}{<z_0^2>} \tag{2.128}$$

We are interested in a small focal spot size and a large Rayleigh range. This leads us to the following ratio for the definition of laser beam quality:

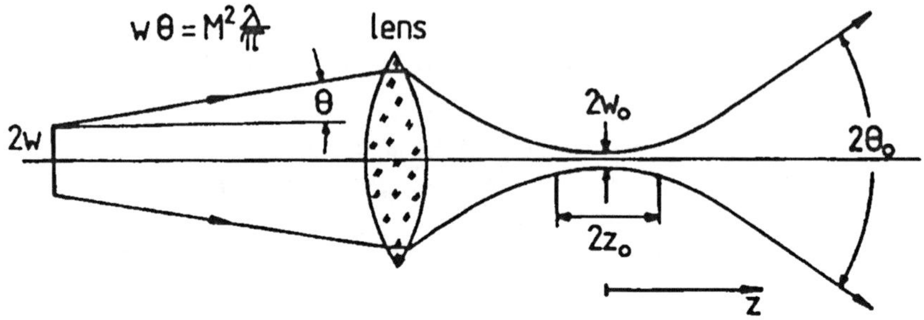

Fig. 2.21 Focusing of a laser beam with propagation factor M^2 by means of lens. Good beam quality is characterized by a small beam cross sectional area in the focal plane and a large Rayleigh range z_0.

$$\frac{w_0^2}{z_0} = w_0\theta_0 = w\theta = M^2\frac{\lambda}{\pi} \tag{2.129}$$

A measure for the beam quality is a low beam parameter product $w\theta$ which is equivalent to a low beam propagation factor M^2. The knowledge of the beam propagation factor is sufficient to characterize the beam quality of the laser. Once M^2 is measured for the laser beam (see Part VI for experimental techniques), the beam quality is well defined no matter what kind of optics are used for beam transformation and focusing (M^2 stays constant in ABCD-optical systems!).

Size and position of the beam waist w_0 can be calculated by using the Gaussian beam imaging conditions (2.69) and (2.70). Both equations also apply to partially coherent fields if the beam radii and the divergence angles are defined by the second intensity moments. To a good approximation, w_0 is determined by the divergence angle θ of the beam in front of the lens and the focal length f:

$$w_0 \approx f\,\theta \tag{2.130}$$

Since we have defined all beam dimensions via the second intensity moments, Eqs. (2.103)-(2.106) exactly hold for all laser beams. Definition and measurement of laser beam quality was recently standardized in ISO-procedures using this second order intensity moment approach [1.86]. Unfortunately, the second moments do not provide consistent information in regards to the power contained within the beam diameter. For Gaussian beams, the second order intensity moment provides a beam diameter that is equal to the Gaussian beam diameter with a power content of 86.5%. For other beams, the power contained in the second order intensity diameter is in general higher. Therefore, for some beams, the beam quality definition via the second order intensity moments may not reflect the focusing properties in materials processing applications. This is especially true for unstable resonators, whose beams exhibit side lobes in the far field (Fig. 2.22). If a hard edged output coupler is used, the second intensity moment in the far field is not defined (integral is infinite).

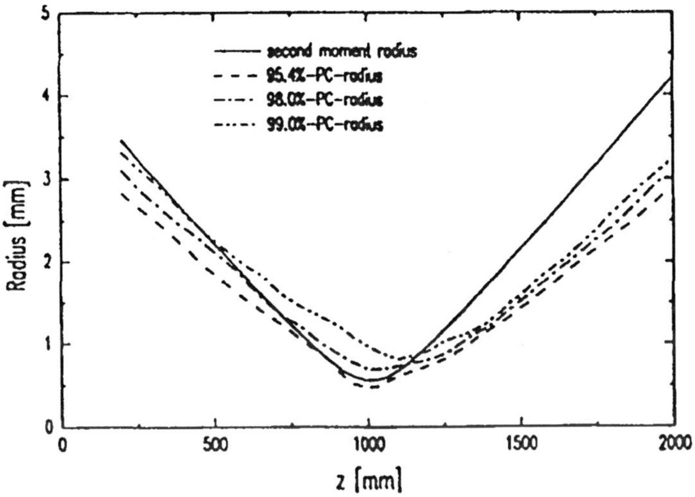

Fig. 2.22 Calculated beam radii of a focused unstable resonator laser beam The position of the beam waist is changed and the beam caustic becomes non-hyperbolic if the beam radii are defined by power contents (PC-radii) [1.70] (© SPIE 1994).

2.7 Coherence

A common definition of coherence is: "Coherence is a fixed relationship between the phase of waves in a beam of single-frequency radiation. Two waves of light are coherent, when the phase difference between their waves is constant; they are non-coherent if there is a random or changing phase relationship. Stable interference pattern are formed only by radiation emitted by coherent (or partially coherent) light". A monochromatic beam with a beam propagation factor of $M^2=1$ is a TEM_{00}-Gaussian beam and always coherent. But a beam with $M^2>1$ can be a coherent higher order mode, or a superposition of several modes with different frequencies. A coherent mode with $M^2>1$ can, in principle, be converted into a fundamental mode by a suitable phase plate, as was shown theoretically and experimentally [1.75,1.91,1.101], whereas a superposition of modes cannot. Therefore, the characterization of a beam only by M^2 is not sufficient. At least one additional number, the global degree of coherence K is required.

Coherence is discussed extensively in the well known book by Born and Wolf [1.2], starting from first principles, but neglecting the photon character of light. A more fundamental approach, taking into account the quantized field, can be found in [1.74]. In the following section a simplified version, adapted to the requirements of laser radiation, will be presented. For an easier understanding, temporal coherence will be briefly summarized.

2.7.1 Temporal Coherence

Normal light sources emit a statistical sequence of light bursts of duration τ with no fixed phase relation between each other. This is also true for high power lasers, with the only difference that the duration τ is much larger. Even the emission of stabilised single frequency-lasers is not a precise sinusoidal wave since the phase fluctuates statistically with a time constant in the range of milliseconds to seconds. Coherence characteristics of some lasers are shown in Tab. 2.4.

Table 2.4 Temporal coherence parameters of some light sources. All numbers depend on the special mode of operation and are rough estimates. ($\Delta t_{coh}=\ell_{coh}/c$, $\Delta t_{coh}\,\Delta\omega=0.5$).

Source	temporal coherence length ℓ_{coh} [m]	coherence time Δt_{coh} [s]	bandwidth $\Delta\omega$ [Hz]
torch light	5×10^{-7}	1.5×10^{-15}	3.34×10^{14}
solid-state laser	10^{-2}	3×10^{-11}	1.67×10^{12}
CO_2 laser	1	3×10^{-9}	1.67×10^{8}
He-Ne laser	10	3×10^{-8}	1.67×10^{7}
He-Ne laser (stabilised)	1.5×10^{8}	0.5	1

A wave train of limited duration τ can be described by:

$$E(\boldsymbol{r},t) = A(\boldsymbol{r})\, f(t/\tau)\, \exp[i\omega t - ikz] \tag{2.131}$$

The time τ is a measure for the duration of the wave. If such a field is superimposed with the same field delayed by Δt, an interference pattern will appear only if the time delay is in the range of τ or smaller. An example for time-limited radiation is the spontaneous emission of atoms. The amplitude decays exponentially in time:

$$f(t) = \begin{bmatrix} 0 & \textit{for } t<t_0 \\ \exp[-(t-t_0)/\tau + i\phi_n] & \textit{for } t \geq t_0 \end{bmatrix}$$

where amplitude A, phase ϕ_n and time of emission τ are random. A gas discharge of high temperature emits Gaussian-like wave-trains, due to Doppler-broadening:

$$f(t) = \exp\left[-\left(\frac{t-t_n}{\tau}\right)^2 + i\phi_n\right]$$

The emission of a cw-laser is nearly stable in amplitude, but fluctuates in phase:

$$f(t) = \exp[i\phi_n(t/\tau)]$$

where the time constant τ is roughly the inverse bandwidth. For a more quantitative description of temporal coherence, an experimental set-up, the Michelson-interferometer has to be discussed.

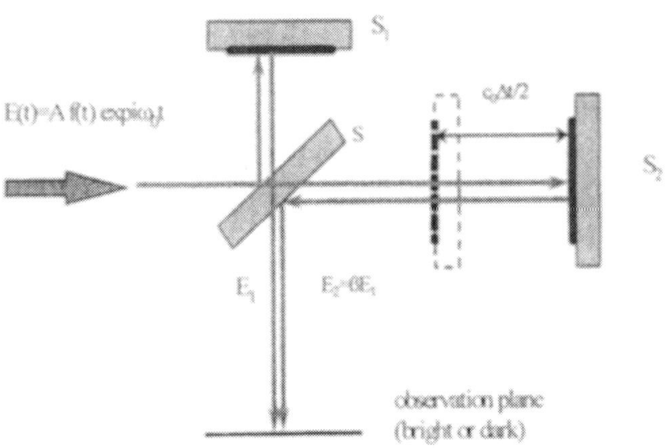

Fig. 2.23 The Michelson-interferometer to determine temporal coherence.

The Michelson Interferometer

In Fig. 2.23 a schematic Michelson-interferometer is shown. The incoming beam is split by the mirror S into two beams. Each beam is reflected by mirrors S_1, S_2, respectively, and both are recombined in the observation plane. By moving mirror S_2, a time delay Δt between the two beams can be generated. Using expression (2.131), the resulting field in the observation plane reads with (2.131):

$$E_{obs}(t) = E_1(t) + E_2(t) = A_1 f(t)\exp[i\omega_0 t] + A_2 f(t+\Delta t)\exp[i\omega_0(t+\Delta t)] \qquad (2.132)$$

The intensity is recorded with a detector (CCD-camera, film), which squares and integrates $E_{obs}(t)$. The resulting signal I is:

$$I = \frac{1}{2}c_0\epsilon_0\frac{1}{T}\int_0^T EE^* dt = \frac{1}{2}c_0\epsilon_0 \langle EE^* \rangle \qquad (2.133)$$

with T being the observation time. It is assumed that T is large compared to the time constant τ of the fluctuations. The bracket denotes the time average. Insertion of (2.132) into (2.133) results in:

$$I = \frac{1}{2}c_0\epsilon_0\left[|A_1|^2\langle f(t)f^*(t)\rangle + |A_2|^2\langle f(t+\Delta t)f^*(t+\Delta t)\rangle + A_1A_2^*\exp[i\omega_0\Delta t]\langle f(t)f^*(t+\Delta t)\rangle + cc.\right]$$

$$\Leftrightarrow \qquad I = I_1 + I_2 + \Gamma_{11} + \Gamma_{11}^* \qquad (2.134)$$

I_1, I_2 are the intensities of the individual wave trains and Γ_{11} is called the auto-correlation function, because the field is correlated with itself. The normalised value of Γ_{11} is the complex degree of coherence $\gamma_{11}(\Delta t)$:

$$\gamma_{11} = \frac{\Gamma_{11}(\Delta t)}{\sqrt{I_1 I_2}} = \frac{A_1A_2^*\exp[-i\omega_0\Delta t]\langle f(t)f^*(t+\Delta t)\rangle}{\sqrt{I_1 I_2}} \qquad (2.135)$$

Using Equation (2.135), (2.134) can be written as:

$$I = I_1 + I_2 + 2|\gamma_{11}|\sqrt{I_1 I_2}\cos(\omega_0\Delta t + \phi) \qquad (2.136)$$

The intensity is modulated with the delay-time Δt. The intensity contrast is referred to as visibility V which is derived from (2.136):

$$V = \frac{I_{max} - I_{min}}{I_{max} + I_{min}} = 2|\gamma_{11}|\frac{\sqrt{I_1 I_2}}{I_1 + I_2} \tag{2.137}$$

For a coherent beam the complex degree of coherence will be modulated, but the modulus is constant and equal one. In all other cases $|\gamma_{11}|$ will decrease with increasing delay Δt. An example is shown in Fig. 2.24.

Coherence Time

The parameter τ in (2.131) is a measure for the coherence time. If the delay Δt is large compared with τ, the contrast will be low and vice versa. It is convenient to define the coherence time by an average value, weighted with the auto-correlation function. By convention, the following definition is used for the coherence time [1.2]:

$$\Delta t_{coh}^2 = \frac{\int \Delta t^2 |\Gamma_{11}(\Delta t)|^2 d(\Delta t)}{\int |\Gamma_{11}(\Delta t)|^2 d(\Delta t)} \tag{2.138}$$

Some examples are given in Tab.2.5. Note that the coherence time is not equal the pulse duration, as will be shown in the next section.

Table 2.5 Characteristic parameters of different pulses

amplitude $E(t)/A_0$	spectrum $E(\omega)/A_0\tau$	coherence function $\Gamma(\Delta t)/\|A_0\|^2\tau$	coherence length $\Delta t_{coh}/\tau$	time-bandwidth product $\Delta t_{coh}\Delta\omega$
rectangular pulse				
$\begin{cases} \exp(i\omega_0 t) & for\ \|t\|<\tau \\ 0 & for\ \|t\|\geq\tau \end{cases}$	$\sqrt{\frac{2}{\pi}}\frac{\sin[(\omega-\omega_0)\tau]}{(\omega-\omega_0)\tau}$	$\left[2-\frac{\|\Delta t\|}{\tau}\right]\exp(i\omega_0\Delta t)$ $\|\Delta t\| \leq 2\tau$	$\sqrt{0.4}$	0.91
exponential decay				
$\begin{cases} \exp(i\omega_0 t - t/\tau) & for\ t>0 \\ 0 & for\ t\leq 0 \end{cases}$	$\frac{1/\sqrt{2\pi}}{1-i(\omega-\omega_0)\tau}$	$\frac{1}{2}\exp(i\omega_0\Delta t - \|\Delta t\|/\tau)$	$\frac{1}{\sqrt{2}}$	0.71
Gaussian pulse				
$\exp\left(i\omega_0 t - (t/\tau)^2\right)$	$\exp-[(\omega-\omega_0)\tau/2]^2$	$\sqrt{\frac{\pi}{2}}\exp\left[i\omega_0\Delta t - \frac{1}{2}(\Delta t/\tau)^2\right]$	$\frac{1}{\sqrt{2}}$	0.5

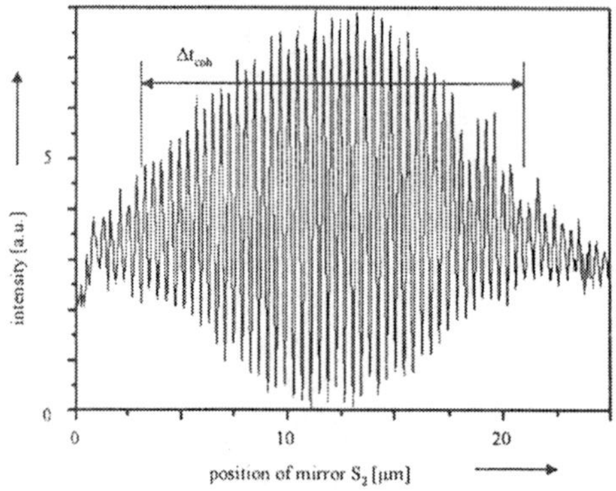

Fig. 2.24 Measurement of the temporal coherence of a beam generated by a short-pulse Ti-Sapphire-laser with $\Delta t_{coh} = 5.8\text{x}10^{-14}$ s [1.85].

Spectrum

The amplitude spectrum $E(\omega)$ and the power spectrum $G(\omega)$ of the field are given by the Fourier transform of the field:

$$E(\omega) = \frac{1}{\sqrt{2\pi}} \int_{-\infty}^{+\infty} E(t)\exp[i\omega t]dt \qquad G(\omega) = E(\omega)E^*(\omega) \tag{2.139}$$

A very useful relation between the auto-correlation function and the power spectrum is the Wiener-Khinchin theorem:

$$G(\omega) = \int \Gamma_{11}(\Delta t)\exp[i\omega\Delta t]d(\Delta t) \cdot \tag{2.140}$$

which is easy to prove [1.2]. The Fourier spectrum of the auto-correlation function is equal to the power spectrum. With the Wiener-Khinchin theorem, it is possible to determine the power spectrum of a light source using a Michelson-interferometer.

Spectral width

The power spectrum can be used to define the center frequency $\hat{\omega}$ and the spectra width $\Delta\omega$ of the field by its first order and second order moments:

$$\hat{\omega} = \int_{-\infty}^{+\infty} \omega G^2(\omega)d\omega \Big/ \int_{-\infty}^{+\infty} G^2(\omega)d\omega \tag{2.141}$$

$$\Delta\omega^2 = \int_{-\infty}^{+\infty} (\omega-\bar{\omega})^2 G^2(\omega)d\omega \Big/ \int_{-\infty}^{+\infty} G^2(\omega)d\omega \tag{2.142}$$

Fourier Relation

Spectral width $\Delta\omega$ and coherence time Δt_{coh} are related by the Fourier relation:

$$\Delta t_{coh}\ \Delta\omega \geq \frac{1}{2} \tag{2.143}$$

This relation holds for arbitrary fields, if the two parameters are defined according to (2.138) and (2.142) and the two weighting function Γ, G are related by the Fourier transform. The equality in (2.143) holds for Gaussian shaped fields [1.2].

Pulse width and Chirp

The width of a single pulse is defined by its second intensity moment:

$$\Delta t_p^2 = 4 \int_{-\infty}^{+\infty} t^2 I^2(t)dt \Big/ \int_{-\infty}^{+\infty} I^2(t)dt \tag{2.144}$$

The factor 4 is arbitrary, but guarantees that the pulse width is larger than the coherence length. In general, the pulse width will differ from the coherence length. A well known example is the chirped Gaussian pulse, given by:

$$E = A\ \exp[i\omega_0 t - \delta t^2] \qquad\qquad \delta = a - ib$$

where δ is a complex constant. This special pulse is produced by non-linear interaction of high intensity pulses with Kerr-active media (glass, liquids). The instantaneous frequency is the time-derivative of the phase and increases (in this case) linearly with time. The constant b is called the chirp:

$$\omega_{inst} = \frac{d}{dt}(\omega_0 t + bt^2) = \omega_0 + 2bt \tag{2.145}$$

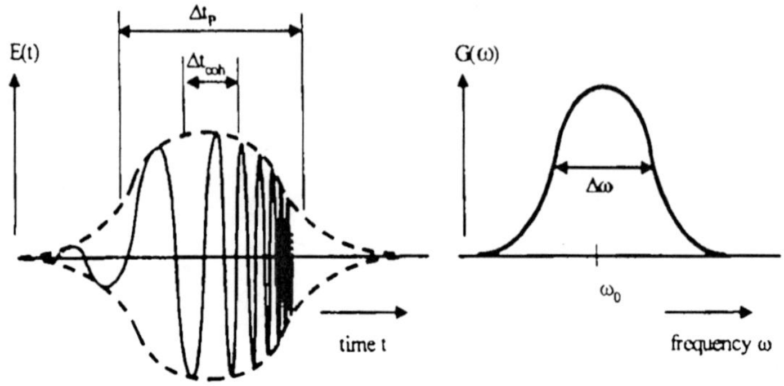

Fig. 2.25 Field $E(t)$ and power spectrum $G(\omega)$ of a chirped Gaussian pulse. The oscillation in the left figure is not to scale.

amplitude spectrum
$$E(\omega) = \frac{A}{\delta\sqrt{2}}\exp\left[-\frac{(\omega-\omega_0)^2}{4\delta}\right]$$

power spectrum
$$G(\omega) = \frac{|A|^2}{2(a^2+b^2)}\exp\left[-\frac{(\omega-\omega_0)^2}{2}\,\frac{a}{a^2+b^2}\right]$$

center frequency
$$\hat{\omega} = \omega_0$$

spectral width
$$\Delta\omega = \sqrt{\frac{a^2+b^2}{2a}}$$

coherence time
$$\Delta t_{coh} = \sqrt{\frac{a}{2(a^2+b^2)}}$$

pulse duration
$$\Delta t_p = \frac{1}{\sqrt{2a}}$$

The following relations hold for a chirped Gaussian pulse:

$$\Delta t_{coh} = \frac{\Delta t_p}{\sqrt{1+b^2/a^2}} \le \Delta t_p \qquad\qquad \Delta t_{coh}\,\Delta\omega = \frac{1}{2}$$

It is important to know that the Michelson-interferometer can measure the spectrum or the temporal coherence length, but not the pulse width. To determine the pulse width correctly, non-linear methods have to be used such as frequency conversion or two-photon fluorescence [1.90].

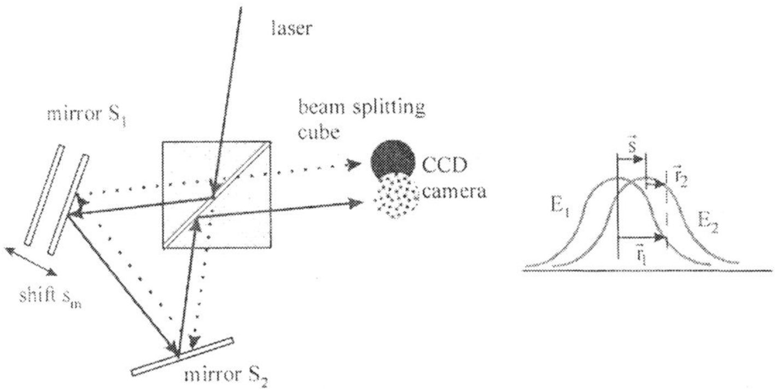

Fig. 2.26 The shear interferometer to determine spatial coherence [1.87].

2.7.2 Spatial Coherence

Spatial coherence is of interest for many applications such as holography, lithography, or laser etching of gratings. Whether high or low spatial coherence is required depends on the application. Whereas temporal coherence is determined via a temporal shift of the field, transverse coherence is measured by a transverse shift between the two fields.

The Shear Interferometer

It was originally used to measure lens aberrations [1.22,1.33]. A schematic set-up is shown in Fig. 2.26. The beam under investigation is divided by a beam splitter into two beams, which are recombined by the mirrors S_1, S_2 in the observation plane. By shifting mirror S_2, a transverse shift $\bar{s}$ of the two beams is produced. The resulting field reads.

$$E = E_1(\boldsymbol{r}_1,t) + E_2(\boldsymbol{r}_2,t)$$

For quasi-monochromatic fields the time difference Δt between the two beams can be neglected. The intensity is the time average, squared value. After some manipulation, the same relation as in the case of temporal coherence (2.134) is obtained:

$$I(\boldsymbol{r}_1,\boldsymbol{r}_2) = I_1(\boldsymbol{r}_1) + I_2(\boldsymbol{r}_2) + \Gamma_{12}(\boldsymbol{r}_1,\boldsymbol{r}_2) + \Gamma_{12}^*(\boldsymbol{r}_1,\boldsymbol{r}_2)$$

where Γ_{12} now denotes the cross-correlation function, which in general is a four-dimensional function.

$$\Gamma_{12}(\boldsymbol{r}_1,\boldsymbol{r}_2) = \frac{1}{2}c_0\epsilon_0\langle E_1(\boldsymbol{r}_1)E_2^*(\boldsymbol{r}_2)\rangle \tag{2.146}$$

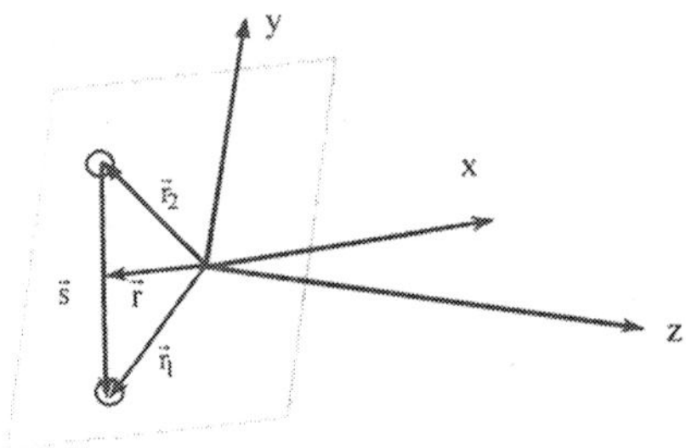

Fig. 2.27 The vectors r,s identifying the two points of the field, which interfere.

Often the variables $\boldsymbol{r},\boldsymbol{s}$ are used instead of $\boldsymbol{r}_1$, $\boldsymbol{r}_2$ (Fig. 2.27). These are defined as:

$$\boldsymbol{r} = (\boldsymbol{r}_1 + \boldsymbol{r}_2)/2 \qquad \boldsymbol{s} = \boldsymbol{r}_1 - \boldsymbol{r}_2 \tag{2.147}$$

The bracket again represents the time average. It can be expanded into a series of coherent functions, which is called modal-expansion [1.42,1.45,1.93,1.97]. For $\boldsymbol{s}=0$ and $\boldsymbol{r}_1=\boldsymbol{r}_2$, the intensity and the power are directly related to Γ_{12}:

$$I(\boldsymbol{r}) = \Gamma_{12}(\boldsymbol{r}, \boldsymbol{s}=0) \tag{2.148}$$

$$P = \iint \Gamma_{12}(\boldsymbol{r}, \boldsymbol{s}=0)\, d\boldsymbol{r} \qquad with \quad d\boldsymbol{r} = dxdy \tag{2.149}$$

The complex degree of coherence is the normalized cross-correlation function:

$$\gamma_{12}(\boldsymbol{r}_1,\boldsymbol{r}_2) = \frac{\Gamma_{12}(\boldsymbol{r}_1,\boldsymbol{r}_2)}{\sqrt{I_1(\boldsymbol{r}_1)I_2(\boldsymbol{r}_2)}} = |\gamma_{12}(\boldsymbol{r}_1,\boldsymbol{r}_2)|\exp[i\Phi_{12}(\boldsymbol{r}_1,\boldsymbol{r}_2)] \tag{2.150}$$

where $\Phi_{12}(\boldsymbol{r}_1,\boldsymbol{r}_2)$ represents the mutual phase of the two fields (phase difference) in the two points $\boldsymbol{r}_1,\boldsymbol{r}_2$. The intensity can now be written as:

$$I = I_1 + I_2 + 2|\gamma_{12}|\sqrt{I_1 I_2}\cos[\Phi_{12}] \tag{2.151}$$

which has a from analogous to the temporal coherence case (2.136). If the field is completely coherent the modulus of γ_{12} is equal to one everywhere. The modulation of the intensity is given by the mutual phase Φ_{12}. If the two fields in the observation plane have a plane wave front, the phase Φ_{12} is constant. For non-planar waves the phase difference depends on the shape of the wave front. An example is given in Fig. 2.28, which shows the shear interferogram of a fundamental mode (Gaussian beam).

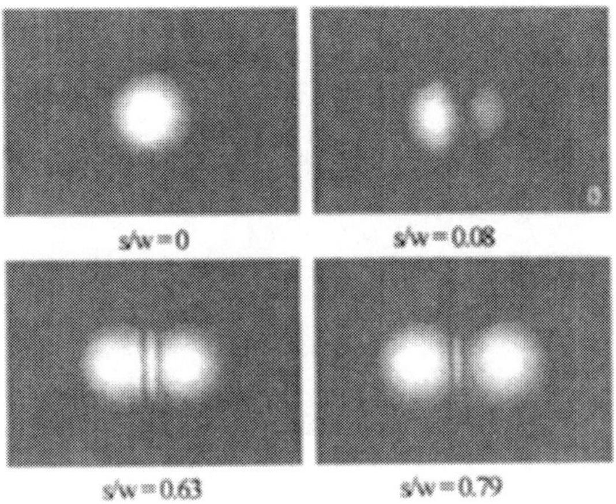

Fig. 2.28 Shear interferogram of a fundamental mode with $M^2=1.2$ of a He/Ne-laser [1.87]. s/w: shear normalized to the beam radius.

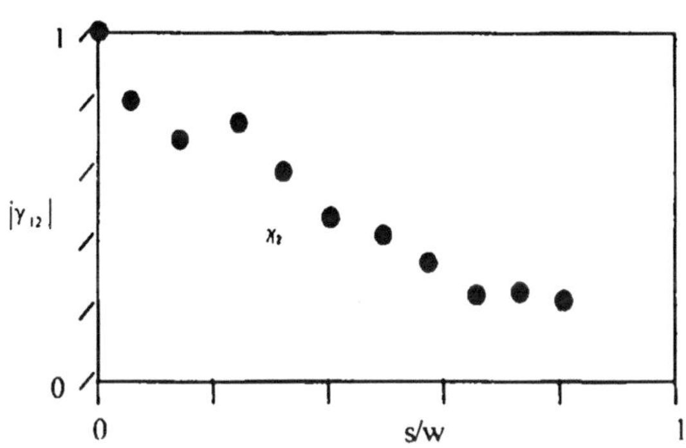

Fig. 2.29 $|\gamma_{12}|$ vs the normalized shear s for a He/Ne-laser with $M^2=1.2$ [1.87]. w: beam radius.

Outside the waist the mode exhibits a spherical wave front, and consequently interference patterns appear. From the intensity pattern the wave front can be reconstructed, using (2.147) and (2.148). If the beam is not completely coherent ($|\gamma_{12}|<1$), two measurements are necessary to determine $|\gamma_{12}|$ and Φ_{12}. This can be done by inserting a quarter wave plate in one beam and performing a second measurement. Equation (2.151) then reads:

$$I = I_1 + I_2 + 2|\gamma_{12}|\sqrt{I_1 I_2}\sin[\Phi_{12}] \tag{2.152}$$

The result of such a measurement is shown in Fig. 2.29. The degree of coherence $|\gamma_{12}|$ is plotted versus the shear for a beam with $M^2=1.2$. To obtain reliable results, the optical elements must be very homogeneous and free of stress induced birefringence. The shear-interferometer has the advantage that for each value s, the CCD-camera delivers the complete set of r parameters. A useful device to measure the two-dimensional spatial coherence is shown in Fig. 2.30, where a rotating Dove-prism is used to scan the full transverse plane. Some experimental results are compiled in Fig. 2.31.

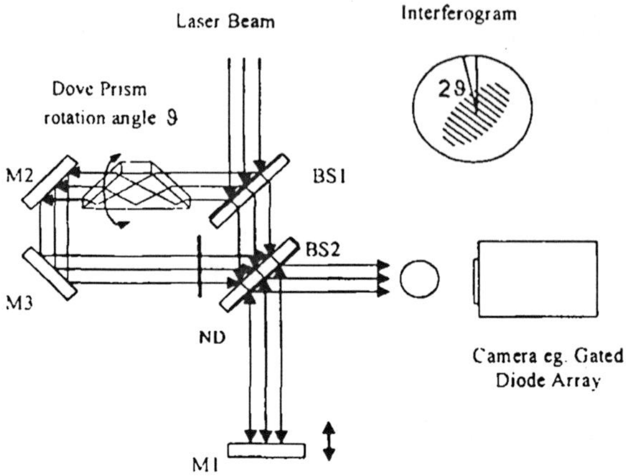

Fig. 2.30 A shearing interferometer with a rotating Dove prism [1.79] (© OSA 1998).

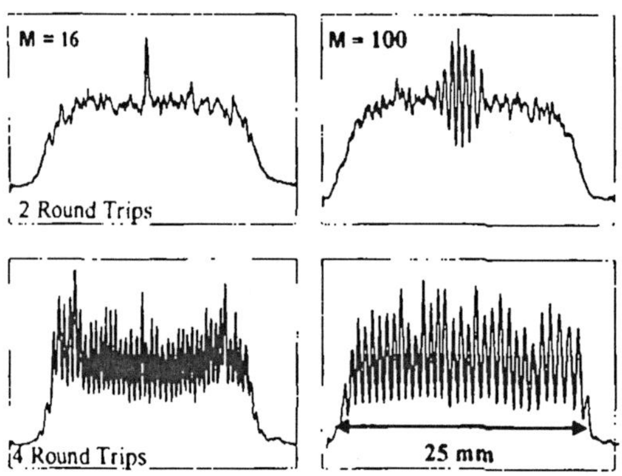

Fig. 2.31 Shear interferogram of a copper-vapour laser beam. Parameters are the magnification M of the unstable resonator and the number of round trips [1.79] (© OSA 1998).

Young's Double-slit Interferometer

One of the most famous experiments in quantum optics is Young's double slit experiment, because it proved the wave theory of light. The set-up is in principle very simple (Fig. 2.32). A light source is illuminating a double slit, or two pin-holes. From the two pin holes separated by the distance s two spherical fields E_1, E_2 emanate and are collimated with a lens. An interference pattern is generated in the observation plane, which is magnified and recorded by a CCD-camera. The two pinholes are realized by two masks with cross slits as shown in Fig. 2.33. The cross-correlation function for this special example is also given in Fig.2.33. Using the cross-correlation function all beam parameters such as phase front, mode-content, beam width, beam divergence, global degree of coherence and transverse coherence length can be evaluated. In an alternate embodiment, the two pin holes can be replaced by two single mode fibers [1.88].

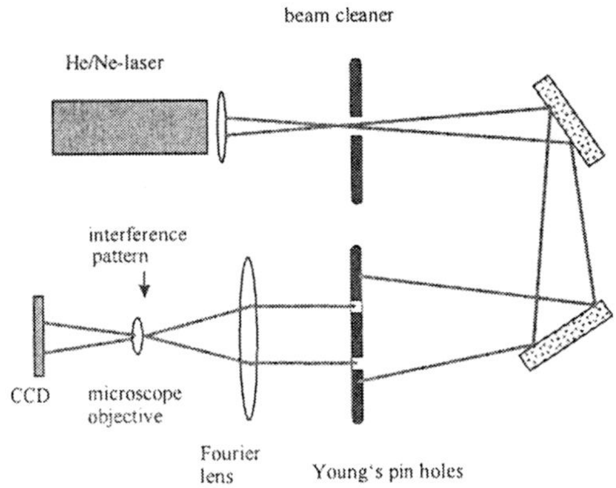

Fig. 2.32 Young's interferometer to determine the cross-correlation function of a multimode He/Ne-laser [1.55] (© Springer-Verlag 1989).

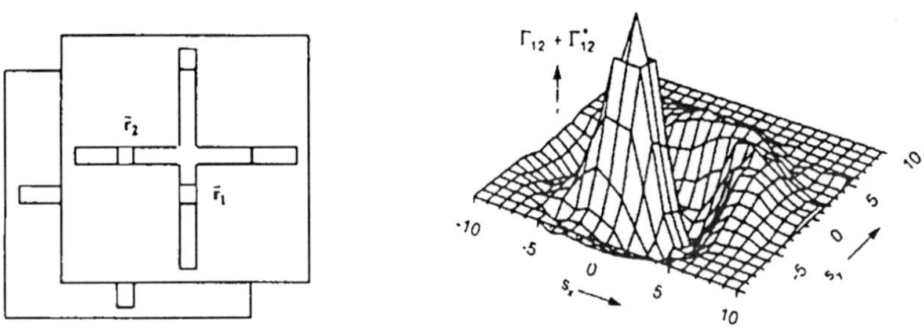

Fig. 2.33 The two pin holes r_1, r_2 of Young's interferometer and a measured cross correlation function [1.55] (© Springer-Verlag 1989).

The determination of the beam parameters by Young's interferometer is very time consuming, because for each value r the parameter s has to be scanned point by point over the complete two-dimensional transverse plane.

Some useful definitions:

Visibility

The visibility or contrast is defined in the same way as for temporal coherence:

$$V = \frac{I_{max} - I_{min}}{I_{max} + I_{min}} = 2|\gamma_{12}|\frac{\sqrt{I_1 I_2}}{I_1 + I_2} \tag{2.153}$$

The decrease of V with increasing shift can be caused by poor coherence or by different intensities of the two beams.

Global Degree of Coherence

The modulus of the cross-correlation function will decay with increasing shear and approaches asymptotically zero. To eliminate the influence of the intensity decay, the cross-correlation function has to be normalised. A convenient definition of the global degree of coherence is [1.102,1.48]:

$$K^2 = \frac{\iiiint |\Gamma_{12}(\boldsymbol{r}_1,\boldsymbol{r}_2)|^2 d\boldsymbol{r}_1 d\boldsymbol{r}_2}{\iint \Gamma_{12}(\boldsymbol{r}_1,\boldsymbol{r}_1) d\boldsymbol{r}_1 \iint \Gamma_{12}(\boldsymbol{r}_2,\boldsymbol{r}_2) d\boldsymbol{r}_2} = \frac{1}{P^2}\iiiint |\Gamma_{12}(\boldsymbol{r}_1,\boldsymbol{r}_2)|^2 d\boldsymbol{r}_1 d\boldsymbol{r}_2 \qquad (2.154)$$

If the field is completely coherent with:

$$E_{1,2} = A_{1,2}(\boldsymbol{r}_{1,2}) \exp[i(\omega t - kz)]$$

the cross-correlation function becomes

$$\Gamma_{12} = 0.5\ c_0 \epsilon_0 A_1 A_2^*$$

and (2.154) results in $K^2=1$. In all other cases, K^2 will be smaller than unity. A relation between the global coherence and the beam propagation factor exists [1.90,1.48]:

$$\sqrt{K^2}\ M^2 \geq 8/9 \qquad (2.155)$$

This relation is useful to check on experimental results. If the cross-correlation function can be factorized, e.g. in the case of rectangular symmetry:

$$\Gamma_{12}(\boldsymbol{r}_1,\boldsymbol{r}_2) = \Gamma_x(x_1,x_2)\Gamma_y(y_1,y_2) \quad or \quad \Gamma_{12}(\boldsymbol{r}_1,\boldsymbol{r}_2) = \Gamma_x(x,s_x)\Gamma_y(y,s_y)$$

the global degree of coherence can be simplified to yield (with $P^2 = P_x^2\, P_y^2$):

$$K^2 = \frac{1}{P_x^2}\iint |\Gamma_x(x_1,x_2)|^2 dx_1 dx_2\ \frac{1}{P_y^2}\iint |\Gamma_y(y_1,y_2)|^2 dy_1 dy_2 = K_x^2 K_y^2 \qquad (2.156)$$

Transverse Coherence Length

For a fixed position r_1 the degree of coherence will decrease with increasing s, but it can be modulated, depending on the beam structure. Therefore, the local coherence length is defined as an average value, weighted with the modulus of the cross-correlation function:

$$\ell^2_{loc}(\boldsymbol{r}_1) = 8\frac{\iint (\boldsymbol{r}_1-\boldsymbol{r}_2)^2|\Gamma_{12}(\boldsymbol{r}_1,\boldsymbol{r}_2)|^2 d\boldsymbol{r}_2}{\iint |\Gamma_{12}(\boldsymbol{r}_1,\boldsymbol{r}_2)|^2 d\boldsymbol{r}_2}$$

Using (2.147), the global coherence length now reads:

$$\ell^2_{glob} = 8\frac{\iiiint (\boldsymbol{r}_1-\boldsymbol{r}_2)^2|\Gamma_{12}(\boldsymbol{r}_1,\boldsymbol{r}_2)|^2 d\boldsymbol{r}_1 d\boldsymbol{r}_2}{\iiiint |\Gamma_{12}(\boldsymbol{r}_1,\boldsymbol{r}_2)|^2 d\boldsymbol{r}_1 d\boldsymbol{r}_2} = \frac{8}{K^2P^2}\iiiint \boldsymbol{s}^2|\Gamma_{12}(\boldsymbol{r},\boldsymbol{s})|^2 d\boldsymbol{s}d\boldsymbol{r} \tag{2.157}$$

Similar to the temporal case, the factor 8 is arbitrary, but guarantees that the coherence length is always smaller or equal the beam diameter, defined by (2.93). It is important to notice that the coherence length is a measure for the decay of the contrast, which can be caused by different intensities or by poor coherence. Even a coherent beam has a limited coherence length given by the beam diameter. In the case of rectangular symmetry the transverse degree of coherence reads:

$$\ell^2_{glob} = \ell^2_{glob,x} + \ell^2_{glob,y} \tag{2.158}$$

with:

$$\ell^2_{glob,x} = \frac{8}{K_x^2P_x^2}\iint s_x^2|\Gamma_x(x,s_x)|^2 dx ds_x\ , \quad \ell^2_{glob,y} = \frac{8}{K_y^2P_y^2}\iint s_y^2|\Gamma_y(y,s_y)|^2 dy ds_y \tag{2.159}$$

The far field angle of coherence is defined by

$$\Theta^2_{coh,x,y} = \lim_{z\to\infty}\left[\frac{\ell_{glob,x,y}}{z}\right]^2 \tag{2.160}$$

Examples
To demonstrate the practical aspects of these definitions, let us discuss some examples [1.102], without going into the mathematical details. In the cases below, beam radius w_{Rx} and coherence length $\ell_{glob,x}$ are normalized to the fundamental mode radius w_{0x}, which propagates according to (2.107).

Pure TEM$_{m0}$
In rectangular symmetry, the field is given by (5.8) and the cross-correlation function reads:

$$\Gamma_x(x_1,x_2) = \frac{1}{2}c_0\epsilon_0|E_0|^2 \exp\left[-\frac{x_1^2+x_2^2}{w_x^2}\right] H_0^2\; H_m\left(\frac{\sqrt{2}x_1}{w_x}\right)\; H_m\left(\frac{\sqrt{2}x_2}{w_y}\right)$$

Beam radius and divergence are defined by the second moments (2.93, 2.94, 5.12):

$$w_{Rx}(z) = w_{0x}(z)\sqrt{2m+1} = w_{0x}(0)\sqrt{1+z^2/z_0^2}\Big/\sqrt{2m+1}$$
$$\Theta_{Rx} = \Theta_{0x}\sqrt{2m+1}$$

The beam radius scales according to (2.107). The coherence parameters result from the above equations:

beam propagation factor: $M_x^2 = 2m+1$

degree of coherence: $|\gamma_{12}| = 1$

global degree of coherence: $K_x = 1$

coherence length: $\ell_{glob,x}(z) = 2w_{0x}(z)\sqrt{2m+1}$

The coherence length is the same as the beam diameter and scales hyberbolically with z.

b) Two oscillating modes TEM$_{00}$ +TEM$_{01}$
The fundamental mode and the next higher order mode may oscillate in a rectangular one-dimensional resonator. Then (5.8) holds and the resulting field reads:

$$E = E_0 \exp\left[-\frac{x^2}{w_{0x}^2}\right]\left[b_0 \exp(i\omega_{00}t) + b_1 \exp(i\omega_{10}t)\; H_1\left(\frac{\sqrt{2}x}{w_{0x}}\right)\right]$$

In non-degenerated resonators the eigenfrequencies ω_{00}, ω_{10} are different and the cross-terms vanish when averaged over time. The cross-correlation function reduces to:

$$\Gamma_x(x_1,x_2) = \frac{1}{2}c_0\epsilon_0|E_0|^2 \exp\left[-\frac{x_1^2+x_2^2}{w_{0x}^2}\right]\left(a_0 + a_1\; H_1\left(\frac{\sqrt{2}x_1}{w_{0x}}\right)\; H_1\left(\frac{\sqrt{2}x_2}{w_{0x}}\right)\right)$$

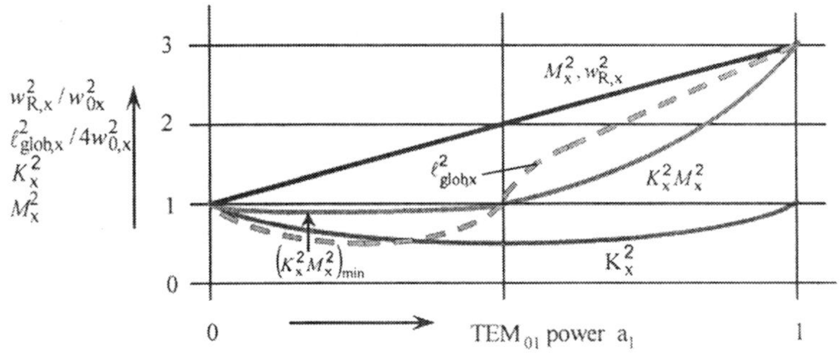

Fig. 2.34 Coherence parameters of the two-mode oscillation vs the power a_1 of the TEM$_{10}$-mode.

Equation (2.149) yields a condition for the mode expansion coefficients:

$$a_0 + a_1 = |b_0|^2 + |b_1|^2 = 1 \qquad 0 \leq a_0, a_1 \leq 1$$

The beam parameters can be evaluated analytically and depend on the fractional powers a_0, a_1 of the modes:

beam radius: $$w_{Rx}(z) = w_{0x}(z)\sqrt{1+2a_1}$$

beam propagation factor: $$M_x^2 = 1+2a_1$$

degree of coherence: $$|\gamma_{12,x}| = \frac{|a_0+4a_1x_1x_2/w_{0x}^2|}{\sqrt{a_0+4a_1x_1^2}\sqrt{a_0+4a_1x_2^2}}$$

global coherence: $$K_x^2 = a_0^2+a_1^2$$

coherence length: $$\ell_{glob,x}(z) = 2w_{0x}(z)\sqrt{1+2a_1(a_1-a_0)/K_x^2}$$

It is interesting to note that the coherence length has a minimum for $a_1=1/3$ (Fig. 2.34).

The Gauss-Schell beam [1.47,1.68,1.74]
This is the most general partially coherent beam with respect to the second order moments and a Gaussian intensity distribution. The general Gauss-Schell-beam, which is often used to describe multimode lasers, is characterized by ten independent parameters [1.68]. We will

restrict the discussion to the one-dimensional beam with its two parameters w_{Rx}, σ, which describe beam radius and degree of coherence, respectively. The Gauss-Schell beam is only defined by its cross-correlation function. It reads in the one-dimensional case:

$$\Gamma_x(x,s_x,z) = \frac{1}{2}c_0\epsilon_0|E_0|^2 \exp\left[-\frac{2x^2}{w_{Rx}^2} - \frac{s_x^2}{2w_{Rx}^2\sigma^2} - \frac{2ixs_x}{w_{Rx}^2\sigma}\frac{z}{z_{Rx}}\right] \tag{2.161}$$

$$z_{Rx} = \sigma z_0 \quad : \quad \textit{Rayleigh range}$$

The beam intensity is given by (2.148) and results from the above equation for $s_x=0$:

$$I(x,z) = \frac{1}{2}c_0\epsilon_0|E_0|^2 \exp\left[-\frac{2x^2}{w_{Rx}^2}\right]$$

with $$w_{Rx}(z) = w_{Rx}(0)\sqrt{1+(z/z_{Rx})^2}$$

The intensity is Gaussian and remains Gaussian when the beam is propagating in free space or through parabolic optical systems. The divergence, however, is higher for the same beam waist and the Rayleigh range shorter compared to a Gaussian TEM_{00}-mode. The parameter σ characterizes the degree of coherence with $0 \leq \sigma \leq 1$. For complete coherence σ is equal to 1.0, an incoherent beam exhibits $\sigma = 0$. The local transverse coherence length of this specific beam does not depend on the position x. Some relevant parameters are:

beam radius:	$w_{Rx}(z)$
beam divergence:	$\Theta_{Rx} = \Theta_{0x}/\sigma$
beam propagation factor:	$M_x^2 = 1/\sigma$
degree of coherence:	$\lvert\gamma_{12,x}(x_1,x_2)\rvert = \exp\left[-\frac{1}{2}\left(\frac{x_1-x_2}{w_{Rx}}\right)^2\left(\frac{1}{\sigma^2}-1\right)\right]$
global coherence:	$K_x^2 = \sigma$
coherence length:	$\ell_{glob,x}(z) = 2w_{0x}(z)\sigma$

with w_{0x} and θ_{0x} being the values of the fundamental mode.
For coherent light ($\sigma = 1$), the parameters of the fundamental mode are obtained. The Gauss-Schell beam can be realized by an incoherent superposition of squared Gauss-Hermite polynomials with suitable coefficients [1.42,1.48].

Determination of Phase and Coherence by Non-interferometric Measurements

For sake of simplicity, fields with only one transverse coordinate $E(x,z)$ are discussed. The two-dimensional case is not straight forward, as will be briefly pointed out in the next section. The cross correlation function of $E(x,z)$ is a two-dimensional function $\Gamma_x(x_1,x_2)$ or $\Gamma_t(x,s_x)$, using (2.147). By measuring Γ_x in one transverse plane (e.g. $z=0$) all parameters of interest such as beam width, divergence, phase front, and coherence can be evaluated. On the other hand, the propagation of the field and the intensity depend on the phase and the amplitude distribution in a z-plane and is also determined by the cross-correlation function Γ_x. The full information on the field is contained in Γ_x as well as in the intensity structure. It is possible to reconstruct the two-dimensional cross-correlation function $\Gamma_t(x,s_x)$ from the two-dimensional intensity $I(x,z)$ by using the ambiguity function Z_A , as will be demonstrated.

The Ambiguity-function Z_A

The propagation of the cross-correlation function $\Gamma_x(x_1,x_2,z)$ in free space is given by the Fresnel-integral in the paraxial approach (2.23). However, Γ_x is the product of two fields resulting in a complicated diffraction integral. It is much easier to use the ambiguity-function, which is the Fourier-transform of the cross-correlation function with respect to x [1.25,1.38]:

$$\square_{A,x}(s_x,\alpha_x) = \frac{1}{\lambda}\int_{-\infty}^{+\infty}\Gamma_x(x,s_x)\exp[ik\alpha_x x]dx \tag{2.162}$$

The ambiguity function is a representation of the field in the s_x,α_x space. The propagation of this function is determined by the ABCD-law [1.38], which for propagation in free space reads:

$$Z_{Ax}(s_x,\alpha_x,z) = Z_{Ax}(s_x-\alpha_x z,\ \alpha_x,\ 0) \tag{2.163}$$

The propagating function remains the same, if s_x is replaced by s_x-$\alpha_x z$. This coordinate transformation means a shearing of the function by the angle φ, as plotted in Fig. 2.35, with $\tan\varphi = z/z_0$ where z_0 is the Rayleigh length and φ the Guoy-shift (see (5.39) and (5.40)). Let us discuss some properties of this new function. For $\alpha_x = s_x = 0$, (2.162) and (2.149) immediately deliver:

$$Z_{Ax}(0,0) = \frac{1}{\lambda}\int_{-\infty}^{+\infty}\Gamma_x(x,s_x=0)dx = \frac{1}{\lambda}P_x$$

This special value of the function is equal the square root of the power. According Parseval's theorem, the following relationship holds for $|Z_{A,x}|^2$:

$$\iint |Z_{A,x}(s_x,\alpha_x)|^2 ds_x d\alpha_x = \frac{1}{\lambda}\iint |\Gamma_x(x,s_x)|^2 dx ds_x = \frac{1}{\lambda}\iint |\Gamma_x(x_1,x_2)|^2 dx_1 dx. \tag{2.164}$$

which according to (2.136) is the x-part of the global degree of coherence.

$$K_x^2 = \frac{\lambda}{P_x^2}\iint |Z_{A,x}(s_x,\alpha_x)|^2 ds_x d\alpha_x$$

The transverse coherence length is given by (2.159). Replacing Γ_x by $Z_{A,x}$ and using (2.162) results in:

$$\ell^2_{glob,x} = \frac{8}{K_x^2 P_x^2 \lambda}\iint s_x^2 |Z_{A,x}(s_x,\alpha_x)|^2 ds_x d\alpha_x \tag{2.165}$$

The second moment of s_x, weighted with the squared modulus of the ambiguity function is the transverse coherence length as shown in Fig.2.35. The second moment of α_x does not depend on z and is the far field angle of coherence as defined by (2.160):

$$\Theta^2_{coh,x} = \frac{8}{K_x^2 P_x^2 \lambda}\iint \alpha_x^2 |Z_{A,x}(s_x,\alpha_x)|^2 ds_x d\alpha_x \tag{2.166}$$

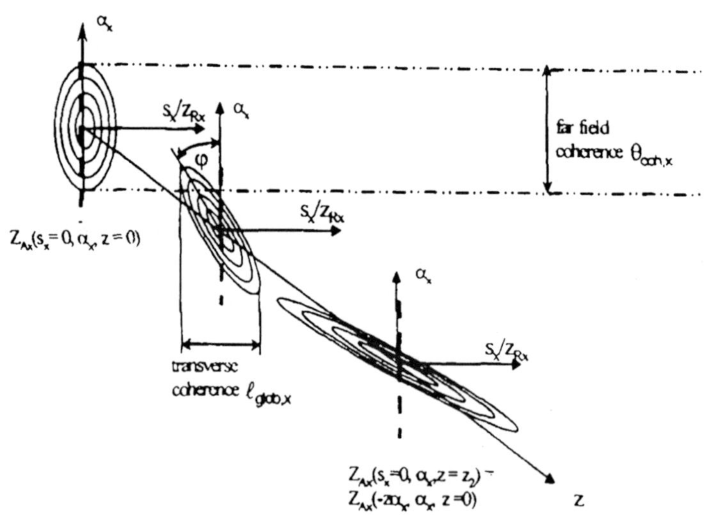

Fig.2.35 The ambiguity function is sheared by the angle $\tan\varphi = z/z_0$, when propagating in free space. It is convenient to normalise the s_x-parameter to the corresponding Rayleigh length z_{Rx}. The Fourier transform of the intensity is given by the dotted lines.

The transverse coherence length is given by (2.159). Replacing Γ_x by $Z_{A,x}$ and using (2.162) results in:

$$\ell^2_{glob,x} = \frac{8}{K_x^2 P_x^2 \lambda} \int\int s_x^2 |Z_{A,x}(s_x,\alpha_x)|^2 ds_x d\alpha_x \tag{2.165}$$

The second moment of s_x, weighted with the squared modulus of the ambiguity functior is the transverse coherence length as shown in Fig. 2.35. The second moment of α_x does not depend on z and is the far field angle of coherence as defined by (2.160):

$$\Theta^2_{glob,x} = \frac{8}{K_x^2 P_x^2 \lambda} \int\int \alpha_x^2 |Z_{A,x}(s_x,\alpha_x)|^2 ds_x d\alpha_x \tag{2.166}$$

The relevant coherence parameters can be evaluated, if $Z_{A,x}$ is known. To understand how it is obtained from intensity measurements, let us start with (2.162) at z=0, which for s_x = 0 together with (2.148) delivers:

$$Z_{A,x}(s_x=0,\alpha_x) = \int I(x,z=0)\ \exp[ik\alpha_x x]\ dx \tag{2.167}$$

Propagation by a distance z is described by (2.163):

$$Z_{A,x}(s_x,\alpha_x) = \int I(x,z)\ \exp[ik\alpha_x x]\ dx \qquad with \quad z = -s_x/\alpha_x \tag{2.168}$$

In a measurement, the one-dimensional intensity distribution $I(x,z)$ is Fourier-transformed to generate the ambiguity function with s_x=-$z\alpha_x$ z is running from -∞ to +∞. In reality it is sufficient to determine the intensity in 10 to 20 positions around the focal range, where it is varying strongly. This simplifies the determination of the coherence parameters, because the transverse intensity structure can be easily measured with a CCD-camera. Details of the measurement technique were published in [1.80]. The procedure is as follows

- measure $I(x,z)$ at 10-20 different positions around the focal region
- calculate $Z_{A,x}$ using (2.168) and replace z by $-s/\alpha$
- With $Z_{A,x}$, global coherence, coherence length and far-field divergence can be calculated according to (2.165) and (2.166)
- Fourier transform of $Z_{A,x}$ generates the cross-correlation function (2.162), from which amplitude and phase front can be calculated.
- beam width, divergence, and beam propagation factor can be calculated from $I(x,z)$ using the second order intensity moments (see Sec. 2.6)

Figure 2.36 shows experimental results for an off-axis diode-end-pumped Nd:YAG laser. By moving the pump spot laterally with respect to the optical axis of the stable resonator, Gauss-Hermite modes of a desired order can be excited. The measured global coherence versus the diode adjustment is shown in Fig. 2.36. The global degree of coherence always has a maximum when a pure TEM_{om} mode oscillates, but never reaches $K_x^2 = 1$, which means that other modes oscillate as well. Table 2.6 compares the results of this non-interferometric measurement with the ones obtained with a Young-interferometer. The theoretical values are based on the assumption that only two modes are oscillating, using the formulas of the two-mode-case.

Table 2.6 Comparison of beam propagation factor M^2, global degree of coherence K_x^2, and global transverse coherence length $\ell_{coh,x}$, obtained by Young's interferometer and by non-interferometric measurements [1.108]. a_0, a_1 are the relative power contents of TEM_{00} and TEM_{01} mode, respectively.

$a_0 : a_1$	M_x^2		K_x^2			$\ell_{glob,x} / w_{Rx}$		
	theory	exp.	theory	Young	non-interf.	theory	Young	non-interf.
1.00 : 0.00	1.0	1.01	1.0	0.96 - 1.0	1.0	1.0	0.99	1.01
0.75 : 0.25	1.75	1.71	0.79	0.70 - 0.75	0.74	0.63	0.63	0.60
0.50 : 0.50	2.0	2.04	0.71	0.66 - 0.73	0.71	0.71	0.74	0.71
0.25 : 0.75	2.5	2.39	0.79	0.72 - 0.77	0.75	0.94	0.89	0.87
0.00 : 1.00	3.0	3.04	1.0	0.89 - 0.99	0.95	1.0	0.97	0.96

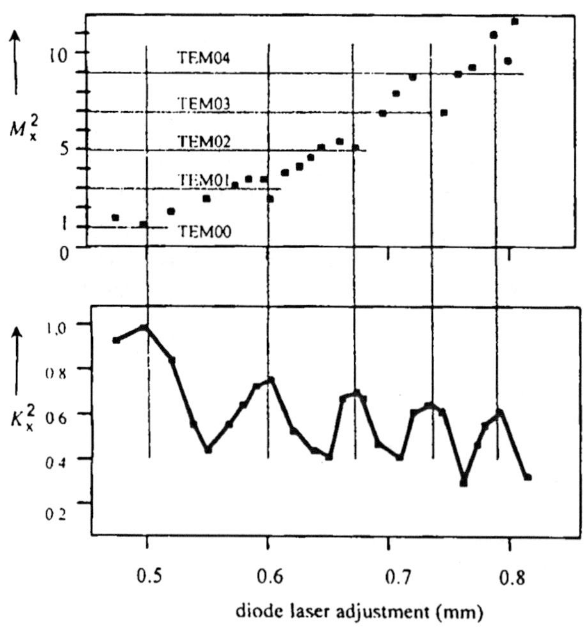

Fig. 2.36 Beam propagation factor M_x^2 and global degree of coherence K_x^2. The modes were generated in a diode end-pumped Nd-YAG-laser, by laterally shifting the diode with respect to the optical axis of the resonator [1.76] (© Elsevier Science 1996).

If the field cannot be separated according to $E(x,y) = E_0 \cdot f(x) \cdot g(y)$, it exhibits two transverse dimensions, resulting in a four-dimensional cross-correlation function. The corresponding ambiguity functions are also four-dimensional. However, since the intensity $I(x,y,z)$ is only a three-dimensional function, the missing information is hidden in the azimuthal phase structure. By inserting a cylinder lens and rotating it by an angle β, the additional information can be recovered. Now the intensity becomes four-dimensional, $I(x,y,\beta,z)$ and can be used to determine all coherence characteristics and beam parameters, but it is a time consuming method.

Wigner-function

Another function often used in Optics is the Wigner-function $h(\boldsymbol{r},\boldsymbol{\theta})$, which is defined as the Fourier transform of the cross-correlation function with respect to $\boldsymbol{s}=\boldsymbol{r}_1-\boldsymbol{r}_2$:

$$h(\boldsymbol{r},\theta) = \frac{1}{\lambda^2}\int \Gamma_{12}(\boldsymbol{r},\boldsymbol{s})\ \exp[-ik\theta\boldsymbol{s}]\ ds \tag{2.169}$$

where $\boldsymbol{k}$ is the wave vector (k_x,k_y), $\boldsymbol{r}$ is the position vector and $\boldsymbol{\theta}$ the far field divergence vector:

$$\boldsymbol{r} = \begin{pmatrix} x \\ y \end{pmatrix} \qquad \theta = \begin{pmatrix} \theta_x \\ \theta_y \end{pmatrix}$$

This function was first introduced by Wigner [1.24] for the description of quantum mechanical phenomena in the phase space and later used by Walther [1.30] and Wolf [1.45]. The properties of this function and its application to beam propagation is summarized in several papers of Bastiaans [1.39,1.40,1.44]. The special feature of coherence is discussed in detail by Eppich [1.78,1.80].

The Wigner-function as well as the ambiguity-function and the cross-correlation function completely characterize a radiation field and all three functions contain the same information. The cross-correlation function is obtained directly from the interference experiments as discussed in the beginning of this section. Ambiguity and Wigner-function have to be evaluated from intensity measurements, which are easier to perform but require sophisticated numerical approaches. The Wigner-function is related to the radiance of geometrical optics, being approximately (but not correctly) the power per area and angle [1.30]. The advantage of using the Wigner-function is its simple propagation law through first order optical elements. The same relation holds as for the ambiguity function:

$$h(\boldsymbol{r}_2,\theta) = h(\boldsymbol{M}^{-1}\boldsymbol{r}_2\ ,\ z) \tag{2.170}$$

where M is the ABCD-matrix of Sec. 2.4.2. For propagation in free space, h is sheared the same way as shown for the ambiguity function in Fig. 2.35.

Without mathematical derivations, some results are summarized:

beam power
$$P = \int h(\boldsymbol{r},\theta)\,dr\,d\theta \tag{2.171}$$

near field intensity
$$I(\boldsymbol{r}) = \int h(\boldsymbol{r},\theta)\,d\theta \tag{2.172}$$

far field intensity
$$I(\theta) = \int h(\boldsymbol{r},\theta)\,dr \tag{2.173}$$

beam width
$$\left.\begin{matrix} \langle x^2\rangle \\ \langle y^2\rangle \\ \langle xy\rangle \end{matrix}\right\} = \frac{1}{P}\int \begin{pmatrix} x^2 \\ y^2 \\ xy \end{pmatrix} h(\boldsymbol{r},\theta)\,dr\,d\theta \tag{2.174}$$

beam divergence
$$\left.\begin{matrix} \langle \theta_x^2\rangle \\ \langle \theta_y^2\rangle \\ \langle \theta_x\theta_y\rangle \end{matrix}\right\} = \frac{1}{P}\int \begin{pmatrix} \theta_x^2 \\ \theta_y^2 \\ \theta_x\theta_y \end{pmatrix} h(\boldsymbol{r},\theta)\,dr\,d\theta \tag{2.175}$$

A look at Section 2.6.2 shows that the Wigner-function delivers all beam parameters. Unfortunately, it can not be measured directly, but for one-dimensional transverse fields it can be evaluated from the intensity by a suitable transformation [1.78]. For two-dimensional fields it is also possible to retrieve the Wigner-function from intensity-measurements, but the procedure is much more complicated [1.78]. Therefore h is more of a mathematical tool, that can be used to calculate the propagation of light, also through non-parabolic optical systems, including slits and arbitrary phase plates. In order to gain a better feeling for this concept, let us apply the Wigner-function theory to a one-dimensional Gaussian beam with a field distribution at the waist position of:

$$E(x,z=0) = E_0 \exp\left[-\frac{x^2}{w_0^2}\right]$$

The cross-correlation function is given by (2.150/2.151):

$$\Gamma(x_1,x_2,z=0) = \frac{1}{2}c_0\epsilon_0|E_0| \exp\left[-\frac{x_1^2+x_2^2}{w_0^2}\right] = \frac{1}{2}c_0\epsilon_0|E_0| \exp\left[-\frac{2x^2+s^2/2}{w_0^2}\right] \tag{2.176}$$

Application of the Fourier transform (2.169) results in the Wigner-function at $z = 0$:

$$h_0(x,\theta,z=0) = \sqrt{2\pi}\frac{w_0}{\lambda}\,\frac{1}{2}c_0\epsilon_0|E_0|^2\,\exp\left[-2\frac{x^2}{w_0^2} - \frac{k^2\theta^2 w_0^2}{2}\right] \tag{2.177}$$

Free space propagation of the Wigner-function is performed by applying the ABCD-law (2.170), replacing x by $x-\theta z$:

$$h_0(x,\theta,z) = \sqrt{2\pi}\frac{w_0}{\lambda}\,\frac{1}{2}c_0\epsilon_0|E_0|^2\,\exp\left[-2\frac{(x-\theta z)^2}{w_0^2} - \frac{k^2\theta^2 w_0^2}{2}\right] \tag{2.178}$$

which again is a Gaussian shaped function in x- and θ- direction. This function and its propagation in free space are shown in Fig. 2.37. Starting at the waist, the projection onto the x-axis is the near field intensity with the waist-radius $\langle x^2\rangle = w_0^2/4$.The projection on the θ -axis is the far field intensity with the divergence $\langle\theta^2\rangle = \theta_0^2/4$. With increasing distance z, the beam radius increases according to (2.57), whereas the far field divergence remains constant.

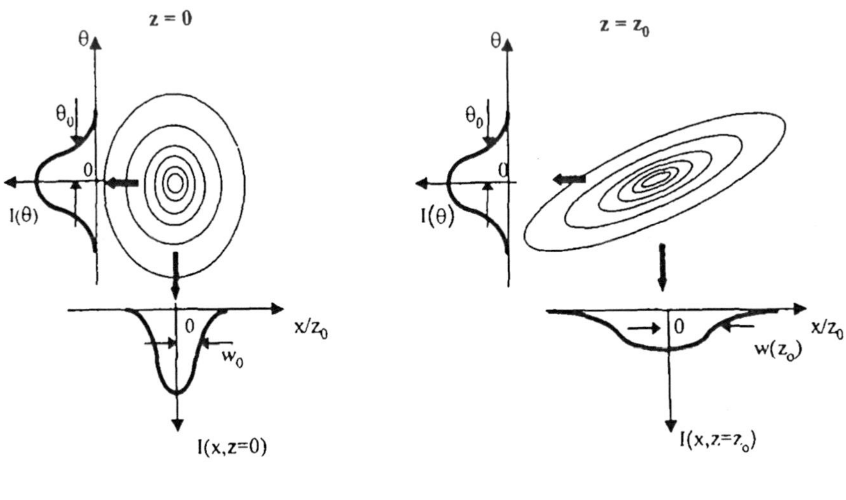

Fig. 2.37 The Wigner-function of the Gaussian beam and its propagation in free space. The x coordinate is normalized to the Rayleigh-length $z_0=\pi w_0^2/\lambda$.

2.8 Diffraction Theory of Optical Resonators

2.8.1 Integral-Equation for the Electric Field Distribution

By using the formalism of the Collins integral introduced previously, it is now a straightforward task to develop a diffraction theory for optical resonators [1.26,1.29, 1.31]. We are basically interested in the steady-state electric field distribution on the two resonator mirrors. Let us consider an optical resonator with rotational symmetry, as depicted in Fig. 2.38, with a field distribution $E_1(x,y)$ on mirror 1. What does the electric field look like after one round trip inside the resonator? In Sec. 1.3 we calculated the ray transfer matrix for a round trip, assuming an empty resonator:

$$M = \begin{pmatrix} G & 2Lg_2 \\ \dfrac{G^2-1}{2Lg_2} & G \end{pmatrix}, \quad G = 2g_1g_2-1 \tag{2.179}$$

By inserting the matrix elements into the Collins-Integral (2.29) we get the field after the round trip:

$$E_2(x_2,y_2) = i\frac{\exp[-i2kL]}{2Lg_2\lambda}\iint E_1(x_1,y_1)\exp\left[\frac{-i\pi}{2Lg_2\lambda}\left(G(x_1^2+y_1^2+x_2^2+y_2^2)-2(x_1x_2+y_1y_2)\right)\right]dx_1dy_1 \tag{2.180}$$

whereby the integration is performed over the surface of mirror 1.

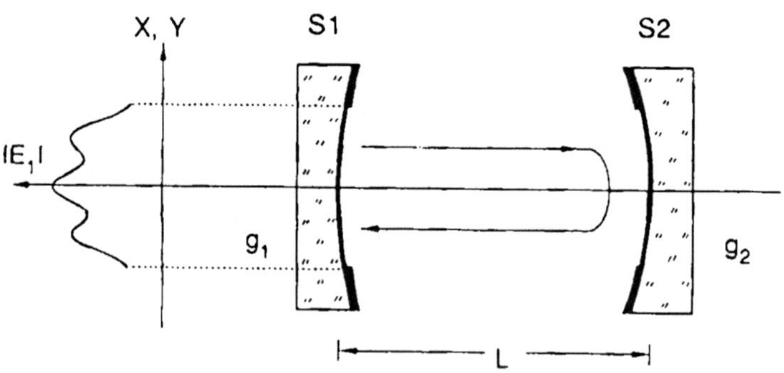

Fig. 2.38 After one round trip inside the resonator, the field distribution $E_1(x,y)$ has reproduced itself. In steady state operation the field amplitude can change but the shape of the field distribution has to remain constant.

The steady state oscillation is defined by the constancy of the field distribution: field E_2 is identical to the starting field E_1 except for a proportionality factor γ. The following relation must therefore hold:

$$E_2(x_2,y_2) = \gamma \; E_1(x_2,y_2) \tag{2.181}$$

The complex proportionality constant γ is called the eigenvalue of the field distribution. Replacing the left hand side of (2.180) with (2.181) yields an integral equation for the field distribution $E_1(x,y)$. Once this distribution is calculated (in general this is done numerically), the electric field on any other plane inside or outside the resonator can be determined by applying the corresponding Collins integral [1.26,1.29,1.31].

It should be kept in mind that the Collins-integral can only be used strictly to calculate the beam propagation between two apertures. If both of the resonator mirrors are limited in size, as is the case in Fig. 2.38, one must not use the Collins integral for the complete round trip. The two transits between the mirrors have to be dealt with separately by using two Collins integrals with the corresponding ray transfer matrices for each transit.

The physical meaning of the eigenvalue γ becomes clear when we compare the total power P_2 hitting the mirror after the round trip with the initial power P_1. We get:

$$P_2 = \frac{1}{2}c\epsilon_0 \iint E_2(x_2,y_2)E_2^*(x_2,y_2)dx_2dy_2 = \gamma\gamma^* \frac{1}{2}c\epsilon_0 \iint E_1(x_2,y_2)E_1^*(x_2,y_2)dx_2dy_2$$

$$= \gamma\gamma^* \; P_1 \tag{2.182}$$

where we have used (2.181). The integration is again performed over the surface of mirror 1. It is obvious that the factor $\gamma\gamma^*$ represents the power fraction that hits the mirror surface again after one round trip. This factor is referred to as the loss factor or the diffraction loss factor V. In Fig. 2.38 only the fraction V of the initial power falls into the aperture and gets reflected by the mirror, whereas the remainder $1\text{-}V$ hits the aperture and gets absorbed or scattered. The fractional power loss is called the diffraction loss $\Delta V = 1\text{-}V$ of the field distribution. For a laser in steady-state oscillation, the diffraction loss as well as other losses generated by output coupling and scattering are exactly compensated by the gain of the active medium. In general an infinite number of self-reproducing field distributions that fulfill (2.180) and (2.181) can be found for a giv en resonator. Each of these field distributions represents an eigenmode of the resonator with a well defined eigenvalue γ. Which of these eigenmodes will oscillate inside a laser resonator depends on the boundary conditions like aperture location and size and the gain of the active medium. All of those eigenmodes whose losses are compensated by the gain can oscillate. Since different eigenmodes experience different loss, the number of oscillating eigenmodes is decreased as the aperture size or the gain of the active medium are reduced. By using this mode selection technique one can force only the lowest-loss mode, also referred to as the fundamental mode, to oscillate.

2.8.2 The Gaussian Beam as a Fundamental Resonator Mode

One special solution of the eigenvalue problem (2.180/2.181) is the Gaussian beam, which represents the fundamental mode in a stable resonator. In the following, we will discuss the integral equation for an arbitrary optical resonator by using the Collins integral (2.29) expressed in terms of a general round trip ray transfer matrix $\boldsymbol{M}$. The mirror surfaces are assumed to extend to infinity and no apertures are located inside the resonator (Fig. 2.39). In reality, this means that the transverse dimension of the field is small compared to the sizes of all apertures.

We consider a resonator with an arbitrary number of optical elements located inside and the ray transfer matrix $\boldsymbol{M}$ represents the matrix for a resonator round trip starting on a given, but arbitrarily chosen, reference plane. The field distribution on the reference plane is a solution of the Collins integral equation:

$$\gamma E(x_2,y_2) = i\frac{\exp[-ik2L]}{\lambda B}\int_{-\infty}^{+\infty}\int_{-\infty}^{+\infty} E(x_1,y_1)\exp\left[\frac{-i\pi}{\lambda B}\left(A(x_1^2+y_1^2)+D(x_2^2+y_2^2)-2(x_1x_2+y_1y_2)\right)\right]dx_1dy_1$$

We know already that a Gaussian beam with beam parameter q_1 will transform into a Gaussian beam with beam parameter q_2 in this integral equation, with:

$$q_2 = \frac{Aq_1 + B}{Cq_1 + D} \tag{2.183}$$

Since we are looking for a steady-state solution, $q_2=q_1$ must hold. By replacing the left hand side of (2.183) with q_1, we can determine q_1 to be:

$$\frac{1}{q_1} = \frac{D-A}{2B} - \frac{i}{2B}\sqrt{4-(A+D)^2} \tag{2.184}$$

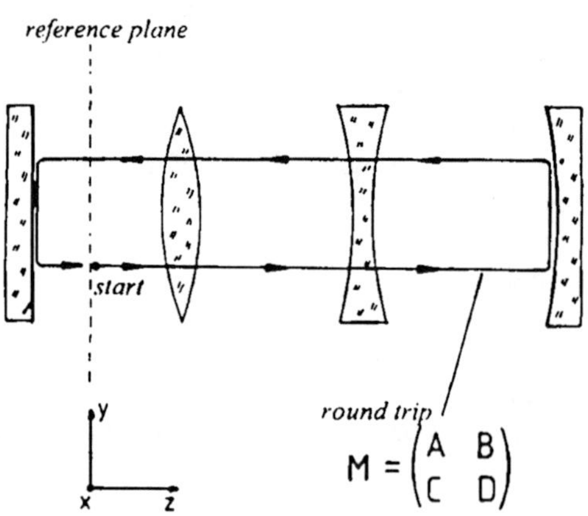

Fig. 2.39 Determination of the self-consistent Gaussian beam parameter on the reference plane for a round trip in an optical resonator.

Comparison with (2.58) yields the radius of curvature and the beam radius of the Gaussian beam on the reference plane:

$$R_1 = \frac{2B}{D-A} \quad , \quad w_1^2 = \frac{\lambda}{\pi}\frac{2B}{\sqrt{4-(A+D)^2}} \tag{2.185a,b}$$

Equation (2.185b) reveals that a Gaussian beam can only be a solution if:

$$|A+D| < 2 \tag{2.186}$$

holds. Optical resonators for which (2.186) holds are called *stable resonators*. It should be added that this result always holds no matter which reference plane in the resonator is chosen. The Gaussian beam represents the fundamental mode inside a stable optical resonator.

Example: Linear two-mirror resonator without internal optical elements
As already discussed in Sec. 1.3 the ray transfer matrix for the roundtrip starting on spherical mirror 1 reads:

$$M = \begin{pmatrix} G & 2Lg_2 \\ \dfrac{G^2-1}{2Lg_2} & G \end{pmatrix} \qquad \textit{with} \quad G = 2g_1g_2-1$$

The stability condition (2.186) now reads:

$$|G| < 1 \quad \Rightarrow \quad 0 < g_1g_2 < +1 \tag{2.187}$$

By using (1.185a,b), the radius of curvature and the beam radius on mirror 1 are:

$$R_1 = \infty \quad , \quad w_1^2 = \frac{2Lg_2\lambda}{\pi\sqrt{1-G^2}}$$

An infinite radius of curvature on the mirror means that the mirror surface represents the wave front of the Gaussian beam (remember that in Sec. 1.3 we have replaced the mirror by two lenses and our reference plane is located in between). The Gaussian beam parameters on mirror 2 can be obtained by switching the indices in (2.187).

Beam radii and radii of curvature at any other plane inside or outside the resonator can be easily obtained by applying the ABCD law for Gaussian beams. However, for propagation outside the resonator one has to incorporate one 'replacement lens' into the corresponding ray transfer matrix.

2.9 Plane Wave Representation of Diffraction

An exact solution of the wave equation (2.1) is the infite, monochromatic plane wave, which, omitting the frequency term, reads in the complex notation:

$$E(x,y,z) = A_0 \exp[-i\boldsymbol{kr}] \tag{2.188}$$

with $E=(E_x, E_y, E_z)$: electric field vector

$\boldsymbol{k}=(k_x, k_y, k_z)$: wave vector with $|\boldsymbol{k}| = 2\pi/\lambda=\omega/c$

$\boldsymbol{r}=(x,y,z)$: position vector

Any superposition of plane waves travelling in different directions $\boldsymbol{k}$ is also a solution of (2.1):

$$E(x,y,z) = \int_{-\infty}^{+\infty}\int_{-\infty}^{+\infty} A(k_x,k_y)\, \exp[-i\boldsymbol{kr}]\, dk_x dk_y \tag{2.189}$$

$A(k_x,k_y)$ is called the angular spectrum of the field $\boldsymbol{E}$. For a single plane wave propagating in direction $\boldsymbol{k}=(k_{x0},k_{y0})$, the spectrum is a delta peak:

$$A(k_{x0},k_{y0}) = A_0\delta(k_x-k_{x0},k_y-k_{y0}) \qquad with \int_{-\infty}^{+\infty}\int_{-\infty}^{+\infty} \delta(k_x-k_{x0},k_y-k_{y0})dk_x dk_y = 1$$

Since $|\boldsymbol{k}|$ is constant, (2.189) can be rewritten into the form: (2.190)

$$E(x,y,z) = \int\int_{k_x^2+k_y^2\le k^2} A(k_x,k_y)\, \exp\left[-i\left(k_x x+k_y y+z\sqrt{k^2-k_x^2-k_y^2}\right)\right] dk_x dk_y$$

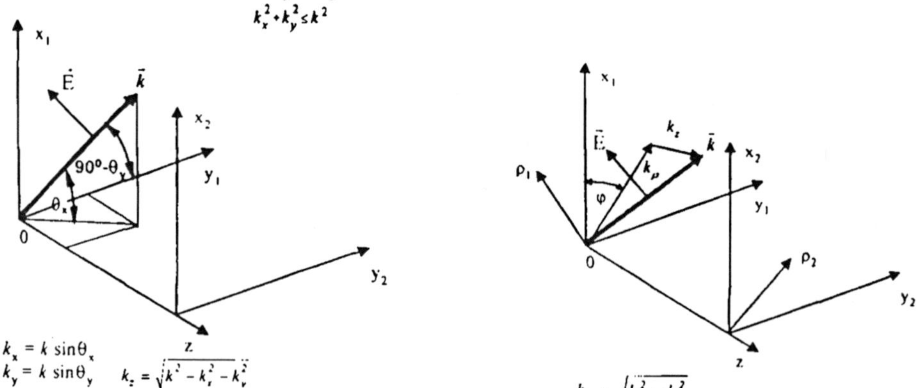

Fig. 2.40 A plane wave propagation in the direction of $\boldsymbol{k}$. Left: rectangular symmetry, right: circular symmetry.

The exponent in (2.190) becomes real for $k_x^2+k_y^2>k^2$ which means that the field amplitude decays very fast with $exp[-2\pi z/\lambda]$. This evanescent term does not contribute to the propagating field and can be neglected. The integration is limited to values $k_x^2+k_y^2\leq k^2$ or angles $|\theta_x|,|\theta_y|\leq\pi/2$ and (2.190) holds for $z>\lambda$. If the field $E(x,y,z)$ is given at the plane z=0, (2.190) results in:

$$E(x_1,y_1,0) = \iint A(k_x,k_y)\ \exp\left[-i(k_x x_1+k_y y_1)\right]\ dk_x dk_y$$

$$\Leftrightarrow \qquad E(x_1,y_1,0) = F^{-1}[A(k_x,k_y)]$$

$E(x_1,y_1,0)$ is the inverse Fourier transform $\mathbf{F}^{-1}$ of the angular spectrum, as discussed in Sec. 2.4.2, and accordingly, the angular spectrum $A(k_x,k_y)$ is the Fourier transform of the field $\boldsymbol{E}$. Now (2.190) can be rewitten as:

$$E(x_2,y_2,z) = \iint F[E(x_1,y_1,0)]\ \exp\left[-i\left(k_x x_2+k_y y_2+z\sqrt{k^2-k_x^2-k_y^2}\right)\right]\ dk_x dk \qquad (2.191)$$

If the field is given at the plane $z=0$, its propagation in free space is determined by (2.191).This so called plane wave representation is a better approach to diffraction than the Kirchhoff/Fresnel-integrals [1.2,1.77]. In addition, it automatically incorporates the vector character of the electric field. The plane wave presentation is useful for the evaluation of beam propagation in anisotropic media, higher order approximation of beam propagation and for the evaluation of diffraction free beams.

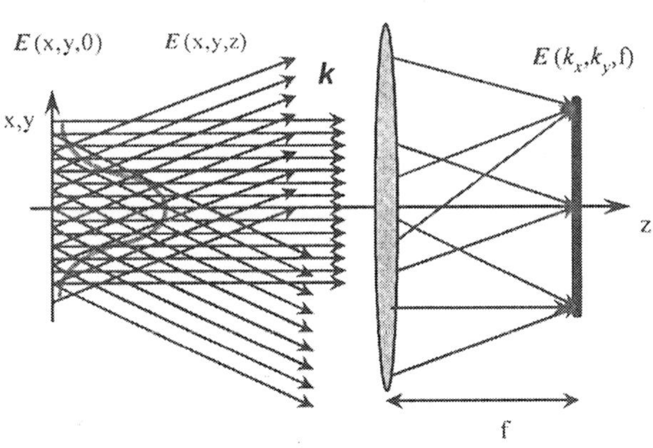

Fig. 2.41 The field distribution in the plane z=0 is decomposed into plane waves. The plane wave spectrum is the Fourier transform of the field in a given plane (z=0) and it can be observed in the far field or in the focal plane of a lens. The propagation of the plane waves is described by (2.188), and their superposition delivers the propagating field $\boldsymbol{E}$(x,y,z).

Equation (2.191) automatically predicts that laterally confined electric field distributions will develop a longitudinal field component in direction of the beam propagation. The field of the single plane wave, travelling with an angle with respect to the z-axis, has a longitudinal component E_z, and the superposition of all fields will also have a resulting longitudinal component. This is a general property of fields with a transverse structure propagating in charge free media. Only the infinite plane wave, propagating in z-direction has no longitudinal component, which is why light is often described as a transverse electro-magnetic field. In the paraxial approach with $k_x, k_y \ll k$, the square root can be expanded. In first order approximation, (2.191) can then be transformed into:

$$E(x_2,y_2,z) \approx \exp[-ikz] \int\!\!\int_{k^2+k^2 \le k^2} F[E(x_1,x_2,0)] \exp\left[-i\left(k_x x_2 + k_y y_2 + \frac{z}{2k}(k_x^2+k_y^2)\right)\right] dk_x dk_y \tag{2.192}$$

By inserting the Fourier-integral:

$$F[E(x_1,y_1,0) = \int_{-\infty}^{+\infty}\!\!\int_{-\infty}^{+\infty} E(x_1,x_1,0)] \exp\left[i(k_x x_1 + k_y y_1)\right] dx_1 dy_1 \tag{2.193}$$

and integrating (2.192) with respect to k_x, k_y, the well known Fresnel-integral is obtained:

$$E(x_2,y_2,z) = i\frac{k}{2\pi z}\exp[-ikz] \int_{-\infty}^{+\infty}\!\!\int_{-\infty}^{+\infty} E(x_1,x_1,0)] \exp\left[-\frac{ik}{2z}\left[(x_2-x_1)^2+(y_2-y_1)^2\right]\right] dx_1 dy_1 \tag{2.194}$$

Example:
A Gaussian field distribution with the waist radius w_0 is given at the plane z=0:

$$E(x,y,0) = E_0 \exp\left[-\frac{x^2+y^2}{w_0^2}\right]$$

The Fourier Transform reads:

$$A(k_x,k_y,0) = F[E(x,y,0)] = E_0 \int\!\!\int \exp\left|-\frac{x^2+y^2}{w_0^2}\right| \exp\left[i(k_x x + k_y y)\right] dx dy$$

The integral can be solved solved and delivers the angular spectrum of the Gaussian shaped field:

$$A(k_x,k_y,0) = E_0 \frac{w_0^2}{4\pi} \exp\left[-\frac{w_0^2}{4}(k_x^2+k_y^2)\right]$$

The paraxial approximation of the plane wave presentation (2.192) then yields for the electric field at the plane z:

$$E(x,y,z) = \frac{E_0}{1+iz/z_0} \exp\left[-\frac{x^2+y^2}{w_0^2(1+iz/z_0)}\right] \qquad k_x,k_y < k \tag{2.195}$$

with $z=\pi w_0^2/\lambda$ being the Rayleigh range. For large values of k_x,k_y, or large angles θ_x,θ_y the correct equation (2.189) has to be solved. For an x-polarized beam, an approximate second order solution reads [1.38,1.41]:

$$E(x,y,z) = \frac{E_0}{1+iz/z_0} \exp\left[-\frac{x^2+y^2}{w_0^2(1+iz/z_0)}\right] \left(1-2i\frac{z/z_0}{kz_0(1+iz/z_0)}\ ;\ 0\ ;\ -i\frac{x/z_0}{1+iz/z_0}\right) \tag{2.196}$$

The x-component is corrected by an additional term. In the far field ($z\to\infty$), this term approaches $-2/kz_0=-\theta_0^2$ where θ_0 is the half angle of divergence of the beam. This term becomes relevant for highly divergent fields. Moreover, a z-component appears, as already discussed. The real part of E_z is the projection of the tilted **E**-vector, which is tangential to the curved wave front, as shown in Fig. 2.42.

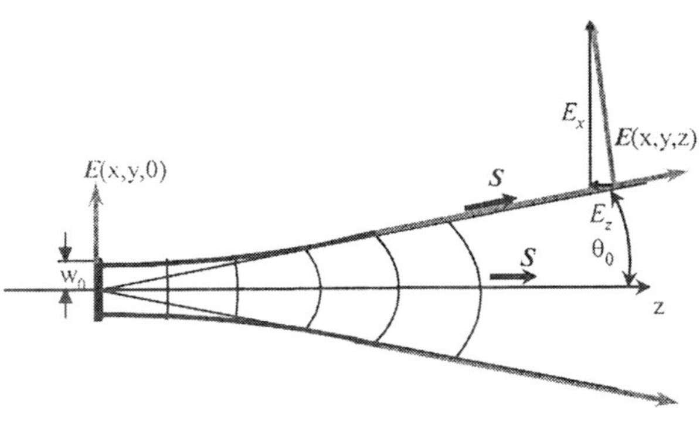

Fig. 2.42 The orientation of the electric field vector of a Gaussian beam

The z-component can be written as:

$$Re\left(\frac{E_z}{E_x}\right) = -\frac{x}{R(z)}$$

with $R(z)$ being the radius of curvature of the phase front. The longitudinal field component vanishes in the waist, where the field is plane and perpendicular to the optical axis. The imaginary part of E_z is shifted by 90° with respect to E_x, which means that the field is slightly elliptically polarized in the x-z plane.

Example:
At the output of a diode laser, the waist radius is about 1 μm for a wavelength of $\lambda=0.8\ \mu m$. The half angle of divergence is $\theta_0=\lambda/(\pi w0) = 0.25\ mrad$, the Rayleigh range is $z_0 = \pi w_0^2/\lambda = 3.9\ \mu m$ and the z-component of the electric field E_z at the waist $x=y=w_0$ becomes $E_z(w_0,w_0,0) = -\theta_0/e^2 = -i\ 0.025\ E_0$. In the far field, the correction of the x-component is $-0.0625\ E_0$. Even for diode lasers, the correction terms are small.

Circular Symmetry
If the electric field exhibits circular symmetry and does not depend on the azimuthal angle φ, the same procedure as shown in Sec. 2.2.2 can be applied. The plane wave presentation in circular symmetry is given by:

$$E(r_2,z) = 2\pi\int_0^k A(k_r)\ J_0(k_r r_2)\ \exp\left[-iz\sqrt{k^2-k_r^2}\right] k_r dk_r \tag{2.197}$$

Replacing the angular spectrum by its Fourier Transform results in:

$$E(r_2,z) = \int_0^k\int_0^\infty E(r_1,0)\ J_0(k_r r_2) J_0(k_r r_1)\ \exp\left[-iz\sqrt{k^2-k_r^2}\right] r_1 dr_1 k_r dk_r \tag{2.198}$$

J_0 is the zero-order Bessel function. In the paraxial approach ($|k_r| \ll |k|$), the Fresnel-integral for circular symmetry of Sec. 2.2.2. is obtained.

2.10 Diffraction-Free Beams

Normally the structure of the electromagnetic field changes while propagating in free space. These changes are described by the wave equation (2.1) or by one of the diffraction integrals (2.12) and (2.191). However, special solutions of the diffraction integral exist, the eigensolutions, for which the shape of the field remains constant. Only the amplitude and the transverse dimension vary in such a way that the total power remains constant. One eigensolution is the Gaussian beam; others are discussed in Sec. 5.1. All of these beams experience diffractive spreading which means that their amplitude decreases during propagation. Fortunately, diffraction-free solutions of the wave equation exist [1.51,1.52,1.106]. One is the infinite plane wave as discussed in the previous Section. Normally, the field distribution generated by a superposition of plane waves will change during propagation. But there are special superpositions that do not change their field distributions.

A set-up for the realization of a beam generated by a superposition of plane waves is shown in Fig. 2.43. In the front focal plane of a lens with focal length f is a coherently illuminated ring with radius R and width ΔR. The field starting from the differential area $R\Delta R\Delta\phi$ is collimated by the lens resulting in a plane wave:

$$\Delta E(x,y,z,t) = E_0 \exp[-i(k_x x + k_y y + k_z z - \omega t)]\ \Delta\phi \qquad (2.199)$$

with the wave vector components:

$$k_x = \frac{2\pi}{\lambda} \sin\phi \sin\theta_P \qquad k_y = \frac{2\pi}{\lambda} \cos\phi \sin\theta_P \qquad k_z = \frac{2\pi}{\lambda} \cos\phi \cos\theta_P$$

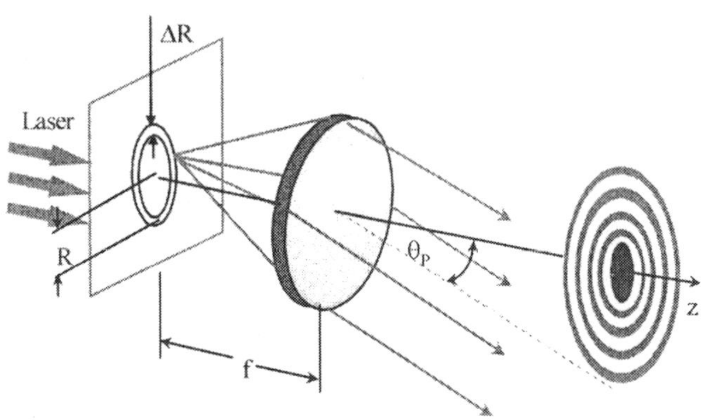

Fig. 2.43 Schematic set-up for generating a diffraction-free Bessel beam. In order to illuminate the lens completely, the width ΔR of the ring must be smaller than $\lambda f/w$., with w being the lens radius.

The resulting field behind the lens is obtained by integrating over all waves with ϕ varying between 0 and 2π. Equation (2.199) yields:

$$E(x,y,z,t) = E_0 \exp[-i(k_z z - \omega t)] \int_0^{2\pi} \exp[-ia(x \sin\phi + y \cos\phi)] d\phi \tag{2.200}$$

with: $a = \frac{2\pi}{\lambda} \sin\theta_P$

The angle θ_P is determined by the focal length f of the lens and the radius R of the ring:

$$\tan\theta_P = R/f \tag{2.201}$$

The above integral can be solved analytically and yields the Bessel function of zero order:

$$E(r,z,t) = E_0 \exp[i(k_z z - \omega t)] \, J_0(ar) \tag{2.202}$$

and the intensity distribution:

$$I(r) = I_0 \, J_0^2(ar) \tag{2.203}$$

with r being the radial coordinate (Fig. 2.44). This is an exact solution of the wave equation, as long as the field is not limited transversally. The intensity distribution does not depend on z, meaning the structure of the beam remains constant during propagation (Fig. 2.45). Be aware, that (2.202) only holds directly behind the lens. Due to the diameter $2w$ of the lens, the plane wave is limited and will spread due to diffraction. However, diffraction effects can be neglected for large Fresnel numbers $N=w^2/(\lambda L)$.

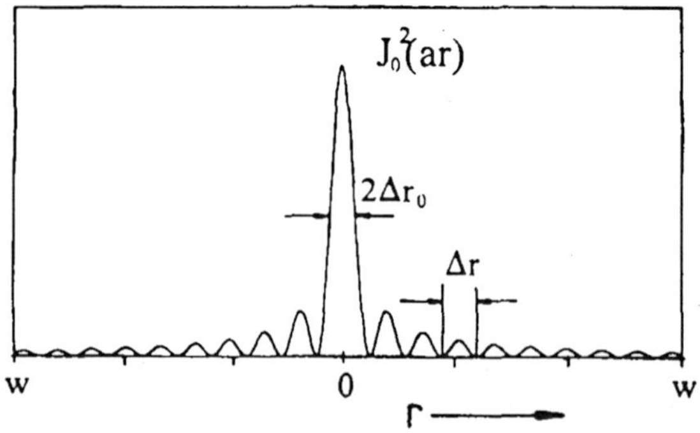

Fig. 2.44 Radial intensity distribution of the diffraction-free Bessel beam.

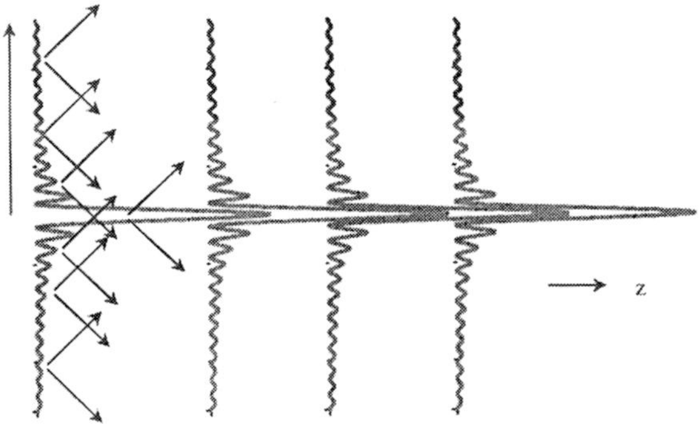

Fig. 2.45 The intensity distribution of a Bessel beam remains constant during propagation, even though power is constantly flowing transversally as indicated by the arrows.

The transverse structure of the intensity is plotted in Fig. 2.44. A central peak with a half maximum radius of:

$$\Delta r_0 \simeq \frac{1}{a} \tag{2.204}$$

is surrounded by side lobes with peak intensities decaying with $1/r$. The distance between two consecutive intensity zeroes is almost constant:

$$\Delta r \simeq \frac{\pi}{a} \tag{2.205}$$

The power in each individual ring is equal to that in the central peak. By using these approximations, the power P_0 in the central peak can be written as:

$$P_0 = P_{tot} \frac{\Delta r}{w} \tag{2.206}$$

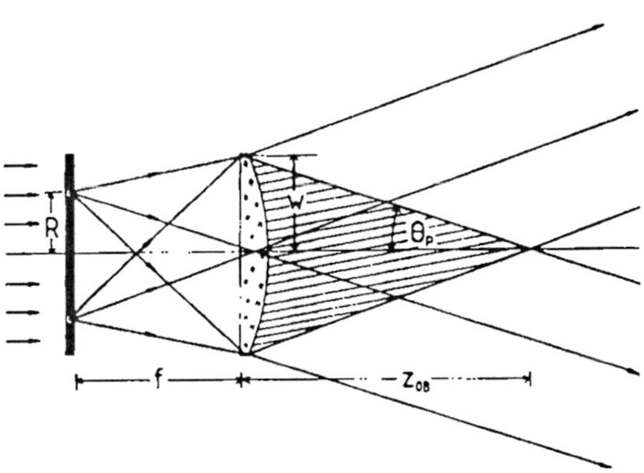

Fig. 2.46 The Rayleigh range z_{0B} of the Bessel beam. The shaded area is the interference range of the plane waves.

with P_{tot} being the total beam power at the lens. The sharp central peak is generated by the interference of the plane waves. It will propagate in the z-direction without broadening as long as the plane waves overlap. This diffraction-free propagation distance z_{0B} is given by (Fig. 2.46):

$$z_{0B} = \frac{w}{\tan\theta_p} \tag{2.207}$$

which again is the well-known Rayleigh range, now for large angles θ_p.

The set-up in Fig. 2.43 exhibits low efficiency, because only a small portion of the laser beam is used. A more efficient set-up using an axicon with radius R is shown in Fig. 2.47 [1.106]. Again the resulting field is a superposition of plane waves which all have the same slope $\sin\theta_p = k_{r0}/k$, $\tan\theta_p = (n-1)\gamma$, where n is the index of refraction of the axicon and γ its apex angle. Let us apply the plane wave presentation of diffraction in circular symmetry (2.197) to calculate the resulting intensity distribution. The angular spectrum is a delta peak:

$$E(k_{r0},0) = E_0\ \delta(k_r - k_{r0}) \qquad with \quad \int_0^\infty \delta(k_r - k_{r0})\ k_r dk_r = 1$$

and (2.197) right away delivers for field and intensity::

$$E_B(r,z) = E_0\ J_0(ar)\ \exp[-ikz] \qquad with \quad a = k\ \sin\theta_p \tag{2.208}$$

$$I(r) = I(0)\ J_0^2(ar) \tag{2.209}$$

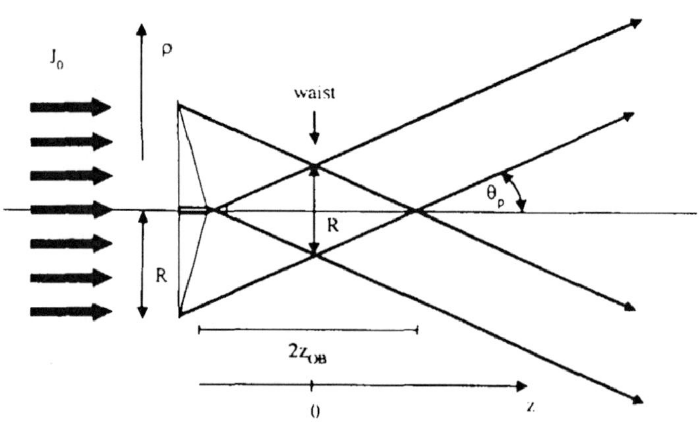

Fig. 2.47 Bessel beam generation using an axicon. The radius of the waist of the Bessel beam is $w=R/2$ at $z=0$.

Again a central peak with a half maximum radius of:

$$\Delta r_0 \approx \frac{1}{a} \tag{2.210}$$

and a power content of:

$$P_0 = \pi \, \Delta r_0^2 \, I(0) \tag{2.211}$$

is surrounded by side lobes. For $ar>5\pi/2$, the Bessel function is approximated by:

$$J_0(ar) = \sqrt{\frac{2}{\pi ar}}\cos(ar-\pi/4)$$

which delivers for the distance between two consecutive intensity zeroes:

$$\Delta r \simeq \frac{\pi}{a} \tag{2.212}$$

The power in each ring is equal to that in the central peak. By using these approximations, the power P_0 in the central peak for the set-up in Fig. 2.46 can be written as:

$$P_0 = P_{tot} \frac{\pi}{Ra} \tag{2.213}$$

with P_{tot} being the total beam power incident on the axicon with radius R. The sharp central peak is generated by the interference of the plane waves. It will propagate in the z-direction without broadening as long as the plane waves overlap. This diffraction-free propagation distance z_{0B} is approximately half the overlap-region in Fig. 2.47:

$$z_{0B} = \frac{R}{2\tan\theta_p} = \frac{R\pi}{\lambda a} \tag{2.214}$$

which is consistent with the usual definition of the Rayleigh range, the ratio of waist radius to divergence. Compared to the Rayleigh range of a Gaussian beam with the same waist radius Δr_0:

$$z_{0G} = \frac{\pi \Delta r_0^2}{\lambda} = \frac{\pi}{\lambda a^2} = \frac{z_{0B}}{Ra} \tag{2.215}$$

the Rayleigh range of the Bessel beam is larger by a factor Ra. Unfortunately, the power content in the main peak is reduced by the same factor.

The beam parameter product of a diffraction-free beam can only be evaluated numerically, but a rough estimation is easily done. The waist is located in the center of the overlap region at $z=0$. The waist radius containing 86.5% of the total power is given by:

$$w_0 = 0.865 \; R/2$$

The far field divergence is θ_P since plane waves propagating at an angle θ_P are considered. For infinite plane waves, the far field is given by a delta ring $\delta(\theta-\theta_P)$. Since the near field is limited to a radius of $R/2$, a broadening occurs with $\Delta\theta_P \approx 2\lambda/R$. The ring, containing 86.%% of the power has a divergence of approximately $\theta_P + 0.865\Delta\theta_P \approx \theta_P$. The beam parameter product thus reads:

$$w_0 \theta_P = \frac{0.865}{4} \frac{Ra\lambda}{\pi}$$

and the beam propagation factor results in:

$$M^2 = w_0 \theta_P \frac{\pi}{\lambda} \approx 0.216 \; Ra$$

which holds for $Ra \gg 1$.

Example 1 (annular apperture):
A circular slit with radius $R=2mm$ and width 20μm is placed in front of a lens with focal length $f=200mm$ and diameter $2w=20mm$. A 1 W laser beam with a beam radius of 3mm (we assume that the intensity profile is homogeneous) and a wavelength of $\lambda=1\,\mu m$ generates a Bessel beam with the following parameters:

Divergence angle of the plane waves:	θ_P	= 10 mrad
Width of central peak:	$2\Delta r_0$	= 31.8 μm
Total power at the lens:	P_{tot}	= 8.89 mW
Power in central peak:	P_0	= 42.4 μW
Rayleigh range:	z_{0B}	= 1 m

A Gaussian beam with the same waist radius w_0 has a Rayleigh range of 794 μm.

This example clearly indicates that there are some disadvantages to the generation of diffraction-free beams:

- only a small part of the laser beam can be transformed into a Bessel beam using the simple annular aperture. This drawback can be overcome by using an axicon, by amplifying the Bessel beam, directly or holographically [1.59,1.106] or by using special optical resonators [1.82,1.106].

- generation of a small spot radius Δr_0 with a Rayleigh range that is $R/\Delta r_0$ larger than the conventional Rayleigh range requires a total beam power that is larger by the same factor.

- the enhancement of the Rayleigh range by the factor $R/\Delta r_0$ reduces the power in the central peak by the same factor.

- if the central peak is used only by cutting off the outer diffraction peaks with a pinhole, the Rayleigh range will be immediately reduced to the normal value. However, the central peak can be used to attain higher efficiencies in nonlinear optics and spectroscopy [1.89,1.92,1.95,1.96]. In this case, the central peak is reduced during propagation, but the Rayleigh range is not reduced in the same way.

Example 2 (axicon) [1.106]:
A ZnSe axicon (n=2.4) with radius $R=9.35mm$ and apex angle of 0.5° is homogeneously illuminated with a 1kW CO_2 laser beam at a wavelength of $\lambda=10.6\mu m$ The generated Bessel beam exhibits the following parameters:

Divergence angle of the plane waves:	θ_P	= 12.2 mrad
Width of central peak:	$2\Delta r_0$	= 277 µm
Power in central peak:	P_0	= 45.7 W
Rayleigh range:	z_{0B}	= 383 mm

A Gaussian beam with the same waist radius w_0 has a Rayleigh range of 5.69 mm. This example indicates, that it is possible to get a considerable fraction of the total power into the central peak of the Bessel beam.

2.11 Beam Propagation in Anisotropic Crystals

In the previous sections of this chapter the diffraction and the propagation of scalar fields was discussed. This is correct for isotropic media, but not for birefringent media and anisotropic crystals. Now the situation becomes more complicated, and we have to deal with uni- and biaxial crystals. The physics of anisotropic crystals are well treated in the textbooks of Optics [1.2,1.6]. Here, only a very simple example will be discussed qualitatively.

A uniaxial crystal is characterized by two principal refractive indices, n_o and n_{eo}. They depend on the orientation of the polarization with respect to the c-axis of the crystal, as shown in Fig. 2.48. Inside the crystal the beam is characterized by the field $\boldsymbol{E}$ and the electric displacement $\boldsymbol{D}$. In the most general case, the correlation between these vectors is determined by the dieletric tensor ϵ:

$$\begin{pmatrix} D_x \\ D_y \\ D_z \end{pmatrix} = \epsilon_0 \; \epsilon \; \boldsymbol{E} = \epsilon_0 \begin{pmatrix} \epsilon_{xx} & \epsilon_{xy} & \epsilon_{xz} \\ \epsilon_{yx} & \epsilon_{yy} & \epsilon_{yz} \\ \epsilon_{zx} & \epsilon_{zy} & \epsilon_{zz} \end{pmatrix} \begin{pmatrix} E_x \\ E_y \\ E_z \end{pmatrix} \tag{2.216}$$

For loss-free crystals, the dielectric tensor is symmetric and can be diagonalized by choosing the three principal axes for which $\boldsymbol{D}$ and $\boldsymbol{E}$ are parallel. For a uniaxial crystal, (2.216) is then reduced to:

$$\begin{pmatrix} D_x \\ D_y \\ D_z \end{pmatrix} = \epsilon_0 \begin{pmatrix} n_o^2 E_x \\ n_o^2 E_y \\ n_{oe}^2 E_z \end{pmatrix} \tag{2.217}$$

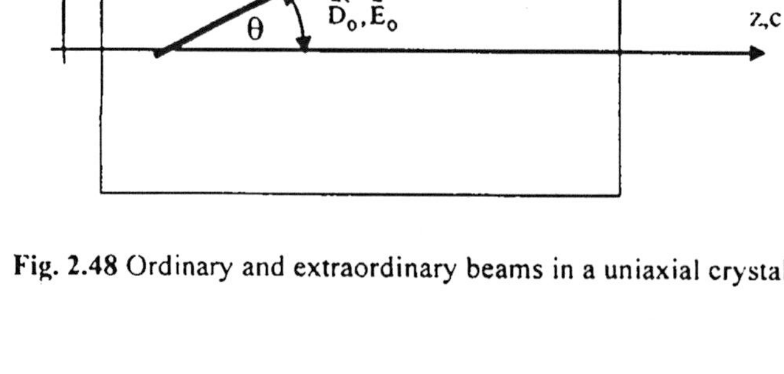

Fig. 2.48 Ordinary and extraordinary beams in a uniaxial crystal.

For uniaxial crystals, the wave equation delivers two polarized waves, which propagate inside the crystal without changing the polarization:

- the ordinary beam $\boldsymbol{D}_o$, $\boldsymbol{E}_o$ with $\boldsymbol{D}_o$ perpendicular to the c-axis of the crystal and perpendicular to the propagation vector $\boldsymbol{e}$ of the phase fronts.

- the extraordinary beam $\boldsymbol{D}_{eo}$, $\boldsymbol{E}_{eo}$ with $\boldsymbol{D}_{eo}$ perpendicular to $\boldsymbol{D}_o$ and e. $\boldsymbol{D}_{eo}$ and $\boldsymbol{E}_{eo}$ are in general not collinear because of the anisotropy.

The two beams are described by:

$$\boldsymbol{D}_o = \boldsymbol{D}_{o,0} \exp[i(\omega t - n_o k_o \boldsymbol{er})] \tag{2.218}$$

$$\boldsymbol{D}_{eo} = \boldsymbol{D}_{eo,0} \exp[i(\omega t - n_{eo}(\theta) k_o \boldsymbol{er})] \tag{2.219}$$

where the refractive index n_{eo} of the extraordinary beam depends on the angle θ:

$$\frac{1}{n_{eo}^2(\theta)} = \frac{\cos^2\theta}{n_o^2} + \frac{\sin^2\theta}{n_{eo}^2} \tag{2.220}$$

The ordinary and the extraordinary beam propagate with different phase velocities.

Now let us discuss the situation of a Gaussian beam, propagating in z-direction, parallel to the crystal axis c, and polarized in x direction. The $\boldsymbol{E}$ field in the y-z plane (shaded in Fig. 2.49) is perpendicular to the c-axis and to the propagation vector $\boldsymbol{e}$. It is an ordinary beam $\boldsymbol{E}_o$, $\boldsymbol{D}_o$. But the field in the x-z-plane is perpendicular to $\boldsymbol{D}_o$ and $\boldsymbol{e}$, and therefore an extraordinary beam. Inside the crystal ordinary and extraordinary parts of the Gaussian beam propagate with different phase velocities.

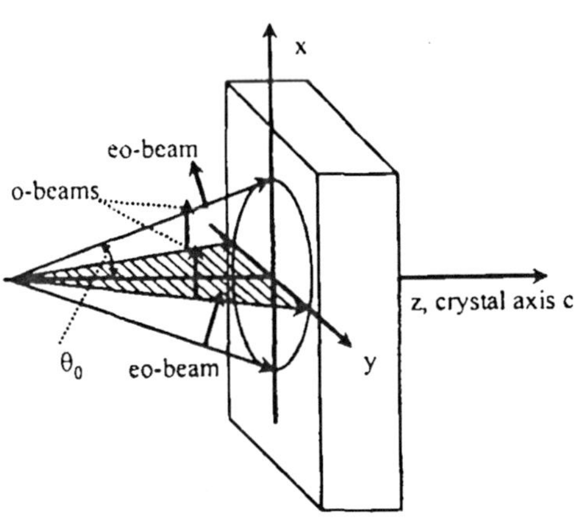

Fig. 2.49 The Gaussian beam incident on a uniaxial crystal consists of ordinary and extraordinary parts [1.104] (© OSA 2002).

These phase shifts, depending on the transverse coordinates, produce interference pattern and distort the Gaussian beam. Additionally the x-polarized beam is partly converted into y-polarization. The theory is straight forward, but tiring. The plane wave method in the paraxial approach is used as discussed in Sec. 2.9 [1.103,1.104]. Some results are presented in Figs. 2.50 and 2.51. The theory can be extended to higher order modes and modes of cylindrical symmetry [1.105].

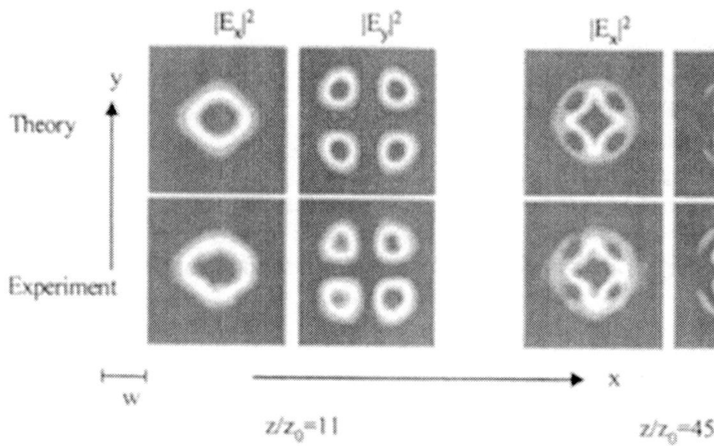

Fig. 2.50 The distortion of an x-polarized TEM_{00} beam propagating in a Calcite-crystal for two different propagation distances z/z_0 with z_0 being the Rayleigh range. The transverse coordinates are normalized to the beam radius *w(z)* [1.104] (© OSA 2002).

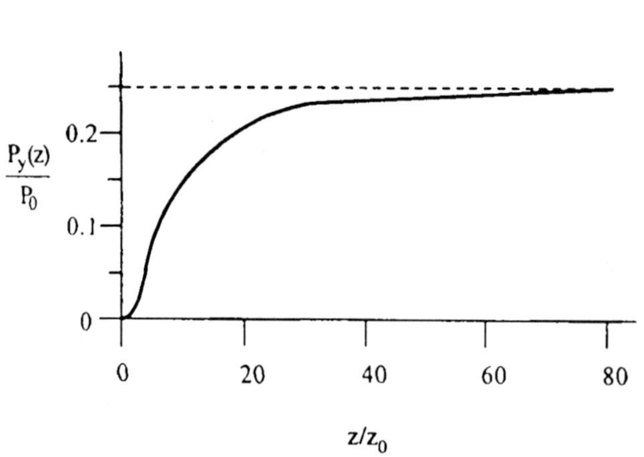

Fig. 2.51 The power $P_y(z)$ of the y-component versus the normalized propagation length z/z_0. In the limit of $z >> z_0$, 25% of the total power is converted into the y-component [1.104] (© OSA 2002).

Chapter 3

Polarization

3.1 General Aspects

An infinite plane wave in free space is transversally polarized which means that the electric field vector is always perpendicular to the direction of wave propagation z. The electric field is thus characterized by its components in the x-direction and the y-direction. For natural light and for the majority of lasers, the field vectors change their orientations randomly and in a short time interval compared to the detection time. Light sources with these properties are called unpolarized. In the following we discuss completely polarized light which means that the electric field vector either points in a fixed direction or changes its orientation periodically. The polarization of an electromagnetic field is completely characterized by the components of the electric field $\boldsymbol{E}$ in the x- and the y-direction. The field vector can be written such that common phase terms in the x- and the y-direction are extracted:

$$\boldsymbol{E} = \begin{pmatrix} E_{0x} \\ E_{0y}\,\exp[i\phi] \end{pmatrix} \exp[i(kz-\omega t+\psi)] \tag{3.1}$$

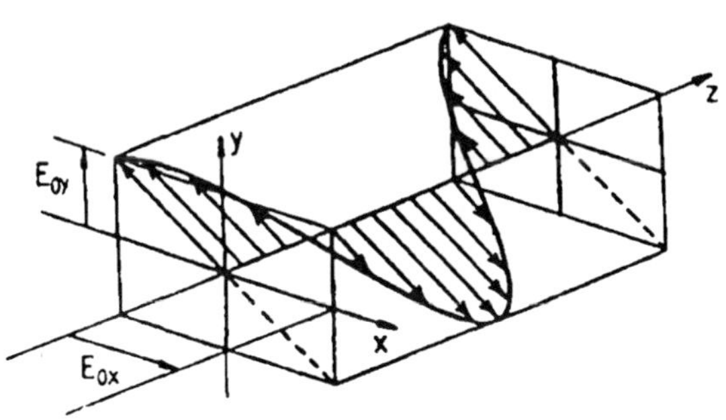

Fig. 3.1 The polarization of the electric field is characterized by the x-y-components field vector with the z-axis pointing into the direction of propagation. The polarization state is visualized in the x-y-plane by the projected curve traversed by the tip of the field vector. The field depicted is linearly polarized.

The field amplitudes E_{0x} and E_{0y} are both real numbers and the time averaged intensity of the wave is given by:

$$I = \frac{1}{2} c_0 \epsilon_0 \left(E_{0x}^2 + E_{0y}^2\right) \tag{3.2}$$

with: c_0: speed of light in vacuum
ϵ_0: vacuum permittivity

For the description of the polarization state the common time-dependent phase factor in (3.1) has no influence. In the following we will omit this term and deal only with the field vector:

$$\boldsymbol{E} = \begin{pmatrix} E_{0x} \\ E_{0y} \exp[i\phi] \end{pmatrix}$$

Special Polarization States

1) Linear Polarization, $\phi=0$ or $\phi=\pi$, (Fig. 3.2)
The two field components are either in phase or have opposite signs. The electric field vector oscillates at an angle α with respect to the x-axis with:

$$\alpha = \tan^{-1}\left(\frac{E_{0y}}{E_{0x}}\right) \qquad \textit{if} \;\; \phi = 0$$

$$\alpha = \pi - \tan^{-1}\left(\frac{E_{0y}}{E_{0x}}\right) \qquad \textit{if} \;\; \phi = \pi$$

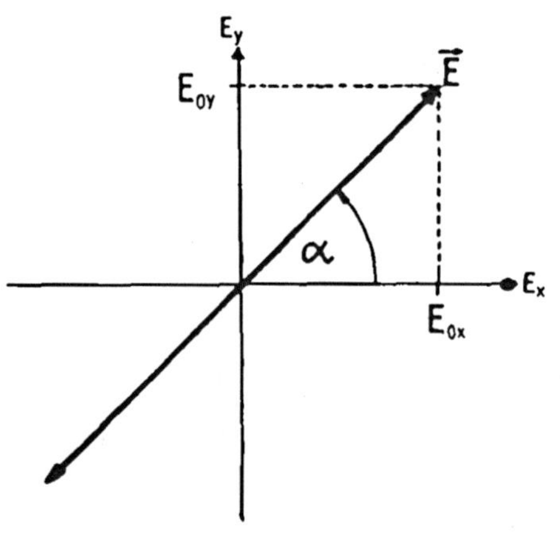

Fig. 3.2 For a linearly polarized electromagnetic wave the electric field vector oscillates only in one direction. The inclination is determined by the field amplitudes in the x- and y-direction.

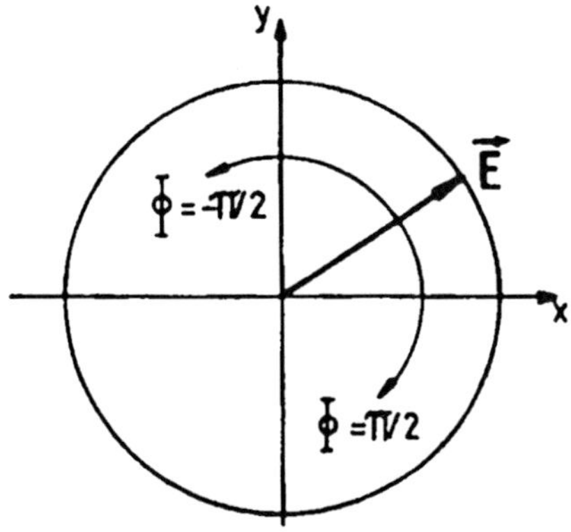

Fig. 3.3 For circularly polarized light, the tip of the field vector rotates on a circle.

2) Circular Polarization, $\phi=\pm\pi/2$, $E_{0x}=E_{0y}=E_0/\sqrt{2}$.
By using (3.1) we get the following field vectors for circularly polarized light:

$$\boldsymbol{E} = \frac{E_0}{\sqrt{2}}\begin{pmatrix}1\\ -i\end{pmatrix} \qquad \text{if } \phi = -\frac{\pi}{2}$$

$$\boldsymbol{E} = \frac{E_0}{\sqrt{2}}\begin{pmatrix}1\\ +i\end{pmatrix} \qquad \text{if } \phi = +\frac{\pi}{2}$$

The temporal behavior of the field vector is depicted in Fig. 3.3. The two field components are oscillating with a phase difference of 90°. The electric field vector sweeps out a circle with radius E_0 as time evolves. If we look back at the source, the field vector rotates in clockwise direction for $\phi=+\pi/2$ (*right circularly polarized*) and in counterclockwise direction for $\phi=-\pi/2$ (*left circularly polarized*).

3) Elliptical Polarization, E_{0x}, E_{0y}, and ϕ arbitrary.
If we assume that both field amplitudes are equal ($E_{0x}=E_{0y}=E_0$) the field vector is given by:

$$\boldsymbol{E} = E_0\begin{pmatrix}1\\ \exp[i\phi]\end{pmatrix}$$

Projected onto the x-y plane, the tip of the field vector describes an ellipse whose semi-axes are rotated by 45° with respect to the x-axis. Linearly and circularly polarized light are special cases of elliptical polarization (Fig. 3.4).

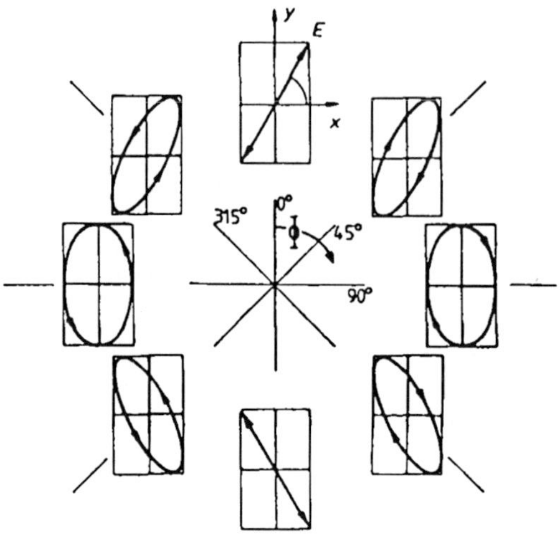

Fig. 3.4 Polarization states of light as a function of the phase difference ϕ between the electric field components in the x- and y-directions. In this graph the amplitudes of both components are different and the view is with the light propagation away from the observer.

Even if the two field amplitudes have different magnitudes, the field vector still traverses an ellipse with the semimajor axis rotated with respect to the x-axis by the angle α with:

$$\tan\alpha = \frac{E_{0y}^2 - E_{0x}^2 + \sqrt{4\cos^2\phi\ E_{0x}^2 E_{0y}^2 + (E_{0x}^2 - E_{0y}^2)^2}}{2\cos\phi\ E_{0x} E_{0y}}$$

Polarized light is called *right elliptically polarized* (clockwise rotation when looking back at the source) for $0<\phi<\pi$ and *left elliptically polarized* light is obtained for $\pi<\phi<2\pi$ (clockwise rotation when viewed in propagation direction).

3.2 Jones Matrices

3.2.1 Definition

Similarly to ray matrix theory in geometrical optics one can describe the changes in polarization generated by optical elements by means of a 2x2 matrix $\boldsymbol{M^P}$. If $\boldsymbol{E_0}$ denotes the field vector in front of the optical element, the new field vector at the output plane is given by:

$$\boldsymbol{E_1} = \boldsymbol{M^P}\ \boldsymbol{E_0} \tag{3.3}$$

The matrix $\boldsymbol{M^P}$ is referred to as the Jones matrix of the optical element [1.111,1.113]. In the following we present the Jones matrices for common polarizing optics.

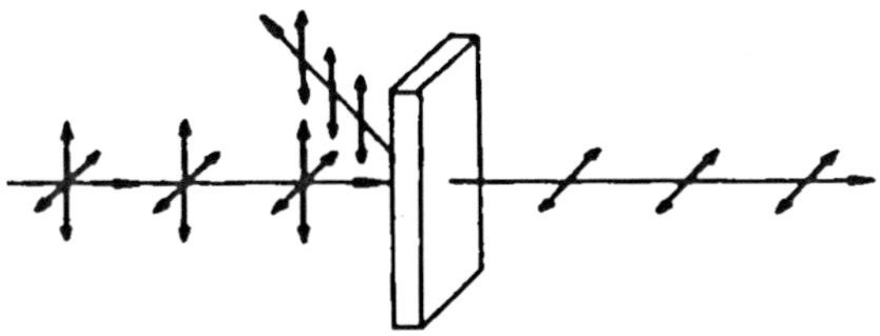

Fig. 3.5 Polarizer with a horizontal pass direction.

1) Polarizer

A polarizer transmits only electric fields oscillating in one direction called the pass direction. In case arbitrary polarized or unpolarized light is incident on the polarizer, only the field components in the pass direction will be transmitted. For an ideal polarizer, the light becomes linearly polarized. Thus, the Jones matrix of a polarizer can be written as:

$$M_P^P = t\begin{pmatrix}1 & 0\\ 0 & 0\end{pmatrix} \qquad \textit{if the pass direction is the x-axis}$$

$$M_P^P = t\begin{pmatrix}0 & 0\\ 0 & 1\end{pmatrix} \qquad \textit{if the pass direction is the y-axis}$$

The factor t with $0 \le t \le 1.0$ denotes the amplitude transmission of the polarizer. If the pass direction does not coincide with the x- or the y-direction, the Jones matrix can be found by rotation, as shall be discussed later.

In reality, polarizers do not provide 100% linear polarization. To a certain degree, field components oscillating perpendicular to the pass direction are also transmitted. If $I_{\parallel}$ and $I_{\perp}$ denote the intensities of the transmitted light being linearly polarized parallel and perpendicular to the pass direction respectively, the quality of the polarizer is characterized by the degree of polarization P:

$$P = \frac{I_{\parallel} - I_{\perp}}{I_{\parallel} + I_{\perp}} \tag{3.4}$$

For the ideal polarizer $I_{\perp}=0$ holds, and the degree of polarization is $P=1$. The Jones matrix for a real polarizer with the pass direction along the y-axis reads:

$$M_P^P = t\begin{pmatrix}\dfrac{\sqrt{1-P}}{\sqrt{1+P}} & 0\\ 0 & 1\end{pmatrix} \tag{3.5}$$

High quality polarizers provide degrees of polarization of $P=0.999$ and greater.

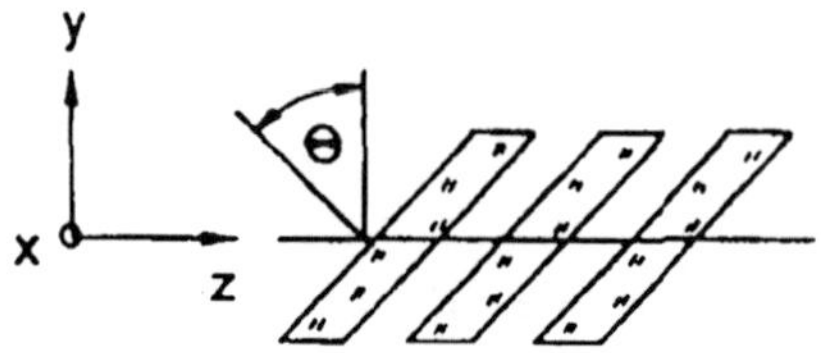

Fig. 3.6 Stack plate polarizer with three plates tilted by the Brewster angle Θ.

A simple way of generating a polarizer is by using an array of dielectric plates with refractive index *n*, arranged at the Brewster angle $\Theta = atan(n)$ (Brewster-plates), as shown in Fig. 3.6. This so called stack plate polarizer, used as early as the 19th century, has the advantage of transmitting one oscillation direction without reflection loss. Light polarized in the plane defined by the surface normal and the wave (propagation) vector of the light (p-polarized, along the y-axis in Fig. 3.6) passes a Brewster plate without loss. Field components oscillating in the perpendicular direction (s-polarized, along the x-axis) experience an amplitude transmission of:

$$t = \frac{2n}{n^2+1}$$

at each of the two interfaces. For a stack plate polarizer with *N* Brewster plates, as shown in Fig. 3.6, the Jones matrix reads:

$$M_{SPP}^{P} = \begin{pmatrix} \left[\frac{2n}{n^2+1}\right]^{2N} & 0 \\ 0 & 1 \end{pmatrix} \tag{3.6}$$

The degree of polarization is given by:

$$P = \frac{1 - t^{4N}}{1 + t^{4N}} \tag{3.7}$$

The intensity goes as the square of the field, and we can define the intensity transmission $T = t^{4N}$. For three glass plates ($n = 1.5$) the intensity transmission for s-polarized light is 0.383 resulting in a degree of polarization *P* of 0.446. In low power gas lasers (HeNe, Ar^+) the discharge tubes are sealed off with Brewster glass plates ($N=2$). Note that the Brewster plates are inside the resonator - the resonator mirrors are typically located a few cm behind the plates. A resonator round trip will generate a loss ($=1-T^2$) of 0.72. Since the gain of these lasers is generally low and can therefore not compensate for combined losses of more than a couple of percent, the output beam is p-polarized.

2) Retardation Plates

Retardation plates are made of birefringent material that exhibit different indices of refraction along two perpendicular axes, denoted as the principal axes. If light incident on the retardation plate is oscillating along one of those principal axes, it experiences a phase shift of ϕ_1 or ϕ_2, depending on which principal axis lies along the oscillation direction. If the principal axes coincide with the axes of the coordinate frame, the Jones matrix of a retardation plate reads:

$$M_R^P = \begin{pmatrix} \exp[i\phi_1] & 0 \\ 0 & \exp[i\phi_2] \end{pmatrix}$$

$$= \begin{pmatrix} 1 & 0 \\ 0 & \exp[i(\phi_2 - \phi_1)] \end{pmatrix} \exp[i\phi_1]$$

The phase term outside the matrix has no influence upon the polarization and therefore the Jones matrix can be written in the form:

$$M_R^P = \begin{pmatrix} 1 & 0 \\ 0 & \exp[i\delta] \end{pmatrix} \qquad \textit{with } \delta = \phi_2 - \phi_1 \tag{3.8}$$

Special Cases:

a) $\delta = \pm\pi/2$, 'Quarter Wave Plate'

The Jones matrix (3.8) now reads:

$$M_{\lambda/4}^P = \begin{pmatrix} 1 & 0 \\ 0 & \pm i \end{pmatrix}$$

For linearly polarized light incident at an angle of 45° (field vector $E_0 = E_0(1,1)$), the output field vector becomes:

$$E_1 = \begin{pmatrix} 1 \\ \pm i \end{pmatrix}$$

Depending on the sign we get right circularly polarized or left circularly polarized light (Fig. 3.7).

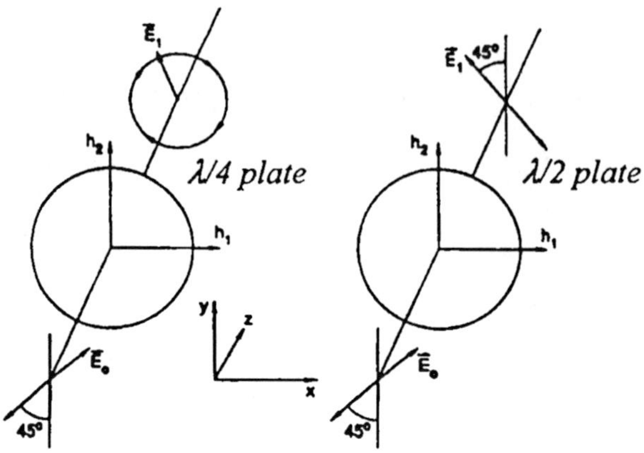

Fig. 3.7 Influence of a quarter wave plate and a half wave plate on the polarization of a linearly polarized beam oscillating at 45°. The principal axes of the plates are denoted by h_i.

b) $\delta = \pm\pi$, 'Half Wave Plate'

According to (3.8), the Jones matrix for the half wave plate is:

$$M^{P}_{\lambda/2} = \begin{pmatrix} 1 & 0 \\ 0 & -1 \end{pmatrix}$$

After passage through the half wave plate, light linearly polarized under 45° will still be linearly polarized but the oscillation direction has rotated by 90°. The same effect can be generated by a series of two quarter wave plates.

3) Faraday Rotator

When inserted into the beam path, a Faraday rotator exhibits the unique property of rotating the plane of polarization for a linearly polarized wave regardless of the inclination angle at the entrance plane. This rotation, called the Faraday effect, can be generated in dielectric materials by applying a static magnetic field $\boldsymbol{B}$ along the propagation direction of the electromagnetic wave. The rotation angle β is proportional to the length L of Faraday rotator and to the magnetic field component in the direction of the wave vector $\boldsymbol{k}$:

$$\beta = V\,L\,\frac{\boldsymbol{k}\boldsymbol{B}}{|\boldsymbol{k}|} \tag{3.9}$$

The Verdet constant V determines the strength of the Faraday effect. Faraday rotators used in laser systems use doped glass or crystal rods with the magnetic field either generated electrically, as shown in Fig. 3.8, or by means of strong permanent magnets (Table 3.1).

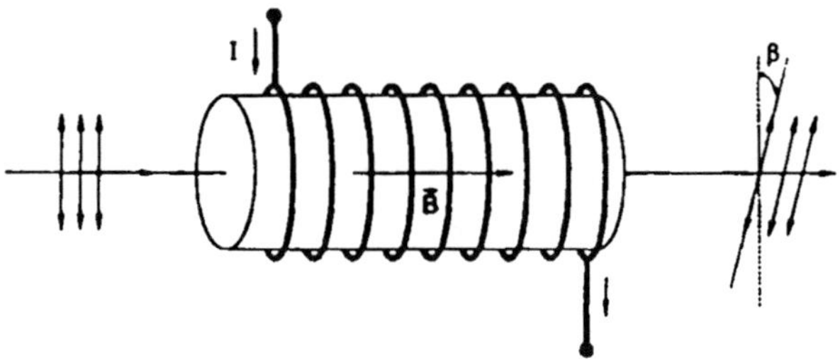

Fig. 3.8 The Faraday rotator rotates the plane of polarization of linearly polarized light.

Since the rotation of the polarization plane does not depend on the inclination of the input field vector, the Jones matrix of a Faraday rotator has the form of a rotation matrix (rotation in the counterclockwise direction when looking along the wave vector):

$$M_{FR}^{P} = \begin{pmatrix} \cos\beta & -\sin\beta \\ \sin\beta & \cos\beta \end{pmatrix} \tag{3.10}$$

Table 3.1 Verdet constants of different materials.

Material	V[degree/(cm Tesla)]
water	2.2
phosphate glass	2.7
quartz	2.8
flint glass	5.3
Terbium doped glass	40.0
Terbium doped GGG	76.8

3.2.2 Matrices for Rotated Polarizing Optics

If the Jones matrix M^P for the aligned optics is known, we can calculate the Jones matrix $M^P(\alpha)$ for a rotation by an angle α by using (see Sec. 1.2.5):

$$M^{P}(\alpha) = \begin{pmatrix} \cos\alpha & -\sin\alpha \\ \sin\alpha & \cos\alpha \end{pmatrix} M^{P} \begin{pmatrix} \cos\alpha & \sin\alpha \\ -\sin\alpha & \cos\alpha \end{pmatrix} \tag{3.11}$$

whereby the optics are rotated in the counterclockwise direction as we look towards the optics in the propagation direction of the electric field (Fig. 3.9).

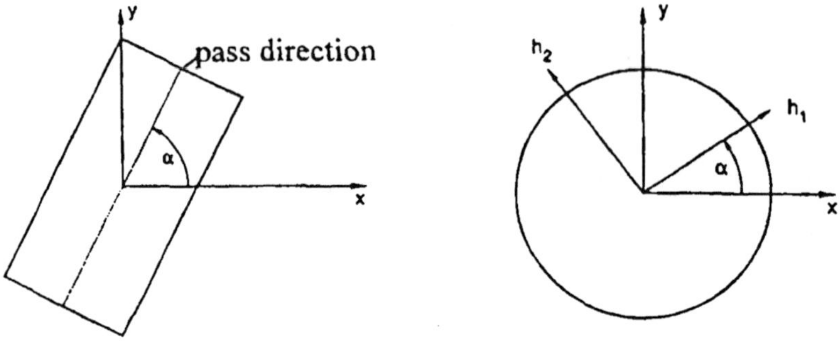

Fig. 3.9 Rotated polarizer and rotated retardation plate (viewed along the propagation direction of the field).

a) Rotated Polarizer (Fig. 3.9)
We assume that the pass direction of the aligned polarizer (with $t=1$) is along the x-axis. Application of (3.11) yields:

$$M_P^P(\alpha) = \begin{pmatrix} \cos^2\alpha & \sin\alpha\cos\alpha \\ \sin\alpha\cos\alpha & \sin^2\alpha \end{pmatrix} \tag{3.12}$$

b) Rotated Retardation Plate (Fig. 3.9)
By inserting the Jones matrix of the aligned retarder (3.8) into the transformation law (3.11) one gets:

$$M_{RP}^P(\alpha) = \begin{pmatrix} \cos^2\alpha + \exp[i\delta]\sin^2\alpha & \sin\alpha\cos\alpha(1-\exp[i\delta]) \\ \sin\alpha\cos\alpha(1-\exp[i\delta]) & \sin^2\alpha + \exp[i\delta]\cos^2\alpha \end{pmatrix} \tag{3.13}$$

For a $\lambda/2$-plate ($\delta=\pi$), the matrix (3.13) reads:

$$M_{RP}^P(\alpha) = \begin{pmatrix} \cos^2\alpha - \sin^2\alpha & 2\sin\alpha\cos\alpha \\ 2\sin\alpha\cos\alpha & \sin^2\alpha - \cos^2\alpha \end{pmatrix} = \begin{pmatrix} \cos2\alpha & \sin2\alpha \\ \sin2\alpha & -\cos2\alpha \end{pmatrix} \tag{3.14}$$

Light that is polarized linearly in x- or in y-direction will be rotated by an angle 2α.

c) Rotated Faraday Rotator

The Jones matrix of the Faraday rotator is invariant under rotation:

$$M_{FR}^{P}(\alpha) = M_{FR}^{P}(0) \tag{3.15}$$

This is easy to understand since the Faraday effect does not depend on the inclination of the polarization plane at the entrance face of the rotator.

3.2.3 Combination of Several Polarizing Optics

When several polarizing optics are located in the beam path, the Jones matrices of the individual components have to be multiplied to find the resulting Jones matrix of the total optical system. If M_i^P denotes the Jones matrix of the i-th optical element passed by the beam, the resulting Jones matrix for N elements is given by:

$$M^{P} = M_{N}^{P}\, M_{N-1}^{P} \dots M_{2}^{P}\, M_{1}^{P} \tag{3.16}$$

Similar to the ray matrices in geometrical optics, the optical element passed first stands at the right hand side of the matrix product. Note that (3.16) can only be used if no more than one Jones-matrix depends on spatial coordinates (such as for radially and radial-azimuthally birefringent materials). If the optical system comprises more than one of these elements, the ray propagation between the elements has to be taken into account.

When the light is reflected off a mirror and travels through the same polarizing optics again, but in the opposite direction, the same Jones matrices for these elements are used. As already discussed in geometrical optics, the observer always rides with the beam which means that the coordinate frame is reflected by the mirror such that the z-axis points again in the propagation direction. There is, however, one exception to this rule: For polarization rotators based on optical activity (such as crystalline quartz, sugar solution, or milk acid), the negative angle $-\beta$ has to be used for the reverse direction in the Jones matrix (3.10). This is the mathematical description of the fact that for an optically active medium, the polarization is completely unaffected if the beam propagates through the medium twice, once in the forward and once in the reverse direction.

A high-reflecting mirror generally does not influence the polarization properties of the beam since both field components experience the same phase shift of π and the same reflectivity. Partially reflecting mirrors with non-normal incidence and all reflecting devices based on total internal reflection (porro prisms, corner cubes), however, affect the polarization state since the reflectivities and the phase shifts are different for p-polarized and s-polarized light. The general Jones matrix for a mirror thus reads:

$$M_M^P = \begin{pmatrix} r_x & 0 \\ 0 & r_y \exp[i\delta] \end{pmatrix} \tag{3.17}$$

with r_x,r_y are the amplitude reflectivities in the x- and the y-direction, respectively. For a 90° roof top prism with index of refraction n, the phase shift δ is given by:

$$\delta = 4 \tan^{-1}\sqrt{1-2/n^2} - \pi \qquad \textit{roof edge along y-axis} \tag{3.18}$$

$$\delta = \pi - 4 \tan^{-1}\sqrt{1-2/n^2} \qquad \textit{roof edge along x-axis} \tag{3.19}$$

A roof top prism acts like a $\lambda/4$ plate for an index of refraction of n=*1.5538*.

Examples:

1) The Optical Diode

A combination of a polarizer and a quarter wave plate with an angle of 45° between the pass direction of the polarizer and the principal axes of the plate is called an optical diode (Fig. 3.10). Light reflected off a mirror or a target (e.g. work piece in laser material processing) cannot pass the polarizer in backward direction. The reflected light is prevented from entering the laser system which can cause unwanted feedback effects. The optical diode is also used to physically separate the signal beam from the source beam path in laser measurement systems that make use of light scattered or reflected back into the source (remote sensing, LIDAR).

The principle of operation of an optical diode is based on the fact that the combination of forward and backward propagation through the quarter wave plate will rotate the polarization plane by 90°. Any back propagating portion of the beam will be completely absorbed or reflected by the polarizer. This is, of course, only true if the mirror or the target does not change the polarization state.

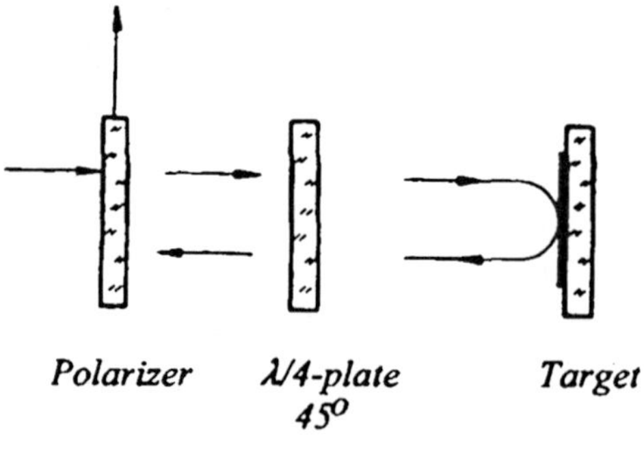

Fig. 3.10 The optical diode. The quarter wave plate rotates the polarization plane by 90°. For best results, the target must not affect the polarization.

By calculating the resulting Jones matrix with (3.16) we find that the resulting Jones matrix has indeed only zero components:

$$M_{OD}^{P} = M_{P}^{P}\, M_{\lambda/4}^{P}(45^\circ)\, M_{\lambda/4}^{P}(45^\circ)\, M_{P}^{P} = \begin{pmatrix} 0 & 0 \\ 0 & 0 \end{pmatrix} \tag{3.20}$$

If the target affects the polarization of the beam, the quarter wave plate has to be replaced by a Faraday rotator with rotation angle β=45° and a polarizer attached to its end face (such that the rotated beam will be completely transmitted). If the pass direction of the first polarizer is in the y-direction, the resulting Jones matrix for this system reads:

$$M_{OD}^{P} = \begin{pmatrix} 0 & 0 \\ 0 & 1 \end{pmatrix}\begin{pmatrix} 1/\sqrt{2} & -1/\sqrt{2} \\ 1/\sqrt{2} & 1/\sqrt{2} \end{pmatrix}\begin{pmatrix} 1/2 & -1/2 \\ -1/2 & 1/2 \end{pmatrix}\begin{pmatrix} m_{11} & m_{12} \\ m_{21} & m_{22} \end{pmatrix}\begin{pmatrix} 1/2 & -1/2 \\ -1/2 & 1/2 \end{pmatrix}\begin{pmatrix} 1/\sqrt{2} & -1/\sqrt{2} \\ 1/\sqrt{2} & 1/\sqrt{2} \end{pmatrix}\begin{pmatrix} 0 & 0 \\ 0 & 1 \end{pmatrix}$$

Pol.(y) FR(45°) Pol.(45°) Target Pol(45°) FR(45°) Pol.(y)

$$M_{OD}^{P} = \begin{pmatrix} 0 & 0 \\ 0 & 0 \end{pmatrix}\begin{pmatrix} m_{11} & m_{12} \\ m_{21} & m_{22} \end{pmatrix}\begin{pmatrix} 0 & -1/\sqrt{2} \\ 0 & 1/\sqrt{2} \end{pmatrix} \tag{3.21}$$

Equation (3.21) clearly shows that this system always blocks the light in the backwards direction, no matter what the Jones matrix elements m_{ij} of the target are.

2) Rotatable Retardation Plate between Crossed Polarizers

Two crossed polarizers cannot transmit light. Only if the polarization is changed between the polarizers can field components in the pass direction of the second polarizer be generated. This effect can be used to visualize stress in transparent materials. Since stress induces birefringence, the intensity distribution behind the second polarizer contains information on the area and the magnitude of stress in materials. The retardation plate is a model for such a stressed material. The resulting Jones matrix for the system depicted in Fig. 3.11 reads:

$$M^{P} = \begin{pmatrix} 0 & 0 \\ 0 & 1 \end{pmatrix}\begin{pmatrix} \cos^2\alpha + \exp[i\delta]\sin^2\alpha & \sin\alpha\cos\alpha(1-\exp[i\delta]) \\ \sin\alpha\cos\alpha(1-\exp[i\delta]) & \sin^2\alpha + \exp[i\delta]\cos^2\alpha \end{pmatrix}\begin{pmatrix} 1 & 0 \\ 0 & 0 \end{pmatrix}$$

$$= \begin{pmatrix} 0 & 0 \\ \sin\alpha\cos\alpha(1-\exp[i\delta]) & 0 \end{pmatrix} \tag{3.22}$$

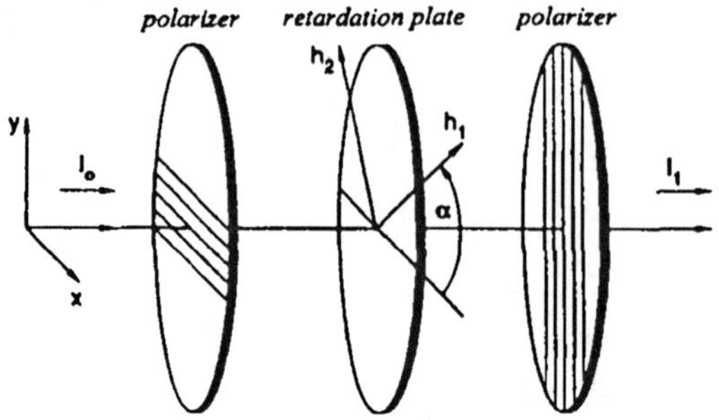

Fig. 3.11 Rotatable retardation plate between crossed polarizers

An unpolarized input beam with intensity I_0 results in an output intensity I of:

$$I = I_0 \; 2\sin^2\alpha\cos^2\alpha(1-\cos\delta) \tag{3.23}$$

We get zero transmission when the pass directions of the polarizers coincide with the principal axes of the retardation plate (α=0°,90°,180°,270°), and maximum transmission is attained for α=45°,135°, 225°, and 315°. The phase shift δ can be determined from two intensity measurements, performed at different angles.

3) Two half-wave plates offset by 45°

The first half wave plate has its principal axes rotated by an angle α, the second plate is rotated by an additional 45° (angle: $\alpha+\pi/4$). By using (3.14), the resulting Jones-matrix can be easily calculated:

$$M^P_{2\cdot\pi}(\alpha) = \begin{pmatrix} \cos2(\alpha+\pi/4) & \sin2(\alpha+\pi/4) \\ \sin2(\alpha+\pi/4) & -\cos2(\alpha+\pi/4) \end{pmatrix} \begin{pmatrix} \cos2\alpha & \sin2\alpha \\ \sin2\alpha & -\cos2\alpha \end{pmatrix}$$

$$= \begin{pmatrix} -\sin2\alpha & \cos2\alpha \\ \cos2\alpha & \sin2\alpha \end{pmatrix} \begin{pmatrix} \cos2\alpha & \sin2\alpha \\ \sin2\alpha & -\cos2\alpha \end{pmatrix}$$

$$= \begin{pmatrix} 0 & -1 \\ 1 & 0 \end{pmatrix}$$

A comparison with the matrix of a Faraday rotator (3.10) indicates that this combination of two half-wave plates is equivalent to a 90° rotator with counter- clockwise rotation (viewed along the propagation direction). If the second wave plate were rotated by an angle $\alpha+\pi/4$, a 90° rotation in the clockwise direction would result.

3.3 Eigenstates of Polarization

For any combination of polarizing optics we can find states of polarization which remain unaffected after passage. These polarization states are called the eigenstates of the optical system and they play an important role in resonator physics since they represent the steady state solutions. If M^P denotes the Jones matrix of the optical system, the field vector E_1 must be proportional to the input field vector E_0, with:

$$E_1 = \mu^P E_0 = M^P E_0 \tag{3.24}$$

The eigenstates of polarization are given by the two eigenvectors of the Jones matrix M^P. The physical meaning of the eigenvalues μ^P is revealed if we compare the intensities at the input plane I_0 and at the output plane I_1:

$$I_1 = \frac{1}{2}c\epsilon_0 E_1 E_1^* = \mu^P \mu^{P*} \frac{1}{2}c\epsilon_0 E_0 E_0^* = |\mu^P|^2 I_0 \tag{3.25}$$

The factor $|\mu^P|^2$ represents the intensity fraction remaining in the beam after passage through the polarizing optics. This factor is called the loss factor V. The power fraction getting lost due to reflection or absorption is given by $\Delta V = 1 - V$. The two eigenvalues for the general Jones matrix given by:

$$M^P = \begin{pmatrix} m_{11} & m_{12} \\ m_{21} & m_{22} \end{pmatrix}$$

can be calculated by using:

$$\mu_{1,2}^P = \frac{m_{11}+m_{22}}{2} \pm \sqrt{\left(\frac{m_{11} - m_{22}}{2}\right)^2 + m_{12}m_{21}} \tag{3.26}$$

The corresponding eigenvectors read:

$$E_i^P = \begin{pmatrix} 1 \\ \dfrac{\mu_i^P - m_{11}}{m_{12}} \end{pmatrix} \; ; \quad i=1,2 \qquad if\ m_{12} \neq 0 \tag{3.27}$$

$$E_1^P = \begin{pmatrix} 1 \\ 0 \end{pmatrix} , \; E_2^P = \begin{pmatrix} 0 \\ 1 \end{pmatrix} ; \qquad if\ m_{12} = m_{21} = 0 \tag{3.28}$$

Examples:

1) Retardation Plate

The Jones matrix of the rotated retardation plate is:

$$M_{RP}^{P}(\alpha) = \begin{pmatrix} \cos^2\alpha + \exp[i\delta]\sin^2\alpha & \sin\alpha\cos\alpha(1-\exp[i\delta]) \\ \sin\alpha\cos\alpha(1-\exp[i\delta]) & \sin^2\alpha + \exp[i\delta]\cos^2\alpha \end{pmatrix}$$

By using (3.26) and (3.27) we get the following eigenvectors and eigenvalues:

$$E_1^P = \begin{pmatrix} \cos\alpha \\ \sin\alpha \end{pmatrix} \quad , \quad \mu_1^P = 1$$

$$E_2^P = \begin{pmatrix} -\sin\alpha \\ \cos\alpha \end{pmatrix} \quad , \quad \mu_2^P = \exp[i\delta]$$

The eigenstates are linear polarization along the two principal axes. Since the retardation plate does not induce any losses (surface reflections are not accounted for in the Jones matrix!), both loss factors V are equal to 1.

2) Brewster Plate

The Jones matrix for a Brewster plate with index of refraction n and the surface normal in the y-z plane reads:

$$M_{BR}^{P} = \begin{pmatrix} \left[\dfrac{2n}{n^2+1}\right]^2 & 0 \\ 0 & 1 \end{pmatrix}$$

The eigenvectors and eigenvalues are:

$$E_1^P = \begin{pmatrix} 1 \\ 0 \end{pmatrix} \quad , \quad \mu_1^P = \left[\frac{2n}{n^2-1}\right]^2$$

$$E_2^P = \begin{pmatrix} 0 \\ 1 \end{pmatrix} \quad , \quad \mu_1^P = 1$$

The s-polarized wave E_1^P exhibits a loss of $1-|\mu_1^P|^2$ due to reflection at the two interfaces.

3) Faraday Rotator

It can be easily shown that left circular and right circular polarizations are the eigenstates of a Faraday rotator, with both loss factors being equal to 1.

3.4 Polarization in Optical Resonators

3.4.1 Eigenstates of the Round trip Jones Matrix

The Jones matrix formalism enables us to determine the polarization of the electric field in laser resonators [1.112]. If polarizing optical elements are inserted into the resonator, the polarization will reproduce itself after every round trip as soon as a steady state electric field has been established. The steady state polarization is thus determined by the polarization eigenstates of the round trip Jones matrix. The reference plane from which the round trip is started can be chosen arbitrarily since the steady state condition must hold everywhere inside the resonator.

A polarized laser beam, however, can only be generated if the loss factor for one polarization eigenstate is lower than that for the other one. In this case, the polarization with the lowest loss is preferred. This is due to the fact that the threshold condition is reached first for the lowest loss polarization. After the onset of laser oscillation in this polarization eigenstate, gain saturation will prevent the second polarization eigenstate from reaching the threshold. In case both loss factors are equal, the electric field will be unpolarized. A laser resonator with an internal Brewster plate (Fig.3.12) will therefore emit a linearly polarized beam. The round trip Jones matrix starting on mirror 1 reads:

$$M^{P} = \begin{pmatrix} \left[\dfrac{2n}{n^2+1}\right]^4 & 0 \\ 0 & 1 \end{pmatrix}$$

The polarization eigenstates are linear polarizations along the x- and y- axis, respectively. The field will, however, be polarized along the y-axis (p-polarization), since this oscillation mode will not generate any reflection losses at the Brewster plate. The second eigenvector representing the s-polarized beam exhibits a round trip loss of:

$$\Delta V = 1 - \left[\frac{2n}{n^2+1}\right]^8$$

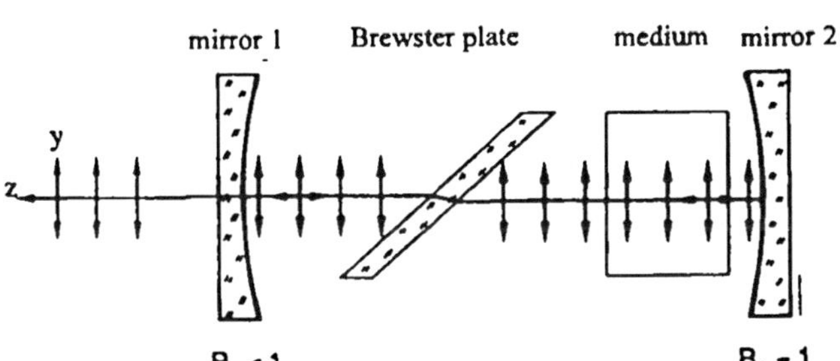

Fig. 3.12 A laser resonator with an internal Brewster plate generates a linearly polarized beam (p-polarized).

3.4.2 Polarization and Diffraction Integrals

In Sec. 2.8 we presented the integral equations for the calculation of the electric field distributions at the resonator mirrors and at any other plane inside the resonator. For the derivation of the Collins integral it was assumed that the electric field is a scalar quantity, neglecting the polarization. Fortunately, as far as optical resonators are concerned, it is not necessary to repeat a similar derivation of the diffraction integrals when the polarization is taken into account. Since the scalar wave equation holds for both the x- and the y-component of the field vector, we can apply the Collins integral in the x- and y-directions separately. If $\boldsymbol{K}E$ denotes the Collins integral for a round trip inside the resonator applied to the scalar field E, we can write the integral equation for the vector field in the form:

$$\begin{pmatrix} \gamma_x E_x \\ \gamma_y E_y \end{pmatrix} = \begin{pmatrix} \boldsymbol{K} & 0 \\ 0 & \boldsymbol{K} \end{pmatrix} \begin{pmatrix} E_x \\ E_y \end{pmatrix} \tag{3.29}$$

The two field components do not mix since we have no polarizing element inside the resonator. We are dealing with two scalar field integral equations that can be solved separately.

If a polarizing optical element with Jones matrix M^p is located inside the resonator, as depicted in Fig. 3.13, the two components will interact. We can still calculate the four propagations between the mirrors and the polarizing elements using diagonal integral matrices as in (3.29), but we now have to mix the x- and y-components of the field at the plane of the polarizing element according to its Jones matrix. The integral equation for a resonator round trip will then read:

$$\begin{pmatrix} \gamma_x E_x \\ \gamma_y E_y \end{pmatrix} = \begin{pmatrix} \boldsymbol{K}_4 & 0 \\ 0 & \boldsymbol{K}_4 \end{pmatrix} \begin{pmatrix} m_{11} & m_{12} \\ m_{21} & m_{22} \end{pmatrix} \begin{pmatrix} \boldsymbol{K}_3 & 0 \\ 0 & \boldsymbol{K}_3 \end{pmatrix} \begin{pmatrix} \boldsymbol{K}_2 & 0 \\ 0 & \boldsymbol{K}_2 \end{pmatrix} \begin{pmatrix} m_{11} & m_{12} \\ m_{21} & m_{22} \end{pmatrix} \begin{pmatrix} \boldsymbol{K}_1 & 0 \\ 0 & \boldsymbol{K}_1 \end{pmatrix} \begin{pmatrix} E_x \\ E_y \end{pmatrix}$$

(3.30)

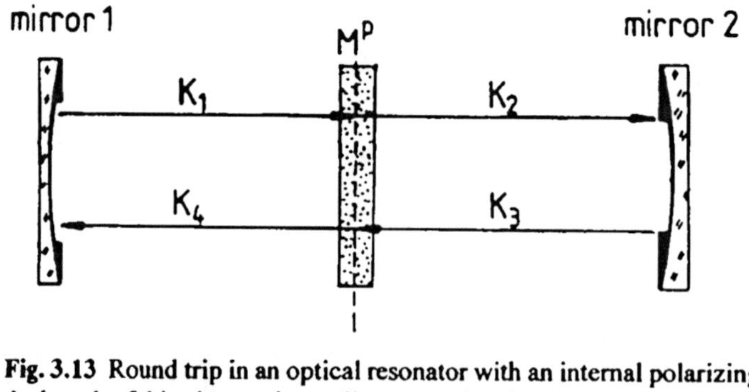

Fig. 3.13 Round trip in an optical resonator with an internal polarizing element. It is assumed that the length of this element is small compared to the resonator length. The field is propagated to and from the dotted plane by means of Collins integrals .

which can be written as:

$$\gamma_x E_x = [K_4 m_{11} K_3 K_2 m_{11} K_1 + K_4 m_{12} K_3 K_2 m_{21} K_1]\ E_x + [K_4 m_{11} K_3 K_2 m_{12} K_1 + K_4 m_{12} K_3 K_2 m_{22} K_1]\ E_y$$

$$\gamma_y E_y = [K_4 m_{12} K_3 K_2 m_{11} K_1 + K_4 m_{22} K_3 K_2 m_{21} K_1]\ E_x + [K_4 m_{21} K_3 K_2 m_{12} K_1 + K_4 m_{22} K_3 K_2 m_{22} K_1]\ E_y$$

If the Jones matrix has non-vanishing elements m_{12}, m_{21}, the two field components become coupled and we have to solve both integral equations simultaneously. Note that the swapping of the integral operators and the Jones matrix elements in the above equations is prohibited if the Jones matrix elements depend on the spatial coordinates. Once the solutions of these coupled integral equations are found (in most cases numerically), the field distribution E and the loss factor V are given by:

$$E = \begin{pmatrix} E_x(x,y) \\ E_y(x,y) \end{pmatrix}, \quad V = \frac{|\gamma_x E_x|^2 + |\gamma_y E_y|^2}{|E_x|^2 + |E_y|^2} \tag{3.31}$$

3.5 Depolarizers

In the preceding sections we have only discussed polarizing optics that either change the polarization state of polarized light, like the retardation plate, or generate a well-defined polarization from unpolarized light, like the polarizer. In addition to this class of polarizing optics, optical elements that transform polarized light into unpolarized light also exist. The latter type of polarizing optics are called *depolarizers*. Whereas the generation of polarized light out of unpolarized light cannot be accomplished without loss of energy, the depolarization can be achieved in a lossless way. The depolarization of polarized laser beams is most often applied to decrease measurement errors in power and intensity measurements since the polarization interacts with the measurement apparatus. Furthermore, as far as material processing is concerned, an unpolarized beam provides a more flexible tool since the interaction with the work piece is less affected by changes in the processing geometry (e.g. angle of incidence, direction of focal spot movement). Depolarization can be attained by reflection off or transmission through a scattering screen. This technique will, however, generate considerable loss and will also spoil the beam quality. A more suitable technique is to generate a continuum of different polarization states across the beam. The beam will then behave like an unpolarized beam. One commonly used depolarizer working on this principle is the Cornu depolarizer, as shown in Fig. 3.14.

This depolarizer for monochromatic light consists of two crystalline quartz prisms attached such that they form a cube. The first quartz prism (left handed quartz) acts like a retardation plate whereby the induced phase shift δ_l is proportional to the distance over which the beam has to propagate within the prism. At the interface, the phase shift of a collimated beam is a function of the entrance height y of the beam:

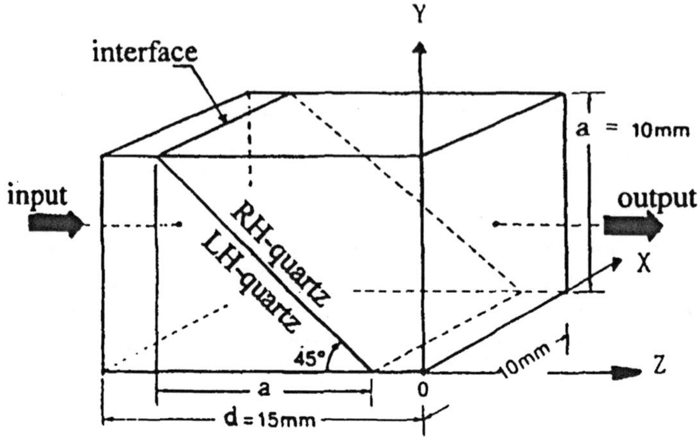

Fig. 3.14 Cornu Depolarizer.

$$\delta_1(y) = \frac{2\pi}{\lambda}[n_2 - n_1]\left(\frac{d-a}{2} + a - y\right) \tag{3.32}$$

Without the second prism, the beam would be deflected from the optical axis due to refraction. The second prism must be arranged such that the phase shift has a negative sign compared to the first one, otherwise the total phase shift would not depend on the y-coordinate. This is accomplished by switching the principal axes (right handed quartz). The phase shift induced by the second prism is given by:

$$\delta_2(y) = \frac{2\pi}{\lambda}[n_1 - n_2]\left(y + \frac{d-a}{2}\right) \tag{3.33}$$

Addition of the two phase shifts results in the total phase shift:

$$\delta(y) = \frac{2\pi}{\lambda}[n_1 - n_2](2y - a) \tag{3.34}$$

The beam emerging from the Cornu depolarizer exhibits different polarization states at different heights. Hence, the beam is not truly unpolarized, but by averaging over the total beam cross section it will simulate the behavior of unpolarized radiation. It is advantageous to focus the exit beam to a smaller spot size to more closely mimic unpolarized light.

3.6 Momentum and Angular Momentum of a Beam

The electromagnetic field carries momentum and angular momentum, which depend on the Poynting vector, the structure of the beam and its polarization. This is easy to understand, if we remember that in Quantum Optics the beam can be represented by a stream of photons. Each of them with a momentum $\hbar k$ and an angular momentum $\hbar$, which is related to the polarisation. If these photons are absorbed by a target (a small piece of matter), momentum and angular momentum are transferred and the target will be pushed away and starts to rotate. However, this is not a quantum effect, and it has been already well known in classical electrodynamics [1.116]. The interaction between field and matter occurs by the Coulomb/Lorentz-force $\boldsymbol{F}$ acting on the electrons [1.117]. This force depends on the electric field $\boldsymbol{E}$ and the magnetic induction $\boldsymbol{B}$. For dielectric media with $\boldsymbol{B} = \mu_0 \boldsymbol{H}$ the Coulomb/Lorentz force reads:

$$\boldsymbol{F} = e\,\boldsymbol{E} + \frac{e}{\epsilon_0 c_0^2}\,\boldsymbol{v} \times \boldsymbol{H} \tag{3.35}$$

with e : charge of the electron ($1.6021 \cdot 10^{-19}$ As)
v : velocity of the electron

The Coulomb force is parallel to the electric field, the Lorentz force perpendicular to the magnetic field and the velocity of the electron. The oscillating electric field forces the electron to oscillate in the same direction as the electric field vector. The oscillating magnetic field interacts with the oscillating electron and generates a force in direction $\boldsymbol{k}$ of the propagating field, as shown in Fig. 3.15. Momentum is transferred to the target, but only if the field is partly absorbed or reflected. This force can be used to deflect atoms [1.118] or to balance small glass beads in the gravitational field [1.119, 1.120].

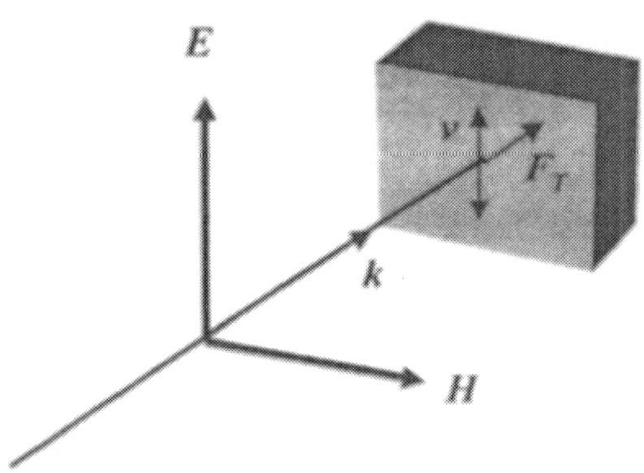

Fig. 3.15 A plane wave is transmitting a dielectric target and partly reflected or absorbed. The interplay of Coulomb and Lorentz force generates a force $\boldsymbol{F_T}$ on the target in the direction $\boldsymbol{k}$ of beam propagation.

An application is the optical tweezer [1.128] a tool for handling microscopic particles. The correct calculation of this force is given in text books [1.117,1.121, 1.130]. A laser pulse with an energy E_p consists of $N = E_p/\hbar\omega$ photons. Each photon carries the momentum $\hbar \boldsymbol{k}$ which results in the total momentum $\boldsymbol{P}$ in direction of propagation:

$$\boldsymbol{P} = \frac{E_p}{c_0} \frac{\boldsymbol{k}}{|\boldsymbol{k}|} \tag{3.36}$$

The momentum flux is the momentum per time and is equivalent to a force. It reads for a continuous field of power $P = dE_p/dt$:

$$\frac{d\boldsymbol{P}}{dt} = \frac{P}{c_0} \frac{\boldsymbol{k}}{|\boldsymbol{k}|} \tag{3.37}$$

This equation holds for plane waves. A more precise expression for arbitrary in non-magnetic media reads [1.121]:

$$\boldsymbol{P} = \frac{1}{c_0^2} \iiint \boldsymbol{E} \times \boldsymbol{H} \, dxdydz = \frac{1}{c_0^2} \iiint \boldsymbol{S} \, dxdydz \tag{3.38}$$

and the momentum flux in z -direction is given by:

$$\frac{dP}{dt} = \frac{1}{c_0} \iint S_z \, dxdy = \frac{P}{c_0} \tag{3.39}$$

This momentum flux can transfer momentum to a target only if power is absorbed. A loss free crystal interacts with the radiation field, demonstrated by the lower speed of light inside the medium. But in average, the Lorentz force is zero and no momentum is transferred, because the velocity of the electron and the magnetic field have a phase shift of 90^0. In order to transfer momentum, absorption is necessary, which results in a slight change the phase shift. If reflection occurs, the incident and the reflected wave generate a standing wave. In this case the electron velocity and the magnetic field are in phase and momentum transfer occurs. The momentum flux absorbed or reflected is equal to the force on the target. Therefore, the force on the target is given by:

$$\boldsymbol{F}_T = \frac{P}{c_0} (1 + R - T) \frac{\boldsymbol{k}}{|\boldsymbol{k}|} \tag{3.40}$$

where P is the incident power, R is the reflectivity and T the transmission of the target.

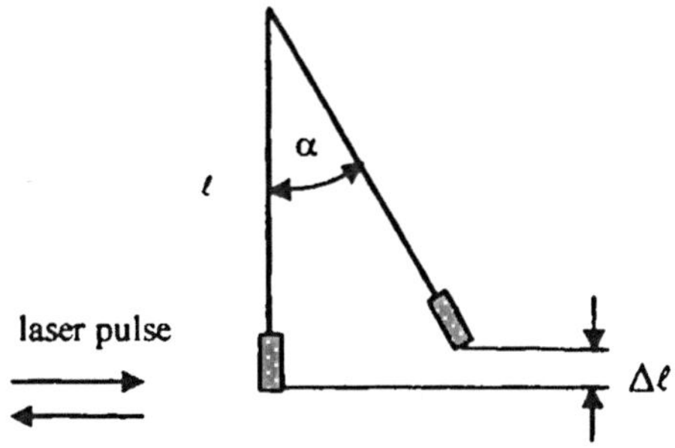

Fig. 3.16 Deflection of a mirror by reflection of a laser pulse.

Example:
A tiny mirror with a mass of $m = 2\times10^{-5}$ kg is suspended by a string with a length of ℓ=0.1 m, as shown in Fig. 3.16. A short laser pulse with energy E_p=1 Joule is completely reflected by the mirror. The momentum of the beam is changed by Δp=$2E_p/c_0$. This momentum has to be absorbed by the mirror $p = mv = 2E_p/c_0$ leading to a recoil. The kinetic energy of the mirror becomes:

$$E_{kin} = \frac{1}{2}mv^2 = \frac{2}{m}\left[\frac{E_p}{c_0}\right]^2$$

The mirror is deflected and lifted by $\Delta\ell$ in the gravitational field, at which point the potential energy equals the initial kinematic energy:

$$E_{pot} = mg\Delta\ell = \frac{2}{m}\left[\frac{E_p}{c_0}\right]^2$$

with g = 9.81 m/s^2. The maximum deflection angle $\Delta\alpha$ is thus given by:

$$\Delta\alpha = \sqrt{\frac{\Delta\ell}{\ell}} = \frac{2}{m}\left|\frac{E_p}{c_0}\right|\frac{1}{\sqrt{g\ell}}$$

For the parameters given above, the resulting angle of 0.3 mrad can be easily observed.

Using the momentum transfer due to reflection, it is even possible to stabilize a small particle in the center of a Gaussian beam [1.121, 1.122], as is shown in Fig.3.17. This can be easily explained by examining the resulting momentum. If the particle is decentred, the conservation of momentum results in a momentum ΔP, which pushes the particle back to the center.

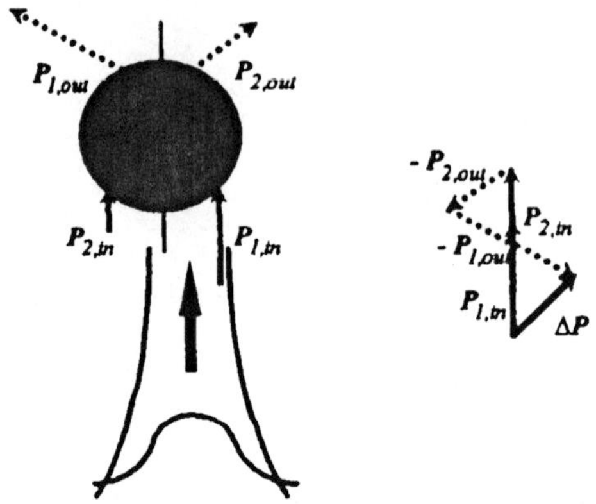

Fig. 3.17 A small particle can be stabilized in the centre of a Gaussian beam (left). If the particle is decentered, the resulting transferred momentum pushes it back to the center (right).

3.6.2 The Poynting Vector of Structured Beams

Before dealing with the angular momentum let us briefly discuss the pointing vector of structured fields. The field of a quasi-plane wave propagating in z-direction, slightly distorted by diffraction, can be described by the slowly varying envelope approximation:

$$E(x,y,z,t) = \boldsymbol{E}_0 \, u(x,y,z) \, \exp[i(\omega t - kz)] \;, \qquad \boldsymbol{E}_0 = \begin{pmatrix} \boldsymbol{E}_T \\ \boldsymbol{E}_z \end{pmatrix} \tag{3.41}$$

where $\boldsymbol{E}_T$ and $\boldsymbol{E}_z$ are the transverse and longitudinal components, respectively. Slowly varying envelope approximation means that

$$\frac{d|u|}{dz} \ll k|u|$$

In the case of elliptically polarized light, the transverse part of the electric field reads:

$$\boldsymbol{E}_z = \frac{E_0}{\sqrt{2}} \begin{pmatrix} 1 \\ \exp[i\Phi] \end{pmatrix} \tag{3.42}$$

The Poynting vector for such a field is again given by Eq.(1.1). It is a bit troublesome, but nonetheless straight forward, to derive the magnetic field by applying Maxwell's equations. Neglecting higher order derivations of u, one obtains [1.123, 1.136] for the Poynting vector:

$$S = \begin{pmatrix} S_T \\ S_z \end{pmatrix}$$

Now a transverse flow of power S_T appears. The width of the beam is increasing when propagating in free space, a result of diffraction and energy flowing radially. The transverse part of the Poynting vector is given by:

$$S_T = \frac{1}{4k}\epsilon_0 c_0 |E_0|^2 \begin{pmatrix} i\left[u^* \frac{\partial u}{\partial x} - u \frac{\partial u^*}{\partial x}\right] & - \sin\Phi \left[u \frac{\partial u^*}{\partial y} - u^* \frac{\partial u}{\partial y}\right] \\ i\left[u^* \frac{\partial u}{\partial y} - u \frac{\partial u^*}{\partial y}\right] & + \sin\Phi \left[u \frac{\partial u^*}{\partial x} - u^* \frac{\partial u}{\partial x}\right] \end{pmatrix} \tag{3.43}$$

The first part is related to the structure of the beam and is called the structural term, the second part depends on the polarisation and is called the polarization term. The z-component of the Poynting vector is the well known plane wave energy flow:

$$S_z = \frac{1}{2}\epsilon_0 c_0 \, |E_0 u|^2 \tag{3.44}$$

Example:
In order to gain a better understanding let us calculate the Poynting vector for a circularly polarized Gaussian beam with $\Phi=\pi/2$. As discussed in detail in Sec. 2.5, the electric field at the distance z from the waist position is given by:

$$E = E_0 \, u = \frac{E_0}{1 - iz/z_0} \exp\left[-ik \, \frac{x^2+y^2}{2q}\right]$$

If the transverse Poynting vector is normalized with respect to the z-component, the two contributions are:

$$\frac{S_{T,struct.}}{|S_z|} = \begin{pmatrix} x/R \\ y/R \end{pmatrix} \quad , \quad \frac{S_{T,pol.}}{|S_z|} = \begin{pmatrix} -y/R \\ x/R \end{pmatrix}$$

with the radius of curvature $R= z(z/z_0+z_0/z)$ (see Sec. 2.5.1). The meaning of these two expressions is easier to understand if cylindrical coordinates are introduced:

$$S = \begin{pmatrix} S_\rho \\ S_\phi \end{pmatrix}$$

Then the two parts of the transverse Poynting vector read:

$$\frac{S_{T,struct.}}{|S_z|} = \frac{\rho}{R}e_\rho \quad , \qquad \frac{S_{T,pol.}}{|S_z|} = \frac{\rho}{R}e_\Phi$$

with $\mathbf{e}_\rho$ and $\mathbf{e}_\Phi$ being the radial and azimuthal unit vectors, respectively. The structural part represents a radial energy flow as discussed in Sec. 2.9, which means that the resulting Poynting vector $\boldsymbol{S}$ is always perpendicular to the phase fronts, as shown in Fig. 2.42. The polarization part has an azimuthal component only. The energy is circulating around the z-axis, a consequence of the circular polarization.

3.6.3 Angular Momentum

The angular momentum $\boldsymbol{L}$ is defined the same way as in classical mechanics as the cross product of the position vector $\boldsymbol{r}$ and the momentum $\boldsymbol{P}$:

$$L = \frac{1}{c_0^2}\iiint r \times E \times H \; dxdydz = \frac{1}{c_0^2}\iiint r \times S \; dxdydz \tag{3.45}$$

In most cases only the angular momentum flow in propagation direction is of interest. This flow is obtained from (3.45) with $dz/dt=c_0$:

$$J_z = \frac{\partial}{\partial t}L_z = \frac{1}{c_0}\iint (r \times S)_z \; dxdy \tag{3.46}$$

The vector of the cross product is perpendicular to both vectors $\boldsymbol{r}$ and $\boldsymbol{S}$. If $\boldsymbol{S}$ has only a z-component, the cross product has no z-component and the angular momentum flow is zero. Due to diffraction, $\boldsymbol{S}$ exhibits x,y-components, as was discussed in the previous section. Inserting (3.43) into (3.46) delivers two terms for the angular momentum [1.124,1.125]:

$$J_{z,struct.} = \frac{1}{4\omega}\epsilon_0 c_0 |E_0|^2 \iint i\left[yu\frac{\partial u^*}{\partial x} - xu\frac{\partial u^*}{\partial y}\right]dxdy \; + \; cc \tag{3.47}$$

$$J_{z,pol.} = \frac{1}{4\omega}\epsilon_0 c_0 |E_0|^2 \, \sin\Phi \iint\left[xu\frac{\partial u^*}{\partial x} + yu\frac{\partial u^*}{\partial y}\right]dxdy \; + \; cc \tag{3.48}$$

where cc denotes the complex conjugate. Note that momentum flux is equivalent to a torque.

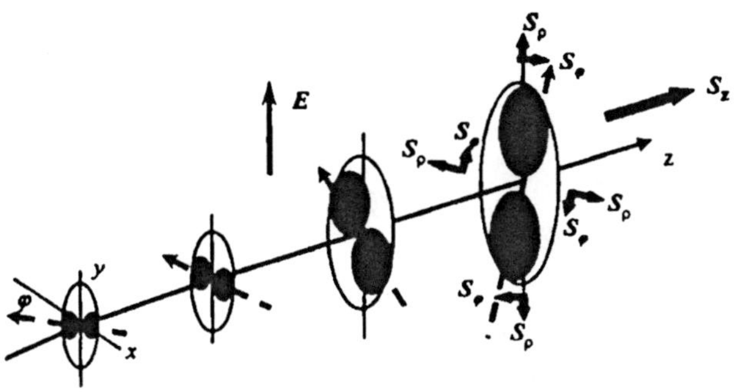

Fig. 3.18 A rotating intensity distribution carries an orbital angular momentum.

The structural (orbital) angular momentum

The structural term is also referred to as the orbital angular momentum. In Sec. 2.6.2, the second intensity moments of abeam were discussed. From the definition of the mixed moments $\langle w_x\theta_y\rangle$ and $\langle w_y\theta_x\rangle$, a relation between these twist parameters and the orbital angular momentum is obtained:

$$J_{z,struct.} = \frac{P}{4c_0}\left[\langle w_x\theta_y\rangle - \langle w_y\theta_x\rangle\right] \tag{3.49}$$

where P is the total power of the beam. The existence of a twist or orbital angular momentum means that the beam is rotating around the z-axis with constant direction of the electric field vector as depicted in Fig. 3.18. An experimental example of such a field are the Gauss-Laguerre eigenmodes, which are discussed in detail in Sec. 5.2.1. The field and the intensity of such an eigenmode of order ℓ for a rotating field read: (see Eq. (5.6)):

$$E(\rho,\phi) = E_0\, f(\rho)\, \exp[i(\omega t - kz) \pm i\ell\phi]$$
$$I(\rho) = I_0\, |f(\rho)|^2 \tag{3.50}$$

For a standing wave, which can be described as the superposition of two counter-rotating modes with equal amplitude, the corresponding expressions read:

$$E(\rho,\phi) = E_0\, f(\rho)\, \exp[i(\omega t - kz)]\, [\exp(-i\ell\phi) + \exp(+i\ell\phi)]$$
$$I(\rho) = 4I_0\, |f(\rho)|^2 \cos^2(\ell\phi) \tag{3.51}$$

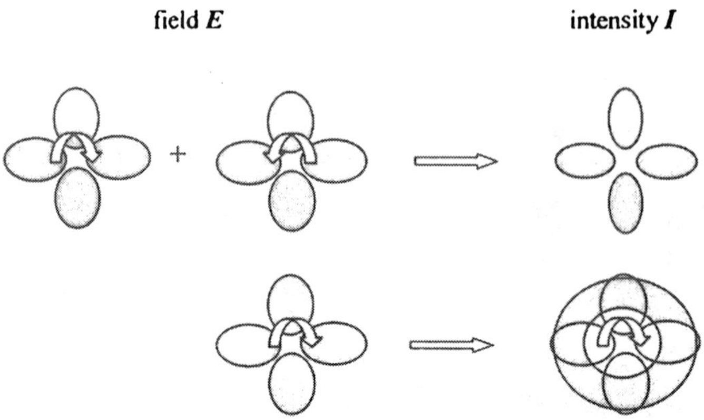

Fig. 3.19 The superposition of two counter-rotating Gaus-Laguerre modes with mode order p=0 and ℓ=2 generate an intensity distribution with an azimuthal structure (top). A single mode exhibits a ring-like structure (bottom).

The standing wave produces an azimuthal intensity pattern, whereas the rotating field intensity (3.50) exhibits a ring-like structure as shown in Fig. 3.19. Therefore, only the ring mode has an angular orbital momentum. Equations (3.50) and (3.47) deliver for the flux:

$$J_{z,struct.} = \pm \ell \frac{P}{\omega} \tag{3.52}$$

which means that each photon of the beam has an orbital angular momentum of $\ell\hbar$.

Generation of modes with orbital angular momentum (twist)

Gauss-Laguerre modes can be generated in laser systems of perfect circular symmetry, which is difficult to realize. It is much easier to transform Gauss-Hermite modes into Gauss-Laguerre modes by using astigmatic optical systems [1.125-1.127]. An experimental setup is shown in Fig. 3.20. At $z = 0$ is the waist of a Gauss-Hermite mode with order $(m,0)$. This mode is transformed by a system of three cylinder lenses with focal lengths $f/2, f$ and $f/2$ equally spaced by a distance $f/2$. The lens system is rotated by 45^0 with respect to the mode axis and the focal length is equal to the Rayleigh range z_0. If the Rayleigh ranges are different in the x- and the y- direction, they have to be equalized by a suitable bifocal lens system [1.134]. In the distance f/2 behind the third lens, the waist of a Gauss-Laguerre mode will appear. Let us first evaluate the variance matrices of the beam.

If the beam has its waist in the plane z=0, the second order moment or variance matrix reads in the x_1,y_1 reference system:

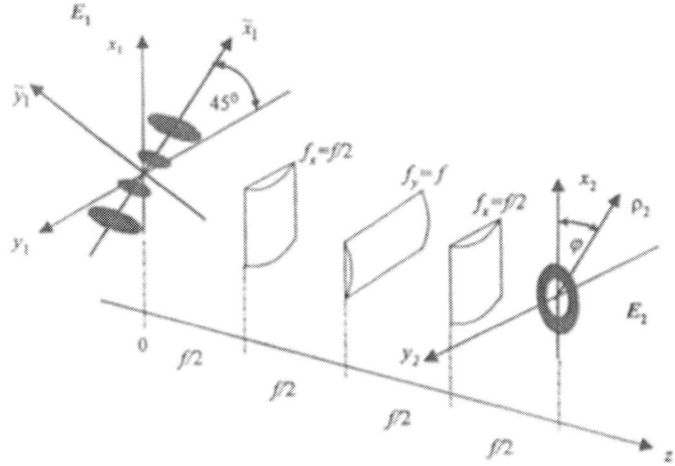

Fig. 3.20 Transformation of a Gauss-Hermite mode with order (m,0) into a Gauss-Laguerre mode of order p=0 and ℓ=2.

$$P_1 = \begin{pmatrix} w^2_{x,1} & 0 & 0 & 0 \\ 0 & w^2_{y,1} & 0 & 0 \\ 0 & 0 & \theta^2_{x,1} & 0 \\ 0 & 0 & 0 & \theta^2_{y,1} \end{pmatrix} \tag{3.53}$$

It is a simple astigmatic beam with different waist sizes and divergence angles in the x_1-and the y_1-directions. For a Gauss-Hermite mode of order (m,0) the following relations hold:

$$w_{x,1} = w_0\sqrt{2m+1} \quad , \qquad \theta_{x,1} = \theta_0\sqrt{2m+1}$$
$$w_{y,1} = w_0 \quad , \qquad \theta_{y,1} = \theta_0 = \frac{\lambda}{\pi w_0} \quad , \qquad z_0 = k w_0^2/2 \tag{3.54}$$

The variance matrix P_2 in the x_2,y_2 plane is obtained by applying Eq. (2.126):

$$P_2 = M P_1 M^T$$

The matrix M is the four-dimensional ray transfer matrix of the complete system and includes all distances and elements as well as the 45° rotation:

$$M = R^{-}(45^{\circ}) M_{lens}(f/2) M_{distance}(f/2) M_{lens}(f) M_{distance}(f/2) M_{lens}(f/2) M_{distance}(f/2) R^{+}(45^{\circ})$$

The different matrices are compiled in Sec. 1.2.4. Instead of going straight ahead with the matrix multiplication, we use an easier method to obtain the resulting matrix. First we calculate the 2x2 matrices of the system in the x- and in the y-direction without taking the rotation into account:

$$M_x = \begin{pmatrix} 1 & f \\ 0 & 1 \end{pmatrix} \begin{pmatrix} 1 & 0 \\ -1/f & 1 \end{pmatrix} \begin{pmatrix} 1 & f \\ 0 & 1 \end{pmatrix} = \begin{pmatrix} 0 & f \\ -1/f & 0 \end{pmatrix}$$

$$M_y = \begin{pmatrix} 1 & f/2 \\ 0 & 1 \end{pmatrix} \begin{pmatrix} 1 & 0 \\ -2/f & 1 \end{pmatrix} \begin{pmatrix} 1 & f \\ 0 & 1 \end{pmatrix} \begin{pmatrix} 1 & 0 \\ -2/f & 1 \end{pmatrix} \begin{pmatrix} 1 & f/2 \\ 0 & 1 \end{pmatrix} = \begin{pmatrix} -1 & 0 \\ 0 & -1 \end{pmatrix}$$

Combination of both matrices results in the four-dimensional ray transfer matrix of the non-rotated system:

$$M_{non-rotated} = \begin{pmatrix} 0 & 0 & f & 0 \\ 0 & -1 & 0 & 0 \\ -1/f & 0 & 0 & 0 \\ 0 & 0 & 0 & -1 \end{pmatrix}$$

By using the rotation matrix (1.68):

$$R^{\pm}(45^{\circ}) = \frac{1}{\sqrt{2}} \begin{pmatrix} 1 & \pm 1 & 0 & 0 \\ \mp 1 & 1 & 0 & 0 \\ 0 & 0 & 1 & \pm 1 \\ 0 & 0 & \mp 1 & 1 \end{pmatrix}$$

the complete matrix of the system is finally obtained:

$$M = R^{-}M_{non-rotated}R^{+} = \frac{1}{2}\begin{pmatrix} -1 & 1 & f & f \\ 1 & -1 & f & f \\ -1/f & -1/f & -1 & 1 \\ -1/f & -1/f & 1 & -1 \end{pmatrix} = \begin{pmatrix} A & B \\ C & D \end{pmatrix} \tag{3.55}$$

Equations (3.53) and (3.55) deliver the new variance matrix P_2 of the field in the x_2,y_2-plane:

$$P_2 = \begin{pmatrix} w^2_{x,1}+w^2_{y,1} & 0 & 0 & \frac{(w^2_{x,1}-w^2_{y,1})}{z_0} \\ 0 & w^2_{x,1}+w^2_{y,1} & \frac{(w^2_{x,1}-w^2_{y,1})}{z_0} & 0 \\ 0 & -\frac{(w^2_{x,1}-w^2_{y,1})}{z_0} & \theta^2_{x,1}+\theta^2_{y,1} & 0 \\ \frac{(w^2_{x,1}-w^2_{y,1})}{z_0} & 0 & 0 & \theta^2_{x,1}+\theta^2_{y,1} \end{pmatrix} \tag{3.56}$$

This is a field of rotational symmetry with a waist in this plane. Waist radii and angles of divergence, respectively, are equal in the two directions:

$$w^2_{x,2} = w^2_{y,2} = \frac{w^2_{x,1}+w^2_{y,1}}{2} \quad , \quad \theta^2_{x,2} = \theta^2_{y,2} = \frac{\theta^2_{x,1}+\theta^2_{y,1}}{2}$$

The astigmatic beam has been transformed into a beam of rotational symmetry. However, it is not a stigmatic beam since new non-diagonal term appear in the variance matrix. The can be identified as a twist:

$$<w_x\theta_y> = -<w_y\theta_x> = \frac{w^2_{x,1}-w^2_{y,1}}{2z_0}$$

or as an orbital angular momentum:

$$J_{z,struct.} = \frac{P}{4c_0}\frac{w^2_{x,1}-w^2_{y,1}}{z_0} \tag{3.57}$$

This equation holds for any astigmatic beam with variance matric P_l, which is transformed into a beam of rotational symmetry by a suitable optical system. By using the parameters (3.54) of the Gauss-Hermite modes of order (m,0), the angular momentum reads:

$$J_{z,struct.} = m \frac{P}{\omega} \tag{3.58}$$

n order to calculate the field distribution in the x_2,y_2-plane, the two-dimensional Collins-ntegral (Sec.2.3.2) has to be solved using the ABCD-parameters of the system matrix (3.55). In the x_1,y_1-plane, the field of a Gauss-Hermite mode of order (m,2) is given by:

$$E_1(x_1,y_1) = E_0\, H_m(\sqrt{2}x_1/w_0)\, \exp\left[-\frac{x_1^2+y_1^2}{w_0^2}\right]$$

After propagation through the optical system, the output field in the x_2,y_2-plane reads [1.125-1.127]:

$$E_2(x_2,y_2) = E_0\, i^m \sqrt{-i}\, \sqrt{2^{m-1}} \left(\frac{\sqrt{2}\rho_2}{w_0}\right)^m \exp\left[-\frac{\rho_2^2}{w_0^2}\right] \exp[im\phi]$$

with $\rho_2^2=x_2^2+y_2^2$. An experimental example is shown in Fig. 3.21. Other combinations of spherical and cylindrical lenses, as proposed by several authors [1.128-1.130], can also be used to transform astigmatic beams into beams with rotational symmetry. This is of particular interest for fiber coupling of diode lasers.

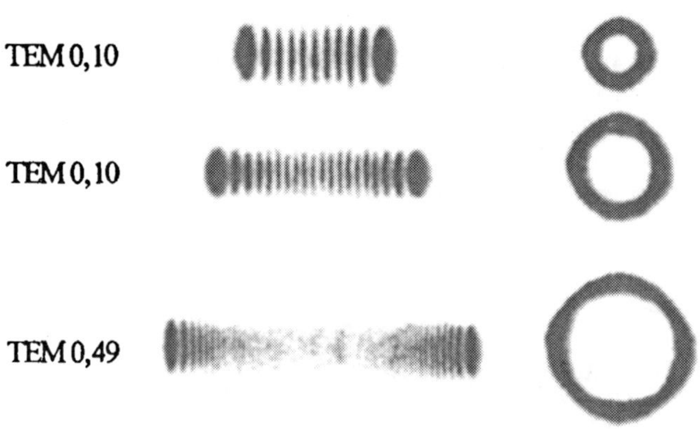

Fig. 3.21 Transformation of Gauss-Hermite modes into Gauss-Laguerre modes using the optical system depicted in the previous figure. A diode-pumped 1064nm Nd:YAG laser was used to generate the Gauss-Hermite modes. The recorded intensity distributions at the input plane (left) and the output plane (right) are shown [1.125,1.127].

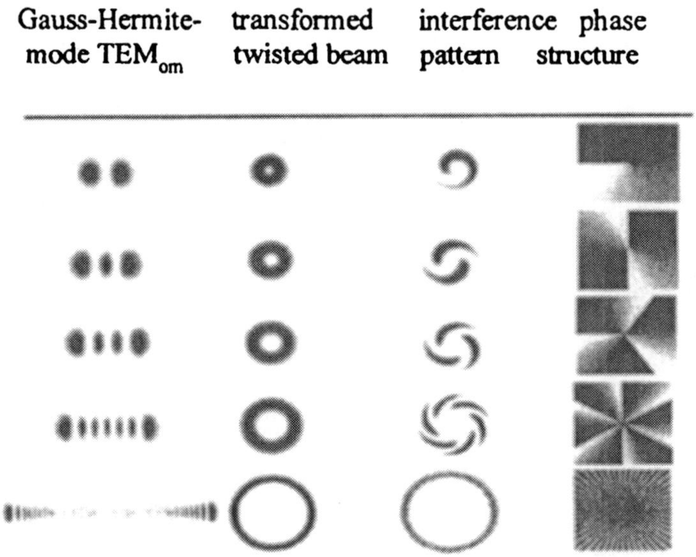

Fig. 3.22 Interference pattern of a plane wave with a twisted Gauss-Laguerre mode and its phase structure [1.125].

Example:
A Nd:YAG laser emits a Gauss-Hermite mode of order (m,n)=(10,0) with an output power of 10W at 1064nm. After transformation into a Gauss-Laguerre mode, the beam, according to (3.58), exhibits torque of $J_{z,struct.}= 5.4 \cdot 10^{-15}$ Nm. This is certainly a low value, but it is sufficient to make microscopic particles rotate.

The polarization angular momentum
The angular momentum flux due to the polarization term given by Eq. (3.48) can be written as:

$$J_{z,pol.} = \frac{\epsilon_0 c_0}{4\omega}|E_0|^2 \sin\Phi \iint \left| x\frac{\partial |u|^2}{\partial x} + y\frac{\partial |u|^2}{\partial y} \right| dxdy$$

Integration of this equation yields a very simple result:

$$J_{z,pol.} = -\sin\Phi \; \frac{P}{\omega}$$

This means, that for counterclockwise or clockwise circular polarization ($\Phi=\mp\pi$), each photon carries an angular momentum of $\pm\hbar$ [1.132]. The relation between circular polarization and spin of the photon was proven experimentally by Beth in 1936 [1.134]. The set-up he used is depicted in Fig. 3.23. The radiation of a conventional light source was collimated, left circularly polarized with a combination of a polarizer and a retarder plate and chopped with a shutter. The light was incident on a $\lambda/2$–plate, which was suspended on a thin quartz fibre. Each photon that goes through the plate, changes its angular momentum by $2\hbar$. This angular momentum is transferred to the plate, which results in an oscillation with the shutter frequency. To enhance the sensitivity the shutter frequency is equal the resonance frequency of the suspended plate.

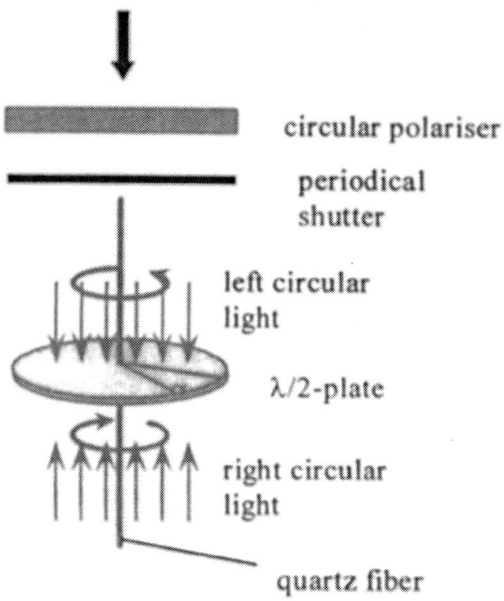

Fig. 3.23 Set-up of Beth's experiment to measure the intrinsic angular momentum of circularly polarized light [1.134].

Part II

Basic Properties of Optical Resonators

Chapter 4

The Fabry Perot Resonator

4.1 General Aspects

Whether a steady state radiation field can be established in an optical resonator depends on the wavelength of the radiation and on the mirror spacing. Steady state means that both the amplitude and the phase reproduce themselves after one round trip. It is easy to understand that both conditions can only be accomplished if the resonator length is an integral multiple of half the wavelength. Only in this case can we obtain standing waves inside the resonator with nodal intensity points on the mirror surfaces (Fig. 4.1). The preceding statement is always true as long as the field is not confined laterally by means of apertures. Hence, for a given mirror spacing L_0 and a medium with index of refraction n between the mirrors, we will find steady state field distributions for all wavelengths λ_q for which the following condition holds:

$$\lambda_q = \frac{2\,L}{q} \tag{4.1}$$

with:

	λ_q	:	wavelength in vacuum
	$L=L_0 n$	:	optical path length between mirrors
	L_0	:	geometrical path length between mirrors
	n	:	index of refraction

The integral number q, called the axial mode order, represents the number of intensity maxima. A resonator thus exhibits a periodic sequence of resonance frequencies $\nu_q=c_0/\lambda_q$ which are separated by:

$$\Delta\nu = \frac{c_0}{\lambda_q} - \frac{c_0}{\lambda_{q+1}} = \frac{c_0}{2\,L} \tag{4.2}$$

with c_0 being the speed of light in vacuum. For an optical path length of $L=1m$ - a typical length for laser resonators - the frequency gap is 150 MHz, which is extremely small compared to the frequency of visible light (ν=6x10^8 MHz for λ=500nm). The axial mode order q, which represents the number of half wavelengths fitting between the mirrors, is very high: for λ=500nm, we will find 4x10^6 intensity maxima.

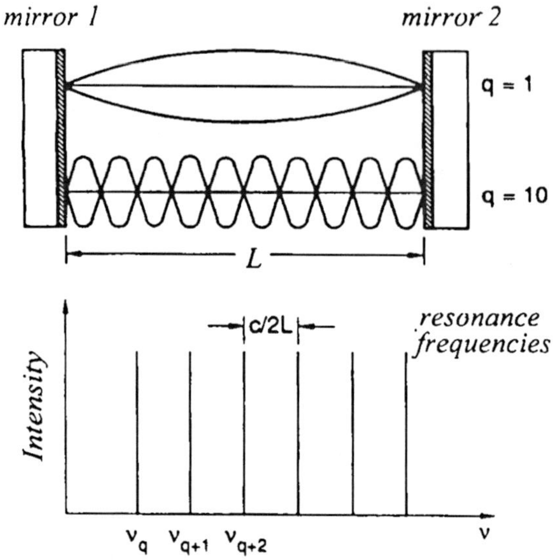

Fig. 4.1 Steady state field distributions in an optical resonator can only be established if the distance between the mirrors is an integral multiple *q* of half the wavelength. Due to the interference between the two fields moving in opposite directions, a standing wave pattern is created. The corresponding resonance frequencies are separated by a gap of *c/2L*, where c is the speed of light inside the resonator.

Standing waves of the kind shown in Fig. 4.1 can only be found if the resonator does not experience losses due to output coupling. If the combined mirror reflectivity R_1R_2 is less than 100%, traveling wave components will be generated too. The intensity minima will then not be equal to zero anymore. The resonance condition (4.1), which determines the wavelengths generating the maximum intracavity power density, is also valid for lossy resonators, but the resonance peaks depicted in Fig. 4.1 will become broader as the loss in the resonator is increased. That means that even wavelengths that do not exactly match the resonance condition (4.1) will induce a resonant behavior inside the resonator.

In the following we shall discuss the basic resonance properties of optical resonators without taking the transverse field structure into account. This is a reasonable approach since the transverse mode structure will only induce variations to the resonance conditions rather than changing the overall physical behavior of the resonator. As will be presented in Part III, the transverse field distribution splits each axial resonance frequency ν_q into a set of frequencies, each corresponding to a transverse mode structure.

We will start with the most basic optical resonator, the plane-parallel Fabry Perot interferometer (FPI), which has many applications in laser engineering and optical measurement techniques. The influence of the mirror reflectivities on the resonance frequencies and on the bandwidth will be presented. The incorporation of an active medium into the FPI will automatically lead to the basic properties of lasers. In Sec 4.3 we will apply the knowledge about FPIs to optical coatings whose characteristics can be investigated by treating them as a series of FPIs. The goal of this section is to make the reader familiar with the physical properties of resonator mirror coatings. A knowledge of coating designs and techniques is essential to understand the impact of the coating on the resonator design and on the beam characteristics of a laser.

4.2 The Fabry Perot Interferometer

4.2.1 Passive Fabry Perot Interferometer

A Fabry Perot interferometer (FPI) consists of two mirrors with reflectivities R_1, R_2, respectively, and are separated by a distance L (Fig. 4.2). For convenience, we assume that both mirrors are planar and are not confined by an aperture. The diffraction of the light can then be neglected and beam propagation inside the FPI can be described by means of geometrical optics. Losses generated by scattering and absorption on the mirror surfaces are incorporated into our resonator model by placing a medium with loss factor V between the mirrors. The loss factor represents the fraction of incident light intensity that is transmitted by the medium [2.14, 2.15].

Monochromatic light at a wavelength λ entering the FPI with an intensity I_0 (electric field E_0) will be partially reflected by mirror 1. The transmitted intensity portion is then reflected back and forth between the two mirrors. With each reflection intensity is coupled out of the FPI in both the forward and the backward direction. Summation of all waves coupled out yields the transmission T and the reflectance R of the FPI. If r_1, r_2 denote the reflection coefficients and t_1,t_2 are the corresponding transmission coefficients of the two mirrors, the reflected field E_r and the transmitted field E_t are given by (with v being the amplitude loss factor):

$$E_r = E_0 \left[r_1 - t_1^2 v^2 r_2 \exp[i2kL] \sum_{n=0}^{\infty} (v^2 r_1 r_2 \exp[i2kL])^n \right] \tag{4.3}$$

$$E_t = - E_0 \, t_1 t_2 v \exp[ikL] \sum_{n=0}^{\infty} (v^2 r_1 r_2 \exp[i2kL])^n \tag{4.4}$$

Note that a transmission phase factor of $\exp(i\pi/2)$ has to be taken into account to get the minus signs [2.1].

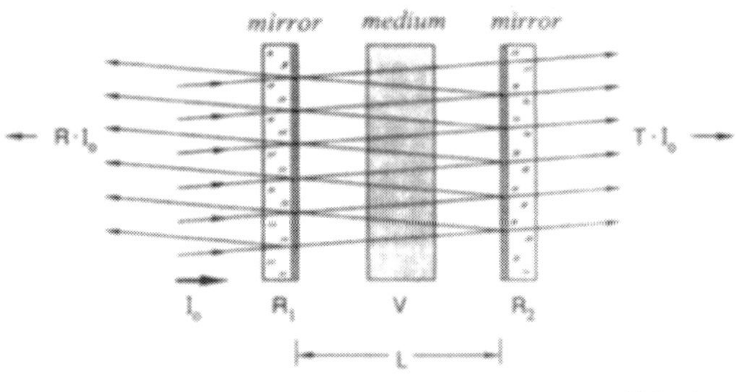

Fig. 4.2 Plane-parallel Fabry Perot interferometer (FPI). Incoming light with intensity I_0 is oscillating between the two mirrors and is partially transmitted with each reflection at a mirror. The rays are shown tilted to show the propagation in a more convenient way.

By calculating the intensities from (4.3) and (4.4), we can determine the fraction of the incident intensity I_0 being transmitted and reflected by the FPI:

$$R = \frac{(\sqrt{R_1}-\sqrt{R_2 V})^2 + 4\sqrt{R_1 R_2 V}\sin^2(kL)}{(1-\sqrt{R_1 R_2 V})^2 + 4\sqrt{R_1 R_2 V}\sin^2(kL)} \tag{4.5}$$

$$T = \frac{(1-R_1)(1-R_2)V}{(1-\sqrt{R_1 R_2 V})^2 + 4\sqrt{R_1 R_2 V}\sin^2(kL)} \tag{4.6}$$

with:

$R_i = |r_i|^2$: reflectivity of mirror i
$V = |v|^2$: loss factor per transit
$k = 2\pi/\lambda$: wave number
λ : wavelength in vacuum
L : optical mirror spacing

Transmission T and reflectance R as a function of the phase kL are shown in Fig. 4.3 for a loss free FPI with $R_1=R_2$. The transmission becomes maximum if the condition

$$\sin(kL) = 0 \quad \Rightarrow \quad \lambda = \frac{2\,L}{q} \quad , \; q=1,2,3,... \tag{4.7}$$

holds. The transmission T thus exhibits a periodic sequence of maxima whose frequencies are separated by $c_0/2L$, called the *free spectral range*. As discussed in Sec. 4.1, standing waves are generated inside the FPI at these resonance frequencies characterized by the condition that the mirror spacing equals an integral multiple of half the wavelength. The following relations hold:

Maximum Transmission:

$$T_{max} = \frac{(1-R_1)(1-R_2)V}{(1-\sqrt{R_1 R_2 V})^2} \tag{4.8}$$

$$T_{max} = 1 \qquad \text{for } R_1=R_2,\ V\ 1$$

Maximum Energy Density:

$$\rho_{max} = \frac{I_0}{c_0}\,\frac{1-R_1^2}{(1-\sqrt{R_1 R_2 V})^2} \tag{4.9}$$

$$\rho_{max} = I_0(1+R)/[c_0(1-R)] \qquad \text{for } R_1=R_2,\ V-1$$

Bandwidth (FWHM):

$$\delta\nu = |\ln(\sqrt{R_1 R_2 V})|\frac{c_0}{2\pi L} \tag{4.10}$$

Free Spectral Range:

$$\Delta\nu = \frac{c_0}{2L} \tag{4.11}$$

Light will be transmitted by the FPI if its frequency lies within the frequency width $\delta\nu$ of one of the resonance peaks. The resonance frequencies, determined by a maximum of the transmission T, exhibit a frequency width that decreases as the mirror reflectivities and the loss factor are increased. The reflectance R of the FPI is minimum at the resonance frequencies, since the waves traveling out of the FPI in the backwards direction partially cancel due to destructive interference. If both mirror reflectivities are equal, complete destructive interference occurs and consequently the reflectance of the FPI goes to zero at the resonance frequencies (if $V=1$). For the loss free FPI, the relation $R+T=1$ always holds.

It seems surprising that the transmission of a loss free FPI at a resonance frequency is equal to one, which means that the incident beam intensity is completely transmitted even though both mirrors exhibit reflectances of higher than zero. Even for mirror reflectances of $R_1=R_2=0.9999$, no intensity is reflected off the FPI. This puzzling phenomenon is caused by the interference of all waves bouncing back and forth inside the FPI and being transmitted through the two mirrors. In resonance, the waves inside the FPI exhibit constructive interference leading to an increase of the intracavity intensity I such that the intensity transmitted through mirror 2 equals the initial beam intensity I_0 . For $R_1=R_2=0.9999$ the intracavity intensity is 10,000 times higher than the intensity I_0 of the beam incident on the FPI. Figure 4.4 depicts the resonant behavior of the intracavity intensity I as a function of kL.

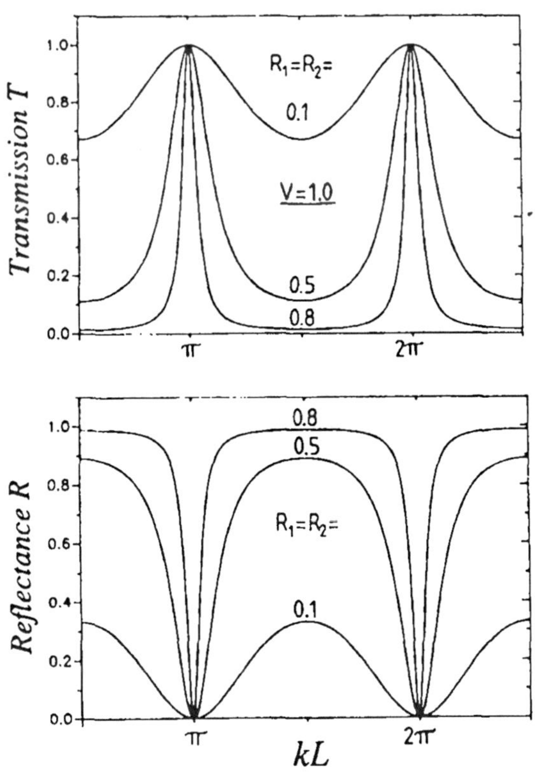

Fig. 4.3 Transmission and reflectance of a loss free FPI ($V=1$) as a function of kL. Transmission maxima are observed when the optical mirror spacing L equals an integral multiple of half the vacuum wavelength. Energy conservation requires $R+T=1$.

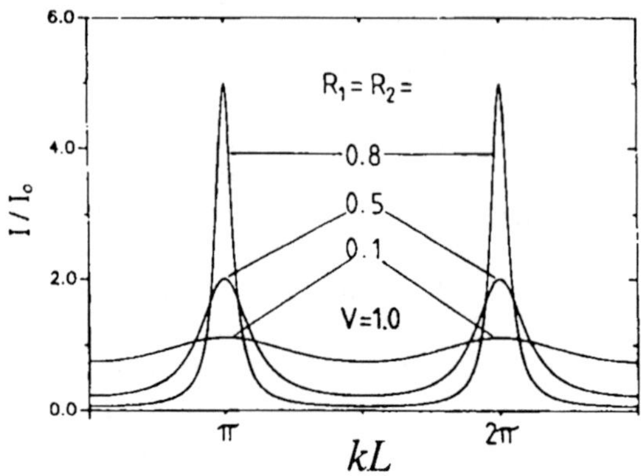

Fig. 4.4 Intensity I inside the FPI (normalized to the incident intensity I_0) as a function of kL for a loss free symmetric FPI ($V=1$).

If the light source is shut off, the energy stored inside the FPI will show a temporal decay due to the output coupling losses experienced on both mirrors. The decay time τ after which the fraction of 1/e (≈ 0.368) of the initial energy has remained inside the FPI is given by:

$$\tau = \frac{L}{c_0} \frac{1}{|\ln(\sqrt{R_1 R_2 V})|} = \frac{1}{2\pi\delta\nu} \tag{4.12}$$

The decay time also determines the time it takes the FPI to be fully operational after the light source is switched on again. As in all linear resonant systems, the decay time and the resonance bandwidth are inversely proportional.

Two quantities, the finesse F and the cavity quality Q, are commonly used to characterize the quality of a FPI. They are defined as:

$$F = \frac{\Delta\nu}{\delta\nu} = \frac{\textit{free spectral range}}{\textit{bandwidth}} = \frac{\pi}{|\ln(\sqrt{R_1 R_2 V})|} \tag{4.13}$$

$$Q = 2\pi\Delta\nu\tau = \frac{\nu}{\delta\nu} \tag{4.14}$$

High quality FPIs exhibit both a high finesse and a high cavity quality. This is accomplished by choosing the mirror reflectivities R_1, R_2 and the loss factor V as close to 1.0 as possible. Table 4.1 compares the resonance properties of a typical FPI used in atomic physics for the measurement of spectral lines with those of a spring oscillator. FPIs used for the active stabilization of laser frequencies have finesses of up to 30,000.

Table 4.1 Resonance properties of an FPI and a spring oscillator.

	FPI	spring oscillator
Resonance frequency	$\nu=4.94\times10^{14}$ Hz (λ=607nm)	2 Hz
Axial mode number	q=33,000	-
Bandwidth	$\delta\nu=3.8\times10^{8}$ Hz	0.1Hz
Free spectral range	$\Delta\nu=1.5\times10^{10}$ Hz	-
Max. transmission	T=0.13	-
Decay time	$\tau=3.61\times10^{-10}$ s	1.6 s
Cavity quality	$Q=1.3\times10^{6}$	20
Finesse	F=39.5	-

If the loss factor of the FPI is less than one which means that the light bouncing back and forth inside the FPI experiences losses due to absorption or scattering, the transmission T of the FPI cannot reach its maximum value of $T_{max}=1.0$ at the resonant frequencies. We can calculate the loss ΔV^* of the FPI by inserting Eqs. (4.5) and (4.6) into the condition for energy conservation:

$$R + T + \Delta V^* = 1$$

This results in:

$$\Delta V^* = (1-V)\frac{(1-R_1)(1+R_2V)}{(1-\sqrt{R_1R_2}V)^2 + 4\sqrt{R_1R_2}V\sin^2(kL)} \tag{4.15}$$

where $1-V$ is the loss per transit inside the FPI. Before we visualize this equation in a graph, we should try to understand what to expect. As already seen in Fig. 4.4, the intracavity intensity is amplified at the resonance frequencies to values which, for FPIs with high cavity quality Q (high mirror reflectivities), can exceed the initial beam intensity by several orders of magnitude . A higher intracavity intensity will, however, generate a higher loss ΔV^* of the FPI. If, for instance, the intracavity intensity is 20 times higher than the initial intensity I_0 a loss of 1% per transit will generate a loss in intensity of 20% in terms of the initial intensity. The loss of the FPI will therefore show maxima at the resonance frequencies and the maximum loss will increase as the mirror reflectivities are increased. One can also use the decay time, τ, to discuss the loss of an FPI. At the resonance frequencies the light remains inside the FPI for a long time. The increased number of reflections induce a high loss since for each transit the amount $(1-V)$ of the intensity is absorbed or scattered. A long decay time (which is equivalent to having high mirror reflectances R_1, R_2 or a small bandwidth $\delta\nu$), therefore, generates a high loss ΔV^* of the FPI. It is for this reason that FPIs with a high transmission T_{max} and a small bandwidth are difficult to realize.

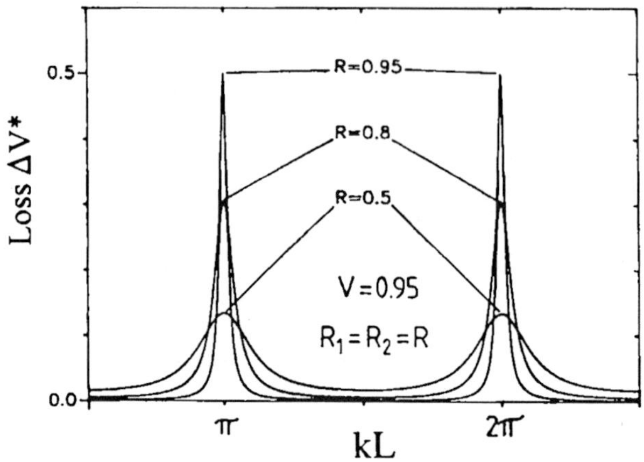

Fig. 4.5 Loss of an FPI with mirror reflectances $R_1=R_2=R$ as a function of kL. The loss per transit is 5% ($V=0.95$).

Figure 4.5 shows the loss of a symmetric FPI as a function of kL for a loss per transit of 5%. As discussed, the maximum loss occurs at the resonance frequencies. Note that the loss of the FPI is usually higher than the loss per transit. For high mirror reflectivities, in particular, the loss can exceed 50% even though the loss per transit may be only 1%. This interesting result indicates how sensitive the properties of an optical resonators are to small losses induced inside the resonator. In Chapter 10 we will present models for the calculation of the output power of laser resonators. We will see that small losses in the few percent range induced by scattering or absorption in the active medium (or on the mirror surfaces) can decrease the output power by up to one order of magnitude, depending on the gain of medium. With the above discussion of the FPI this is not surprising: the high number of bounces inside the resonator magnifies the loss.

It is worth noting that the FPI can be designed such that its reflectance R vanishes at the resonance frequencies even if losses occur between the mirrors. By using (4.5), the condition to meet is found to be:

$$R_1 = R_2 V^2$$

A FPI for which this relation holds is called *impedance-matched*. This is not the case for the FPIs in Fig. 4.6 in which the maximum loss ΔV^*_{max} and the maximum transmission T_{max} are plotted as a function of the mirror reflectances $R=R_1=R_2$. This is why in this figure the loss of the FPI vanishes as the mirror reflectivities approach unity: the incoming beam is 100% reflected by the FPI.

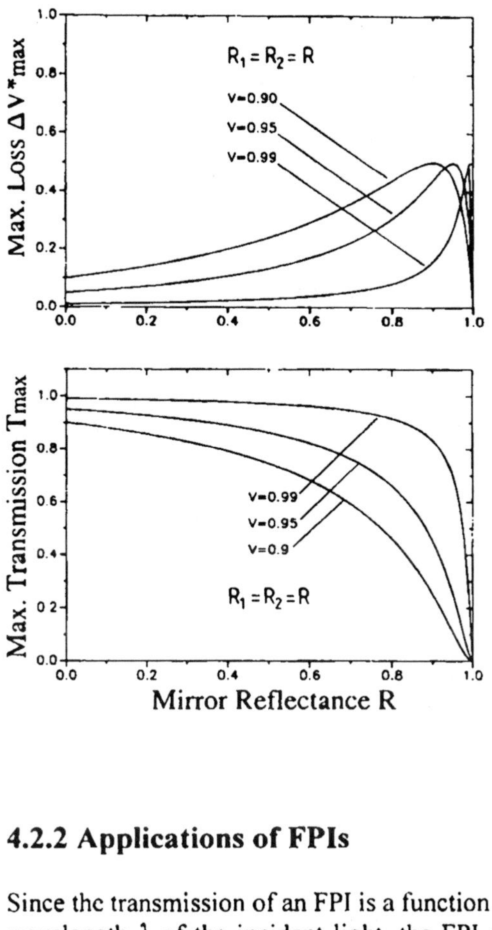

Fig. 4.6 Maximum loss and maximum transmission of an FPI at the resonance frequencies as a function of the mirror reflectances $R=R_1=R_2$. In each graph the three curves correspond to a loss of 1%, 5%, and 10% per transit. Even with small losses inside the FPI, high transmission with a small bandwidth (equivalent to high mirror reflectivities) is difficult to attain.

4.2.2 Applications of FPIs

Since the transmission of an FPI is a function of both the optical mirror spacing L and the wavelength λ of the incident light, the FPI can be used to measure small variations in lengths, or to determine the spectral characteristics of a light source, Furthermore, it can serve as a filter to decrease the spectral width of the light source [2.7-2.9, 2.11]. An example demonstrating the length measurement capabilities of an FPI, Fig. 4.7 depicts an experimental set-up to measure the expansion characteristics of a piezo ceramic. The piezo ceramic is driven by an AC voltage which results in an oscillating movement of one mirror of the FPI. The FPI used here has concave mirrors, preferably with the center of curvature located on the opposite mirror (confocal FPI) This geometry provides the lowest sensitivity to mirror tilt [2.13]. The transmitted intensity of a laser beam with known wavelength λ is measured with a photodiode as a function of the drive voltage. A sequence of transmission maxima is observed corresponding to a piezo movement of $\lambda/2$ from maxima to maxima. By using these data, the variation in piezo length as a function of the applied voltage can be determined. With such a set-up it is possible to resolve length differences on the order of nanometers (10^{-9}m).

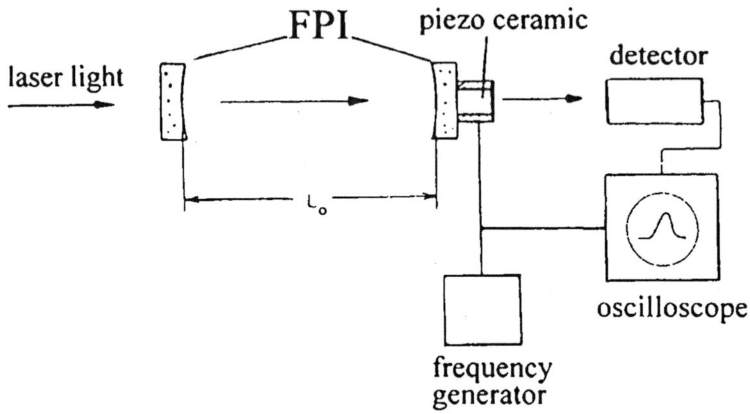

Fig. 4.7 Experimental set-up for measuring the expansion characteristic of a piezo ceramic [2.9].

This arrangement can also be used to measure the spectral characteristics of a laser beam. However, the bandwidth of the laser must be smaller than the free spectral range of the FPI, otherwise we will get maximum transmission for several wavelengths simultaneously. Furthermore, the bandwidth of the FPI should be as small as possible to get high frequency resolution. It is for these reasons that FPIs with small mirror spacings (10 mm) and high mirror reflectivities are used for laser frequency measurements. Typical free spectral ranges are on the order of 10 GHz with bandwidths around 10 MHz (for spectral properties of lasers see Table 4.2). The laser with known wavelength λ is used as a calibration standard to relate each drive voltage of the piezo to a certain wavelength. After replacing the laser with the unknown laser, the spectral distribution of the latter can be observed on the oscilloscope. The spectrum of a HeNe laser shown in Fig. 4.11 was recorded using this technique.

The spectral filtering properties of the FPI are used in **etalons**, which are fixed spaced FPIs with cavity lengths varying from 10 mm down to 0.1 mm. They are either made of two coated plano-plano glass plates (air gap etalon) or they consist of one plate coated on both sides. The latter technique provides higher transmission and higher damage threshold. Etalons exhibit free spectral ranges of up to 1,000 GHz with a Finesse of 50. The resonance wavelengths λ_q can also be changed to a certain degree by tilting the etalon. If θ denotes the tilt angle, the new resonance wavelengths are given by $\lambda'_q = \lambda_q / \cos\theta$.

Higher free spectral ranges are realized with **interference filters**. They can be considered as a stack of several FPIs with mirror spacings on the order of µm. Technically, these filters are generated by coating multilayer dielectric films on top of each other with dielectric spacers in between representing the mirror gap. Bandwidths are on the order of 10^4 GHz (this corresponds to about 10nm for visible wavelengths) and free spectral ranges of 10^5 GHz (100nm) are obtained. In addition, higher order transmission peaks can be suppressed by adding colored filter glass. At the desired wavelength, the typical transmission of interference filters is 50%.

4.2.3 Fabry Perot Interferometer with Gain - Laser Resonator

We now consider the case that the medium placed inside the FPI amplifies the light oscillating between the two mirrors. The gain factor G with $G>1$ denotes the factor by which the intensity has been amplified after a transit through the medium. In the passive FPI discussed in the last section, the intensity I incident on the medium was decreased to VI after the transit. With light amplification taken into account, the intensity at the other side of the medium is given by GVI. We can therefore use all equations derived in Sec. 4.2 and take the gain into account by replacing V with GV. Equations (4.8)-(4.11) then yield:

Resonance frequencies: $$\nu_q = q\frac{c_0}{2L} \qquad q = 1,2,3, \ldots$$

Free spectral range: $$\Delta\nu = \frac{c_0}{2L}$$

Bandwidth: $$\delta\nu = |\ln(G\sqrt{R_1R_2}V)|\frac{c_0}{2\pi L} \approx \frac{\Delta\nu}{\pi}(1-G\sqrt{R_1R_2}V)$$

Decay time: $$\tau = \frac{L}{c_0}\frac{1}{|\ln(G\sqrt{R_1R_2}V)|} \approx \frac{L}{c_0}\frac{1}{1-G\sqrt{R_1R_2}V}$$

Maximum transmission: $$T_{max} = \frac{(1-R_1)(1-R_2)GV}{(1-G\sqrt{R_1R_2}V)^2}$$

The resonance frequencies are not affected by the gain of the medium. A more detailed theory shows that the resonance frequencies are shifted by the gain, but the shift is negligible for most applications (see Sec. 9.6). The bandwidth of the active FPI is decreased as the gain of the active medium is increased. This is understandable because the losses that the light experiences per round trip are now partially compensated by the active medium. This results in an increase of both the decay time and the transmission of the FPI. The insertion of an active medium is a well established technique to enhance the spectral resolution of an FPI. However, the gain of the medium should not overcompensate the losses generated by output coupling at the mirrors, absorption, and scattering. In other words, the total gain loss product should not exceed 1.0:

$$G\sqrt{R_1R_2}V < 1$$

If the gain factor G is chosen so that all losses are exactly compensated:

$$G\sqrt{R_1 R_2}V = 1 \tag{4.16}$$

our mathematical model predicts that the decay time, the transmission, and the energy density inside the FPI diverge. The wave summation used to derive (4.3) and (4.4) is not applicable in this case, especially since the gain factor G is affected by the increasing intensity inside the FPI. When the threshold condition $GRV=1$ holds, the spontaneous emission of the active medium leads to the generation of standing light waves inside the FPI, even without an external light source. The FPI has now become a laser resonator [2.6].

A laser resonator thus emits light at the resonance frequencies ν_q if (4.16) holds for the gain factor G at these frequencies. The corresponding electromagnetic waves - these are standing waves if both mirror reflectivities are equal - are referred to as the axial modes of the resonator. They are characterized by the mode number q, which determines the number of half wavelengths fitting into the optical resonator length.

The gain factor G depends on the frequency and the intensity of the light inside the resonator and it can be increased by transferring more power into the active medium via the pump process (e.g. flashlamp pumping or laser diode pumping for solid state lasers, gas discharge for gas lasers). Without the pumping process, the gain is less than 1.0, which means that the light is absorbed as it propagates through the medium. In general, the gain factor G can exceed 1.0 only at distinct frequencies which are characteristic for the active material used. By using special mirror coating designs, the laser emission is usually restricted to one of these possible emission wavelengths (the mirror reflectivities are chosen too low at the other wavelengths to reach the threshold condition (4.16)). Figure 4.8 shows the gain profile at the preferred laser emission wavelength of ruby for four different pump powers. A threshold pump power is required to reach a gain factor G of more than 1.0.

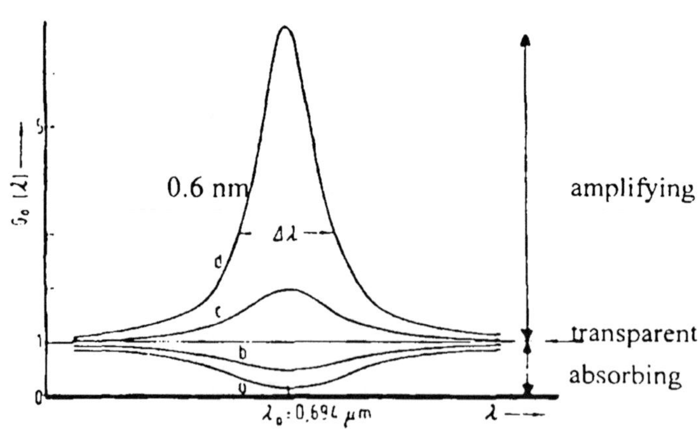

Fig. 4.8 Laser materials exhibit gain profiles only at characteristic wavelengths. This example shows gain profiles for a ruby crystal at $\lambda=694nm$ for four different pump powers (a-d). The axial modes whose wavelengths are within the bandwidth can oscillate in the laser resonator.

Laser emission can only be generated within the bandwidth of the gain profile. The number of axial modes, to a good approximation, is given by the ratio of the gain bandwidth to the free spectral range of the resonator. The bandwidth of the laser emission strongly varies with the active material used. Some gas lasers have gain bandwidths as low as 100 MHz, whereas dye lasers can exceed this value by more than a factor of 10,000. Accordingly, the number of axial modes varies considerably with the type of laser. Table 4.2 gives an overview of the spectral emission characteristics of common lasers including the number of axial modes that may oscillate in a resonator with an optical length of 1m. In homogeneously broadened lasers, the number of observed axial modes is much smaller. Alexandrite and Ti:sapphire lasers, for instance, exhibit laser bandwidths of less than 10nm.

Table 4.2 Center wavelength λ_0, gain bandwidth $\Delta\lambda$, and number of axial modes n within the gain bandwidth (resonator with optical length 1m) for different lasers.

Medium	Type of Laser	λ_0[nm]	$\Delta\lambda$[nm]	n
KrF	excimer	249	20	646,000
Ar^+	gas	488	0.003	3
HeNe	gas	632.8	0.002	10
CO_2	gas	10,600	2	36
Rhodamin 6G	dye	600	60	334,000
GaAlAs	semiconductor	804	2	6,200
Alexandrite	solid state	760	70	240,000
Nd:glass	solid state	1,054	10	18,000
Nd:YAG	solid state	1,064	0.5	885
Nd:YLF	solid state	1,047	1.0	1,825
Yb:YAG	solid state	1,030	20	38,000
Ti:sapphire	solid state	800	200	635,000

What happens if the gain is so high that GRV is greater than 1.0? The consequence would be that the intensity inside the resonator is increased more by the gain than it is decreased by the losses. The intensity would therefore become infinite! This catastrophic behavior is prevented by the saturation of the gain. The gain factor is decreased as the intensity of the light is increased. The intensity dependence of the gain factor is given by (see Sec. 9.5):

$$G = \exp\left[\frac{g_0 l}{(1 + I/I_S)^x}\right] \tag{4.17}$$

The *small-signal gain* g_0l is proportional to the inversion, which for four-level lasers increases linearly with the pump power. The saturation intensity I_S is a characteristic of the laser material. The quantity x depends on the type of medium used. The two limits are $x=1$ for homogeneously broadened media and $x=0.5$ for inhomogeneously broadened media (see Sec. 9.5 on line broadening). How does the gain saturation influence the oscillation of axial modes in a laser ? Let us assume that we cover one resonator mirror to prevent laser oscillation and we pump the active medium. Since we have negligible intensity inside the resonator (the losses are too high to start laser oscillation) the gain factor (4.17) is given by:

$$G_0 = \exp(g_0 l) \tag{4.18}$$

The small-signal gain factor G_0 determines how light with an intensity much smaller than the saturation intensity gets amplified by the medium in one transit. As soon as we remove the cover on the resonator mirror, all axial modes for which the relation $G_0RV \geq 1$ holds will oscillate. The intensity of the modes will now rapidly increase. According to (4.17), the gain factor will decrease resulting in a continuous decrease of the net amplification of the light. Finally, a steady state intensity is reached for which $GRV=1$ holds. By using (4.17), we find the steady state intensity to be:

$$I = I_S \frac{(g_0 l)^{1/x} - |\ln(RV)|^{1/x}}{|\ln(RV)|^{1/x}} \tag{4.19}$$

When the laser is inhomogeneously broadened all axial modes which started to oscillate will reach this steady state intensity (as shown in Fig. 4.9). For truly homogeneously broadened lasers, only one axial mode will oscillate in steady state, since competition between modes occurs (we shall discuss this phenomenon in detail in Sec. 9.5). Note that it is the gain saturation that makes a laser work! The intensity dependence of the gain ensures that we get a steady state solution for the intensity and thus a well defined output power from the laser.

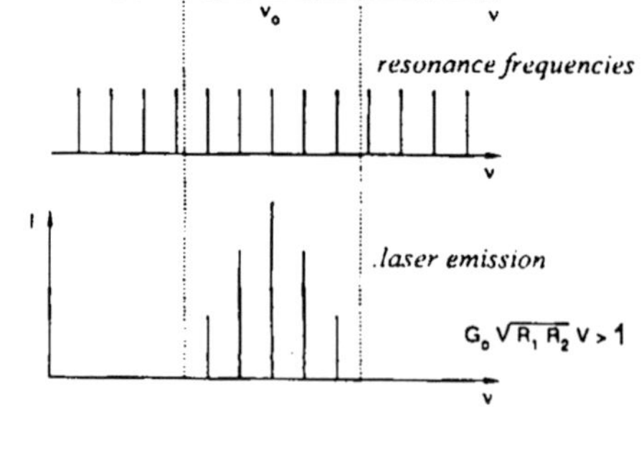

Fig. 4.9 Only axial resonator modes for which $G_0RV>1$ holds can oscillate. Due to saturation, the gain factor will decrease as the mode intensity increases. In steady state the intensity in each axial mode is determined by $GRV=1$, with G given by (4.17).

Figure 4.9 presents the spectral emission characteristics of a laser resonator. Only the axial modes of the resonators for which the small signal gain factor G_0 is high enough to overcome the losses generated by absorption and output coupling will oscillate. The steady state intensity is highest for the modes that experienced the highest small signal gain. For an inhomogeneously broadened active medium, oscillation of all these axial modes is observed because each mode depletes the gain only at its resonance frequency without interacting with other modes. The envelope of the axial mode intensity spectrum is then determined by the gain profile of the medium (Figs. 4.10 and 4.11). For homogeneously broadened lasers, however, only the highest gain mode survives due to the fact that all axial modes have access to the whole gain profile (see Sec. 9.5). This holds as long as no thermal or mechanical distortions occur and spatial hole-burning is prevented.

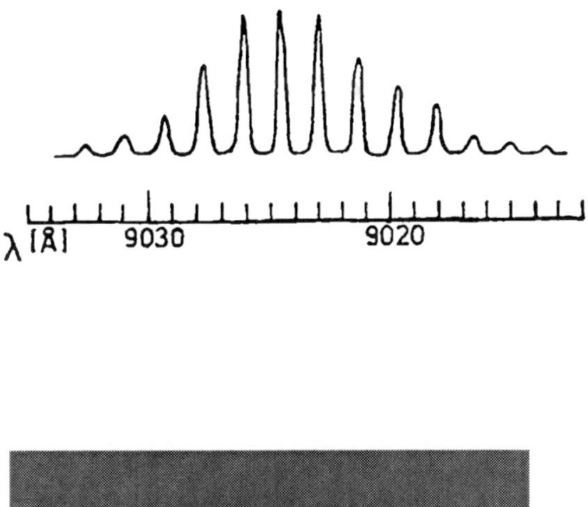

Fig. 4.10 Axial mode intensity spectrum of a GaAs diode laser (measured with a scanning FPI). The polished endfaces of the crystal serve as the resonator mirrors. The short resonator length (L_0=*1mm*) generates a large free spectral range. This is the reason why only 13 axial modes are observed even though the gain bandwidth is quite large.

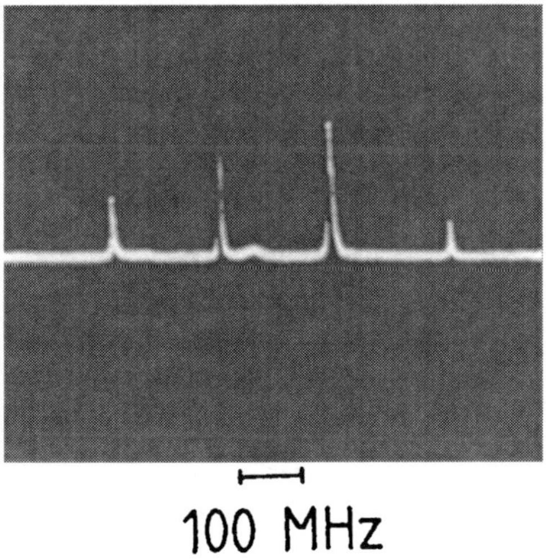

Fig. 4.11 Spectrum of a HeNe laser ($\lambda_0 = 632.8 nm$) measured with a scanning FPI. The optical resonator length is $L = 0.7 m$. The small gain bandwidth allows only four axial modes to oscillate.

With increasing pump power the gain profile increases resulting in a wider frequency range within which the gain is high enough to overcome the resonator losses. For an inhomogeneously broadened laser, new axial modes will be added at the outer areas of the gain profile. The mode numbers n shown in Table 4.2 represent upper limits achievable at high pump powers. In a laser, the bandwidth $\delta\nu$ of each axial mode can assume very low values. According to the theory of the active FPI, the bandwidth of a laser in steady state oscillation should be zero. In reality, however, the bandwidth is much larger due to spontaneous emission which generates a noise background to the axial mode build up by induced emission. The noise determines the lower bandwidth limit theoretically attainable in a laser (see the Schawlow-Townes formula in Sec. 9.7). Furthermore, the resonator set-up experiences small scale optical length variations due to vibrations and temperature changes. The effect of the length variations on the bandwidth is generally several orders of magnitude higher compared to the influence of the noise.

By using special stabilizing and feedback techniques, bandwidths as low as 100 mHz can be realized. Without active or passive frequency stabilization, laser modes exhibit bandwidths on the order of MHz. In general, many axial modes are found in the spectral emission of a laser resonator (see Table 4.2) since true homogeneously broadened lasers which operate at one axial mode are not very common. Without additional frequency selecting elements inside the resonator, the bandwidth of inhomogeneously broadened lasers is determined by the bandwidth of the gain profile. The technical realization of narrow bandwidths or single axial mode operation in these lasers requires the incorporation of etalons into the resonator. In homogeneously broadened lasers, oscillation at one axial mode can be forced by using resonator designs that prevent spatial hole burning. A detailed discussion of single mode lasers shall be given in Chapter 21.

4.3 Optical Coatings

4.3.1 A Coating Design Matrix Method

The principle of the FPI can be used to generate optical elements that exhibit spectral reflectances suitable for a desired application. We have seen that an FPI has its highest reflectance R if the mirror spacing is an odd multiple of quarter wavelengths. A preferred technique to realize such an FPI is to coat a substrate with a dielectric material whose optical thickness nL equals one quarter wavelength (Fig. 4.12a). The two mirror reflectances R_1 and R_2 are now given by the reflectances of the two dielectric interfaces, air-coating and coating-substrate. Unfortunately, this design provides only limited control over the amplitude and the spectral profile of the reflectance R. In order to become more flexible in the design of the reflection properties, multi-layer coatings of different dielectric materials are used as depicted in Fig. 4.12b. This is the preferred way to produce resonator mirrors and anti-reflection coatings of optical elements [2.4,2.5,2.16,2.17]. In the following we will derive a matrix method to calculate the reflectance of such a stack of layers deposited onto a substrate [2.1].

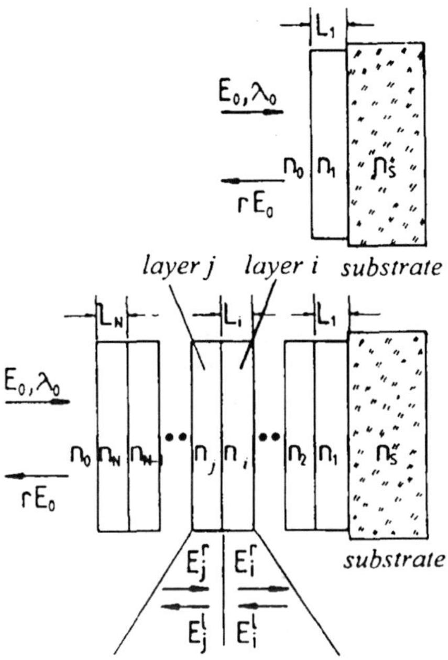

Fig. 4.12 a) An FPI can be realized by coating a substrate with dielectric material. If the thickness L_1 is an odd multiple of $\lambda_0/(4n)$, the reflectance becomes a maximum at the wavelength λ_0. b) A multi-layer coating enhances the flexibility in designing spectral reflectance profiles $r(\lambda)$.

In order to keep the discussion as general as possible we assume that the thickness of each layer can be chosen arbitrarily. We will, however, restrict ourselves to normal incidence. Incorporation of a tilt of the surface normal with respect to the wave propagation vector can be accomplished by modifying the final result.

Let a beam with field amplitude E_0 and wavelength λ_0 be incident on the stack of layers as shown in Fig. 4.12b. Reflection and transmission at each interface inside the stack leads to the generation of two counterpropagating waves in each layer. Our goal is to calculate the amplitude of the wave leaving the stack in the direction opposite to the incident wave. We first consider one interface in the stack between layer i and layer j and determine how the electric field amplitudes propagating in opposite directions are related to one another. If r_{ij}, r_{ji} denote the reflection coefficients for a transition from layer i to layer j and from layer j to layer i, respectively, and t_{ij},t_{ji} are the corresponding transmission coefficients, we obtain:

$$E_j^l = t_{ij}E_i^l + r_{ji}E_j^r \tag{4.20}$$

$$E_i^r = r_{ij}E_i^l + t_{ji}E_j^r \tag{4.21}$$

Since we want to express the fields in layer j as a function of the fields in layer i we rewrite (4.21) as:

$$E_j^r = \frac{1}{t_{ji}} E_i^r - \frac{r_{ij}}{t_{ji}} E_i^l \tag{4.22}$$

The second equation is obtained by inserting (4.22) into (4.20):

$$E_j^l = \frac{r_{ji}}{t_{ji}} E_i^r + \frac{t_{ij}t_{ji} - r_{ji}r_{ij}}{t_{ji}} E_i^l \tag{4.23}$$

By using the Fresnel relations $t_{ij}=t_{ji}$ (the transmission cannot depend on the propagation direction) and $r_{ij}=-r_{ji}$ (a phase shift of π is induced when the light is reflected off the high index material) together with $r_{ij}^2+t_{ij}^2=1$ (energy conservation), we can write Eqs. (4.22) and (4.23) in the form of a matrix equation:

$$\begin{pmatrix} E_j^l \\ E_j^r \end{pmatrix} = \frac{1}{t_{ij}} \begin{pmatrix} 1 & -r_{ij} \\ -r_{ij} & 1 \end{pmatrix} \begin{pmatrix} E_i^l \\ E_i^r \end{pmatrix} = T_{ij} \begin{pmatrix} E_i^l \\ E_i^r \end{pmatrix} \tag{4.24}$$

Hence, we found the transition matrix T_{ij} connecting the field amplitudes on both sides of the interface between layer i and layer j. Now we have to incorporate the propagation of the fields inside each layer from right to left. We take layer j and mark the field after propagation with an asterisk. The propagation will induce a phase shift on both field components as well as an amplitude loss factor V (note that for the field propagating to the right we have to subtract the phase and divide by the loss factor to get the amplitude on the left side of the layer!). The propagation matrix P_j can be written as:

$$\begin{pmatrix} E_j^{*l} \\ E_j^{*r} \end{pmatrix} = \begin{pmatrix} V_j \exp[i\frac{2\pi}{\lambda_0} n_j L_j] & 0 \\ 0 & \frac{1}{V_j} \exp[-i\frac{2\pi}{\lambda_0} n_j L_j] \end{pmatrix} \begin{pmatrix} E_j^l \\ E_j^r \end{pmatrix} = P_j \begin{pmatrix} E_j^l \\ E_j^r \end{pmatrix} \tag{4.25}$$

In the substrate we have only one field amplitude E_S propagating to the right since there is no reflecting interface to generate a left-propagating wave. Starting at the substrate, we combine all transition matrices T and propagation matrices P until we reach the first interface on the left side of the stack:

$$\begin{pmatrix} rE_0 \\ E_0 \end{pmatrix} = T_{NA} P_N T_{(N-1)N} P_{N-1} \cdots T_{23} P_2 T_{12} P_1 T_{S1} \begin{pmatrix} 0 \\ E_S \end{pmatrix} \tag{4.26}$$

where T_{NA} and T_{SI} are the transition matrices from layer N to the outer medium and from the substrate to the first layer, respectively. The whole multi-layer stack can thus be described by the stack matrix S with:

$$\begin{pmatrix} rE_0 \\ E_0 \end{pmatrix} = \begin{pmatrix} S_{11} & S_{12} \\ S_{21} & S_{22} \end{pmatrix} \begin{pmatrix} 0 \\ E_S \end{pmatrix} = S \begin{pmatrix} 0 \\ E_S \end{pmatrix} \tag{4.27}$$

The reflection coefficient r of the stack can now be calculated to be:

$$r = \frac{S_{12}}{S_{22}} \tag{4.28}$$

In general, the reflection coefficient r is a complex number which means that the field experiences a phase shift. The reflectance R and the phase shift ϕ read:

$$R = rr* = \left|\frac{S_{12}}{S_{22}}\right|^2 \tag{4.29}$$

$$\phi = atan\left[\frac{Im(r)}{Re(r)}\right] \tag{4.30}$$

Note that this matrix method of calculating the reflectance is equivalent to the wave summation method used for the FPI! All the field components in (4.24) represent the sums of the partial waves which means that the interference is taken into account.

To verify the equivalency of the two methods let us try to derive the reflectance R of an FPI. We use only one layer of thickness L and refractive index n with reflection coefficients r_1 and r_2 at the front and the rear interface, respectively. The stack matrix S then reads:

$$S = T_1 P_1 T_2 = \frac{1}{t_1 t_2} \begin{pmatrix} 1 & -r_1 \\ -r_1 & 1 \end{pmatrix} \begin{pmatrix} v\,\exp[i\frac{2\pi}{\lambda_0}Ln] & 0 \\ 0 & \frac{1}{v}\,\exp[-i\frac{2\pi}{\lambda_0}Ln] \end{pmatrix} \begin{pmatrix} 1 & -r_2 \\ -r_2 & 1 \end{pmatrix}$$

Equation (4.29) yields for the reflectance of the FPI:

$$R = \left| \frac{r_1 - r_2 v^2 \exp[i4\pi nL/\lambda_0]}{1 - r_1 r_2 v^2 \exp[i4\pi nL/\lambda_0]} \right|^2 \tag{4.31}$$

This is equivalent to the square of the reflection coefficient determined by the wave sum (4.3), with $k=2\pi n/\lambda_0$.

The calculation of the reflectance R for multilayer coatings using the stack matrix method can become quite a task if it is done manually. It is much more convenient (especially because one gets the same result when the calculation is repeated!) to use a computer to perform the algebra. Input parameters of such a coating design program are the layer thicknesses, their indices of refraction and the angle of incidence of the light. The reflection coefficients r_{ij} are given by the Fresnel equations [1.1,1.2]. For a transition from layer i to layer j with normal incidence, the reflection coefficient does not depend on the polarization of the incident beam. The Fresnel equations yield:

$$r_{ij} = \frac{n_i - n_j}{n_i + n_j} \tag{4.32}$$

Note that r_{ij} is negative if the beam is reflected off the higher index material. If the normal of the substrate is tilted by an angle θ_A with respect to the wave vector of the incident beam (Fig. 4.13), we first have to calculate the angle of refraction θ_j for each layer j by using Snell's law. If n_A denotes the index of refraction of the outside medium (usually this is air), the angles θ_j are determined by:

$$\cos\theta_j = \sqrt{1 - \frac{n_A^2}{n_j^2}\sin^2\theta_A} \tag{4.33}$$

The propagation matrix P_j for each layer then has the form:

$$P_j = \begin{pmatrix} v_j \exp[i\frac{2\pi}{\lambda_0}n_j L_j \cos\theta_j] & 0 \\ 0 & \frac{1}{v_j}\exp[-i\frac{2\pi}{\lambda_0}n_j L_j \cos\theta_j] \end{pmatrix} \tag{4.34}$$

The reflection coefficients r_{ij}, to be inserted into the transition matrix T_{ij}, are different for s-polarized and p-polarized light.

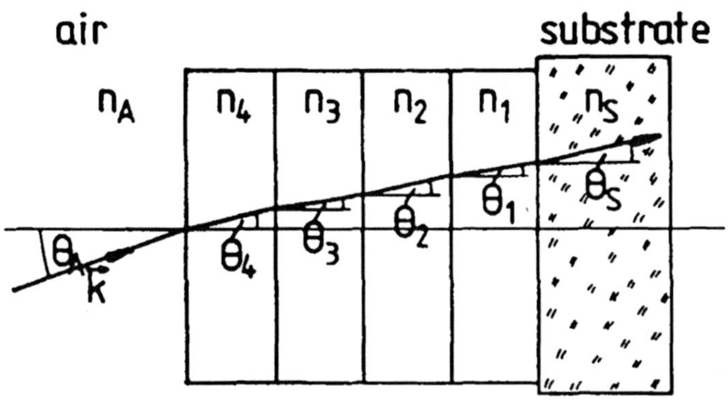

Fig. 4.13 Angles of refraction in each layer for incidence at an angle θ_A.

The coefficients are given by the Fresnel equations:

$$s\text{-}polarized: \qquad r_{ij} = \frac{n_i \cos\theta_i - n_j \cos\theta_j}{n_i \cos\theta_i + n_j \cos\theta_j} \tag{4.35}$$

$$p\text{-}polarized: \qquad r_{ij} = \frac{n_j \cos\theta_i - n_i \cos\theta_j}{n_j \cos\theta_i + n_i \cos\theta_j} \tag{4.36}$$

The stack matrices for the two polarizations have to be calculated separately. Tilted multilayer coatings generally exhibit different reflectances R for the two polarizations.

4.3.2 Quarter Wavelength Systems

Single λ/4 Layer

We consider one dielectric layer with index of refraction n_l and geometrical thickness L_l. The index of refraction of the substrate and the outer medium are n_S and n_A, respectively. According to (4.32), the reflection coefficients are given by:

$$r_{Sl} = \frac{n_S - n_1}{n_S + n_1} \quad , \qquad r_{1A} = \frac{n_1 - n_A}{n_1 + n_A}$$

Application of (4.29) to the stack matrix results in the reflectance:

$$R = \frac{n_1^2(n_s-n_A)^2\cos^2\delta + (n_1^2-n_Sn_A)^2\sin^2\delta}{n_1^2(n_s+n_A)^2\cos^2\delta + (n_1^2+n_Sn_A)^2\sin^2\delta} \tag{4.37}$$

with: $$\delta = \frac{2\pi}{\lambda_0}n_1L_1$$

If we choose an optical thickness of a quarter wavelength for the design wavelength λ_0 ($L_l=\lambda_0/(4n_l)$), the reflectance assumes its maximum of:

$$R_{max} = \left[\frac{n_1^2 - n_An_S}{n_1^2 + n_An_S}\right]^2 \tag{4.38}$$

For such a $\lambda/4$ layer, the reflectance as a function of the wavelength varies between this maximum value (at the wavelengths $\lambda_0/(2q+1)$) and the reflectance of the bare substrate (at the wavelengths $\lambda_0/(2q)$) given by:

$$R_{min} = \left[\frac{n_S - n_A}{n_S + n_A}\right]^2 \tag{4.39}$$

For the latter wavelengths, the layer represents a $\lambda/2$ coating which does not affect the reflectance of the substrate. Due to dispersion (the refractive indices increase as the wavelength is decreased), the reflectance maxima are shifted towards longer wavelengths. Figure 4.14 shows the transmission spectrum for a single layer of Ta_2O_5 with a geometrical thickness of 1050nm ($n_l=2.1$), coated on a BK7 glass substrate ($n_S=1.5$). The outside medium is air ($n_A=1.0$). This coating represents a quarter wavelength layer for λ_0=8,820nm. If there were no dispersion, we would also expect the maximum reflectance of $R_{max}=0.242$ at the wavelengths ($q=2$-5): 1,764nm, 1,260nm, 980nm, and 802nm. Considering the fact that the transmission curve includes the reflectance of the rear substrate interface as well as absorption losses in the substrate, the measured maximum reflectance is in good agreement with (4.38). Note the spectral shifts of the transmission minima at shorter wavelengths due to dispersion. Measured dispersion curves for different coating materials and coating conditions are presented in Fig. 4.15.

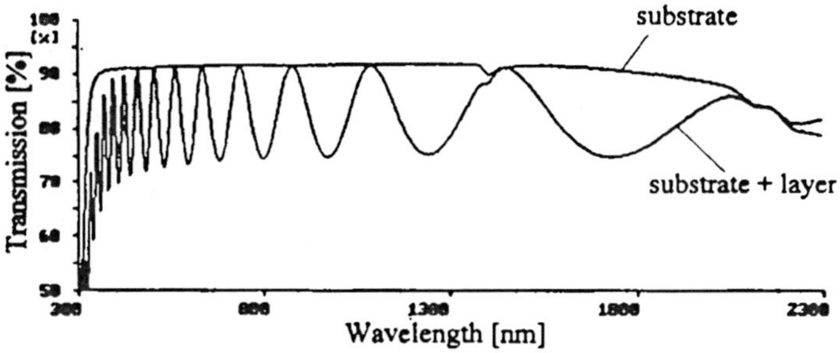

Fig. 4.14 Transmission spectrum for a single layer of Ta_2O_5 (n=*2.1*) on a BK7 substrate (n_S=*1.5*). The geometrical layer thickness is *1,050nm*. The transmission spectrum of the plain substrate is also shown for comparison. The layer was deposited using ion beam sputtering with argon [S.1].

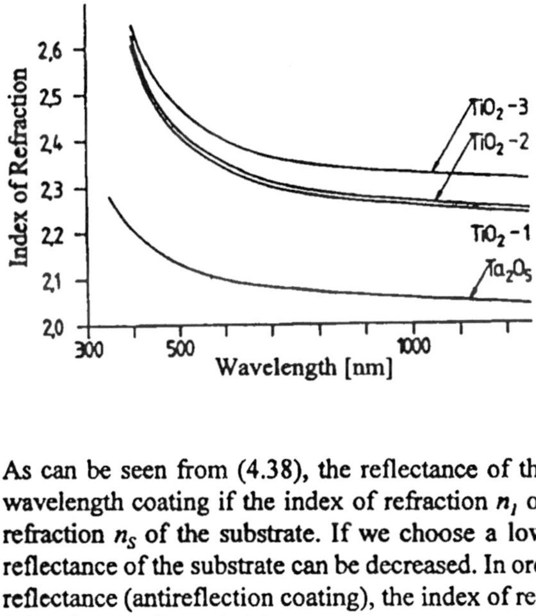

Fig. 4.15 Measured index of refraction of single layers of TiO_2 and Ta_2O_5 with an optical thickness of 2,120nm at a wavelength of λ_0=1,060nm (deposited with ion beam sputtering). Equation (4.37) was used to determine the refractive index from the spectral transmission minima. The three TiO_2 layers were coated using different source materials and different O_2 atmospheres [S.1].

As can be seen from (4.38), the reflectance of the substrate is increased by the quarter wavelength coating if the index of refraction n_l of the layer is greater than the index of refraction n_S of the substrate. If we choose a lower index material for the coating, the reflectance of the substrate can be decreased. In order to completely suppress the substrate reflectance (antireflection coating), the index of refraction n_l must be chosen such that the following condition holds:

$$n_1 = \sqrt{n_A n_S} \tag{4.40}$$

For a glass substrate (n_S=*1.5*) in air, we thus need an index of refraction of n_l=*1.225*. The coating material exhibiting an index close to this value is magnesium fluoride (n_l=*1.37 at 600nm*). A MgF_2 coating lowers the reflectance of a glass/air interface from 4% to about 1%.

Multiple λ/4 Layers

Let k be the number of λ/4 layers with geometrical thickness:

$$L_j = \frac{\lambda_0}{4 n_j(\lambda_0)}$$

deposited onto the substrate. Note that in the notation we use, the layer with refractive index n_1 and thickness L_1 is always the one in contact with the substrate. The reflectance R at the wavelength λ_0 is given by:

$$R = \left[\frac{(n_1 n_3 n_5 \ldots n_{k-1})^2 n_A - n_S (n_2 n_4 \ldots n_k)^2}{(n_1 n_3 n_5 \ldots n_{k-1})^2 n_A + n_S (n_2 n_4 \ldots n_k)^2}\right]^2 \qquad \text{for } k \text{ even},\ k \geq 2 \tag{4.41}$$

$$R = \left[\frac{(n_1 n_3 n_5 \ldots n_k)^2 - n_S n_A (n_2 n_4 \ldots n_{k-1})^2}{(n_1 n_3 n_5 \ldots n_k)^2 + n_S n_A (n_2 n_4 \ldots n_{k-1})^2}\right]^2 \qquad \text{for } k \text{ odd},\ k \geq 3 \tag{4.42}$$

Usually, alternating layers of two coating materials - a high index material ($n_H = 1.9\text{-}4.2$) and a low index material ($n_L = 1.35\text{-}1.46$) with $n_L < n_S < n_H$ - are evaporated onto the substrate. The above equations can then be simplified:

$$R = \left[\frac{n_1^k n_A - n_S n_2^k}{n_1^k n_A + n_S n_2^k}\right]^2 \qquad \text{for } k \text{ even},\ k \geq 2 \tag{4.43}$$

$$R = \left[\frac{n_1^{k+1} - n_S n_A n_2^{k-1}}{n_1^{k+1} + n_S n_A n_2^{k-1}}\right]^2 \qquad \text{for } k \text{ odd},\ k \geq 3 \tag{4.44}$$

For a multilayer antireflection coating, the first layer deposited onto the substrate has to be the high index material (with refractive index $n_1 = n_H$). From (4.43) and (4.44) we get the conditions for zero reflectance:

$$\left(\frac{n_1}{n_2}\right)^k = \frac{n_S}{n_A}, \quad k \text{ even} \tag{4.45}$$

$$\left(\frac{n_1}{n_2}\right)^{k+1} = \frac{n_A n_S}{n_2^2}, \quad k \text{ odd} \tag{4.46}$$

These conditions are much easier to meet than the single layer condition (4.40). If a large number of multiple quarter wavelength layers are used, the reflectance of the substrate will be enhanced and will eventually reach 100%. In order to attain the desired reflectance with a minimum number of layers, it is advantageous to choose coating materials which exhibit a large difference in their index of refraction. Resonator mirrors for lasers emitting in the visible and near infrared spectral range are usually produced by applying up to 15 alternating dielectric layers on glass or quartz substrates. Figure 4.16 shows how the reflectance of the substrate at the design wavelength λ_0 varies with an increasing number of layers for different dielectric materials. Spectrally resolved reflectance curves for case c) of this figure are presented in Fig. 4.17 for a center wavelength $\lambda_0 = 460nm$.

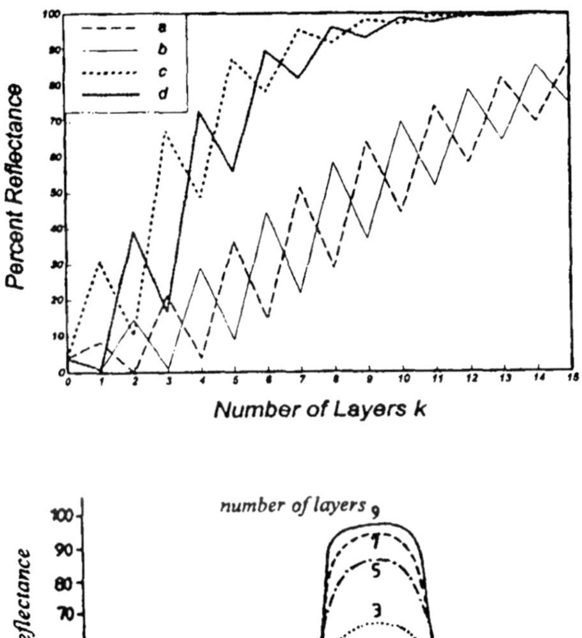

Fig. 4.16 Reflectivity of multilayer quarter wavelength coatings as a function of the number of layers, calculated with (4.43) and (4.44). Alternating coating materials are deposited on a glass substrate ($n_S = 1.5$): a) $n_1 = 1.65$, $n_2 = 1.35$; b) $n_1 = 1.35$, $n_2 = 1.65$; c) $n_1 = 2.3$, $n_2 = 1.35$; d) $n_1 = 1.35$, $n_2 = 2.3$.

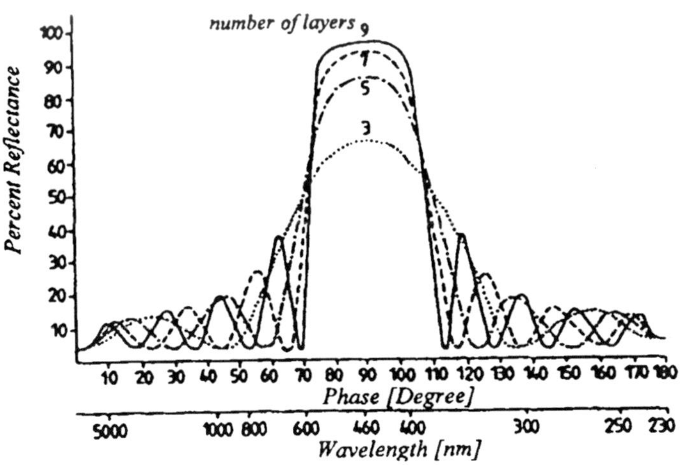

Fig. 4.17 Calculated reflectance as a function of the phase $2\pi Ln/\lambda$ for odd numbers of quarter wave layers with $n_1 = 2.3$, $n_2 = 1.35$, and $n_S = 1.5$ (high index material deposited first). The lower axis shows the corresponding wavelengths assuming a design wavelength of $\lambda_0 = 460nm$.

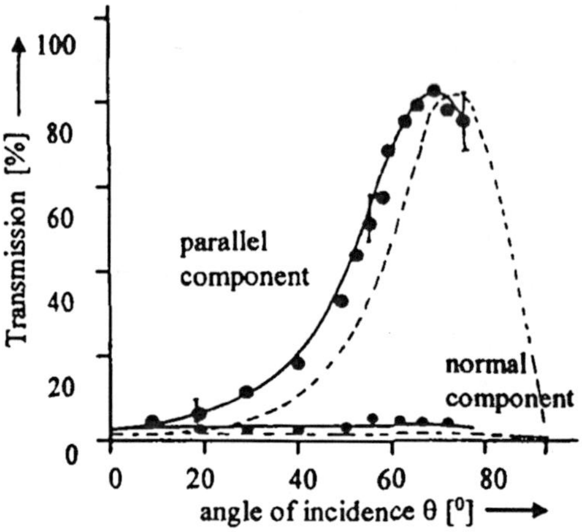

Fig. 4.18 Transmission of a dielectric polarizer vs. the angle of incidence for λ=534nm. Layer structure: five layers of low and five layers of high index material. The broken line represents the calculated transmission.

Dielectric plates can be used to polarize light, as was discussed in Sec. 3.2.1. The degree of polarization increases with the number of stacked plates. The stackplate polarizer can be replaced by a system of thin layers resulting in very efficient and compact polarizers, The calculation of the reflectance of such a thin film polarizer can be done numerically using Eqs. (4.33-4.36). An example is shown in Fig. 4.18.

4.3.3 Coating Methods and Materials

There are different methods to deposit dielectric coatings on a substrate. The most common one is the **evaporation** of the dielectric material onto the substrate in a vacuum chamber at typical pressures of 10^{-6} torr (Fig.4.19). The substrates are mounted on a carousel that rotates on top of the chamber to achieve a homogeneous deposition. The vapor of the coating material is generated at the bottom of the chamber by heating the source material (in the form of a granulate) with an electron beam or, in older machines, by inductance or resistance heating. The thickness of the coating is monitored during the process by analyzing the spectral reflectance of a test substrate. Since evaporation is a thermal process, the low kinetic energy (about 50meV) of the atoms is not high enough to generate a homogeneous coating. The coatings exhibit a certain porosity which leads to a lower refractive index compared to the bulk material and also to a decreased damage threshold. By heating the substrate to temperatures of about 300°C the atoms get additional energy to move on the substrate and thus distribute themselves more evenly. Electron-beam evaporation generates optical coatings with the highest damage threshold to laser radiation. High quality, high reflectivity mirrors can withstand fluences of over $4J/cm^2$ at 355nm and over $25J/cm^2$ at 1064nm (for 10ns long laser pulses). In cw-operation at 1064nm, high reflectors can withstand power densities of more than $5MW/cm^2$.

A higher homogeneity of the coatings can be attained by providing additional energy to the coating atoms. In **ion assisted deposition** techniques, the substrate is bombarded during the deposition with an ion beam transferring additional energy on the order of 100eV per ion to the growing coating. A process of similar energy is the **ion beam sputtering**. A beam of inert ions is aimed at the target and the target atoms are kicked out of the material and land on the substrate with a high energy. Both of these energetic processes generate layers with bulk-like indices of refraction, resulting in a better durability of the coating. The damage threshold, however, is lower as compared to e-beam evaporation, due to intrinsic stress generated in the optical layers.

Table 4.3 Physical properties of coating materials used for laser mirrors. The indices of refraction are those obtained in thermal deposition processes (electron beam evaporation). The transparent spectral range is defined by an absorption coefficient of less than 10^3 per cm [2.4].

Material	index of refraction (at λ_0=500nm)	transparent spectral range
Na_3AlF_6	1.35	0.20 - 14μm
MgF_2	1.37	0.22 - 2 μm
SiO_2	1.46	0.20 - 8 μm
HfO_2	1.95	0.22 - 12μm
ZrO_2	2.10	0.34 - 12μm
Ta_2O_5	2.16	0.30 - 10μm
TiO_2	2.25	0.35 - 12μm
ZnS	2.55	0.38 - 25μm

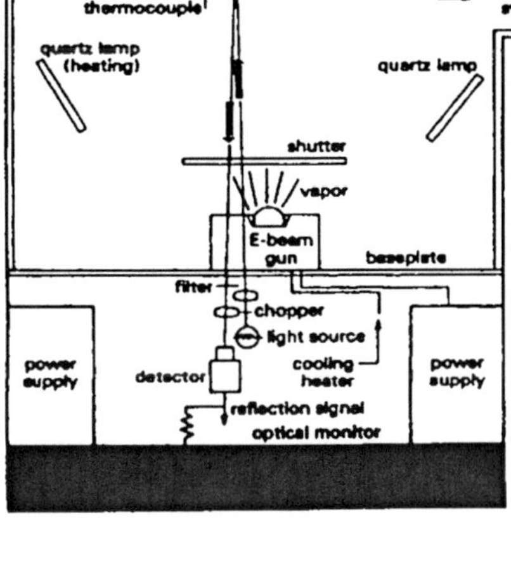

Fig. 4.19 Schematic view of a typical vacuum deposition chamber with electron beam heating [S.2] (courtesy of Melles-Griot, Irvine, CA). The electron beam is aimed at the coating material contained in water-cooled crucibles. The crucibles are located to the side of the electron beam gun (not shown).

Part III

Passive Open Resonators

Chapter 5

Stable Resonators

5.1 General Aspects

In this part the basic properties of both stable and unstable optical resonators with spherical mirrors are explained. We neglect the presence of an active medium inside the resonator and assume that both resonator mirrors exhibit 100% reflectance at the wavelengths considered. Resonators that do not provide amplification of the light are called passive resonators. Although the active medium is required to generate laser emission, the concept of the passive resonator is applicable to the investigation of the physics of laser radiation. The inclusion of the gain only modifies the resonator properties. The influence of the amplification upon the radiation characteristics of resonators will be discussed in Part IV.

A stable optical resonator generally consists of two mirrors with radii of curvature ρ_1 and ρ_2 separated by an optical distance $L=nL_0$ (L_0: geometrical mirror spacing, n: index of refraction inside the resonator). The range of L within which a resonator is stable is determined by the condition that a ray launched inside the resonator parallel to the optical axis remains inside the resonator after an infinite number of bounces. Equivalent to this definition is the lack of self-reproducing spherical waves inside the resonator (eigensolutions of the ABCD law, see Sec. 1.3).

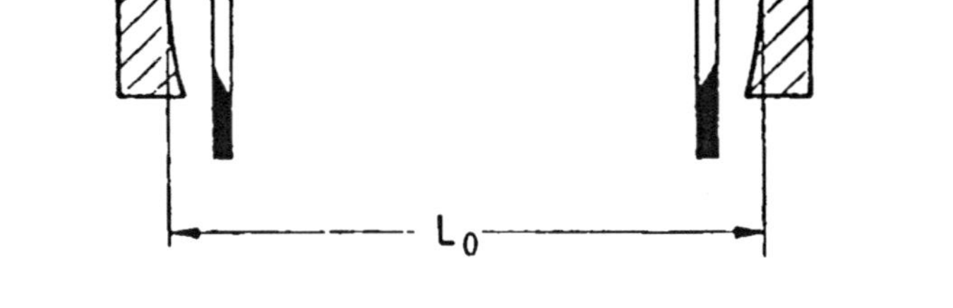

Fig. 5.1 The general optical resonator is determined by the g-parameters of the mirrors, the mirror spacing L_0 and the radii a_i of internal apertures.

By introducing the g-parameters of the resonator mirrors:

$$g_i = 1 - \frac{L}{\rho_i}, \quad i=1,2 \tag{5.1}$$

the condition for a stable resonator reads:

$$0 < g_1 g_2 < 1 \tag{5.2}$$

As discussed in Sec. 2.8.2, stable resonators exhibit a Gaussian beam as the fundamental eigenmode. Note that the radius of curvature is positive for a concave mirror and negative for a convex mirror. In Fig. 5.1 both radii of curvature are positive. It is convenient to visualize optical resonators in the g-diagram, also referred to as the stability diagram, in which a resonator is determined by a point in the $g_1 g_2$ plane (Fig. 5.2). The area of stable resonators is limited by the coordinate axes and the hyperbolas $g_2 = \pm 1/g_1$. The resonators on the stability limits, represent a unique class of resonators since the Gaussian beam is not an eigensolution of the electric field. The exception is the confocal resonator with $g_1 = g_2 = 0$ which is usually considered a stable resonator. However, since it exhibits some properties that are quite unique compared to common stable resonators, we will discuss the stable confocal resonator as well as resonators on the stability limits in a later section.

An optical resonator is generally defined by the g-parameters of the mirrors, the mirror spacing, and the dimensions of apertures that might be located inside the resonator. The goal of this chapter is to achieve a detailed understanding of how the mode structure and the diffraction losses are affected by these resonator parameters. We will first assume that there is no aperture inside the resonator and both mirrors have infinite lateral dimensions.

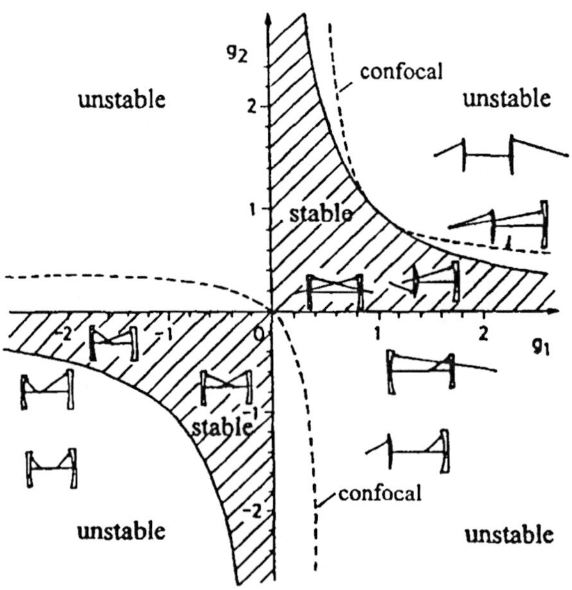

Fig. 5.2 The stability diagram of optical resonators with spherical mirrors. The hatched area indicates the region of stable resonators.

5.2 Unconfined Stable Resonators

We first want to know which field distributions on the resonator mirrors represent the steady-state solutions of the stable resonator. Such an eigensolution $E_i(x,y)$ on mirror i will reproduce itself after each round trip. The round trip in the resonator is described mathematically by the Kirchhoff integral equation derived in Sec. 2.7.1 [3.7,3.10,3.16-3.20,3.24,3.25]:

$$\gamma E_i(x_2,y_2) = i\frac{\exp[-ikL]}{2Lg_j\lambda_0}\int\int E_i(x_1,y_1)\exp\left[\frac{-i\pi}{2Lg_j\lambda_0}\left(G(x_1^2+y_1^2+x_2^2+y_2^2)-2(x_1x_2+y_1y_2)\right)\right]dx_1dy_1$$

with: $G = 2g_1g_2-1;\ \ i,j=1,2;\ \ i\neq j$ (5.3)
L : optical resonator length $=nL_0$ (L_0: geometrical length)
λ_0: vacuum wavelength, $k=2\pi/\lambda_0$: wave number

The solutions to this integral equation represent the eigenmodes of the optical resonator. In general an infinite number of eigenmodes exist. The field distributions of the eigenmodes do not change their shapes but they might experience a decrease in amplitude due to diffraction losses. This is taken into account by the complex eigenvalue γ. The loss factor,

$$V = |\gamma|^2 , \qquad (5.4)$$

represents the fraction of the initial power hitting the mirror after the round trip. The loss factor V is related to the loss ΔV via $\Delta V=1-V$. If both resonator mirrors are unconfined and perfectly reflecting, no power is lost during the round trip and the loss factor, therefore, is equal to 1.0. The condition $\gamma=1.0$, referred to as the *resonance condition* of the optical resonator, yields the resonance frequencies.

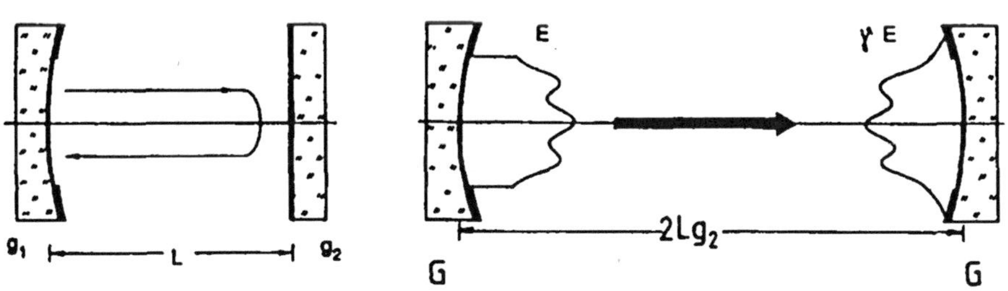

Fig. 5.3 A steady-state field distribution *E(x,y)* must reproduce itself after each round trip. On the right side the round trip is presented as a transit in the equivalent resonator (see Sec. 5.3.2).

5.2.1 Transverse Mode Structures

Equation (5.3) can be solved analytically yielding an infinite number of eigensolutions. Which of these eigensolutions will actually be observed in the resonator depends on the geometry and the size of the mirrors. In reality, the mirrors will have a finite size with a shape that usually is round or rectangular. This boundary condition is taken into account by choosing those eigensolutions that exhibit circular or rectangular symmetry, respectively. The solutions to (5.1) for the two symmetries are described below (the index i denotes the mirror) [3.1,3.5,3.7]:

a) Circular Symmetry (Gauss-Laguerre Modes):

$$E_{p\ell}^{(i)}(r,\Phi) = E_0 \left[\frac{\sqrt{2}r}{w_i}\right]^{\ell} L_p^{(\ell)}\left(\frac{2r^2}{w_i^2}\right) \exp\left[\frac{-r^2}{w_i^2}\right] \cdot \left\{\begin{matrix}\cos(\ell\Phi)\\ \sin(\ell\Phi)\end{matrix}\right. \tag{5.6}$$

$$\gamma = \exp\left[ik\left(2L - \frac{\lambda_0}{\pi}(2p+\ell+1)\, acos\sqrt{g_1 g_2}\right)\right] \tag{5.7}$$

with	$L_p^{(\ell)}[t]$	:	Laguerre polynomial of order p, ℓ; p, ℓ: integer
	r, Φ	:	radial and azimuthal coordinate
	k	:	wave number = $2\pi/\lambda_0$; λ_0: wavelength in vacuum
	L	:	optical mirror spacing = nL_0

The Laguerre polynomials can be found in mathematical handbooks [3.13,3.14]. For low orders p,ℓ they read:

$$L_0^{(\ell)}(t) = 1$$

$$L_1^{(\ell)}(t) = \ell+1 - t$$

$$L_2^{(\ell)}(t) = 0.5(\ell+1)(\ell+2) - (\ell+2)t + 0.5t^2$$

$$L_3^{(\ell)}(t) = (\ell+1)(\ell+2)(\ell+3)/6 - 0.5(\ell+2)(\ell+3)t + 0.5(\ell+3)t^2 - t^3/6$$

b) Rectangular Symmetry (Gauss-Hermite Modes):

$$E_{mn}^{(i)}(x,y) = E_0 \exp\left[\frac{-(x^2+y^2)}{w_i^2}\right] H_m\left(\frac{\sqrt{2}x}{w_i}\right) H_n\left(\frac{\sqrt{2}y}{w_i}\right) \tag{5.8}$$

$$\gamma = \exp\left[ik\left(2L - \frac{\lambda_0}{\pi}(m+n+1)\, acos\sqrt{g_1 g_2}\right)\right] \tag{5.9}$$

The Hermite polynomials H_m can also be looked up in mathematical handbooks. For low order numbers *m,n* they read:

$$H_0(t) = 1 \qquad H_1(t) = 2t$$

$$H_2(t) = 4t^2 - 2 \qquad H_3(t) = 8t^3 - 12t$$

$$H_4(t) = 16t^4 - 48t^2 + 12 \qquad H_5(t) = 32t^5 - 160t^3 + 120t$$

In both symmetries the loss factor V is equal to 1.0 since the mirrors are not limited by an aperture and, consequently, no power can leak out of the resonator. Figure 5.4 presents intensity distributions as a function of the order numbers p, ℓ and m, n, calculated with (5.6) and (5.8), respectively.

The steady-state electric field distribution is characterized by the indices *plq* and *mnq*, whose meaning becomes apparent by looking at Fig. 5.4. In rectangular symmetry the first two indices represent the number of nodal lines of the intensity distribution in the corresponding direction. In circular symmetry the intensity distributions exhibit p radial and ℓ azimuthal nodes at which the intensity is equal to zero. The index q was already discussed in Chapter 4; it represents the number of half wavelengths fitting into the mirror spacing.

A steady-state field distribution oscillating inside the resonator is called an eigenmode of the resonator. The eigenmodes are characterized by the transverse mode structure (transverse mode index p, ℓ or m,n) and the axial mode order q. The notations for the eigenmodes are

$$TEM_{p\ell q} \qquad \text{and} \qquad TEM_{mnq}$$

where the abbreviation *TEM* represents the fact that the electric and the magnetic field vectors are perpendicular to each other and to the wave vector k (Transverse Electro Magnetic). This is not entirely true since the diffraction generates small field components in the direction of the wave propagation(see Sec. 2.9). Only in the limit of large beam radii (large Fresnel numbers) is the field truly transverse. The notation TEM was adopted from the modes of waveguides which in fact are transversal. Although not physically correct, it is customary to refer to the modes of open resonators as TEM modes as well. In general, the axial mode index q is not used and the mode structure is specified by the order numbers p, ℓ and m, n (as we have seen, q can be very large).

The lateral extent of the eigenmodes on mirror i is determined by the beam radius w_i of the TEM_{00} mode. The beam radii w_1 and w_2 on the two mirrors depend on the mirror spacing and the g-parameters of the resonator:

$$w_i^2 = \frac{\lambda L}{\pi} \sqrt{\frac{g_j}{g_i(1 - g_1 g_2)}} \; ; \quad i,j = 1,2 \; ; \quad i \neq j \tag{5.10}$$

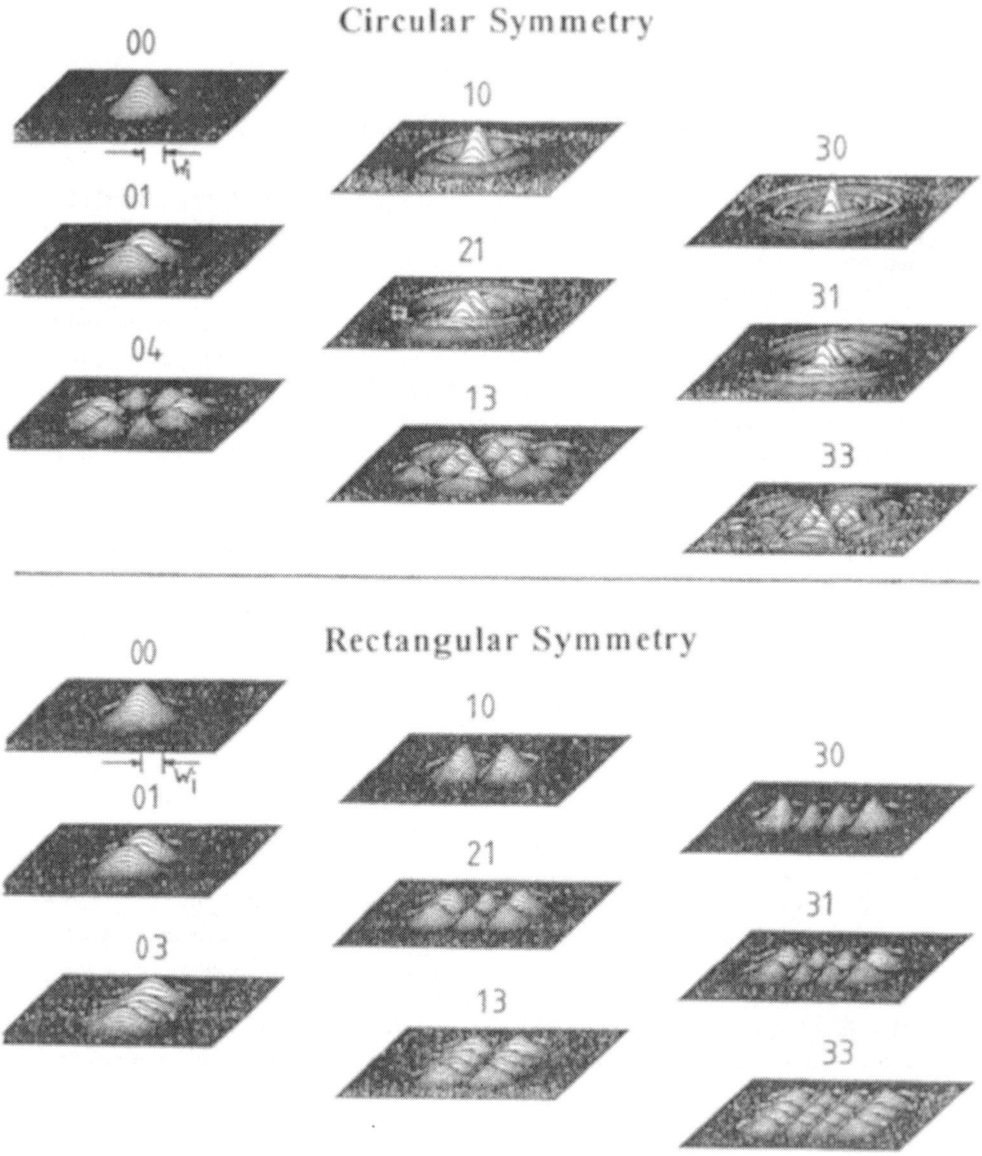

Fig. 5.4 Intensity distributions on the mirrors of stable resonators calculated with (5.6) and (5.8) for different mode orders $p\ell$ and mn. The radius w_I is the same for all distributions [S.3].

As can be seen in Fig. 5.4, the size of the intensity distributions increases as the mode order is increased. In both symmetries the TEM_{00} mode has the same shape (Fig. 5.5); the intensity distribution is Gaussian. At a distance $r=w_i$ from the center of gravity, the intensity has decreased by a factor of $1/e^2$ and 86.5% of the beam power is contained within the corresponding circle. The lateral extent of the TEM_{00} mode is therefore defined by the beam radius w_i, also referred to as the Gaussian beam radius. The TEM_{00} mode is the mode with the smallest size that can oscillate in a stable resonator. It is generally referred to as the *fundamental mode* or the Gaussian beam.

For transverse modes of higher order $p\ell$ or mn, the beam radii are defined via the second intensity moments introduced in Chapter 2.6 (see Eqs. (2.93) and (2.97)). This definition of the beam radii enables one to calculate the propagation of arbitrary field distributions through ABCD-type optics by applying the generalized ABCD law (2.111) (see Sec. 2.6).

The beam radii on mirror i read [3.31,3.32]:

Circular Symmetry: $$w_{pl}^{(i)} = w_i\sqrt{2p + l + 1} \tag{5.11}$$

Rectangular Symmetry: $$w_{mn}^{(i)} = w_i\sqrt{2m + 1} \quad x\text{-direction} \tag{5.12}$$

$$w_{mn}^{(i)} = w_i\sqrt{2n + 1} \quad y\text{-direction} \tag{5.13}$$

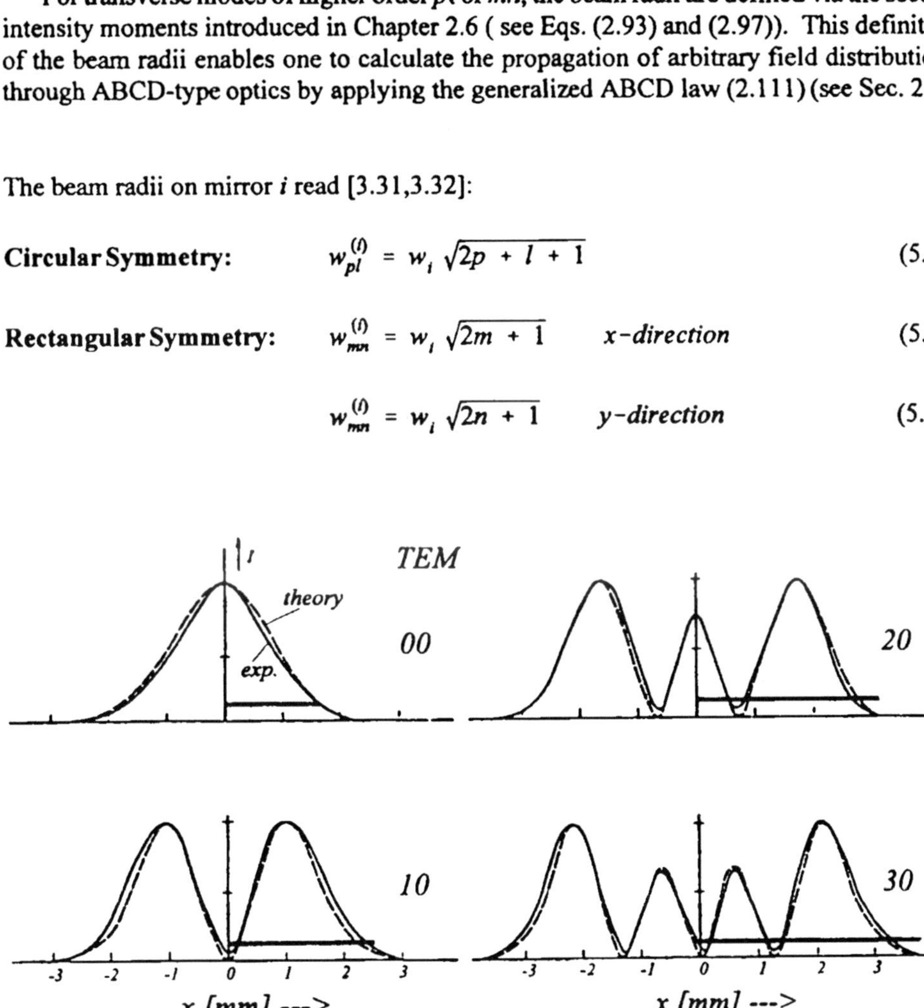

Fig. 5.5 Measured (solid line) and calculated (dashed line) one-dimensional intensity distributions of transverse modes TEM_{mn} in rectangular symmetry. The horizontal lines indicate the beam radii w_{m0} defined by the second intensity moments.

Note that, as far as the power content is concerned, the second intensity moments yield beam radii that are too large, especially for modes in rectangular symmetry. For higher order modes the power fraction contained within the radius $w_{pl}^{(i)}$ is greater than 86.5%. Table 5.1 shows the difference between the beam radii and the 86.5% power content radii for different mode orders p,ℓ and m,n. For some laser applications it may be more applicable to define the beam radii via the 86.5% energy content. However, since the beam propagation can be calculated for arbitrary field distributions with the ABCD law (2.111), the definition of the beam radius using the second intensity moments was standardized by ISO for the measurement of beam quality, beam diameter and beam divergence [1.86]. As far as the modes of stable resonators are concerned, the generalized ABCD law can also be applied to the power content radii (see Fig. 2.20) provided that the far field angle is defined the same way.

In order to get a better agreement with the power content radii, we modify the beam radii for rectangular symmetry modes by dividing by √2:

$$w_{mn}^{(i)} = w_i \sqrt{m+1/2} \qquad x\text{-direction}$$

Table 5.1 Calculated ratio of 86.5% power content beam radii $r_{86.5\%}$ to the beam radii w_{pl} defined via the second intensity moment for transverse modes of stable resonators in circular and rectangular symmetry (Eqs. (5.11),(5.12)). For comparison: a homogeneous circular intensity profile with radius R yields a second moment radius of $w_{pl}=R$ and an 86.5% power content radius of $r_{86.5\%}=0.93R$ (ratio of 0.93). The radius ratio for a homogeneous intensity profile of width $2a$ is $(0.865a)/(2a/\sqrt{3})=0.749$.

p	ℓ	$r_{86.5\%}/w_{pl}$	p	ℓ	$r_{86.5\%}/w_{pl}$
0	0	1.000	0	7	0.853
1	0	0.960	0	10	0.813
4	0	0.960	0	100	0.746
7	0	0.960	0	∞	1/√2
∞	0	0.960	1	10	0.865
0	1	0.944	4	10	0.920
0	4	0.857	7	10	0.941

m	$x_{86.5\%}/w_{mn}$	m	$x_{86.5\%}/w_{mn}$
0	0.740	3	0.667
1	0.678	4	0.665
2	0.670	5	0.664

If we use the modified beam radii in rectangular symmetry, the ratio of the 86.5% power content radius to the modified beam radius is increased to about *0.94* for higher order modes and to *1.05* for the fundamental mode. However, we have to apply the same correction factor of $1/\sqrt{2}$ to the angles of divergence in order to preserve the validity of the generalized ABCD law (2.111). The modified beam radius gives the position of the outermost inflection point of the intensity distribution. To be consistent in our notation for the beam radii, we will use $w_{00}^{(i)}$ instead of w_i for the beam radius of the fundamental mode (Gaussian beam) on mirror *i*. Let us first look at some examples to get a feeling for the transverse mode size of stable resonators.

Examples:

a) HeNe laser, circular symmetry, λ=632.8nm, L=1m, $\rho_1=\rho_2$=2m.

$$w_{00}^{(1)} = w_{00}^{(2)} = 0.4823\ mm$$

$$w_{10}^{(1)} = w_{10}^{(2)} = 0.8354\ mm$$

$$w_{11}^{(1)} = w_{11}^{(2)} = 0.9646\ mm$$

b) HeNe laser, circular symmetry, λ=632.8nm, L=1m, ρ_1=5m, $\rho_2=\infty$ (flat).

$$w_{00}^{(1)} = 0.7092\ mm \qquad w_{00}^{(2)} = 0.6437\ mm$$

$$w_{10}^{(1)} = 1.2284\ mm \qquad w_{10}^{(2)} = 1.0992\ mm$$

$$w_{11}^{(1)} = 1.4184\ mm \qquad w_{11}^{(2)} = 1.2692\ mm$$

c) CO_2 laser, circular symmetry, λ=10,600 nm, L=1m, ρ_1=5m, $\rho_2=\infty$m (flat).

$$w_{00}^{(1)} = 2.9026\ mm \qquad w_{00}^{(2)} = 2.5972\ mm$$

$$w_{10}^{(1)} = 5.0275\ mm \qquad w_{10}^{(2)} = 4.4984\ mm$$

$$w_{11}^{(1)} = 5.8052\ mm \qquad w_{11}^{(2)} = 5.1944\ mm$$

The cross sectional area of the Gaussian beam radius scales linearly with the wavelength and the resonator length. It is for this reason that CO_2 lasers with their large emission wavelength of 10.6µm exhibit relatively large beam radii of the fundamental mode. On the other hand, diode lasers have very small Gaussian beam radii on the order of µm due to their short mirror spacing in the sub-mm range. Furthermore, at a fixed wavelength and a fixed resonator length, the Gaussian beam radius will decrease as the resonator design is chosen closer to a stability limit in the g-diagram. This fact is visualized in Fig. 5.6 which presents curves of constant cross sectional area of the Gaussian beam at mirror 1. The beam radius at mirror 2 can be obtained from this graph by switching g_1 and g_2.

The transverse modes that are actually observed in a laser resonator, are mainly determined by the size of the mirrors. The resonator mirrors are generally limited either by apertures or by the active medium itself. Only those transverse modes whose beam radii are smaller than the radii of the mirror apertures can be observed. This means that only modes up to a certain transverse order are allowed to oscillate. Modes with higher transverse orders exhibit losses that are too high to be compensated by the gain medium. If the gain of the medium is increased, more transverse modes will, however, reach the laser threshold. As a rule of thumb, a higher order transverse mode will oscillate if the radius of the aperture a is greater than 0.9-1.0 times the beam radius. Fundamental mode operation requires an aperture radius a of 0.9-1.3 times the Gaussian beam radius. Upper values and lower values correspond to low gain and to high gain active media, respectively.

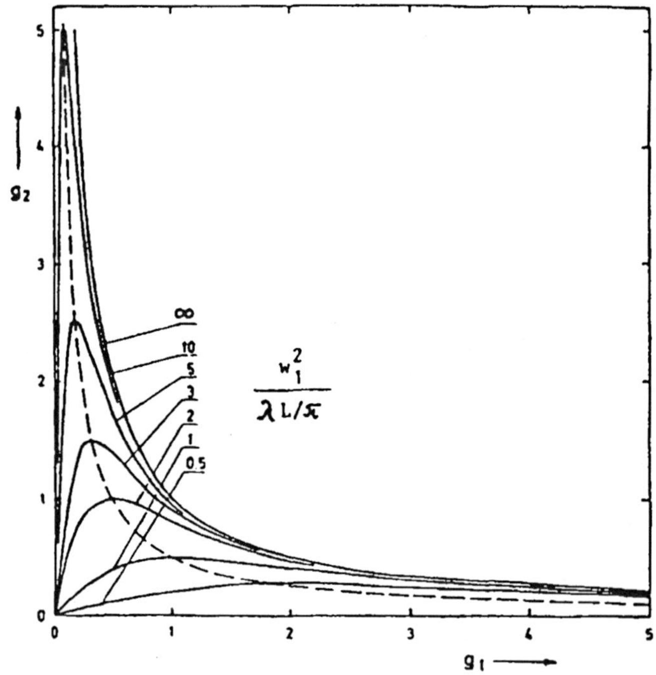

Fig. 5.6 Curves of constant cross sectional area of the Gaussian beam at mirror 1 in the stability diagram. The cross sectional area is normalized to $\lambda L/\pi$. The beam radius at mirror 2 can be obtained by switching g_1 and g_2.

Example:
HeNe laser, circular symmetry, λ*=632.8nm, L=0.3m,* ρ_1*=5m,* ρ_2*=3m.* The mirrors are close to the endfaces of the discharge tube with inner diameter *of 2a=3mm.*

According to (5.10), the Gaussian beam radii on the mirrors are $w_{00}^{(1)}=0.388mm$ and $w_{00}^{(2)}=0.397mm$. Only those transverse modes are observed for which the beam radii at both mirrors are smaller than the tube radius:

$$a \geq 1.0\ w_{00}^{(i)} \sqrt{2p + \ell + 1}$$

This is equivalent to the condition $2p+\ell+1 \leq 14$. The transverse modes with the highest radial and highest azimuthal order oscillating in the HeNe laser are TEM_{60} and TEM_{012}, respectively.

Lasers can be forced to oscillate only in the fundamental mode by inserting apertures into the resonator with a diameter close to the diameter of the Gaussian beam at the aperture plane. If the apertures are considerably larger than the Gaussian beam, all higher transverse modes fitting into the aperture will oscillate. It is for this reason that in multimode lasers one does generally not observe the characteristic intensity distributions of Fig. 5.4. Since all of these modes exhibit virtually no loss, they are oscillating simultaneously resulting in a more or less homogenous beam profile. The more modes that participate in this process the more homogenous the laser beam becomes. This multimode behavior is supported by the fact that different modes exhibit their intensity maxima in different areas of the active medium. The gain at the nodal lines of the intensity distribution is not depleted and can then be used by a different mode that has its intensity peaks in these vacant areas. Figure 5.7 presents a photograph of the intensity distribution at the output coupling mirror of a Nd:YAG rod laser. The large rod diameter of 10mm enables all transverse modes with $2p+\ell+1<60$ to oscillate. The high number of modes results in a good homogeneity of the beam.

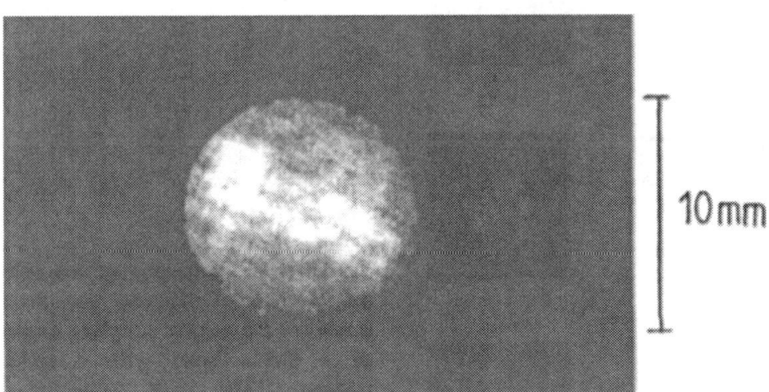

Fig. 5.7 Photograph of the beam profile of a Nd:YAG rod laser ($\lambda=1.064\mu m$) with a stable resonator in multimode operation (maximum of $2p+\ell+1$ is about 60).

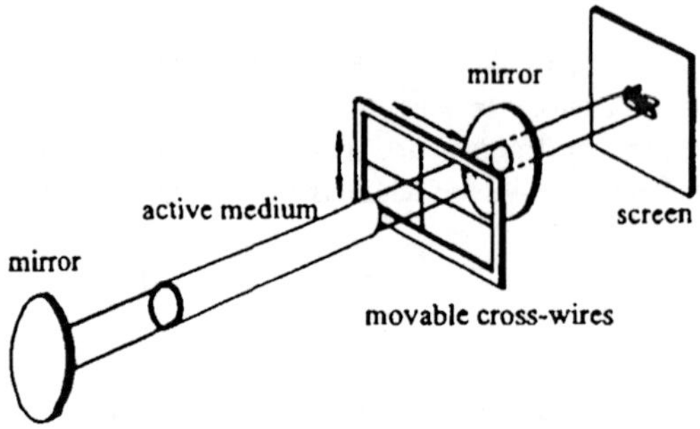

Fig. 5.8 Set-up to observe the transverse intensity distributions of individual transverse modes in a multimode laser.

It is possible to observe the intensity structure of individual transverse modes by generating losses for all other modes. One technique to accomplish this is to insert cross-wires into the resonator as depicted in Fig. 5.8. The mode having nodal lines along the wires is preferred since it experiences lower losses than all other modes. By moving the cross wires, different transverse modes can be selected. Due to the symmetry of the obstruction, only modes with rectangular symmetry are observed. The intensity distributions of the modes of a HeNe laser shown in the photographs of Fig. 5.9 were generated this way.

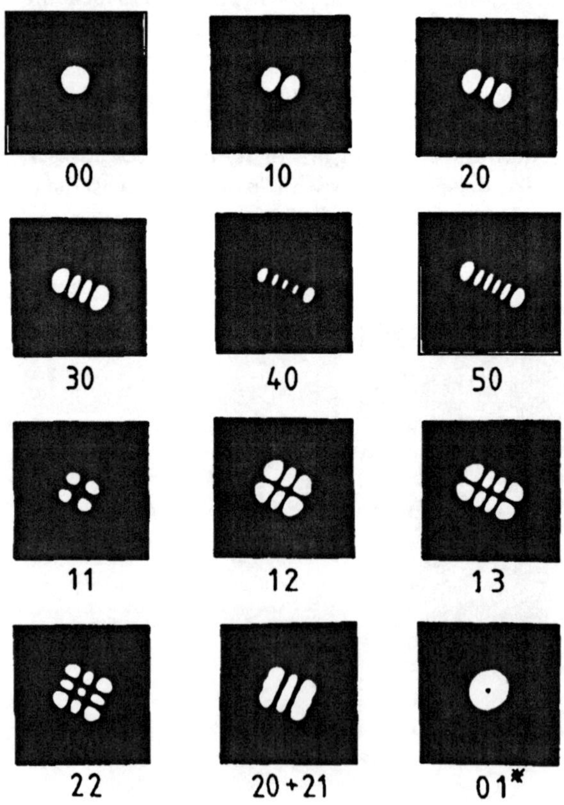

Fig. 5.9 Photographs of intensity distributions of different transverse modes at the output coupling mirror of a HeNe laser with a stable resonator. The cross- wires select modes with rectangular symmetry. The lower right photograph shows the lowest order donut mode (without cross-wires).

Hybrid Modes

In circularly symmetric laser resonators one can quite often observe intensity distributions of modes that exhibit an annular intensity profile with almost zero intensity in the center. These beam profiles are generated by a superposition of two circularly symmetric transverse modes of the same order $p\ell$ which are both linearly polarized and oscillate rotated by an angle of 90° with respect to each other (Fig. 5.10). There are four different ways to combine two linearly polarized modes resulting in different polarization states of the sum mode. The superposition always yields the same annular intensity profile with $p+1$ rings with maximum intensity and a characteristic hole in the center. It is quite obvious why these modes are called donut modes. Sometimes they are also referred to as hybrid modes. Hybrid modes are marked by an asterisk next to the mode order numbers. Due to the different polarizations, the two transverse modes do not interfere and the intensity distribution of the sum is given by the sum of the individual mode intensity profiles. Application of (5.6) yields for the radial intensity distribution of hybrid modes at mirror i:

$$I_{p\ell^*}(t) = I_0\ t^{\ell} \exp[-t]\ [L_{p\ell}(t)]^2\ , \tag{5.14}$$

with $$t = 2\left[r/w_{00}^{(i)}\right]^2 .$$

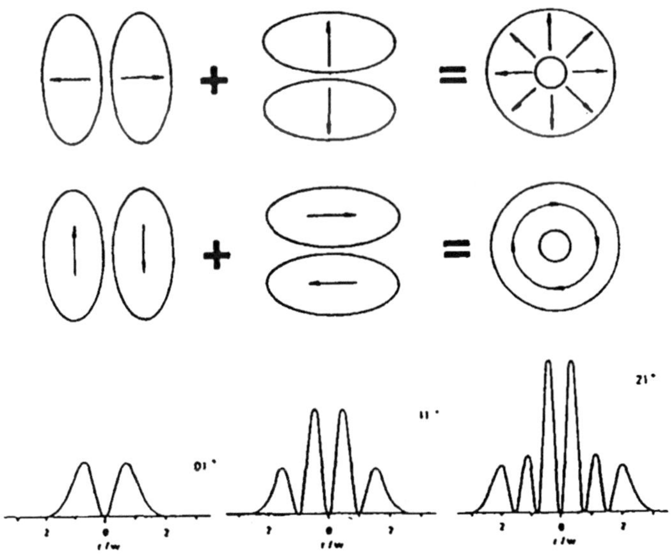

Fig. 5.10 Hybrid modes are generated by the superposition of two linearly polarized circularly symmetric transverse modes of the same order $p\ell$ which oscillate rotated by 90° with respect to each other. The graph shows two of the four possible ways to superimpose the mode structures. The lower graphs present the radial intensity distributions of the three lowest order hybrid modes TEM_{01^*}, TEM_{11^*}, and TEM_{21^*}.

Hybrid modes are quite often observed if an aperture inside the resonator is continuously varied. As an example, Fig. 5.11 presents the development of the beam profile of an Nd:YAG laser as the diameter of an aperture inside the stable resonator is increased. Although these intensity distribut ions represent the superposition of more than two transverse modes, they often exhibit the center hole characteristic of hybrid modes.

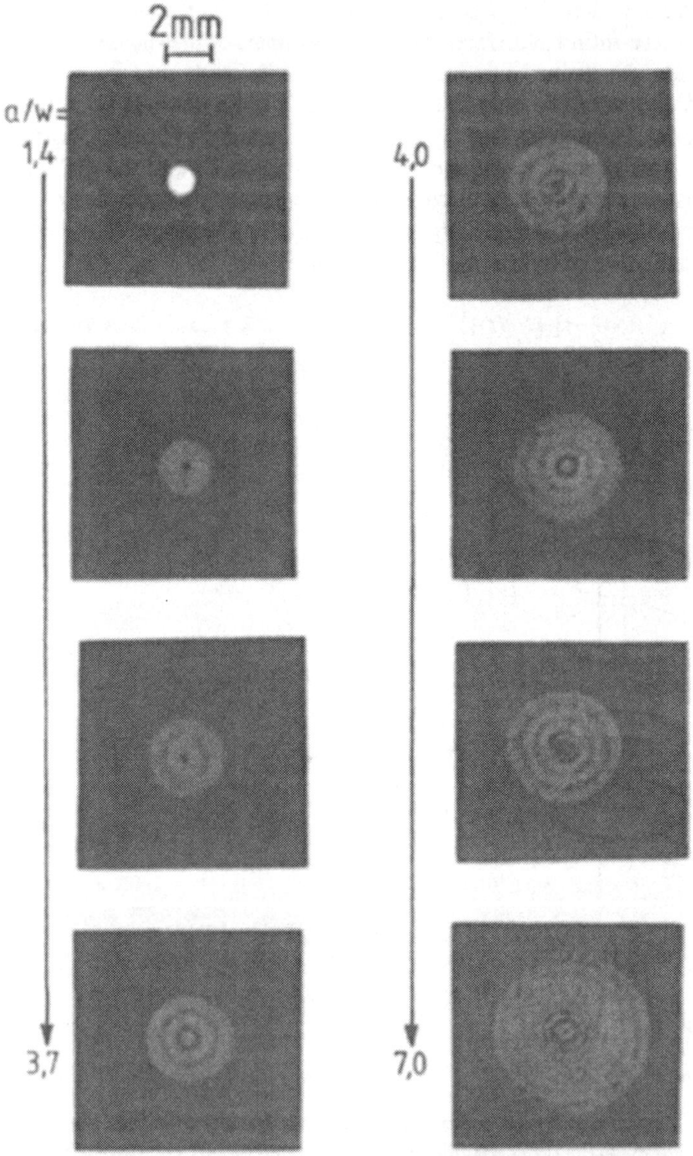

Fig. 5.11 Photographs of intensity distributions at the output coupling mirror of an Nd:YAG rod laser. The radius *a* of an aperture located inside the stable resonator is increased; *w* denotes the Gaussian beam radius at the aperture ($g_1=1$, $g_2=0.5$, $L=0.5m$, $\lambda=1.064\mu m$).

5.2.2 Resonance Frequencies

The development of steady-state field distributions in any optical resonator requires that both the amplitude and the phase of the electric field $E(x,y)$ are reproduced after each round trip. According to (5.3) this requirement is met if the eigenvalue γ is equal to 1.0, which is referred to as the *resonance condition*. Together with (5.7) and (5.9), the resonance condition yields the resonance frequencies ν of the eigenmodes [3.7]:

Circular Symmetry:

$$\nu_{plq} = \frac{c_0}{2L}\left[q + \frac{2p+\ell+1}{\pi}\,acos\sqrt{g_1 g_2}\right] \tag{5.15}$$

Rectangular Symmetry:

$$\nu_{mnq} = \frac{c_0}{2L}\left[q + \frac{m+n+1}{\pi}\,acos\sqrt{g_1 g_2}\right] \tag{5.16}$$

with c_0 : speed of light in vacuum
L : optical mirror spacing = nL_0 (L_0: geometrical spacing, n: index of refraction inside the resonator)

The resonance frequencies thus depend on both the axial and the transverse mode order. In contrast to the plane-parallel FPI ($g_1=g_2=1$, Sec. 4.1) whose resonance frequencies are only determined by the axial mode index, each axial mode of stable resonators is subdivided into a sequence of frequencies corresponding to different transverse modes. This separation is controlled by the g-parameters of the resonator mirrors (Fig. 5.12). As the origin of the g-diagram is approached, the frequency gap between different transverse modes having the same axial mode order becomes wider. In the limit of the stable confocal resonator ($g_1=g_2=0$), the frequency gap equals $c_0/(4L)$ which is half the axial mode distance. The confocal resonator exhibits frequency degeneracy which means that all modes meeting the conditions $2q + 2p + \ell + 1 = k$ and $2q + m + n + 1 = k$, with k : integer, oscillate at the same resonance frequency $kc_0/(4L)$.

The difference in resonance frequencies plays an important role in the temporal stability of the laser emission. If several transverse modes oscillate, the emission is modulated with the difference frequencies of the modes, called the beat frequency (Fig. 5.13). Chaotic laser emission can occur if more than two transverse modes oscillate simultaneously. Lasers featuring a stabilized temporal output operate at one single transverse mode (usually the fundamental mode) and preferably at one single axial mode. Single transverse mode operation is usually achieved by aperture-limiting the Gaussian beam inside the resonator. Single axial mode operation can be attained with one or a combination of the following techniques (depending on the type of laser): a) reduction of the gain bandwidth with intracavity etalons, gratings, or Lyot filters, b) increase of the axial mode spacing by using longer resonators or coupled resonators, c) prevention of standing waves inside the active medium (unidirectional ring resonator, twisted mode resonator) (see also Chapter 21).

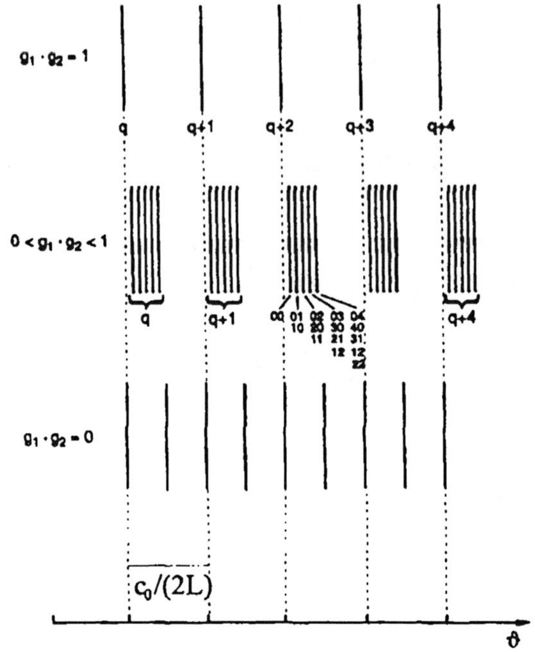

Fig. 5.12 Resonance frequencies of the modes of stable resonators with rectangular symmetry for different g-parameter products. The indices *mn* of some transverse modes are shown for the axial mode of order *q+2*.

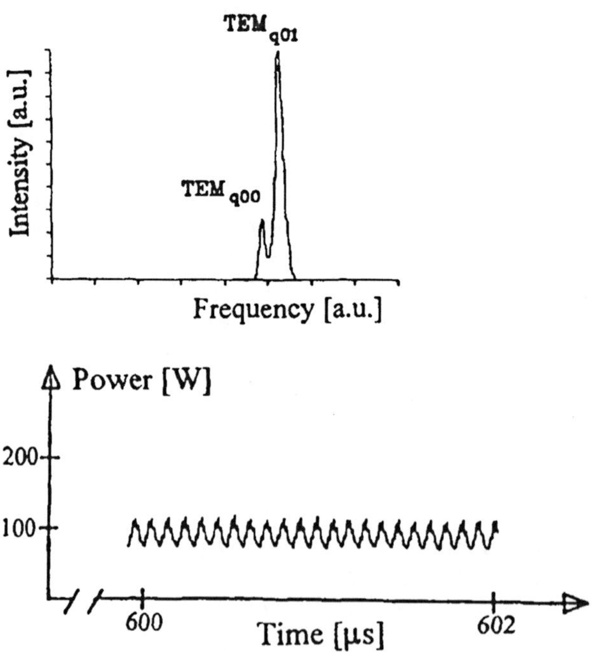

Fig. 5.13 Measured frequency spectrum of a pulsed single axial mode Nd:YAG ring laser ($L=1m$, $g_1= g_2=0.5$) operating at two transverse modes and the temporal emission [S.4]. The beat frequency is 45MHz. The additional modulation at 12MHz is caused by the nonlinear interaction between the modes in the active medium (sloshing of power between the modes).

5.2.3 The TEM_{00} Mode

So far we have only discussed the intensity distributions of the modes at the resonator mirrors and how the mode structure depends on the resonator length and the g-parameters of the mirrors. The Gaussian beam radius $w_{00}^{(i)}$ at mirror i determines the lateral dimension of the mode. For stable resonators, the intensity distributions at any plane inside or outside the resonator exhibit the same shape as the distributions on the mirrors; only the beam radius changes with the propagation.

For the fundamental mode, or Gaussian beam, the beam radius as a function of the distance can be calculated by using the ABCD law (2.51). If z denotes the distance along the optical axis from the position of the beam waist with radius w_0, the Gaussian beam radius $w_{00}(z)$ inside the resonator reads (Fig. 5.14):

$$w_{00}(z) = w_0 \sqrt{1 + \left(\frac{z}{z_0}\right)^2} \tag{5.17}$$

with:

waist radius $$w_0^2 = \frac{\lambda L}{\pi} \frac{\sqrt{g_1 g_2 (1 - g_1 g_2)}}{|g_1 + g_2 - 2 g_1 g_2|} \tag{5.18}$$

Rayleigh range $$z_0 = \frac{\pi w_0^2}{\lambda} \tag{5.19}$$

waist position $$L_{01} = L \frac{(1 - g_1) g_2}{|g_1 + g_2 - 2 g_1 g_2|} \tag{5.20}$$

The waist position L_{0l} is the distance of the minimum beam radius (beam waist) from mirror 1. If L_{0l} is positive the waist is located to the right of mirror 1 (as shown in Fig. 5.14), for a negative L_{0l} the waist is found to the left of mirror 1 (see Fig. 5.15).

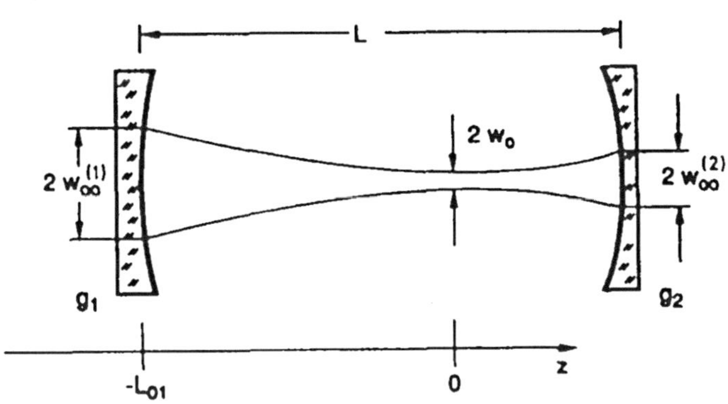

Fig. 5.14 Propagation of a Gaussian beam in a stable resonator.

At the distance of one Rayleigh range z_0 from the position of the beam waist the beam radius has increased by a factor of $\sqrt{2}$, which means that the cross sectional area of the beam has doubled. The physical meaning of the Rayleigh range will be discussed in Sec. 5.2.5. If we insert the values $z=L_{01}$ and $z=L-L_{01}$ into (5.17) we obtain expression (5.10) for the Gaussian beam radius on mirror 1 and mirror 2, respectively.

Example: Stable Resonator with L=1.5m, ρ_1=−1m, ρ_2=2m, λ=632.8nm (Fig. 5.15)
The g-parameters of this resonator are $g_1=2.5$ and $g_2=0.25$. With (5.17)-(5.20) we obtain:
$w_0 = 0.3123mm$, $z_0=485mm$, $L_{01}=-375mm$, $w_{00}^{(1)}=0.395$ mm, $w_{00}^{(2)}=1.249mm$

The beam waist is located at a distance of 375mm to the left of mirror 1. It is for this reason that the beam radius on mirror 1 is much smaller than the beam radius on mirror 2, since the latter mirror is farther away from the beam waist.

The Gaussian beam is fully determined by the beam radius, the Rayleigh range, and the location of the beam waist. The divergence angle (half cone angle) θ_0 is obtained from these quantities by using the relation:

$$\theta_0 = \frac{\lambda}{\pi w_0} = \frac{w_0}{z_0} \tag{5.21}$$

The beam parameter product $w_0\theta_0$ is a constant of the Gaussian beam as long as the beam propagates through ABCD-type optical systems.

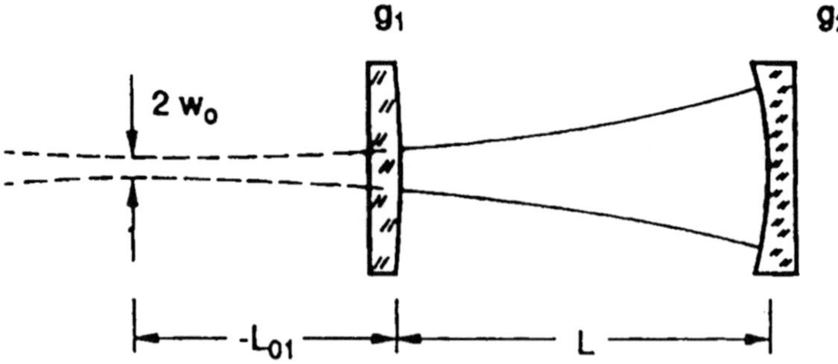

Fig. 5.15 A stable resonator with the beam waist being outside the resonator. According to (5.20), the distance L_{01} is negative.

The radius of curvature $R^{(i)}$ of the Gaussian beam at mirror i is always equal to the radius of curvature ρ_i of the mirror. The general expression for the radius of curvature R as a function of the propagation distance z reads:

$$R(z) = z_0 \left[\frac{z}{z_0} + \frac{z_0}{z} \right] \tag{5.22}$$

If we set $z=L_{0i}$ and $z=L-L_{0i}$ we can easily verify that the mirror surfaces indeed represent surfaces of constant phase of the Gaussian beam.

The general expression for the electric field of a Gaussian beam as a function of the distance z from the waist is given by:

$$E(x,y,z) = \frac{E_0}{\sqrt{1 + (z/z_0)^2}} \exp\left[\frac{-(x^2+y^2)}{w_{00}^2(z)} - \frac{ik(x^2+y^2)}{2R(z)} \right] \exp\left[-i\, atan\left(\frac{z}{z_0} \right) \right] \tag{5.23}$$

The last term is referred to as the Gouy phase shift. This additional phase is the reason why the resonance frequency (5.15/5.16) has an additional term that depends on the g-parameters of the resonator. By introducing the q-parameter of the Gaussian beam:

$$\frac{1}{q(z)} = \frac{1}{R(z)} - \frac{i\lambda}{\pi w_{00}^2(z)} \tag{5.24}$$

the field distribution can be written as (see Eqs. 2.50/2.59):

$$E(x,y,z) = \frac{E_0}{\sqrt{1 + (z/z_0)^2}} \exp\left[-\frac{ik(x^2+y^2)}{2q(z)} \right] \exp\left[-i\, atan\left(\frac{z}{z_0} \right) \right] \tag{5.25}$$

Propagation of the Gaussian beam through ABCD-type optics can be accomplished by using the ABCD-law (2.51) for the q-parameter. For the propagation of the Gaussian beam outside the resonator the imaging properties of the resonator mirrors have to be taken into account. In general, the mirror substrates have a planar rear surface and, therefore, act as lenses transforming the Gaussian beam into a new Gaussian beam (Fig. 5.16). The new divergence angle behind mirror i with refractive index n is given by:

$$\theta_{0i}^{*2} = \theta_0^2 \, \frac{g_i(1+g_1g_2(n^2-1)) + n^2(g_j-2g_1g_2)}{|g_1+g_2-2g_1g_2|} \tag{5.26}$$

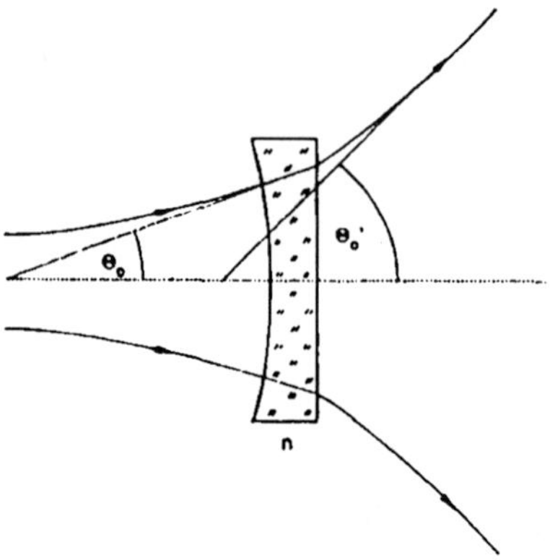

Fig. 5.16 Propagation through a mirror substrate changes the angle of divergence θ_0 of the Gaussian beam.

The two divergence angles are different for nonsymmetric resonators ($g_1 \neq g_2$) and usually greater than the intracavity divergence angle θ_0. The new beam waist radius of the Gaussian beam behind mirror i can be determined with $w_{0i}=\lambda/(\pi\theta_{0i})$.

Resonator Schemes, Beam Radii, and Angles of Divergence of the TEM$_{00}$ Mode
The beam radii and the angles of divergence of common stable resonators are presented in Table 5.2 and in Fig. 5.17. The resonators on the stability limits have been included in this table to show that the beam waist radius goes to zero or infinity and, accordingly, the angle of divergence assumes values of $\pi/2$ or 0 if the stability limit is approached. This behavior is caused by the fact that a Gaussian beam is not an eigensolution of these resonators. The only exception is the confocal resonator located at the origin of the g-diagram.

Table 5.2 Stable resonators and the properties of their Gaussian beam.

Resonator	$(w_{00}^{(1)})^2$	$(w_{00}^{(2)})^2$	w_0^2	θ_0^2
symmetric ($g_1=g_2=g$)	$\frac{\lambda L}{\pi\sqrt{1-g^2}}$		$\frac{\lambda L\sqrt{1+g}}{2\pi\sqrt{1-g}}$	$\frac{2\lambda\sqrt{1-g}}{\pi L\sqrt{1+g}}$
plane-plane ($\rho_1=\rho_2=\infty$, $g_1=g_2=1$)	∞	∞	∞	0
semi-confocal ($g_1=1$, $g_2=0.5$)	$\frac{\lambda L}{\pi}$	$\frac{2\lambda L}{\pi}$	$\frac{\lambda L}{\pi}$	$\frac{\lambda}{\pi L}$
symmetric confocal ($g_1=g_2=0$)	$\frac{\lambda L}{\pi}$		$\frac{\lambda L}{2\pi}$	$\frac{2\lambda}{\pi L}$
general concentric ($\rho_1+\rho_2=L$, $g_1g_2=1$)	∞	∞	0	$\pi^2/4$

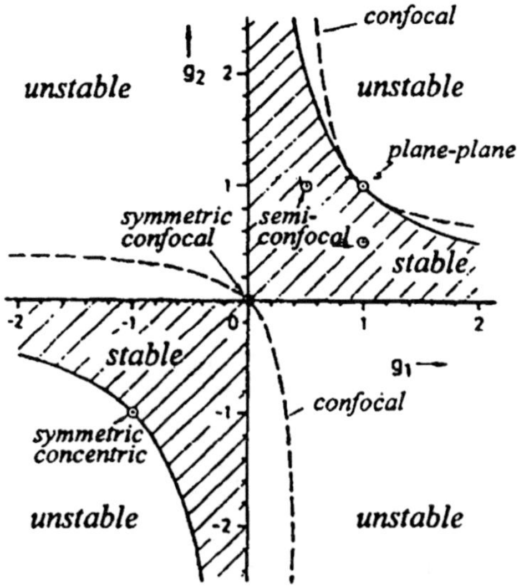

Fig. 5.17 The g-diagram of optical resonators and the location of common resonators.

Figure 5.18 presents the dependence of the angle of divergence on the resonator parameters. By combining (5.18) and (5.21), the angle of divergence θ_0 for a general stable resonator is given by:

$$\theta_0^2 = \frac{\lambda}{\pi L} \frac{|g_1 + g_2 - 2g_1 g_2|}{\sqrt{g_1 g_2 (1 - g_1 g_2)}} \tag{5.27}$$

The normalized angle $\theta_0 \sqrt{(L/\lambda)}$ depends only on the g-parameters of the resonator. For a constant resonator length L, the angle of divergence becomes very small if the resonator is chosen close to the hyperbola in the first quadrant. The corresponding beam waist radius w_0 is, of course, very large for these resonators since the beam parameter product $w_0\theta_0$ is a constant. The largest angles of divergence are found for negative g-parameters at the stability limit $g_2=1/g_1$, where the concentric resonators are located. Since the large angles of divergence are generated by extremely small intracavity beam waists, these resonators have a small mode volume and have therefore found only limited application in laser systems.

Examples:

1) L=1m, λ=1064nm, semi-confocal:	w_0=0.582mm,	θ_0=0.582mrad
2) L=1m, λ=1064nm, confocal:	w_0=0.412mm,	θ_0=0.823mrad
3) L=1m, λ=1064nm, ρ_1=-1.5m, ρ_2=1.5m:	w_0=0.435mm,	θ_0=0.778mrad
4) L=1m, λ=1064nm, $\rho_1=\infty$ m, ρ_2=5m:	w_0=0.823mm,	θ_0=0.412mrad
5) L=1m, λ=1064nm, $\rho_1=\rho_2$=0.55m:	w_0=0.232mm,	θ_0=1.463mrad

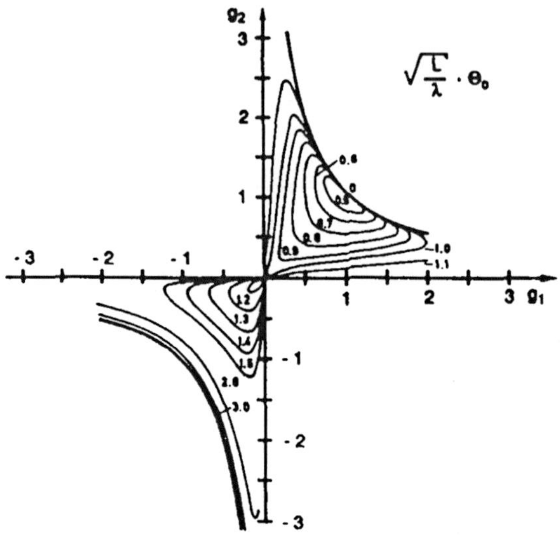

Fig. 5.18 Curves of constant angle of divergence θ_0 of the Gaussian beam in the g-diagram. The parameter is the normalized angle of divergence $\theta_0\sqrt{(L/\lambda)}$ which is 1.0 for the confocal resonator.

Mode Volume of the TEM$_{00}$ Mode

The output power of a laser resonator, among other parameters, is determined by the volume of the active medium that is filled by the mode. Only in this area can the stored energy be extracted by the mode via induced emission. Let us calculate the volume of the Gaussian beam between the resonator mirrors assuming that the mode volume is determined by the Gaussian beam radius $w(z)$. As far as the output power is concerned, we consider an active medium that fills the whole resonator. Integration of the square of the beam waist $w(z)$ given by (5.17)-(5.19) from $z=-L_{01}$ to $z=L-L_{01}$ yields the mode volume of the Gaussian beam:

$$V_{00} = \pi\lambda L^2 \left[\frac{\sqrt{g_1 g_2(1-g_1 g_2)}}{|g_1+g_2-2g_1 g_2|}\right]\left[1 + \frac{(g_2-g_1)^2 + (1-g_1)(1-g_2)g_1 g_2}{3g_1 g_2(1-g_1 g_2)}\right] \tag{5.28}$$

The normalized mode volume $V_{00}/(\pi\lambda L^2)$ as a function of the g-parameters is shown in Fig. 5.19. It is apparent from this graph that the highest volume of the fundamental mode is achieved for resonators close to the stability limit at positive g-parameters and one g-parameter being much lower than 1. These resonators are formed by a concave and a convex mirror (see Fig. 5.15) and, therefore, are referred to as concave-convex resonators. A concave-convex resonator is the preferred means to achieve high output power in fundamental mode operation, even though they are more sensitive to mirror tilt than other resonators. Figure 5.20 presents measured output energies of a pulsed Nd:YAG laser in fundamental mode operation for different resonator configurations.

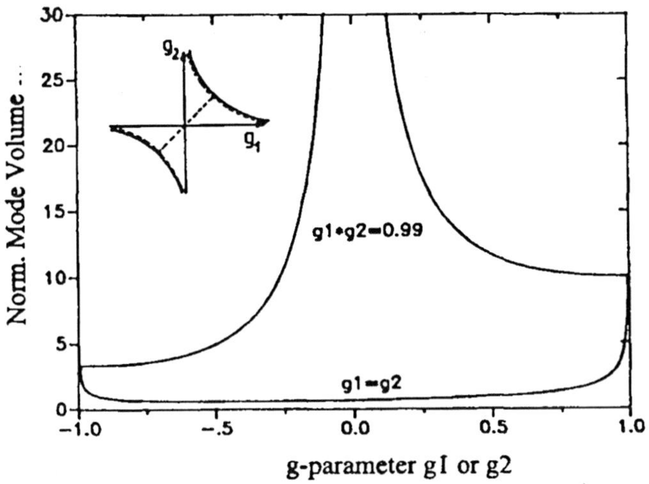

Fig. 5.19 Normalized mode volume $V_{00}/(\pi\lambda L^2)$ of the fundamental mode of stable resonators as a function of one of the g-parameters. The corresponding resonators are indicated in the g-diagram by dotted lines.

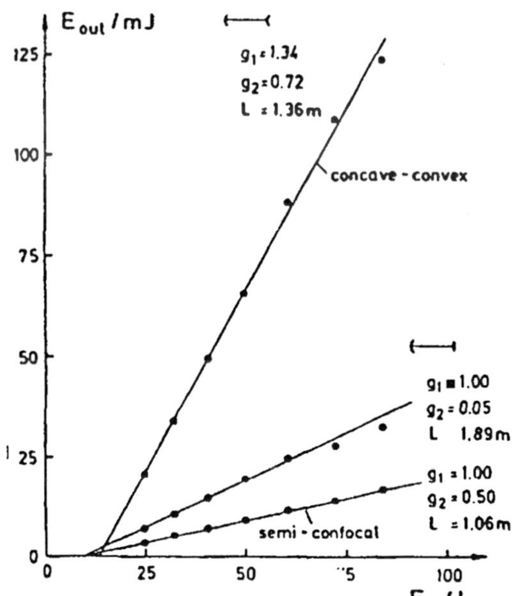

Fig. 5.20 Measured output energy of a pulsed 1.064μm Nd:YAG rod laser (rod diameter: 6.35mm, rod length: 76mm) in fundamental mode operation for different resonator configurations as a function of the electric energy supplied to the flashlamp.

5.2.4 Higher Order Modes

The treatment of the propagation of higher order modes in the resonators becomes quite simple given the preceding detailed discussion of the propagation of Gaussian beams. In both circular and rectangular symmetries the propagation of the fundamental mode and of the higher order modes is similar (Fig. 5.21). If w_0 and θ_0 denote the waist radius and the angle of divergence (half angle) of the fundamental mode, the beam radius as a function of the propagation distance from the waist position reads as follows:

a) in circular symmetry:

beam radius at location z: $$w_{p\ell}(z) = w_0\,\sqrt{2p+\ell+1}\,\sqrt{1+\left(\frac{z}{z_0}\right)^2} \qquad (5.29)$$

waist radius: $$w_{p\ell} = w_0\,\sqrt{2p+\ell+1} \qquad (5.30)$$

angle of divergence: $$\theta_{p\ell} = \theta_0\,\sqrt{2p+\ell+1} \qquad (5.31)$$

b) in rectangular symmetry:

beam radius at location z: $$w_m(z) = w_0\,\sqrt{2m+1}\,\sqrt{1+\left(\frac{z}{z_0}\right)^2} \qquad (5.32)$$

waist radius: $$w_m = w_0\,\sqrt{2m+1} \qquad (5.33)$$

angle of divergence: $$\theta_m = \theta_0\,\sqrt{2m+1} \qquad (5.34)$$

The Rayleigh range is the same as for the Gaussian beam:

$$z_0 = \frac{w_m}{\theta_m} = \frac{w_0}{\theta_0} = \frac{\pi w_0^2}{\lambda} \qquad (5.35)$$

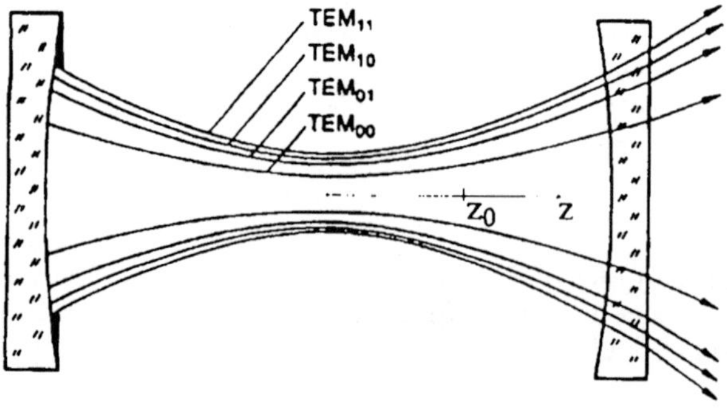

Fig. 5.21 Beam propagation of higher order modes. All modes have the same Rayleigh range. At any plane the beam radii are $\sqrt{2p+\ell+1}$ larger than the Gaussian beam radius.

The beam parameter product $w\theta$ is a constant of the beam with:

$$w_{p\ell}\theta_{p\ell} = (2p+\ell+1)\frac{\lambda}{\pi} = M^2\frac{\lambda}{\pi} \tag{5.36}$$

$$w_m\theta_m = (2m+1)\frac{\lambda}{\pi} = M^2\frac{\lambda}{\pi} \tag{5.37}$$

The term M^2 is referred to as the propagation factor of the mode. In a laser resonator, the maximum propagation factor corresponding to the highest order transverse mode determines the beam quality of the laser beam. Let us assume that N apertures with radii a_i are located inside the resonator at positions $z_1..z_N$ and $w_{00}(z_i)$ are the beam radii of the Gaussian beam at the apertures. The maximum propagation factor, to a good approximation, is given by:

$$M^2_{max} = int\ \min\left(\left[\frac{a_1}{w_{00}(z_1)}\right]^2, \left[\frac{a_2}{w_{00}(z_2)}\right]^2, \ldots, \left[\frac{a_N}{w_{00}(z_N)}\right]^2\right) \tag{5.38}$$

where *min* is the minimum term in the group and *int* rounds to the next integer value.

The electric field distribution of individual transverse modes as a function of the propagation distance z reads:

a) in circular symmetry: (5.39)

$$E(r,\Phi,z) = \frac{E_0}{\sqrt{1+(z/z_0)^2}}\exp\left[\frac{-r^2}{w_{00}^2(z)} - \frac{ikr^2}{2R(z)}\right]\left[\frac{\sqrt{2}r}{w_{00}(z)}\right]^{\ell} L_{p\ell}\left[\frac{2r^2}{w_{00}(z)^2}\right]\begin{Bmatrix}\cos(\ell\Phi)\\ \sin(\ell\Phi)\end{Bmatrix}\exp\left[-i(2p+\ell+1)\,atan\left(\frac{z}{z_0}\right)\right]$$

b) in rectangular symmetry: (5.40)

$$E(x,y,z) = \frac{E_0}{\sqrt{1+(z/z_0)^2}}\exp\left[\frac{-(x^2+y^2)}{w_{00}^2(z)} - \frac{ik(x^2+y^2)}{2R(z)}\right]H_m\left[\frac{\sqrt{2}x}{w_{00}(z)}\right]H_n\left[\frac{\sqrt{2}y}{w_{00}(z)}\right]\exp\left[-i(m+n+1)\,atan\left(\frac{z}{z_0}\right)\right]$$

The beam radius $w_{00}(z)$ and the radius of curvature $R(z)$ are those of the Gaussian beam (see (5.17) and (5.22)). The Gouy phase shift (far right term) generates the frequency difference between modes of different transverse order. By introducing the q-parameter:

$$\frac{1}{q(z)} = \frac{1}{R(z)} - \frac{i\lambda M^2}{\pi w_{p\ell}^2(z)} \tag{5.41}$$

the propagation of higher order modes from a plane 1 to a plane 2 through ABCD-type optics can be calculated by using the ABCD law:

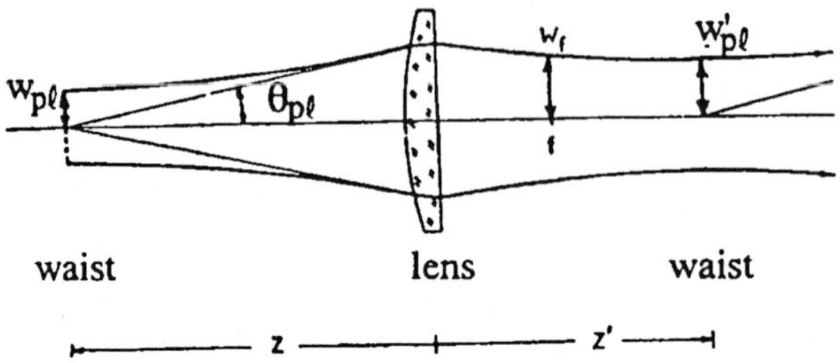

Fig. 5.22 Focusing of transverse modes TEM_{pl}. The beam waist is imaged at a distance z' from the lens.

$$q_2 = \frac{Aq_1 + B}{Cq_1 + D} \tag{5.42}$$

Note that the q-parameter is the same for all modes including the fundamental mode since the square of the waist is larger by M^2, which cancels the numerator term in (5.41)! We can use the ABCD law to calculate the focusing properties of the modes. If plane 1 is at the beam waist with radius w_{pl} and a focusing lens with focal length f is located at a distance z from the waist, the position z' (measured from the lens) of the focal beam waist with radius w'_{pl} are given by (see Fig. 5.22 and Sec. 2.5.1):

$$\frac{1}{z} + \frac{1}{z'} = \frac{1}{f} + \frac{z_0^2}{z[z^2+z_0^2-zf]} \tag{5.43}$$

$$w'_{pl} = w_{pl} \frac{f}{\sqrt{z_0^2 + (z-f)^2}} \tag{5.44}$$

Example (Fig. 5.23):
CO_2 laser, $\lambda=10.6\mu m$, tube diameter $d=30mm$, tube length $l=300mm$, $\rho_1=-5m$, $\rho_2=+5m$, $L=0.5m$. The gas tube is in the middle of the resonator. A focusing lens with $f=100mm$ is located at a distance of *50mm* from mirror 2. The g-parameters of the mirrors are: $g_1=1.1$ and $g_2=0.9$. With (5.10) and (5.17)-(5.20) we get the following properties of the Gaussian beam:

beam radius at mirror 1	$w_{00}^{(1)}$	= 3.906mm
beam radius at mirror 2	$w_{00}^{(2)}$	= 4.319mm
beam waist	w_0	= 2.897mm
waist position	L_{01}	=-2,250mm
Rayleigh range	z_0	= 2,488 mm
beam radius at left tube end	$w_{00}(z_1)$	= 3.985 mm
beam radius at right tube end	$w_{00}(z_2)$	= 4.233 mm

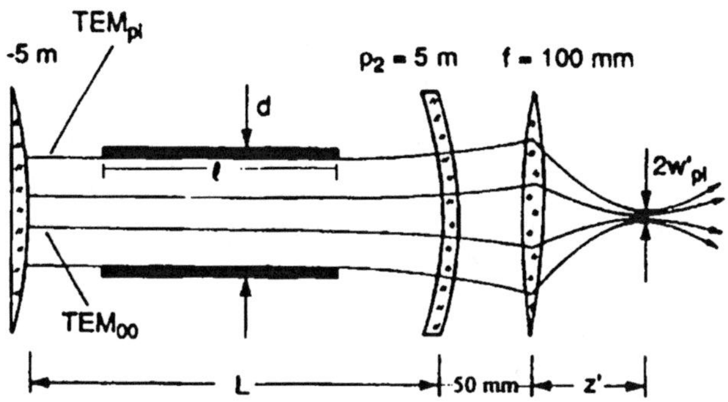

Fig. 5.23 CO_2 laser with concave-convex resonator.

The Gaussian beam radius is largest at the right end of the tube. The maximum propagation factor (5.38) is given by:

$$M^2_{max} = (2p+\ell+1)_{max} = int\left(\left[\frac{d/2}{w_{00}(z_2)}\right]^2\right) = 13$$

All modes with transverse mode order $2p+\ell+1 \le 13$ will oscillate simultaneously. The beam radii at any plane inside or outside the resonator are determined by the highest order mode and therefore are $\sqrt{13}$ larger than those of the Gaussian beam. The lens is at a distance $z=|L_{01}|+L+50mm=2,800mm$ from the beam waist. With Eqs. (5.38) and (5.39) we obtain:

position of focal spot: $z' = 101.9$ *mm*

spot radius: $w'_{p\ell} = 0.079$ *mm*

Transverse Multimode Operation

All transverse modes whose beam radii fit into the apertures located inside the resonator will oscillate simultaneously. The resulting output beam is a superposition of these transverse modes. Although each individual mode is a steady state solution of the resonator, the superposition is not since the modes exhibit different resonance frequencies $\nu_{qp\ell}$. The frequency difference leads to temporal oscillations of the intensity distributions. To get a better understanding of this phenomenon let us investigate a superposition of a TEM_{00} mode with amplitude E_0 and a TEM_{10} mode with amplitude E_1 (in rectangular symmetry and one dimension, w is the Gaussian beam radius). The time-dependent electric field is given by:

$$E(x,z,t) = E_0 \exp\left[\frac{-x^2}{w^2}\right] \exp[i(2\pi\nu_{q00}t-\Delta)] + E_1\left[\frac{2\sqrt{2}x}{w}\right]\exp\left[\frac{-x^2}{w^2}\right] \exp[i(2\pi\nu_{q10}t-2\Delta)]$$

with: $\Delta = atan(z/z_0)$ (5.45)

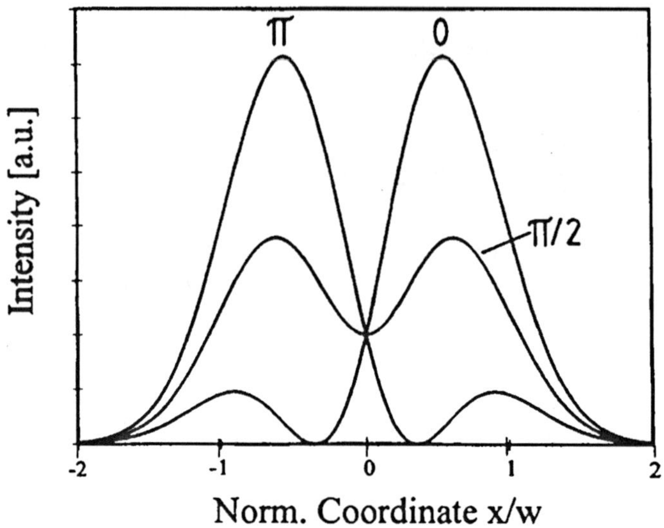

Fig. 5.24 Superposition of a TEM_{00} and a TEM_{10} mode according to (5.41) with $E_0=E_1$. The different curves depict various phase delays $\phi=\Omega t-\Delta$ between the modes. The curve for $\phi=\pi/2$ represents the time-averaged intensity distribution.

This results in an intensity distribution: (5.46)

$$I(x,z,t) = const.\ \exp\left[\frac{-2x^2}{w^2}\right]\left(E_0^2 + E_1^2\frac{8x^2}{w^2} + E_0E_1\frac{4\sqrt{2}x}{w}\cos(\Omega t-\Delta)\right)$$

with: $$\Omega = 2\pi(\nu_{q10}-\nu_{q00}) = \frac{c_0}{L}\ a\cos\sqrt{g_1g_2}$$ (see Eq. (5.16))

The intensity distribution is the sum of the intensity distributions of the individual modes plus an oscillating term generated by the frequency difference. The oscillation frequency Ω typically is on the order of 100 MHz. Figure 5.24 presents the temporal change of the intensity distribution during one oscillation cycle. If the intensity pattern is recorded, the instrumentation (e.g. CCD camera) will average over the mode oscillations and only the sum of the intensity distributions of the individual modes is observed. However, during short time intervals the intensity can be much higher at certain areas, especially if a high number of modes with varying amplitudes is oscillating. These so-called *hot spots* can induce damage on the surfaces of optical components.

The calculation of the intensity distribution for a high number of modes generally has to be done numerically. For an infinite number of modes, however, the superposition can, in special cases, be performed analytically. In rectangular symmetry, the infinite sum of eigenmodes yields the field distribution:

$$E(x,z,t) = \exp\left[\frac{-x^2}{w^2}\right] \sum_{m=0}^{\infty} E_m \, H_m\left[\frac{\sqrt{2}x}{w}\right] \exp[i(m+1)(\Omega t - \Delta)] \tag{5.47}$$

If we assume that the field amplitudes are given by $E_m = E_0/m!$, we can use the generating function of Hermite polynomials,

$$\exp[2yt - t^2] = \sum_{m=0}^{\infty} \frac{H_m(y) t^m}{m!} ,$$

to calculate the intensity distribution. The final expression then reads:

$$I(x,z,t) = const.\ E_0^2 \exp\left[-\left(\frac{\sqrt{2}x}{w} - 2\cos(\Omega t - \Delta)\right)^2\right] \tag{5.48}$$

The intensity distribution is Gaussian with a beam radius of $w/\sqrt{2}$ and the whole profile oscillates back and forth in the x-direction with an amplitude of $\sqrt{2}w$, as shown in Fig. 5.25. This effect is referred to as *transverse modelocking*. If we recorded this intensity distribution with a CCD camera we would observe the time-averaged intensity indicated by the dotted line. At any time, however, the peak intensity is more than twice as high!

In contrast to the cases discussed above, the amplitudes E_m of the transverse modes in laser resonators will exhibit temporal variations caused by their interaction in the active medium. Since different modes make use of different gain areas, the gain unused by one mode will allow a different mode to oscillate with an increased amplitude. This mode competition may be such that no steady-state solution for the mode amplitudes is found. Depending on the type of active material and the boundary conditions of the system (e.g. resonator set-up and pumping conditions), this may lead to a periodic, a quasiperiodic, or a chaotic behavior of the laser emission.

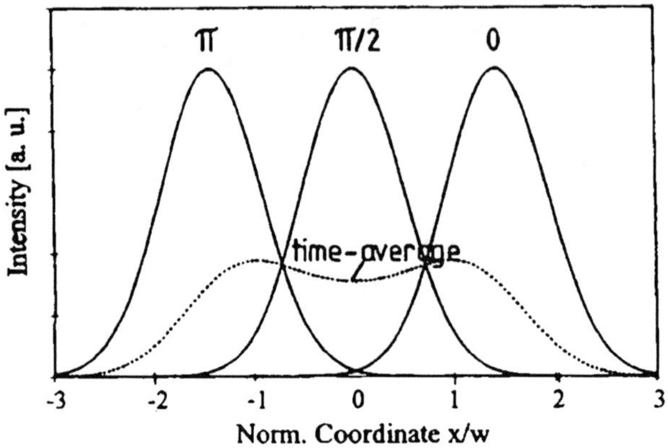

Fig. 5.25 The superposition of an infinite number of Gauss-Hermite modes with Gaussian beam radius w according to Eq. (5.48). The resulting Gaussian intensity profile exhibits temporal oscillations in x-direction with an amplitude of $\sqrt{2}w$. The parameter of the curves is the phase $\phi = \Omega t - \Delta$.

Orthogonality of Transverse Modes

Any field distribution inside or outside of an unconfined stable resonator can be written as a sum of the field distributions of the transverse eigenmodes [3.1,3.30]. Let us consider Gauss-Hermite eigenmodes in one dimension with field distributions $f_m(x,z)$ at a distance z from the beam waist, given by (5.8). Any field distribution $E(x,z)$ can then be expressed as:

$$E(x,z) = \sum_{m=0}^{\infty} c_m f_m(x,z) \tag{5.49}$$

where c_m is the amplitude (complex number) of each eigenmode. The power of the field distribution is given by:

$$P = const. \int |E(x,z)|^2 dx = const. \left[\sum_{m=0}^{\infty} \int |c_m|^2 |f_m(x,z)|^2 dx + \sum_{m=0}^{\infty} \sum_{\substack{n=0 \\ n \neq m}}^{\infty} \int c_m c_n^* f_m(x,z) f_n^*(x,z) dx \right] \tag{5.50}$$

For Gauss-Hermite modes the right-most integral is zero, since:

$$\int f_m f_n^* dx = \int_{-\infty}^{+\infty} H_m\left[\frac{\sqrt{2}x}{x}\right] H_n\left[\frac{\sqrt{2}x}{w}\right] \exp\left[\frac{-2x^2}{w^2}\right] \exp[i(n-m)\Delta] dx = \sqrt{\pi}\, 2^n n! \exp[i(n-m)\Delta]\, \delta_{mn} \tag{5.51}$$

with $\delta_{mn} = 1$ for $m=n$ and $\delta_{mn} = 0$ otherwise. The Gauss-Hermite modes thus form an orthogonal set and the total power is the sum of the powers of each mode:

$$P = const.\left[\sum_{m=0}^{\infty}\int |c_m|^2 |f_m(x,z)|^2 dx\right] = \sum_{m=0}^{\infty} P_m \tag{5.52}$$

All modes contain the same power as the fundamental mode, if the amplitudes are given by:

$$c_m = \frac{c_0}{\sqrt{2^m m!}} \tag{5.53}$$

For Gauss-Laguerre modes the orthogonality relation reads: (5.54)

$$\int_0^{2\pi}\int_0^{\infty} f_{p\ell} f_{qm}^* r dr d\phi = 2\pi \int_0^{\infty} L_{p\ell}[t] L_{qm}[t]\ t^{\ell+m}\ \exp[-t]\ \exp[i2(q-p)\Delta]\ r dr = 2\pi \frac{(p+\ell)!}{p!}\delta_{pq}\delta_{\ell m}$$

with: $t = \left[\frac{2r^2}{w^2}\right]$

Any field distribution $E(r, \Phi)$ can be expressed as a sum of eigenmodes:

$$E(r,\phi,z) = \sum_{p=0}^{\infty}\sum_{\ell=0}^{\infty} c_{p\ell} f_{p\ell}(r,\phi,z) \tag{5.55}$$

Similar to rectangular symmetry, the power is given by the sum of the powers of each mode and the modes exhibit equal power if the following relation holds for the amplitudes:

$$c_{p\ell} = c_{00}\sqrt{\frac{p!}{(p+\ell)!}} \tag{5.56}$$

The expansions of a field distribution into the transverse eigenmodes (5.49) and (5.55) are only possible because the Gauss-Hermite and the Gauss-Laguerre modes form a complete set of orthogonal functions. These modes represent the eigenmodes of passive stable resonators with unconfined mirrors. If we insert apertures or an active medium into the resonator, the field distributions of the transverse eigenmodes are changed. The new eigenmodes are generally not orthogonal which means that the total power cannot be expressed as a sum of individual mode powers (the cross integral (5.51) is not equal to zero). We can still expand a field distribution in the resonator as a series of Gauss-Hermite or Gauss-Laguerre modes, but it is mathematically not guaranteed that the power can be expressed as the sum of the powers of the resonator eigenmodes.

5.2.5 Focusability and BeamQuality

The focusing of a laser beam can be considered as the imaging of the beam waist by means of transformation optics such as a lens or a telescope. The position and the beam radius of the focal spot can be calculated by using the Gaussian imaging conditions (5.43) and (5.44). Let w be the beam waist radius of a circularly symmetric beam with Rayleigh range z_0 and θ the corresponding angle of divergence (Fig. 5.26). Two quantities are preserved when the beam is focused no matter what kind of focusing optics are used: the beam parameter product $w\theta$ (we drop the mode indices for simplicity) and the ratio of the cross sectional area in the waist πw^2 to the Rayleigh range z_0. The following relation holds:

$$\frac{\pi w^2}{z_0} = \pi\, w\theta \tag{5.57}$$

In order to attain a small focal spot and a large Rayleigh range (remember that this is the distance from the waist at which the beam area has doubled), a low beam parameter product is required. The ratio of the beam area in the focal plane to the Rayleigh range is a characteristic of the laser beam and proportional to the beam parameter product. The ratio cannot be changed by transformation optics. This is shown in Fig. 5.26 in which two different optics are used to focus the same laser beam. It is for this reason that the beam quality is defined via the beam parameter product. The general expression for the beam parameter product, with M^2 being the propagation factor, is given by:

$$w\theta = M^2 \frac{\lambda}{\pi}\ , \quad M^2 \geq 1 \tag{5.58}$$

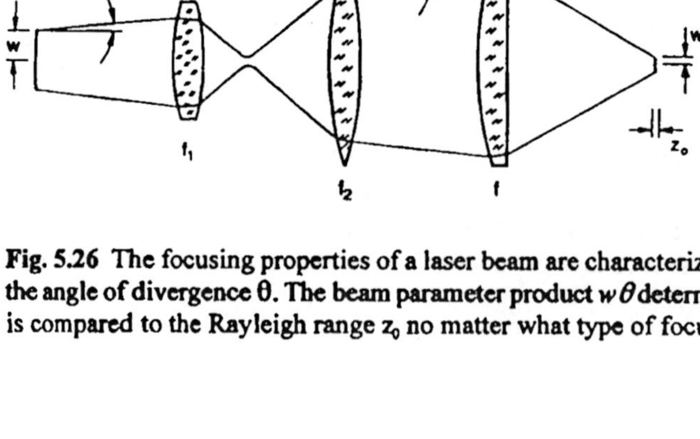

Fig. 5.26 The focusing properties of a laser beam are characterized by the beam waist radius w and the angle of divergence θ. The beam parameter product $w\theta$ determines how small the focal spot area is compared to the Rayleigh range z_0 no matter what type of focusing optics are used.

The propagation factor is equal to 1.0 for the fundamental mode. Both the beam radius and the angle of divergence are defined via the second order intensity moments (see Sec. 2.6). Note that for the same beam, the propagation factor M^2 assumes different values in (5.58) and (5.59) (except for the Gaussian beam where $M^2=1$ in both cases). In order to specify the beam quality it is necessary to measure the beam waist diameter and the angle of divergence separately. This is generally done by generating a waist with a focusing lens and the waist diameter d is determined by the second intensity moment. The corresponding far field with divergence angle can be measured in the focal plane of a second lens (see Chapter 23 for different beam quality measurement techniques). An alternate quantity occasionally used to specify the focusability of a beam is the beam quality factor K defined by:

$$K = \frac{1}{M^2} \tag{5.60}$$

Focusing of Arbitrary Beams to Equal Spot Size

Let us investigate the beam propagation of different quality beams which exhibit the same focal spot radius w_T (Fig. 5.27). For a pure Gaussian beam the following relation holds for the beam radius in the vicinity of the focus:

$$w(z) = w_T \sqrt{1 + \left(\frac{z}{z_0}\right)^2} \tag{5.61}$$

with: $$z_0 = \frac{\pi w_T^2}{\lambda}$$

For a transverse multimode beam being focused to the same spot size, the corresponding, "embedded", Gaussian beam radius is M times smaller. This results in the propagation law:

$$w(z) = w_T \sqrt{1 + M^4\left(\frac{z}{z_0}\right)^2} = w_T \sqrt{1 + \left(\frac{z}{z_M}\right)^2} \tag{5.62}$$

with: $$z_0 = \frac{\pi w_T^2}{\lambda}, \quad z_M = \frac{z_0}{M^2}$$

The Rayleigh range is now shorter by a factor $1/M^2$. This is not in contradiction with (5.35) and Fig. 5.21 where we assumed that higher order modes exhibit an increased beam radius and, therefore, the same Rayleigh range as the Gaussian beam.

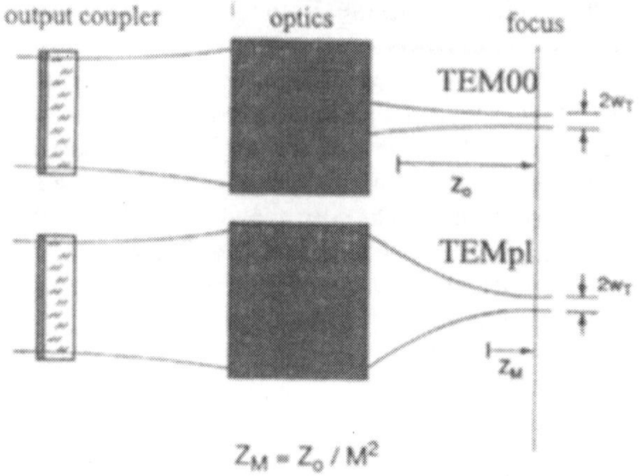

Fig. 5.27 Beam propagation around the focal spot for a fundamental mode and a multimode beam.

Since we now have a fixed beam diameter and more modes have to fit within this diameter, the Rayleigh range will decrease. This behavior is typical for solid-state lasers exhibiting thermal lensing. Attaining a constant focal spot size when the beam propagation factor M^2 changes will be discussed later.

The propagation laws (5.17) and (5.43/5.44) of Gaussian beams can be applied to arbitrary beams if the following transformations are made:

$$w_0 \text{ ---------------> } w_T$$
$$z_0 \text{ ---------------> } z_M = z_0/M^2$$
$$\theta_0 \text{ ---------------> } \theta$$

The following relations are useful for the experimental determination of the beam propagation factor:

$$w_T\theta = M^2\frac{\lambda}{\pi} = \frac{w_T^2}{z_M} \tag{5.63}$$

$$z_M = \frac{\pi w_T^2}{\lambda M^2} \tag{5.64}$$

$$\theta = \frac{w_T}{z_M} \tag{5.65}$$

Note that according to (5.43), the beam propagation factor M^2 also affects the position z' of the image beam waist (Fig. 5.28).

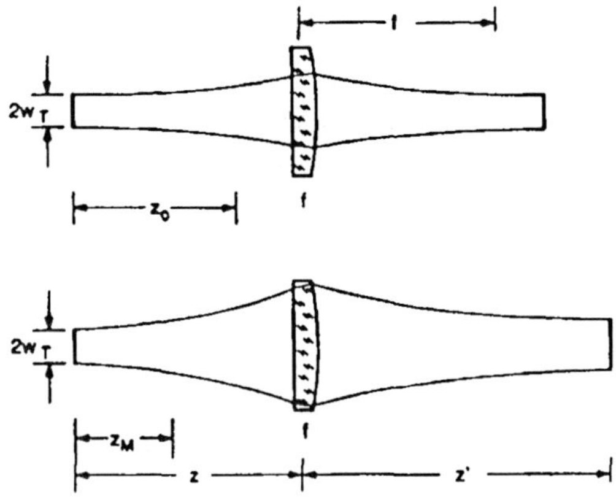

Fig. 5.28 The imaging of a laser beam with beam waist w_T depends on the beam propagation factor M^2.

Beam Quality and Resonator Parameters

The magnitude of the beam parameter product of a stable laser resonator depends on the number of transverse modes oscillating. The beam propagation factor is determined by the radius and the location of the smallest aperture (with respect to the Gaussian beam radius) inside the resonator. To a very good approximation, the beam propagation factor can be calculated by using (5.38). If b denotes the radius of the active medium with length ℓ and the distances from the endfaces to the nearest mirror are ℓ_1 and ℓ_2, the beam propagation factor as a function of the resonator parameters reads (Fig. 5.29):

$$M^2 = \frac{\pi b^2}{\lambda L}\frac{|g_1+g_2-2g_1g_2|}{\sqrt{g_1g_2(1-g_1g_2)}}\left[1+\frac{[[g_1+g_2-2g_1g_2][X - Lg_2(1-g_1)/(g_1+g_2-2g_1g_2)]]^2}{L^2(g_1g_2(1-g_1g_2))}\right]^{-1} \quad (5.66)$$

with
$X = \ell_l + \ell$ if $L_{0l} < \ell_l + \ell/2$ (see Fig. 5.29 and (5.20))
$X = \ell_l$ otherwise
$L = \ell_l + \ell_2 + \ell$

The same expression holds for rectangular media if the radius b is replaced by half the thickness t of the medium. Once the beam propagation factor M^2 is determined, we can calculate, in addition to the beam parameter product, the beam radius $w(z)$ as a function of the distance z from the waist w, the angle of divergence θ, and the mode volume V inside the resonator by using the corresponding expressions for the fundamental mode (5.17), (5.21), and (5.28):

$$w(z) = w_{00}(z)\, M$$
$$\theta = \theta_0\, M$$
$$V = V_{00}\, M^2$$
$$w\theta = w_0\theta_0\, M^2$$

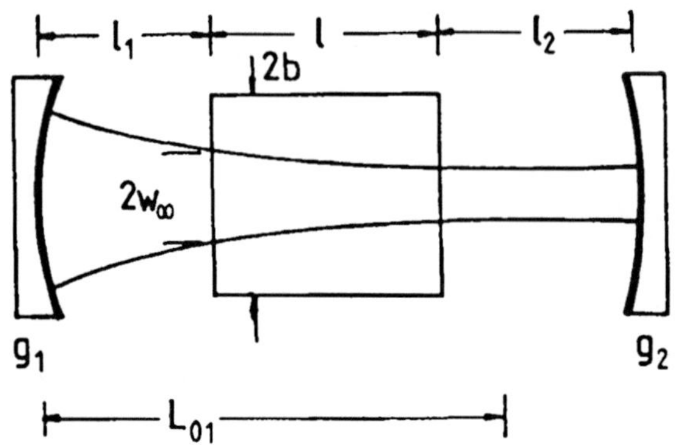

Fig. 5.29 The size of the active medium determines the number of transverse modes. The ratio of the radius b to the largest Gaussian beam radius inside the medium w_{00} determines the beam propagation factor. L_{0l} is the distance of the beam waist from mirror 1, given by (5.20).

Figure 5.30 presents the beam parameter product as a function of the g-parameters for stable resonators of equal length $L=1m$ and different locations of the active medium with radius $b=5mm$. The best beam quality is attained near the stability limits at positive g-parameters since the Gaussian beam radii go to infinity in this area. If the active medium is placed in the middle of the resonator, the concentric resonator generally exhibits the worst beam quality. Note that the beam parameter product in multimode operation does not depend on the wavelength! This means that the size of the focal spot does not depend on the type of laser material used. This is due to the fact that a smaller wavelength will generate a smaller radius Gaussian beam which in turn increases the number of transverse modes fitting into the medium. Therefore, the product $M^2\lambda$ is a constant of the resonator (see (5.66)).

Beam Quality of Laser Systems

We have discussed the beam parameter product and its dependence on the resonator parameters. Now let us generate a small focal spot size for a given laser beam (Fig. 5.31). We focus a beam with a diameter d_0 at the waist and a full angle of divergence Φ, both defined by the second intensity moments (2.93/94) or (2.97/98), by means of a lens with focal length f. According to (5.44), the beam diameter d'_0 in the focus is given by:

$$d_0' = d_0 \frac{f}{\sqrt{z_M^2 + (z-f)^2}} \tag{5.67}$$

with: $$z_m = \frac{d_0}{\Phi}$$

and z being the distance of the beam waist d_0 from the lens.

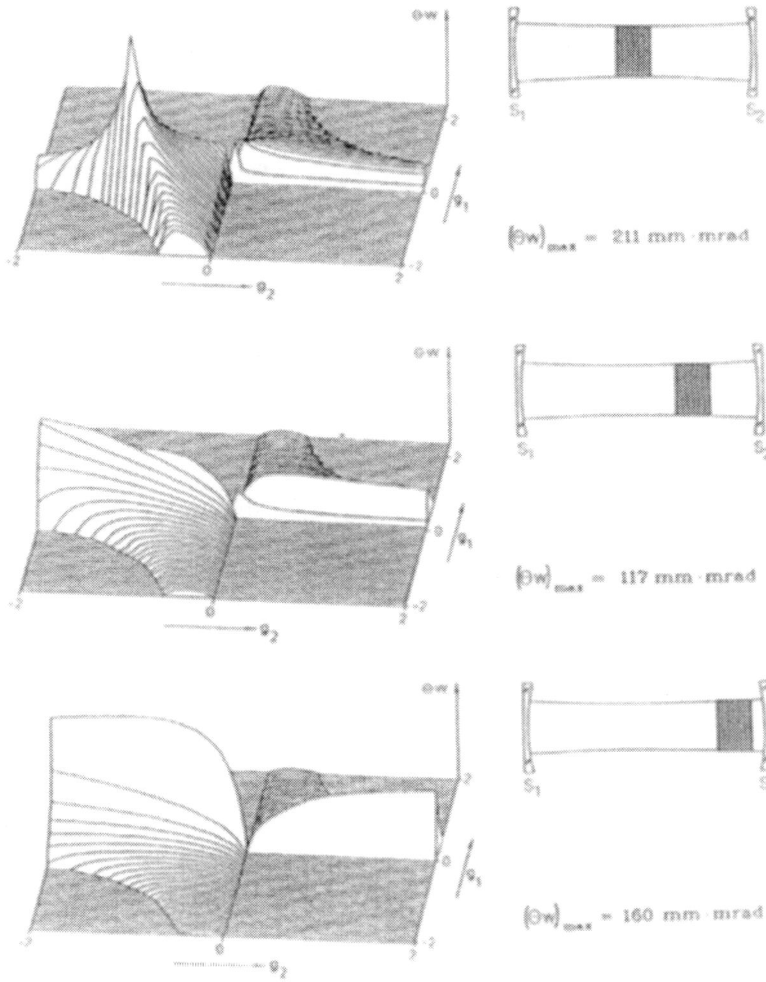

Fig. 5.30 Beam parameter products in the stability diagram for different positions of the active medium ($b=5mm$, $\ell=100mm$). The resonator length is always $L=1m$. For each graph the maximum beam parameter product is given [S.5].

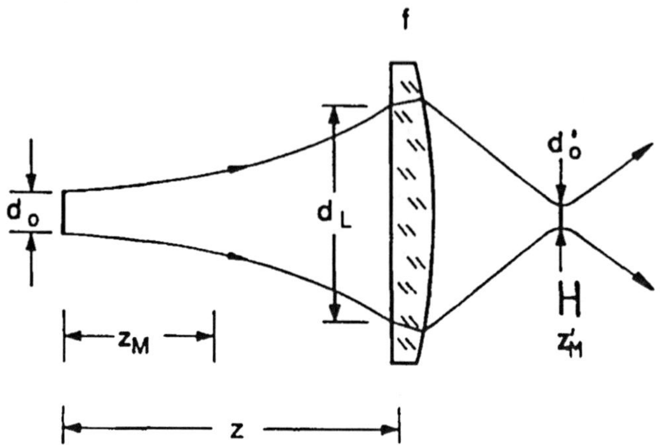

Fig. 5.31 Focusing of a laser beam with beam waist diameter d_0 and angle of divergence $\Phi = d_0/z_M$.

In general, the exact waist position z is not known. This is not a serious problem if we know the beam diameter d_L at the lens. The propagation law (5.62) yields:

$$d_L = \frac{4M^2\lambda}{\pi d_0}\sqrt{z_M^2 + z^2} \tag{5.68}$$

By inserting (5.68) into (5.67) and making the assumption $z \gg f$, we get:

$$d_0' = \frac{4M^2\lambda f}{\pi d_L} \tag{5.69}$$

The new Rayleigh range z'_M can be calculated with:

$$z_M' = \frac{\pi d_0'^2}{4\lambda M^2} \tag{5.70}$$

Since the beam propagation factor is a constant of the beam, the spot size can be decreased by either using a shorter focal length lens or by increasing the beam diameter at the lens. This is the reason laser beams are generally expanded by means of a telescope in front of the focusing lens. But keep in mind that this will also decrease the Rayleigh range z'_M because the beam parameter product is preserved. Table 5.3 presents typical beam propagation factors M^2, Rayleigh ranges z'_M, and spot diameters d'_0 for different laser systems. A beam diameter at the focusing lens of $d_L = 10mm$ (calculated using (5.69) and (5.70)).

Table 5.3 Focusing properties of different laser systems at several output powers (focal length $f=100mm$, beam diameter at lens $d_L=10mm$).

Laser	P_{out}[W]	M^2	d'_0[mm]	z'_M[mm]
HeNe (λ=0.633μm)	0.005	1	0.008	0.079
Nd:YAG (λ=1.064μm)	20	1	0.0136	0.136
(lamp-pumped rod)	400	50	0.677	6.77
	1,500	60	0.813	8.13
CO_2 (λ=10.6μm)	500	1	0.135	1.35
	3,000	2	0.270	2.70
	10,000	3	0.405	4.05

Special Focusing Optics

1) Focusing of beams with constant beam divergence
For some resonators with an internal variable lens (thermal lensing) the beam waist w is changed as the pump power is increased, whereas the angle of divergence θ remains constant. Despite this behavior a constant spot size can be attained by placing the beam waist in the front focal plane of the focusing lens (Fig. 5.32). In this case the focal spot is found in the back focal plane with $w'=f\theta$.

2) Focusing of beams with constant waist radius
If the beam waist radius stays constant and the angle of divergence varies, focusing optics can also be designed so that the spot size is preserved. This beam behavior is typical for resonators with thermal lensing. Furthermore, the radiation exiting a fiber falls into this category since the core radius of the fiber defines the beam waist. Focusing is accomplished by imaging the beam waist with a telescope with magnification $f_2/f_1<1$, as shown in Fig. 5.33. The position z' of the focus and the spot radius w' are given by:

$$z' = -\left(\frac{f_2}{f_1}\right)^2 z + f_2\left(1+\frac{f_2}{f_1}\right) \quad , \quad w' = w\frac{f_2}{f_1}$$

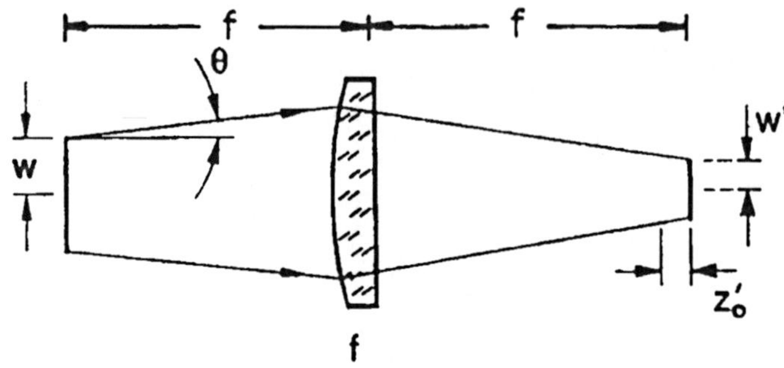

Fig. 5.32 If the beam waist is located at the front focal plane of a focusing lens, the spot radius w' is only a function of the angle of divergence θ and the focal length *f*.

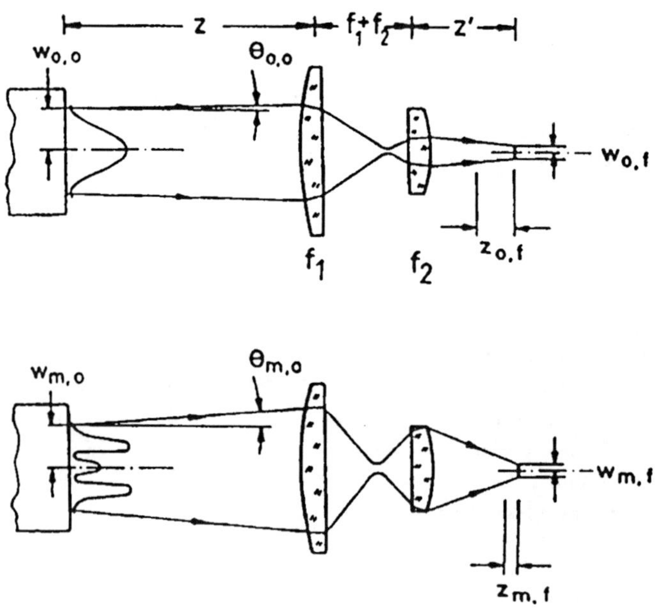

Fig. 5.33 Focusing of laser beams having a constant beam waist by means of a telescope. The higher order mode (lower graph) with its increased angle of divergence exhibits the same spot size as the fundamental mode. The Rayleigh range, however, is decreased.

5.3 Aperture Limited Stable Resonators

In the preceding section we have calculated the transverse eigenmodes for stable resonators without any limiting apertures inside the resonator. We have only used an aperture to determine the maximum number of transverse modes that can oscillate. A transverse mode represented an eigensolution of the resonator if its beam radius was smaller than the aperture radius. However, the aperture will also generate diffraction losses which are the result of a changed transverse mode structure. For instance, if we decrease the aperture radius so that it becomes smaller than the fundamental mode beam radius, the angle of divergence will increase resulting in an increased power fraction hitting the aperture after each round trip (Fig. 5.34). The intensity distribution at the mirrors is no longer Gaussian and the beam propagation inside the resonator does not follow the Gaussian beam propagation rules. The losses of the fundamental mode will decrease as we open the aperture again and the intensity profile will approach that of a Gaussian beam. If the aperture is much larger than the Gaussian beam radius, the diffraction losses go to zero and we will observe the Gaussian beam again. The next higher transverse mode has by then also decreased its diffraction losses far enough to reach the laser threshold. The output beam will then consist of a superposition of the two lowest order transverse modes.

If we plotted the diffraction loss as a function of the aperture radius we would expect a graph as shown in Fig. 5.35 (for circular symmetry). With increasing aperture radius the losses of an increased number of transverse modes will go below a certain loss threshold required to maintain a steady-state oscillation. Since the beam radius is proportional to $\sqrt{2p+\ell+1}$, it is a reasonable assumption that the diffraction losses of the TEMpℓ mode will exhibit the highest slopes if the aperture radius is varied around a value of about $\sqrt{2p+\ell+1}$ times the Gaussian beam radius.

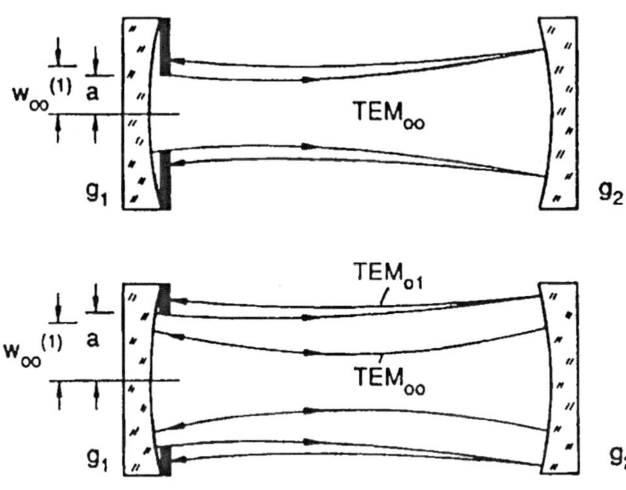

Fig. 5.34 Influence of an aperture on the propagation of the fundamental mode and the next order transverse mode.

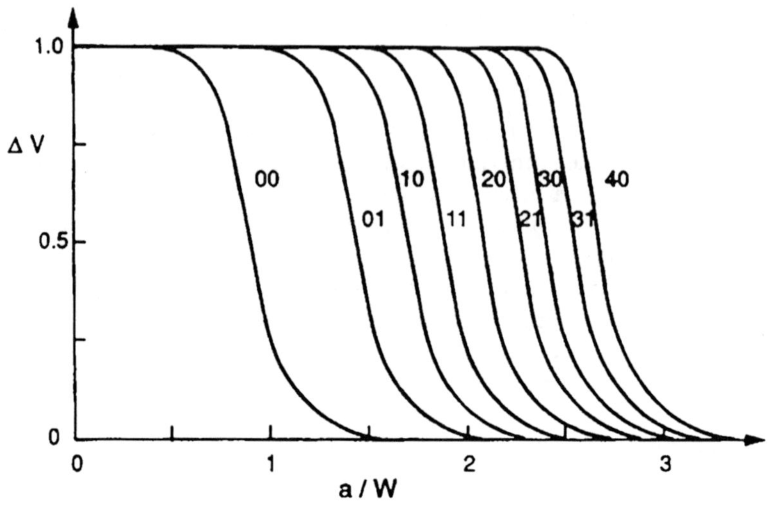

Fig. 5.35 Qualitative dependence of the diffraction losses per round trip of different circularly symmetric modes $TEM_{p\ell}$ on the ratio of aperture radius to Gaussian beam radius a/w.

The detailed knowledge of the diffraction losses of the transverse modes is crucial for the optimized design of a laser resonator, especially for lasers with low gain media. The pump power required to reach threshold is determined by the losses the radiation experiences in a round trip. The gain factor per transit G_0 has to compensate the losses generated by output coupling and diffraction. The threshold condition reads:

$$G_0 \geq \frac{1}{\sqrt{R_1 R_2 V}}$$

where R_i is the reflectance of mirror i, and V is loss factor per round trip (=1-loss). Furthermore, the efficiency of a laser is strongly affected by internal losses. As discussed for the FPI resonator, even low losses in the percent range can decrease the output power by orders of magnitude if the gain of the medium is low. A typical HeNe laser, for instance, would stop laser emission if the losses per round trip were increased only by 1-2%. In this case the resonator design has to be carefully optimized to attain lossless fundamental mode operation while preventing higher order modes from oscillating.

In the following, we will discuss the dependence of the diffraction losses of transverse modes on the g-parameters, the aperture radii, and the number and location of the apertures inserted into the resonator. Our main attention is on the fundamental mode since its losses determine the behavior of the laser at threshold. The time it takes to build up the threshold gain after onset of the pumping, is only affected by the losses of the fundamental mode, even in lasers operating in multiple transverse modes.

5.3.1 One Aperture Limited Mirror

Let us first consider the case that only one aperture is placed inside the resonator directly in front of mirror 1 [3.6,3.7,3.10,3.16-3.20] (Fig. 5.36). The field distributions at mirror 1 are eigensolutions of the Kirchhoff integral equation (5.3) for the round trip. However, the integral limits are not infinite, but given by the size of the aperture. In rectangular geometry with an aperture width of *2a* and an aperture height of *2b*, the integral equation for the fields of the transverse eigenmodes reads:

$$\gamma_{mn}E_{mn}(x_2,y_2) = i\frac{\exp[-ikL]}{2Lg_2\lambda_0}\int_{-b}^{b}\int_{-a}^{a}E_{mn}(x_1,y_1)\exp\left[\frac{-i\pi}{2Lg_f\lambda_0}\left(G(x_1^2+y_1^2+x_2^2+y_2^2)-2(x_1x_2+y_1y_2)\right)\right]dx_1dy_1$$

with: $G = 2g_1g_2-1$; (5.71)
L : optical resonator length $=nL_0$ (L_0: geometrical length)
λ_0: vacuum wavelength, $k=2\pi/\lambda_0$: wave number

In circular symmetry with a round aperture with radius a, the corresponding equation is given by:

$$\gamma_{pl}E_{pl}(r_2,\phi_2) = i\frac{\exp[-ikL]}{2Lg_2\lambda_0}\int_{0}^{2\pi}\int_{0}^{a}E_{pl}(r_1,\phi_1)\exp\left[\frac{-i\pi}{2Lg_f\lambda_0}\left(G(r_1^2+r_2^2)-2r_1r_2\cos(\phi_1-\phi_2)\right)\right]r_1dr_1d\phi_1$$

(5.72)

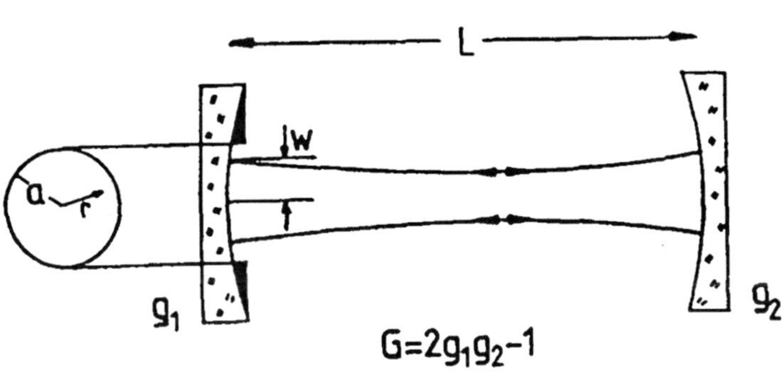

Fig. 5.36 Stable resonator with aperture limited mirror 1 (circular symmetry).

In the following we restrict the discussion to the circularly symmetric resonator. By separating the radial and the azimuthal field components according to:

$$E_{p\ell}(r,\Phi) = u_p(r)\exp[i\ell\Phi]$$

the integration over the azimuthal coordinate can be performed, resulting in:

$$\gamma_{p\ell}u_p(r_2) = (-i)^\ell 2\pi N_{eff}\exp[-ikL]\int_0^1 u_p(r_1)J_\ell(2\pi N_{eff}r_1r_2)\exp\left[-i\pi N_{eff}G(r_1^2+r_2^2)\right]r_1dr_1$$

with: r_i : normalized radial coordinate (5.73)
J_ℓ: : Bessel function of order ℓ
N_{eff}: effective Fresnel number (see below)

In contrast to the unconfined resonator, the loss factors $V=|\gamma_{p\ell}|$ are now lower than 1.0 since the eigenmodes exhibit diffraction losses at the aperture. Eq. (5.68) indicates that the losses and the mode structure depend only on two parameters:

the equivalent G-parameter $G=2g_1g_2-1$
the effective Fresnel number $Neff = a^2/(2Lg_2\lambda_0)$

The effective Fresnel number is proportional to the area of the aperture. Its relationship to the Gaussian beam radius $w_{00}^{(1)}$ at mirror 1 is given by:

$$\left[\frac{a}{w_{00}^{(1)}}\right]^2 = \pi\, N_{eff}\sqrt{1-G^2} \tag{5.74}$$

We see that stable resonators in fundamental mode operation with an aperture radius that is 1.2-1.5 times larger than the Gaussian beam radius exhibit an effective Fresnel number around 1.0. All resonators having the same absolute value of the equivalent g-parameter and the same absolute value of the effective Fresnel number exhibit the same eigenmodes and loss factors. These resonators are referred to as equivalent resonators. If we knew the functional relationship between the losses and the resonator parameters $|G|$, $|N_{eff}|$, we could determine the mode properties of any stable resonator provided that only one mirror is aperture limited.

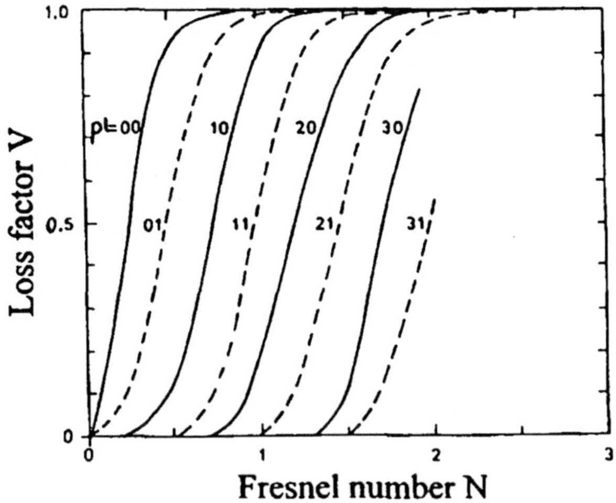

Fig. 5.37 Loss factor per round trip of different transverse modes in circular symmetry for $G=0.0$ as a function of the effective Fresnel number N. The aperture radius is equal to the Gaussian beam radius for $N_{eff}=1/\pi\approx 0.318$.

Examples of equivalent resonators:

1) $\rho_1=\infty$, $\rho_2=2$m, L=1m, a=0.8mm, $\lambda_0=1.064\mu$m
------------> $g_1 = 1.0$, $g_2 = 0.5$, $G = 0.0$, $N_{eff} = 0.602$, $a/w_{00}^{(1)} = 1.376$
2) $\rho_1=2.5$m, $\rho_2=1.333$m, L=0.5m, a=0.633mm, $\lambda_0=1.064\mu$m
------------> $g_1 = 0.8$, $g_2 = 0.625$, $G = 0.0$, $N_{eff} = 0.602$, $a/w_{00}^{(1)} = 1.376$
3) $\rho_1=0.2777$m, $\rho_2=0.3077$m, L=0.5m, a=0.633mm, $\lambda_0=1.064\mu$m
------------> $g_1 = -0.8$, $g_2 = -0.625$, $G = 0.0$, $N_{eff} = -0.602$, $a/w_{00}^{(1)} = 1.376$

The integral equation (5.73) generally cannot be solved analytically. Figure 5.37 presents the numerically calculated loss factors of different low order transverse modes as a function of the effective Fresnel number for resonators with $G=0.0$. The curves agree quite well with the expected dependence of Fig. 5.35. Since the losses are now plotted versus the square of the aperture radius, the curves are almost equally spaced.

In order to determine the losses of all transverse modes of a stable resonators, we have to make graphs like the one shown in Fig. 5.37 for all values of the G-parameter between 0.0 and 1.0. This time-consuming procedure is not really necessary since only the properties of the fundamental mode are of practical interest. The loss of the fundamental mode determines both the laser threshold and the efficiency, whereas the losses of higher transverse modes affect only the output power. Figure 5.38 shows calculated loss factors per round trip in fundamental mode operation as a function of the equivalent G-parameter, a comparison with experimental data is shown in Fig. 5.39. As to be expected, the losses are symmetric at $G=0$ since resonators having the same absolute value of G are equivalent. Note that the diffraction losses for a constant ratio of aperture radius to Gaussian beam radius depend on the G-parameter and go to zero as the G-parameter approaches 1.0. This means that resonators close to the stability limits experience the lowest loss for the fundamental mode. In order to minimize the fundamental mode loss this graph encourages us to increase the aperture radius as much as possible. However, to prevent the next transverse mode from oscillating it is necessary to choose an aperture radius of 1.2-1.4 times the Gaussian beam radius. The larger $|G|$ and the higher the gain of the medium, the closer the aperture radius can approach the Gaussian beam radius.

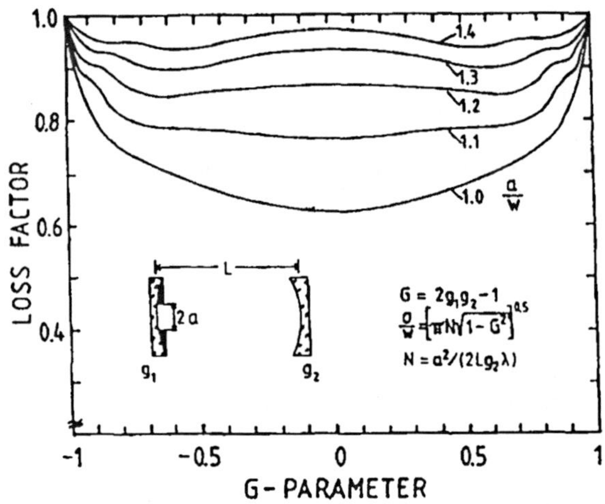

Fig. 5.38 Calculated loss factors per round trip of the fundamental mode in stable resonators with a circular aperture of radius a as a function of the equivalent G-parameter $G=2g_1g_2-1$. The aperture radius is shown normalized to the Gaussian beam radius w.

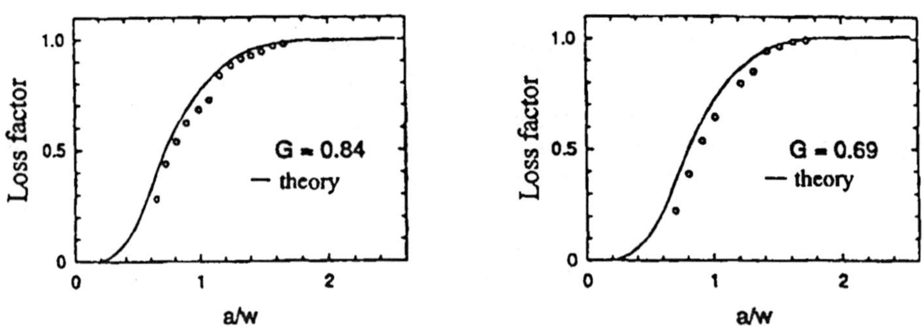

Fig. 5.39 Measured loss factors per round trip as a function of the ratio of the aperture radius a to the Gaussian beam radius w for two different G-parameters. The solid line represents the numerically calculated loss factor [S.6].

The loss per round trip of the fundamental mode and the next transverse mode TEM_{01} can be approximated by using the empirical relation [3.39,3.41]:

$$\Delta V = \exp\left[-\alpha N_{eff}^{\beta}\right] \tag{5.75}$$

The parameters α and β read:

G	0.0	0.2	0.4	0.5	0.6	0.7	0.8	0.85	0.9	0.95	0.99
TEM_{00}:											
α	8.4	6.9	4.9	4.4	3.8	3.5	2.9	2.6	2.3	2.0	1.83
β	1.84	1.66	1.38	1.34	1.34	1.27	1.16	1.08	1.01	0.86	0.59
TEM_{01}:											
α	5.1	4.3	2.84	2.46	2.18	1.86	1.5	1.34	1.18	1.05	1.02
β	2.69	2.46	1.91	1.83	1.84	1.81	1.58	1.46	1.35	1.12	0.84

Diffraction Losses in Multimode Operation

If the aperture radius is much larger than the Gaussian beam radius, all transverse modes fitting into the aperture oscillate simultaneously. Only those modes whose beam radii are close to the aperture radius exhibit noticeable diffraction losses. As the aperture is increased the loss of the highest order mode will decrease until the next mode starts oscillating and the loss increases again. The loss will thus show an oscillating behavior as a function of the aperture radius. A numerically calculated example is presented in Fig. 5.40. As soon as the next transverse mode reaches the laser threshold, the loss factor decreases again whereas the mode volume becomes higher due to the larger beam radius of this mode. The oscillation depth will, however, become smaller with increasing aperture radius and the loss factor will slowly approach 1.0. In multimode lasers the number of transverse modes is on the order of 100 resulting in a round trip loss of 1.0-1.5% (Fig. 5.41). This loss has to be taken into account if the output power is calculated (see Chapter 10). Although a loss of 1% seems to be small, it has a considerable effect on the design of high power lasers. For a laser with an output power of 1kW, the intracavity power is typically 2kW which means that a power of 20W falls onto the aperture. Therefore, it is necessary to cool the aperture in order to prevent damage.

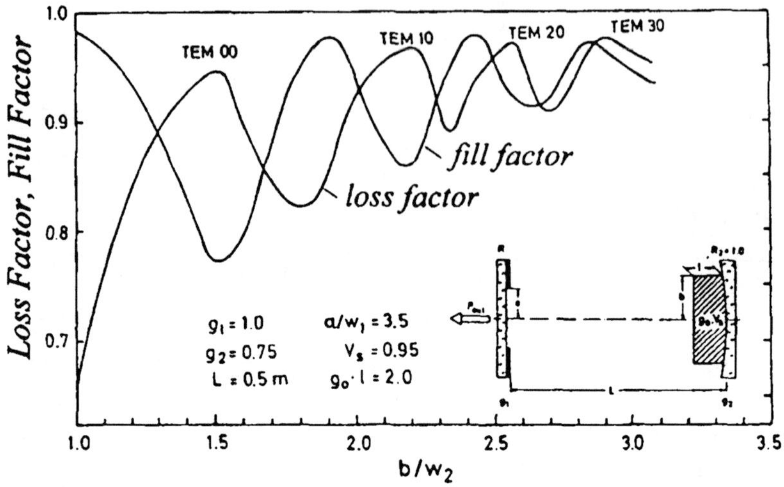

Fig. 5.40 Calculated loss factor per round trip and fill factor (mode volume normalized to the volume of the active medium) for a stable resonator as a function of the ratio of the radius b of the active medium to the Gaussian beam radius w_2. The aperture radius a at mirror 1 is held constant at $a=3.5w_1$. w_1, w_2 refer to the Gaussian beam radii at mirrors 1 and 2, respectively. The oscillations of both the loss factor and the fill factor become less pronounced as the transverse mode order increases. Only modes without azimuthal structure ($\ell=0$) were calculated (small signal gain $g_0l= 2.0$, loss factor per transit through the medium $Vs=0.95$).

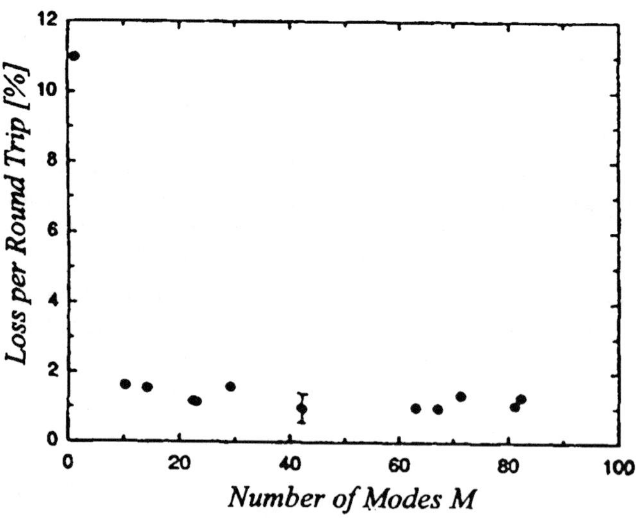

Fig. 5.41 Measured loss per round trip of an Nd:YAG laser in multimode operation as a function of the number of transverse modes M. The loss was determined by measuring the power hitting the aperture with respect to the total intracavity power.

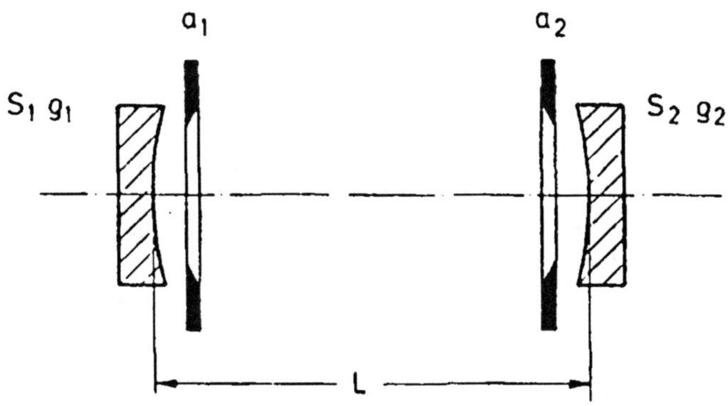

Fig. 5.42 Stable resonator with two aperture limited mirrors.

5.3.2 Both Mirrors Aperture Limited

This case is much more complicated to deal with since the mode properties depend on three parameters rather than only on two parameters. If the aperture radii at mirror 1 and 2 are a_1 and a_2, respectively (Fig. 5.42), the loss per round trip of each transverse mode is a function of the modified g-parameters g_1a_1/a_2, g_2a_2/a_1, and the Fresnel number N with:

$$N = \frac{a_1 a_2}{\lambda L} \tag{5.76}$$

It is, therefore, difficult to present a general overview of the mode properties of these resonators. Furthermore, the number of publications dealing with double aperture resonators is very limited [3.16,3.18,3.35,3.36].

In order to calculate the loss factor and the mode structure we have to apply the Kirchhoff integral to both transits inside the resonator. First we propagate the field from mirror 1 to mirror 2 and then back again. We thus get two coupled integral equations:

$$u_{2p}(r_2) = (-i)^l 2\pi N \exp[-ikL] \int_0^1 u_{1p}(r_1) J_l(2\pi N r_1 r_2) \exp\left[-i\pi N\left(g_1\frac{a_1}{a_2}r_1^2 + g_2\frac{a_2}{a_1}r_2^2\right)\right] r_1 dr_1$$

$$\gamma_{pl} u_{1p}(r_1) = (-i)^l 2\pi N \exp[-ikL] \int_0^1 u_{2p}(r_2) J_l(2\pi N r_1 r_2) \exp\left[-i\pi N\left(g_2\frac{a_2}{a_1}r_2^2 + g_1\frac{a_1}{a_2}r_1^2\right)\right] r_2 dr_2$$

with $u_{ip}(r_i)$: radial field distribution at mirror i (5.77)
r_1, r_2 : normalized radial coordinates
N : Fresnel number = $a_1 a_2/(\lambda L)$

The loss per round trip is given by $\Delta V_R = 1-|\gamma_{pl}|^2$. It is common to define the loss per transit via $\Delta V_T = 1-\sqrt{|\gamma_{pl}|^2}$, which represents the average loss per transit since the losses are generally different for the two directions. Figure 5.43 presents calculated losses per transit of the TEM_{00} mode and the TEM_{01} mode for different resonators as a function of the Fresnel number N for symmetric apertures ($a_1=a_2=a$).

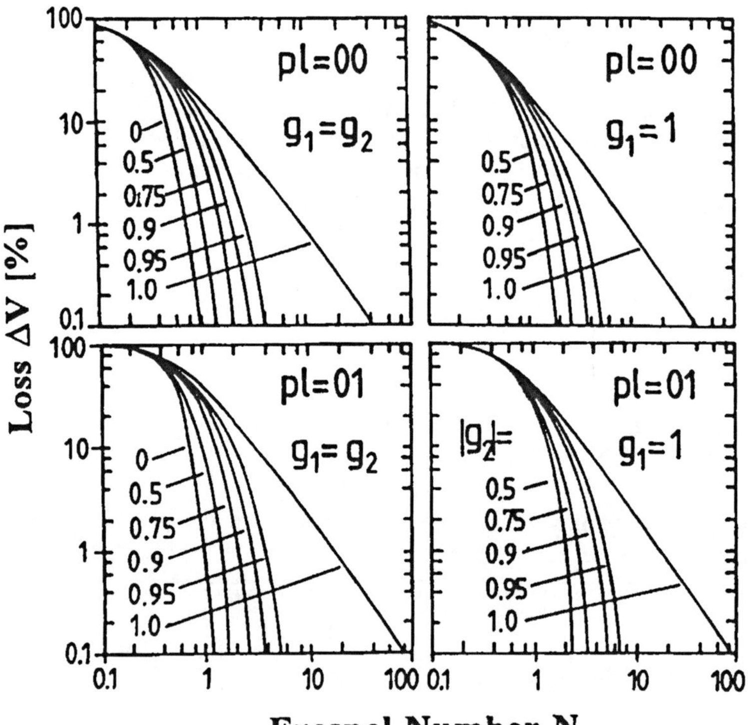

Fig. 5.43 Diffraction losses per transit of stable resonators in circular symmetry with both mirrors limited by apertures with radius $a_1=a_2=a$ as a function of the Fresnel number $a^2/(\lambda L)$. The losses for the fundamental mode and the next order transverse mode are shown. The curve parameter is the g-parameter of mirror 2 [after 3.16].

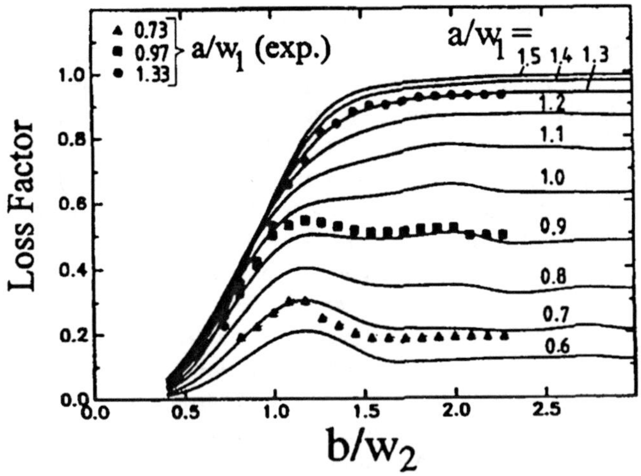

Fig. 5.44 Calculated and measured loss factor per round trip of a stable resonator in fundamental mode operation as a function of the aperture radius b at mirror 2. The curve parameter is the aperture radius a at mirror 1. The Gaussian beam radii at mirror 1 and mirror 2 are w_1 and w_2, respectively. Note that the losses decrease when the aperture with radius b truncates the Gaussian beam. Resonator parameters: $g_1=1.0$, $g_2=0.5$, $L=0.5m$ (pulsed Nd:YAG rod laser, λ=1.064μm).

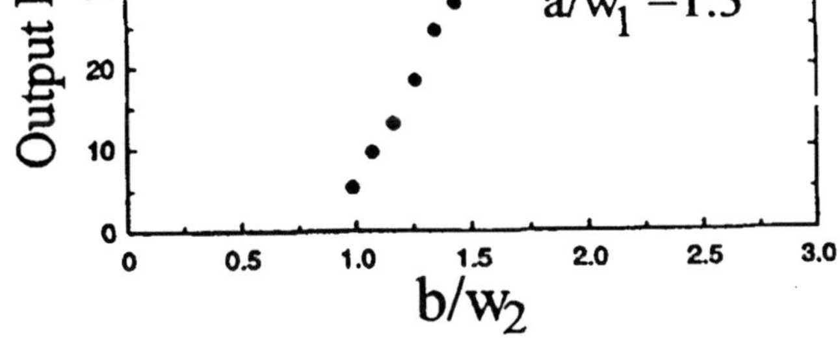

Fig. 5.45 Measured output power per pulse of a stable resonator in fundamental mode operation as a function of the ratio of the aperture radius b to the Gaussian beam radius w_2 at mirror 2. The output coupling mirror 1 is limited by an aperture with radius $a=1.3w_1$. Resonator parameters: $g_1=1.0$, $g_2=0.56$, effective length $L=0.66m$ (6x3/8" Nd:YAG rod, small-signal gain $g_0\ell=2.2$, reflectance $R=0.7$).

The losses of stable resonator with one aperture limited mirror do not necessarily increase if a second aperture is inserted in front of the second mirror. This is due to the fact that diffraction increases the on-axis intensity. Diffraction at the second aperture, therefore, may decrease the power fraction hitting the first aperture. This effect is shown in Figs. 5.44 and 5.45 for a stable resonator in fundamental mode operation. In both figures, the aperture radius at mirror 1 is remained fixed and the losses and the output power are shown as a function of the radius of the second aperture. If both apertures truncate the Gaussian beam, the loss factor and, consequently, the output power may be higher than with one aperture alone. However, this does not mean that the maximum efficiency in fundamental mode operation is attained with two apertures. In general, the highest output power is achieved if only one aperture is located inside the resonator (see Chapter 11).

The special case of symmetric resonators with both the aperture radius and the radius of curvature being equal for the two mirrors can be discussed by using the properties of an equivalent resonator with one aperture [3.12,3.37]. Any resonator with one aperture limited mirror (mirror 1) and a non-vanishing g-parameter of mirror 2 can be transformed into a symmetric resonator having the same aperture at both mirrors (Fig. 5.46). The transit in this equivalent symmetric resonator is equivalent to the round trip in the resonator with one aperture. The imaging properties of the unconfined mirror 2 are taken into account by changing the resonator length and the mirror curvatures. Since we get the same ray transfer matrix, the loss per round trip in the original resonator with g-parameters g_1,g_2 and length L is the same as the loss per transit in the equivalent symmetric resonator with g-parameters $G=2g_1g_2-1$ and length $2Lg_2$. The loss per transit of a resonator with $g_1=g_2=g$ and aperture radii $a_1=a_2=a$ can thus be determined by using Fig. 5.38 (and Eq. (5.75) with $G=g$ and $N_{eff}=a^2/(\lambda L)$).

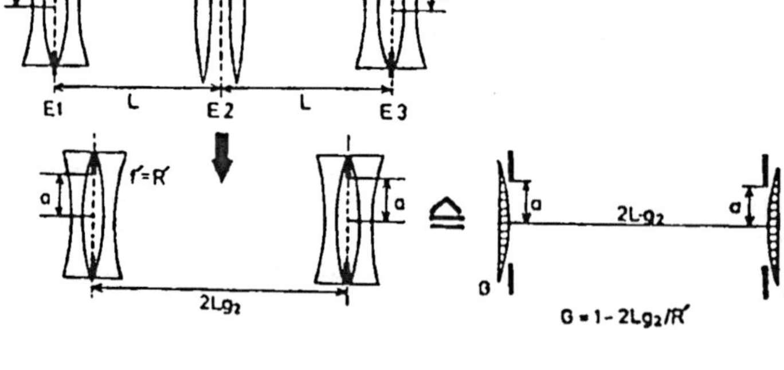

Fig. 5.46 The round trip in a resonator with one aperture is equivalent to a transit in the equivalent symmetric resonator. The transformation between the two resonators is shown with each mirror being replaced by two lenses. The reference planes for the ray transfer matrix are indicated by broken lines. R' denotes the mirror curvature of the equivalent resonator.

5.4 Misalignment Sensitivity

The misalignment sensitivity of a resonator is defined as the sensitivity with which the diffraction losses or the output power are changed due to mirror tilt. In this section we will only discuss the influence of the mirror misalignment on the resonator losses. The resulting variation of the output power will be dealt with in Chapter 14.

The geometrical effect of a mirror tilt is shown in Fig. 5.47 for a resonator with one aperture-limited mirror. Rotation of mirror j by an angle α_j results in a rotation of the optical axis by an angle θ_j with the center of curvature of mirror i being the pivot point. As in the aligned resonator, the optical axis is defined by the line going through the centers of curvature of the mirrors. The angle of rotation of the optical axis θ_j, also referred to as the pointing stability, and the shifts Δ_{jj}, Δ_{ij} of the intersecting points of the optical axis on mirror j and mirror i, respectively, read:

$$\theta_j = \alpha_j \frac{1-g_i}{1-g_1 g_2} \tag{5.78}$$

$$\Delta_{jj} = \alpha_j \frac{L g_i}{1-g_1 g_2} \tag{5.79}$$

$$\Delta_{ij} = \alpha_j \frac{L}{1-g_1 g_2} \tag{5.80}$$

where g_i is the g-parameter of mirror i, and $i,j=1,2$ with $i \neq j$.

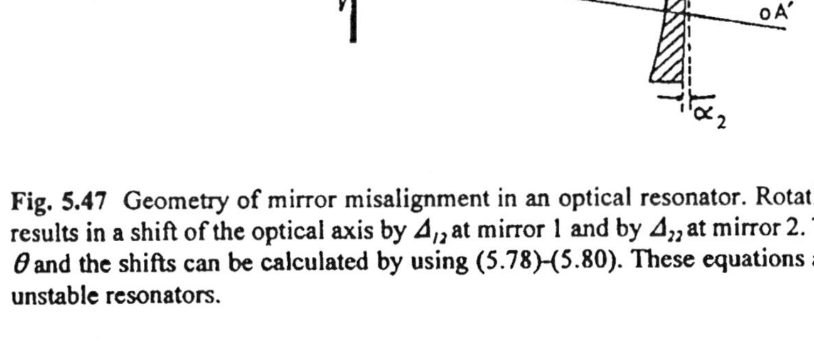

Fig. 5.47 Geometry of mirror misalignment in an optical resonator. Rotation of mirror 2 by α_2 results in a shift of the optical axis by Δ_{12} at mirror 1 and by Δ_{22} at mirror 2. The pointing stability θ and the shifts can be calculated by using (5.78)–(5.80). These equations are also applicable to unstable resonators.

The transverse eigenmodes will keep oscillating parallel to the optical axis whether the resonator is aligned or not. Furthermore, as far as stable resonators are concerned, the mode structure also stays almost symmetric to the optical axis. Since the optical axis comes closer to one side of the aperture if one or both mirrors are tilted, additional diffraction losses are generated. The effective aperture radius is decreased by the shift of the optical axis. In Fig. 5.47 this means that the aperture radius now is $a-\Delta_{12}$ rather than a.

If Δ denotes the total shift of the optical axis at the aperture, to a good approximation, we can describe the properties of the tilted resonator by using an effective Fresnel number $N_{eff}(\alpha)$ which takes the reduction of the aperture into account:

$$N_{eff}(\alpha) = \frac{(a-\Delta)^2}{2Lg_2\lambda_0} \approx N_{eff}\,(1 - 2\Delta/a) \tag{5.81}$$

The Fresnel number will decrease as the angle of rotation α is increased resulting in an increase of the diffraction losses and a corresponding decrease of the loss factor. For all linear resonators, stable as well as unstable ones, the loss factor decreases parabolically with the angle of rotation for small mirror tilts:

$$V(\alpha) = V(0)\left[1 - 0.1\left(\frac{\alpha}{\alpha_{10\%}}\right)^2\right] \tag{5.82}$$

with $V(0)$ being the loss factor for the aligned resonator (Fig. 5.48). The angle $\alpha_{10\%}$ denotes the angle of misalignment at which the loss factor has decreased by 10% and consequently the losses have increased by 10%. This angle is used to define the misalignment sensitivity of optical resonators. A low misalignment sensitivity is equivalent to a small 10%-angle $\alpha_{10\%}$.

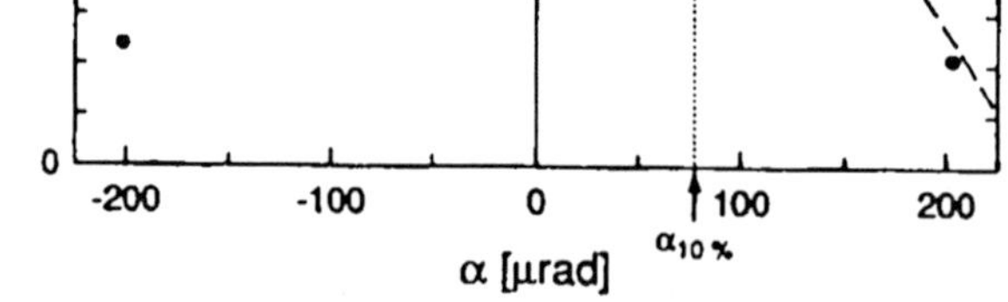

Fig. 5.48 Measured loss factors per round trip of a stable resonator as a function of the angle of misalignment of mirror 2. Mirror 1 is limited by an aperture with radius a ($G=0.34$, $L=0.7m$, $a=0.55mm$, $\lambda_0=1.064\mu m$).

Typical values of $\alpha_{10\%}$ for a stable resonator in fundamental mode operation are on the order of 50 μrad. The exact value of the 10%-angle depends on the g-parameters of the mirrors, the Fresnel number, the resonator length, and the aperture radii. All resonators exhibit two 10%-angles, each corresponding to the tilt of one mirror. It is customary to define an average misalignment sensitivity by taking the geometrical mean value of the two 10%-angles:

$$\alpha_{10\%} = \frac{1}{2}\sqrt{\alpha^2_{10\%,1} + \alpha^2_{10\%,2}} \tag{5.83}$$

with the additional indices indicating the corresponding mirror. This mean angle defines the average angle by which both mirrors can be rotated simultaneously before a 10% increase in diffraction losses is generated.

A theoretical investigation of the Kirchhoff integral for misaligned resonators shows that the 10%-angle is proportional to the resonator length and inversely proportional to the aperture radius. In order to compare the misalignment sensitivity of different resonators it is, therefore, advantageous to introduce the misalignment parameter D_i for the mirrors:

$$D_i = \frac{L}{a}\alpha_{10\%,i} \quad , \qquad i=1,2 \tag{5.84}$$

The misalignment parameters of the two mirrors depend only on the g-parameters and on the effective Fresnel number.

5.4.1 One Aperture Limited Mirror

Let us consider a resonator with mirror 1 being limited by an aperture with radius a and with the unconfined mirror 2 being misaligned by an angle α_2, as shown in Fig. 5.47. The mode structures and the diffraction loss can be calculated as a function of the angle of misalignment by using a two-dimensional Kirchhoff integral. Starting at mirror 1 with the field $E(r,\phi)$, the round trip in the resonator reads:

$$\gamma E(r_2,\phi_2) = -i\,\exp[ikL]\; N_{eff}\,\exp\left[\frac{-8\pi i D_2^2}{G-1}\right] \cdot \tag{5.85}$$

$$\cdot \int_0^{2\pi}\int_0^1 E(r_1,\phi_1)K(r_1,r_2,\phi_1,\phi_2)\;\exp\left[i4\pi D_2 N_{eff}(r_1\cos\phi_1 + r_2\cos\phi_2)\right]\, r_1 dr_1 d\phi_1$$

where K is the kernel of the aligned resonator according to (5.67) and r_i being the normalized radial coordinates.

The resonator properties depend on the equivalent G-parameter $G=2g_1g_2-1$, the effective Fresnel number N_{eff}, and the misalignment parameter D_2. Figure 5.49 presents the calculated misalignment parameter D_2 for stable resonators in fundamental mode operation as a function of the effective Fresnel number N_{eff} for different equivalent G-parameters G. The comparison with experimental data is shown in Fig. 5.50.

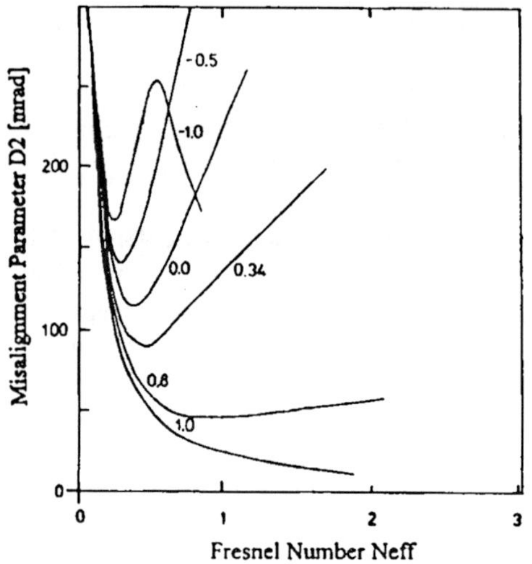

Fig. 5.49 Calculated misalignment parameter for stable resonators in fundamental mode operation as a function of the effective Fresnel number $N_{eff}=a^2/(2Lg_2\lambda_0)$. Mirror 1 is limited by the aperture and the unconfined mirror is tilted (circular symmetry).

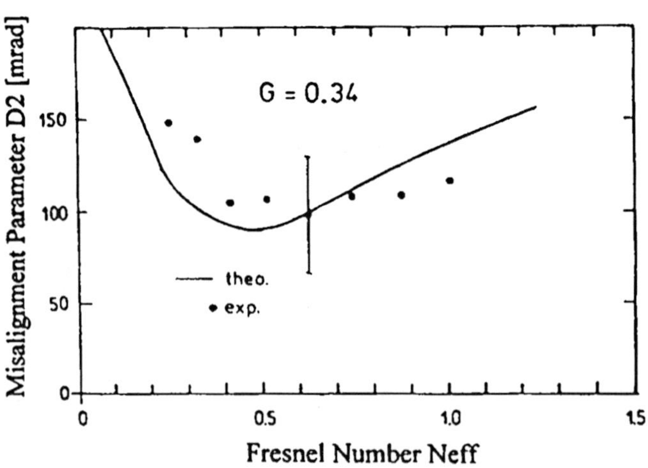

Fig. 5.50 Calculated and measured misalignment parameter D_2 for resonators in fundamental mode operation with $G=0.34$ as a function of the effective Fresnel number. Mirror 1 is aperture limited and the unconfined mirror 2 is misaligned (circular symmetry, Nd:YAG laser, $\lambda_0=1.064\mu m$).

Of particular interest are resonators in fundamental mode operation with the aperture a being adapted to the Gaussian beam radius w_{00}. Figure 5.51 shows the misalignment parameter of the unconfined mirror for stable resonators with $a/w_{00}=1.3$. This figure indicates how strongly the misalignment sensitivity depends on the location of the resonator in the stability diagram. Note that the misalignment parameter is not a function of the wavelength λ_0 because the ratio of aperture radius to Gaussian beam radius is only a function of N_{eff} and G (see (5.74)). However, the Gaussian beam radius, and therefore the aperture radius a, are proportional to $\sqrt{\lambda_0 L}$ (see (5.10)) which means that the 10%-angle is larger for longer wavelength lasers. The lowest misalignment sensitivity is attained near the axes of the stability diagram ($G=-1$), whereas concave-convex resonators with G being close to +1.0 exhibit a sensitivity to mirror tilt. A high mode volume of the fundamental mode (see Fig. 5.19) and a low misalignment sensitivity, therefore, can generally not be achieved simultaneously.

So far we have discussed only the misalignment of the unconfined mirror 2 which results in a shift Δ_{12} at the aperture limited mirror 1. If mirror 1 is tilted we get a shift Δ_{11} at mirror 1. If both mirrors are rotated by the same angle the two shifts are related to each other via:

$$\Delta_{11} = \Delta_{12}\, g_2 \tag{5.86}$$

A rotation of the unconfined mirror 2 by an angle α is thus equivalent to the rotation of the aperture limited mirror 1 by the angle α/g_2. The misalignment parameter D_1 can be calculated by dividing the misalignment parameter D_2 by the g-parameter of mirror 2. We can now determine the sensitivity of all resonators in fundamental mode operation to tilt of one of the two mirrors by using Figs. 5.49-5.51.

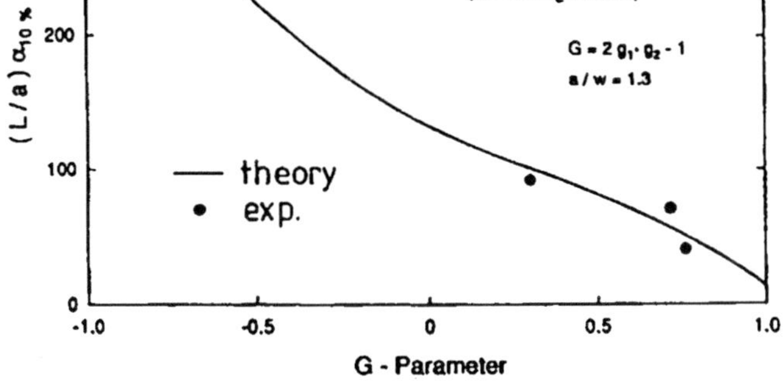

Fig. 5.51 Measured and calculated misalignment parameter D_2 of stable resonators in fundamental mode operation with the aperture radius a at mirror 1 being adapted to the Gaussian beam radius ($a/w_{00}=1.3$). The unconfined mirror 2 is misaligned.

The mean 10%-angle is given by:

$$\alpha_{10\%} = \frac{a}{2L}\sqrt{D_2^2 + D_1^2} = \frac{a}{2L} D_2 \sqrt{1 + (1/g_2)^2} \tag{5.87}$$

For reliable operation as well as safe shipping and handling of the laser system, each mirror should not be misaligned by more than this mean angle. This defines the tolerances required for the stability of the mirror mounts. Resonators with $g_2=0$ exhibit the lowest misalignment sensitivity. Misalignment of mirror 1 does not increase the losses at all (in the geometrical model used we do not get a shift of the optical axis intersection point on mirror 1 since the center of curvature of mirror 2 is located at this mirror) and the sensitivity of mirror 2 to tilt is low compared to other stable resonators (see Fig. 5.51).

Example: Misalignment Sensitivity of a CO_2 Laser Resonator
Resonator geometry: $\rho_1=3m$, $\rho_2=-4m$, $L=1m$, $\lambda_0=10.6\mu m$, $g_1=0.667$, $g_2=1.25$. The cylindrical gas tube with diameter 2a=8.75mm is positioned close to mirror 1. The Gaussian beam radius at mirror 1 is 3.37mm which means that the ratio of the tube radius to the Gaussian beam radius is 1.3. With $G=2g_1g_2-1=0.666$, Fig. 5.51 yields a misalignment parameter of about D_2=75mrad. The 10%-angles for the misalignment of the mirrors are:

$$\alpha_{10\%,2} = \frac{a}{L}D_2 = 0.328\ mrad\ , \quad \alpha_{10\%,1} = \frac{\alpha_{10\%,2}}{g_2} = 0.263\ mrad$$

These values result in a mean 10%-angle of 0.21 mrad.

Multimode Operation
If the aperture radius is much larger than the Gaussian beam radius, higher order modes will also oscillate in the misaligned resonator. The diffraction losses, however, will only increase if the optical axis is rotated so far that the Gaussian beam gets limited by the aperture. For smaller tilts, the misalignment only results in a decrease in the number of transverse modes without noticeably increasing the diffraction losses (Fig. 5.52). However, the output power will show a decrease since the mode volume becomes smaller as the mirrors are tilted.

If $K=1/M^2$ denotes the beam quality factor of the aligned resonator (see (5.60) and (5.66)), the 10%-angle in multimode operation, to a good approximation, is given by:

$$\alpha_{10\%,pl} = \frac{1}{\sqrt{K}}\,\alpha_{10\%,00} \tag{5.88}$$

with $\alpha_{10\%,00}$ being the 10%-angle for fundamental mode operation with adapted aperture radius $a/w_{00}=1.3$.

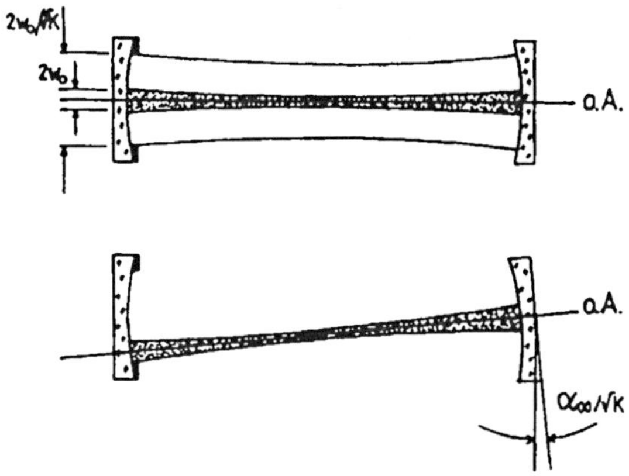

Fig. 5.52 Mirror misalignment in a multimode stable resonator results in an increase in diffraction losses as soon as the fundamental mode gets clipped by the aperture. Since the Gaussian beam radius is $\sqrt{K}$ times smaller than the multimode beam radius (*K:* beam quality factor, $K<1$), the multimode 10%-angle is $1/\sqrt{K}$ times larger than the 10%-angle for fundamental mode operation.

5.4.2 Two Aperture Limited Mirrors

If both mirrors are limited by apertures, the mirror misalignment generates additional diffraction losses at both apertures (Fig. 5.53). Similar to the aligned resonators, the theoretical investigation is more complicated and beyond the scope of this book. In the following we will summarize the main results for fundamental mode operation [3.27,3.33]. Let both aperture radii a_1,a_2 be adapted to the Gaussian beam radii $w_{00}^{(1)}, w_{00}^{(2)}$ at the mirrors:

$$a_i = s\, w_{00}^{(i)} \tag{5.89}$$

If mirror i is misaligned by an angle α_i, a first order perturbation analysis of the diffraction integral yields for the loss factor per round trip [3.33]:

$$V(\alpha_i) = V(0)\left[1-\alpha_i^2\,\frac{s^2}{\exp[2s^2]-1}\,D_i^2\right] \tag{5.90}$$

with:
$$D_i^2 = \frac{\pi L^*}{\lambda_0}\sqrt{\left(\frac{g_j}{g_i}\right)}\,\frac{1+g_1g_2}{(1-g_1g_2)^{1.5}}$$

$L^* =$	$L_0-(n-1)\ell/n$: effective resonator length
L_0 :	geometrical resonator length
n:	index of refraction of active medium
ℓ :	length of active medium.

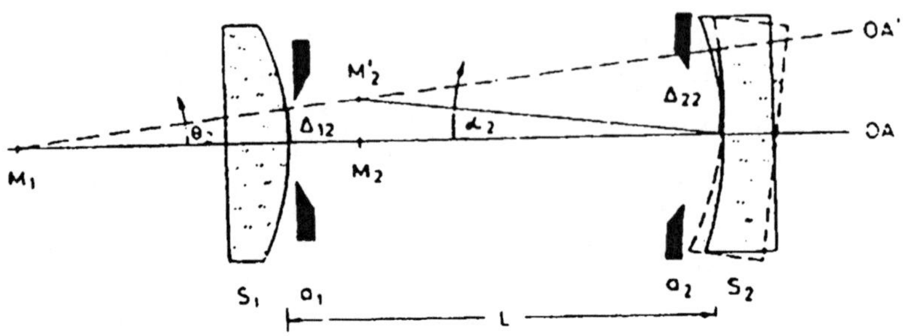

Fig. 5.53 Misalignment of a resonator with both mirrors aperture limited [3.33] (© OSA 1980).

The 10%-angle for the misalignment of mirror i thus reads:

$$\alpha_{10\%,i} = \sqrt{0.1}\,\frac{\sqrt{\exp[2s^2]-1}}{sD_i} \tag{5.91}$$

If both mirrors are misaligned, the 10%-angle is given by:

$$\alpha_{10\%} = \sqrt{0.1}\,\frac{\sqrt{\exp[2s^2]-1}}{s\sqrt{D_1^2+D_2^2}} = \sqrt{0.1}\,\frac{\sqrt{\exp[2s^2]-1}}{sD} \tag{5.92}$$

with: $$D = \sqrt{\left(\frac{\pi L}{\lambda_0}\right)\frac{1+g_1g_2}{(1-g_1g_2)^{1.5}}\frac{|g_1+g_2|}{\sqrt{g_1g_2}}}$$

We define D as the misalignment sensitivity of the resonator since a small value of D results in a large 10%-angle. Figure 5.54 presents calculated misalignment sensitivities D of stable resonators in fundamental mode operation with both apertures adapted to the Gaussian beam radii at the corresponding mirror. The misalignment sensitivity in this graph is normalized to the sensitivity D_0 of the symmetric confocal resonator with $g_1=g_2=0$, given by:

$$D_0 = \sqrt{\frac{2\pi L}{\lambda_0}} \tag{5.93}$$

The misalignment sensitivity is only a function of the resonator length and the g-parameters.

Lowest misalignment sensitivities again are attained for resonators near the axes of the stability diagram. Figure 5.55 shows a comparison of measured 10%-angles with the theoretical prediction given by (5.91).

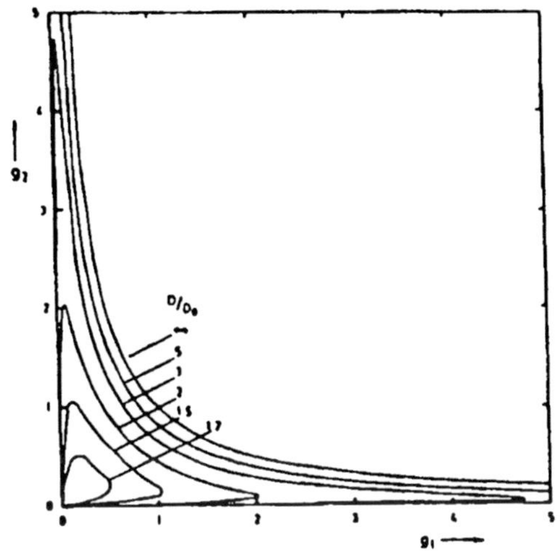

Fig. 5.54 Curves of constant misalignment sensitivity D for resonators with both apertures adapted to the Gaussian beam radius. D_0 is the sensitivity of the confocal resonator [3.33] (© OSA 1980).

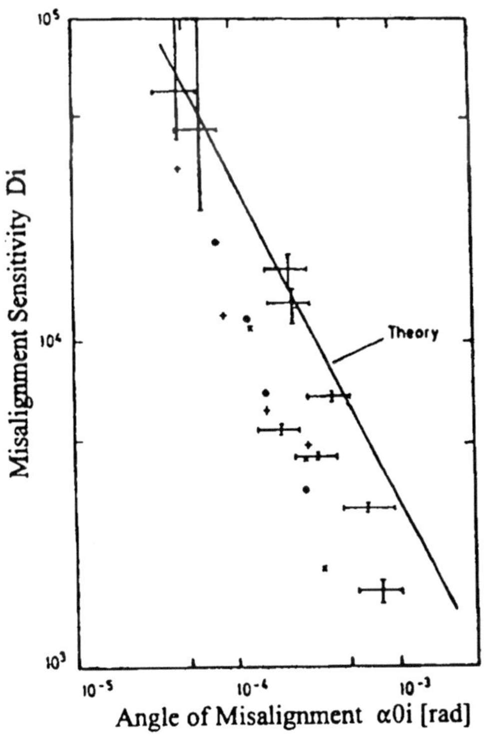

Fig. 5.55 Measured angles of misalignment $\alpha_{0i}=\alpha_{10\%,i}/\sqrt{0.1}$ and the corresponding misalignment sensitivities D_i of the resonators according to (5.90). The solid line represents relation (5.91) [3.33] (© OSA 1980).

Chapter 6

Resonators on the Stability Limits

6.1 Resonators with g_1g_2=1

If the g-parameters of the stable resonator approach the stability boundaries at $g_1g_2=0$ or $g_1g_2=1$, the beam radii at the mirrors of the Gaussian beam go to zero or infinity. On the stability boundaries, the Gaussian beam is not an eigensolution of the resonator. The field distributions of the eigenmodes are still solutions of the diffraction integral equation (5.66), but the mode structure and the diffraction losses have to be calculated numerically. A helpful property of resonators on the stability limits is their equivalency since they all exhibit the same absolute value of the equivalent g-parameter $G=2g_1g_2-1=\pm 1$. If the properties are known for one resonator as a function of the aperture radius, the mode structure and the losses of all other resonators on the stability limits can be determined.

Let us first discuss resonators that are located on the two hyperbolas $g_2=1/g_1$ in the stability diagram. They either have two plane mirrors or the resonator length is given by the sum of the radii of curvature of the two mirrors:

$$L = \rho_1 + \rho_2 \tag{6.1}$$

The centers of curvature are on top of each other.

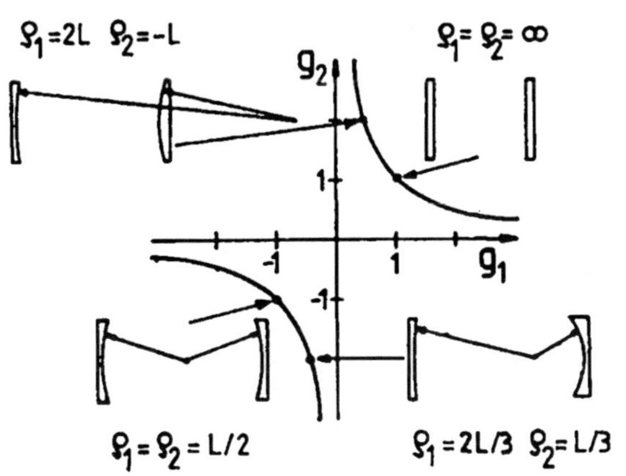

Fig. 6.1 Resonators with $g_1g_2=1$ in the stability diagram.

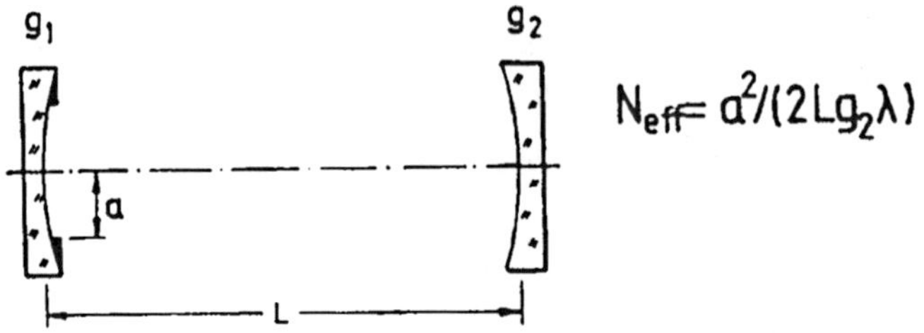

Fig. 6.2 Resonator with $g_1g_2=1$ and one aperture limited mirror. Loss and mode structure depend only on the absolute value of the effective Fresnel number N_{eff}.

If one of the mirrors is limited by the aperture (we choose mirror 1 again), the intensity distribution at this mirror and the diffraction loss per round trip is obtained by applying the integral equation used in Sec. 5.3 to calculate the properties of stable resonators with one aperture. The mode structure and the losses depend only on the absolute values of both the equivalent g-parameter G and the effective Fresnel number N_{eff} (Fig. 6.2). Since the equivalent g-parameter is $G=1$ for all resonators on the two stability hyperbolas, we only have to investigate the resonator properties as a function of the effective Fresnel number to deal with all resonators having $g_1g_2=1$.

Similar to stable resonators, these resonators exhibit an infinite set of transverse eigenmodes whose beam radii increase with the mode indices $p\ell$ and mn. This leads to the oscillation of transverse modes of higher order as the aperture radius is increased, resulting in a decrease of the beam quality and a decrease of the diffraction losses. Figures 6.3-6.5 show the intensity distributions of the TEM_{00} mode at the aperture limited mirror, the diffraction loss factor per round trip, and the beam parameter product as a function of the absolute value of the effective Fresnel number, respectively. By using these graphs, the properties of any resonator with $g_1g_2=1$ can be determined. The beam radius of the TEM_{00} mode at mirror 1 can, to a good approximation, be calculated by using the relation:

$$w_{00} = 2\sqrt{2Lg_2\lambda} \tag{6.2}$$

In order to realize fundamental mode operation the effective Fresnel number N_{eff} has to be less than 3. The beam propagation factor M^2 as a function of the effective Fresnel number can be approximated by (see Fig. 6.5 with $N_{eff}=N/2$):

$$M^2 = \frac{w\theta}{\lambda/\pi} \approx 2\sqrt{N_{eff}} \tag{6.3}$$

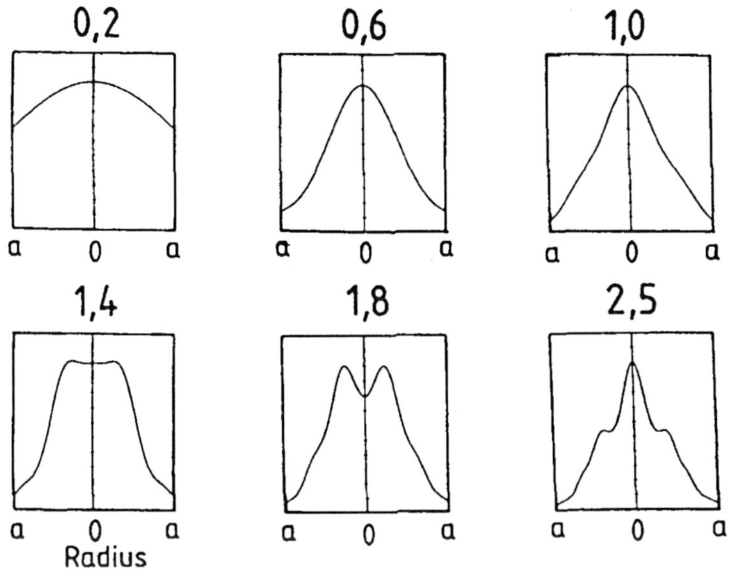

Fig. 6.3 Radial intensity distributions of the TEM_{00} mode for resonators with $|G|=1$ as a function of the effective Fresnel number N_{eff}.

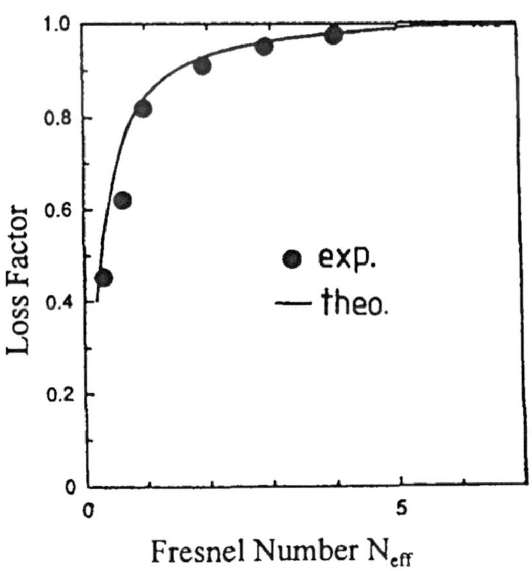

Fig. 6.4 Calculated and measured loss factor per round trip for resonators with $g_1g_2=1$ as a function of the absolute value of the effective Fresnel number N_{eff} (round aperture with radius a).

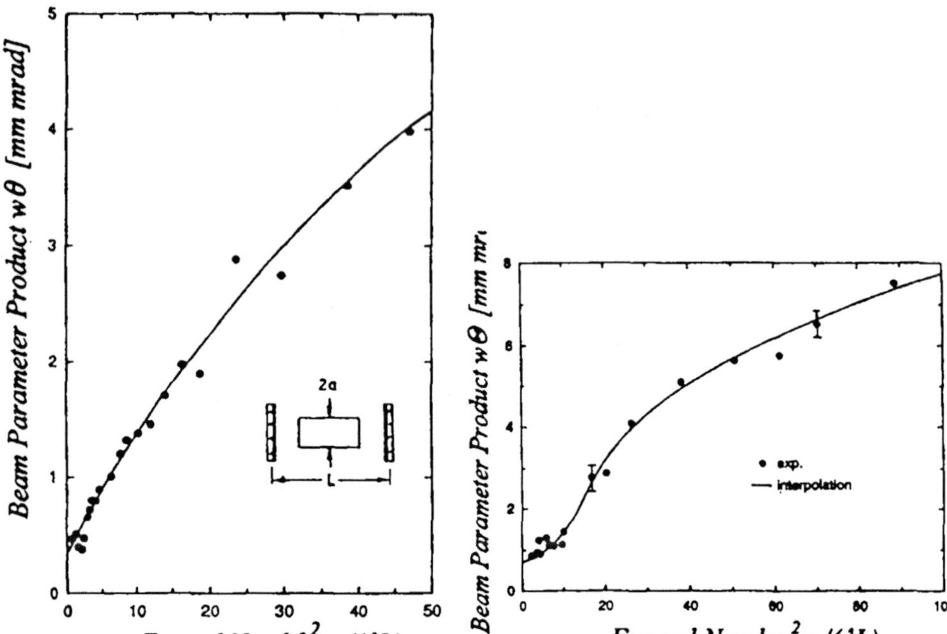

Fig. 6.5 Measured beam parameter products wθ of symmetric plane-plane resonators ($g_1=g_2=1$, *w:* waist radius, *θ:* half angle of divergence) as a function of the Fresnel number $N=a^2/(\lambda L)$. Left: round aperture with radius *a*, right: square aperture with side *2a*, pulsed Nd:YAG laser (λ=1.064μm) in single shot operation. Note that the effective Fresnel number is given by $N_{eff}=N/2$.

Examples:

1) Nd:YAG laser (λ=1.064μm) with rod radius of 5mm, ρ_1=2m, ρ_2=-1m, L=1m
The rod is placed close to mirror 1 which means that the aperture radius *a* is 5mm. Figs. 6.3-6.5 provide the following properties:

Resonator parameters:	g_1=0.5, g_2=2.0, N_{eff}=5.8
Loss of the TEM_{00} mode per round trip:	ΔV = 1%
Beam propagation factor:	M^2 = 4.0
Beam parameter product:	wθ= 1.33 mm mrad

2) CO_2 laser (λ=10.6μm) with tube radius of 10mm, ρ_1=1m, ρ_2=1m, L=2m
Again, the gas tube represents the aperture that limits one mirror.

Resonator parameters:	g_1=g_2=-1, N_{eff}= -2.358
Loss of the TEM_{00} mode per round trip:	ΔV = 7%
Beam propagation factor:	M^2 = 2.5
Beam parameter product:	wθ= 8.35 mm mrad

6.2 Resonators with One Vanishing g-Parameter

If one of the two g-parameters is equal to zero, which means that the radius of curvature of the corresponding mirror is equal to the mirror spacing, the resonator is located on an axis of the stability diagram. If the axis is approached from the stable region, the Gaussian beam radius at the mirror with the vanishing g-parameter becomes infinite, whereas the beam radius goes to zero at the other mirror. This means that the Gaussian beam is not an eigensolution. However, we can conclude that the beam propagation in a resonator on an axis of the stability diagram exhibits beam radii at the mirrors that are different by several orders of magnitude. In the following we choose mirror 1 as the mirror with a zero g-parameter, as shown in Fig. 6.6. In order to attain the best beam quality possible, it is advantageous to place the aperture or the active medium as close as possible to mirror 1. We are thus dealing with resonators on the axis $g_1=0$ with mirror 1 being limited by an aperture with radius a.

Since the resonators on the g-axes are equivalent to the resonators on the hyperbolas $g_2=1/g_1$, we can use Figs 6.3-6.5 and Eq. (6.3) to determine their mode properties. Again, fundamental mode operation requires an effective Fresnel number N_{eff} of less than 3. The beam radius at mirror 1 is given by the aperture radius a, on mirror 2 the beam radius can be calculated to a good approximation with:

$$w^{(2)} = \frac{\lambda\rho_1}{\pi a}\sqrt{M^2} \tag{6.4}$$

with the propagation factor M^2 according to (6.3). In fundamental mode operation, this beam radius is very small (typically on the order of 100μm for λ=1μm). This can lead to damage of the mirror's surface.

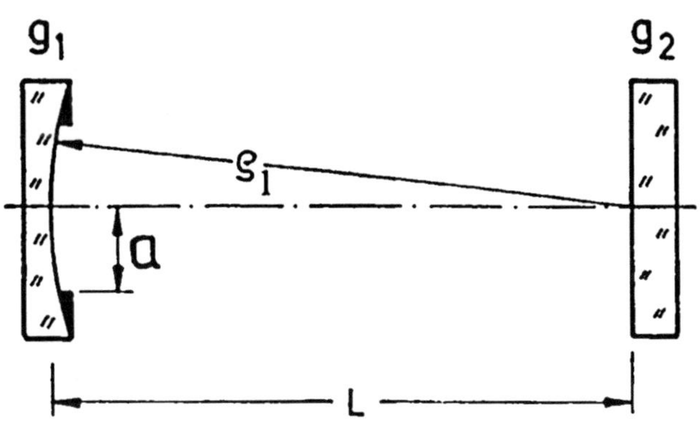

Fig. 6.6 Resonators on one axis of the g-diagram (here $g_1=0$) have one mirror with a radius of curvature equal to the mirror spacing.

Example:
A resonator with a length of $L=0.5m$ has to be designed for a pulsed Nd:glass laser ($\lambda=1.054\mu m$) with a 50mm long and 6mm thick rod. The laser should provide a high output energy in fundamental mode operation, and a low sensitivity to mirror tilt is required. The repetition frequency is well below 0.3 Hz which means that thermal effects can be neglected.

What we need is a resonator that exhibits a large TEM_{00} mode with a radius on the order of 2-2.5mm. A stable concave-convex resonator (see Sec. 5.2.3) provides the large Gaussian beam radius but it exhibits a high sensitivity to mirror misalignment. A much better choice is a resonator on an axis of the g-diagram. We choose a radius of curvature of mirror 1 of $\rho_1=0.5m$ ($g_1=0$) and place the glass rod as close as possible to this mirror. For fundamental mode operation we need an effective Fresnel number of less than 3. The g-parameter of mirror 2 is given by:

$$g_2 = \frac{a^2}{N_{eff} 2L\lambda}$$

which yields $g_2=3.4155$ for $N_{eff}=2.5$. The radius of curvature of mirror 2 has to be chosen to be $\rho_2=-0.207m$. According to (6.4) with $M^2=1$, the beam radius at this mirror is $w^{(2)}=56\mu m$. The 10%-angle of the misalignment sensitivity (10% loss increase due to mirror tilt, see Sec. 5.4.1) are:

$$\alpha_{10\%,1} = \infty \quad , \quad \alpha_{10\%,2} = 2.4\ mrad$$

This example indicates that resonators with one vanishing g-parameter provide a good means to simultaneously achieve a high fundamental mode volume and a low sensitivity to mirror misalignment. However, the small beam radius on the unconfined mirror may cause damage in high power lasers. This problem can be avoided if the unconfined mirror is chosen as the output coupler since the damage threshold of optical coatings decreases with the reflectivity.

In practice, it is difficult to set-up the resonator so that it ends up on the g-axis right away. Due to tolerances of the mirror curvatures and the resonator length it is useful to first set the resonator length shorter than intended. The resonator will then operate in the stable region which makes the initial alignment much easier. By monitoring the beam radius at mirror 2 or the beam quality while slowly backing off one mirror, the resonator can be tuned to a vanishing g-parameter.

6.3 The Confocal Resonator

The symmetric confocal resonator has g-parameters that lie at the origin of the stability diagram and is characterized by the fact that both radii of curvature are equal to the resonator length [3.44,3.45]. This means that the centers of curvature are located on the mirror surfaces (Fig. 6.7). In contrast to the resonators on the axes of the stability diagram, the confocal resonator exhibits a Gaussian beam as an eigensolution. It is for this reason that the confocal resonator is sometimes referred to as a stable resonator. In the limit $g_1,g_2 \to 0$, Eqs. (5.10) and (5.18) yield for the beam radii:

waist radius:
$$w_0 = \sqrt{\frac{\lambda L}{2\pi}} \tag{6.5}$$

beam radii at the mirrors:
$$w_{00}^{(1)} = w_{00}^{(2)} = \sqrt{\frac{\lambda L}{\pi}} \tag{6.6}$$

The waist is located in the middle of the resonator and the Rayleigh range is given by:

$$z_0 = \frac{L}{2} \tag{6.7}$$

If the resonator is limited by apertures with radii $a_1=a_2=a$ at both mirrors, the beam propagation factor can be calculated using the relation:

$$M^2 \cong \pi N = \pi \frac{a^2}{\lambda L} \tag{6.8}$$

where N is the Fresnel number.

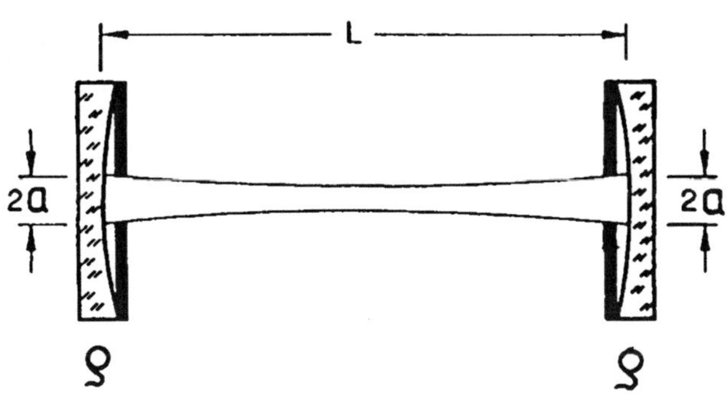

Fig. 6.7 The symmetric aperture-limited confocal resonator.

Fundamental mode operation is obtained for Fresnel numbers N of less than 0.5. Unfortunately, the mode volume of the fundamental mode is quite low resulting in a low output power. For a wavelength λ of 600nm and an effective resonator length L of 1m, for instance, the waist radius w_0 is only 0.31mm. However, the confocal resonator exhibits some interesting properties that make it possible to increase the fundamental mode volume by several orders of magnitude by choosing different aperture sizes at the mirrors.

Let us first consider a confocal resonator with only mirror 1 limited by an aperture (Fig. 6.8). Any field distribution with radial symmetry starting at mirror 1 will be reproduced after the round trip inside the resonator. The aperture is imaged onto itself since the ray transfer matrix for the round trip reads (see Sec. 1.3):

$$\boldsymbol{M} = \begin{pmatrix} -1 & 0 \\ 0 & -1 \end{pmatrix}$$

All field distributions that are circularly symmetric or exhibit symmetry along the x-axis and the y-axis are eigensolutions of the resonator. As a consequence, any Gaussian beam with q-parameter q at the aperture limited mirror is an eigensolution. For a given aperture radius a, we can thus always find a modified Gaussian beam with beam radius $w_{00}^{(1)}=a$ and radius of curvature ρ at mirror 1 that is reproduced after each round trip. If the aperture radius a, and thus the Gaussian beam radius $w_{00}^{(1)}$, is decreased the beam radius $w_{00}^{(2)}$ is increased in such a way that the following relation holds (Fig. 6.9):

$$w_{00}^{(1)}\, w_{00}^{(2)} = \frac{\lambda L}{\pi} \tag{6.9}$$

The distance z of the beam waist from mirror 1 is given by:

$$z = \frac{L}{1 + \left[\dfrac{\lambda L}{\pi [w_{00}^{(1)}]^2}\right]} \tag{6.10}$$

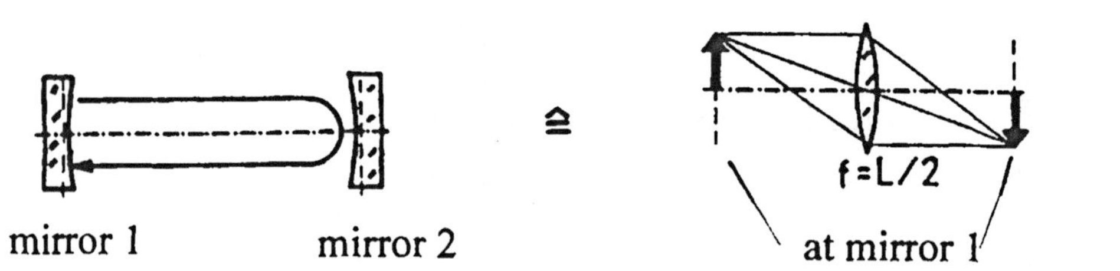

Fig. 6.8 In a confocal resonator each point at mirror 1 is imaged by the round trip into a point opposite the optical axis.

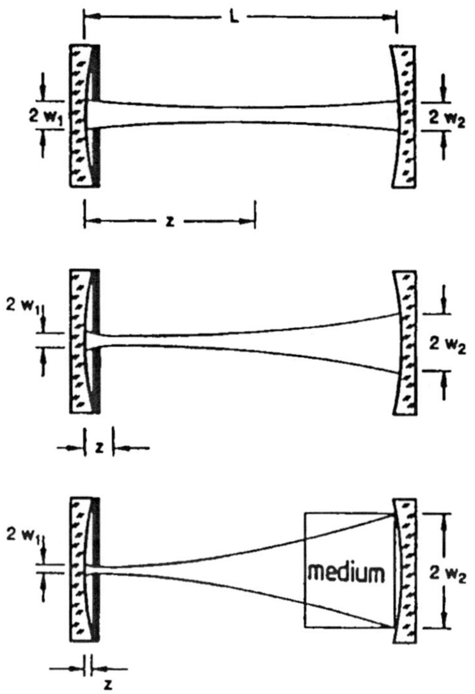

Fig. 6.9 For any aperture radius at mirror 1, one can find a Gaussian beam that exhibits a radius of curvature equal to that of the mirrors and a beam radius that is adapted to the aperture radius. z denotes the distance of the beam waist from mirror 1.

If the active medium is positioned at the infinite mirror 2, a high fundamental mode volume in the medium can be attained by decreasing the aperture radius *a*.

Example:
The symmetric Gaussian beam in a confocal resonator with $L=1m$ and $\lambda=500nm$ exhibits beam radii at the mirrors of $w_{00}^{(1)}=w_{00}^{(2)}=0.4mm$. If mirror 1 is limited by an aperture with $a_l=0.1mm$, the beam radius at mirror 2 is increased by a factor of four to be $w_{00}^{(1)}=1.6mm$. We can thus fill a cylindrical active medium with a diameter of about 4mm with the fundamental mode.

This special property of the confocal resonator is generated by the fact that the field distributions at the two mirrors are related to each other via a Fourier transform. The field at mirror 2 is the Fourier transform of the field at mirror 1. Decreasing the aperture radius a_l will therefore lead to an increase of the beam radius at mirror 2. This becomes clearer if we apply the diffraction integral to the field propagation inside the resonator (Fig. 6.10). By assuming a circularly symmetric field distribution on mirror *i* :

$$E_i(r,\Phi) = u_{pl}(r)\,\exp[il\Phi] \tag{6.11}$$

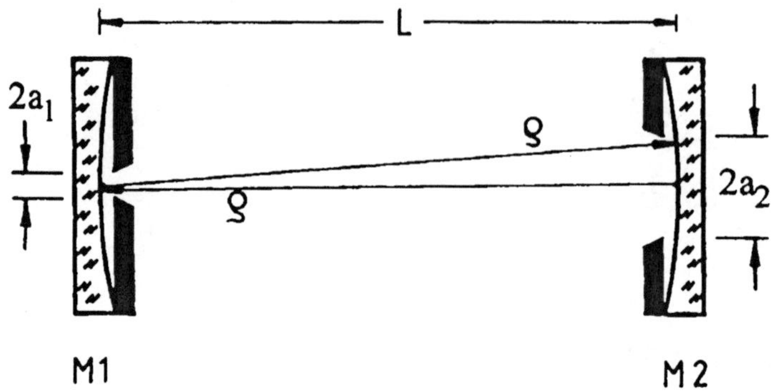

Fig. 6.10 Confocal resonator with different aperture radii at the mirrors.

and scaling the fields with the square root of the Fresnel number $N_i = a_i^2/(\lambda L)$:

$$E'_{pl}(r,\Phi) = E_{pl}(r,\Phi)\sqrt{N_i} \tag{6.12}$$

we find the following relations between the radial parts u'_{p1} and u'_{p2} of the fields on the two mirrors:

$$u'_{p2}(r_2) = (-i)^l\, 2\pi\, N \exp[-ikL] \int_0^1 u'_{p1}(r_1) J_l(2\pi N r_1 r_2) r_1 dr_1 \tag{6.13}$$

$$\gamma_{pl} u'_{p1}(r_1) = (-i)^l\, 2\pi\, N \exp[-ikL] \int_0^1 u'_{p2}(r_2) J_l(2\pi N r_1 r_2) r_2 dr_2 \tag{6.14}$$

with	J_l	:	Bessel function of order l
	$N = \sqrt{N_1 N_2}$	:	Fresnel number
	r_i	:	normalized radial coordinate on mirror i
	$V_{pl} = \gamma_{pl}\gamma^*_{pl}$	:	loss factor per round trip of the TEM_{pl} mode

The similarity of (6.13) and (6.14) indicates that the field distributions on the two mirrors have the same shape since the two integrals are identical. This means that if one aperture is completely filled by the oscillating modes, the same filling is obtained at the second one. The mode structure and the losses depend only on the Fresnel number N. By decreasing the Fresnel number, the number of oscillating modes will decrease until we get fundamental mode operation at low Fresnel numbers N. Figure 6.11 presents loss factor and intensity distributions of the fundamental mode as a function of the Fresnel number N, calculated with (6.13) and (6.14). This figure indicates that the Fresnel number has to be chosen lower than 0.8 to prevent higher modes from oscillating.

If we place the active medium at mirror 2 (the aperture radius a_2 now represents the radius of the active medium), we can decrease the aperture at mirror 1 until the fundamental mode condition $N<0.8$ is fulfilled. Maximum output power is attained for a Fresnel number of about 0.6. The optimum aperture radius a_1 is thus given by:

$$a_1 = 0.6 \frac{\lambda L}{a_2} \tag{6.15}$$

As the aperture is decreased a continuous increase in beam quality is observed but the output power will remain fairly constant since the active medium is always filled by the modes. If Fresnel numbers lower than 0.5 are reached, the increased diffraction losses experienced by the fundamental mode will result in a sudden decrease of the output power.

Figures 6.12 and 6.13 show measured output energies per pulse and the corresponding beam parameter products of an Nd:YAG laser with a 1m long confocal resonator. These graphs indicate that the asymmetric confocal resonator is capable of providing high output power and excellent beam quality simultaneously. Furthermore, the misalignment sensitivity is lower compared to other stable resonators in fundamental mode operation (see Sec. 5.4.2). The misalignment sensitivity is characterized by the tilt angle at which the diffraction losses have increased by 10%. For the resonator set-up shown, misalignment of the unconfined mirror 2 resulted in a 10%-angle of 150μrad at a ratio of aperture radius to Gaussian beam radius of $a_1/w=0.7$. For the tilt of mirror 1, a 10%-angle of more than 5mrad was observed.

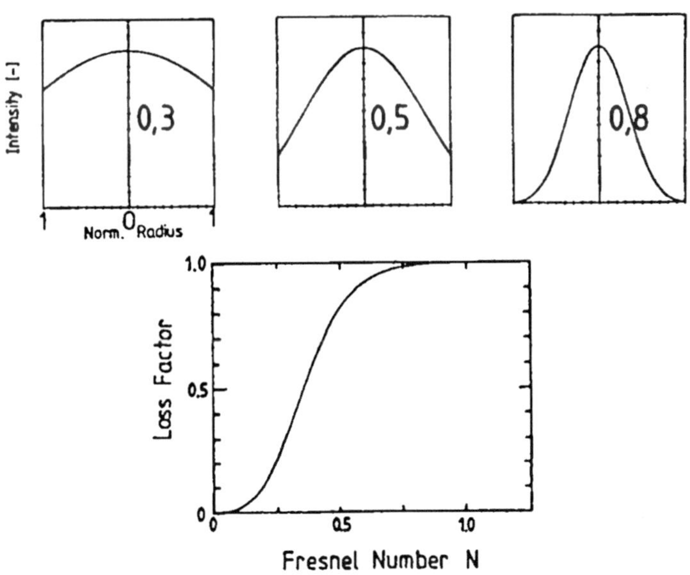

Fig. 6.11 Calculated loss factor per round trip of the fundamental mode as a function of the Fresnel number $N=a_1a_2/(\lambda L)$; The radial intensity distributions in the aperture for different Fresnel numbers N are shown as well (top).

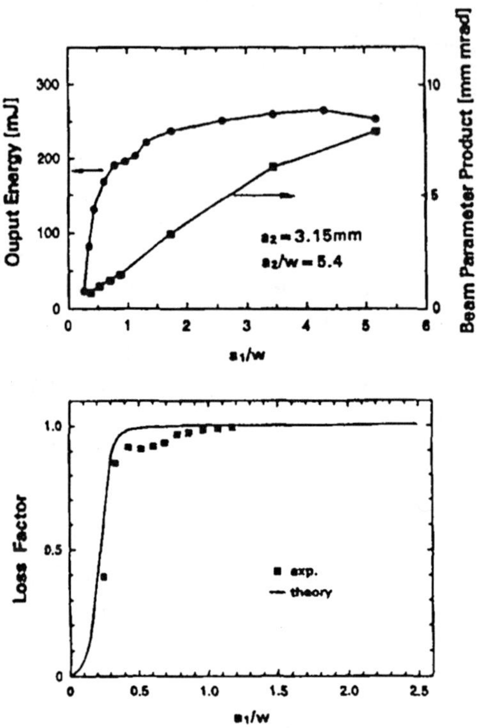

Fig. 6.12 Measured output energy and beam parameter product for a confocal resonator as a function of the aperture radius a_1. The Gaussian beam radius on the mirror w is *0.582mm*; small-signal gain $g_0l=1.0$; $R=80\%$; $L=1.0m$; $a_2=3.15mm$ (Nd:YAG laser, $\lambda=1.064\mu m$). With a stable resonator in multimode operation, the maximum output energy was 320mJ for optimum output coupling [3.53] (© OSA 1993).

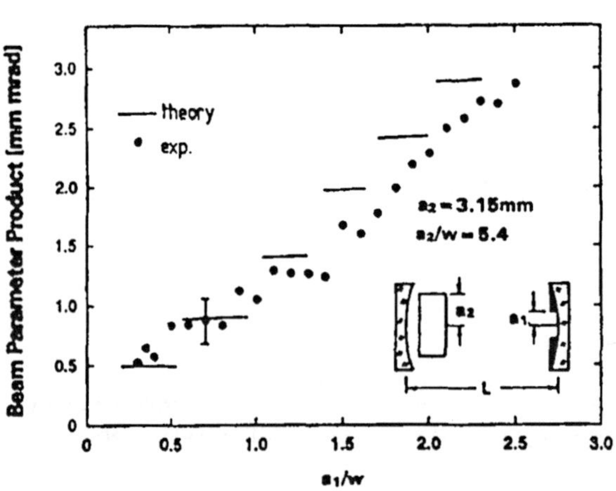

Fig. 6.13 Measured beam parameter products for the experiments presented in Fig. 6.12. The solid lines mark the theoretical values obtained by solving the diffraction integrals (6.13) and (6.14) for $l=0$ including the amplification of the field by the active medium [3.53] (© OSA 1993).

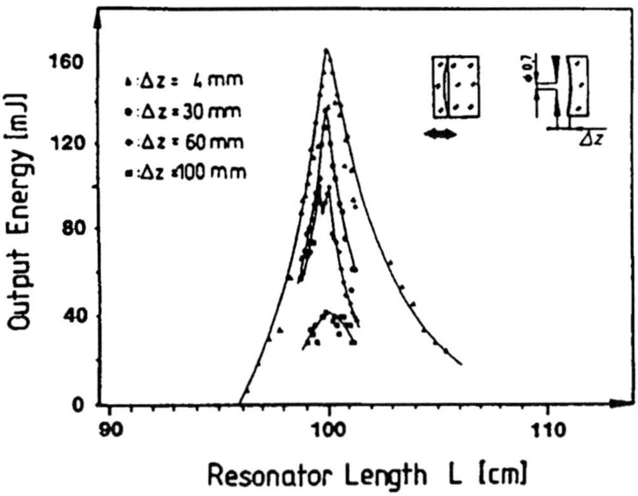

Fig. 6.14 Measured output energy for the resonator of Fig. 6.12 as a function of the resonator length for a Fresnel number of $N=1.0$. The curve parameter is the distance Δz between the aperture and mirror 1 [3.53] (© OSA 1993).

Unfortunately, the price we have to pay for the low misalignment sensitivity and the high fundamental mode volume is a fairly high sensitivity to variations in the resonator length. Small length variations will destroy the Fourier transform relationship between the fields and the resonator will go stable. Only one Gaussian beam is an eigensolution of the stable resonator and this Gaussian beam exhibits a beam radius at mirror 1 that is much larger than the aperture radius a_l. A slight mismatch of the resonator length will therefore lead to a sudden increase of the diffraction losses. As the experimental data in Fig. 6.14 indicate, the length of the confocal resonator has to be controlled with an accuracy of ±1% in order to maintain an output power of at least 90%. In a single-shot operation, this required length stability does not cause any problems, but as far as high power operation is concerned, the confocal resonator can be used only for one fixed input power. Most active materials exhibit a lensing effect which is brought about by a combination of heat generation due to the absorption of pump and laser radiation and the flow of heat to the outer periphery due to cooling. This thermal effect is small in gas lasers but it poses a serious problem in solid state lasers where refractive powers on the order of several diopters are observed. The refractive power increases linearly with the average pumping power. In order to operate the resonator at the confocal point, the refractive power has to be compensated by an appropriate choice of the mirror curvatures. This requires a constancy of the refractive power of better than 1%, and any decrease in the pumping efficiency due to degradation will considerably decrease the output power. Furthermore, any nonparabolic refractive index profile that cannot be compensated for by the mirror curvatures will prevent the resonator from working efficiently.

Chapter 7

Unstable Resonators

7.1 General Aspects

If the g-parameters of the resonator mirrors fulfill the relations $g_1g_2<0$ or $g_1g_2>1$, the radiation inside the resonator cannot be characterized by a Gaussian beam. Equation (5.10) indicates that in these cases the Gaussian beam radii become complex quantities. The steady state field distributions are not given by Gauss-Hermite or Gauss-Laguerre polynomials. These resonators are referred to as ***unstable resonators*** [3.54,3.67], whereby the term "unstable" accounts for the fact that a Gaussian beam launched into the resonator will increase its beam radius after each round trip and, therefore, does not represent a "stable" eigensolution of the resonator (Fig. 7.1). Note that the term "unstable" does not mean that the resonator is more sensitive to mirror misalignment as compared to stable resonators. In fact, unstable resonators are generally less sensitive to mirror tilt than stable resonators in fundamental mode operation. By using the equivalent G-parameter $G=2g_1g_2-1$ we can characterize the different resonator schemes as follows:

stable resonators	:	$0 < \|G\| < 1$
resonators on the stability limits	:	$\|G\| = 1$
unstable resonators	:	$\|G\| > 1$

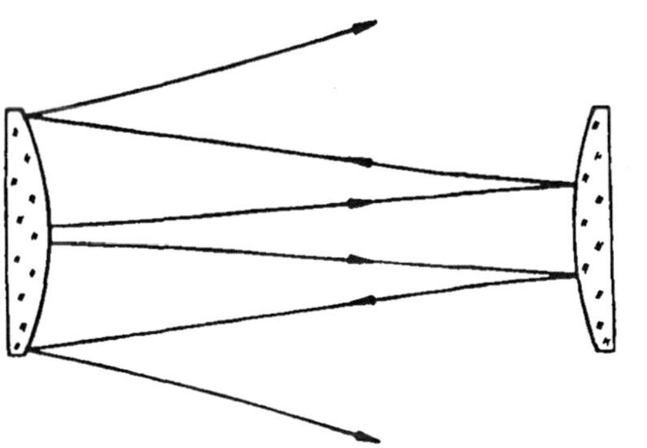

Fig. 7.1 No Gaussian beam can reproduce itself in an unstable resonator. The beam radii at the mirrors increase with each round trip.

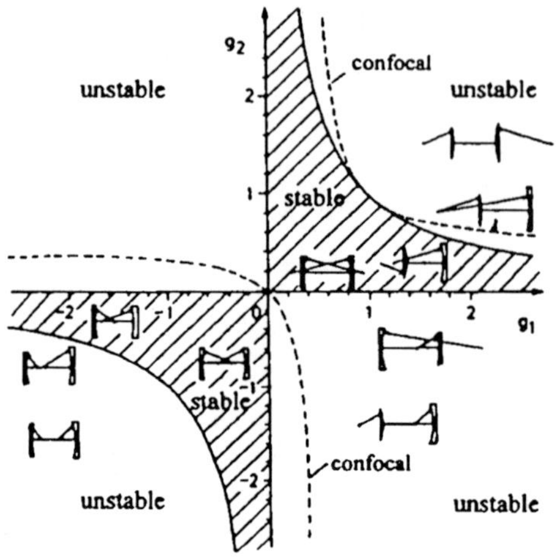

Fig. 7.2 The stability diagram of optical resonators. Unstable resonators are located in the non-hatched areas.

Furthermore, one is able to distinguish *positive branch* unstable resonators ($G>1$) and *negative branch* unstable resonators ($G<-1$). The steady state field distributions on the mirrors of unstable resonators, similar to stable resonators, are solutions of the Kirchhoff integral equation (see (5.71),(5.72)). In contrast to stable resonators, the beam propagation inside unstable resonators, to a good approximation, can be described by the propagation laws of geometrical optics. As already discussed in Sec. 1.3, unstable resonators are characterized by the presence of spherical waves whose radii of curvature reproduce themselves after each round trip. Before we investigate the mode structures and the diffraction losses by applying diffraction theory, we will discuss the basic properties of unstable resonators by analyzing the propagation of these spherical waves. Although this geometric-optical treatment does not provide information on the field distributions, the geometrical description will provide the reader with a better understanding of the basic principles of unstable resonators.

7.2 Geometric Optical Description of Unstable Resonators

7.2.1 Beam Propagation

Unstable resonators are characterized by the presence of spherical waves inside the resonator that reproduce themselves after each round trip [3.54,3.55,3.67] (see Sec. 1.3). A spherical wave starting at mirror 1 with a radius of curvature R_1 is transformed by mirror 2 into a spherical wave with radius of curvature R_2 (Fig. 7.3). After hitting mirror 1 again, the initial radius of curvature R_1 is reproduced. The resonator mirrors thus image the centers of curvatures of the wave fronts Z_1 and Z_2 onto each other. In every unstable resonator we can find two spherical waves whose radii of curvature at any plane inside the resonator are

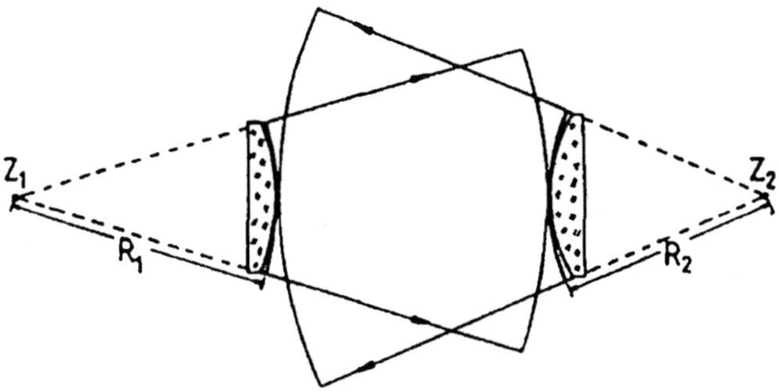

Fig. 7.3 **In unstable resonators spherical waves can be found whose radii of curvatures are reproduced after each round trip. The beam propagation of the diverging wave is shown.**

are reproduced after a round trip (note that in Fig. 7.3 only one spherical wave is shown!). If R_+ and R_- denote the radii of curvature at mirror 1 (after being reflected off the mirror) of the two spherical waves (Fig. 7.4), the following relation holds:

$$R_{\pm} = \frac{\pm\, 2Lg_2}{|G| \pm \sqrt{G^2-1} - 2g_2 + 1} \tag{7.1}$$

If mirror 1 is limited by an aperture with radius a, which means that the beam starting at mirror 1 has a diameter of $2a$, the radii of curvature R_+ and R_- reproduce themselves after the round trip, but the beam diameter is magnified by a factor M_+ and M_-, respectively, with:

$$M_{\pm} = |G| \pm \sqrt{G^2 - 1} \qquad and \quad |M_+ M_-| = 1 \tag{7.2}$$

The spherical wave with radius R_+ increases the beam radius after each round trip by the factor $|M_+|$, called the *magnification*. Since $|M_+|>1$ holds, the corresponding spherical wave is referred to as the *diverging wave*. If the power starting at mirror 1 (inside the aperture) is given by P_0, only the power

$$P_1 = \frac{1}{M_+^2}\, P_0 \tag{7.3}$$

hits the mirror inside the aperture after the round trip, provided that the intensity profile is homogeneous. The loss ΔV per round trip and the loss factor V per round trip are thus given by:

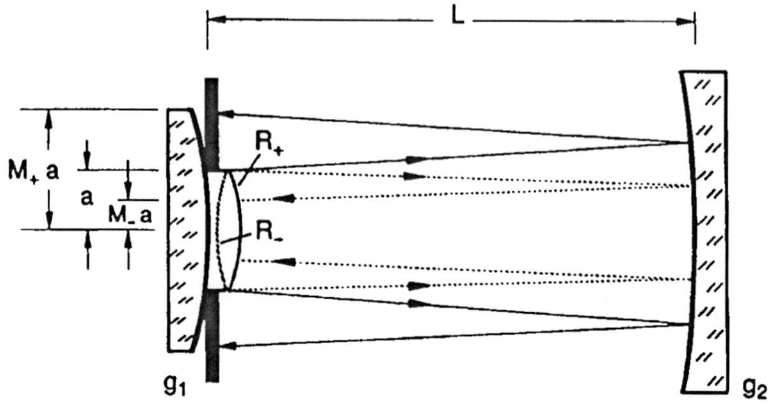

Fig. 7.4 The two self-reproducing spherical waves with radii of curvature R_+ and R_- at mirror 1 (after the reflection). After one round trip the beam diameter is changed by a factor M_+ (>1) and $|M_-|$ (M_-<1), respectively.

$$\Delta V = 1 - \frac{1}{M_+^2}\ , \qquad V = 1 - \Delta V = \frac{1}{M_+^2} \tag{7.4}$$

The loss factor represents the power fraction that stays inside the resonator after the round trip; the remainder of the power is absorbed by or reflected off the aperture material.

In contrast to the diverging wave, the *converging wave* (R_-) decreases the beam diameter after each round trip by a factor $|M_-|$ with $|M_-|<1$. The power P_0 starting at mirror 1 is conserved, but with every round trip the beam radius at mirror 1 is continuously decreased by $|M_-|$ so that no steady state beam radius can be established on the mirror. After a few round trips in the resonator, the beam radius of the converging wave reaches its minimum value given by the diffraction limit and will then expand again. The converging wave transforms itself into a diverging wave due to diffraction. It is for this reason that the beam propagation in an unstable resonator is characterized by the divergent wave only. However, the convergent wave may have an influence on the mode properties of unstable resonators, if it is continuously excited by reflection off apertures or the endfaces of the active medium [3.67,3.125]. In the following we will deal only with the diverging wave and we will drop the index + in both the magnification and the radius of curvature for convenience.

Figure 7.5 depicts the beam propagation in an unstable resonator. Instead of limiting mirror 1 by an aperture, the size of the highly reflecting area now defines the beam size on the mirror. The laser beam is generated by output coupling around the reflective spot on the mirror. In circular symmetry, the near field exhibits the shape of an annulus with inner radius a and outer radius Ma. The size of mirror 2 is chosen such that no power is coupled out at this side of the resonator. The coatings on both mirrors are highly reflecting for the desired wavelength of laser emission.

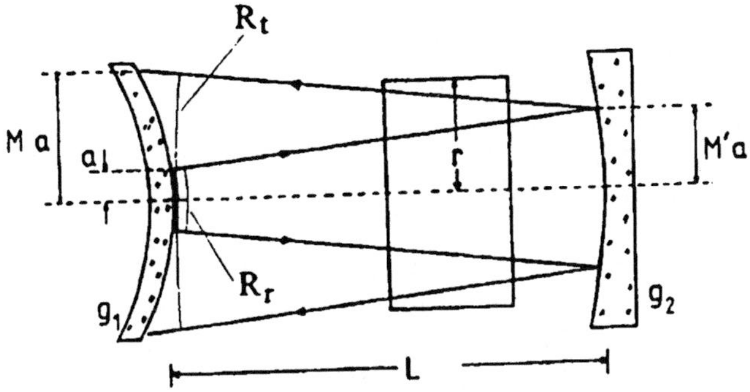

Fig. 7.5 Beam propagation inside an unstable resonator (diverging wave).

The following relations hold (see Fig. 7.5):

g-parameters: $$g_i = 1 - \frac{L}{\rho_i}\ , \quad i=1,2 \tag{7.5}$$

Equivalent G-parameter: $$G = 2g_1g_2 - 1 \tag{7.6}$$

Magnification (round trip): $$M = |G| + \sqrt{G^2 - 1} \tag{7.7}$$

Magnification (transit): $$M' = g_1 + \frac{\sqrt{G^2 - 1}}{2g_2} \tag{7.8}$$

Radius of curvature of the spherical wave at mirror 1:

- travelling towards the mirror: $$R_t = \frac{L}{M' + 1 - 2g_1} \tag{7.9}$$

- reflected off the mirror: $$R_r = \frac{2Lg_2}{M + 1 - 2g_2} \tag{7.10}$$

Loss factor per round trip: $$V = \frac{1}{M^2} \tag{7.11}$$

Note that in contrast to stable resonators, the loss of the resonator represents the power fraction coupled out of the resonator.

Example: $\rho_1 = -0.5m$, $\rho_2 = 2m$, $L = 0.75m$
---------> $g_1 = 2.5$, $g_2 = 0.625$, $G = 2.125$
$M = 4$, $M' = 4$, $R_t = \infty m$, $R_r = 0.25$ m
Loss factor per round trip : $V = 0.0625$

Three different ways to accomplish this special output coupling scheme in unstable resonators exist (Fig. 7.6). The high reflecting, confined mirror can be held in place by thin pins or mounted in the bore of a highly transmitting substrate (b). More common is the application of a high reflecting coating on an AR coated substrate as depicted in (a). For CO_2 lasers and other lasers emitting in the wavelength range on the order of 10μm, the output coupling by means of a scraper is a well established technique (c). The mirror is confined by an aperture (usually made of copper and polished to optical quality) that reflects the beam out of the resonator in a direction perpendicular to the optical axis of the resonator.

The loss of an unstable resonator is generated by the output coupling, which means that a higher loss might generate a higher output power. This probably sounds strange to readers that are more familiar with stable resonators, since the output power of stable resonators always decreases as the losses are increased. It takes time to adjust to the fact that the losses of unstable resonators are determined by the power fraction coupled out of the resonator and that the output coupling can be changed by varying the geometrical dimensions (mirror curvatures and length) of the resonator. It is definitely helpful to keep in mind that an unstable resonator with diffraction loss factor V per round trip provides the same output coupling as a stable resonator with a reflectivity of the output coupling mirror of $R=V$. Those readers who are not familiar with unstable resonators should therefore associate the loss factor with a mirror reflectivity.

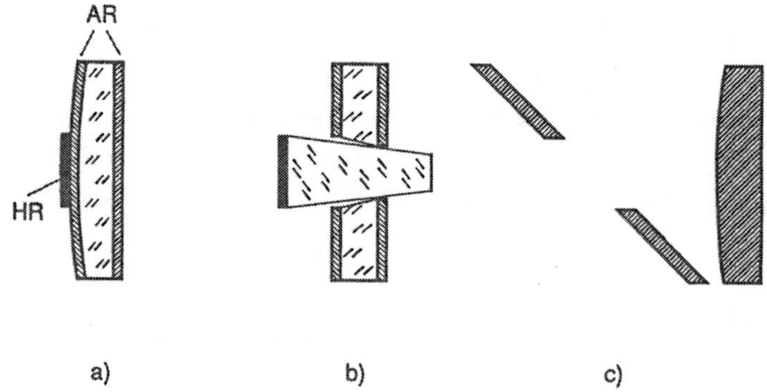

Fig. 7.6 Technical realization of the output coupling mirror of unstable resonators.

Resonator Schemes
Unstable resonators can be subdivided into two different classes:

1) $g_1g_2 > 1$, positive branch
These resonators exhibit either zero or two focal points inside the resonator. The centers of curvatures of the spherical waves at the mirrors (see Fig. 7.3) are located either outside ($g_1>0$, $g_2>0$) or inside ($g_1<0$, $g_2<0$) the resonator. Unstable resonators exhibiting these properties are referred to as *positive branch unstable resonators.*

2) $g_1g_2 < 0$, negative branch
One center of curvature of the spherical wave is located inside the resonator. The *negative branch unstable resonators* therefore exhibit a focal spot in the resonator. Due to possible damage of the active medium by the high intracavity intensities, these resonators are only used in high gain gas lasers. Compared to positive branch resonators they exhibit a much lower misalignment sensitivity.

Resonators for which the relation:

$$g_1 + g_2 = 2g_1g_2 \tag{7.12}$$

holds, are referred to as confocal resonators (Fig. 7.7). For confocal resonators, the focal points of the two resonator mirrors are on top of each other, which means that the unstable resonator acts like a telescope. This special mirror arrangement provides several advantages:

a) the beam radius stays constant as the beam propagates from the unconfined mirror to the output coupler (as long as diffraction is neglected). The mode volume can thus be easily adapted to the volume active medium yielding optimum fill factors.

b) the beam is coupled out in the form of a plane wave which saves transformation optics for beam handling.

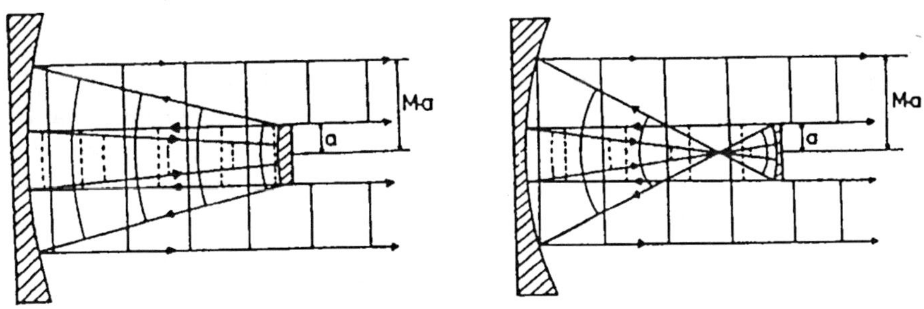

a) b)

Fig. 7.7 Beam propagation in confocal unstable resonators with magnification |M|=2. a) positive branch, b) negative branch.

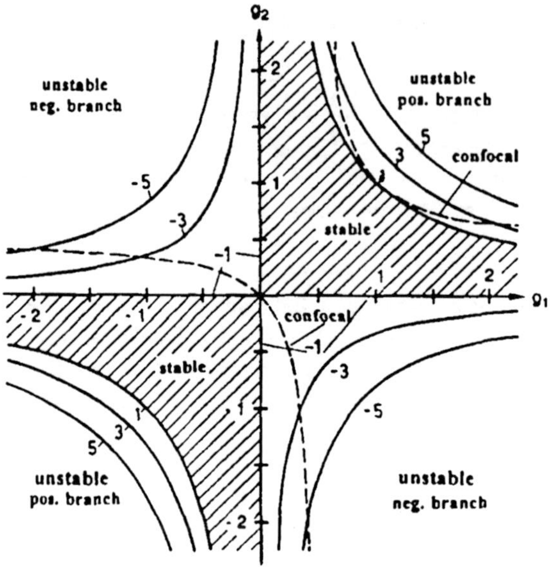

Fig. 7.8 Curves of constant magnification in the stability diagram. The broken lines indicate the confocal resonators. The curves with magnification $|M|=1$ are the stability limits.

So far the geometrical description indicates that unstable resonators having the same absolute value of the equivalent g-parameter $G=2g_1g_2-1$ exhibit the same absolute value of the magnification. Consequently, they have the same loss factor per round trip. Unstable resonators with equal losses are thus located on hyperbolas in the stability diagram, as shown in Fig. 7.8.

Unstable resonators can, of course, also be realized in rectangular symmetry. If the confined mirror has a rectangular shape and both mirrors are spherical, the beam radii in the x- and in the y-direction are both magnified by the magnification M (Fig. 7.9). The decision as to whether rectangular or circular symmetry is chosen depends on the symmetry of the active medium. In general, the resonator geometry has to be designed so that the beam travelling towards the output coupler fills the entire active medium, and optimum output coupling is provided. The geometry of the output coupler is thus determined by the geometry of the active medium. Rectangular geometry active media (slab technology) are used in CO_2 lasers and Nd:YAG lasers. The extraction of the power from the active medium is optimum for confocal resonators since the beam radius of the back travelling wave stays constant. It is for this reason that confocal positive branch unstable resonators play an important role in practical applications. The magnification is determined by the small-signal gain of the active medium. Higher gain media tolerate higher magnifications as far as the laser efficiency is concerned. Since the focusability and the misalignment sensitivity are enhanced as the magnification is increased (see later sections), unstable resonators are generally utilized in medium to high gain lasers. Typical values of the magnification are around $M=2$.

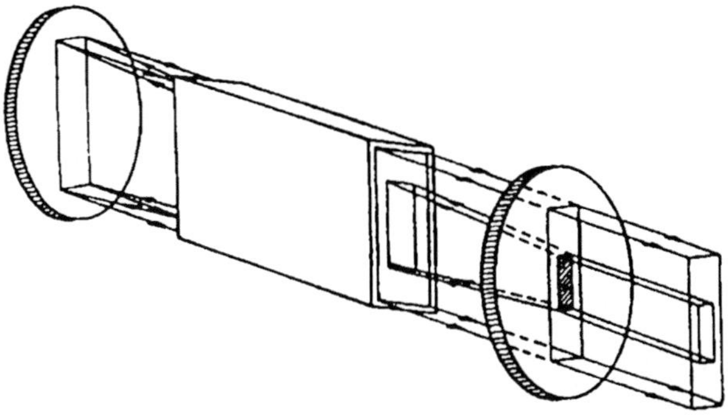

Fig. 7.9 Confocal positive branch unstable resonator in rectangular symmetry.

7.2.2 Focusability

The laser beam emerges from the unstable resonator in the form of an annular ring with inner radius a, outer radius Ma, and a radius of curvature of R_t (see (7.9)). The beam thus exhibits an angle of divergence θ_g (Fig. 7.10), with:

$$\theta_g = \frac{Ma}{R_t} \tag{7.13}$$

This angle, which can assume very high values, is generated by the geometrical set-up of the resonator (for a confocal unstable resonator $\theta_g=0$ holds), and it does not affect the beam quality of the laser. This is due to the fact that the angle of divergence can be varied by using a lens right behind the output coupler without changing the beam quality. It is advantageous to use a focal length that transforms the emerging beam into a plane wave ($\theta_g=0$), as shown in Fig. 7.11.

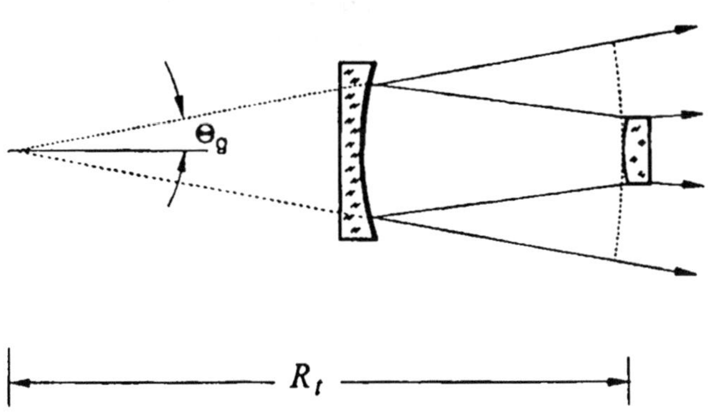

Fig. 7.10 The geometrical angle of divergence θ_g of an unstable resonator.

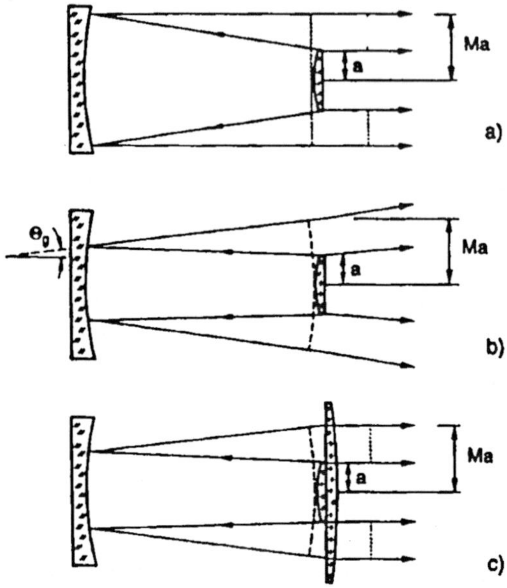

Fig. 7.11 The angle of divergence θ_g can be varied by changing the design of the resonator or by collimating the beam with a focusing lens. All three resonators shown exhibit the same beam quality. The beam quality depends on the magnification *M* and the mirror radius *a* only.

All unstable resonators having the same magnification and the same mirror radius a exhibit the same beam quality, no matter how large the geometrical angle of divergence θ_g is. The theoretical reason for this is the fact that the beam parameter product remains constant. Therefore, the beam waist w_0, which is located at the center of curvature of the spherical wave with radius R_t (virtual beam waist), is increased as the angle of divergence is decreased. The geometrical divergence angle, however, determines the position of the focus if the beam is focused by means of a lens (Fig. 7.12). Only for confocal resonators is the focal spot at the focal plane of the lens. The distance *z* of the focus from the lens, to a good approximation, can be calculated by using the imaging condition of geometrical optics. If *x* denotes the distance of the focusing lens from the output coupling mirror, the distance *z* reads:

$$z = \frac{f}{1 - f/(R_t+x)} \tag{7.14}$$

If *w* is the beam radius at the focusing lens, this relation can be written as:

$$z = \frac{f}{1 - f\theta_g/w} \tag{7.15}$$

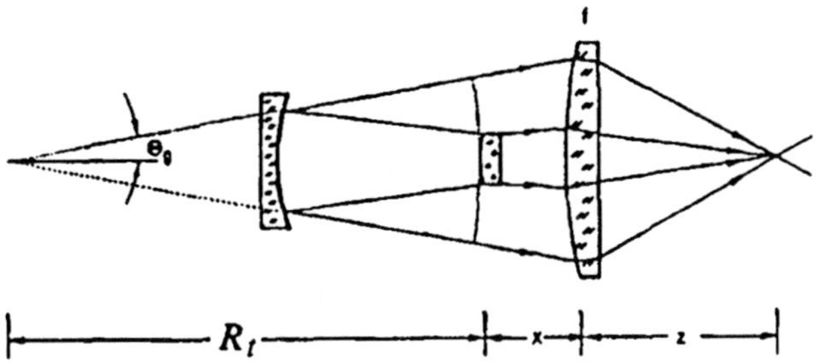

Fig. 7.12 The position of the focus of unstable resonators, to a good approximation, can be calculated by using the imaging condition of geometrical optics, as long as $w/\theta_g \ll f$.

Example: ρ_1=-0.5m, ρ_2=1.5m, L=0.7m, a=3mm, x=100mm, f=150mm (Fig. 7.12)

equivalent G-parameter	:	G	= 1.56
magnification	:	M	= 2.76
radius of curvature	:	R_f	= -2.52 m
angle of divergence	:	θ_g	= 3.28 mrad
beam radius at lens	:	w	= 8.61 mm
position of focus	:	z	= 159.1 mm

The beam quality of unstable resonators is determined by the diffraction at the confined output coupling mirror. If we assume that the outcoupled beam exhibits a plane phase front, the intensity distribution in the focal spot is given by the intensity distribution in the far field of a homogeneously illuminated annular ring with inner radius a and outer radius Ma. Application of the diffraction integral in the Fraunhofer approximation yields the far field intensity distribution:

$$I(\theta) = I(0)\,\frac{M^2}{M^2-1}\left[\frac{J_1(2\pi Ma\theta/\lambda)}{\pi Ma\theta/\lambda} - \frac{1}{M^2}\frac{J_1(2\pi a\theta/\lambda)}{\pi a\theta/\lambda}\right] \qquad (7.16)$$

where J_1 is the Bessel function of order 1. Figure 7.13 presents far field intensity distributions calculated with (7.16) as a function of $z=2\pi Ma\theta/\lambda$. For high magnifications ($M \gg 1$) we obtain the far field intensity distribution of a round aperture (see Sec. 2.2.2) with a full width half maximum (FWHM) diameter of the central peak of:

$$\Delta\theta = 0.51\frac{\lambda}{Ma} \qquad (7.17)$$

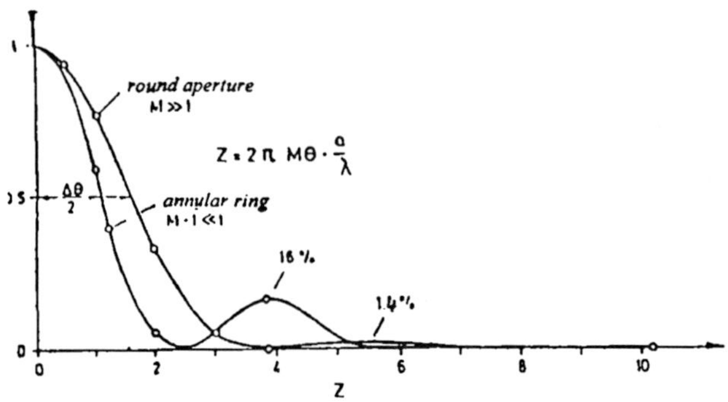

Fig. 7.13 Far field intensity distributions of a homogeneously illuminated ring with inner radius a and outer radius Ma in the limits of low and high magnification M (calculated with (7.16)).

In the limit of low magnification ($M-1 \ll 1$), we get a distribution with a narrower center peak ($\Delta\theta = 0.35\lambda/a$), but now the power fraction in the first side lobe is increased. No matter what the radius of curvature R_f of the outcoupled beam is, the intensity distribution at the focus is always given by the distribution (7.16). The radial extent of the focal spot can be calculated by using the distance z of (7.14):

$$r = z\ \Delta\theta/2 \tag{7.18}$$

The power fraction in the side lobe and the spot radius r decrease as the magnification of the resonator is increased. It is for this reason that the magnification of unstable resonators is generally chosen as high as possible with an upper limit given by the gain of the medium (a high magnification requires a high gain in order to operate the resonator at optimum output coupling). Unstable resonators are mostly applied to active media with sufficiently high small-signal gain ($g_0\ell > 1.5$). For low gain media the optimum magnification would be too low to have the major power fraction going into the central peak.

In our geometrical model, the beam waist is located at the plane of the output coupling mirror. We can thus make a first estimate of the beam parameter product by multiplying the near field diameter d_0 with 86.5% power content:

$$d_0 = 2a\sqrt{0.865M^2 + 0.135} \tag{7.19}$$

with the numerically calculated full angle of divergence Φ with 86.5% power content of the far field distribution (7.16) to get the beam parameter product:

$$\frac{d_0\Phi}{4} = \frac{a}{2}\sqrt{0.865M^2 + 0.135}\ \Phi = \frac{1}{K}\frac{\lambda}{\pi} \tag{7.20}$$

where K is the beam quality factor with $K<1$.

Figure 7.14 shows numerically calculated beam parameter products as a function of the magnification M and the power fractions in the central peak and in the first side lobe. This graph indicates that the beam quality is enhanced as the magnification is increased. In the limit of high magnifications, the beam parameter product approaches the value for a round beam with a homogeneous intensity distribution ($K=0.375$). Compared to a Gaussian beam ($K=1$), the beam parameter product of unstable resonators is at least 2.5 times higher. For typical magnifications between $M=2$ and $M=4$, the beam quality of unstable resonators is about three times worse as compared to a Gaussian beam.

Keep in mind that the model used to calculate the beam quality assumes that the near field is a perfect annular ring. In reality the mode structure in the near field exhibits variations in both the amplitude and the phase. This diffraction effect leads to a lower power fraction in the side lobe and to slightly lower beam parameter products (see Fig. 7.14a). This is mainly due to a less steep slope in the outer area of the beam. Furthermore, with the incorporation of the mode structure the beam quality will become dependent on the mirror radius a, as will be discussed in a later section.

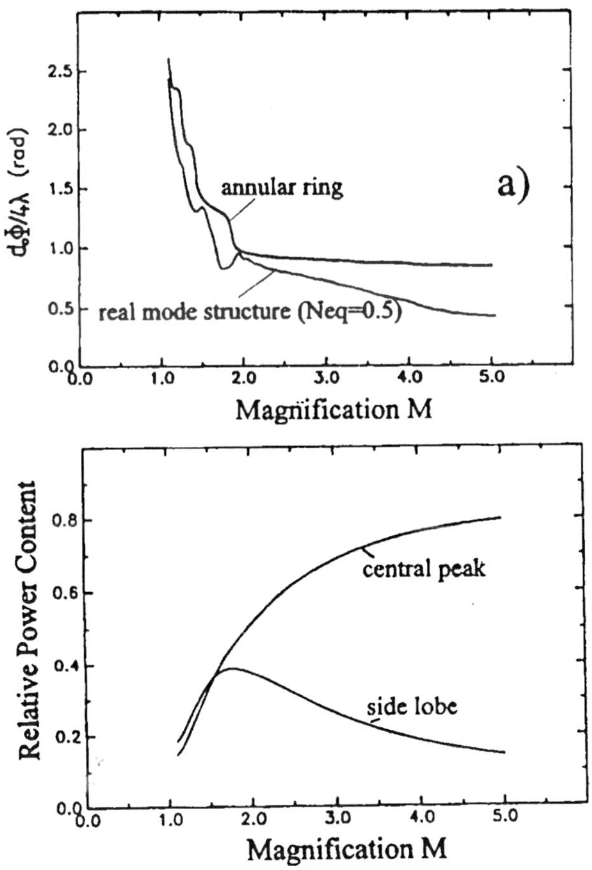

Fig. 7.14 Calculated beam parameter products (86.5% power content, upper graph) and power fractions in the center peak and in the side lobe as a function of the magnification (using the far field distribution of a homogeneously illuminated ring with inner radius a and outer radius Ma).

A similar geometrical treatment can be performed for unstable resonators in rectangular symmetry. The one-dimensional far field is given by the Fourier transform of a double slit with a slit width of *(M-1)a* and a slit separation (center-to-center) of *(M+1)a*. The angular intensity distribution in one dimension reads:

$$I(\theta) = I(0)\left[\frac{\sin[\pi\theta(M-1)a/\lambda]\ \cos[\pi\theta(M+1)a/\lambda]}{\pi\theta(M-1)a/\lambda}\right]^2 \tag{7.21}$$

For the same magnification, the intensity distribution at the focus of unstable resonators in rectangular symmetry exhibits much higher side lobes as compared to circular symmetric unstable resonators. Figure 7.15 presents measured focal intensity distributions for both symmetries in comparison with the theoretical distributions calculated numerically with taking the near field mode structure into account. This graph clearly indicates the relation between the height of the side lobes and the magnification. Measured beam parameter products for an Nd:YAG slab laser with unstable resonators in rectangular symmetry as a function of the magnification are shown in Fig. 7.16. Similar to circular symmetry unstable resonators, the beam parameter products are 3-6 times higher as compared to a Gaussian beam.

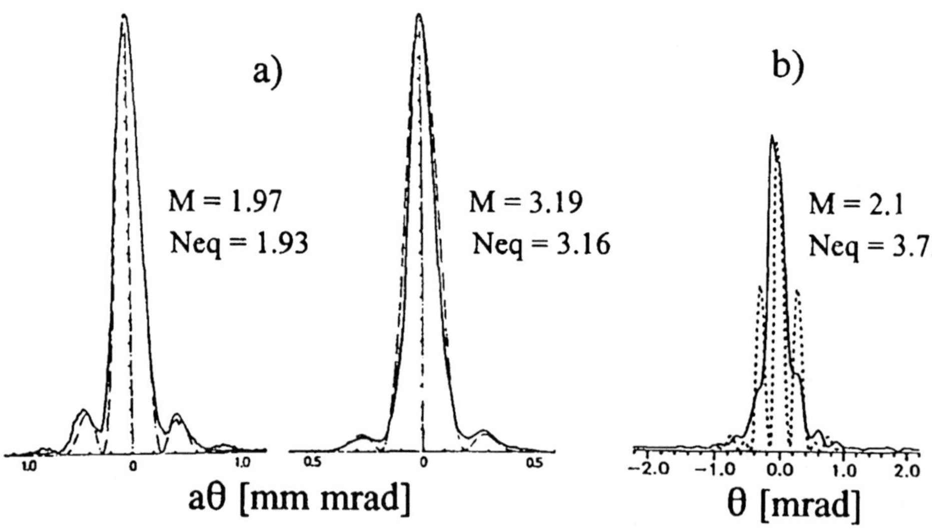

Fig. 7.15 Measured (solid lines) and calculated (broken lines) intensity distributions at the focus of unstable resonators for Nd:YAG lasers ($\lambda = 1.064\mu m$). a) circular symmetry, $a = 1.5mm$, b) rectangular symmetry, $a = 2.0mm$ [S.7].

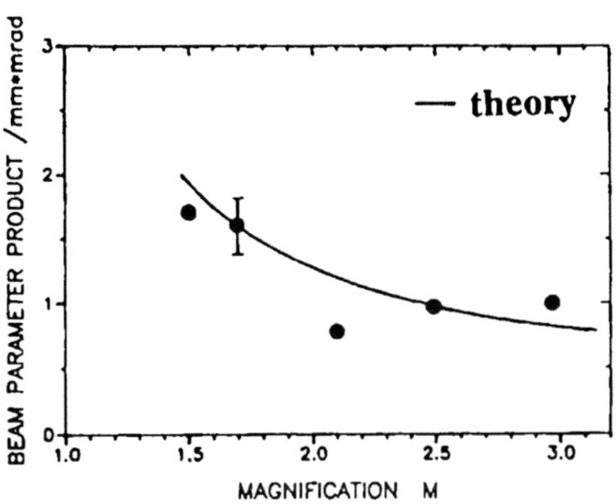

Fig. 7.16 Measured beam parameter products (86.5% power content) of an Nd:YAG slab laser (λ=1.064μm) with unstable resonators in rectangular symmetry. The beam parameter product of a Gaussian beam is 0.25mm mrad (note that this is less than λ/π due to the symmetry) [3.129] (© Chapman and Hall 1992).

It is important to understand that for unstable resonators, the intensity distribution in the far field and the intensity distribution at the focus are generally different. The two distributions are only similar if the geometrical angle of divergence θ_g is equal to zero. For a nonvanishing geometrical divergence, the far field has an annular shape whereas the intensity distribution in the focal spot is still determined by the distributions (7.16) and (7.21) and exhibits a central intensity peak. The measurement of the far field in the focal plane of a lens will thus not provide any information on the focusability of the beam. This is in contrast to stable resonator modes where identical distributions are observed in the far field and in the focus.

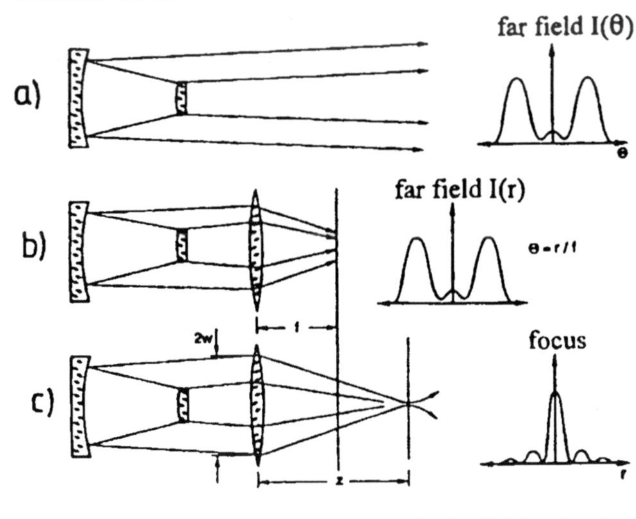

Fig. 7.17 The intensity distributions in the far field and at the focus are identical only for confocal resonators. The far field is observed in the focal plane of the lens (b) whereas the focus is found at a distance z from the lens (c).

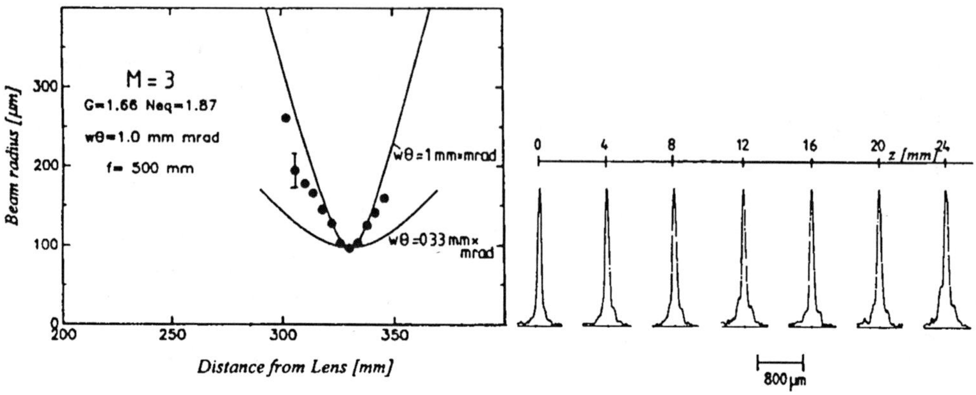

Fig. 7.18 Measured beam radii (86.5% power content) and radial intensity distributions in the vicinity of the focal spot for an unstable resonator (Nd:YAG laser, $\lambda=1.064\mu m$, $M=3, f=500mm$). The curves represent the caustics for a Gaussian beam (lower curve) and a higher order stable resonator mode with the same beam parameter product as that of the unstable resonator. In the right hand graph, the focus is at $z=0$.

The reason for this behavior is the fact that the mode structures of unstable resonators do not represent eigensolutions of the free space diffraction integral. This means that the mode structure changes as the beam propagates; an annular ring in the near field is transformed into a centered intensity distribution at the focus. The power fraction in the center peak is continuously transferred to the sidelobe as the observation plane is moved from the focal plane back to the output coupler. In the vicinity of the focal spot, unstable resonators thus exhibit an increase in the side lobe intensity in addition to the increase in the beam size (Fig. 7.18). This graph indicates that despite the changing intensity structure, the beam radii follow, to a good approximation, the same parabolic propagation law as a higher order stable resonator mode. The knowledge of the beam parameter product thus enables us to calculate the Rayleigh range in a way similar to stable resonators:

$$z_0 = \frac{w_0^2}{d_0\Phi/4} \tag{7.22}$$

where w_0 is the beam radius at the focus and $d_0\Phi/4$ is the beam parameter product.

So far we have assumed that the intensity distribution inside the unstable resonator is represented by a rectangular profile. Unfortunately, this is only a fair approximation of the real mode structure. Due to diffraction at the output coupler, the field distributions inside and outside the resonator exhibit a more complicated phase and amplitude structure which influence both the losses and the beam quality. The geometrical loss factor (7.11) represents only a lower bound for the real loss factor in unstable resonators. In order to get a detailed understanding of unstable resonators the mode structure has to be calculated by using diffraction integrals. However, the geometrical discussion presented so far is not worthless, since the beam propagation can always be described by geometrical optics. Furthermore, the mode structure and the losses will converge to the geometrical quantities if the Fresnel number is chosen high enough. This is a consequence of the fact that for Fresnel numbers greater than 100, the field distributions do not change significantly during propagation and can therefore be propagated by using the laws of geometrical optics.

7.3 Diffraction Theory of Unstable Resonators

7.3.1 Mode Structures, Beam Quality, and Losses

The steady state field distributions on the mirrors of unstable resonators can be calculated similarly to those of stable resonators by applying the integral equations (5.71) in rectangular symmetry and (5.73) in circular symmetry. For both stable and unstable resonators the same integral equations can be used, the only difference is that the absolute value of the G-parameter is now greater than 1.0. Similar to the treatment of stable resonator modes, the integral equations can be simplified by separating the coordinates. The intensity distributions of the eigenmodes and the loss factor $V=|\gamma|^2$ again depend on the absolute value of the equivalent g-parameter G and the absolute value of the effective Fresnel number $N_{eff}=a^2/(2Lg_2\lambda)$. Unstable resonators exhibit an infinite set of TEM modes with mode indices $p\ell$ and mn, but the mode properties are completely different as compared to the modes of stable resonators. Figure 7.20 presents calculated loss factors per round trip of unstable resonators with magnification $M=2$ for some transverse modes in circular symmetry with $\ell=0$ and $\ell=1$. In contrast to stable resonator modes, the loss factor does not always increase as the aperture radius is increased leading to extreme values of the loss factor.

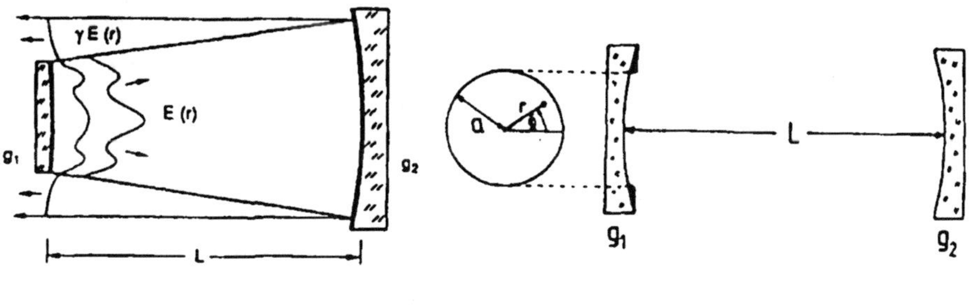

Fig. 7.19 Calculation of the field distribution at the plane of the output coupler for unstable resonators in circular symmetry. After the round trip the shape of the field at mirror 1 is still the same.

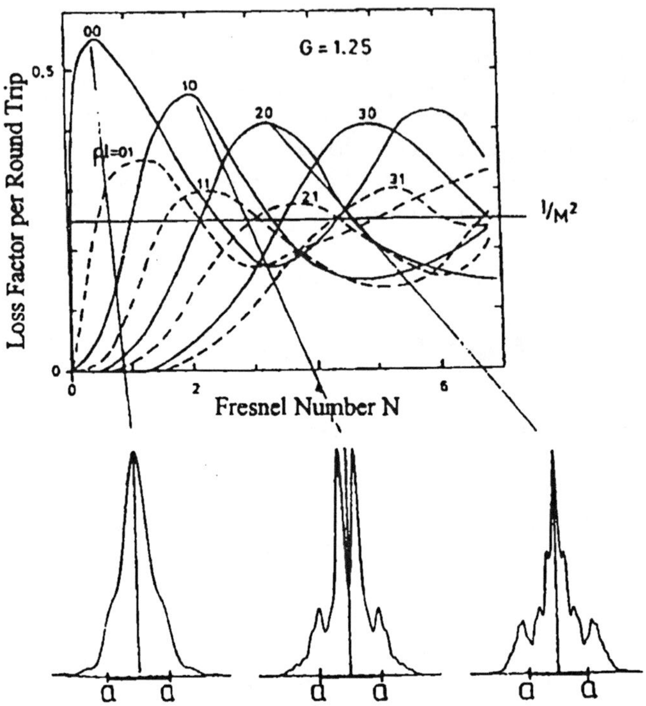

Fig. 7.20 Calculated loss factor per round trip as a function of the effective Fresnel number N for unstable resonators in circular symmetry with magnification $M=2$. The radial intensity distributions at the plane of the output coupler are shown for the first three loss factor maxima.

Since different transverse modes exhibit their maxima at different Fresnel numbers, the loss factor curves cross, which means that at this point two transverse modes have the same diffraction loss. For transverse modes without azimuthal structure ($\ell=0$) the loss factor is higher as compared to other modes and also higher than the geometrical loss factor of $1/M^2$. The reason for the lower loss becomes apparent if we investigate the intensity profiles of the modes at the loss factor maxima. Since the profiles are more centered as compared to the homogeneous profile assumed in the preceding section, a higher power fraction hits the output coupler again after the round trip resulting in a higher loss factor.

It is important to note that the beam radii of different transverse modes are the same. This property, together with the difference in diffraction loss between transverse modes, determines the special oscillation behavior of unstable resonators. The mode with the lowest loss will start oscillating first and deplete the gain in the same area of the active medium that might be used by other transverse modes. These modes, however, exhibit too high of a loss for the leftover gain to reach the laser threshold. Except for the operation at mode crossing points, unstable resonators thus oscillate in a single transverse mode with no azimuthal structure ($\ell=0$).

For practical applications, therefore, it is sufficient to determine the loss and the mode structure only for the lowest loss modes, which means that we only have to calculate the envelope of the curves in Fig. 7.20. The calculated loss factor per round trip of the lowest loss mode for unstable resonators in circular and rectangular symmetry are presented in Fig. 7.21. In these graphs, the loss factor is plotted versus the equivalent Fresnel number N_{eq}, which is related to the effective Fresnel number via:

$$N_{eq} = N_{eff}\sqrt{G^2 - 1} \tag{7.23}$$

By using the equivalent Fresnel number as a mode parameter, the mode crossing points are characterized by integral values of N_{eq}. As the intensity distributions in Fig. 7.22 indicate, the mode structure suddenly changes if the Fresnel number is slightly increased around an integral value of N_{eq}. Since the two modes have complementary intensity profiles and nearly the same diffraction losses, they can oscillate simultaneously. In order to attain true single transverse mode operation in an unstable resonator, it, therefore, is recommended to choose half integral values of the equivalent Fresnel number because the loss difference between modes is a maximum in these areas (Fig. 7.23).

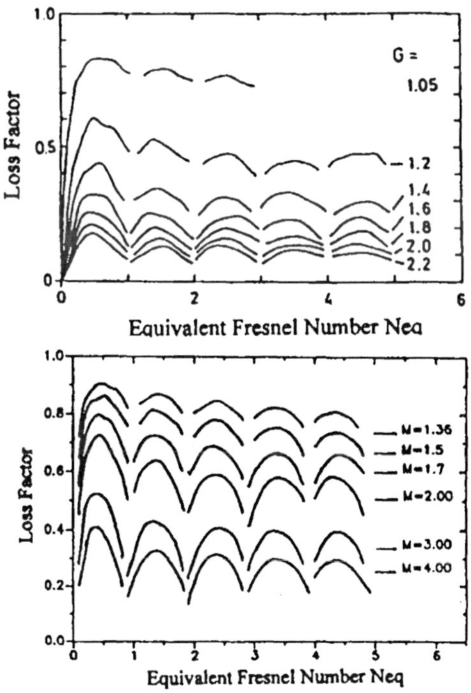

a)

b)

Fig. 7.21 Calculated loss factor per round trip of the lowest loss modes as a function of the equivalent Fresnel number N_{eq}. a) circular symmetry with mirror radius a, b) rectangular symmetry with mirror width $2a$, one-dimensional loss factor V_x, the total loss factor is $V=V_xV_y$ [3.129] (Chapman & Hall 1992).

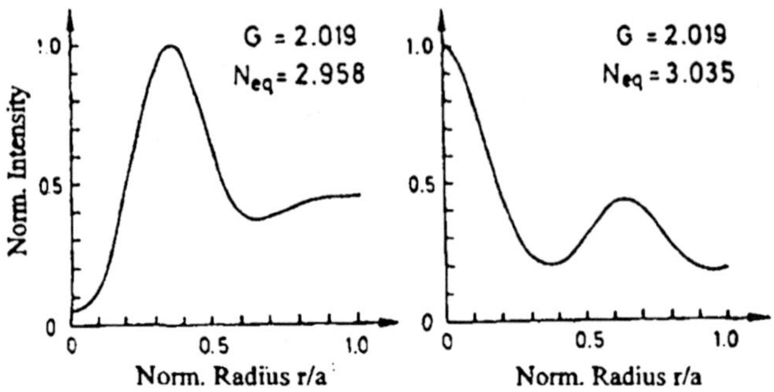

Fig. 7.22 Calculated radial intensity distributions on the confined mirror at an equivalent Fresnel number $N_{eq}=3$ for a magnification of $M=3.75$ [3.116] (© AIP 1988).

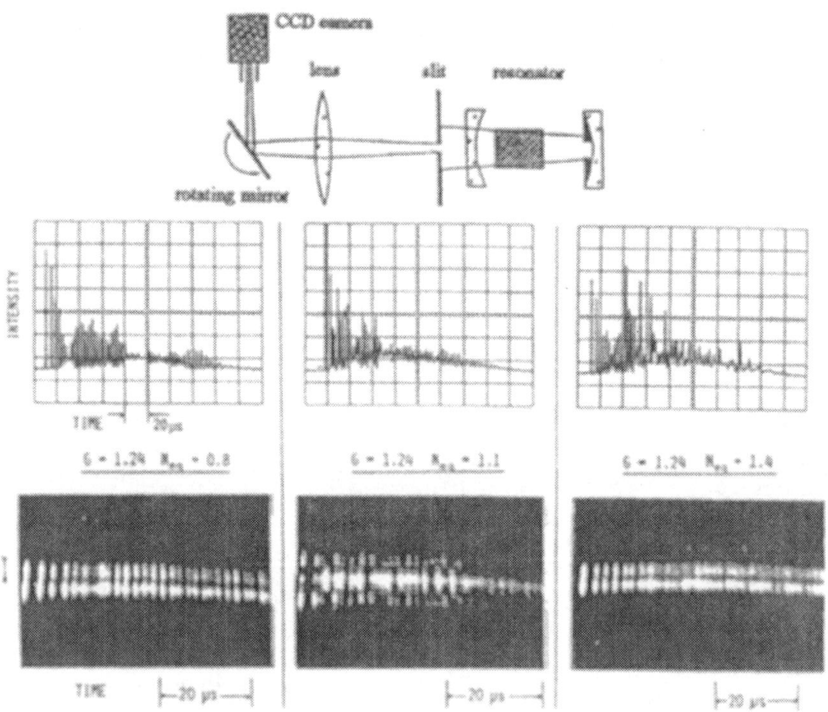

Fig. 7.23 Time resolved measurement of the radial mode structure of an Nd:YAG laser with an unstable resonator. The drawing depicts the experimental set-up. The upper row represents the temporal laser emission, the photographs below show the corresponding radial intensity distributions. The simultaneous oscillation of two transverse modes is observed at $N_{eq}=1.1$ (photograph in the middle) [3.116] (© AIP 1988).

Figure 7.24 shows measured radial intensity distributions at the plane of the output coupler in comparison to the theoretical profiles. The radius *a* of the output coupler and the geometrical beam radius *Ma* are indicated. With increasing Fresnel numbers the mode structures exhibit more side lobes and will approach a homogeneous profile for very large Fresnel numbers. Note that the geometrical beam radius *Ma* is quite a good approximation for the lateral extent of the modes. Similar to the mode structures, the loss factors converge to the geometrical loss factor for large Fresnel numbers (Fig. 7.25).

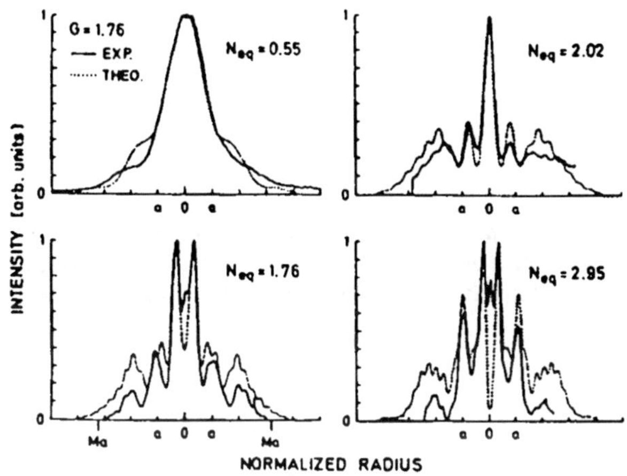

Fig. 7.24 Measured radial intensity distributions at the plane of the output coupler of unstable resonators in circular symmetry with *G=1.76* (*M=3.21*) and different equivalent Fresnel numbers. The profile within the radius *a* is reflected by the mirror, the outer portion of the mode is coupled out. The dotted lines are the numerically calculated distributions (pulsed Nd:YAG laser) [3.116] (© AIP 1988).

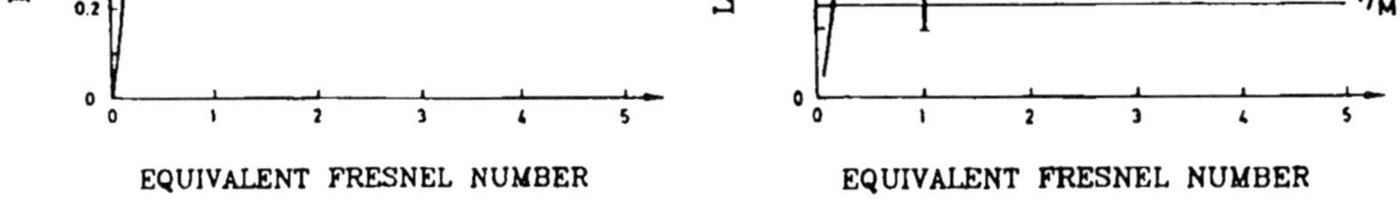

Fig. 7.25 Measured and calculated loss factors per round trip of unstable resonators in circular symmetry for different magnifications *M* as a function of the equivalent Fresnel number (pulsed Nd:YAG laser) [3.116] (© AIP 1988).

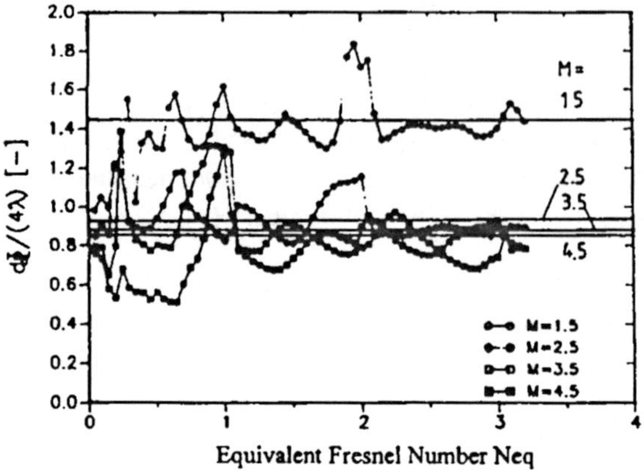

Fig. 7.26 Calculated normalized beam parameter products $d_0\Theta/(4\lambda)$ (86.5% power content) of unstable resonators in circular symmetry as a function of the equivalent Fresnel number for different magnifications *M*. The corresponding values for a homogeneous annular ring (Fig. 7.14) are indicated by the horizontal lines.

The mode structure at the output coupler not only influences the loss factor, but also affects the beam quality of the unstable resonator. Figure 7.26 shows beam parameter products (86.5% power content) as a function of N_{eq} and M, taking into account the mode structure in the near field. The comparison with the beam parameter products obtained by approximating the near field by a homogeneous ring profile indicates that the incorporation of the mode structure at the output coupler generally improves the beam quality (except close to integral values of N_{eq}). Furthermore, the power content in the central peak is increased as compared to the far field of an annular ring (Fig. 7.27, cf. Fig. 7.14).

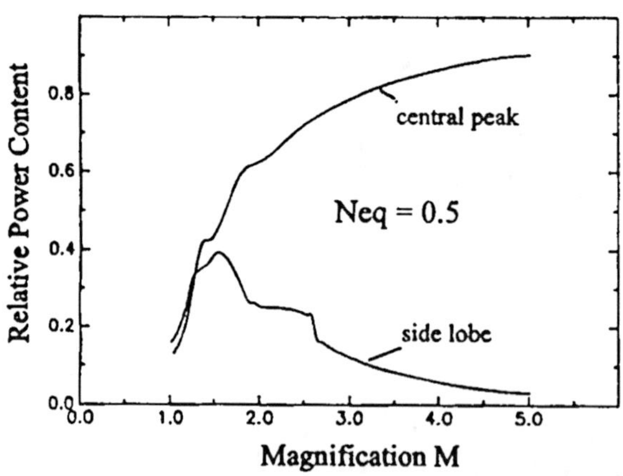

Fig. 7.27 Calculated power content in the central peak and in the side lobe of the far field intensity distribution as a function of the magnification (circular symmetry, $N_{eq}=0.5$).

7.3.2 Applications of Unstable Resonators

The main advantage of unstable resonators is the adaptation of the beam radius inside the resonator by changing the mirror radius a. Since single transverse mode operation is not linked to the size of the mirror, as is the case for stable resonators, beam qualities close to the diffraction limit can be attained even for gain media with very large cross sections. In contrast to stable resonators, a high mode volume and a high beam quality can be realized simultaneously. This sounds too good to be true. So what are the drawbacks ?

First of all, the power content in the side lobes of the far field might cause problems in specific applications. In order to minimize the power content in the side lobes the magnification has to be chosen as high as possible. This, however, requires an active medium with a high gain. It therefore makes no sense to use an unstable resonator in a HeNe laser, since the gain is too low to obtain a good laser efficiency. Secondly, the output power of unstable resonators generally is 20%-30% lower as compared to a stable resonator in multimode operation. This is due to a lower fill factor in combination with additional diffraction losses induced by the rim of the active medium. If one is more interested in the output power than in attaining an excellent beam quality, it is therefore more advantageous to use a stable resonator in multimode operation. This preference of multimode stable resonators is also due to their easier alignment and their lower sensitivity to mirror misalignment.

Unstable resonators have found application in high power CO_2 lasers 3.60,3.64,3.66,3.69,3.83,3.122] and excimer lasers [3.92,3.102,3.119], both of which generally exhibit a high gain and a large cross section of the active medium. Unstable resonators are also used in pulsed solid state lasers [3.85,3.106,3.111,3.125-3.129,3.137,3.146,3.152,3.156,3.163-3.169]. Unfortunately, the thermal lensing of the solid state laser materials deteriorates the beam quality and makes beam handling quite difficult, as will be discussed in Chapter 12. It is for this reason that most applications of unstable resonators in solid state laser engineering are limited to low power Q-switch systems. The utilization of unstable resonators in diode lasers [3.124], dye lasers [3.171], and free electron lasers [3.100,3.107,3.114] has also been reported.

7.4 Misalignment Sensitivity

All geometrical relations presented in Sec. 5.4 for stable resonators also apply to unstable resonators. Since mirror 2 is always unconfined, the additional losses occur at the output coupling mirror 1 due to a shift Δ_{1j} of the optical axis with respect to the center of mirror 1. If mirror 2 is tilted by an angle α_2 and the output coupling mirror exhibits a radius a, the relative shift is given by (Fig. 7.28, see also Fig 5.47):

$$\frac{\Delta_{12}}{a} = \frac{L\alpha_2}{a} \frac{2}{|G-1|} \tag{7.24}$$

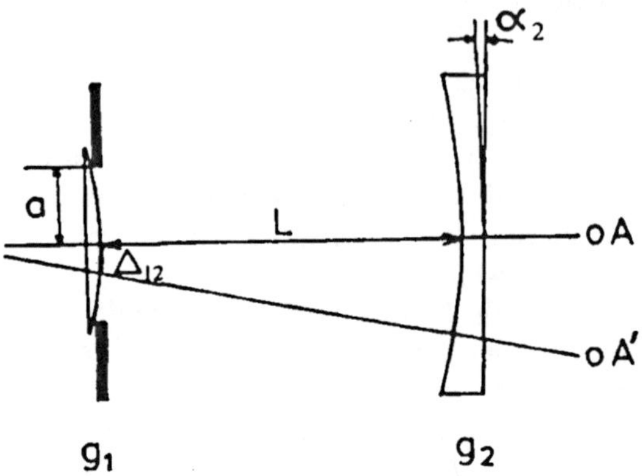

Fig. 7.28 Misalignment of mirror 2 induces a shift Δ of the optical axis (oA) on mirror 1. Additional diffraction losses are generated at the confined mirror 1.

If mirror 1 is tilted by an angle α_1, the corresponding relation reads:

$$\frac{\Delta_{11}}{a} = \frac{Lg_2\alpha_1}{a}\frac{2}{|G-1|} \tag{7.25}$$

We see that a tilt of the unconfined mirror by the angle α_2 is equivalent to the misalignment of the output coupling mirror by the angle $\alpha_1=\alpha_2/g_2$. These two geometrical relations already provide us with valuable information on the tilt sensitivity of unstable resonators:

- the misalignment sensitivity gets lower if the equivalent g-parameter G or the magnification M is increased.
- negative branch resonators ($G<-1$) are less sensitive to mirror tilt than positive branch resonators. For a magnification of $|M|=2$, we can expect a nine times lower sensitivity in the negative branch (assuming equal length L and equal mirror radius a).

Similar to stable resonators, unstable resonators experience a parabolic increase of the losses if one of the resonator mirrors is tilted [3.60,3.68,3.79,3.89,3.118,3.129]. However, the diffraction losses of unstable resonators can also increase at higher tilt angles as shown in Fig. 7.29. The resulting loss factor minima can be related to a switching of transverse modes similar to the mode switching of the aligned resonator at integer equivalent Fresnel numbers N_{eq}. Using the shift Δ of the optical axis, we can define two new equivalent Fresnel numbers $N_{eq}^{\pm}$ with:

$$N_{eq}^{\pm} = \frac{(a\pm\Delta)^2}{2Lg_2\lambda}\sqrt{G^2-1} \tag{7.26}$$

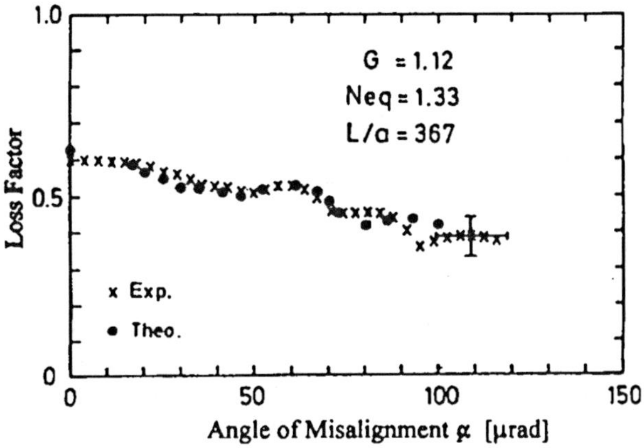

Fig. 7.29 Measured loss factor per round trip of a circularly symmetric unstable resonator for misalignment of the unconfined mirror 2 (Nd:YAG rod laser). The dots represent a numerical calculation using diffraction integrals [3.118] (© Taylor & Francis 1988).

The tilt angles at which the mode crossings occur can be related to integer values of this newly defined equivalent Fresnel number. For the misalignment of the unconfined mirror, the loss factor minima occur at the angles:

$$\alpha_{2,p} = \frac{(G-1)a}{2L}\left[\mp 1 \pm \sqrt{1 - \frac{|N_{eq}-p|}{|N_{eq}|}}\right] \tag{7.27}$$

with p being an integer. The upper sign holds for $p>|N_{eq}|$, the lower sign for $p<|N_{eq}|$. For the measurement shown in Fig. 7.29, Eq. (7.27) yields 21.7μrad ($p=1$), 37 μrad ($p=2$), and 82μrad ($p=3$), which is in good agreement with the experimental data.

In general, the tilt angle at which the first mode crossing occurs is greater than the tilt angle $\alpha_{10\%}$ at which the losses have increased by 10%. Similar to stable resonators, we can therefore define a parabolic dependence of the loss factor V on the angle of misalignment:

$$V(\alpha) = V(0)\left[1 - 0.1\left(\frac{\alpha}{\alpha_{10\%}}\right)^2\right] \tag{7.28}$$

As was already done in Sec. 5.4 for the misalignment of stable resonators, it is convenient to introduce the misalignment parameter $D=L\,\alpha_{10\%}/a$ to quantify the misalignment sensitivity of unstable resonators. The misalignment parameter depends only on the equivalent g-parameter G and the absolute value of the equivalent Fresnel number N_{eq}. Figure 7.30 presents calculated misalignment parameters for the misalignment of the unconfined mirror 2 as a function of the equivalent Fresnel number for both positive branch and negative branch unstable resonators. The calculations were performed by numerically solving integral equation (5.85). These graphs enable us to determine the 10% angles of any unstable resonator. Note that the 10% tilt angle of the confined mirror 1 can also be determined by applying the above mentioned equivalency relation between the tilts of the two mirrors. As we already found in our preliminary geometrical result (7.24), the misalignment sensitivity becomes lower for high magnifications and small ratios of resonator length L to mirror radius a. Furthermore, a low Fresnel number is advantageous for the stability of an unstable resonator. Figure 7.30 also confirms the lower sensitivity of negative branch resonators predicted by our geometrical analysis!

How does this compare to the misalignment sensitivity of stable resonators ? The misalignment parameters for stable resonators in fundamental mode operation with adapted aperture radius ($a=1.3w$, w: Gaussian beam radius at the aperture) and for positive branch unstable resonators with $N_{eq}=0.5$ are presented in Fig. 7.31. This graph indicates that the misalignment parameters are comparable. However, in order to get a high mode volume of the Gaussian beam in the stable resonator, an equivalent g-parameter G between 0.95 and 1.0 or a large resonator length is required. Considering this, the 10%-angle of stable resonators in fundamental mode operation, therefore, is several times smaller as compared to unstable resonators.

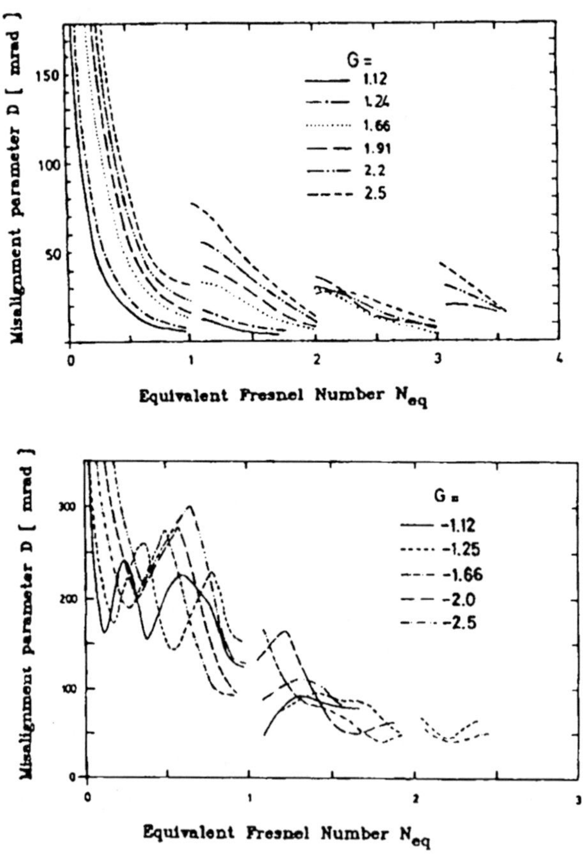

a)

b)

Fig. 7.30 Calculated misalignment para-meter $L\alpha_{10\%}/a$ as a function of the equivalent Fresnel number for misalignment of the unconfined mirror of unstable resonators in circular symmetry. a) positive branch, b) negative branch [3.118] (© Taylor & Francis 1988).

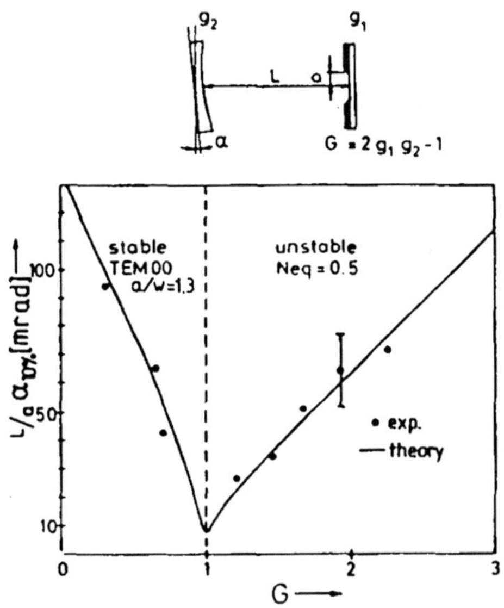

Fig. 7.31 Measured and calculated misalignment parameter of stable resonators in fundamental mode operation with adapted aperture radius a and positive branch unstable resonators with equivalent Fresnel number N_{eq} as a function of the equivalent g-parameter G. The unconfined mirror 2 is tilted (circular symmetry).

Example:

Confocal resonator with M=2 (g_1=1.5, g_2=0.75, G=1.25, L=0.5m, λ=1.064μm)
In order to fill an active medium of diameter 6.35mm we need a mirror radius a of 1.58mm. This results in an equivalent Fresnel number of $N_{eq}=2.36$. For this resonator the misalignment parameter in Fig. 7.30 is 20 mrad. This results in a 10%-angle of 63 μrad for the unconfined mirror. If the output coupling mirror is misaligned the 10%-increase of the loss occurs at a tilt angle of 84μrad (=63μrad/g_2).

Now let us design a stable resonator in fundamental mode operation having a comparable mode volume and the same resonator length. The Gaussian beam radius w_i at mirror i is given by (see Sec. 2.8.2):

$$w_i^2 = \frac{2Lg_j\lambda}{\pi\sqrt{1-G^2}} \quad , \quad i,j=1,2 \ ; \ i \neq j$$

We place the active medium close to mirror 1. In order to completely fill the medium, the Gaussian beam radius w_i has to be 1.3 times smaller than the radius of the medium, resulting in w_i=2.44mm (the radius of the medium is the aperture radius a). We choose a mirror radius of mirror 2 of -0.25m. With this choice all other resonator parameters are defined: $g_2=3.0$, $G=0.985$, $g_1=0.331$. According to Fig. 7.31 (or Fig. 5.49), the misalignment parameter for mirror 2 is about 8mrad. This results in 10%-angles of 51μrad for mirror 2 and 17μrad for mirror 1. This resonator would be several times more sensitive to misalignment than the unstable resonator!

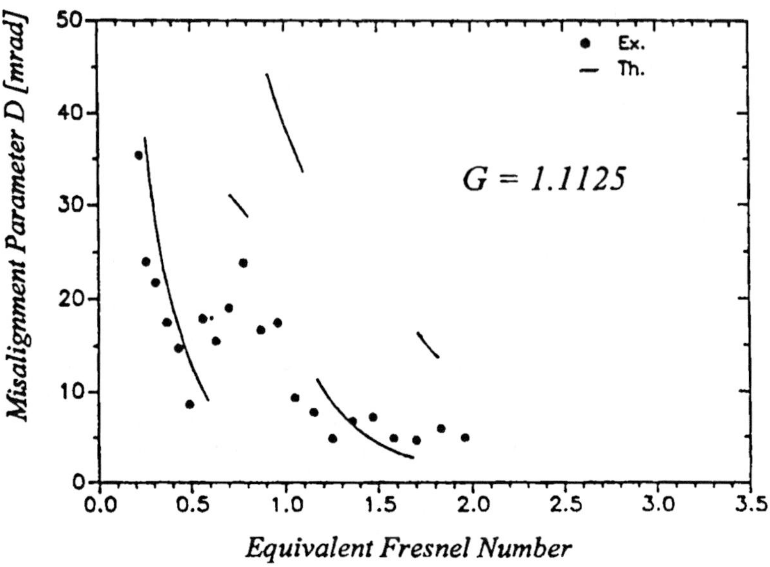

Fig. 7.32 Measured and calculated misalignment parameter as a function of the equivalent Fresnel number for unstable resonators in rectangular symmetry with magnification $M=1.6$ (Nd:YAG slab laser) [S.7].

In rectangular symmetry, the misalignment sensitivity of unstable resonators is comparable to the circularly symmetric ones as a comparison of Fig. 7.32 and Fig. 7.30a indicates.

The reader should keep in mind that the misalignment of unstable resonators increases the output coupling. This is in contrast to stable resonators in which the output coupling remains constant and the increase in loss is due to clipping of the beam at an aperture. It is for this reason that the influence of mirror misalignment on the output power is quite different for the two resonator types. In contrast to stable resonators in which a mirror tilt is equivalent to a decrease of output power, unstable resonators may exhibit an increase of power. This increase can be observed if the aligned resonator is undercoupled, which means that the output coupling is too low for the given small-signal gain of the active medium. This interesting property will be discussed in more detail in Chapter 14. We will see that the increase in output coupling results in a misalignment sensitivity of the output power that is twice as low as compared to the sensitivity of the power of stable resonators in fundamental mode operation, assuming the same 10%-angle for the loss for both resonators. A similar statement holds for the shape of the far field intensity distribution. Even for tilt angles much higher than the 10%-angle, the shape stays almost constant, as the measured intensity distributions in Fig. 7.33 indicate. For stable resonators in fundamental mode operation, the far field structure already exhibits a noticeable asymmetry at the 10%-angle.

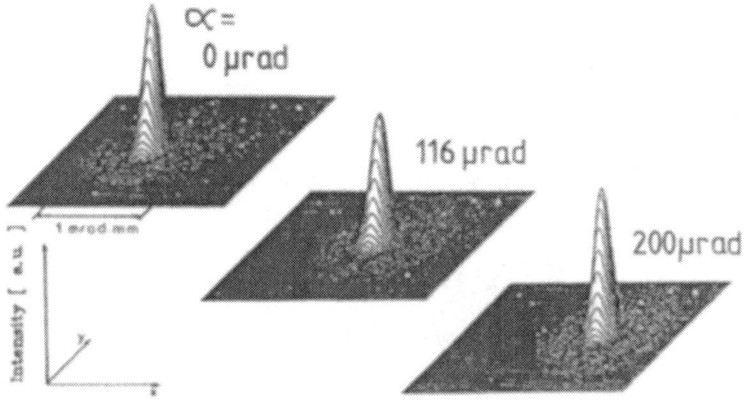

Fig. 7.33 Measured, normalized intensity distributions in the far field of a positive branch unstable resonator for different tilt angles of the unconfined mirror ($G=1.75$, $N_{eq}=2.85$, $L=0.32m$, $a=1mm$, $\lambda=1.064\mu m$, Nd:YAG rod laser). The 10%-angle $\alpha_{10\%}$ is 50μrad [3.118] (© Taylor & Francis 1988).

7.5 Unstable Resonators in Off-Axis Geometry

Unstable resonators in rectangular geometry exhibit high side lobes in the far field, even for high magnifications ($M>2$). This is due to the fact that the far field is determined by diffraction at a double slit which leads to a strong modulation of the intensity distribution in the far field (see Fig. 7.15b). Fortunately, the side lobes can be decreased considerably if the radiation is coupled out at one side of the high reflecting output coupling mirror [3.1,3.129]. This can be accomplished by misaligning both resonator mirrors in such a way that the optical axis is positioned along a corner of the active medium rather than choosing the symmetry axis of the medium as the optical axis (Fig. 7.34). If spherical mirrors are used the beam is magnified along the x- and the y-axis during a round trip but only in one direction. Accordingly, the beam is coupled out at one side of the output coupler only. The one-dimensional intensity distribution in the far field is now given by the Fourier transform of an illuminated single slit (see Sec. 2.2.1). The effect of this special output coupling scheme on the mode structure and the far field intensity distribution is presented in Fig. 7.35 for a confocal unstable resonator with magnification $M=1.5$. The power content in the side lobes of the far field is clearly reduced resulting in an improvement of the beam quality by a factor of 2. Note that in off-axis geometry the near field experiences diffraction losses at the edge of the active medium (hatched area in the upper right graph). These additional losses can be minimized by adjusting the mirror alignment so that the optical axis is slightly moved outside the active medium. Since the off-axis unstable resonator can be viewed as a misaligned on-axis resonator, the diffraction losses are higher as compared to on-axis unstable resonators in rectangular symmetry (Fig. 7.36, see Fig. 7.21b for a comparison).

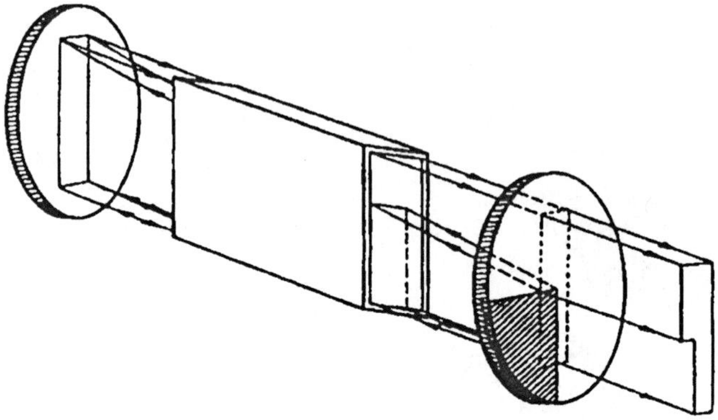

Fig. 7.34 Confocal unstable resonator with spherical mirrors in off-axis geometry. The optical axis is positioned along the lower left corner of the active medium. The hatched area marks the HR coating of the output coupling mirror [3.129] (© Chapman and Hall 1992).

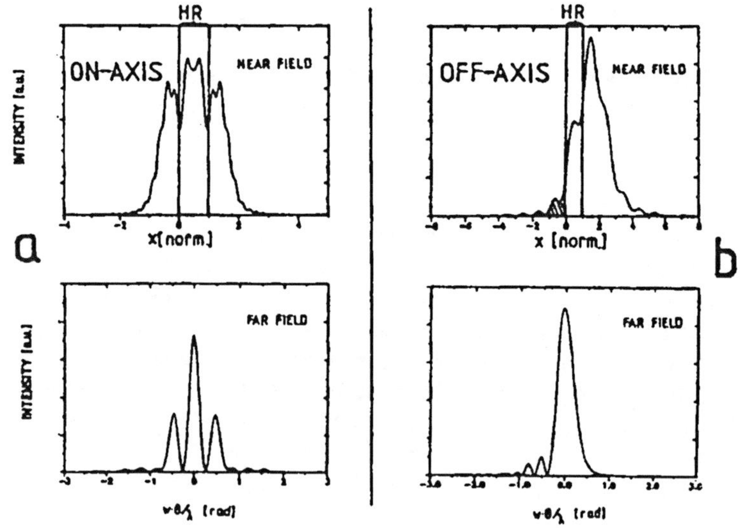

Fig. 7.35 Calculated intensity distributions in the near field and in the far field of an unstable resonator with magnification $M=2$, both in on-axis and in off-axis geometry. The near field inside the resonator at the plane of the output coupler, is shown (HR indicates the high reflecting area) [3.129] (© Chapman and Hall 1992).

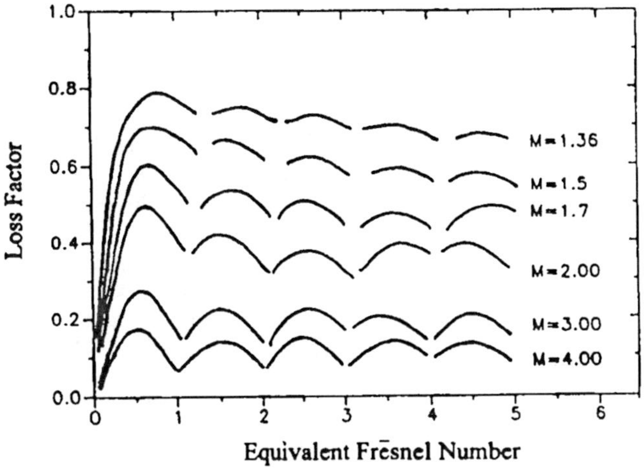

Fig. 7.36 Calculated one-dimensional loss factors of unstable resonators in off-axis geometry as a function of the equivalent Fresnel number $N_{eq} = a^2\sqrt{G^2-1}/(2Lg_2\lambda)$ (a: width of the HR coating on the output coupler hit by the beam). The curve parameter is the magnification M (see also Fig. 7.21b).

In spite of the asymmetry of the L-shaped intensity distribution coupled out of the resonator, the intensity distribution in the far field is symmetric with respect to the x-axis and the y-axis. The slight asymmetry in the far field intensity distribution in Fig. 7.35 is a result of the near field mode structure; if the near field were a perfect L with homogeneous illumination, the focus would be completely symmetric. Figure 7.37 presents photographs of the near field and the far field of a Nd:YAG slab laser with an unstable resonator with magnification $M=1.4$ in off-axis geometry. The comparison of the beam quality of on-axis and off-axis unstable resonators shown in Fig. 7.38 clearly indicates the advantages of the off-axis geometry.

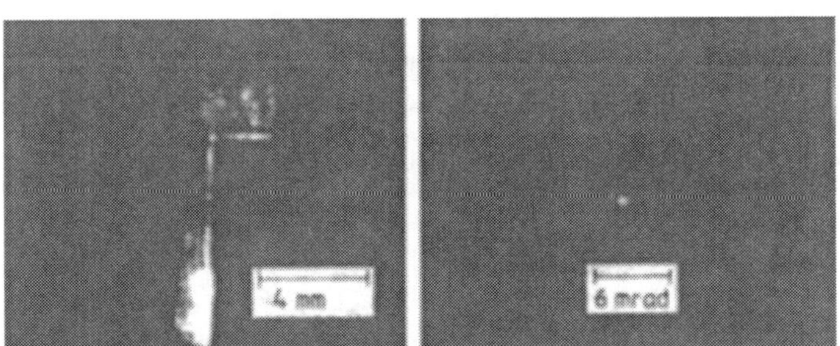

Fig. 7.37 Photographs of the near field and the far field intensity distributions of a confocal unstable resonator in off-axis geometry with $M=1.4$ (Nd:YAG slab laser, $\lambda=1.064\mu m$, cross sectional area of the slab: 4x12 mm^2) [3.129] (© Chapman and Hall 1992).

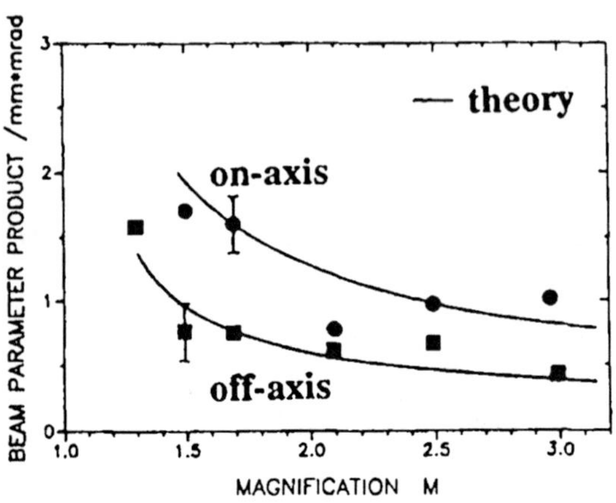

Fig. 7.38 Measured and calculated beam parameter products $d\Phi/4$ (d: waist width, Φ: full angle of divergence, both defined by the 86.5% power content) of unstable resonators in on-axis and in off-axis geometry versus the magnification M (Nd:YAG laser, $\lambda=1.064\mu m$) [3.129] (© Chapman and Hall 1992).

For active media with rectangular cross section the off-axis geometry is the preferred means to attain good beam quality. Since the aspect ratio of the medium (ratio of height to width) is generally high (for solid state slab lasers and CO_2 slab lasers the aspect ratio is 4-6 and 5-10, respectively) it is common to apply the off-axis unstable resonator scheme only in the direction of the larger dimension. In the perpendicular direction (width) a stable resonator (solid state) or a waveguide resonator (CO_2) is usually chosen [5.80,5.104]. Since the width is on the order of millimeters it is possible to obtain beam qualities near the diffraction limit by using the latter two resonator concepts. A detailed discussion on these so called hybrid resonators is given in Chapter 19. Figures 7.39 and 7.40 show layout and performance of a diode-pumped Nd:YAG slab laser using an off-axis unstable resonator along the slab width and a stable resonator along the thickness of the slab.. By using a variable reflectivity mirror (VRM) as the output coupler, a beam propagation factor M^2 of less than 1.2 was attained in the unstable direction [3.179]. Similar results have been reported at even higher power levels [3.173-175] and also for Vanadate slabs [3.177,3.178].

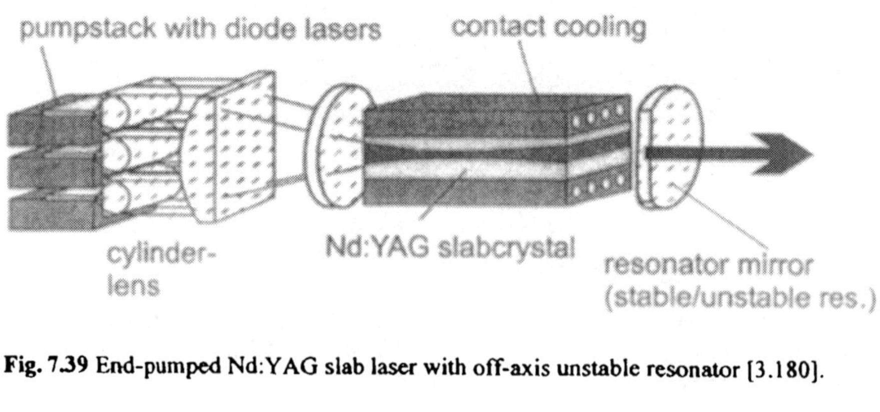

Fig. 7.39 End-pumped Nd:YAG slab laser with off-axis unstable resonator [3.180].

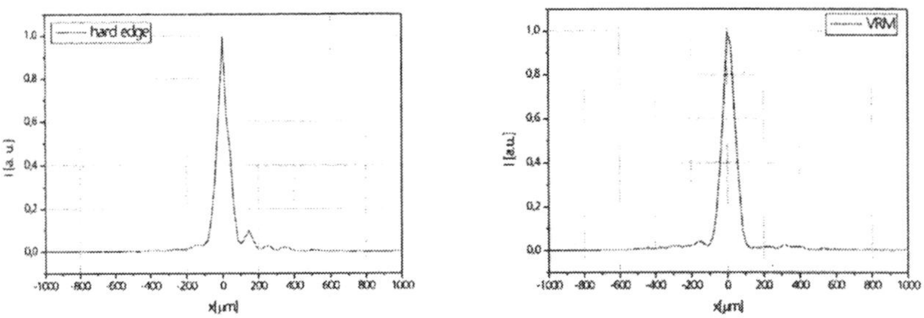

Fig. 7.40 Measured focus intensity distribution for the slab laser of Fig. 7.39 in the unstable resonator direction, with a hard edge (left) and a variable reflectivity output coupler (VRM, right). Output power in both cases was 60W, magnification :1.7 for hard edge and 1.2 for VRM [3.180].

It is also possible to apply the off-axis concept to circularly symmetric media (Fig. 7.41). Unfortunately, it is very difficult to manufacture glass substrates exhibiting the aspheric shapes required for this resonator concept. It is for this reason that circularly symmetric off-axis resonators have only been realized in CO_2 lasers for which copper mirrors of any shape can be made using diamond milling [3.130]. The basic principle of the resonator set-up can be understood if we imagine a one-dimensional off-axis unstable resonator in rectangular symmetry whose cylindrical mirrors are bent to form a tube. The output coupling mirror now exhibits an annular shape and the laser beam emerges from the resonator through the central aperture. Thus, the far field is similar to the Fraunhofer diffraction pattern of a circular aperture having much lower side lobes as compared to an on-axis unstable resonator (Fig. 7.42). Note that for a given magnification *M*, the output coupling losses are lower in the off-axis geometry. Instead of $1/M^2$ for the on-axis resonator, the geometric loss factor is now given by (Fig. 7.43):

$$V = \frac{2}{M} - \frac{1}{M^2} \tag{7.29}$$

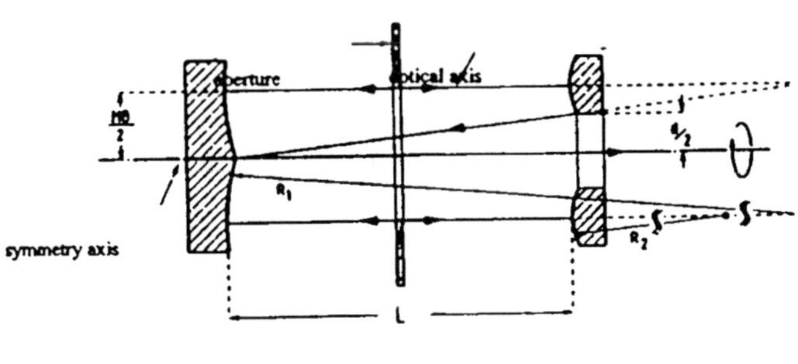

Fig. 7.41 The off-axis unstable resonator in circular geometry [S. 8].

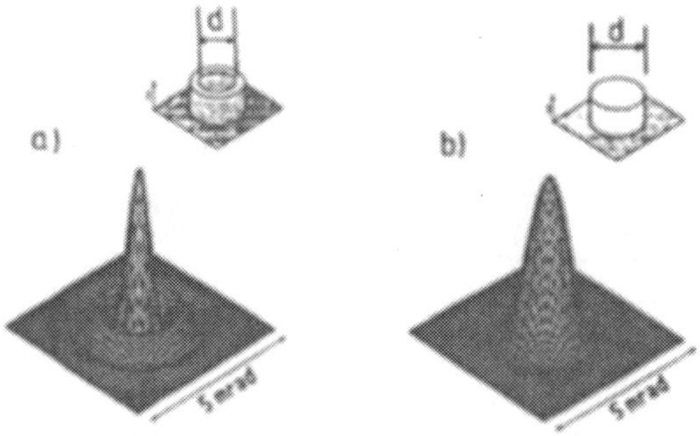

Fig. 7.42 Calculated far field intensity distributions for confocal on-axis (a) and off-axis (b) unstable resonators in circular symmetry. The magnification is $M=1.5$ ($d=20mm$, $\lambda=10.6\mu m$) [S.8].

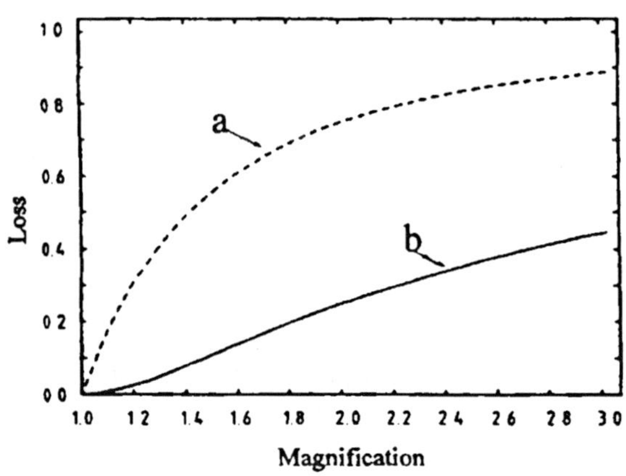

Fig. 7.43 Geometric loss per round trip for circularly symmetric unstable resonators in on-axis (a) and off-axis geometry (b) as a function of the magnification [S.8].

7.6 Unstable Resonators with Homogeneous Output Coupling

The side lobes in the far field of unstable resonators are caused by diffraction off the edge of the output coupling mirror in combination with the annular shape of the near field. A method to overcome this disadvantage is to transmit the laser beam through a partially reflecting unconfined mirror (Fig. 7.44). The beam size is defined by an aperture in front of the high reflecting mirror which replaces the output coupler described in previous sections. Similar to the off-axis resonators, the far field distribution is given by the Fraunhofer diffraction pattern of a round aperture which exhibits smaller side lobes. However, due to the mode structure at the aperture, which for low equivalent Fresnel numbers exhibits a Gaussian-like field distribution, the power content in the side lobes is even smaller.

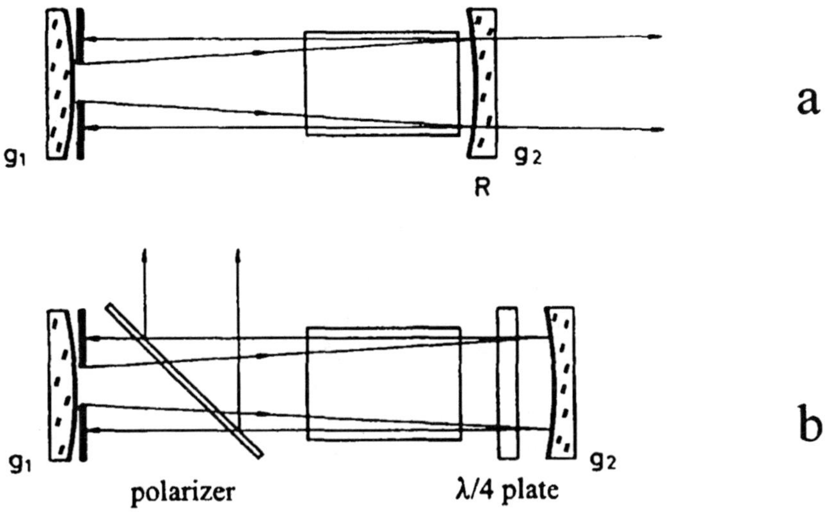

Fig. 7.44 Unstable resonators with homogeneous output coupling. a) using a partially reflecting mirror 2, b) using a polarizer with a rotatable quarter wave plate to vary the output coupling.

The laser beam can be described as a Gaussian beam to a good approximation. Unfortunately, the price to be paid for this improvement is a considerable decrease of output power because the diffraction loss of the unstable resonator represents a true loss. It is recommended to use a low magnification to keep this loss as low as possible. This resonator concept, therefore, is only suitable for applications in which a perfect beam quality and a Gaussian intensity distribution is more important than the optimization of the laser efficiency (e.g. as a master oscillator to feed the beam into an amplifier chain) [5.29]. Unstable resonators with homogeneous output coupling are used for active media with high gain and small emission wavelengths (Q-switched solid state lasers and excimer lasers). For these lasers no other resonator concept exists that combines near diffraction limited focusability with a high efficiency. A common stable resonator in fundamental mode operation might not provide the output power required. Note that replacing the aperture by a high reflecting flat mirror in order to decrease the losses will not work since the two mirrors will then form a stable resonator depleting the gain in the outer area of the medium.

7.7 Unstable Resonators with Variable Reflectivity Mirrors

7.7.1 Resonator Properties

A better way than using homogeneous output coupling techniques to decrease the power content in the side lobes of the far field of unstable resonators is the utilization of output coupling mirrors with a variable reflectivity profile [3.131-3.171]. The reflectivity profile which generally exhibits a maximum at the center improves the beam quality by two effects. The diffraction of the power into the side lobes is considerably decreased by replacing the hard edge by a continuous transition from high to low reflectance. Furthermore, the center reflectivity and the shape of the profile can be varied to generate a specified intensity profile in the near field (e.g. flat top beams). Figure 7.45 shows the effect of the reflectivity profile on the near field and the far field intensity distributions of a confocal unstable resonator in circular symmetry with magnification $M=2$. The parabolic and the Gaussian profile both decrease the power content in the first side lobe from 25% to about 3-4%. The specific shape of the reflectivity profile is only of secondary importance, especially if the center reflectivity is chosen lower than 100%. However, in order to maximize the fill factor, steeper profiles are preferred since the mode with its correspondingly steeper slopes can be better adapted to the active medium. From 1985 to 1995 much effort was spent to develop coating techniques for generating variable reflectivity mirrors (VRM) exhibiting Gaussian, super-Gaussian, and parabolic reflectivity profiles [3.140,3.151,3.154,3.160,3.162, 3.166, 3.168].

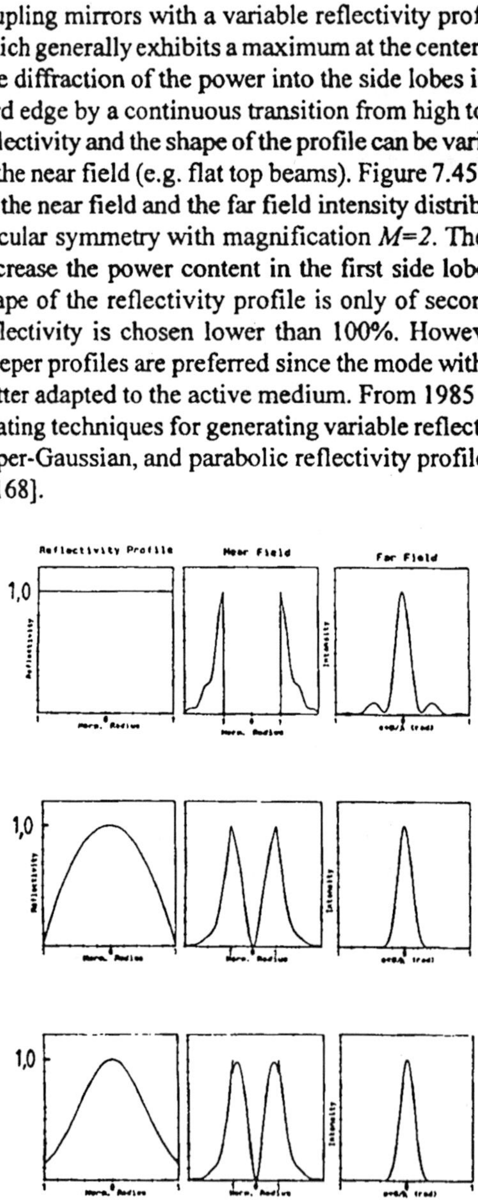

Fig. 7.45 Calculated intensity distributions in the near field and the far field of a confocal unstable resonator with $M=2$ and $N_{eq}=0.5$ for different reflectivity profiles of the output coupler.

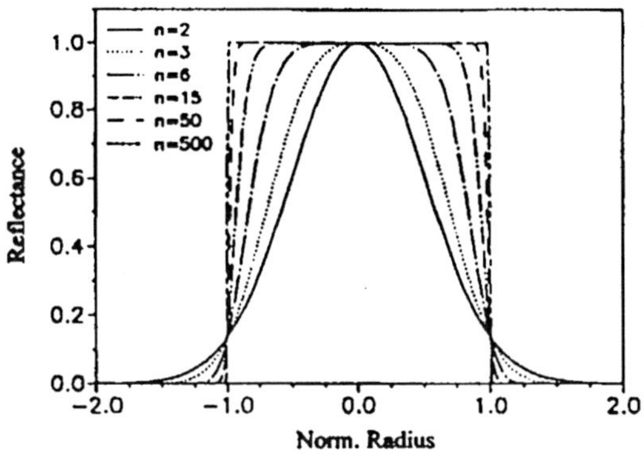

Fig. 7.46 Super-Gaussian reflectivity profiles for different indices n as a function of the radius [S.9].

Super-Gaussian Mirrors

The VRMs most commonly used exhibit a super-Gaussian reflectivity profile which in circular symmetry reads:

$$R(r) = R_0 \exp\left[-2\left(\frac{r}{w}\right)^n\right] \tag{7.30}$$

with:
- R_0: center reflectivity
- w: profile radius
- n: super-Gaussian index
- r: radial coordinate

The super-Gaussian index n determines the shape of the reflectivity profile. For $n=2$ the profile is Gaussian and with increasing n the slopes become steeper until the hard edge profile is approached in the limit $n \to \infty$ (Fig. 7.46).

The mode properties can be evaluated by using the stationary condition for the electric field. If $E_1(r)$ denotes the electric field hitting the output coupling mirror, the electric field after one round trip reads (we treat the round trip by magnifying the field by the magnification M):

$$E_2(r) = \frac{1}{M} E_1(r/M) \sqrt{R(r/M)} \tag{7.31}$$

The stationary condition $E_2(r)=\gamma E_1(r)$ with γ being the eigenvalue yields:

$$\gamma E_1(r) = \frac{1}{M} E_1(r/M) \sqrt{R(r/M)} \tag{7.32}$$

After insertion of the super-Gaussian reflectivity profile (7.30), the stationary electric field can be calculated. The corresponding intensity profile is given by:

$$I(r) = I_0 \exp\left[-2\left(\frac{r}{w_M}\right)^n\right] \tag{7.33}$$

with: $$w_M = w\,(M^n - 1)^{1/n} \qquad M > 1 \tag{7.34}$$

The total loss factor per round trip reads:

$$V = \gamma\gamma^* = \frac{R_0}{M^2} \tag{7.35}$$

Thus, the intensity distribution at the output coupler inside the resonator is also super-Gaussian with a beam radius w_M. The intensity profile of the near field can be obtained by multiplying (7.33) with the intensity transmission *1-R(r)* of the VRM. Note that the loss factor per round trip is the same for all super-Gaussian indices and equal to the geometrical loss factor! This result can be verified if diffraction integrals are used to propagate the field inside the resonator. Figure 7.47 presents numerically calculated loss factors per round trip of unstable resonators with *M=2* as a function of the equivalent Fresnel number for different super-Gaussian indices. Only for low equivalent Fresnel numbers can the beam propagation not be described by geometrical optics. This is to be expected since the purpose of the VRMs is the reduction of diffraction effects.

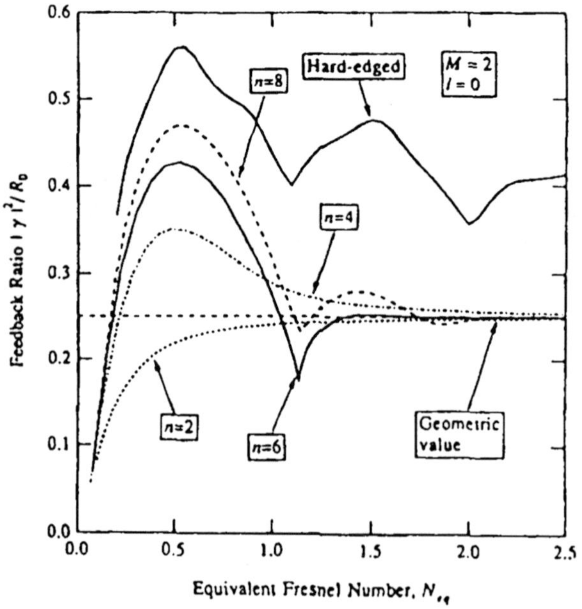

Fig. 7.47 Calculated loss factor per round trip of the lowest loss mode (ℓ=0) as a function of the equivalent Fresnel number for circularly symmetric unstable resonators with magnification *M=2* using different super-Gaussian indices *n* [3.158] (© OSA 1992).

Ray Transfer Matrix for Gaussian Mirrors
For a mirror with a Gaussian reflectivity profile it is possible to define a ray transfer matrix which can be used to determine the beam properties by applying the ABCD law. Let us assume that a Gaussian beam with beam radius w_0 and infinite radius of curvature is incident on a Gaussian mirror with profile radius w and radius of curvature ρ. The beam parameters q_1,q_2 before and after the reflection read:

$$q_1 = -\frac{i\lambda}{\pi w_0^2} \quad , \quad q_2 = -\frac{2}{\rho} - \frac{i\lambda}{\pi}\left[\frac{1}{w_0^2} + \frac{1}{w^2}\right] \qquad (7.36)\ (7.37)$$

We know that the beam parameters transform according to the ABCD law:

$$q_2 = \frac{Aq_1 + B}{Cq_1 + D} \qquad (7.38)$$

with A,B,C,D being the elements of the ray transfer matrix of the mirror. Since $B=0$ (no propagation) and $AD=1$, we find the ray transfer matrix of the Gaussian VRM to be:

$$M_{VRM} = \begin{pmatrix} 1 & 0 \\ -\frac{2}{\rho} - \frac{i\lambda}{\pi w^2} & 0 \end{pmatrix} = \begin{pmatrix} 1 & 0 \\ -\frac{2}{\rho} & 1 \end{pmatrix}\begin{pmatrix} 1 & 0 \\ -\frac{i\lambda}{\pi w^2} & 1 \end{pmatrix} \qquad (7.39)$$

By using this matrix we can calculate the radius of curvature and the beam radius of the self-consistent Gaussian beam. Starting in front of the VRM (mirror 1), the ray transfer matrix for a round trip is given by:

$$M_{RT} = \begin{pmatrix} 4g_1g_2-1-2g_2 & 2Lg_2 \\ \frac{2}{L}[2g_1g_2-g_1-g_2] & 2g_2-1 \end{pmatrix}\begin{pmatrix} 1 & 0 \\ -\frac{i\lambda}{\pi w^2} & 1 \end{pmatrix} \qquad (7.40)$$

Application of the ABCD law yields the stationary q-parameter (see Sec. 2.7.2). By using the assumption that the effective Fresnel number $N_{eff}=w^2/(2Lg_2\lambda)$ and the equivalent G-parameter $G=2g_1g_2-1$ are related by:

$$G > \frac{1}{2\pi N_{eff}} \qquad (7.41)$$

the radius of curvature R and the beam radius w_M of the Gaussian beam in front of the VRM result. Note that this assumption usually holds since in most lasers the effective Fresnel numbers are greater than one (see also Fig. 7.47 for the validity of (7.41)). A straightforward calculation yields the final results:

$$R = \frac{2Lg_2}{\sqrt{G^2-1} - G + 2g_2 - 1} \tag{7.42}$$

$$w_M = w\sqrt{M^2 - 1} \tag{7.43}$$

The radius of curvature R of the Gaussian beam represents the expected value for unstable resonators given by (7.9). The beam parameters of the Gaussian beam at any plane inside and outside the resonator can now be determined using the ABCD law for Gaussian beams. Although a Gaussian beam is an eigensolution, this resonator is not a stable resonator!

7.7.2 Production of VRMs

Variable reflectivity mirrors are produced by coating an AR-coated substrate with one or several layers having a decreasing thickness along the radius. The optical thickness in the center of each layer is preferably a quarter wavelength. The graded profile is generated by using a mask with diameter d located at a certain distance h on top of the substrate. The two mask parameters control the shape of the reflectivity profile. The deposition rate of the coating particles continuously decreases with increasing distance from the substrate's center (Fig. 7.48). With only one graded layer it is difficult to attain high center reflectances due to the limited availability of coating materials exhibiting a high enough refractive index (see Sec. 4.3.2).

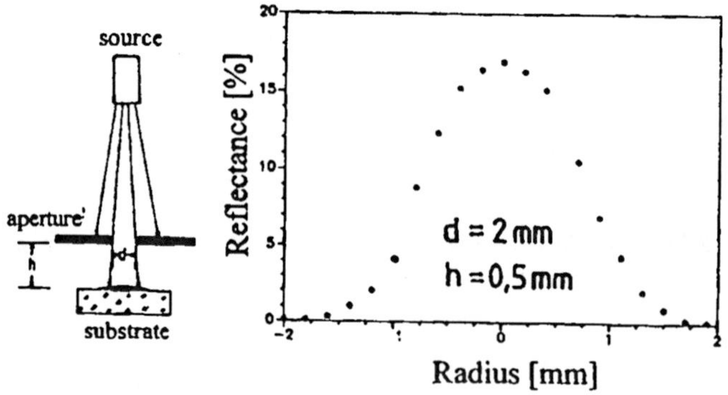

Fig. 7.48 Production of VRMs by coating through an aperture. The number of layers and the coating materials determine the center reflectivity. The aperture diameter d and the aperture distance h control the profile. The right figure shows the measured reflectivity profile of a single TiO_2 $\lambda/4$ layer on an AR-coated BK7 substrate.

For TiO_2(n=2.25), commonly used as high index material for lasers in the 1μm wavelength range, the maximum reflectance is about 50%. Laser mirrors with higher reflectance consist of multiple λ/4 layers, alternating high index and low index coating materials. If *N* denotes the number of high index layers and the thicknesses of all layers go to zero as we move away from the substrate center, the reflectivity profile exhibits *N-1* radial side lobes (Fig. 7.49). The phase shift associated with these maxima deteriorates the beam quality considerably. It is possible to get rid of the maxima by holding the layer thicknesses constant for radii greater than the radius of the first reflectance minimum. This can be accomplished by using a specially shaped rotating mask. Unfortunately, the rotation restricts this technique to mirrors with circular geometry.

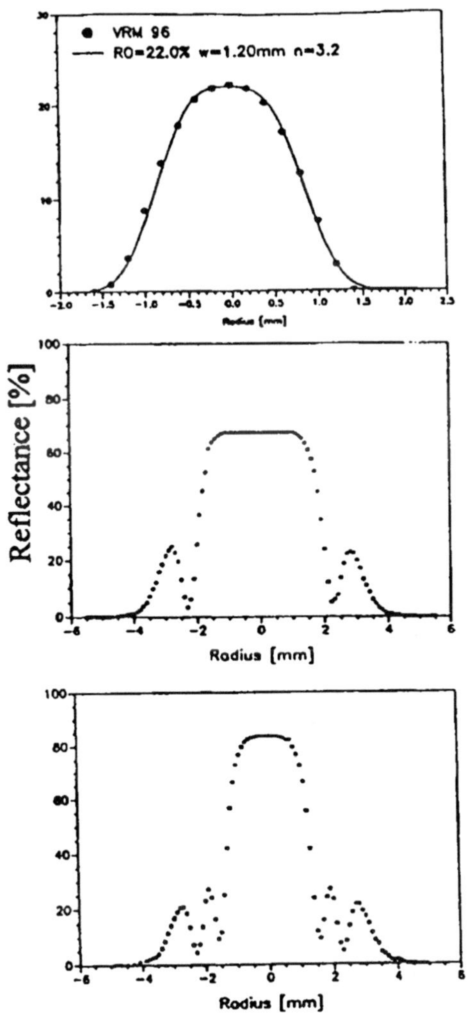

Fig. 7.49 Measured reflectivity profiles of VRMs with one, three, and five λ/4 layers using SiO_2 as the low index material (L) and TiO_2 as the high index material (H) on a substrate S. Coating designs: a) air vH AR S, b) air vH vL vH AR S, c) air vH vL vH vL vH AR S [S.9].

A better method to generate VRMs with high center reflectances is to use a radially variable Fabry Perot interferometer made of dielectric layers (Fig. 7.50). A radially varying layer (etalon layer) is situated between two identical stacks of quarter wave layers, which determine the maximal possible reflectance. In the center, the maximum reflectivity is achieved due to the quarter wavelength spacing between the stacks, and at the positions where the etalon layer thickness goes to zero, the residual reflectance is that of the AR coating. By using this technique it is possible to generate VRMs with center reflectances as high as 95% [3.168]. Figure 7.51 shows a measured reflectivity profile of an FPI mirror for a Nd:YAG rod laser. VRMs in circular and rectangular geometry can be produces using this technique (Fig. 7.52).

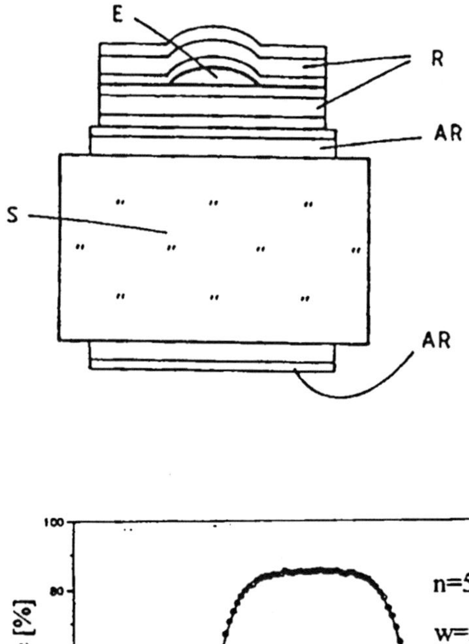

Fig. 7.50 Coating design of a variable Fabry Perot interferometer mirror (FPI mirror). S: substrate, AR: AR-coating, R: reflectance enhancing stack, E: etalon layer [S.9].

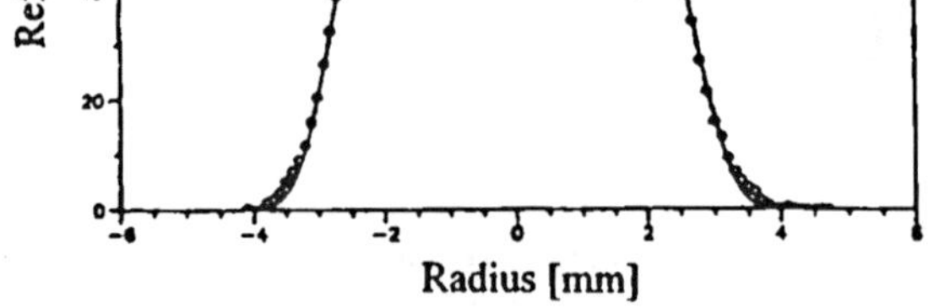

Fig. 7.51 Measured reflectance profile of an FPI mirror for a Nd:YAG laser at $\lambda=1.064\mu m$. The solid line represents a super-Gaussian profile with $n=5.5$, $w=3.2mm$, and $R_0=85\%$. Coating design: quarter wavelength layers (air H L vH L H AR S), L:SiO_2, H:TiO_2 [S.9].

Fig. 7.52 Three VRMs for Nd:YAG lasers. The two 1" substrates in the front are used in rod lasers. The left mirror is an FPI mirror, the reflectance profile of the right mirror is shown in the middle graph of Fig. 7.49. The 2" substrate with the rectangular reflectivity profile is used in a slab laser [S.9].

7.7.3 Laser Properties of VRM Unstable Resonators

Owing to their excellent beam qualities, unstable resonators with variable reflectivity mirrors have found application in both gas and solid state lasers. VRMs are commercially available for a variety of laser wavelengths ranging from the visible to the infrared at 10.6μm. In solid state lasers, these resonators are mostly applied to low to medium power Q-switch systems [3.146,3.147,3.177,3.178], but successful operation in free running lasers with output powers in the kW range has also been reported [3.156,3.157,3.163-3.165,3.167,3.172-176]. Figure 7.53 presents results obtained for a high power Nd:YAG slab laser system using one oscillator and two amplifier stages. By using an unstable resonator with VRM this system provides a maximum output power of 2 kW at a repetition rate of 22 Hz. The VRM of the negative branch confocal unstable resonator exhibits a rectangular shape with profile radii w of 2mm and 7mm in the x- and y-direction, respectively. The beam quality is about ten times diffraction limited. The deterioration of the beam quality is caused by stress induced aberration in the active medium. With a flat-flat resonator an output power of 3 kW is obtained, but the beam quality along the y-direction is decreased by a factor of 25 (Fig. 7.53b). Near diffraction limited beam was generated with a diode side-pumped thin Nd:YAG slab laser with positive branch confocal unstable resonator in the wide direction (Fig.7.45). In the perpendicular direction, the spherical HR mirror generated a stable resonator with a Gaussian beam diameter inside the slab of about 0.4mm. The output coupler was a VRM with a Gaussian profile with 67% center reflectivity and a profile radius of w=3.5mm. A cw output power of 220W was reported at beam propagation factors of less than 1.5 in both directions (Fig. 7.55) [3.176].

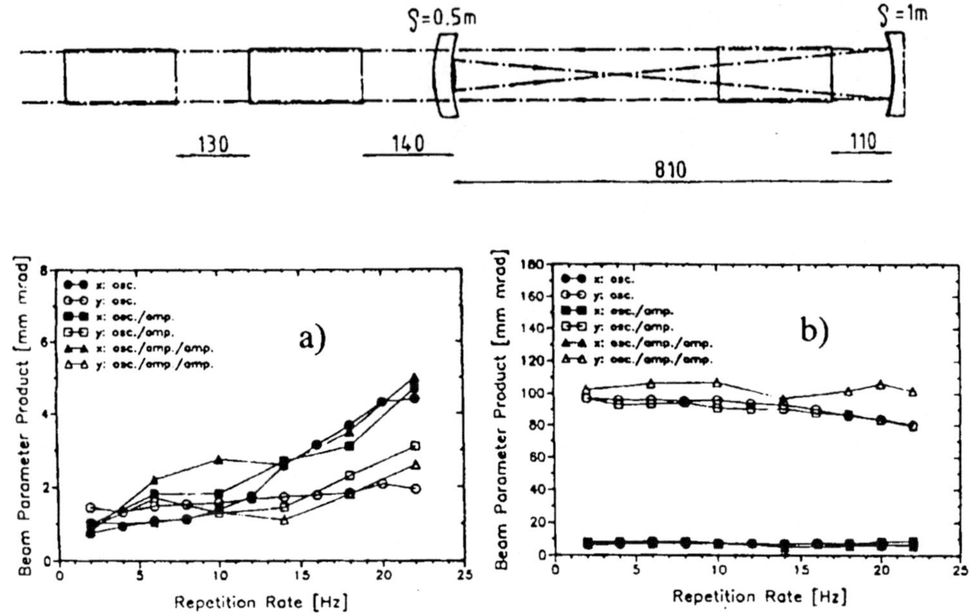

Fig. 7.53 Measured beam parameter products $d\Phi/4$ (d: beam width, Φ: full angle of divergence, both defined via 86.5% power content) in the x- (width) and the y-direction (height) for a pulsed Nd:YAG slab laser system with one oscillator slab and two amplifier slabs of dimensions 7x26x179mm^3. a) unstable resonator with VRM having profile radii of 2mm and 7mm in the x- and y-direction, respectively; b) stable resonator with length 550mm, flat output coupler and 2m concave HR mirror. The schematic shows the set-up with the unstable resonator. Total pump energy: 3,400J; pulse length: 4ms. Data are shown for one (osc.), two (osc./amp.) and three slabs (osc./amp./amp.) [3.165] (© OSA 1993).

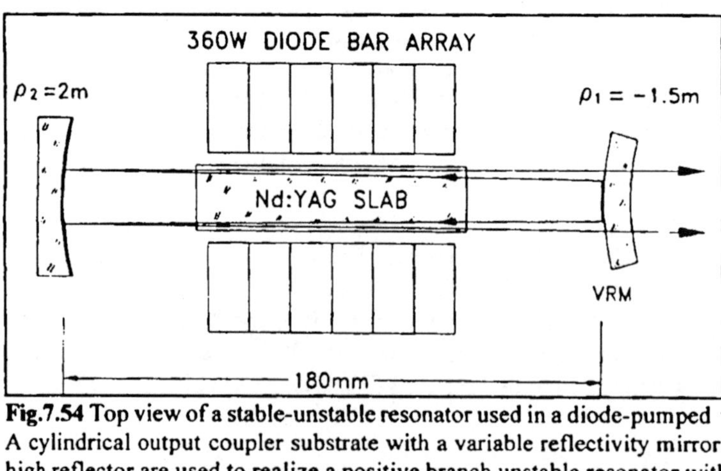

Fig.7.54 Top view of a stable-unstable resonator used in a diode-pumped thin Nd:YAG slab laser. A cylindrical output coupler substrate with a variable reflectivity mirror (VRM) and a spherical high reflector are used to realize a positive branch unstable resonator with magnification M=1.26 in the horizontal direction and a stable resonator in the perpendicular direction [3.176] (© OSA

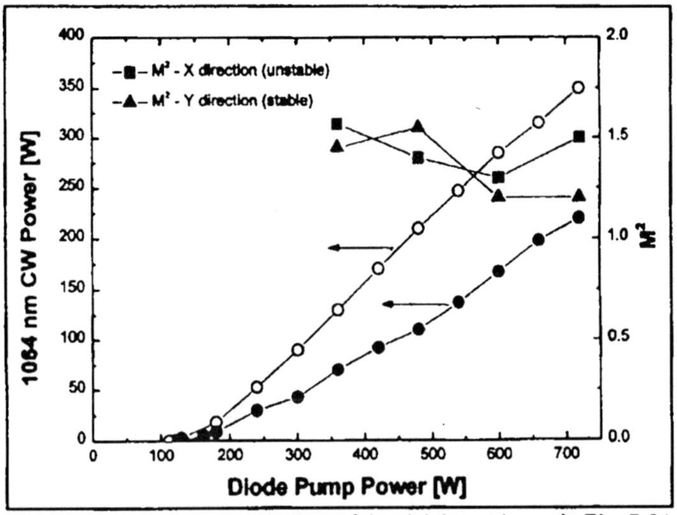

Fig.7.55 Measured output power of the slab laser shown in Fig. 7.54, with a short, stable resonator (open dots) and with the stable-unstable resonator with VRM (filled dots). For the latter resonator, the measured M^2 values in the x-direction (unstable) and the y-direction (stable) are shown as well [3.176] (© OSA 2002).

The lower efficiency of unstable resonators is caused by diffraction losses generated by the edges of the active medium in combination with a lower mode volume. As will be discussed in Chapter 11, unstable resonators generally provide only 70-80% of the extraction efficiency attainable with short stable resonators. The reflectivity profile of the VRM is of lower importance for the beam quality, but it considerably affects the near field distribution and the beam propagation. This is shown in Fig. 7.56 in which measured near field and far field intensity distributions for different VRMs with equal profile radii are compared.

If a flat topped intensity distribution in the near field is required (e.g. for efficient filling of amplifier stages), the center reflectivity and the reflectivity profile can be used as design parameters. In order to attain optimum filling of the active medium by the laser mode an optimization of the reflectivity profile is required. Since the adaptation of the beam width to the width of the active medium is easier to accomplish if the mode exhibits steep slopes, super-Gaussian profiles with high index *n* provide better means to maximize the output power. For an unstable resonator with a Gaussian VRM, a complete filling of the active medium will generate side lobes in the far field. However, for super-Gaussian profiles with $n>2$, the super-Gaussian field distributions at the output coupler are not eigensolutions of the paraxial wave-equation. The intensity distributions inside the resonator, therefore, will change when the field propagates inside the resonator. This may result in ripples and hot spots. The same statement holds for the field transmitted through the VRM. Depending on the type of application and the damage threshold of the optical components used, the beam propagation has also to be considered when the reflectance profile is chosen [3. 159].

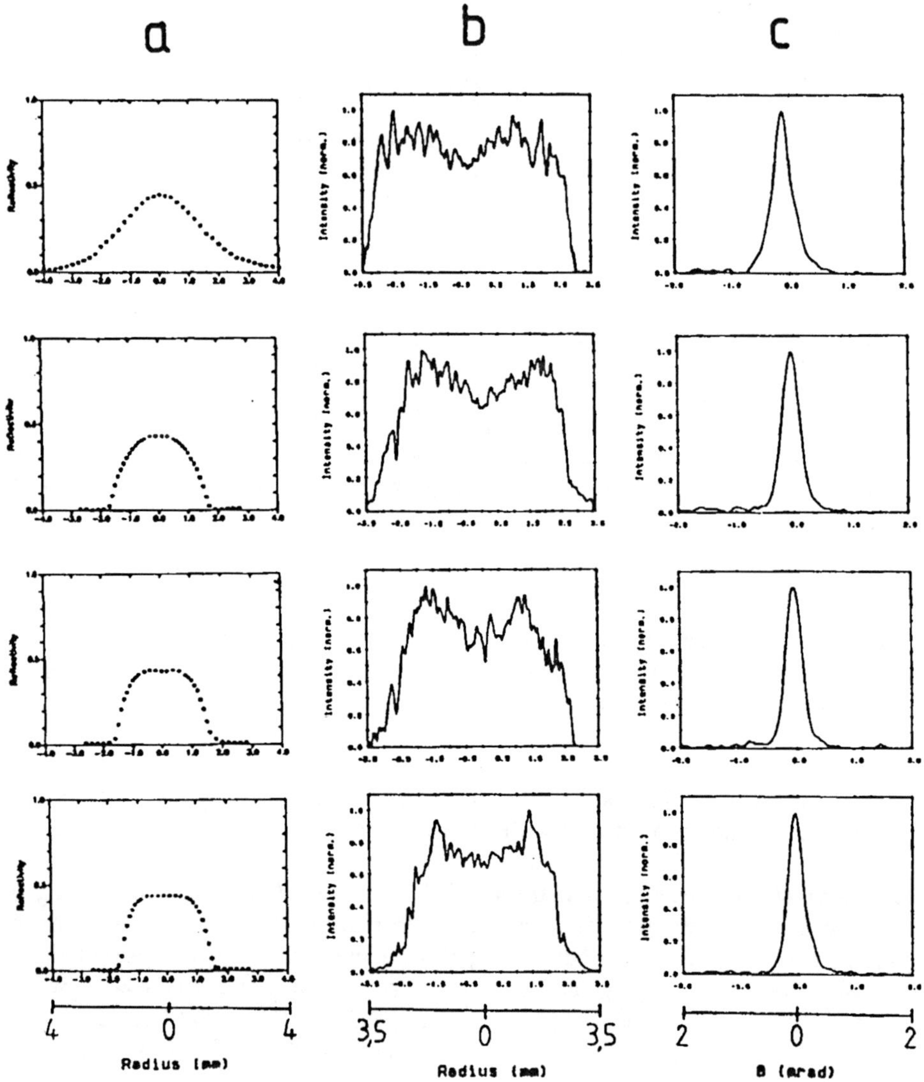

Fig. 7.56 Measured reflectivity profiles of the VRM (a), near field intensity distributions (b), and far field intensity distributions (c) for a confocal unstable resonator with magnification $M=1.5$ (pulsed Nd:YAG rod laser, $\lambda=1.064\mu m$, single shot operation).

Chapter 8

Resonators with Internal Optical Elements

8.1 Resonators with Internal Lenses

We have seen in Sec. 5.1 that stable resonators exhibit large Gaussian beam radii if the resonator length is large or the g-parameters of the two mirrors are chosen close to a stability limit in the g-diagram. In both cases a large fundamental mode volume and, consequently, a high efficiency are obtained. In order to avoid a high misalignment sensitivity or a long resonator set-up, lenses can be used inside the resonator to enlarge the Gaussian beam radius in the active medium. Such a resonator with internal lenses is referred to as a lens resonator [3.182].

Let us first discuss the case in which only one lens with focal length f is located between the resonator mirrors as depicted in Fig. 8.1. If we calculate the ray transfer matrix for the transit from mirror 1 to mirror 2 we get (see Fig. 1.25 for the location of the reference planes):

$$M_D = \begin{pmatrix} g_1^* & L^* \\ \dfrac{g_1^* g_2^* - 1}{L^*} & g_2^* \end{pmatrix} \tag{8.1}$$

with:

$$g_i^* = g_i - Dd_j(1 - d_j/\rho_i) \qquad i,j=1,2;\ i \neq j \tag{8.2}$$

$$L^* = d_1 + d_2 - Dd_1d_2 \tag{8.3}$$

$$D = 1/f \ : \ \text{refractive power}$$

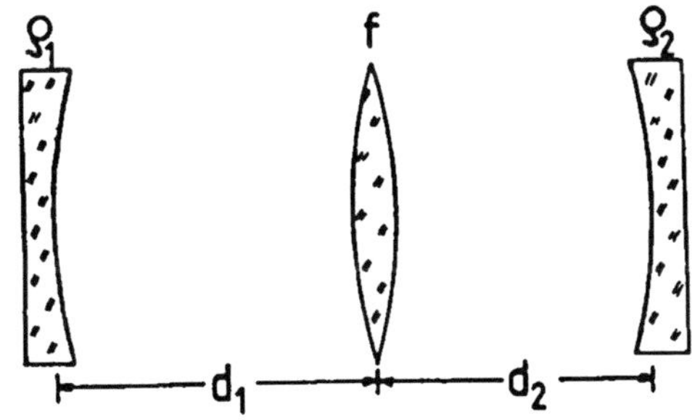

Fig. 8.1 Resonator with one internal lens.

Without the internal lens ($D=0$) the ray transfer matrix M_D becomes equivalent to the matrix M_0 of the empty resonator:

$$M_0 = \begin{pmatrix} g_1 & L \\ \dfrac{g_1 g_2 - 1}{L} & g_2 \end{pmatrix} \tag{8.4}$$

A comparison of the two ray transfer matrices indicates that a resonator with an internal lens exhibits the same ray transfer matrix for the transit as an equivalent, empty resonator with g-parameters g^*_1, g^*_2 and length L^*. Since the Gaussian beam radii at the mirrors are only a function of the ray transfer matrix elements, the lens resonator has the same Gaussian beam radii at the mirrors as the equivalent resonator. However, this equivalency holds only for the beam radii at the mirror planes. The beam caustic inside the lens resonator is different from that inside the equivalent resonator (the resonator in Fig. 8.1 exhibits two beam waists, one on each side of the lens). Since the Gaussian beam still exhibits a constant phase at the mirror surfaces the beam propagation inside the resonator can be easily calculated by using the Gaussian ABCD law (2.51).

Similar to empty resonators, lens resonators can be visualized in the equivalent g-diagram whose axes are defined by the g-parameters g^*_1 and g^*_2 (Fig. 8.2). Without an internal lens the resonator is located in the point (g_1, g_2). With increasing refractive power the location moves along a straight line through the g-diagram. The lens resonator is stable if the condition $0<g^*_1 g^*_2<1$ holds for the equivalent g-parameters. In the stable regions, the Gaussian beam radius at mirror i is given by:

$$w_i^2 = \frac{\lambda L^*}{\pi} \sqrt{\frac{g_j^*}{g_i^*(1-g_1^* g_2^*)}} \qquad i,j=1,2;\ i \neq j \tag{8.5}$$

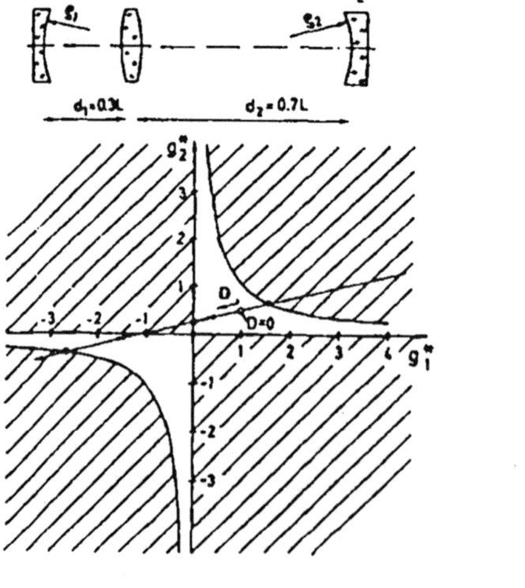

Fig. 8.2 Equivalent g-diagram of lens resonators. The resonator shown moves along the line as the refractive power is varied.

Example: semiconfocal resonator with $\rho_1=\infty$, $\rho_2=2$m, L=1m, $\lambda=500$nm

a) Without an internal lens we obtain:

$g_1=1.0$ and $g_2=0.5$

According to (8.5), the Gaussian beam radii at the mirrors are given by:

$w_1=0.399$mm and $w_2=0.798$mm

b) Insertion of a negative lens with focal length f=-2m at the position $d_1=0.8$m results in the equivalent resonator parameters:

$g^*_1=1.1$, $g^*_2=0.86$, $L^*=1.08$m

The lens resonator is stable since the product of the equivalent g-parameters is 0.946. Equation (8.5) yields for the Gaussian beam radii at the mirrors:

$w_1=0.809$mm and $w_2=0.915$mm

If more than one lens is located inside the resonator the beam radii at the mirrors can be determined in a similar way. After calculating the ray transfer matrix for a resonator transit starting at mirror 1 (the reference plane is the mirror surface):

$$M = \begin{pmatrix} A & B \\ C & D \end{pmatrix}$$

the equivalent g-parameters and the equivalent resonator length can be defined as:

$$g_1^* = A \,, \qquad g_2^* = D \,, \qquad L^* = B$$

and the Gaussian beam radii are again given by (8.5). A commonly used lens resonator with two lenses is the telescope resonator [3.194] (Fig. 8.3). This resonator comprises a telescope of magnification $M=|f_2/f_1|$ and length $\ell=f_1+f_2$, which increases the beam radius at mirror 2 and decreases the beam radius at mirror 1. The equivalent resonator parameters read:

$$g_1^* = M - \frac{L^*}{\rho_1} \,, \qquad g_2^* = \frac{1}{M} - \frac{L^*}{\rho_2} \,, \qquad L^* = \ell + d_1 M + \frac{d_2}{M}$$

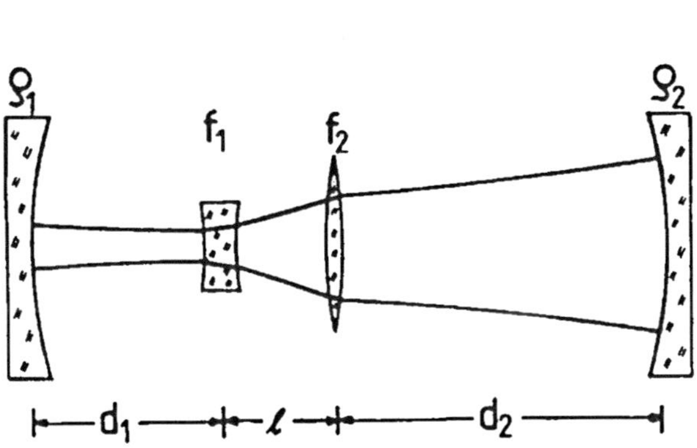

Fig. 8.3 Telescope resonator.

8.2 Resonators with Polarizing Elements

A resonator which consists only of mirrors is degenerated as far as the polarization of the emitted beam is concerned. All oscillation directions of the electromagnetic radiation exhibit the same probability and the resonator will therefore emit unpolarized light. This means that the polarization vector varies statistically with a time constant determined by perturbations of the resonator set-up and the physical properties of the active medium. By inserting polarizing elements into the resonator a well-defined polarization can be generated resulting in a polarized electric field which is reproduced after each round trip. Examples of polarizing intracavity elements are birefringent lenses, retardation plates, and polarizers. In most lasers, the active medium itself will also affect the polarization, either due to inherent or pump induced birefringence, or due to the special geometry of the end faces.

The theoretical treatment of the polarization was already discussed in Chapter 3. The influence of an optical element on the polarization can be mathematically described by using 2x2 matrices which relate the input and the output polarization vector to each other [3.181]. In the following we summarize the main results of the Jones matrix formalism. In optical resonators, the polarization is determined by the resulting Jones matrix M^P for a round trip. The polarization is characterized by the field vector $\boldsymbol{E}$ whose components represent the field amplitudes in the two perpendicular directions. If a cartesian coordinate system is used, the field vector is defined by the amplitudes in the x- and the y-direction and the relative phase shift Φ in between:

$$\boldsymbol{E} = \begin{pmatrix} E_{0x} \\ E_{0y}\ \exp[i\phi] \end{pmatrix} \tag{8.6}$$

For circularly symmetric optical elements it is convenient to express the field vector in radial and azimuthal components:

$$\boldsymbol{E}_r = \begin{pmatrix} E_{0r} \\ E_{0\phi}\ \exp[i\phi] \end{pmatrix} \tag{8.7}$$

In both coordinate systems the two field vectors $\boldsymbol{E}_1, \boldsymbol{E}_2$, which are eigenvectors of the Jones matrix, represent the eigenstates of the polarization with:

$$\mu_i^P \boldsymbol{E}_i = \boldsymbol{M}^P \boldsymbol{E}_i \tag{8.8}$$

The eigenvalue μ_i determines the loss factor per round trip V for the corresponding field vector $\boldsymbol{E}_i$:

$$V = \mu_i \mu_i^* \tag{8.9}$$

In general only the polarization exhibiting the lower loss factor will be observed. If the Jones matrix M^P in the x-y coordinate system is known, the Jones matrix M_r^P in the polar coordinate system is given by:

$$M_r^P = \begin{pmatrix} \cos\theta & \sin\theta \\ -\sin\theta & \cos\theta \end{pmatrix} M^P \begin{pmatrix} \cos\theta & -\sin\theta \\ \sin\theta & \cos\theta \end{pmatrix} \tag{8.10}$$

with $x=r\cos\theta$ and $y=r\sin\theta$. The corresponding transformation rule for the field vectors (8.6) and (8.7) reads:

$$E_r = \begin{pmatrix} \cos\theta & \sin\theta \\ -\sin\theta & \cos\theta \end{pmatrix} E \tag{8.11}$$

The eigenvalue equation (8.8) holds in both coordinate systems if the Jones matrix and the field vector are transformed according to (8.10) and (8.11), respectively. If the resulting Jones matrix M^P for the resonator round trip is given in the general form:

$$M^P = \begin{pmatrix} m_{11} & m_{12} \\ m_{21} & m_{22} \end{pmatrix}$$

the eigenvalues and the eigenvectors can be calculated by using the relations:

$$\mu_{1,2}^P = \frac{m_{11}+m_{22}}{2} \pm \sqrt{\left(\frac{m_{11}-m_{22}}{2}\right)^2 + m_{12}m_{21}} \tag{8.12}$$

$$E_i^P = \begin{pmatrix} 1 \\ \dfrac{\mu_i^P - m_{11}}{m_{12}} \end{pmatrix} \; ; \qquad i=1,2 \quad \mathit{if} \;\; m_{12} \neq 0 \tag{8.13}$$

$$E_1^P = \begin{pmatrix} 1 \\ 0 \end{pmatrix},\; E_2^P = \begin{pmatrix} 1 \\ 0 \end{pmatrix}; \quad i=1,2 \quad \mathit{if} \;\; m_{12} = m_{21} = 0 \tag{8.14}$$

The resulting Jones matrix for the resonator round trip is obtained by multiplying the Jones matrices of all elements in the same sequence as the elements are passed. Thus, the Jones matrix of the first element stands on the right hand side of the matrix product. For a collection of Jones matrices see Chapter 3. In the following sections, common optical resonators with polarizing elements are discussed using the Jones matrix formalism.

8.2.1 The Twisted Mode Resonator

This resonator employs a polarizer and two quarter wave plates whose fast axes are rotated by +45° and -45° with respect to the transmission direction of the polarizer (we chose the y-axis). The resulting Jones matrix for the round trip starting at the left mirror reads:

$$M^P = M_P^P \, M_R^P(45°) \, M_R^P(-45°) \, M_R^P(-45°) \, M_R^P(45°) \, M_P^P = \begin{pmatrix} 0 & 0 \\ 1 & 0 \end{pmatrix}$$

In the y-direction linearly polarized light exhibits no losses in the resonator and the field vector is reproduced after each round trip. Starting at mirror 1 the field becomes right-circularly polarized by the first λ/4 plate, and after passage through the second λ/4 plate the beam is again linearly polarized in the y-direction at the output coupling mirror. On the way back to the HR mirror the beam is transformed into a left-circularly polarized beam between the retardation plates. Due to the different circularity inside the medium of the back and the forth travelling wave, no interference between these waves can occur. The generation of standing waves is thus prevented and the time-averaged intensity is constant along the length of the active medium [3.191]. This lack of spatial hole burning stabilizes the temporal laser emission since the interaction between axial modes is missing. In homogeneously broadened lasers, this resonator provides single axial mode operation. The twisted mode resonator can also be used in frequency doubled Nd:YAG lasers to prevent chaotic laser emission at 532nm due to axial mode competition, often referred to as the "green problem".

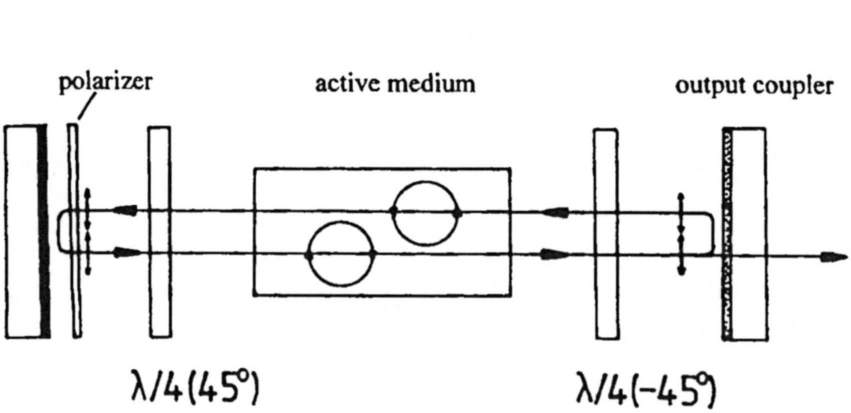

Fig. 8.4 The twisted mode resonator. Two quarter wave plates generate right-circular and left-circular polarization for the forth and the back travelling waves, respectively. At both mirrors the field is linearly polarized in the y-direction.

8.2.2 Resonators with Variable Output Coupling

If only one quarter wave plate is used in the twisted mode resonator of Fig. 8.4, the linearly polarized field vector is rotated by 90° after the round trip. The loss factor is then equal to zero and all the intracavity power is coupled out of the resonator. By rotating the quarter wave plate the loss factor can be continuously varied between 0 and 1 resulting in a variable output coupling (Fig. 8.5). If the rotation angle α is defined by the angle between the fast axis of the quarter wave plate and the transmission direction of the polarizer, no output coupling (loss factor of 1.0) is attained at angles α of 0°, 90°, 180°, and 270°, whereas maximum output coupling is observed at 45°, 135°, 225°, and 315°. Rotation of the $\lambda/4$ plate thus enables one to vary the output power of the laser.

Let us consider the general case that the retardation plate induces an arbitrary phase shift δ between the principal axes (Fig. 8.5). Starting at the left mirror, the resulting Jones matrix reads (transmission direction of the polarizer is the y-direction):

$$M^P = M_P^P \, M_R^P(\alpha) \, M_R^P(\alpha) \, M_P^P$$

$$= \begin{pmatrix} 0 & 0 \\ 1 & [\cos\alpha\sin\alpha(1-\exp[i\delta])]^2 + [\sin^2\alpha + \exp[i\delta]\cos^2\alpha]^2 \end{pmatrix} \tag{8.15}$$

The loss factor per round trip is given by:

$$V = 1 - [\sin\delta\sin 2\alpha]^2 := 1 - R \tag{8.16}$$

with R being the reflectance of the polarizer which corresponds to the reflectance of the output coupling mirror in a standard resonator.

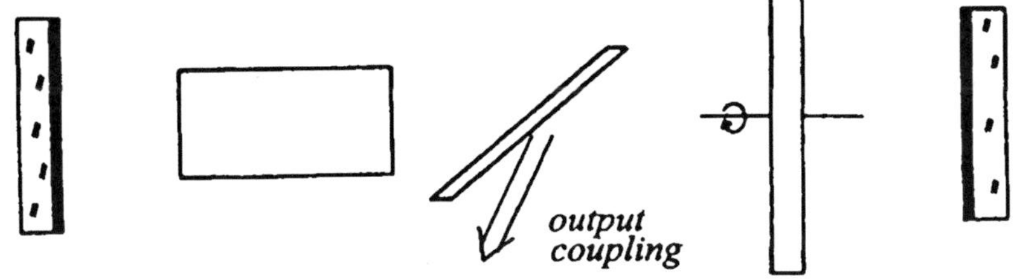

Fig. 8.5 Resonator with internal polarizer and rotatable retardation plate. The polarizer reflects the field component oscillating perpendicular to its transmission direction out of the resonator. The angle of rotation of the retardation plate determines the output coupling.

The upper graph of Fig. 8.6 presents the measured and the calculated loss factor per round trip for a retardation plate with δ=73° as a function of the angle of rotation α. Measured output energies per pulse of a Nd:YAG laser utilizing this retardation plate are shown in the lower diagram. There are two advantages of this resonator concept. First, the output power can be varied without changing the pump power. Variations in mode structure and focusability caused by the change in pump power can thus be avoided. Secondly, the output coupling of the resonator can be adjusted so that the maximum output power is obtained for any value of the pump power. In Fig. 8.6, the rotation angle has to be changed from 12° to 22° as the pump energy is increased to always attain maximum output energy. This is in contrast to resonator schemes with an output coupling mirror which provide optimum output coupling only at one value of the pump power (see Chapter 10).

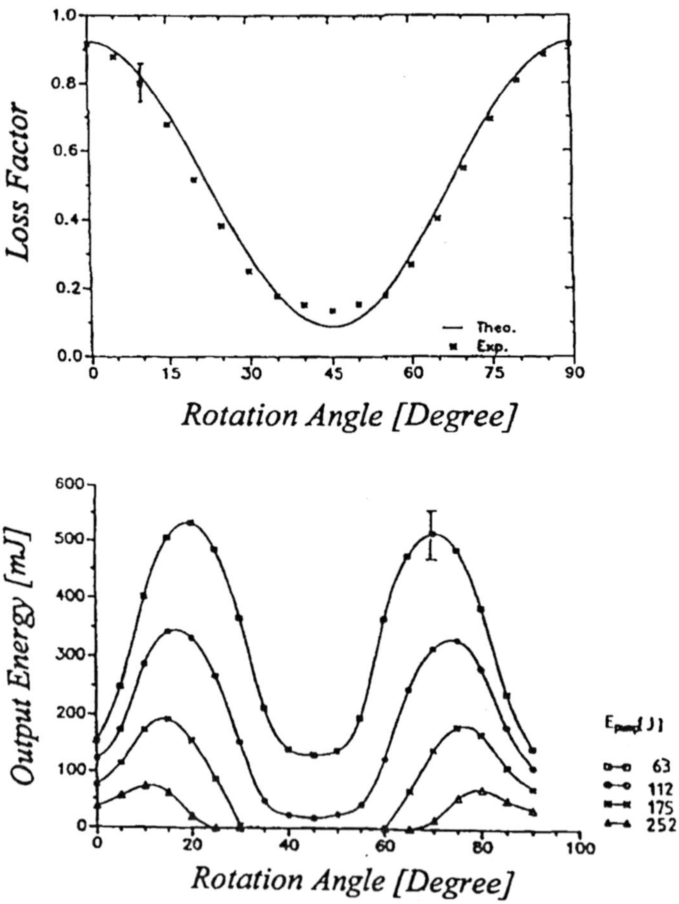

Fig. 8.6 Measured and calculated dependence of the loss factor and the output energy of a pulsed Nd:YAG rod laser (repetition rate: 0.5Hz) as a function of the angle of rotation α of the retardation plate (δ=73°). The curve parameter in the lower diagram is the pump energy [S.10].

8.2.3 Pockels Cell Resonator

Instead of rotating a retardation plate to change the output coupling it is also possible to use a Pockels cell with varying voltage U [3.190] (Fig. 8.7). A Pockels cell consists of a nonlinear crystal to which a voltage in the kV range is applied. A Pockels cell acts as a retardation plate whose principal axes are rotated by 45° with respect to the direction of the applied electric field. The phase shift δ is a linear function of the voltage U:

$$\delta = \frac{U}{U_{\lambda/4}} \frac{\pi}{2} \tag{8.17}$$

The quarter wave voltage $U_{\lambda/4}$ is a characteristic of the Pockels cell and it denotes the voltage required to generate the characteristics of a λ/4 plate (typically in the multi kV range). The resulting Jones matrix for the resonator round trip, starting at the polarizer, is given by:

$$\begin{aligned} M^P &= M_P^P \, M_R^P(45°) \, M_R^P(45°) \, M_P^P \\ &= \begin{pmatrix} 0 & 0 \\ 0 & \cos\delta \end{pmatrix} \end{aligned} \tag{8.18}$$

Therefore, the loss factor per round trip is given by $V=\cos^2\delta$ which results in a reflectance R of the polarizer of:

$$R = 1 - V = \sin^2\left[\frac{\pi U}{2U_{\lambda/4}}\right] \tag{8.19}$$

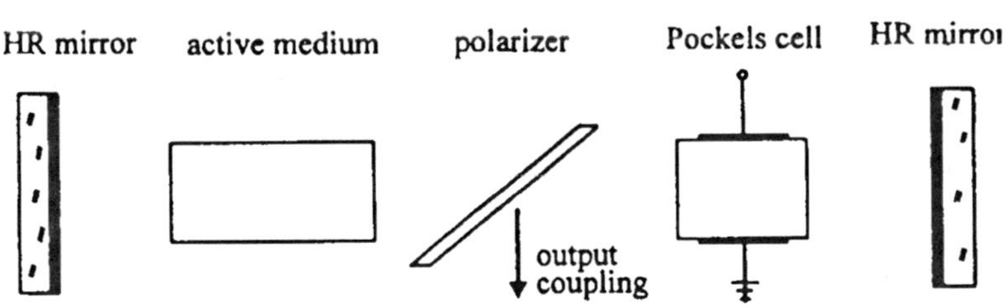

Fig. 8.7 Pockels cell resonator. The Pockels cell acts like a retardation plate whose principal axes are rotated by 45°. The phase shift δ can be varied with the applied voltage.

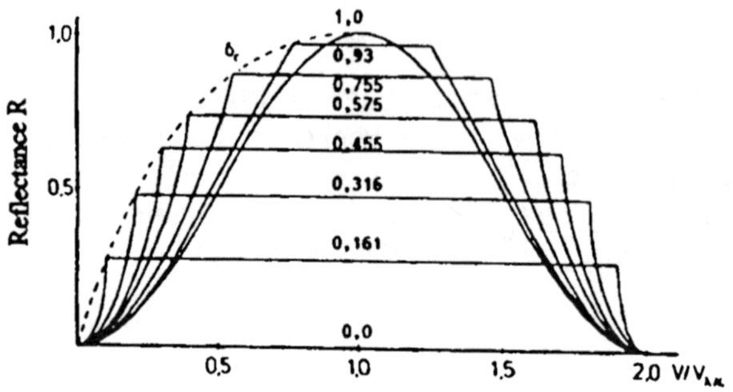

Fig. 8.8 Reflectance R of the polarizer as a function of the Pockels cell voltage. The curve parameter is the degree of polarization P of the polarizer.

The sinusoidal variation of the reflectance R can only be observed for polarizers with an ideal degree of polarization of P=1.0 (see Sec. 1.3). As the degree of polarization is decreased, regions of constant reflectance develop as shown in Fig. 8.8.

The Pockels cell resonator has found widespread application in lasers generating short pulses in the ns-range. The beam is coupled out of the resonator via a standard output coupling mirror and the polarizer is only used to generate a high loss. This loss can be suddenly decreased by switching off the Pockels cell. (Fig. 8.9). If the quarter wave voltage is applied, the linear polarization is rotated by 90° after a round trip resulting in a reflectance of 100% at the polarizer. In this configuration the laser threshold cannot be reached and the pump process keeps building up the inversion inside the active medium. As soon as the steady state inversion is reached the voltage at the Pockels cell is switched off (switching times are on the order of ns) and the reflectance of the polarizer is decreased to a value close to zero. The laser will start oscillating and the high inversion is depleted by a short, intense light pulse with a duration in the 10-100 ns range. Since the cavity Q of the resonator is changed from a very low to a high value, this technique is referred to as Q-switching. In cw pumped lasers the switching frequencies using Pockels cells can be as high as 50kHz. If a pulsed pump source is used, the switching of the Pockels cell is usually in synchronization with the pump frequency (in the range of 1-1000 Hz). In high power lasers, the absorption losses of a Pockels cell are too high resulting in beam distortion and even damage. Therefore, in most commercial lasers acousto-optical elements are used which generate losses due to diffraction by an acoustically excited refractive grating.

Another way to rapidly change the resonator loss is the use of a saturable absorber whose transmission is a function of the light intensity (passive Q-switching). The nonlinear transmission is caused by dye molecules which are either dissolved in a fluid or embedded in a transparent foil or by a doped solid state material (e.g. Cr:YAG). This technique is easier to realize since no switching of high voltages is required. Passive Q-switching is, therefore, used in some commercial systems. However, the Q-switching cannot be actively controlled and the repetition rate is determined by the net round trip gain of the resonator. In addition, absorption in the Q-switch material limits this method to low power lasers (<10W).

Fig. 8.9 Pockels cell resonator for the generation of short pulses via Q-switching. If the quarter wave voltage is applied to the Pockels cell, the laser cannot oscillate due to the high reflectance of the polarizer. After the voltage is switched off the inversion in the active medium is depleted by a short, intense light pulse.

8.2.4 Resonators with Radially Birefringent Elements

A radially birefringent element is a retardation plate whose phase shift δ is a function of the radial distance r from the optical center. The radial dependence is generated by changing the thickness of the retardation plate with increasing radius. Preferably one surface of the plate is curved with a radius of curvature ρ generating a birefringent lens (Fig. 8.10). If n_1 and n_2 denote the indices of refraction along the two principal axes ($n_2 > n_1$), an electric field polarized linearly along the axis i experiences the phase shift:

$$\delta_i(r) = \frac{2\pi}{\lambda} n_i \left(d_0 + \frac{r^2}{2\rho} \right) \tag{8.20}$$

Note that the refractive power of the lens depends on the polarization. However, the difference between the two indices of refraction is usually very small (less than one %) and the polarization effect on the refractive power can therefore be neglected. If the slow axis coincides with the y-axis of the reference frame, the Jones matrix of the radially birefringent lens reads:

$$M_{RB} = \begin{pmatrix} 1 & 0 \\ 0 & \exp[i\delta(r)] \end{pmatrix} \tag{8.21}$$

with: $$\delta(r) = \frac{2\pi}{\lambda}(n_2 - n_1)\left[d_0 + \frac{r^2}{2\rho} \right]$$

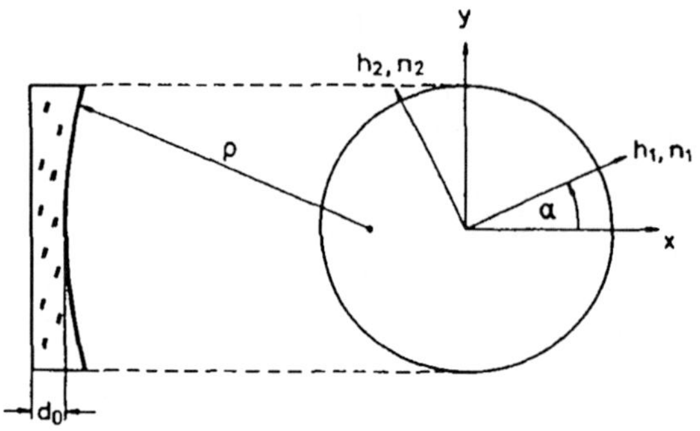

Fig. 8.10 Radially birefringent element. Electric fields that are linearly polarized along the two principal axes h_1 and h_2 experience different indices of refraction n_1 and n_2. The thickness of the element is a function of the radial coordinate.

The Jones matrix is the same as the one for a retardation plate except for the radial dependence of the phase shift. If we replace the retardation plate in the resonator with variable output coupling (Fig. 8.5) by the birefringent lens (Fig. 8.11), the loss factor becomes a function of the radius r:

$$V(r) = 1 - [\sin(\delta(r))\sin 2\alpha]^2 \tag{8.22}$$

Thus, the birefringent lens generates a radially variable output coupling [3.193]. The combination lens-polarizer-mirror simulates a variable reflectivity mirror (VRM) with reflectivity profile $R = V(r)$. This resonator set-up was used for the first demonstration of an unstable resonator with VRM [3.193,3.195]. The angle of rotation α of the lens and its center thickness d_0 can be used to vary the shape of the reflectivity profile. Figure 8.12 shows calculated loss factor profiles $V(r)$ for a quartz lens with a radius of curvature of 1m. The thickness d_0 was chosen such that the loss factor $V(0)$ at the center is equal to 1.0 ($d_0\Delta n/\lambda$ is an integer value). By changing d_0 arbitrary values for the center loss factor can be generated.

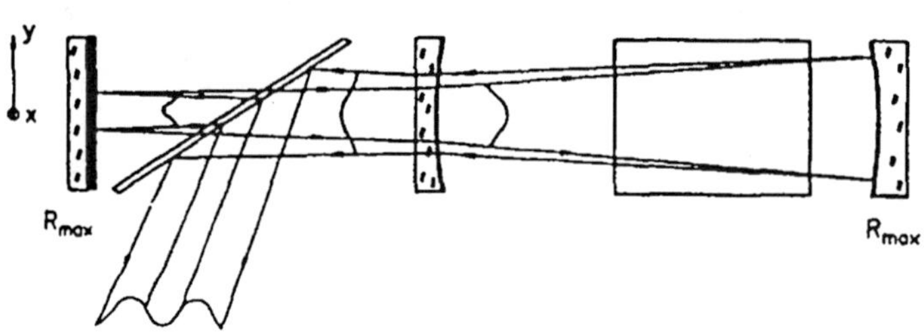

Fig. 8.11 Resonator with radially birefringent lens. The radial dependence of the phase shift generates a radially variable output coupling.

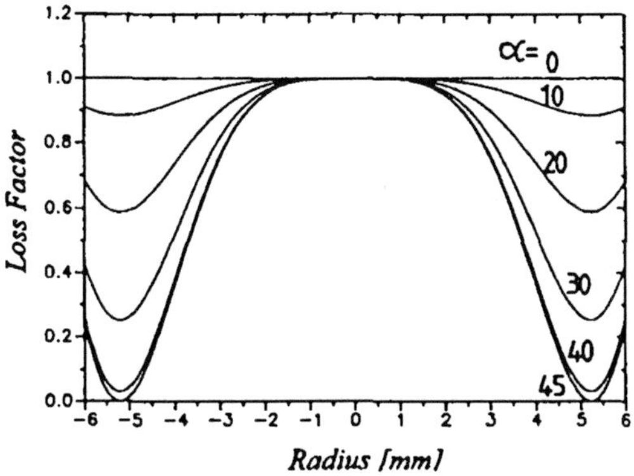

Fig. 8.12 Radial loss factor profiles (8.22) of the resonator shown in the previous figure for a quartz lens (n_3-n_1=0.0092, ρ=1m) and a wavelength of λ=500nm. The curve parameter is the angle of rotation α. The system simulates a VRM with reflectance profile $V(r)$.

8.2.5 Resonators with Azimuthally Birefringent Elements

Azimuthally birefringent materials exhibit different indices of refraction n_r and n_θ for radially and azimuthally polarized light (Fig. 8.13). If in the y-direction linearly polarized light is incident on such an optical element, the field vector has different components in the radial and the azimuthal direction and the relative amplitudes of the two components depend on the azimuthal angle θ. After passage through the birefringent material, the polarization state is a function of the angle θ. Azimuthal birefringence is observed in pumped solid state laser rods [3.2,3.186,3.188].

If $\delta=2\pi\ell(n_r-n_\theta)/\lambda$ denotes the relative phase shift between the r- and the θ-component of the field vector, the Jones matrix for the azimuthally birefringent material reads in polar coordinates:

$$M^P_{ABr} = \begin{pmatrix} \exp[i\delta] & 0 \\ 0 & 1 \end{pmatrix} \tag{8.23}$$

By applying the transformation (8.10), the Jones matrix in the cartesian reference frame is obtained:

$$M^P_{AB} = \begin{pmatrix} \cos\theta & -\sin\theta \\ \sin\theta & \cos\theta \end{pmatrix} M^P_{Abr} \begin{pmatrix} \cos\theta & \sin\theta \\ -\sin\theta & \cos\theta \end{pmatrix}$$

$$= \begin{pmatrix} \exp[i\delta]\cos^2\theta + \sin^2\theta & (\exp[i\delta]-1)\sin\theta\cos\theta \\ (\exp[i\delta]-1)\sin\theta\cos\theta & \exp[i\delta]\sin^2\theta + \cos^2\theta \end{pmatrix} \tag{8.24}$$

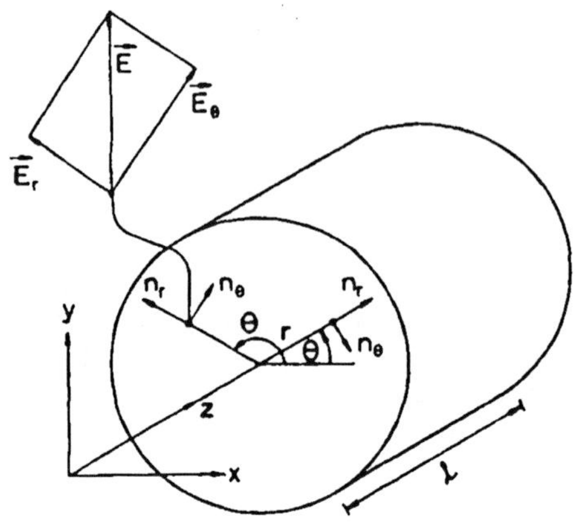

Fig. 8.13 Azimuthally birefringent element. Radially and azimuthally polarized fields experience different indices of refraction n_r and n_θ. After passage through the element, linearly polarized light exhibits a polarization state which depends on the azimuth θ.

The elements of the Jones matrix depend only on the azimuthal angle θ. If the active medium exhibits azimuthal birefringence, output coupling via a polarizer will therefore generate a beam profile with an azimuthal structure. By using the resulting Jones matrix for the resonator round trip, the loss factor can be calculated to be:

$$V(\theta) = 1 - [\sin\delta \sin 2\theta]^2 = 1 - R(\theta) \tag{8.25}$$

The output coupling is maximum (minimum loss factor V and maximum reflectance R) at the angles θ of 45°, 135°, 225°, and 315°. The beam profile thus looks like a Maltese cross rotated by 45°. Note that a rotation of the azimuthally birefringent element does not change the output pattern. The average reflectance R of the polarizer is obtained by integrating the reflectance $R(\theta)$ over the azimuthal angle:

$$R = \frac{1}{2\pi}\int_0^{2\pi} R(\theta) d\theta = \frac{1}{2}\sin^2\delta \tag{8.26}$$

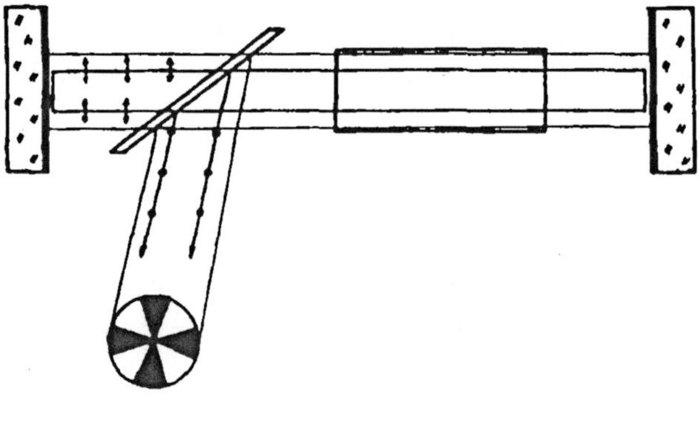

Fig. 8.14 Resonator with polarizer and azimuthally birefringent element. The output coupling depends on the azimuthal angle θ.

8.2.6 Resonators with Radial-Azimuthally Birefringent Elements

In many solid state laser rods, radial and azimuthal birefringence are generated simultaneously by the pump process. The azimuthal intensity pattern coupled out by the polarizer thus exhibits radial rings whose number increase with increasing pump power (Fig. 8.15). The combination of heat generation due to absorption of pump radiation and the flow of heat to the outer periphery due to cooling leads to a parabolic radial temperature profile. The induced stress generates azimuthal birefringence and the temperature profile leads to a radial decrease of both indices of refraction:

$$n_r(r) = n_0\,(1 - \gamma_r r^2)\,, \quad n_\theta(r) = n_0\,(1 - \gamma_\theta r^2) \qquad (8.27)\ (8.28)$$

The shape factors γ_r and γ_θ are proportional to the average pump power. The pumped laser rod acts like a lens with refractive powers D_r and D_θ for the two polarizations. If end effects are neglected, the refractive powers for a rod of length ℓ read in a first order approximation:

$$D_r = 2\gamma_r n_0 \ell\,, \quad D_\theta = 2\gamma_\theta n_0 \ell \qquad (8.29)\ (8.30)$$

The Jones matrix of the laser rod can now be determined by inserting (8.27)-(8.30) into the Jones matrix (8.24) of the azimuthally birefringent element:

$$M_{LR}^{P} = \begin{pmatrix} \exp[i\delta(r)]\cos^2\theta + \sin^2\theta & (\exp[i\delta(r)]-1)\sin\theta\cos\theta \\ (\exp[i\delta(r)]-1)\sin\theta\cos\theta & \exp[i\delta(r)]\sin^2\theta + \cos^2\theta \end{pmatrix} \qquad (8.31)$$

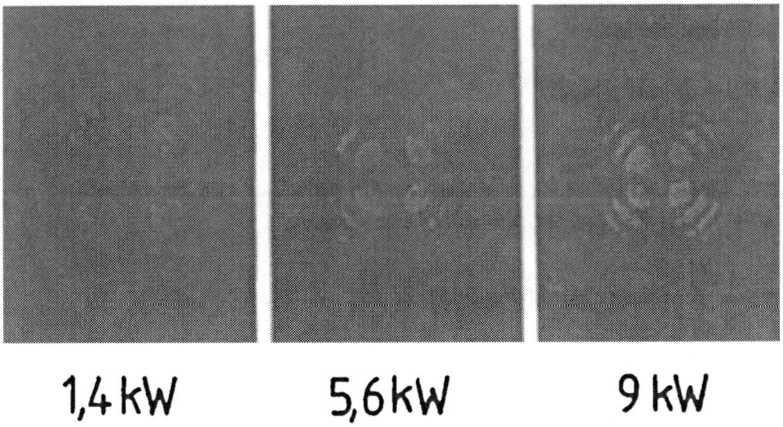

Fig. 8.15 Photographed intensity distributions generated by a collimated HeNe laser beam after passage through a pumped Nd:YAG rod. The rod is placed between crossed polarizers. The parameter is the electrical pump power [S.10].

with $$\delta(r) = -\frac{\pi}{\lambda}(D_r - D_\theta)r^2 = -\frac{\pi}{\lambda}\Delta D r^2 \tag{8.32}$$

For a flashlamp pumped Nd:YAG rod with radius b=5mm typical values for the refractive powers per kW of electrical pump power are D_r=0.3 m^{-1} and D_θ=0.255m^{-1}.
By introducing the thermal lensing coefficient α [3.195] (see Sec. 12.1):

$$\alpha = \frac{(D_r + D_\theta)}{2}\frac{\pi b^2}{P_{pump}} = \frac{D_{ave}\pi b^2}{P_{pump}} \tag{8.33}$$

where we introduced the average refractive power D_{ave}. and the pump power P_{pump} The phase shift (8.32) can be rewritten as:

$$\delta(\eta) = -\frac{\beta}{\lambda}\alpha P_{pump}\,\eta^2 \tag{8.34}$$

where we used the relation $\Delta D = \beta D_{ave}$ ($\beta<1$, $\beta \approx 0.15$ for Nd:YAG) and $\eta = r/b$ is the normalized radial coordinate. Equation (8.34) indicates that for the same pump power, the radial phase shift in the rod is the same, independent of the rod radius. For Nd:YAG, typical values of the thermal lensing coefficient are 0.1-0.15 μm/W for diode pumping (P_{pump}: optical pump power) and 0.02-0.03 μm/W for flashlamp pumping (P_{pump}: electrical pump power), depending on doping concentration and pump light spectrum.

If a polarizer is placed inside the resonator to generate a linearly polarized beam (as in Fig. 8.14), the loss factor per round trip is a function of r and θ. According to (8.25) the reflectance of the polarizer reads:

$$R(\eta,\theta) = [\sin\delta(\eta)\sin 2\theta]^2 \tag{8.35}$$

The total loss R per round trip due to reflection at the polarizer can be calculated by integrating (8.32) over the radial and the azimuthal coordinates:

$$R = \frac{1}{\pi}\int_0^{2\pi}\int_0^1 [\sin\delta(\eta)\sin 2\theta]^2 \eta\, d\eta\, d\theta = \frac{1}{4}\left[1 - \frac{\sin x}{x}\right] \tag{8.36}$$

with $x = 2\delta(\eta=1)$. Figure 8.16 shows the graphic presentation of (8.36). Amazingly, for high pump powers the loss per round trip converges towards a value of 25% which means that only a quarter of the intracavity power is coupled out. An experimental verification of this effect is presented in Fig. 8.17. A Nd:glass rod was placed between crossed polarizers and the transmitted power P of a collimated Nd:glass laser beam was measured as a function of the electrical pump power. The solid line represents the theoretical dependence given by (8.36) with $x = \delta(\eta = 1)$ (the factor 2 is missing because only the loss for a transit is measured). Note that a smaller rod radius b does not reduce the loss since $b^2 \Delta D$ is a constant of the laser head. However, it is possible to decrease the loss by inserting an aperture (radius a) to reduce the beam size. In this case, (8.34) has to be multiplied by a^2/b^2.

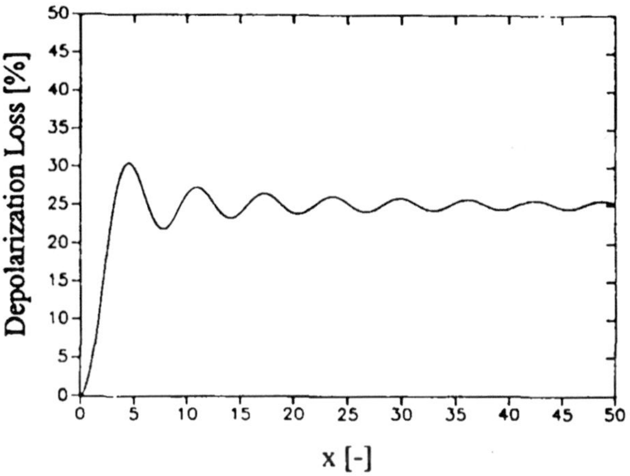

Fig. 8.16 Calculated loss per round trip of resonators with a radial-azimuthally birefringent medium and an intracavity polarizer, according to (8.36). The loss is generated by output coupling at the polarizer.

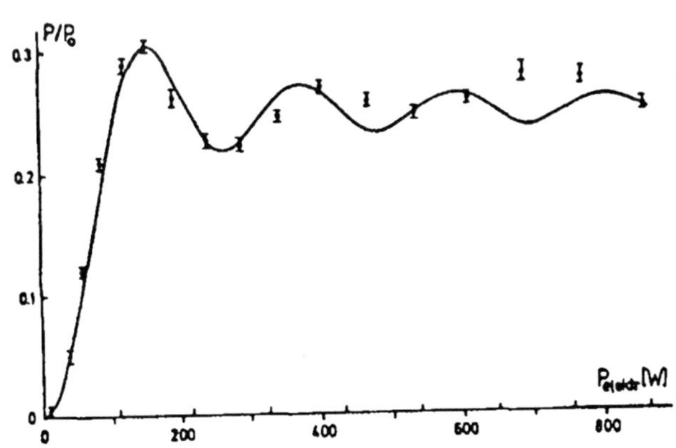

Fig. 8.17 Measured and calculated dependence of the transmitted power P of a Nd:glass laser beam passing through a pumped Nd:glass rod as a function of the electrical pump power. The Nd:glass rod is placed between crossed polarizers. The power P is normalized with respect to the power P_0 obtained for parallel polarizers and no pumping [S.10].

8.2.7 Compensation of Radial-Azimuthal Birefringence

The treatment of radial-azimuthal birefringence is important to estimate the influence of the birefringence on the performance of solid state laser resonators. There are two main effects of birefringence on the laser properties. First, generation of a linearly polarized output by placing a polarizer into the resonator will result in losses and, consequently, a considerable decrease in output power. Secondly, the different refractive powers for radially and azimuthally polarized fields result in different beam propagation for the two fields inside the resonator. This leads to a deterioration of the beam quality [3.198] and a decrease of the mode volume inside the active medium.

It is therefore very important to come up with resonator schemes that provide an intracavity compensation of the thermally induced birefringence [3.185,3.196-199,3.205,3.209]. The basic idea of birefringence compensation is the rotation of the field vector by 90° between transits through the active medium. The rotation switches the r- and the θ-component of the field and the second transit will then equalize the phase shifts experienced by the two components. This means that both polarizations experience the same refractive power of $D_{ave}=(D_r+D_\theta)/2$.

If two rods are used, the rotation can be accomplished by placing a 90° quartz rotator between the rods (Fig. 8.18a). This technique of birefringence compensation was patented in 1969 [3.185] and first demonstrated in 1971 in a dual rod flashlamp pumped TEM_{00} mode Nd:YAG laser [3.187]. Alternatively, the quartz rotator can be replaced by two half wave plates whose principal axes are rotated by 45° with respect to one another (see Sec. 3.2.3). This may be a good alternative if the space between the rods is limited, since a quartz rotator has a length of appr. 14mm at 1064nm.

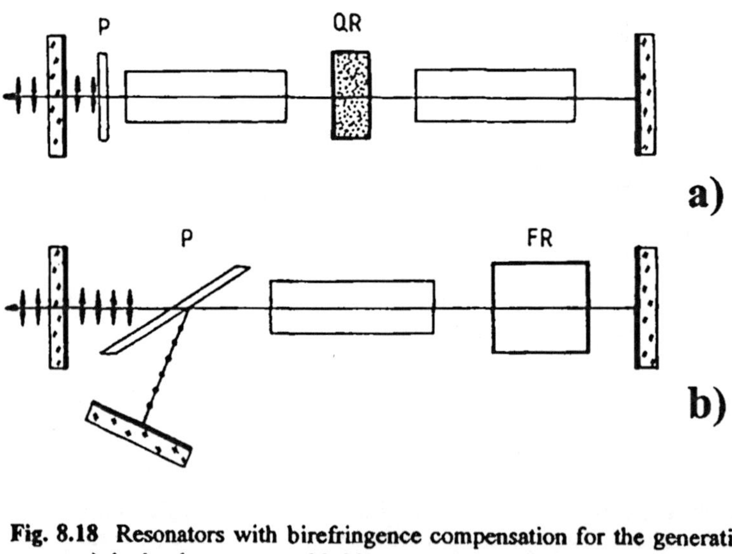

Fig. 8.18 Resonators with birefringence compensation for the generation of linearly polarized output. a) dual rod resonator with 90° quartz rotator (QR), b) single rod resonator with 45° Faraday rotator (FR).

For resonators with a single rod, a 45° Faraday rotator located between the rod and the HR mirror has to be used (Fig. 8.18b). Note that a 45° quartz rotator will not work in this case because the rotation is canceled when the wave passes the rotator the second time on its way back to the rod. After one round trip in the resonator the light is again linearly polarized but with an oscillation direction which is rotated by 90°. The polarizer now totally reflects the beam and an additional mirror feeds the beam back into the resonator. Another round trip will then generate the initial linear polarization state again. This compensation scheme thus requires four transits through the active material. By adding a second 45° Faraday rotator between the laser rod and the polarizer, the reflection at the polarizer can be prevented [3.209].

Unfortunately, Faraday rotators are relatively expensive devices. More affordable is the poor man's version of a Faraday rotator compensation scheme as depicted in Fig. 8.19 [3.214]. The p- and the s- polarized components of the beam are separated by a polarization cube and recombined after rotating each linear polarization component by 90°. To the authors' best knowledge, there has not been an experimental verification of this set-up yet.

A simple technique to partially compensate the birefringence at lower pump powers is depicted in Fig. 8.20 [3.203,3.205,3.207]. The quarter wave plate has its principal axes aligned parallel and perpendicular to the transmission direction of the polarizer. Fig. 8.21 shows calculated loss per roundtrip at the polarizer as a function of pump power for a diode pumped Nd:YAG rod (α=0.15μm/W, β=0.15 in (8.35)) both for a flat-top beam profile and a Gaussian TEM_{00} mode. This compensation technique is useful for low gain, low power lasers with diode pump powers and flashlamp pump powers of up to 100W and 1kW, respectively, where any loss in the cavity has a considerable effect on the output power. Examples are the 946nm, 1319nm and 1444nm transitions in Nd:YAG.

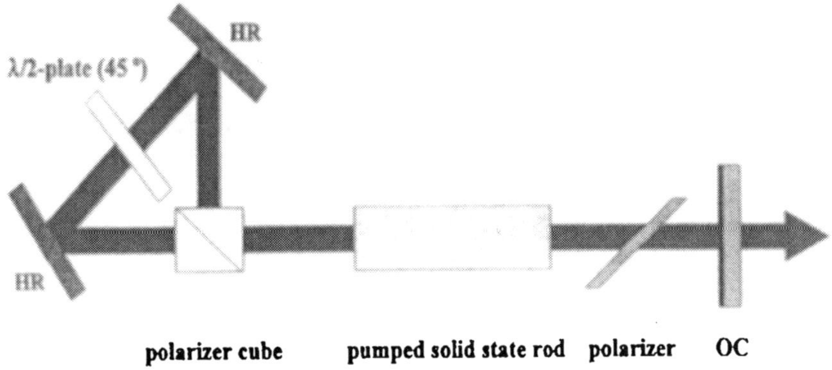

Fig. 8.19 Poor man's Faraday rotator using a half wave plate with its principal axes rotated by 45°.

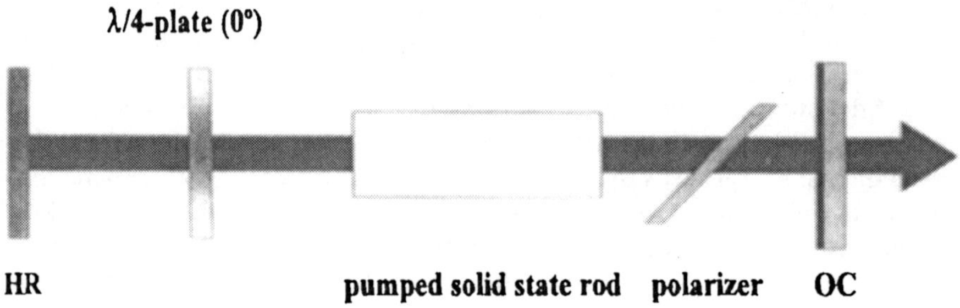

Fig. 8.20 Partial birefringence compensation in a single rod resonator using a quarter wave plate.

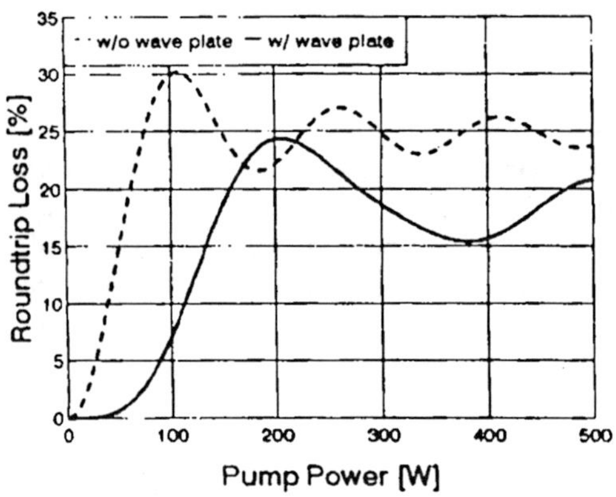

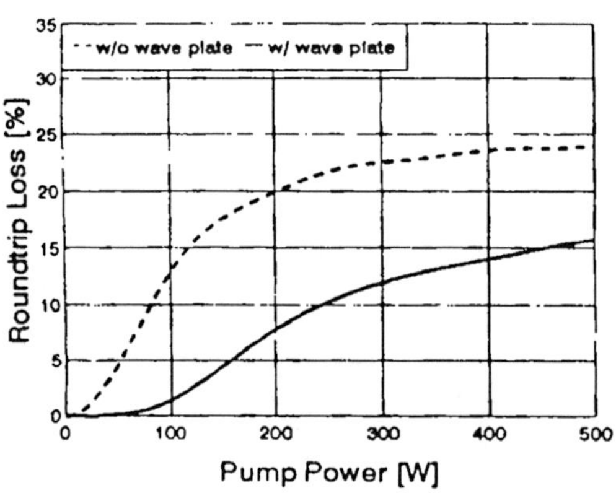

Fig. 8.21 Calculated roundtrip loss for the resonator of Fig. 8.20 with and without the quarter wave plate inserted as a function of the absorbed diode pump power (Nd:YAG, α=0.15µm/W, β=0.15). Top: Flat top beam filling the entire crod cross-section, bottom: Gaussian TEM_{00} mode with beam radius of 0.7 times the rod radius.

Unfortunately, the resonator schemes presented in Figs. 8.18 and 8.19 will not completely compensate for birefringence as the pump power is increased. Due to thermal focusing, the rays propagate through the active medium not parallel to its axis and differently in both rods, or for a single rod resonator, the rays intersect the rod differently during the second transit. Since the phase shift is a function of the radial and the azimuthal coordinate it is necessary that a ray passes through the rod at the same coordinates after the oscillation direction is rotated. By using transformation optics between the rods it is theoretically possible to achieve perfect birefringence compensation by matching the beam propagation in the two rods [3.199] (Fig. 8.22). These compensation schemes can also be applied to single rod resonators, but the imaging property of the rear resonator mirror has to be incorporated.

a) Single Lens Scheme (Fig. 8.22a)

With one focusing lens of focal length f located at distances d_1 and d_2 from the end faces of the rods, perfect birefringence compensation will occur if the following conditions hold [3.199]:

$$d_1 = d_2 = f - \frac{\ell}{2n_0} + \sqrt{f^2 - \left[\frac{\ell}{2n_0}\right]^2} \tag{8.37}$$

where ℓ is the length of the rod and n_0 is the index of refraction at the rod center.

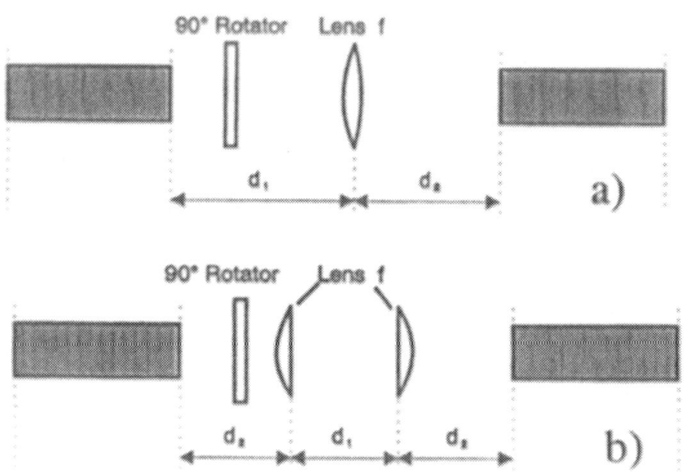

Fig. 8.22 Birefringence-compensated laser confi-gurations using a 90° polarization rotator and transformation optics. a) single lens scheme, b) dual lens scheme [3.199] (© Chapman & Hall 1996).

Note that (8.37), which holds only for $f \geq \ell/(2n_0)$, does not represent an imaging of the principal planes of the rods. This compensation technique is also applicable to the single rod resonator of Fig. 8.18b. If d denotes the distance of the HR mirror to the rod end face, the radius of curvature ρ of the mirror can be determined by setting $f=\rho/2$ in (8.37). The final result for the mirror curvature reads:

$$\rho = d + \frac{\ell}{2n_0} + \left[\frac{\ell}{2n_0}\right]^2 \frac{1}{d+\ell/(2n_0)} \tag{8.38}$$

A similar technique to compensate birefringence in a single rod resonator with two Faraday rotators is described in [3.209]. Here a lens-mirror combination is used to image the rod's principal plane onto itself.

b) Dual Lens Scheme (Fig. 8.22b)

Two of the three parameters d_1,d_2 and f can be chosen freely by using the condition:

$$[f-d_2-\frac{\ell}{2n_0}][d_1f+d_2f-d_1d_2] + \frac{\ell}{2n_0^2}[2n_0f^2-2\ell f+\ell d_1] = 0 \tag{8.39}$$

If we choose $d_1=2f$ (adjusted telescope), (8.39) will be met if:

$$d_2 = f - \ell/2n_0 \tag{8.40}$$

This means that the principal planes of the rods are imaged onto each other in this case.

Figure 8.23 presents measured depolarization losses for different dual Nd:YAG rod configurations as a function of the electrical pump power per rod. The 6x3/8 inch rods were placed between crossed polarizers and the transmitted power fraction of a collimated HeNe laser beam at λ=632.8nm was measured. This graph indicates the importance of using transformation optics in birefringence compensation schemes. With a polarization rotator alone the compensation works only at very low pump powers and the loss quickly rises to the maximum value of 27%. The remaining incomplete compensation for configuration C and D arises mainly as the result of rod end effects.

In recent years, birefringence compensation using one of the techniques described above has gained widespread application in solid state lasers, especially in dual rod Nd:YAG laser systems [3.200,3.201,3.206,3.208,3.210,3.213]. The main reason for this development is the need for high output power at near diffraction limited beam quality. Power scaling of TEM_{00} mode rod lasers beyond 50W can only be realized by increasing the number of rods. Since the TEM_{00} mode is only defined for linear polarization, compensation of the birefringence is mandatory. Average TEM_{00} mode output powers of up to 208W have been demonstrated with diode-pumped dual rod Nd:YAG lasers [3.206,3.210,3.213].

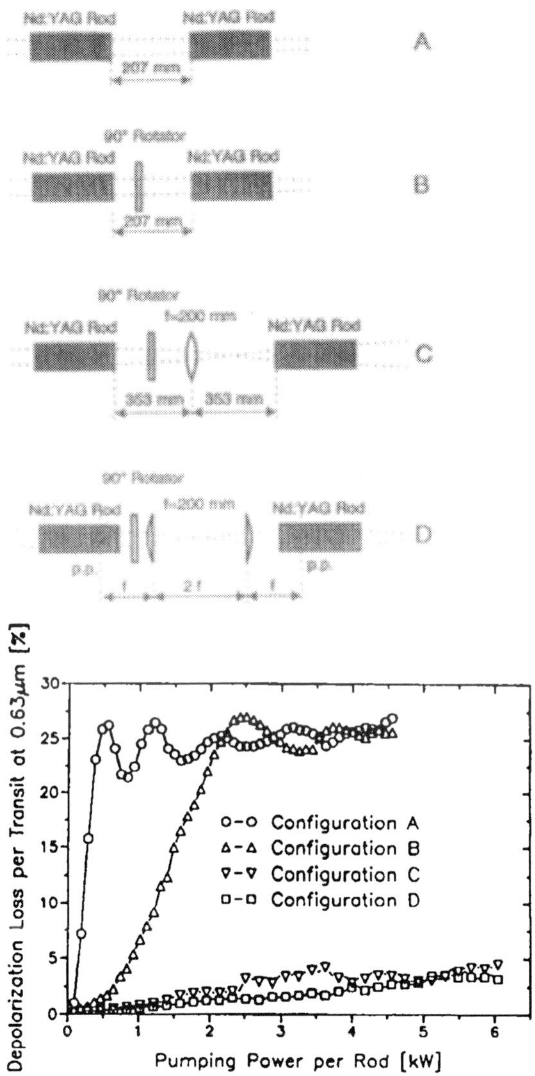

Fig. 8.23 Comparison of measured depolarization losses for four different birefringence compensation schemes of a dual Nd:YAG rod laser system as a function of the average pump power per rod. The 6x3/8 inch Nd:YAG rods were flashlamp pumped with a pump energy of 40J per rod and a pulse duration of 1ms. The pump power was varied with the repetition rate. The configurations were placed between crossed polarizers and the transmitted power fraction of a HeNe laser was measured [3.199] (© Chapman and Hall 1996).

Part IV

Open Resonators with Gain

Chapter 9

The Active Medium

9.1 General Aspects

We have so far neglected the influence of the active medium on the resonator properties and calculated the stationary field distributions inside the empty resonator. We have thereby assumed that the losses of the resonator are compensated by an imaginary medium, but we did not take into account the nonlinear interaction between the atoms of the medium and the electric field.

The active medium affects the electromagnetic wave oscillating inside the resonator in different ways. With every transit through the medium the intensity of the light is amplified by a gain factor G which is a nonlinear function of the intensity itself and depends on the transverse coordinates. This nonlinear interaction gives rise to changes in the mode structure and may also lead to temporal fluctuations of the output power (periodic, quasi-periodic, or chaotic emission). Furthermore, the index of refraction and the geometry of the medium change the propagation characteristics of the beam so that the resonator parameters (g-parameters, length) depend on the dimensions and the material properties of the medium. In addition, as was discussed in the previous chapter, the polarization state is altered if the laser material exhibits pump induced birefringence.

Mode structures and diffraction losses of an active resonator will thus differ from those of the passive resonator. As the gain factor is increased by increasing the pump power, these differences will become more pronounced. Only in the limit of a low gain, which means the laser is operated close to the laser threshold, will the physical properties of active and passive resonators become alike. The detailed treatment of the passive resonator in Part III will help us to better understand the influence of the active medium since the physical properties are only modified rather than completely changed. Besides the physics of the interaction of the laser light with the active medium discussed in this chapter, we shall understand the dependence of the output power on the resonator parameters such as g-parameters, mirror reflectances, and mirror tilts, as well as the changes induced on the mode structure and on the diffraction losses. In addition, the effect of the gain on the temporal emission and its frequency bandwidth will be discussed in more detail. This knowledge will enable the reader to design resonators that exhibit optimum performance as far as output power, beam quality, or misalignment sensitivity are concerned.

9.2 Effective Length of a Resonator

Insertion of the active medium will change the intracavity beam propagation even if the medium does not exhibit any gain. The beam is refracted at the endfaces as it passes through the active medium resulting in a decrease of the resonator length. The origin of a spherical wave is moved towards the active medium due to the beam refraction as shown in Fig. 9.1. If n denotes the index of refraction of the medium and ℓ is its length, the geometrical resonator length L_0 is shorted by the length Δ. The shift Δ can be calculated by determining the B-element in the ray transfer matrix of the medium (Fig. 9.1):

$$\Delta = \frac{(n-1)}{n}\ell \tag{9.1}$$

The mode in an active resonator with geometrical length L_0 exhibits the same beam radii at the resonator mirrors as a passive resonator with effective length $L_{eff}=L_0-\Delta$. This effective length has to be used to calculate the eigenmodes of the resonator. However, as far as the resonance conditions (5.7) and (5.9) are concerned, the optical length $L_{opt}=L_0+(n-1)\ell$ has to be used to determine the eigenfrequencies.

The shortening of the resonator length is negligible for gas lasers, but for solid state lasers and diode lasers which exhibit indices of refraction between 1.5 and 3, the shift Δ might have a considerable impact on the mode properties. For a 150mm long Nd:YAG laser rod ($n=1.82$) the shift is about 68mm, a value that cannot be neglected. The shortening applies not only to the active medium itself, but also to any other optical element placed into the resonator such as thick lenses, cube polarizers, or nonlinear crystals. In any case, the effective resonator length is given by the B-element of the ray transfer matrix for a resonator transit.

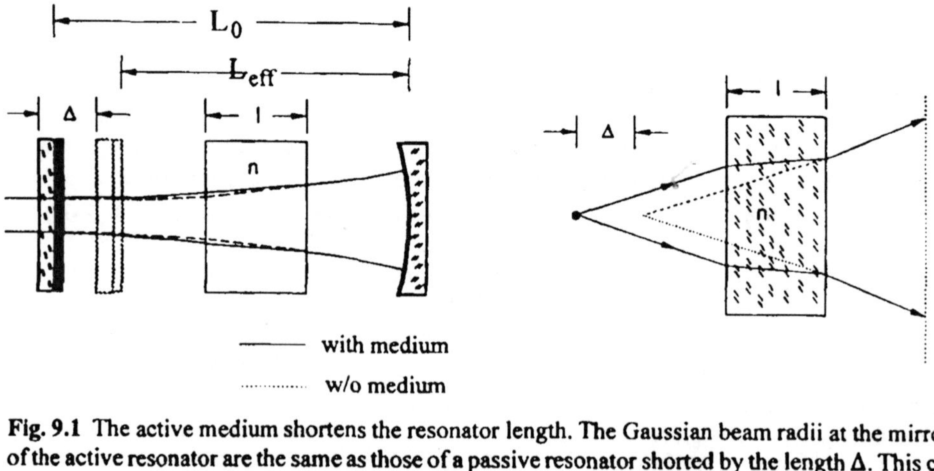

Fig. 9.1 The active medium shortens the resonator length. The Gaussian beam radii at the mirrors of the active resonator are the same as those of a passive resonator shorted by the length Δ. This can be easily understood if the transit of a spherical wave through the medium is considered.

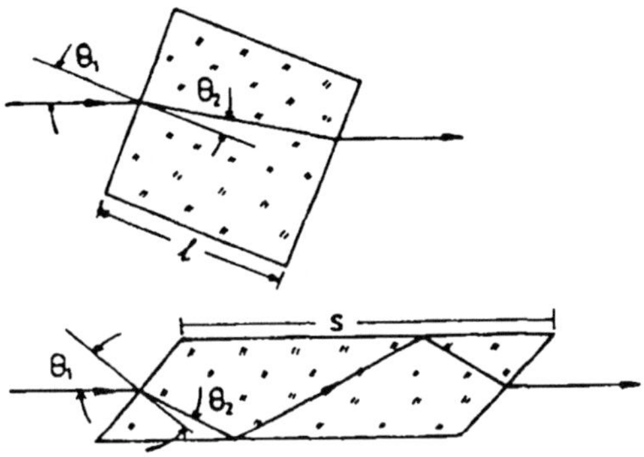

Fig. 9.2 Beam propagation in a tilted plate and in a zig-zag slab. The tangential plane is the paper plane.

Very often the active medium is placed inside the resonator at the Brewster angle to minimize reflection losses for the p-polarization (Fig. 9.2). The shift Δ will then become different for the x- (tangential plane) and the y-direction (sagittal plane) resulting in elliptical beam radii at the resonator mirrors. By using the 4x4 ray transfer matrices (1.82/1.83) of a slab with length ℓ tilted by the angle θ_1, we get for the shifts:

$$\Delta_x = \ell \left[1 - \frac{n \cos^2\theta_1}{n^2 - \sin^2\theta_1} \right] \quad (9.2) \qquad , \qquad \Delta_y = \ell \, \frac{n-1}{n} \quad (9.3)$$

If θ_1 is equal to the Brewster angle ($\tan\theta_1 = n$), Eq. (9.2) can be written as:

$$\Delta_x = \ell \left[1 - \frac{1}{n^3} \right] \qquad (9.4)$$

Example (Fig. 9.2):
A zig-zag slab is considered with side length s and the endfaces cut at the Brewster angle. The index of refraction is $n = 1.82$. Note that we have to use the zig-zag path length ℓ_{zz} as the length ℓ of the medium with:

$$\ell_{zz} = s \left[\frac{\sqrt{1 + n^2}}{n} \right]$$

By using (9.3) and (9.4) the shifts in the x- and the y-direction are found to be:

$$\Delta_x = 0.514 \, s \quad , \quad \Delta_y = 0.189 \, s$$

9.3 Amplification and Efficiencies

Let us first discuss the light amplification on an empirical basis before we treat the interaction of the laser light with the active medium in a physically correct manner. We assume that the light intensity is amplified by a factor G during a transit with:

$$G = \exp[g\ell] \tag{9.5}$$

and attenuated by a loss factor V_S given by:

$$V_S = \exp[-\alpha_0\ell] \tag{9.6}$$

As shall be shown later the gain factor G is a function of the intensity I:

$$G = \exp[g\ell] = \exp\left[\frac{g_0\ell}{(1 + I/I_S)^x}\right] \tag{9.7}$$

with $g_0\ell$ being the small-signal gain and I_S being the saturation intensity, a characteristic of the laser material. The parameter x is equal to 1.0 for homogeneously broadened lasers, and $x=0.5$ holds for inhomogeneously broadened lasers. The difference between the two line broadening mechanisms will be discussed later. The small-signal gain $g_0\ell$, which increases with the pump power, determines the amplification of light whose intensity is small compared to the saturation intensity. The gain will saturate as the intensity is increased. The saturation is necessary for the conservation of energy because otherwise the intensity would become infinite if we passed the light repeatedly through the same medium. If we now consider a round trip in the resonator and start with the intensity I at the high reflecting mirror as shown in Fig. 9.3, the stationary condition for the intensity reads:

$$G^2 R_1 R_2 V_S^2 = 1 \tag{9.8}$$

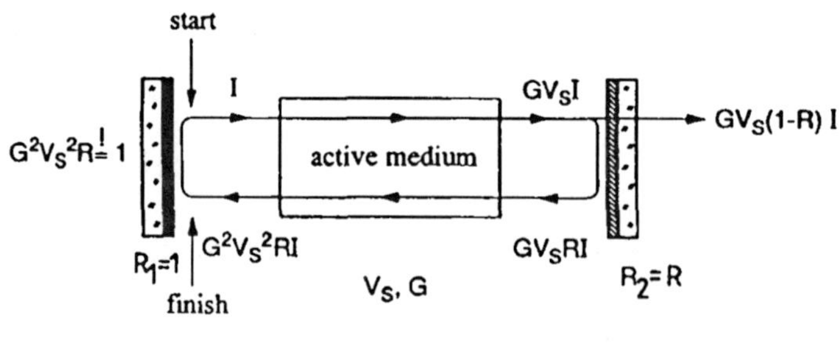

Fig. 9.3 Gain and loss must have compensated each other after a round trip.

By inserting (9.7) into (9.8) and considering the fact that the intensity inside the medium is doubled (both the back and the forth travelling wave contribute to the gain saturation), the average intracavity intensity for one propagation direction is obtained:

$$I = \frac{I_S}{2}\left[\left(\frac{g_0\ell}{|\ln(\sqrt{R_1R_2}V_S)|}\right)^{\frac{1}{x}} - 1\right] \tag{9.9}$$

This holds under the assumption that the z-dependence of the intensity inside the gain medium can be neglected, which is true for high reflectivities (R>0.7). Equation (9.9) indicates that the small-signal gain has to be higher than a certain value to attain laser oscillation:

$$g_0\ell > |\ln(\sqrt{R_1R_2}V_S)| \tag{9.10}$$

The laser threshold condition thus reads:

$$g_0\ell = |\ln(\sqrt{R_1R_2}V_S)| \tag{9.11}$$

We see that the parameters required to describe the active medium are the small-signal gain and the saturation intensity I_S. The functional dependence of both parameters on the material properties and atomic constants are derived in the next section. The final expressions read:

$$g_0\ell = \Delta N\,\sigma\,\ell \tag{9.12}$$

$$I_S = \frac{h\nu}{\sigma\tau} \qquad \textit{for a four level system} \tag{9.13}$$

$$I_S = \frac{h\nu}{2\sigma\tau} \qquad \textit{for a three level system} \tag{9.14}$$

with:

	$h\nu$	: photon energy
	σ	: emission cross section for stimulated emission
	τ	: spontaneous decay time of the upper laser level
	ΔN	: inversion density between upper and lower laser level

Keep in mind that the emission cross section σ is a function of the laser frequency ν. Table 9.1 presents the atomic constants of some common laser materials. If we knew the dependence of the inversion density ΔN on the pump power we would be able to determine the threshold pump power needed to achieve laser emission (Eq. (9.11)) as well as the output power of the laser as a function of the pump power. If A is the cross sectional area of the active medium, the output power is given by:

$$P_{out} = A\,I\,(1-R_2) = \frac{AI_S}{2}(1-R_2)\left[\left(\frac{g_0\ell}{|\ln(\sqrt{R_1R_2V_S})|}\right)^{\frac{1}{x}} - 1\right] \qquad (9.15)$$

Table 9.1 Laser wavelengths, emission cross sections, decay times and saturation intensities of some laser materials. Ruby is the only true three level laser.

Laser medium	$\lambda[\mu m]$	$\sigma[10^{-19}cm^2]$	$\tau[s]$	$I_S\,[W/cm^2]$
CO_2	10.6	1,000	$1x10^{-5}$	20
HeNe	0.63	300,000	$1x10^{-8}$	50
KrF	0.248	1,000	$5x10^{-9}$	$1.6x10^{6}$
Rhodamin 6G in alcohol	0.57	4,000	$5x10^{-9}$	$1.7x10^{5}$
$Cr:Al_2O_3$ (ruby)	0.69	0.2	$3x10^{-3}$	2,350
Nd:YAG	1.06	4.1	$2.3x10^{-4}$	2,000
$Nd:YVO_4$	1.06	15.0	$1.0x10^{-4}$	1,260
Nd:YLF	1.047	1.8	$4.8x10^{-4}$	2,150
Yb:YAG	1.03	0.21	$9.7x10^{-4}$	9,500
Nd:glass	1.05	0.4	$3x10^{-4}$	12,000
Alexandrite	0.76	0.055	$2.6x10^{-4}$	$1.8x10^{5}$
Ti:sapphire	0.79	2.8	$3.2x10^{-6}$	$1.6x10^{5}$
GaAs	0.81	1,000	$4x10^{-9}$	$6.2x10^{5}$

For a homogeneously broadened system ($x=1$) the output power can be approximated for low output coupling by (with $R_1=1.0$ and $1-R_2=2|ln\sqrt{R_2}|$):

$$P_{out} = AI_S\,|\ln\sqrt{R_2}|\left[\frac{g_0\ell}{|\ln(\sqrt{R_2V_S})|} - 1\right] \qquad (9.16)$$

The maximum output power is attained if we have no losses ($V_S=1.0$ and $R_2 \rightarrow 1.0$):

$$P_{out,max} = A\,I_S\,g_0\ell \qquad (9.17)$$

By inserting the expressions for the small-signal gain and the saturation intensity, we see that the maximum output power represents the power P_{UL} transferred into the upper laser level:

$$P_{UL} = \Delta N\,A\ell\,h\nu/\tau = P_{out,max}$$

The output power will generally be lower than this maximum value. The efficiency with which the resonator converts the maximally available power P_{UL} into laser output power is called the extraction efficiency η_{extr} :

$$\eta_{extr.} = \frac{P_{out}}{P_{UL}} = \frac{P_{out}}{AI_S g_0 \ell} \tag{9.18}$$

The extraction efficiency is only a function of the resonator parameters since the output power is proportional to the inversion in the active medium. The efficiency of the pump process itself is given by the excitation efficiency η_{excit} which relates the power in the upper laser level to the pump power P_{pump} (this is the electrical power to the flashlamp, or the gas discharge, or in the case of optical pumping, it is the optical power of the pump laser):

$$\eta_{excit} = \frac{P_{UL}}{P_{pump}} \tag{9.19}$$

The total laser efficiency is then given by:

$$\eta_{tot} = \frac{P_{out}}{P_{pump}} = \eta_{excit}\eta_{extr} \tag{9.20}$$

The excitation efficiencies depend on the pump process and the laser material. Values range from 0.2% for argon lasers, 4-8% for flashlamp pumped solid state lasers, 30% for RF-excited CO_2 lasers, 60-70% for diode pumped solid state lasers to 70-80% for diode lasers and CO lasers. If the excitation efficiency is known, the small-signal gain can be calculated using:

$$g_0 \ell = \frac{\eta_{excit} P_{pump}}{A\ I_S} \tag{9.21}$$

This relation can also be used to experimentally determine the excitation efficiency in combination with (9.11). The pump power P_{pump} required to reach laser threshold is measured for different output coupling reflectances R_2. If the log of R_2 is plotted versus the pump power and the saturation intensity is known, the slope of the curve provides the excitation efficiency (see also Chapter 23). Note that the excitation efficiency is a constant only for true four level systems. In three level systems, the excitation efficiency is a function of the pump power!

The extraction efficiency, on the other hand, is mainly determined by the small-signal gain. At laser threshold, the extraction efficiency is zero and it rises as the small-signal gain is increased. For high gain lasers like diode lasers, extraction efficiencies of up to 80% can be attained. Extraction efficiencies higher than this value are difficult to achieve since the output power is strongly affected by any loss inside the resonator (surface reflections, absorption, scattering). This holds for low- gain lasers in particular, like HeNe lasers, where the extraction efficiency hardly exceeds 20%.

Examples:
1) Flashlamp pumped Nd:YAG laser (cw), rod radius: 5mm, η_{excit}=4%
By using (9.21) the small-signal gain is found to be 0.04 per kW of electrical pump power. In order to reach threshold for R_1=1.0, R_2=0.9, and V_S=0.95, according to (9.11) a small-signal gain of $g_0\ell$=0.104 is required. The necessary electrical pump power thus is P_{pump}=2.6kW. The output power at 10kW ($g_0\ell$=0.4) can now be calculated using (9.15) with x=1.0: P_{out}= 235.6W. The total efficiency of the laser is 2.36% and the extraction efficiency is 59%.
2) Diode-end-pumped Nd:YVO_4 laser (cw), pump spot diameter: 1mm, 95% pump light absorption (at 808nm)
The excitation efficiency is given by the product of pump light absorption (=0.95) and quantum defect (=0.76), resulting in η_{excit}=72%. Per Watt of pump power (9.21) predicts a small-signal gain of 0.073. With R_1=1.0, R_2=0.7, and V_S=0.99, the pump power to reach threshold is 2.6W. For 50W of pump power, the small-signal gain is 3.65, resulting in an output power (9.15) of 27.2W. Total efficiency of this system is 54.4% and its extraction efficiency is 75.6%.
3) CO_2 laser (homogeneously broadened), tube radius: 30mm, η_{excit}=25%
Similar calculations as in example 1) yield:

small-signal gain	: 0.44 per kW of pump power
laser threshold pump power (R_1=1,R_2=0.5,V_S=0.98)	: 828 W
output power at 5kW pump power	: 710 W
extraction efficiency	: 56.8%
total efficiency	: 14.2 %

9.4 The Laser Equations

We will now use the semiclassical description of light amplification in an active medium [4.1,4.3]. The classical part of this description is the utilization of the wave equation to propagate the field through the active medium (Fig. 9.4). The response of the active atoms (the term atoms stands for atom, ion, or molecule depending on the type of laser material used) is characterized by the atomic susceptibility χ. Unfortunately, a classical derivation of the atomic susceptibility cannot account for the quantum mechanical process of stimulated emission. We will, therefore, replace the atomic susceptibility by the corresponding quantum mechanical expression in the final result. In our model the stimulated emission takes place between two energy levels with a spontaneous decay time τ from the upper to the lower level. The center frequency is ν_0 and light can be amplified within a frequency bandwidth $\Delta\nu=1/(2\pi\tau)$.

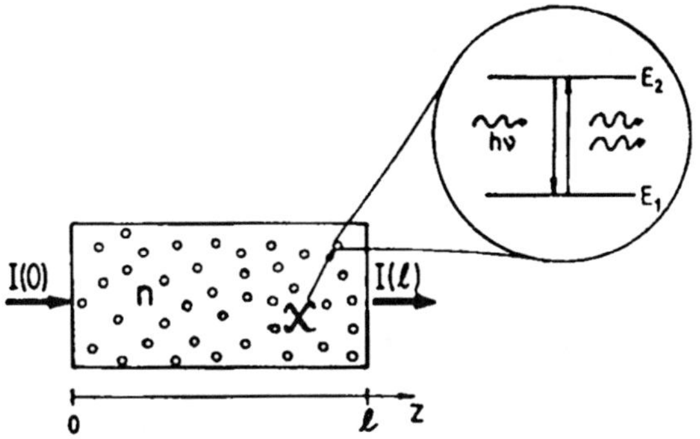

Fig. 9.4 Amplification of an electric field with intensity $I(z)$ by an active medium. The host material with index of refraction n is doped with active atoms that amplify light at a frequency ν via stimulated emission between two energy levels. The response of the atoms to the incident field is described by the atomic susceptibility χ.

Let us consider a host material with index of refraction n which is doped with the active atoms or ions at a density N_0 (atoms per cm^3). The interaction of the electric field E (without lack of generality we use scalar fields) with the active atoms induces an electric polarization P given by:

$$P = \epsilon_0 \epsilon \ \chi \ E = \epsilon_0 \epsilon \ (\chi_1 + i\chi_2) E \tag{9.22}$$

where ε is the permittivity of the host material. As will be shown later, both the real part and the imaginary part of the atomic susceptibility are a function of the frequency ν. Insertion of (9.22) into the one-dimensional wave equation yields:

$$\frac{\delta^2 E}{\delta^2 t} - \frac{1}{c^2}\frac{\delta^2 E}{\delta^2 t} = \frac{1}{\epsilon_0 \epsilon c^2}\frac{\delta^2 P}{\delta^2 t} = \frac{1}{c^2}(\chi_1 + i\chi_2)\frac{\delta^2 E}{\delta^2 t} \tag{9.23}$$

with $c = c_0/n$ being the speed of light in the medium.

For a plane wave with wave number k and frequency $\omega = 2\pi\nu$ in the medium:

$$E(z,t) = E_0(z,t) \exp[-i(kz - \omega t)] \tag{9.24}$$

the wave equation yields in the SVE-approximation (slowly varying envelope: the second derivatives can be neglected) with the assumption $|\chi| \ll 1$:

$$\frac{\delta E_0}{\delta z} + \frac{1}{c}\frac{\delta E_0}{\delta t} = \frac{k\chi_2}{2} E_0 - i\frac{k\chi_1}{2} E_0 \tag{9.25}$$

The field will be absorbed or amplified if the imaginary part χ_2 of the atomic susceptibility is negative or positive, respectively. This can be seen easily if we assume a steady state condition $\delta E_0/\delta t = 0$. Integration of (9.25) then yields:

$$E_0(z) = E_0(0)\ \exp\left[\frac{k}{2}\chi_2 \ell\right]\ \exp\left[-i\frac{k}{2}\chi_1 \ell\right] \tag{9.26}$$

The real part of the susceptibility gives rise to a phase shift of the transmitted field.

A classical analysis using electron oscillators to calculate the atomic susceptibility, however, yields a negative χ_2 which means that the field is partially absorbed by the medium. The concept of light amplification is not known in classical physics because the existence of quantized atomic energy levels is not taken into account. A quantum mechanical derivation indeed shows that the imaginary part of the susceptibility becomes positive if an inversion between two energy levels is created. This means that more atoms are found in the upper level than in the lower one. The quantum mechanical expressions for the real part and the imaginary part of the atomic susceptibility read:

$$\chi_1 = \frac{\sigma_0}{k}\Delta N\ \frac{\nu - \nu_0}{\Delta\nu}\ f(\nu) \tag{9.27}$$

$$\chi_2 = \frac{\sigma_0}{k}\Delta N\ f(\nu) \tag{9.28}$$

with: $f(\nu) = \dfrac{(\Delta\nu/2)^2}{(\nu-\nu_0)^2 + (\Delta\nu/2)^2}$: Lorentzian line function

σ_0 : cross section for stimulated emission at $\nu = \nu_0$

ΔN : inversion density (number of inverted atoms per cm^3)

The function $f(\nu)$ describes the spectral gain profile of the transition. Equations (9.27) and (9.28) are visualized in Fig. 9.5. At the center frequency ν_0 the real part χ_1 of the susceptibility is zero. It is this center frequency, the laser frequency, that we are interested in. Let us for now focus on this resonant case and set χ_1 equal to zero. We will get back to the spectral distribution of the susceptibilities in Sec. 9.6. In this section we will also discuss the effect of the real part of the susceptibility on the laser performance. The combination of (9.28) and (9.25) results in the differential equation for the amplification of light by an active medium (with $\chi_1 = 0$):

$$\frac{\delta E_0}{\delta z} + \frac{1}{c}\frac{\delta E_0}{\delta t} = \frac{\Delta N \sigma_0}{2} E_0 \tag{9.29}$$

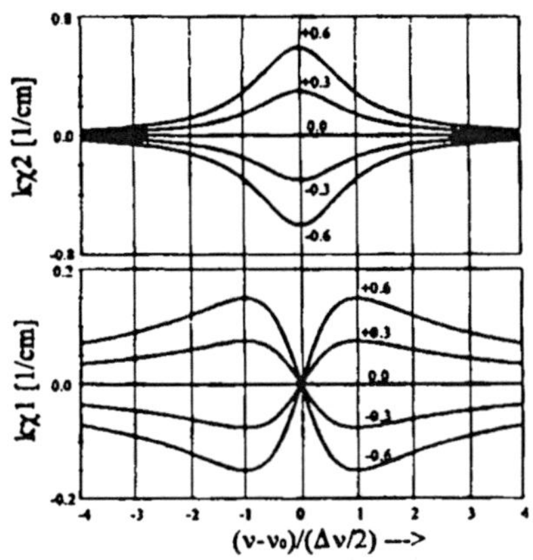

Fig. 9.5 The real part and the imaginary part of the atomic susceptibility (multiplied with the wave number k) in the vicinity of the resonance frequency ν_0. The curve parameter is the small-signal gain coefficient $g_0=\Delta N\sigma_0$ in cm^{-1}.

A positive inversion density ΔN generates amplification of the transmitted light. If we take the losses inside the medium due to scattering and absorption into account by introducing a loss coefficient α_0, the final equation for the intensity I reads:

$$\frac{\delta I}{\delta z} + \frac{1}{c}\frac{\delta I}{\delta t} = (\Delta N\sigma_0 - \alpha_0)I \tag{9.30}$$

Now we have to find a similar equation for the inversion density ΔN. Again, we have to use a quantum mechanical treatment of the atomic energy levels including the quantization of the electromagnetic field (spontaneous decay is a result of the field quantization). The resulting equation is known as the rate equation for the inversion. It describes the temporal change of the inversion due to pumping, spontaneous emission, and induced emission. Let us now discuss the rate equations for different level systems (Fig. 9.6).

a) Four-level system:
The density of active atoms is $N_0=N_1+N_2+N_3+N_4$. The pump process lifts atoms from the ground level to the pump level with a pump rate W. The transition times from the pump level to the upper laser level and from the lower laser level to the ground level are assumed to be zero. This means that the densities N_4 and N_2 are always equal to zero. The atoms decay from level 3 to level 2 under emission of a photon with the frequency ν via spontaneous emission (decay time τ) or induced emission (the incident light intensity is I). If the electric polarization P adiabatically follows the electric field E, the rate equation for the inversion density $\Delta N=N_3-N_2$ reads:

$$\frac{\delta\Delta N}{\delta t} = W(N_0-\Delta N) - \frac{\Delta N}{\tau} - \frac{I}{I_S}\frac{\Delta N}{\tau} \tag{9.31}$$

where $I_S = h\nu/(\sigma_0\tau)$ is the saturation intensity.

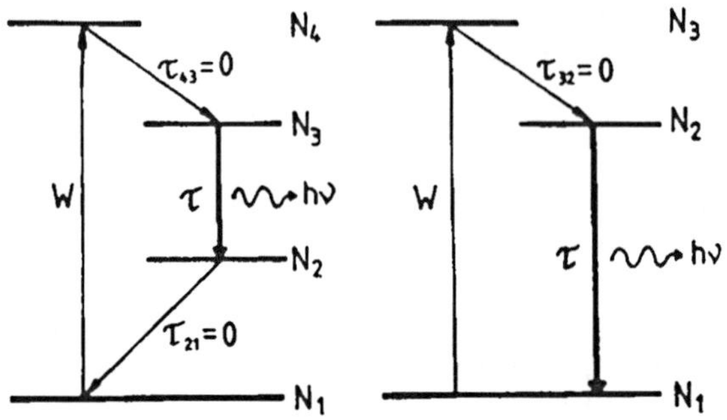

Fig. 9.6 Energy level diagrams of an idealized four-level (left) and a three-level system. The inversion density N_3-N_2 *(4-level)* and N_2-N_1 (3-level) is changed by the pumping of atoms from the ground level with a pumping rate W, spontaneous emission with decay time τ, and induced emission caused by the incident field with intensity I.

b) Three-level system:

The lower laser level and the ground level are identical and the atom density is $N_0=N_1+N_2+N_3$. Assuming $N_3=0$, The rate equation for the inversion $\Delta N=N_2-N_1$ reads:

$$\frac{\delta \Delta N}{\delta t} = (W-\frac{1}{\tau})N_0 - (W+\frac{1}{\tau})\Delta N - 2\frac{I}{I_S}\frac{\Delta N}{\tau} \tag{9.32}$$

The factor 2 in the induced emission term reflects the fact that a transition from level 2 to level 1 reduces the inversion by 2. In the four-level system, a transition from level 3 to 2 reduces the inversion only by 1 since level 2 is assumed to be empty (fast relaxation to the ground level).

Equation (9.30), in combination with (9.31) or (9.32), describes the amplification of light by the active medium. However, the light pulses must not be too short (SVE approximation must hold), and it is assumed that the polarization adiabatically follows the electric field. This means that the relaxation time of the polarization must be much smaller than the time constant with which the electric field changes. Typical polarization relaxation times are on the order of 10^{-9}s for gas lasers and 10^{-12}s for solid state lasers. Thus, for free running laser operation (pulse duration of greater than 10μs) and Q-switched lasers (pulse durations of greater than 1ns), the above shown equations can be used to describe all aspects of the laser performance.

Let us now find the solutions of the laser equations for free running laser operation with pulse widths that are longer than the spontaneous emission lifetime τ. Since the laser pulse duration is large compared to any of the time constants, the time derivatives of the inversion density in (9.31), (9.32), and of the intensity in (9.30) can be set to zero. This results in the steady state inversion density ΔN:

$$\Delta N = \frac{WN_0}{W+1/\tau} \frac{1}{1 + \frac{1}{1+W\tau}\frac{I}{I_S}} \qquad \text{(4-level system)} \qquad (9.33)$$

$$\Delta N = \frac{(W-1/\tau)N_0}{W+1/\tau} \frac{1}{1 + \frac{2}{1+W\tau}\frac{I}{I_S}} \qquad \text{(3-level system)} \qquad (9.34)$$

In four-level lasers, the pump rate W is typically on the order of 10 per second (for diode-end-pumped lasers on the order of 100 per second) and thus negligible compared to $1/\tau$. Insertion into (9.30) yields for both level systems:

$$\frac{\delta I}{\delta z} = \left[\frac{g_0}{1 + I/I_S} - \alpha_0\right] I \qquad (9.35)$$

where for the 3-level system, I_S is replaced by the effective saturation intensity

$$I_S^* = \frac{I_S}{2}(1 + W\tau)$$

The small-signal gain coefficient g_0 is proportional to the inversion density ΔN_0 without an incident field (intensity $I=0$ in (9.33) and (9.34)):

$$g_0 = W\tau\sigma_0 N_0 = \Delta N_0\sigma_0 \qquad \text{(4-level system)} \quad (9.36)$$

$$g_0 = \frac{W\tau-1}{W\tau+1}\sigma_0 N_0 = \Delta N_0\sigma_0 \qquad \text{(3-level system)} \quad (9.37)$$

These are the final equations used for the calculation of the output power of homogeneously broadened lasers (for inhomogeneous line broadening see the next section). The reader should keep in mind that (9.30) determines the increase in intensity due to stimulated emission only! Since the photons created via spontaneous emission are emitted in all directions, only a small number of these photons travel along the wave vector of the amplified beam, and they can be neglected for power calculations. Typically, the relative power content of the spontaneous emission in the beam is on the order of 10^{-9}. However, the total spontaneously emitted power (photons emitted in all directions) is on the same order of magnitude as the power generated via stimulated emission. The photons emitted via spontaneous emission need to be taken into account if the oscillation build-up in a laser cavity, the noise, and the frequency bandwidth are to be evaluated. Equation (9.30) can be modified to include spontaneous emission resulting in:

$$\frac{\delta I}{\delta z} + \frac{1}{c}\frac{\delta I}{\delta t} = \left(\Delta N \sigma_0 \left[1 + \frac{h\nu_0 \Delta\nu}{A\ I} \right] - \alpha_0 \right) I \tag{9.38}$$

where $h\nu_0$ is the photon energy, A is the cross sectional area of the active medium, and $\Delta\nu$ is the bandwidth of emission, which depends on the number of oscillating modes. The contribution of the spontaneous emission is generally very small and can be neglected. This equation can be transformed into a more convenient and commonly used form by introducing the photon number q inside the resonator length L with:

$$I = \frac{h\nu_0 c}{A\ L} q \tag{9.39}$$

Insertion of (9.39) into (9.38) yields for the single mode case with $\Delta\nu = c/L$:

$$c\frac{\delta q}{\delta z} + \frac{\delta q}{\delta t} = \Delta N \sigma_0 c\ (q+1) - \alpha_0 c\ q \tag{9.40}$$

This equation, which holds for single mode operation, is referred to as the rate equation for the photon number. It is commonly used in combination with the rate equation for the inversion density (9.31) to model the temporal emission of free-running and Q-switched lasers. By introducing the loss due to output coupling in a resonator with length L, and considering that at any time, only the fraction ℓ/L of the total number of photons inside the resonator are located inside the gain medium, the final set of rate equations can be written as (neglecting the z-dependence of q):

$$\frac{\delta q}{\delta t} = \frac{\sigma_0 c \ell \Delta N}{L}(q+1) - \frac{q}{\tau_C} \tag{9.41}$$

$$\frac{\delta \Delta N}{\delta t} = W(N_0 - \Delta N) - \frac{\Delta N}{\tau} - \frac{\sigma_0 c \Delta N q}{A\ L} \tag{9.42}$$

where τ_C is the cavity lifetime (4.12) and q is the total number of photons inside the resonator. We will use these two equations in a later chapter to calculate the performance of Q-switched lasers.

Let us now go back to (9.35) and discuss its solution. Given the assumption that the intensity is constant along the length of the active medium (which holds for low gains and high intensities), we can easily integrate the differential equation:

$$I(\ell) = I(0)\exp\left[\frac{g_0\ell}{1+I/I_S} - \alpha_0\ell\right] = I(0)\ GV_S \tag{9.43}$$

The power available in the form of inversion can be written as:

$$P_{UL} = A\ell\ \Delta N_0\ \frac{h\nu}{\tau} = A\ I_S\ g_0\ell \tag{9.44}$$

The pump rate W is proportional to the electrical pump power P_{pump}:

$$W = \gamma\ P_{pump} \tag{9.45}$$

By using this relation in combination with (9.36) and (9.37) we obtain expressions for the excitation efficiency:

$$\eta_{excit.} = \frac{P_{UL}}{P_{pump}} = A\ell\ N_0\ h\nu\ \gamma \qquad \text{(4-level system)} \tag{9.46}$$

$$\eta_{excit} = A\ell\ N_0\ h\nu\ \gamma\ \frac{1-1/(\gamma\tau P_{pump})}{1+\gamma\tau P_{pump}} \qquad \text{(3-level system)} \tag{9.47}$$

In the three-level system, the excitation efficiency is negative for pump rates W of less than $1/\tau$, or $\gamma\tau P_{pump}<1$, because the density of the lower laser level exceeds the upper level density.

In the following we will not distinguish between 3-level and 4-level systems. In both cases, we can use the differential equation (9.35) and the maximally available power (9.44) to calculate output power and extraction efficiency. Keep in mind, however, that in 3-level systems the saturation intensity is a function of the pump power and the small-signal gain does not increase proportionally to the pump power.

Examples:

Nd:YAG rod (four-level system), 6mm diameter, 100mm length

N_0 = 1.5×10^{20} cm^{-3} $\quad \tau$= 230 µs $\quad \sigma_0$= 4.1×10^{-19} cm^2 $\quad h\nu_0$=1.87×10^{-19} J

α_0 = 0.003 cm^{-1}

η_{excit} = 6%

γ = 0.757 kW^{-1} s^{-1}

ΔN_0 = 2.61×10^{16} cm^{-3} per kW of pump power

$g_0\ell$ = 0.106 per kW of pump power (lamp pumped)

CO_2 tube (four-level system), 20 mm diameter, 200 mm length

N_0 = 2.6x10^{19} cm^{-3} τ= 10 µs σ_0= 1x10^{-16} cm^2 $h\nu_0$=1.87x10^{-20} J

α_0	= 0.001 cm^{-1}
η_{excit}	= 25%
γ	= 8.2 kW^{-1} s^{-1}
ΔN_0	= 2.13x10^{15} cm^{-3} per kW of pump power
$g_0\ell$	= 4.25 per kW of pump power

Ruby rod (three-level system), 6 mm diameter, 100 mm length

N_0 = 1.6x10^{19} cm^{-3} τ= 3,000 µs σ_0= 2.5x10^{-20} cm^2 $h\nu_0$=2.86x10^{-19} J

α_0	= 0.004 cm^{-1}
γ	= 4 kW^{-1} s^{-1}
P_{electr}	> 83 kW (for positive inversion)
ΔN_0	= 0.8x10^{19} cm^{-3} at 250 kW of pump power
$g_0\ell$	= 2.0 at 250 kW of pump power (lamp pumped)

9.5 Line Broadening and Hole Burning

9.5.1 Homogeneous and Inhomogeneous Line Broadening

The frequency range within which the laser material exhibits a gain factor G greater than 1.0 does not only depend on the physical properties of the active atoms or ions, but is also affected by the interaction of these atoms with one another and with the host material. If the active atoms could be considered as isolated, the relation between the gain bandwidth and the spontaneous emission lifetime τ would read:

$$2\pi\Delta\nu = 1/\tau \tag{9.48}$$

For the laser transition of Neodymium at λ=1.064µm the bandwidth would be 800Hz, and for the 3p-3s transition of Neon a bandwidth of 16MHz would result. This is in contrast to the experimental observation that the bandwidths are 120GHz for Nd:YAG and 500MHz for HeNe lasers. The reason for the discrepancy is the interaction between the atoms. In a laser material, the density of active atoms is on the order of 10^{20} per cubic centimeter, and the atoms cannot be considered as isolated oscillators. Due to the interaction with the surrounding atoms and the movement of the atoms, the bandwidth of the laser transition becomes broadened. Two line broadening mechanisms are distinguished (Fig. 9.7):

- Homogeneous line broadening

All atoms exhibit the same broadening of the line with a width of $\Delta\nu_{hom}$ and the same center frequency ν_0. The lineshape is Lorentzian (9.28) and the relation between the linewidth and the spontaneous decay time τ is given by:

$$2\pi\Delta\nu_{hom} = \frac{1}{\tau} + \frac{2}{\tau^*} \tag{9.49}$$

where τ^* is the decay time due to the interaction between the atoms. Homogeneous line broadening is caused by collisions between the atoms (for collision broadening in gas lasers, τ^* is on the order of 10^{-7}s) and by the interaction of the active atoms with phonons in solid state lasers ($\tau^* \approx 10^{-12}$s). In gas lasers, the homogeneous linewidth is proportional to the gas pressure p (see Table 9.2):

$$\Delta\nu_{hom} = 64\,\frac{\sigma^*\,p}{\sqrt{mkT/\pi}} \tag{9.50}$$

where T is the temperature, k is the Boltzmann constant, σ^* is the collision cross section, and m is the atom/molecule mass. Note that the collision cross section is a function of the temperature. Typical values for collision broadened line widths are on the order of 100 MHz/Torr.

- Inhomogeneous line broadening

The line width is again the same for all atoms, but now different atoms exhibit different center frequencies ν_0. The Lorentzian transition lines of the individual atoms with homogeneous line width $\Delta\nu_{hom}$ sum up to a line whose bandwidth $\Delta\nu_{inhom}$ is referred to as the inhomogeneous linewidth. Inhomogeneous broadening is very common in gas lasers due to the Doppler-effect. The velocity distribution of the atoms gives rise to a shift of the center frequency which depends on the direction and the magnitude of the velocity. The inhomogeneous linewidth due to Doppler broadening is given by:

$$\Delta\nu_{inhom} = \frac{\nu_0}{c}\sqrt{\frac{8kT\ln 2}{m}} \tag{9.51}$$

and the shape of the gain distribution is Gaussian:

$$g_0(\nu) = g_0(\nu_0)\,\exp\left[\frac{-4\ln 2(\nu-\nu_0)^2}{(\Delta\nu_{inhom})^2}\right] \tag{9.52}$$

In solid state lasers, inhomogeneous broadening is caused by inhomogeneities of the host material. In glass, as an amorphous host material, different doping ions experience slightly different crystal fields.

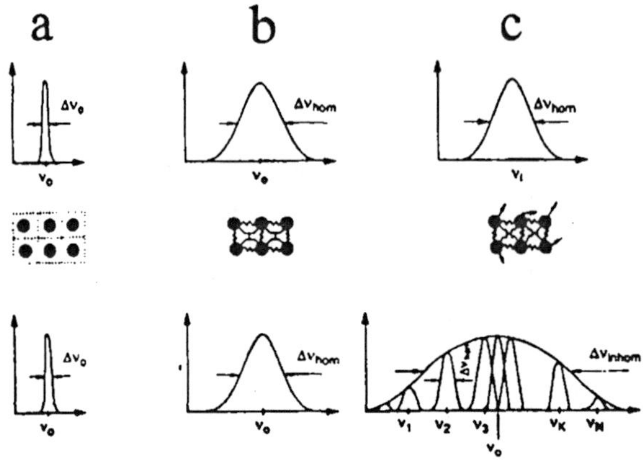

Fig. 9.7 The interaction of the atoms with one another and their movement broadens the line width of the laser transition. The upper row shows the gain coefficient of one single atom, the lower row presents the superposition of the lines of N_0 atoms. a) isolated atoms, b) homogeneously broadened, c) inhomogeneously broadened.

Table 9.2 Homogeneous and inhomogeneous linewidths of different lasers. The first value of the homogeneous linewidth represents the linewidth of the isolated atom/ion/molecule. For gas lasers the homogeneous line broadening is proportional to the pressure (in Torr). See Eq. (9.7) for the definition of the factor x.

Laser	homogeneous linewidth	inhomogeneous linewidth	factor x
HeNe	100 MHz + 150 MHz/Torr	1,500 MHz	0.5
Argon	460 MHz + 1,000 MHz/Torr	5,000 MHz	0.5
CO_2	16 kHz + 6 MHz/Torr	60 MHz	1*,0.5**
Nd:glass	800 Hz + 1,000-2,000 GHz	2,500-3,500 Ghz	1.0
Nd:YAG	800 Hz + 120 GHz	10 MHz	1.0
Nd:YVO_4	800 Hz + 150 Ghz	10 MHz	1.0
Nd:YLF	800 Hz + 200 Ghz	10 MHz	1.0
Cr:Al_2O_3	53 Hz + 1,000 GHz	5 MHz	1.0
Rhodamin 6G	32 MHz + 1,000 GHz	1,000 GHz	1.0

* for high pressure systems (p > 30 Torr),** for low pressure systems (p < 10 Torr)

How is the light amplification affected if the active medium exhibits an inhomogeneously broadened line? Let us assume that we want to amplify light at a certain frequency with a bandwidth that is small compared to the homogeneous linewidth of the material. A perfect example would be the amplification of an axial mode with frequency ν_q in a laser resonator (Fig. 9.8). Only those atoms will contribute to the amplification process whose gain factor is greater than 1 at this frequency. If N_0 denotes the total density of atoms, a good approximation for the density of participating atoms is:

$$N_0^* = N_0 \frac{\Delta\nu_{hom}}{\Delta\nu_{inhom}} \tag{9.53}$$

As in the homogeneously broadened case, the power transferred into inversion is given by:

$$P_{UL} = A\, I_S\, g_0 \ell \tag{9.54}$$

The axial mode will interact only with the portion (9.53) of the total inversion density. Thus the small-signal gain experienced by the mode reads:

$$g_0^* \ell = g_0 \ell \frac{\Delta\nu_{hom}}{\Delta\nu_{inhom}} \tag{9.55}$$

A theoretical analysis (see e.g. [4.22]) indicates that the width of the spectral hole burned into the gain profile becomes larger as the intensity is increased (saturation broadening):

$$\Delta\nu = \Delta\nu_{hom} \sqrt{1 + \frac{I}{I_S}} \tag{9.56}$$

This means that more atoms will interact with the axial mode. By replacing the homogeneous linewidth in (9.55) with the corrected one we obtain the small-signal gain 'seen' by the mode:

$$g'_0 \ell = g_0^* \ell \sqrt{1 + \frac{I}{I_S}} \tag{9.57}$$

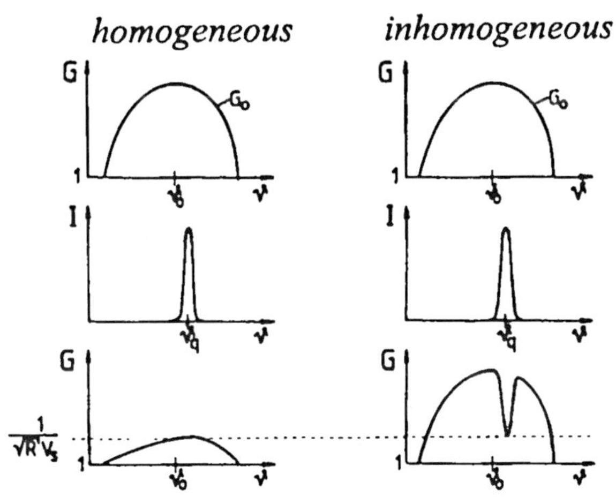

Fig. 9.8 In contrast to a homogeneously broadened laser transition, only the density N_0^* (see (9.53)) of the active atoms will amplify the axial resonator mode. The mode burns a hole into the gain profile and it can extract only a fraction of the total power available in the form of inversion.

Inserting this small-signal gain into the differential equation for the intensity (9.35) results in:

$$\frac{\delta I}{\delta z} = \left[\frac{g_0^{\bullet}}{\sqrt{1 + I/I_S}} - \alpha_0\right] I \tag{9.58}$$

Integration of this equation yields the output power per axial mode. Keep in mind that the axial mode does not have access to the whole inversion density ΔN. If we later wanted to calculate the extraction efficiency of resonators with inhomogeneously broadened media, the following expression for the maximum power available per axial mode would have to be used:

$$P_{UL} = A\, I_S\, g'_0 \ell = A\, I_S\, g_0^{\bullet} \ell \sqrt{1 + I/I_S} \tag{9.59}$$

Unfortunately, the concept of the extraction efficiency is not very useful for inhomogeneously broadened lasers since the output power is related to a maximally available power that depends on the intracavity intensity. The maximum of the output power (optimum output coupling), therefore, does not coincide with a maximum of the extraction efficiency as is the case in homogeneously broadened systems. In the next chapter we, therefore, will optimize the output power and not the extraction efficiency when we deal with inhomogeneously broadened laser resonators.

9.5.2 Spatial Hole Burning

The interference of the back and the forth traveling waves of an axial mode in the laser resonator results in a z-dependent intensity distribution. Only for passive resonators with 100% reflectance of both mirrors is a standing wave generated with a periodical appearance of intensity maxima at a distance of half the wavelength. If the beam is coupled out at one mirror only and the active medium is inserted into the resonator, the two counterpropagating waves exhibit different amplitudes. The intensity in the minima is then not equal to zero. Let us discuss this more quantitatively with the resonator model depicted in Fig. 9.9. The active medium is realized as a thin disk at the high reflecting mirror and the optical resonator length is L. If G denotes the gain factor per transit we get for the sum $E(z,t)$ of the two counterpropagating fields for the axial mode of order q:

$$\begin{aligned} E(z,t) &= E^{+}(z,t) + E^{-}(z,t) \\ &= E_0[G^2\exp[i(kz-\omega t)] + \exp[-i(kz+\omega t+\pi)] \end{aligned} \tag{9.60}$$

with the wave number $k = q\pi/L$, q : *integer*

By using the stationary condition $G\sqrt{R}=1$, the time-averaged intensity distribution can be written as:

$$I(z) = I_0\left[1 - \frac{2R}{1+R^2}\cos(2\pi qz/L)\right] \tag{9.61}$$

where I_0 is a function of the reflectance R. For $R=1$ we get a standing wave and as the output coupling is increased the spatial modulation of the intensity distribution becomes less pronounced (Fig. 9.10). An axial mode of order q exhibits q intensity maxima inside the resonator where the gain becomes more strongly saturated. The axial gain profile will, therefore, show a periodic sequence of holes at a distance of half the wavelength. This effect is referred to as *spatial hole burning*. The effect of the spatial hole burning on the laser performance is different for homogeneously and inhomogeneously broadened lasers. For homogeneous line broadening all axial modes are amplified by the same active atoms. If there were no spatial hole burning only one axial mode (the one with the highest small-signal gain) would oscillate (as was depicted in Fig. 9.8). However, since the location of the intensity minima is different for different mode orders q, several axial modes will share the gain and thus oscillate simultaneously.

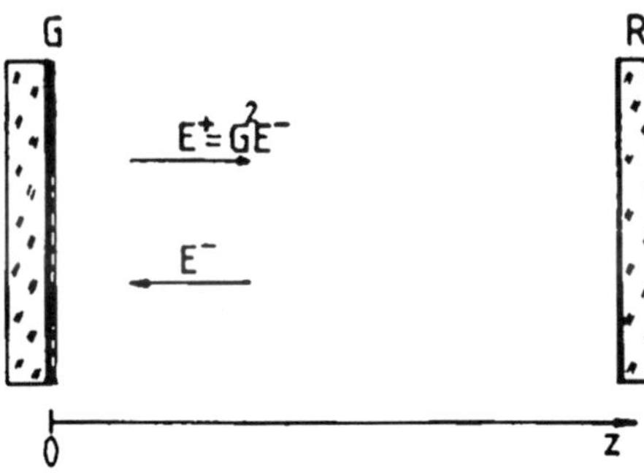

Fig. 9.9 Resonator model for the calculation of the axial intensity distribution generated by a single axial mode.

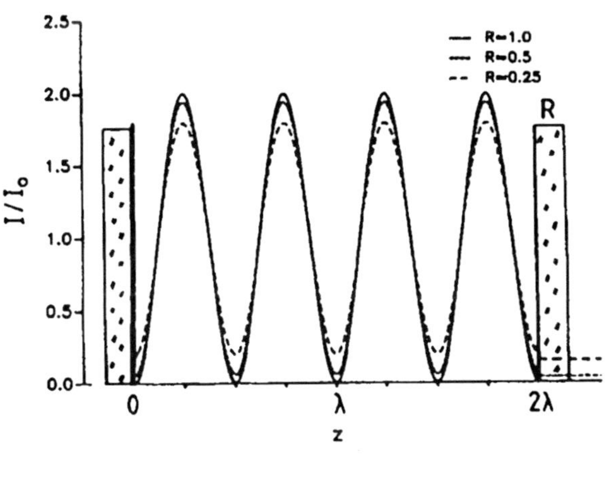

Fig. 9.10 The time-averaged intensity distribution in the resonator of Fig. 9.9 for different reflectances R (axial mode order $q=4$).

The competition between the modes leads to axial mode hopping and to fluctuations of the output power. This problem can be solved by using special resonator schemes which prevent the interference between the forth and the back travelling waves (see the twisted mode resonator in Sec. 8.2) or restrict the wave propagation to one direction (unidirectional ring resonator, see Chapter 21). If the gain profile is inhomogeneously broadened, the axial modes interact with different atoms. Each axial mode will generate its own spatial gain profile without affecting the performance of the other modes. This is why hole burning does not generate intensity fluctuations in inhomogeneously broadened lasers. The lack of the axial mode competition leads to a stable axial mode spectrum.

9.6 Spectral Gain Distribution and Frequency Pulling

So far we have only discussed the laser performance at the center of the transition line ($\nu=\nu_0$) where the small-signal gain is $g_0\ell$. We know already that the laser emission is possible within a linewidth $\Delta\nu$ and the small-signal gain coefficient is a function of the light frequency. For homogeneous broadening the line shape is Lorentzian:

$$g_0(\nu) = \Delta N\sigma_0 \frac{(\Delta\nu/2)^2}{(\nu-\nu_0)^2 + (\Delta\nu/2)^2} = \Delta N\sigma(\nu) \tag{9.62}$$

with ΔN : total inversion density

All axial modes whose frequency fall under this gain curve and have a sufficiently high small-signal gain to reach the laser threshold (9.11) might oscillate (Fig. 9.11). How do we calculate the laser intensity I in this case? Again, we have to distinguish between homogeneous and inhomogeneous line broadening. In a homogeneously broadened laser, only the axial mode at the center frequency will reach the laser threshold since it is capable of saturating the gain at all frequencies. In this case the small-signal gain coefficient at $\nu=\nu_0$ has to be used to calculate the light amplification according to (9.35). For an inhomogeneously broadened system the lack of any axial mode competition results in the simultaneous oscillation of different axial modes at the frequencies ν_q. For each mode, the small- signal gain at its specific resonance frequency has to be determined. Considering that only the fraction (9.53) of all atoms participate in the amplification process, the small-signal gain coefficient to be inserted into (9.58) is given by (for Doppler broadening):

$$g_0^{\bullet} = \frac{\Delta\nu_{hom}}{\Delta\nu_{inhom}} \Delta N\sigma_0 \exp\left[\frac{-4\ln 2(\nu_q-\nu_0)^2}{(\Delta\nu_{inhom})^2}\right] \tag{9.63}$$

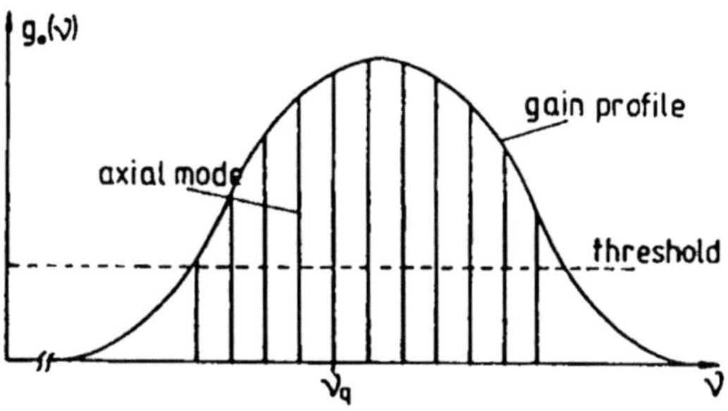

Fig. 9.11 The spectral profile of the gain coefficient of a laser transition and the axial mode frequencies of the optical resonator.

In addition, the frequency dependence of the cross section σ in (9.62) results in a change of the saturation intensity which has to be taken into account:

$$I_S = \frac{h\nu_q}{\sigma(\nu_q)\tau} \tag{9.64}$$

The solution of (9.58) will then yield the intracavity intensity for the axial mode at the frequency ν_q. The total intracavity intensity is obtained by adding the intensities of the individual modes.

The spectral distribution of the real part χ_r of the atomic susceptibility gives rise to a phase shift of the transmitted field if the frequency ν does not coincide with the center frequency ν_0 of the atomic transition. According to (9.26) and (9.27) this phase shift is given by:

$$\Delta\phi = \frac{g_0\ell}{2}\,\frac{\nu-\nu_0}{\Delta\nu}\,\frac{(\Delta\nu/2)^2}{(\nu-\nu_0)^2+(\Delta\nu/2)^2} \tag{9.65}$$

In a laser resonator the additional gain-dependent phase shift will change the axial mode frequencies. Without gain, the axial mode frequencies ν_q are determined by the condition that multiples of half the wavelength fit into the resonator:

$$2k_0L_{opt} = 2\pi\, q \qquad \text{or} \quad \nu_q = q\,\frac{c_0}{2L_{opt}} \tag{9.66}$$

with
k_0, c_0 : wave number and speed of light in vacuum
L_{opt} $= L_0+(n-1)\ell$: optical resonator length
q : axial mode order

When a population inversion is generated in the active medium, the phase shift (9.5) has to be included into the resonance condition (9.66):

$$2k_0 L_{opt} + 2\Delta\phi = 2\pi\, q \tag{9.67}$$

Given the assumption that the additional phase shift is small, we can rewrite this equation to get an expression for the new axial mode frequencies $\nu_q^\bullet$:

$$\nu_q^\bullet = \nu_q \left[1 - \frac{g_0 \ell}{2\pi} \frac{\nu_q^\bullet - \nu_0}{\Delta\nu} \frac{(\Delta\nu/2)^2}{(\nu_q^\bullet - \nu_0)^2 + (\Delta\nu/2)^2} \right] \tag{9.68}$$

It is difficult to solve this equation analytically. However, in a first order approximation we can assume that the shift in the resonance frequency is small and we can replace $\nu_q^\bullet$ by ν_q on the right hand side. If we further assume that the axial mode with order p exhibits its resonance frequency at the center frequency ν_0 and that the gain bandwidth is m times larger than the axial frequency spacing (Fig. 9.12), we obtain a more convenient form of (9.68):

$$\delta\nu = \nu_q^\bullet - \nu_q = \nu_q \frac{g_0 \ell}{2\pi} \frac{p-q}{m} \frac{1}{1 + [2(p-q)/m]^2} \tag{9.69}$$

The frequency shift $\delta\nu$ is zero at the center frequency (q=p), positive for all the axial modes with frequencies lower than the center frequency (q<p), and negative for q>p. The axial mode frequencies thus are pulled towards the line center. This frequency shift is small compared to the axial mode frequency spacing as the following two examples show.

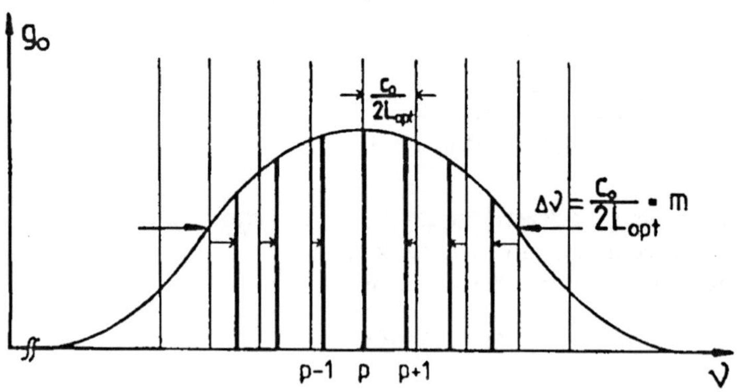

Fig. 9.12 The spectral gain distribution gives rise to a shift of the axial mode frequencies. The axial modes are 'pulled' towards the line center.

Examples:

1) HeNe laser: $\Delta\nu$=750MHz, L_{opt}=1m, $g_0\ell$=0.05
With the axial frequency spacing of the passive resonator of 150MHz we get m=5. The frequency shift of the first axial modes to the left and right of the center frequency (|p-q|=1) is 115kHz. This represents a relative shift of 7.6×10^{-4} with respect to the frequency spacing.

2) Nd:YAG laser: $\Delta\nu$=120 GHz, L_{opt}=1m, $g_0\ell$=2.0
The gain bandwidth is 800 times larger than the frequency spacing of the axial modes of 150MHz. The gain induced frequency shift is 60 kHz for |p-q|=1 and 5.62MHz for |p-q|=100.

9.7 The Spectral Linewidth of Laser Modes

In Sec. 4.2.3 we have discussed the bandwidth $\delta\nu$ of the axial modes in an active Fabry Perot resonator. Applied to the resonator model used in this chapter this bandwidth reads:

$$\delta\nu = \frac{c_0}{2\pi L_{opt}} |\ln(G\sqrt{R}V_S)| \tag{9.70}$$

In a laser resonator the gain factor G becomes saturated so that $G\sqrt{R}V_S=1$ holds. If (9.70) were the correct equation, the bandwidth would then be equal to zero. This is to be expected since steady state laser operation implies that the field amplitude does not decay and we know that an infinite decay time is equivalent to a zero bandwidth. A vanishing linewidth is, of course, a result which does not agree with experimental observations. As long as no special care is taken to stabilize the resonator set-up, small-scale fluctuations of the resonator geometry (mirror jitters) and variations induced by the pump process will result in typical bandwidths on the order of MHz. However, even with sophisticated stabilization techniques laser radiation can never be made perfectly monochromatic. The reason for this lower bound of the bandwidth is the spontaneous emission. The spontaneously emitted photons are not added coherently to the photons generated via stimulated emission, resulting in small field amplitude fluctuations of the laser mode. This noise sets the lower limit to the bandwidth of the laser mode. This lower band width limit was derived by Schawlow and Townes [4.4]. If P_{out} is the output power in a single axial mode with photon energy $h\nu$, the bandwidth is given by:

$$\delta\nu = \pi \frac{h\nu}{P_{out}} (\delta\nu_0)^2 \frac{N_2}{N_{th}} \tag{9.71}$$

with: $\delta\nu_0$: bandwidth of the passive resonator (Eq. (9.70) with $G=1$)
N_2 : population density in the upper laser level
N_{th} $= N_2 - N_1 = |\ln(\sqrt{R}V_S)|/(\sigma_0 \ell)$: threshold inversion density

In an ideal four-level system, the population density N_1 of the lower laser level is always zero. Equation (9.71) then can be simplified by setting N_2 equal to N_{th}, because for a laser in steady-state cw emission, the inversion density is clamped at the value given by the threshold inversion density. The examples presented below indicate that lasers can in principle produce linewidths in the sub-Hertz range. Due to practical problems (environmental fluctuations) the bandwidths that observed in laboratory systems usually exceed the Schawlow-Townes limit by several orders of magnitude. However, bandwidths in the mHz range have been reported for highly stabilized diode pumped solid state lasers [4.23,5.272] (see also Chapter 22).

Examples:

GaAs diode laser (λ=800nm)
$h\nu=2.49\times10^{-19}$ J, L_{opt}=500µm, P_{out}=10mW, R=0.25, V_S=0.99
-------> **$\delta\nu$ = 353 kHz**

HeNe laser (λ=632.8nm)
$h\nu=3.14\times10^{-19}$ J, L_{opt}=0.3m, P_{out}=1mW, R=0.98, V_S=0.99
-------> **$\delta\nu$ = 0.01 Hz**

Nd:YAG laser (λ=1.064µm)
$h\nu=1.87\times10^{-19}$ J, L_{opt}=1.0m, P_{out}=0.5W, R=0.5, V_S=0.95
-------> **$\delta\nu = 4.2\times10^{-4}$ Hz**

CO_2 laser (λ=10.6µm)
$h\nu=1.87\times10^{-20}$ J, L_{opt}=2.0m, P_{out}=50W, R=0.3, V_S=0.98
-------> **$\delta\nu = 2.6\times10^{-7}$ Hz**

Chapter 10

Output Power of Laser Resonators

10.1 Output Power of Stable Resonators

10.1.1 Linear Resonators

The differential equations (9.35) and (9.56) describing the amplification of the intensity inside the active medium are now used to derive the output power of stable resonators. In the resonator model used (Fig. 10.1) it is assumed that both the forward travelling beam with intensity $I^+(z)$ and the backward traveling beam with intensity $I^-(z)$ cover the same area of the active medium. The complete overlap of the two counterpropagating beams is characteristic for stable resonators. During a round trip the intensity is decreased due to diffraction losses (loss factors V_1-V_4), scattering, and absorption inside the medium (loss factor V_S), and by output coupling. In steady state operation, these losses are compensated by the amplification process characterized by the small-signal gain coefficient g_0. The next assumption we make is that no spatial hole burning is present meaning that at any plane inside the medium the intensity $I(z)$ is given by the sum of the two intensities $I^+(z)$ and $I^-(z)$. This is a reasonable approach for most lasers since the effect of spatial hole burning on the output power is smoothed out by atomic motion (gas lasers), energy migration, or axial multimode operation (solid state lasers). Furthermore, the mode is assumed to exhibit a flat-top intensity profile. The incorporation of the real mode structure will be discussed in Chapter 11.

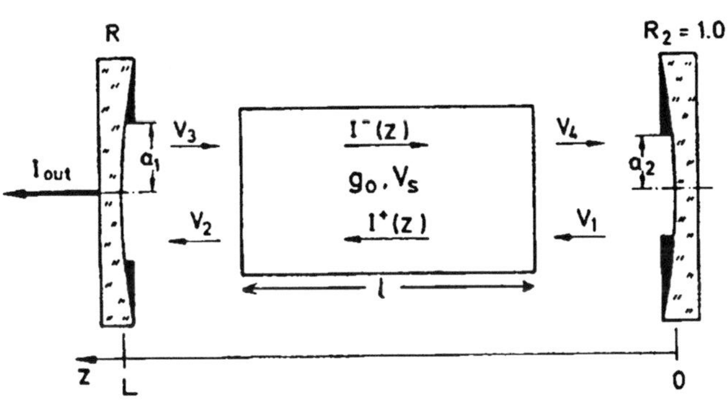

Fig. 10.1 Resonator model used for the calculation of the output power. The intensity is decreased by the diffraction loss factors V_1-V_4 (=1-loss), the loss factor per transit V_S of the medium, and the reflectance R of the output coupling mirror.

However, in single axial mode solid state or liquid dye lasers, spatial hole burning has an impact on the output power (it is lower by up to 30%). In these cases a realistic model for the output power requires the incorporation of the interference between the two counterpropagating fields resulting in the intensity profile:

$$I(z) = I^{+}(z) + I^{-}(z) - 2\sqrt{I^{+}(z)I^{-}(z)}\,\cos[2kz] \tag{10.1}$$

with k being the wave number inside the medium. Without spatial hole burning, the differential equations for the intensities according to (9.35) and (9.56) read [4.1,4.3,4.11]:

$$\frac{dI^{\pm}(z)}{dz} = \pm\left[\frac{g_0}{\left(1 + \dfrac{I^{+}(z) + I^{-}(z)}{I_S}\right)^{X}} - \alpha_0\right] I^{\pm}(z) \tag{10.2}$$

where g_0 is the small-signal gain coefficient, I_S is the saturation intensity, α_0 is the loss coefficient, and X is equal to 1.0 (0.5) for homogeneous (inhomogeneous) broadening.

Before we solve these equations to obtain the intensity $I^{+}(L)$, which is proportional to the output power P_{out}, let us derive an approximate solution by assuming that the intensity sum in (10.2) is constant with $I^{+}(z)+I^{-}(z)=2I$. The mean intensity I now represents the intensity of the beam traveling towards the output coupling mirror. Equation (10.2) then yields for the factor GV_S by which the intensity is amplified in a transit through the medium:

$$GV_S = \exp\left[\frac{g_0\ell}{(1 + 2I/I_S)^{X}}\right]\exp\left[-\alpha_0\ell\right] \tag{10.3}$$

By using the steady state condition:

$$G^2RV_S^2V_1V_2V_3V_4 = 1 \tag{10.4}$$

the intensity I is found to be:

$$I = \frac{I_S}{2}\left[\left(\frac{g_0\ell}{|\ln\sqrt{RV_S^2V_1V_2V_3V_4}|}\right)^{\frac{1}{X}} - 1\right] \tag{10.5}$$

If the cross sectional area of the beam is A_b ($=\pi a_i^2$ in Fig. 10.1), the output power is given by:

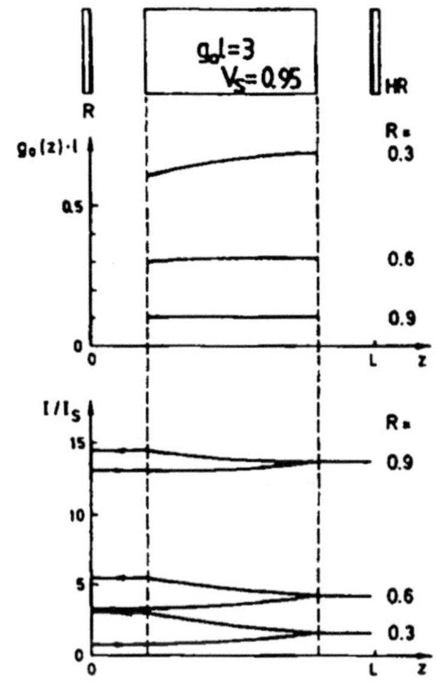

Fig. 10.2 Gain lnG and intensities as a function of the axial coordinate z for different reflectances R of the output coupler, calculated with (10.2) ($g_0\ell=1.5$, $\alpha_0\ell=0.053$, no diffraction losses).

$$P_{out} = A_b\, I\, (1-R)V_2 = A_b\, I_S \frac{(1-R)V_2}{2\left(\left|\ln\sqrt{RV_S^2V}\right|\right)^{\frac{1}{X}}} \left[(g_0\ell)^{\frac{1}{X}} - \left(\left|\ln\sqrt{RV_S^2V}\right|\right)^{\frac{1}{X}}\right] \tag{10.6}$$

with the round trip diffraction loss factor $V=V_1V_2V_3V_4$. This expression for the output power of stable resonators can be used to a very good approximation, if the output coupling is low (high reflectance, $R>0.7$). In this case the sum of the two intensities inside the medium is almost constant as a numerically calculated example in Fig. 10.2 indicates. For lower reflectances, the z-dependence of the intensities has to be taken into account by solving the differential equation (10.2) using the boundary conditions at the mirrors. Unfortunately, the solution can only be found numerically due to the homogeneously distributed loss (loss coefficient α_0) inside the medium [4.11-4.13]. However, for homogeneously broadened lasers ($x=1$) it is possible to derive an analytical solution by setting α_0 equal to zero in (10.2) and taking into account the loss by multiplying the intensity at the high reflecting mirror with the loss factor $V_S^2=exp[-2\alpha_0\ell]$. This method provides an analytical expression for the output power that is very close to the numerical solution (the difference is less than 0.5%). After a lengthy calculation the final expression reads [4.18]:

$$P_{out} = A_b\, I_S \frac{(1-R)V_2}{1 - RV_2V_3 + \sqrt{RV}\,(1/(V_1V_2V_S) - V_S)} \left[g_0\ell - \left|\ln\sqrt{RV_S^2V}\right|\right] \tag{10.7}$$

with: $V = V_1 V_2 V_3 V_4$.

A comparison with (10.6) indicates that the z-dependence of the gain leads to a change in the fraction. For high reflectances R both equations provide similar output powers. In the following we use a simplified version of (10.7) by ignoring the diffraction losses. We thus consider a resonator that exhibits a loss factor per transit V_S. Equation (10.7) then reads:

$$P_{out} = A_b I_S \frac{1-R}{1 - R + \sqrt{R}\,(1/V_S - V_S)} \left[g_0\ell - |\ln\sqrt{RV_S^2}|\right] \tag{10.8}$$

Figure 10.3 presents the normalized output power $P_{out}/(A_b I_S)$ as a function of the small-signal gain $g_0\ell$ which for most laser materials is proportional to the pump power. Starting at the threshold small-signal gain $(g_0\ell)_{th} = |\ln(\sqrt{R}V_S)|$ the power increases linearly with the small-signal gain and the slope of the curve becomes steeper as the reflectance R of the output coupler is decreased. The output power cannot exceed the power P_{OL} that is available in the medium in the form of inversion, with (Sec. 9.3):

$$P_{OL} = A\, I_S\, g_0\ell = \eta_{excit}\, P_{pump} \tag{10.9}$$

with:

A	:	cross sectional area of the medium
η_{excit}	:	excitation efficiency
P_{pump}	:	pump power

Only if the laser resonator exhibits no losses (V_S=*1.0* and R-->*1.0* in (10.8)) and the laser beam fills the whole medium (A_b=A) can all the available power P_{OL} be extracted from the active medium (broken line in Fig. 10.3). Note that (10.9) holds only for homogeneously broadened lasers. For inhomogeneous line broadening, the available power is a function of the laser intensity due to the broadening of the homogeneous line width (see Sec. 9.5).

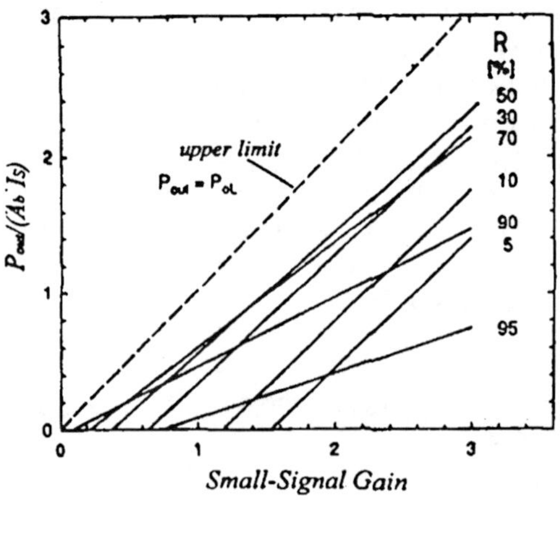

Fig. 10.3 Normalized laser power as a function of the small-signal gain for different reflectances of the output coupler according to (10.8) assuming a loss per transit of 5% (V_S=0.95).

The slope of the output power with respect to the electrical pump power is referred to as the slope efficiency η_{slope} with:

$$\eta_{slope} = \frac{dP_{out}}{dP_{pump}} = \frac{\eta_{excit}}{A\ I_S}\frac{dP_{out}}{d(g_0\ell)}$$

$$= \eta_{excit}\ \gamma\ \frac{1 - R}{1 - R + \sqrt{R}(1/V_S - V_S)} \tag{10.10}$$

where $\gamma = A_b/A$ is the fill factor. The slope efficiency is a function of the loss, the output coupling, and the ratio of the mode volume to the volume of the active medium. This ratio, called the fill factor, can be calculated in a first approach by using the beam diameter of the highest order mode inside the active medium. If the active medium represents the only aperture inside the stable resonator and a high enough number of transverse modes are able to oscillate (let us say more than 10), the maximum fill factor of 1.0 is obtained. In fundamental mode operation with the Gaussian beam radius w being adapted to the radius b of the active medium ($b \approx 1.3w$), the fill factor typically is around 0.9.

Unless the resonator exhibits no losses, only a fraction of the maximally available power P_{OL} can be extracted in the form of a laser beam. This fraction is referred to as the extraction efficiency η_{extr} of the resonator:

$$\eta_{extr} = \frac{P_{out}}{P_{OL}} = \frac{1}{\eta_{excit}}\frac{P_{out}}{P_{pump}} \tag{10.11}$$

The total efficiency of the laser resonator is given by:

$$\eta_{tot} = \frac{P_{out}}{P_{electr}} = \eta_{excit}\ \eta_{extr} \tag{10.12}$$

Example: cw Nd:YAG laser ($I_S = 2kW/cm^2$), rod diameter: 10mm ($A = 0.785cm^2$), multimode operation ($\gamma = 1.0$), electrical pump power : 10 kW. The loss factor per transit and the small-signal gain were determined experimentally (see Chapter 23) to be $V_S = 0.95$ and $g_0\ell = 0.04$ per kW of pump power. According to (10.9), this corresponds to an excitation efficiency of 6.28%. Using (10.8) the following output powers, extraction efficiencies, and total efficiencies are to be expected:

R	0.6	0.7	0.8	0.9	0.95
P_{out} [W]	122	208	255	235	169
η_{extr} [%]	19.5	33.1	40.6	37.5	26.9
η_{tot} [%]	1.22	2.08	2.55	2.35	1.69

Optimum Output Coupling and Maximum Output Power

If the reflectance of the output coupler is varied at a given small-signal gain or pump power, the output power exhibits a maximum at the optimum output coupling. This behavior is easy to understand considering the fact that the output power is zero at low reflectances (laser threshold is not reached) and at a reflectance of 100% (no power is coupled out of the resonator). Thus a maximum of the output power and the extraction efficiency must exist for a certain value of the output coupling. Going back to our approximate power formula (10.6) we can find the optimum reflectance R_{opt} and the maximum output power $P_{out,max}$ by setting the derivative $\delta P_{out}/\delta(\ln R)$ equal to zero. Again, we assume that the losses per transit are represented by the loss factor $V_S = \exp[-\alpha_0 \ell]$, and we consequently set the diffraction loss factor V equal to 1.0. By using the approximation $1-R \approx |\ln R|$, the following expressions are obtained [4.15]:

a) homogeneous line broadening:

Maximum output power:

$$P_{out,max} = A_b\, I_S\, \alpha_0\ell \left[\sqrt{\frac{g_0\ell}{\alpha_0\ell}} - 1\right]^2 \qquad (10.13)$$

$$= \left[\sqrt{\eta_{excit}\, P_{electr}\, A_b/A} - \sqrt{A_b\, I_S\, \alpha_0\ell}\right]^2$$

Maximum extraction efficiency:

$$\eta_{extr,max} = \frac{\alpha_0\ell}{g_0\ell}\left[\sqrt{\frac{g_0\ell}{\alpha_0\ell}} - 1\right]^2 \qquad (10.14)$$

Optimum output coupling:

$$\ln R_{opt} = -2\alpha_0\ell \left[\sqrt{\frac{g_0\ell}{\alpha_0\ell}} - 1\right] \qquad (10.15)$$

Although (10.6) is only an approximate expression for the output power, the extreme values given by (10.13)-(10.15) can, to a very good approximation, be used to optimize the power performance of a laser system. This can be easily verified in Fig. 10.4 in which the correct extreme values (found by solving numerically the differential equation (10.2)) are shown [4.13]. Even for lasers with high small-signal gain ($g_0\ell > 3$) and high loss ($\alpha_0\ell > 0.1$), which require a low reflectance for optimum performance, the difference between the exact extreme values and the ones given by (10.14) and (10.15) is negligible. Note that the graph presented in Fig. 10.4 assumes a fill factor of $\gamma = A_b/A = 1.0$. The optimum parameters for a given laser are represented by the intersecting point of the two curves with constant small-signal gain and constant loss. Again, we find that an extraction efficiciency of 100% is only attainable if the laser exhibits no loss ($\alpha_0\ell = 0$) and the output coupling is close to zero. As the loss is increased, the extraction efficiency decreases rapidly and the optimum mirror reflectance is shifted to lower values (Fig. 10.5). Figure 10.5 clearly indicates how sensitively the output power reacts to slight increases of the losses. A loss of 5% per transit in a low-gain laser with $g_0\ell = 0.4$ results in a reduction of the extraction efficiency by a factor of 2! Half the available power P_{OL} gets lost through spontaneous emission.

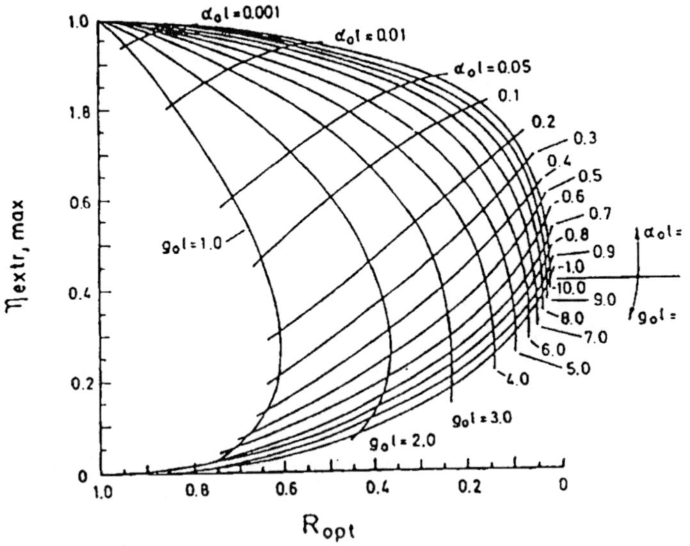

Fig. 10.4 Diagram for the determination of the optimum reflectance R_{opt} and the maximum extraction efficiency $\eta_{extr,max}$ of homogeneously broadened lasers using the small-signal gain $g_0\ell$ and the loss per transit $\alpha_0\ell = -ln V_S$ (fill factor $\gamma = 1.0$) [4.13].

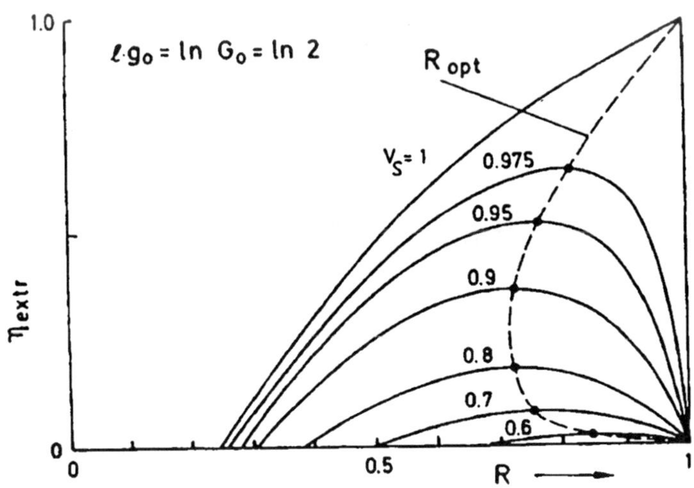

Fig. 10.5 Extraction efficiency as a function of the output coupling reflectance for a small-signal gain of $g_0\ell = ln2$ and different losses per transit. The curve parameter is the loss factor per transit V_S.

The losses have a less dramatic impact on the output power if the small-signal gain of the laser is high. This is why at the same average pump power, lasers provide a higher average output power in pulsed operation than in cw operation. For a repetition rate f and a pulse width Δt, the small-signal gain is higher by the factor $1/(f\Delta t)$ as compared to cw operation at the same average pump power. According to Fig. 10.4, the extraction efficiency is increased leading to a higher average output power. Futhermore, the extraction efficiency becomes less sensitive to changes in output coupling as Fig. 10.6 indicates. The realization of the optimum output coupling is thus much less critical than in a low-gain laser. Figures 10.7 and 10.8 present experimental examples for output power and output energies as a function of the output coupling. In Fig. 10.8 the theoretical curve according to (10.7) is shown too. This example indicates that the expression (10.7) can be used to calculate the output power of a stable resonator if the small-signal gain, the losses, and the fill factor of the resonator are known. The experimental determination of the gain and the loss is easy to accomplish as will be discussed in Chapter 23. The determination of the fill factor, however, is more difficult if only a small number of transverse modes is considered.

In multimode operation the fill factor can, to a good approximation, be calculated by using the cross sectional area of the aperture (see Fig 10.8). If the aperture with radius a is located close to the active medium with radius b, the fill factor is given by:

$$\gamma = \frac{\pi a^2}{\pi b^2} \tag{10.16}$$

For TEM_{00} mode lasers that have the active medium as the limiting aperture, the fill factor is typically around 0.50, meaning that 50% of the multimode power can be extracted in fundamental mode operation. For diode end-pumped TEM_{00} solid state lasers, the fill factor is typically between 0.6 and 0.75, depending on the ratio mode dimaeter to pump spot diameter (typically 0.7-0.9) and aberrations of the thermal lens.

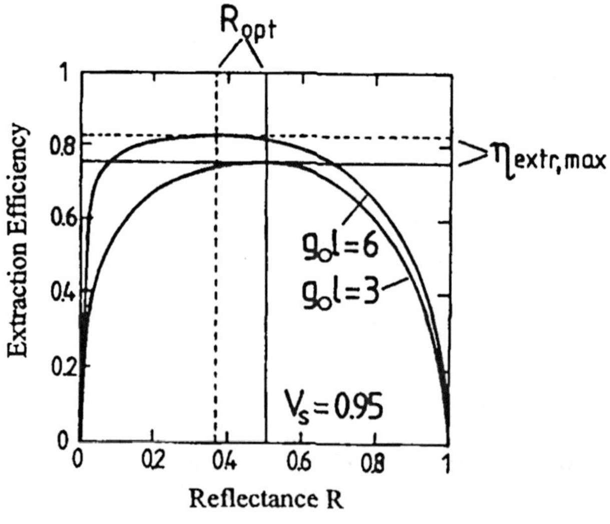

Fig. 10.6 Numerically calculated extraction efficiency as a function of the mirror reflectance R for high small-signal gains. The horizontal and vertical lines mark the analytical values given by (10.14) and (10.15). Loss factor per transit V_S=0.95.

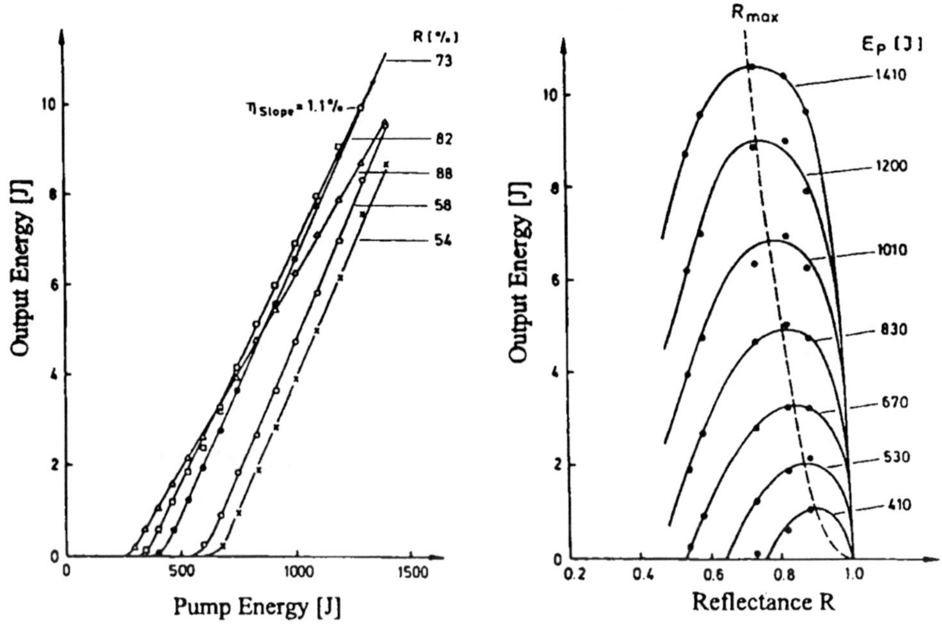

Fig. 10.7 Measured output energy per pulse of Nd:YAG lasers as a function of the pump energy E_p and the output coupling reflectance R [S.11]. The lines represent interpolations.

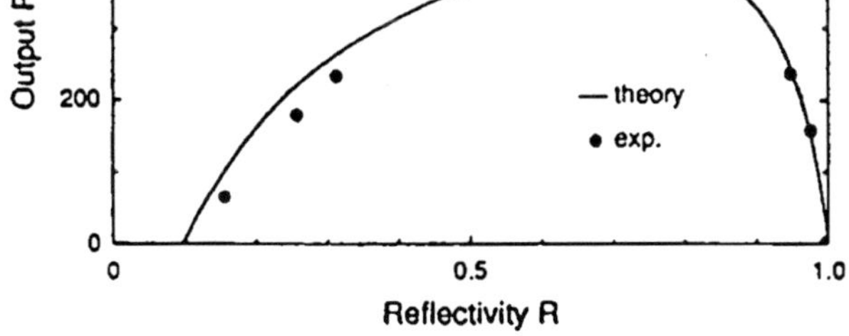

Fig. 10.8 Measured output power of a Nd:YAG laser as a function of the output coupling reflectance. The solid line represents (10.7) with $V_S=0.955$, $V_1=V_4=0.995$, $V_2=V_3=1.0$ ($\rho_1=\infty$, $\rho_2=5m$, $L=1m$, $g_0\ell=1.0$, aperture radius $a=2.8mm$, rod radius $b=3.17mm$).

b) inhomogeneous line broadening:

Maximum output power: $$P_{out,max} = A_b\, I_S\, \frac{|\ln R|^2}{\alpha_0\ell + \ln\sqrt{R}} \tag{10.17}$$

Optimum output coupling: $$[\alpha_0\ell - \ln\sqrt{R_{opt}}]^3 = (g_0\ell)^2[\alpha_0\ell + \ln\sqrt{R_{opt}}] \tag{10.18}$$

Note that the output power (10.17) is always positive since (10.18) implies that $\alpha_0\ell+\ln\sqrt{R}$ is greater than zero. The output power refers to the power of one single axial mode, and the small-signal gain is generated only by those inverted atoms whose resonance frequencies lie within the homogeneous linewidth around the axial mode frequency. In order to obtain the total output power the contributions of the other axial modes have to be added. Also keep in mind that we based our optimization on (10.6) which is valid for high mirror reflectances only.

10.1.2 Folded Resonators without Beam Overlap

For active media with a large cross section it is sometimes advantageous to fold the resonator by means of high reflecting mirrors or roof prisms as depicted in Fig. 10.9. Folding the resonator N times will decrease the beam diameter by a factor $1/(N+1)$ resulting in an enhancement of the beam quality without decreasing the fill factor. Both the small-signal gain and the loss are now higher by a factor of $N+1$. By using (10.14) and (10.15) for homogeneously broadened lasers and assuming a fill factor of 1.0, we get for the maximum extraction efficiency and the optimum mirror reflectance:

$$\eta^{(N)}_{extr,max} = \frac{\alpha_0\ell}{g_0\ell}\left[\sqrt{\frac{g_0\ell}{\alpha_0\ell}} - 1\right]^2 = \eta^{(0)}_{extr,max} \tag{10.19}$$

$$\ln R^{(N)}_{opt} = -2(N+1)\alpha_0\ell\left[\sqrt{\frac{g_0\ell}{\alpha_0\ell}} - 1\right] \tag{10.20}$$

The extraction efficiency does not depend on the number N of folds, and the optimum mirror reflectance is given by the (N+1)th power of the optimum mirror reflectance for the unfolded resonator:

$$R^{(N)}_{opt} = \left[R^{(0)}_{opt}\right]^{N+1} \tag{10.21}$$

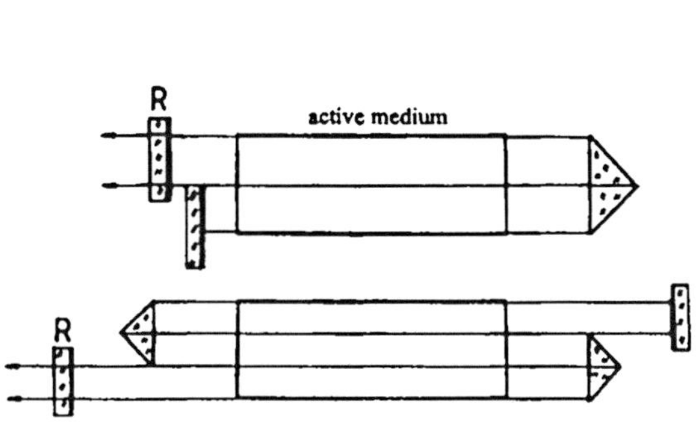

Fig. 10.9 Folded resonators with $N=1$ (top) and $N=2$.

10.1.3 Folded Resonators with Beam Overlap

If the folding is realized such that the different beams are not parallel, they will overlap in certain areas of the active medium. A typical example of such a folding scheme is the tightly folded resonator (TFR) shown in Fig. 10.10 and the slab laser depicted in Fig. 10.11. In order to calculate the output power we first have to determine the influence of the overlapped areas on the extraction efficiency. If the whole beam area is overlapped, the output power will, of course, be decreased by a factor of 2 since the same gain saturation is present for half the laser intensity. As the ratio of the overlapped beam volume to the total beam volume is increased, the normalized extraction efficiency, therefore, will decrease from 100% to 50%. It can be shown numerically that the output power decreases nearly linearly with the ratio of the overlapped mode volume V_{ol} to the total mode volume V_{tot} (Fig. 10.12) [4.19-4.21]. The output power will thus be lower by a factor γ_1 with:

$$\gamma_1 = 1 - 0.5\frac{V_{ol}}{V_{tot}} \tag{10.22}$$

A second correction factor γ_2 takes the increase of the beam cross section from A_b to A_b^*, due to refraction, into account:

$$\gamma_2 = A_b^*/A_b \tag{10.23}$$

Finally, we have to incorporate the longer optical path length ℓ_{zz} (zig-zag) into the expression for the output power. Both the small-signal gain and the loss have to be multiplied by:

$$\gamma_3 = \ell_{zz}/\ell \tag{10.24}$$

with ℓ being the side length of the active medium.

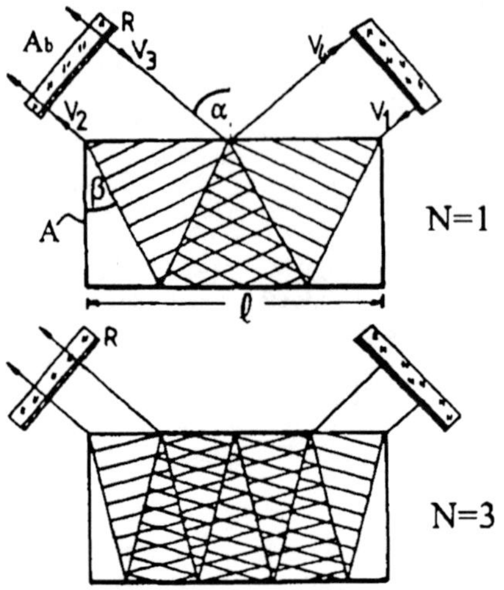

Fig. 10.10 Tightly folded resonators (TFR) with one and three reflections at the bottom surface of the active medium.

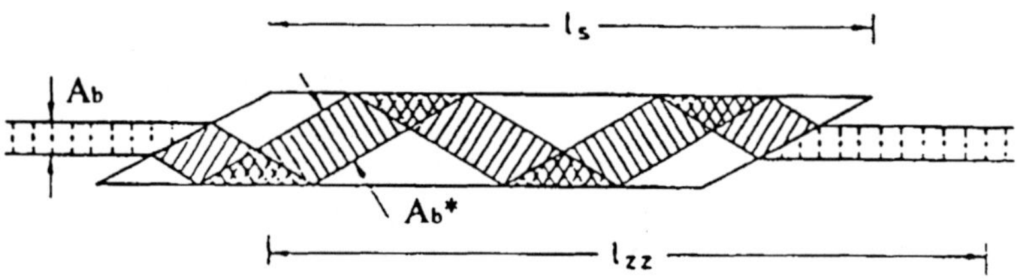

Fig. 10.11 Beam propagation in a solid state slab. The endfaces are cut at the Brewster angle [S.11].

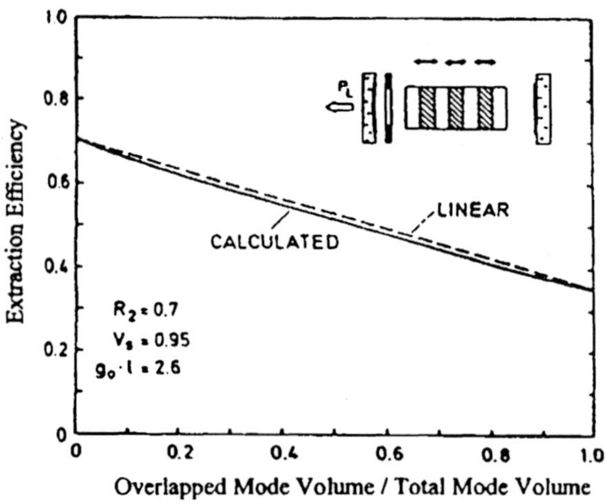

Fig. 10.12 Numerically calculated decrease of the extraction efficiency due to the beam overlap. The hatched areas ma-k the double pass areas. The decrease of the output power is almost linear with the overlapped mode volume [4.20] (© AIP 1988).

Incorporation of these three correction factors into Eq. (10.8) yields the general expression for the output power of folded resonators with a zig-zag path:

$$P_{out} = \gamma_1\,\gamma_2\,A_b\,I_S \frac{1-R}{1-R+\sqrt{R}\,(1/V_S - V_S)} \left[\gamma_3 g_0 \ell - |\ln\sqrt{R V_S^2}|\right]$$

with: $V_S = \exp[-\alpha_0 \ell \gamma_3]$ (12.25)

a) Tightly Folded Resonator

Let us consider a slab with cross sectional area A, length ℓ, and index of refraction n. For complete beam overlap, which means that $(N+1)a = \ell\cos\alpha$ holds as shown in Fig. 10.10, the three correction factors as a function of the number of reflections N at the bottom surface read:

$$\gamma_1 = 1 - \frac{1}{2}\,\frac{2N-1}{2N+1} \tag{10.26}$$

$$\gamma_2 = \frac{A}{A_b}\,\sin\beta \tag{10.27}$$

$$\gamma_3 = \frac{2N}{N+1}\,\frac{1}{\sin\beta} \tag{10.28}$$

with: $\sin\alpha/\sin\beta = n$

The expression for the output power then reads:

$$P_{out} = \left[1-0.5\frac{2N-1}{2N+1}\right]\left[\frac{2N}{N+1}\right] A\,I_S \frac{1-R}{1-R+\sqrt{R}(1/V_S - V_S)} \left[g_0\ell - \alpha_0\ell - \frac{1}{\gamma_3}|\ln\sqrt{R}|\right] \tag{10.29}$$

For a high number of reflections N the product of the two first terms goes to 1.0. We then get the same slope efficiency as compared to a linear resonator whose beam penetrates the medium through the side face with cross section A, but the laser threshold is reduced by the factor $1/\gamma_3$. The lower laser threshold results in a higher extraction efficiency. This is shown in Fig. 10.13 in which extraction efficiencies for different reflection numbers N are compared. For high numbers N the maximum extraction efficiency is higher as compared to the linear resonator.

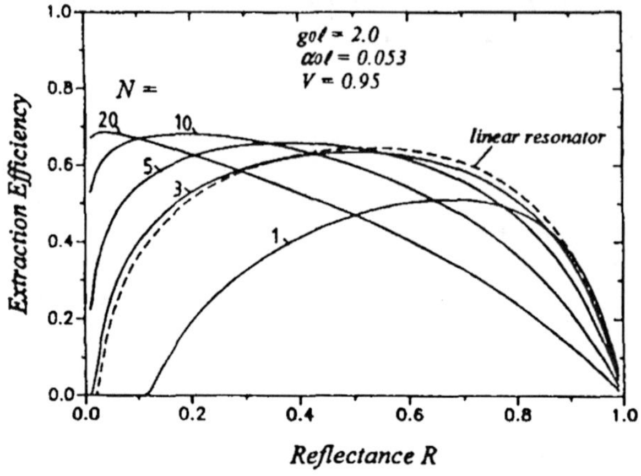

Fig. 10.13 Numerically calculated extraction efficiencies for tightly folded resonators with complete beam overlap as a function of the output coupling reflectance. A round trip diffraction loss factor V of 0.95 was used ($V_1=V_3=\sqrt{0.95}$, $V_2=V_4=1.0$, see Fig. 10.10). The differential equation (10.2) was solved.

b) Slab Resonator

The endfaces of the slab with index of refraction n and side length ℓ are cut at the Brewster angle $\alpha = atan(n)$ (Figs. 10.11 and 10.14). If a_0 denotes the width of the incoming beam and a is the width of the slab, the three correction factors read:

$$\gamma_1 = 1 - \frac{1}{2}\,\frac{1}{(8a/a_0)\cos^2\alpha - 1} \tag{10.30}$$

$$\gamma_2 = \tan\alpha \tag{10.31}$$

$$\gamma_3 = \frac{1}{\sin(2\alpha)} \tag{10.32}$$

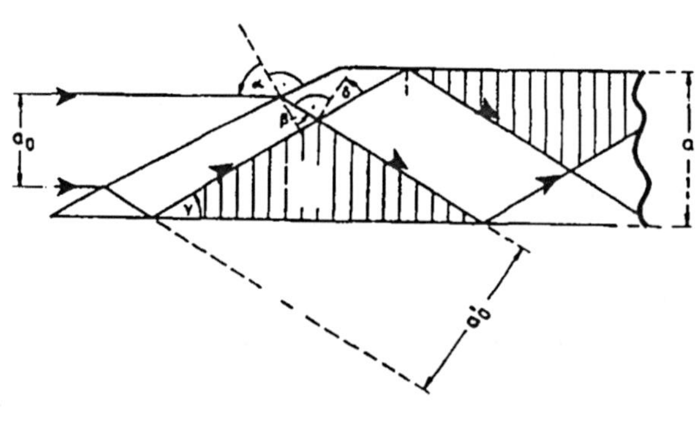

Fig. 10.14 Beam propagation inside a slab active medium. The hatched areas indicate the areas of beam overlap.

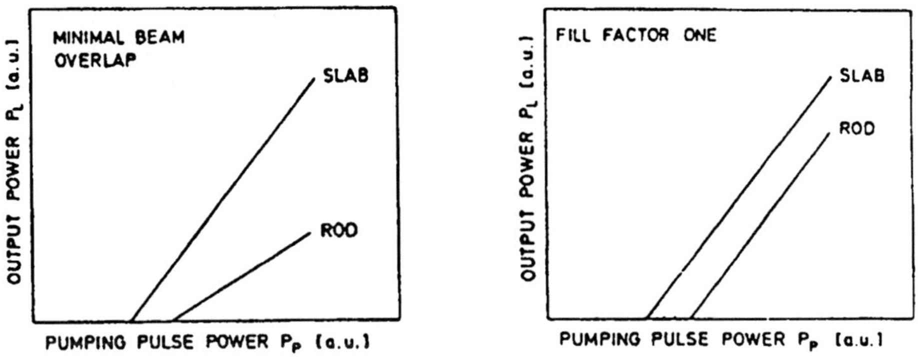

Fig. 10.15 Qualitative comparison of the output power of a Nd:YAG Brewster slab resonator and a linear Nd:YAG rod resonator for minimum and complete beam overlap in the slab (same side length, gain, loss, and cross sectional area of the active medium in both geometries) [4.20] (© AIP 1988).

Insertion into (10.25) yields:

$$P_{out} = \frac{1}{2\cos^2\alpha}\left[1 - \frac{1}{2}\frac{1}{(8a/a_0)\cos^2\alpha - 1}\right] A_b\, I_S\, \frac{1-R}{1-R+\sqrt{R}(1/V_S - V_S)}\left[g_0\ell - \alpha_0\ell - \sin(2\alpha)|\ln\sqrt{R}|\right]$$

with A_b: cross sectional area of the incident beam (10.33)

Again, we get a reduction of the laser threshold due to the longer optical path (factor *sin2α*). If the width of the incident beam is much smaller than the slab width (fundamental mode operation), we get minimal beam overlap resulting in a factor γ_1 close to one. Compared to a linear resonator with active medium with length ℓ (same slab but without zig-zag), the slope is higher by *1/(2cos2α)* due to the increased mode volume inside the slab. For complete beam overlap, the slope is the same as for the linear resonator, but the laser threshold is still reduced by *sin2α*. In any case, the slab configuration provides a higher output power as compared to a linear resonator (Fig. 10.15) [4.19-4.21].

10.1.4 Ring Resonators

Ring resonators are used to prevent spatial hole burning in the active medium. By using an optical isolator inside the resonator, the laser light is forced to travel in one direction only (Fig. 10.16). The missing counterpropagating beam prevents the formation of standing waves. With only one travelling wave, the amplification of the light in the active medium is described by a differential equation that is different from (10.2) for the linear resonator:

$$\frac{dI}{dz} = \left[\frac{g_0}{1 + I/I_S} - \alpha_0\right] I \tag{10.34}$$

Again, we can solve this equation analytically if we concentrate the losses $\alpha_0\ell$ of the medium at one endface of the active medium. Starting at the output coupling mirror with intensity $I(0)$, the intensity $I(L)$ after one round trip must obey the stationary condition:

$$\frac{I(L)}{I(0)} = \frac{1}{V_1 V_2 V_3 V_4 R V_S} \tag{10.35}$$

with $V_S = \exp[-\alpha_0 \ell]$ and V_1-V_4 being diffraction loss factors per transit (=1-loss), as depicted in Fig. 10.16. By using this boundary condition, the solution of (10.34) yields the output power:

$$P_{out} = A_b\ I(L)\ (1 - R) \tag{10.36}$$

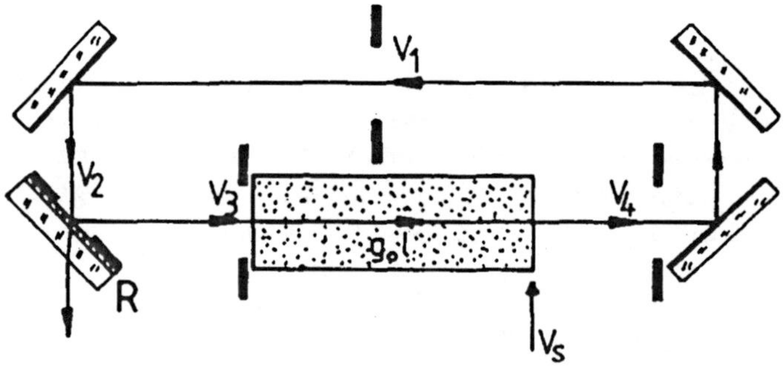

Fig. 10.16 Model of a ring resonator used for the calculation of the output power. The optical isolator generating the unidirectional beam propagation is not shown. Diffraction losses at the apertures are characterized by the loss factors V_1-V_4, and the loss of the medium is taken into account at the end of the medium by the loss factor V_S.

The final result reads:

$$P_{out} = A_b\, I_S\, \frac{(1-R)V_2}{1 - 1/(VV_S R)} \left[g_0\ell - |\ln(VV_S R)|\right] \tag{10.37}$$

with: $V = V_1 V_2 V_3 V_4$

Compared to the corresponding expression for the linear resonator (10.7), we find an increased laser threshold due to the unidirectionality (the light is only amplified once per round trip). However, the maximum extraction efficiencies attainable are similar for both resonator schemes since the impact of the higher threshold on the output power can be compensated by increasing the mirror reflectance R. Figure 10.17 compares maximum extraction efficiencies $\eta_{extr,max}$ and optimum output coupling transmittances $T_{opt}=1-R_{opt}$ for linear resonators and ring resonators (see also Fig. 10.4). These graphs were generated by numerically solving the differential equation (10.2) and (10.34) with the corresponding boundary conditions. Compared to a linear resonator, optimized ring resonators require a lower output coupling to provide a similar output power for a given small-signal gain.

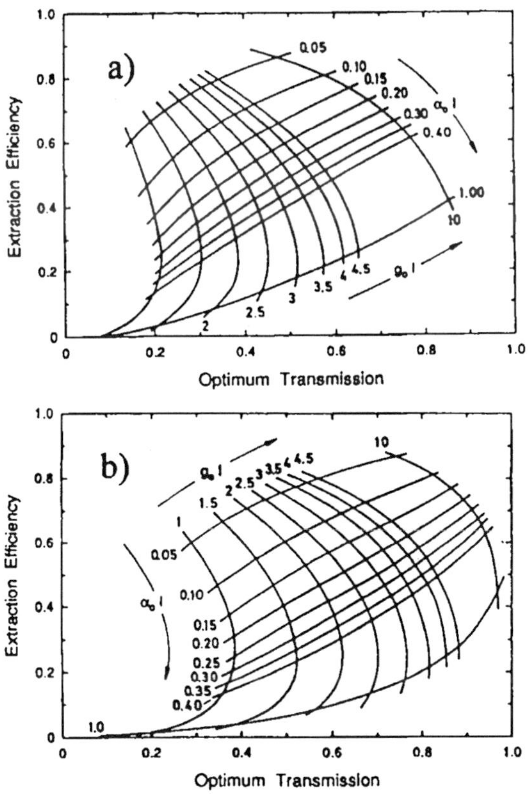

Fig. 10.17 Numerically calculated maximum extraction efficiencies and corresponding optimum mirror transmittances for homogeneously broadened lasers. The curve para-meters are the small-signal gain $g_0\ell$ and the loss per transit $\alpha_0\ell$. a) ring resonator, b) linear resonator.

10.2 Output Power of Unstable Resonators

In unstable resonators the differential equation (10.2) cannot be applied without modifications since the two counterpropagating beams do not overlap completely. Depending on the resonator geometry and the position of the active medium inside the resonator, an outer area exists in which the gain is only saturated by the back travelling wave (see hatched area in Fig. 10.18). Only in the central area of the medium (marked by I) is the inversion depleted in a similar way as in stable resonators, whereas the periphery of the active medium acts as an amplifier for the outcoupled field. Furthermore, the propagation of the diverging wave leads to a longitudinal intensity profile that is different from the one for stable resonators shown in Fig. 10.2. The intensity at the high reflecting mirror, for instance, is lower than the intensity at the output coupler although the light is being amplified by the active medium.

Fortunately, we can take the special beam propagation into account by modifying the differential equation (10.2) [4.9,4.14,4.16,4.18]. For a confocal unstable resonator as shown in Fig. 10.18 the equations read:

$$\frac{dI^+}{dz} = +\left[\frac{g_0}{1+(I^+ + I^-)/I_S} - \alpha_0\right] I^+ - \frac{2I^+}{z+z_0} \qquad (10.38)$$

$$\frac{dI^-}{dz} = -\left[\frac{g_0}{1+(I^+ + I^-)/I_S} - \alpha_0\right] I^- \qquad (10.39)$$

The numerical solution of these equations provides the intensities as a function of z. However, in the outer area marked by II, the intensity I^- also depends on the radial coordinate. This radial dependency has to be incorporated numerically by subdividing the medium into a sequence of disks with each disk having radial points at which the intensity is calculated.

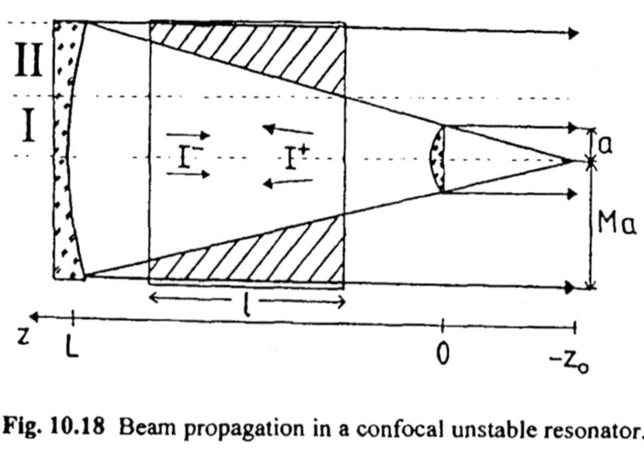

Fig. 10.18 Beam propagation in a confocal unstable resonator.

If the intensity $I^-(z,r)$ at $z=0$ is known as a function of the radius r, the output power P_{out}, extraction efficiency η_{extr}, and the loss factor V (due to output coupling) can be determined using the following relations (for a circular medium with radius Ma):

$$P_{out} = 2\pi \int_a^{Ma} I^-(0,r)\, r dr \tag{10.40}$$

$$\eta_{extr} = \frac{P_{out}}{\pi M^2 a^2 \, I_S \, g_0 \ell} \tag{10.41}$$

$$V = \frac{\pi a^2 I^-(0,0)}{\pi a^2 I^-(0,0) + P_{out}} \tag{10.42}$$

Since the amplification in the outer area is radially dependent, the output coupling loss is increased as the small-signal gain of the medium is increased. As shown in Fig. 10.19, the loss factor equals the geometrical loss factor $1/M^2$ only at the laser threshold and immediately decreases as the gain is increased. This is caused by the lower gain saturation in the periphery of the active medium which in turn leads to a higher intensity in the outer areas of the outcoupled beam profile. Remember that the loss factor V is caused by the output coupling and it therefore corresponds to the mirror reflectance in a stable resonator. The change in output coupling implies that the output power does not increase linearly with the input power, a well-known behavior of unstable laser resonators. The reader may argue that the geometrical model used is too rough since it neglects the true mode structure. However, in Chapter 11 we will incorporate diffraction into our output power model and obtain very similar results.

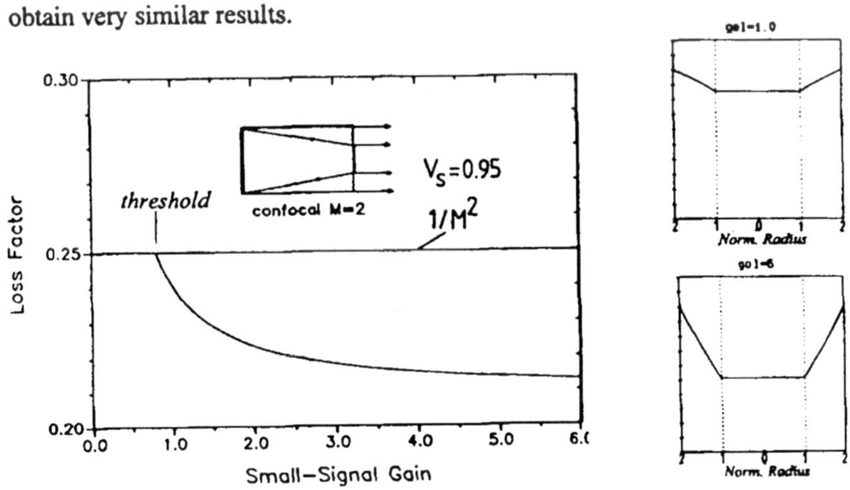

Fig. 10.19 Loss factor as a function of the small-signal gain calculated with Eq. (10.38)-(10.42) for a confocal unstable resonator with magnification $M=2$. The radial intensity profiles $I^-(0,r)$ at the plane of the output coupler are shown on the right side for two different small-signal gains.

The problem we are interested in is, of course, how the unstable resonator compares to the stable resonator as far as the maximum extraction efficiency is concerned. For a given unstable resonator with loss factor V we can calculate the output power of a stable resonator with mirror reflectance $R=V$ using the same active medium. A comparison of the output power will then reveal any influence of the special beam propagation in unstable resonators on the extraction efficiency. Some examples are shown in Fig. 10.20. We see that both resonator geometries provide similar extraction efficiencies as long as the area with one intensity (hatched area in Fig. 10.18) is kept small [4.16,4.18]. The preferred geometry would be to position the medium close to the high reflecting mirror. As soon as the medium is moved closer to the output coupler, the amplifier region becomes larger resulting in a decrease of the output power. However, as far as the maximum extraction efficiency is concerned, the location of the active medium does not really matter and both resonator schemes provide a similar performance. Thus we can apply (10.8) also to unstable resonators if we replace the mirror reflectance R with the loss factor V:

$$P_{out} = A_b \, I_S \, \frac{1-V}{1 - V + \sqrt{V}\,(1/V_S - V_S)} \left[g_0 \ell - |\ln\sqrt{V V_S^2}| \right] \qquad (10.43)$$

with $\qquad A_b = \pi M^2 a^2$ (fill factor of one)

Keep in mind that the loss factor V is a function of the small-signal gain!

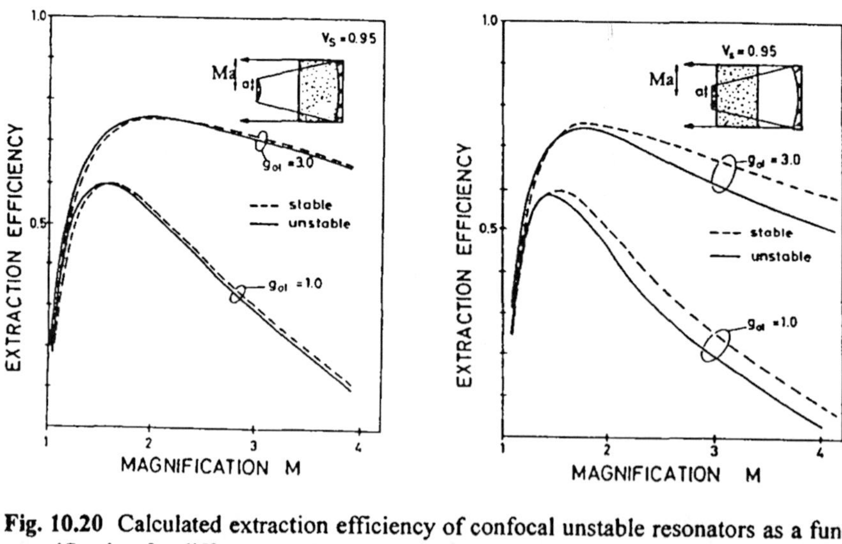

Fig. 10.20 Calculated extraction efficiency of confocal unstable resonators as a function of the magnification for different small-signal gains and different locations of the active medium. The broken line represents the extraction efficiency of a stable resonator exhibiting the same output coupling.

Chapter 11

Influence of Gain on Mode Structure and Loss

11.1 General Aspects

In the previous discussion of the output power we did not take the mode structure into account. The volume of the beam inside the medium was only characterized by an effective cross sectional area A_b resulting in a fill factor $\gamma = A_b/A$, with A being the cross sectional area of the active medium. The question we want to answer in this Chapter is how the beam cross section has to be defined to get a realistic expression for the output power of a laser resonator. In Part III we defined the mode radius w by using the second intensity moments. A first approximation, therefore, would be to use this mode radius for the definition of the mode cross section $A_b = \pi w^2$. For transverse multimode lasers this is a very good approach that provides output powers close to the observed ones.

Unfortunately, in fundamental mode operation or operation with a low number of transverse modes, the above expression yields output powers that are too low. This means that the effective mode cross section generally is larger than expected from the beam radius. Furthermore, the mode cross section is a function of the small-signal gain, an effect caused by the saturation of the gain [4.24,4.27,4.29,4.36]. If the mode exhibits a radial intensity profile $I_1(r)$ in front of the active medium, the steady-state intensity distribution after passage through the medium is given by (we neglect the z-dependence of the gain):

$$I_2(r) = I_1(r) \exp\left[\frac{g_0 \ell}{1 + I_1(r)/I_S}\right] \exp\left[-\alpha_0 \ell\right] \tag{11.1}$$

For a Gaussian beam, the intensity profile $I_2(r)$ will no longer be Gaussian since its slopes are amplified more (higher gain) than the center of the beam (Fig. 11.1). With increasing pump power, the beam will, therefore, spread in size resulting in an increased fill factor. The increasing beam diameter will in turn lead to increased diffraction losses at the intracavity aperture. The expressions for the output power derived in the previous chapter are, of course, still applicable if the functional dependencies of the diffraction losses and the fill factor on the small-signal gain are known and inserted into (10.6). The changes in mode structure and loss are caused by gain saturation, but that does not mean that a high gain is required to observe these effects. In fact, as will be presented later, the induced changes are most pronounced near the laser threshold where the small-signal gain generally is low.

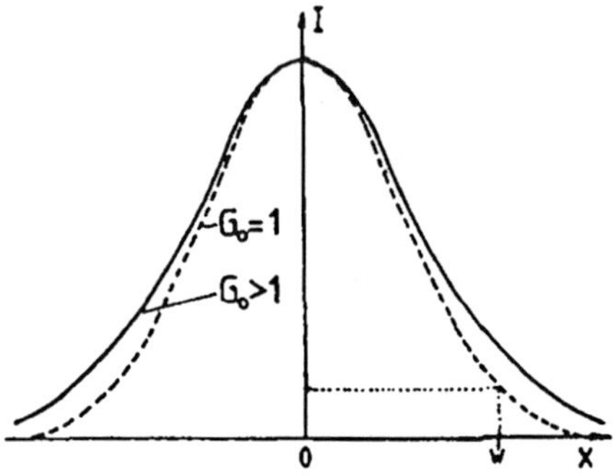

Fig. 11.1 Compared to the intensity distribution in the passive resonator (gain factor $G_0=1$), the mode of the active resonator exhibits a lifting of the slopes due to gain saturation.

This statement may be surprising to those readers who believe that gain induced changes in the mode structure are only noticeable in high gain lasers. No matter which resonator scheme is used and how many transverse modes oscillate simultaneously, the nonlinear amplification will always lead to an increase of the fill factor and of the diffraction losses as the pump power is increased. The impact on the output power, however, is different for single mode and multimode lasers.

11.2 Stable Resonators

11.2.1 Fundamental Mode Operation

Figure 11.2 presents a numerical example of gain saturation in a stable resonator in fundamental mode operation. The calculation was based on the differential equation (10.2) in combination with Kirchhoff integrals to propagate the electric field inside the resonator. The fill factor can be obtained by comparing the calculated output power with the output power predicted by (10.6). This graph indicates that the fill factor increases by more than a factor of two as the small-signal gain is increased. Interestingly, the mode cross section at the laser threshold is determined by the Gaussian beam radius in the active medium w_2. The mode size inside the medium spreads as the gain is increased. This spreading is more pronounced the larger the radius of the active medium is in comparison to the Gaussian beam radius. At the output coupling mirror a higher fraction of the intracavity power will hit the aperture, resulting in an increase of the diffraction loss. As a consequence, the slope of the output power is not a constant anymore. Gain saturation leads to a bending of the output power curve near the laser threshold. This effect is in agreement with experimental data (Fig. 11.3). Note that the change of mode structure and loss is most pronounced close to the laser threshold!

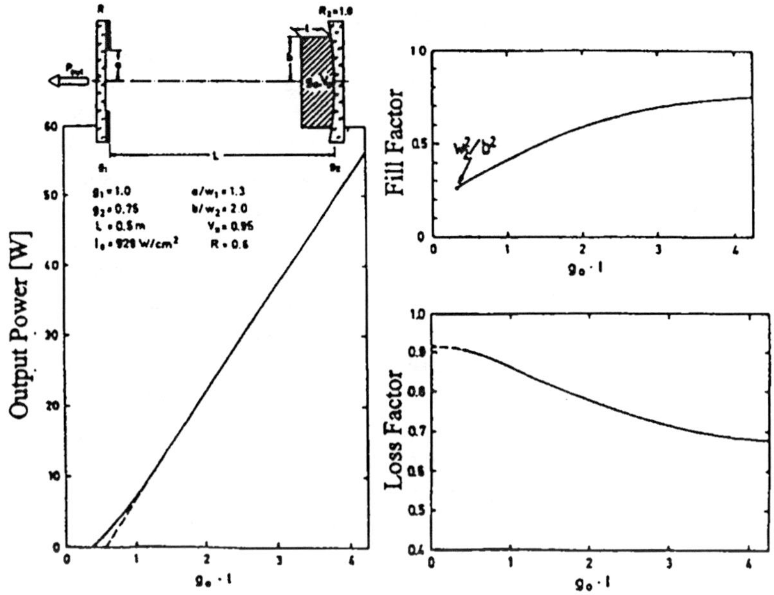

Fig. 11.2 Numerically calculated output power, fill factor, and loss factor per round trip of a stable resonator in fundamental mode operation as a function of the small-signal gain ($\rho_1=\infty$, $\rho_2=2m$, $L=0.5m$, aperture radius $a=1.3w_1$, radius of active medium $b=2w_2$, $w_{1,2}$: Gaussian beam radii at mirror 1,2, output coupling reflectance $R=0.6$, saturation intensity $I_S=929W/cm^2$, loss factor per transit $V_S=0.95$) [4.39] (© SPIE 1988).

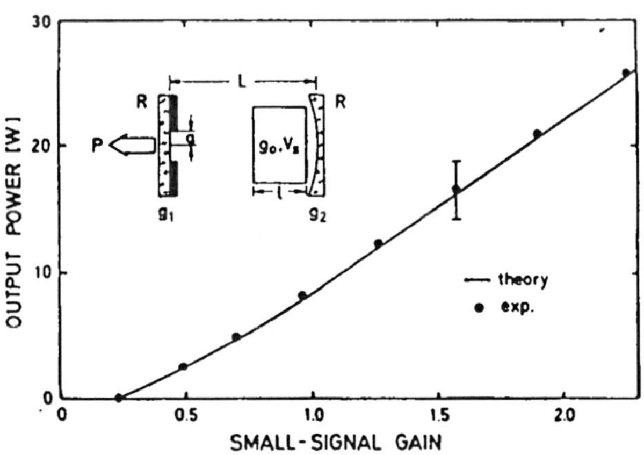

Fig. 11.3 Measured and calculated output power per pulse as a function of the small-signal gain for a stable resonator in fundamental mode operation (pulsed Nd:YAG rod laser, $I_S=2kW/cm^2$, $g_1=1.0$, $g_2=0.5$, effective resonator length $L=0.5$, $\ell=7.5cm$, $R=0.8$, $a=1.22w_1$) [4.39] (© SPIE 1988).

In general, the fill factor will be high if the intracavity intensity is high since gain saturation is more effective in this case. Therefore, we expect that the fill factor is not only increased with higher gain, but will also become higher as the reflectance of the output coupling mirror is increased. This is, in fact, true as shown Fig. 11.4 in which the fill factor and the loss factor per transit are plotted as a function of the mirror reflectance for different small-signal gains. The resonator geometry is the same as in the previous figures. Again, at the threshold we observe the diffraction loss of the passive cavity (V=*0.9*), and the beam size at this point is given by the Gaussian beam radius inside the medium.

Is there an analytical way to incorporate these gain saturation effects into our output power model? Unfortunately, the answer is no since there is no way to evaluate the dependence of fill factor and losses on the laser parameters. However, we can make some corrections to our output power model (10.6) that bring us close to a reasonable result. Let us first take a look at some measured and calculated output power curves for two resonators in fundamental mode operation (Fig. 11.5). For both resonators, graph c) represents the analytical expression (10.6) for the output power assuming a mode cross section of $A_b = \pi w^2$ (w: average Gaussian beam radius inside the medium) and the passive diffraction losses. Graph b) shows the numerical results including diffraction theory. We see that the numerical results differ from the values given by (10.6), but they agree with the experimental data shown in graph a).

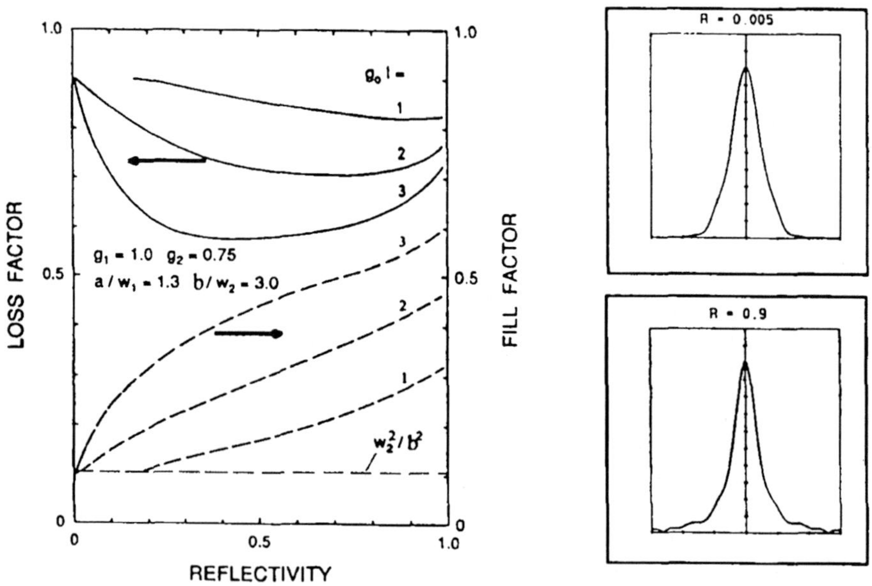

Fig. 11.4 Calculated fill factor and loss factor per round trip of a stable resonator in fundamental mode operation as a function of the output coupling mirror reflectance R. The resonator setup is the one depicted in Fig. 11.2 except for the larger size medium (b/w_2=*3.0*). The curve parameter is the small-signal gain $g_0\ell$. The normalized radial intensity profiles on the right side are the intensity distributions at the high reflecting mirror for $g_0\ell$=*3.0* and reflectances R=*0.005* and R=*0.9*.

For both resonators the slopes of the output power are about twice as high as predicted by the analytical expression (10.6). This leads us to a simple correction to our output power model, which provides reasonable results for high small-signal gains. If the radius of the active medium is at least twice as large as the Gaussian beam radius we can use the cross section of the passive cavity mode $A_b = \pi w^2$ and the diffraction losses of the passive cavity, and multiply the resulting power by a factor of 2.

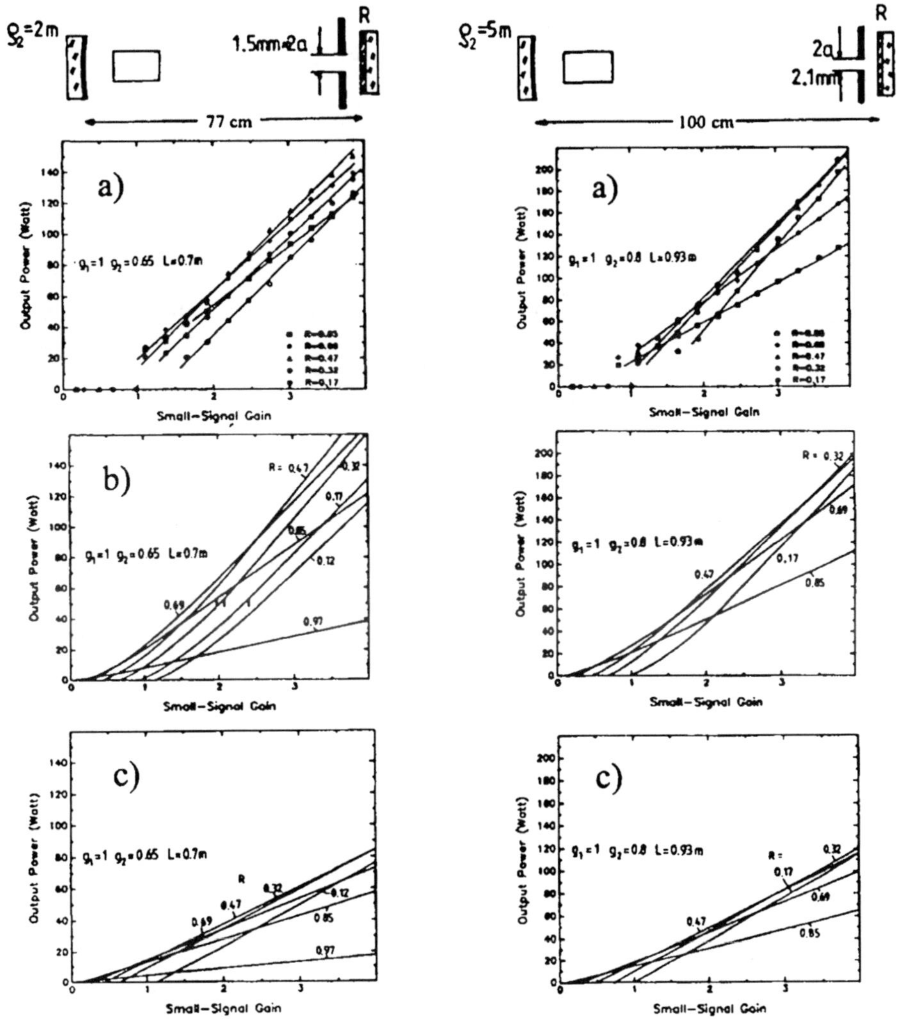

Fig. 11.5 Measured and calculated output power per pulse as a function of the small-signal gain for two stable resonators in fundamental mode operation (pulsed Nd:YAG rod laser, pulse width: 2ms, rod radius: 5mm, rod length ℓ=*150mm*). a) measurement, b) numerical solution including diffraction theory, c) analytical expression (10.6) using the cross sectional area πw^2 of the Gaussian beam inside the active medium and the passive diffraction losses.

The resonators shown in the previous figure , of course, will not provide the optimum extraction efficiency attainable in fundamental mode operation since the size of the active medium is not adapted to the size of the beam. The highest extraction efficiency is generally obtained if the active medium acts as the mode stop. This means that its radius is in the range of the Gaussian beam radius. Let us now discuss how this optimum condition can be achieved in stable resonators and what the maximum extraction efficiency is in fundamental mode operation.

We consider a resonator set-up as depicted in Fig. 11.6 in which the medium with radius b is located at the high reflecting resonator mirror 2. As was discussed in Section 5.1, the Gaussian beam radius at mirror 2 is given by:

$$w_2^2 = \frac{2L_{eff}\, g_1 \lambda}{\pi\sqrt{G^2-1}} \tag{11.2}$$

with $G = 2g_1g_2-1$; $g_{1,2}$: g-parameters of mirror 1,2; λ : wavelength, and L_{eff} : effective resonator length (see Sec. 9.2). For a given radius b of the medium we can always adjust the equivalent g-parameter G and the resonator length L_{eff} to realize a certain radius ratio b/w_2. Figure 11.7 shows the result of the optimization of this resonator scheme. The radius ratio b/w_2 as well as the mirror reflectance R were varied in order to determine the maximum extraction efficiency. The parameters for optimum performance are shown in the graphs. In general, the radius of the active medium should be about 1.3-1.4 times larger than the Gaussian beam radius. A higher gain requires a tighter adaptation of the rod radius to prevent higher order modes from oscillating. The fill factors range between 0.8 for low gain and 0.94 for high gain systems with corresponding diffraction losses per round trip between 5% and 15%. Due to the lower fill factor and the additional diffraction losses, the maximum extraction efficiency is lower as compared to multimode operation in which fill factors close to 1.0 and negligible diffraction losses are attained (as we shall see later). However, it is possible to enhance the extraction efficiency in fundamental mode operation by two techniques: the realization of a Gaussian gain profile and the utilization of a variable reflectivity mirror (VRM) as the output coupling mirror.

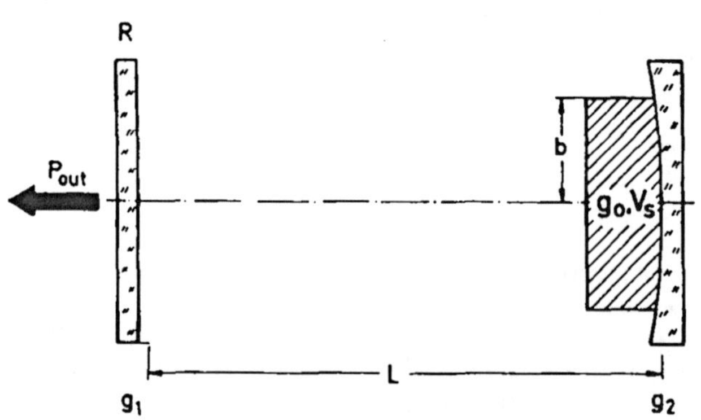

Fig. 11.6 Resonator scheme for attaining maximum extraction efficiency in fundamental mode operation.

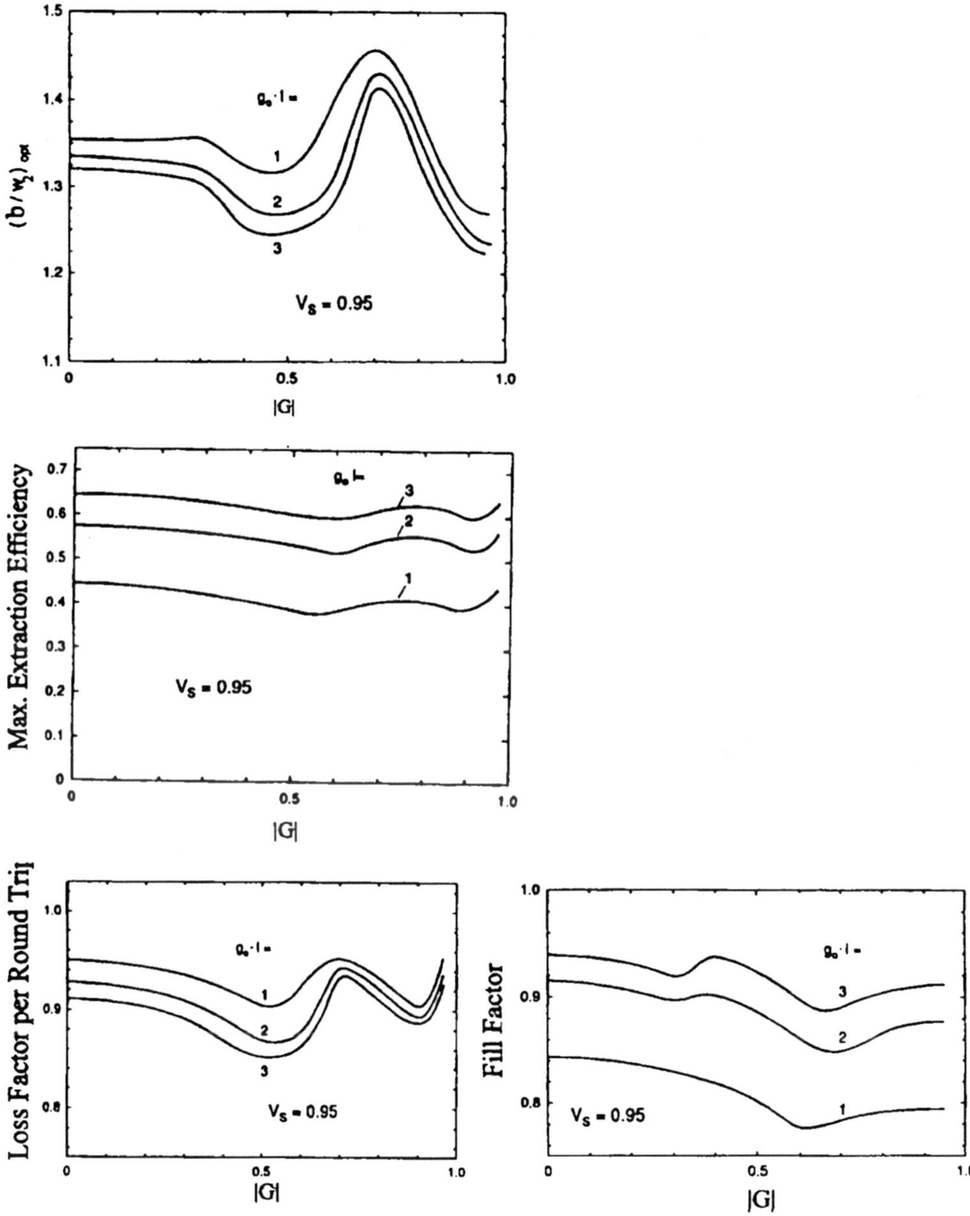

Fig. 11.7 Optimum radius ratios b/w_2 for maximum extraction efficiency of the resonator shown in Fig. 11.6 and corresponding values for extraction efficiency, fill factor, and diffraction loss factor per round trip. The curve parameter is the small-signal gain. The mirror reflectance R and the radius ratio b/w_2 were varied to determine the maximum extraction efficiency (homogeneous line broadening).

Radial Gaussian gain profiles are typical for longitudinally laser pumped active media like diode laser pumped solid state laser materials. Instead of controlling the transverse mode spectrum with an intracavity aperture or the lateral dimension of the active medium, the width of the gain profile can be used to allow only the fundamental mode to oscillate [4.30]. The advantage of this technique is that a higher fill factor can be realized since the mode intensity profile can be better adapted to the gain profile. Furthermore, the diffraction losses can be kept low as long as the pump laser beam is not too tightly focused with respect to the Gaussian beam. Let us consider an active medium with radius b and radial gain coefficient distribution:

$$g_0(r) = \epsilon \exp\left[-\left(\delta\frac{r}{w_2}\right)^2\right] \tag{11.3}$$

where w_2 is the Gaussian beam radius in the medium. The factor ϵ is chosen such that the average small-signal gain coefficient is constant:

$$\int_0^b g_0(r)\, rdr = \frac{1}{2}b^2 g_0 \tag{11.4}$$

$$\epsilon = \frac{\delta\, b^2\, g_0}{w_2\,[1 - \exp[-(\delta b/w_2)^2]]} \tag{11.5}$$

Thus, we always have the same power available in the form of inversion, but the gain profile exhibits a radius ($1/e^2$-definition) of $w_g = w_2/\delta$ which becomes narrower as the shape factor δ is increased. As the pump beam is more tightly focused inside the active medium, the beam radius w_2 of the fundamental mode becomes smaller too since the gain profile acts as an aperture for the resonator mode, a condition that is also referred to as gain-guiding [4.42,4.45]. This is shown in Fig. 11.8 in which the gain profiles and the corresponding intensity profiles of the fundamental mode are shown for a diode pumped Nd:YAG laser. The resulting beam radius w_2^* of the fundamental mode can, to a good approximation, be calculated using the relation [4.44]:

$$\frac{1}{w_2^{*2}} = \frac{1}{w_2^2} + \frac{1}{w_g^2} \tag{11.6}$$

Since the small-signal gain is increased compared to a homogeneously distributed gain, the extraction efficiency is expected to rise with decreasing pump profile radius w_g. Figure 11.9 presents the calculated extraction efficiency for different gain profile parameters δ for the resonator already shown in Fig. 11.2. We obtain a maximum extraction efficiency for $\delta \approx 2.5$ which means that the radius of the gain profile is 2.5 times smaller than the radius of the Gaussian beam. Keep in mind that this graph is not representative for all stable resonators since the optimum gain profile radius also depends on the gain and the Rayleigh ranges of the pump beam and of the laser beam. It is interesting that the extraction efficiency can be higher than the one for a homogeneous gain profile and fill factor 1.0 (dotted line). However, such a tight focusing of the pump beam may lead to side lobes in the far field due to diffraction when an aberrated lens is present [4.41]. For this reason, diode-end pumped solid state lasers in TEM_{00} mode operation generally exhibit a pump spot that is about 10-30% larger than the Gaussian beam radius [4.43].

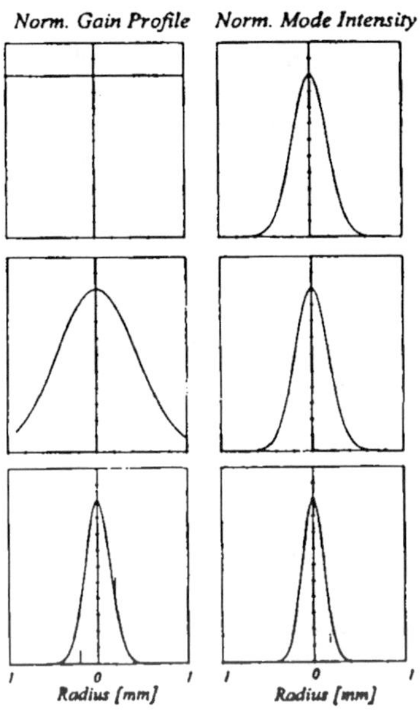

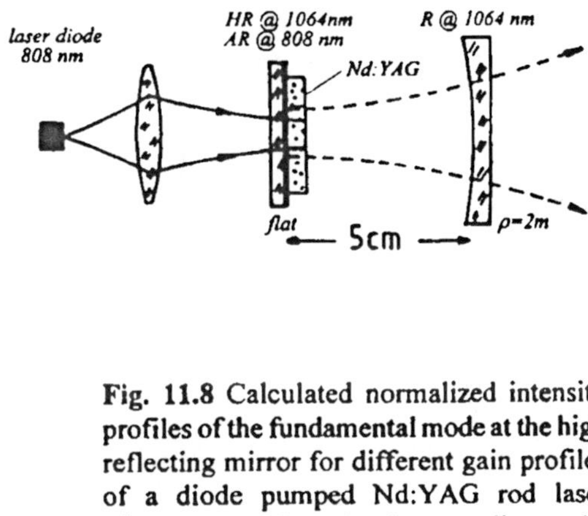

Fig. 11.8 Calculated normalized intensity profiles of the fundamental mode at the high reflecting mirror for different gain profiles of a diode pumped Nd:YAG rod laser ($\lambda = 1.064 \mu m$, Gaussian beam radius at the rod $w_2 = 325 \mu m$, $g_0 \ell = 2$, $R = 0.8$).

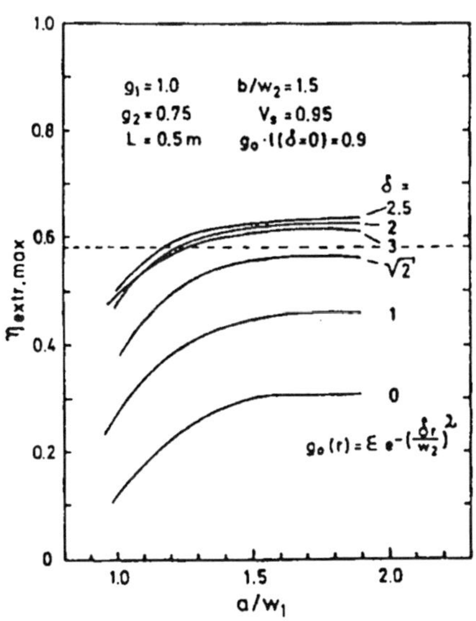

Fig. 11.9 Calculated maximum extraction efficiency of a stable resonator in fundamental mode operation as a function of the aperture radius. The parameter of the curves is the shape factor δ of the radial gain profile (see text). The resonator set-up is the same as in Fig. 11.2. The mirror reflectance R was varied to obtain the maximum extraction efficiency. The dashed curve is the extraction efficiency attainable with flat-top pump and beam profiles of equal diameter.

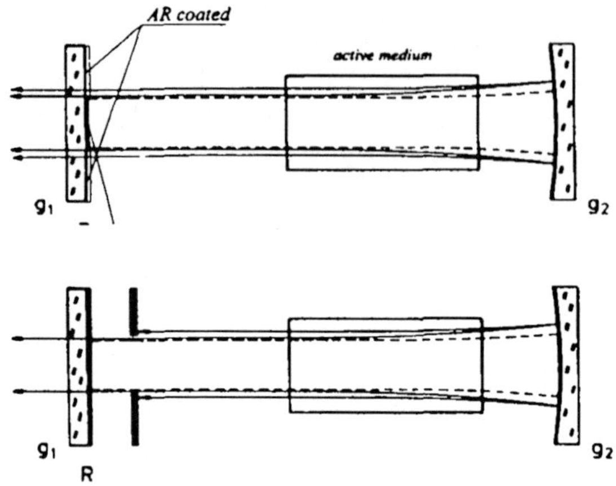

Fig. 11.10 Stable resonator utilizing a VRM as the output coupler (top) in order to use the power fraction that is absorbed by an apertur e in a conventional resonator design (bottom). The dotted lines mark the Gaussian beam propagation in the passive resonator.

A very similar and easier to realize technique is the utilization of a variable reflectivity mirror (VRM) as the output coupler. The width of the reflectivity profile has to be adapted to the Gaussian beam diameter in order to suppress higher order modes. The shape of the profile is of secondary importance. We can use a Gaussian profile or a flat top mirror like the one shown in Fig. 11.10. The basic advantage of this design is that the diffraction losses generated by an intracavity aperture (see Fig. 11.2) now represent useful output power. The power fraction that is hitting the aperture after one round trip is now coupled out of the resonator. This scheme enables one to increase the output power by up to a factor of 2 (a 10% loss at the aperture represents a 100% increase in output power for a mirror reflectance of 0.9). However, there are some constraints regarding the design of the mirror since the phase difference between the field transmitted through the center of the mirror and the field coupled out in the periphery has to be close to a multiple integer of 2π. Fortunately, with the design techniques of VRMs described in Sec. 7.7.2, it is possible to manufacture mirrors that exhibit a phase shift of less than $\pi/10$ over the whole substrate.

11.2.2 Transverse Multimode Operation

In multimode operation the saturation of the gain also leads to an increase of the mode diameter, but the intensity profile of the laser beam stays relatively homogeneous (Figs. 11.11 and 11.12). Similar to stable resonators in fundamental mode operation, the increase of the fill factor generates higher losses at the intracavity aperture. However, the output power is not affected by these changing mode properties, since the same power fraction is always transmitted through the aperture. The additional power generated in the periphery of the active medium does not add to the output power but is blocked by the aperture.

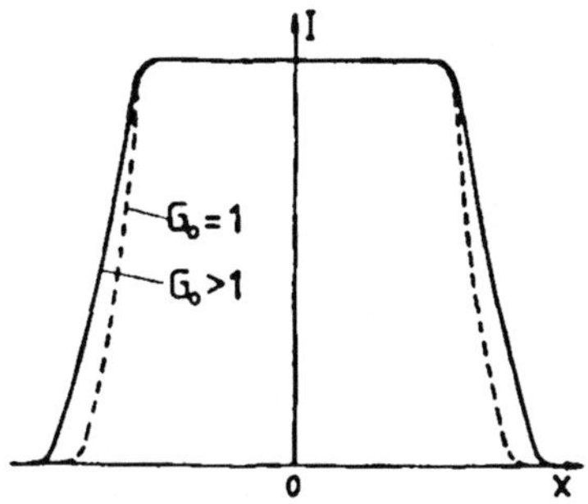

Fig. 11.11 Multimode laser beams increase their diameter as the gain of the active medium is increased. The portion of the intensity profile that passes the aperture, however, stays homogeneous.

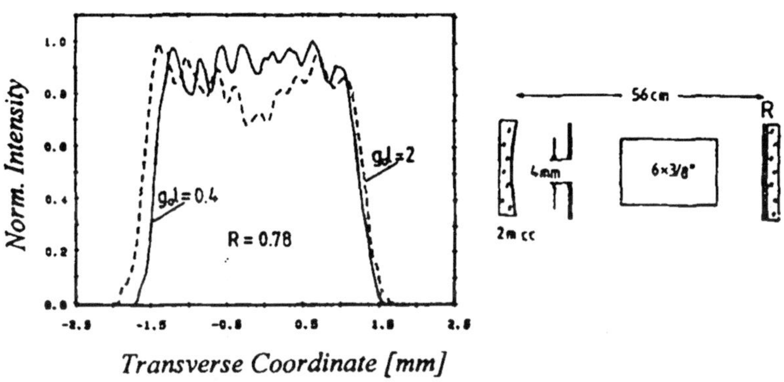

Fig. 11.12 Measured normalized radial intensity profiles at the output coupling mirror for a stable resonator in multimode operation and two different small-signal gains (6x3/8 inch Nd:YAG rod laser, $g_1=1.0$, $g_2=0.75$, $L_{eff}=0.5m$, single pulse operation).

The output power can thus be calculated to a very good approximation with (10.6) by using the cross section of the aperture (or the cross section of the active medium if no aperture is used) and the diffraction losses of the passive resonator of 1-2% per round trip (see Fig. 5.41). The calculated fill factor, loss factor, and beam propagation factor M^2 as a function of the radius b of the active medium are presented in Fig. 11.13 for a stable resonator without additional limiting aperture (the number of transverse modes is controlled by the size of the active medium). This figure clearly indicates that both the fill factor and the loss factor approach 1.0 if more than twenty transverse modes oscillate simultaneously.

If an aperture in the resonator is used to limit the number of modes, the additional losses at the aperture can assume high values. Figure 11.14 presents measured losses at the aperture as a function of the output coupling reflectance R. The losses represent the fraction of the intracavity power that hits the aperture after one round trip (a scraper was used to measure the power at the aperture). In high power lasers, the aperture has to be cooled to prevent damage. Similar to fundamental mode operation, we can make use of this lost power fraction by using a mirror with a variable reflectivity profile instead of an aperture.

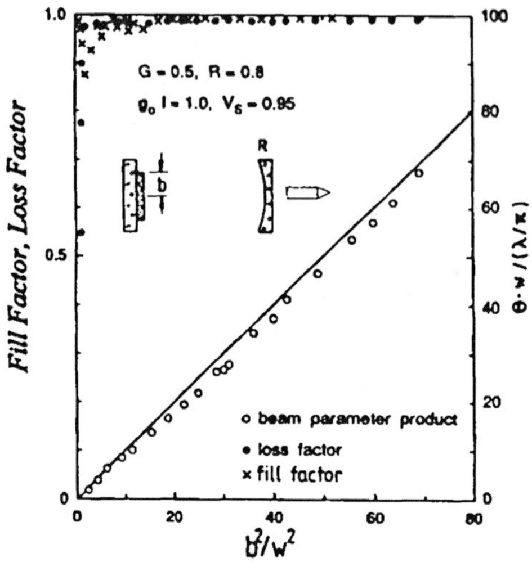

Fig. 11.13 Calculated fill factor, loss factor per round trip, and beam propagation factor $M^2 = w\theta\pi/\lambda$ for a stable resonator as a function of the ratio of the cross sectional area of the active medium (radius b) to the cross sectional area of the Gaussian beam (beam radius w). Resonator parameters: $g_1 = 1.0$, $g_2 = 0.75$, $L_{eff} = 0.5m$, $R = 0.8$, $g_0 \ell = 1.0$, $V_S = 0.95$. The solid line represents the approximation $M^2 = b^2/w^2$ (see also Sec. 5.2.4).

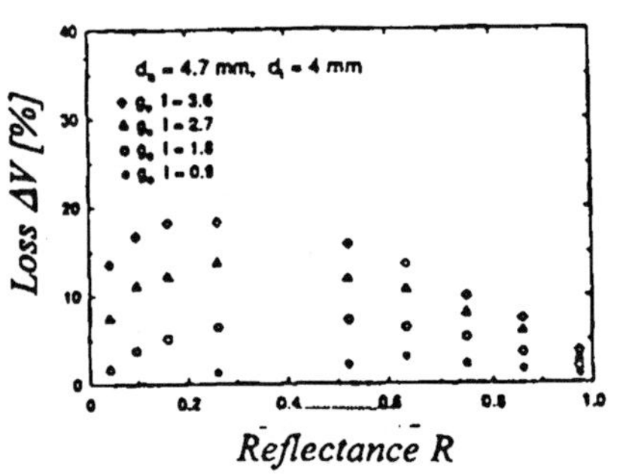

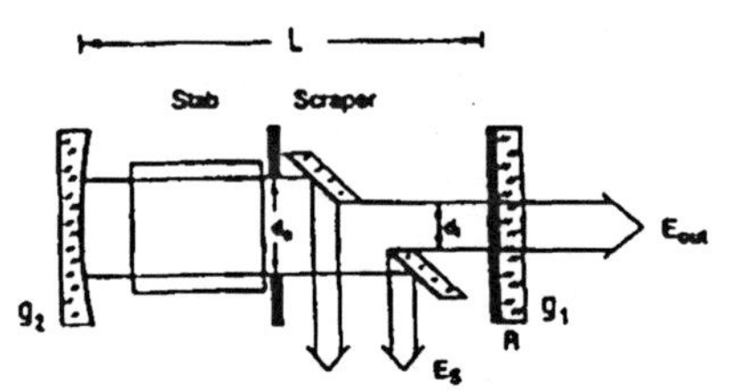

$$\Delta V = \frac{1}{1 + \dfrac{E_{out}}{(1-R)E_S}}$$

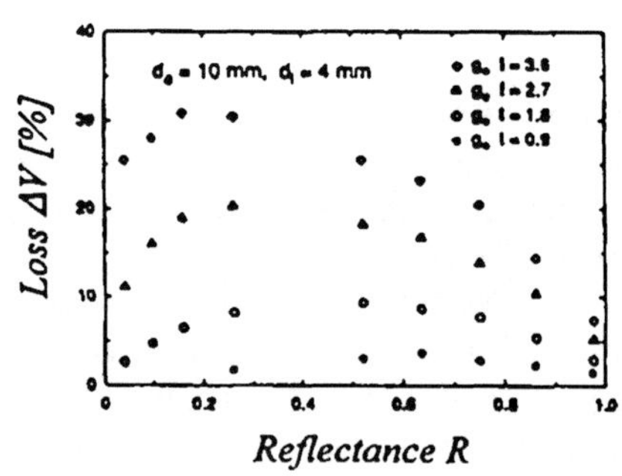

Fig. 11.14 Measured loss at the intracavity aperture (scraper) of a stable resonator in multimode operation as a function of the output coupling reflectance R. The curve parameter is the small-signal gain. The loss represents the fraction of the total intracavity energy that is reflected by the scraper. The additional aperture with diameter d_a is used to perform a spatially resolved measurement of the energy hitting the scraper. The energies E_{out} and E_s are measured to determine the loss using the shown formula ($g_1 = 1.0$, $g_2 = 0.75$, $L_{eff} = 0.5m$, pulsed Nd:YAG rod with length 150mm and diameter 10mm).

11.3 Unstable Resonators

11.3.1 Mode Structure and Loss

The gain saturation effects in unstable resonators are similar to those in stable resonators. The slopes of the intensity profiles are lifted with increasing small-signal gain resulting in a higher fill factor and higher losses. The increase in loss, however, has a different impact on the output power since the loss represents the output coupling of the resonator (Fig. 11.15). We have seen in Chapter 10 that the optimum output power is only attained for a certain output coupling. The change in loss implies that we cannot use the loss factor V of the passive cavity to design an unstable resonator with maximized output power. Remember that the loss factor of unstable resonators corresponds to the output coupling reflectance R of stable resonators as far as output coupling is concerned (see Sec. 7.3). An optimum resonator design requires a passive loss factor that is higher than the optimum loss factor so that the resonator reaches its optimum output coupling with increasing small-signal gain. Figure 11.16 presents a numerical example of gain saturation effects in a confocal unstable resonator for different magnifications M. The radius r of the medium is adapted to the geometrical beam radius $(r=1.05\ Ma)$. The decrease of the loss factor becomes understandable if we take a look at the radial intensity profiles at the mirrors shown for $M=1.5$. The beam sizes at both locations increase resulting in a higher output coupling.

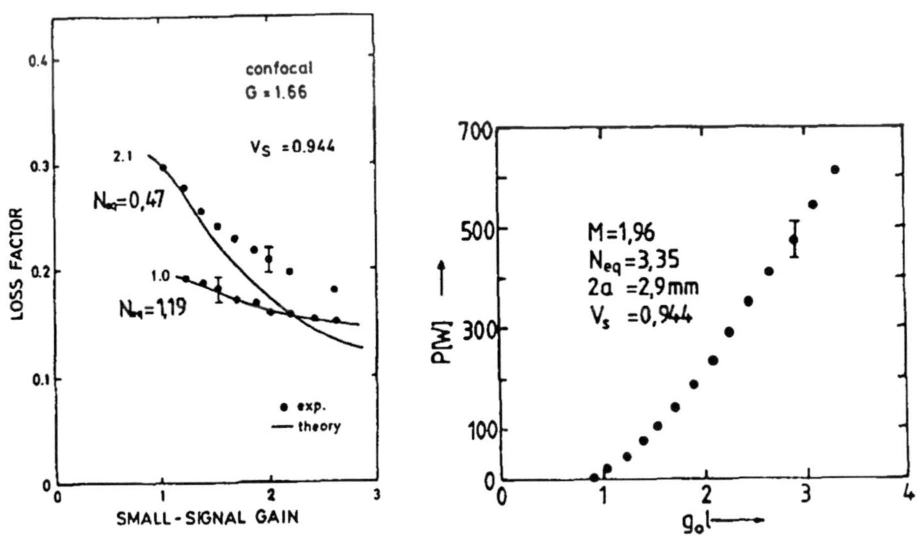

Fig. 11.15 Measured loss factor per round trip (left, for two different radii a of the output coupling mirror) and output power per pulse as a function of the small-signal gain for unstable resonators (pulsed Nd:YAG rod laser, rod diameter: 6.35mm, pulse duration: 1ms).

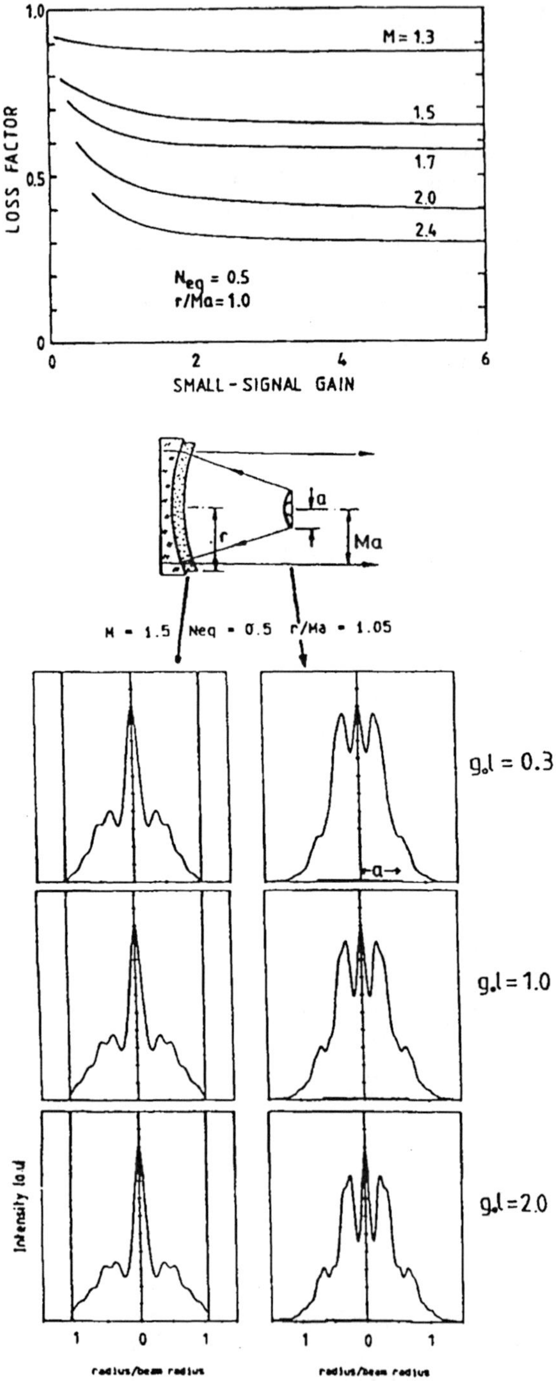

Fig. 11.16 Calculated loss factor per round trip for confocal unstable resonators with different magnifications as a function of the small-signal gain. The equivalent Fresnel number is 0.5. The gain medium is located at the unconfined mirror and the radius r of the medium is adapted to the beam radius ($r=1.05Ma$). Note the change of the mode structure in the periphery as the gain is increased.

11.3.2 Optimum Extraction Efficiency

The realization of maximum extraction efficiency from unstable resonators requires the adaptation of the radius of the beam to the radius of the active medium as well as the optimization of the output coupling. The output coupling is a function of the small-signal gain and the resonator geometry. Unfortunately, for a given small-signal gain several different optimized resonator designs exist. This occurs since the same output coupling can be realized for different values of the magnification *M* and the equivalent Fresnel number N_{eq} (see Sec. 7.3). Therefore, we will restrict ourselves to resonators with $N_{eq}=0.5$. This means that only one value of the magnification exists at which the extraction efficiency exhibits an absolute maximum. The active medium with radius *r* is again located at the unconfined mirror and we consider confocal unstable resonators. The optimization rules can also be applied to non-confocal resonators provided that the corresponding beam radius inside the medium is used.

Figure 11.17 presents calculated extraction efficiencies as a function of the equivalent g-parameter *G* which is related to the magnification via $M=G+\sqrt{G^2-1}$. Generally, the extraction efficiency exhibits two maxima: one at a lower magnification and a radius ratio *r/(Ma)* around 1.05, and a second one at a higher magnification and a radius ratio around 0.95. Although the second maximum is higher for high small-signal gains, operation of the resonator at these parameters will generate additional side lobes in the far field due to diffraction by the edge of the active medium. This medium induced deterioration of the beam quality is a known problem of unstable resonators and in practice the radius ratio is, therefore, chosen higher. Since unstable resonators are usually utilized in high gain laser systems, a higher radius ratio *r/(Ma)* will result only in a slight decrease of the extraction efficiency. As the lower diagram in Fig. 11.17 indicates, a radius ratio of 1.2 will provide a 10% decrease in efficiency, but considering the resulting improvement of the far field this seems to be a small price to pay.

Figure 11.18 presents the optimum equivalent g-parameter G and the optimum magnification *M* as a function of the small-signal gain for the resonator scheme used in the previous figure. If no increase of the losses due to gain saturation were present, the optimum equivalent g-parameter G would be given by the curve marked with 'passive'. Fortunately, for medium to high gain, the domain in which unstable resonators are preferably used, the output power only depends slightly on the output coupling. For unstable resonators, the sensitivity of the output power to changes in the output coupling basically is the same as for stable resonators (see Sec. 10.2) This is demonstrated in Fig. 11.19 in which the optimum loss factor V_{opt} is plotted as a function of the small-signal gain. As long as the loss factor lies somewhere between the curves marked with $V_{99\%}$ and $V_{95\%}$, the output power will be greater than 99% and 95% of its maximum value, respectively. Usually the relative change of the loss factor due to gain saturation is small compared to the 95% tolerance range. This means that for high gain lasers, the loss factor per round trip *V* of the passive resonator (see Sec. 7.3) can be used as a basis of the resonator design.

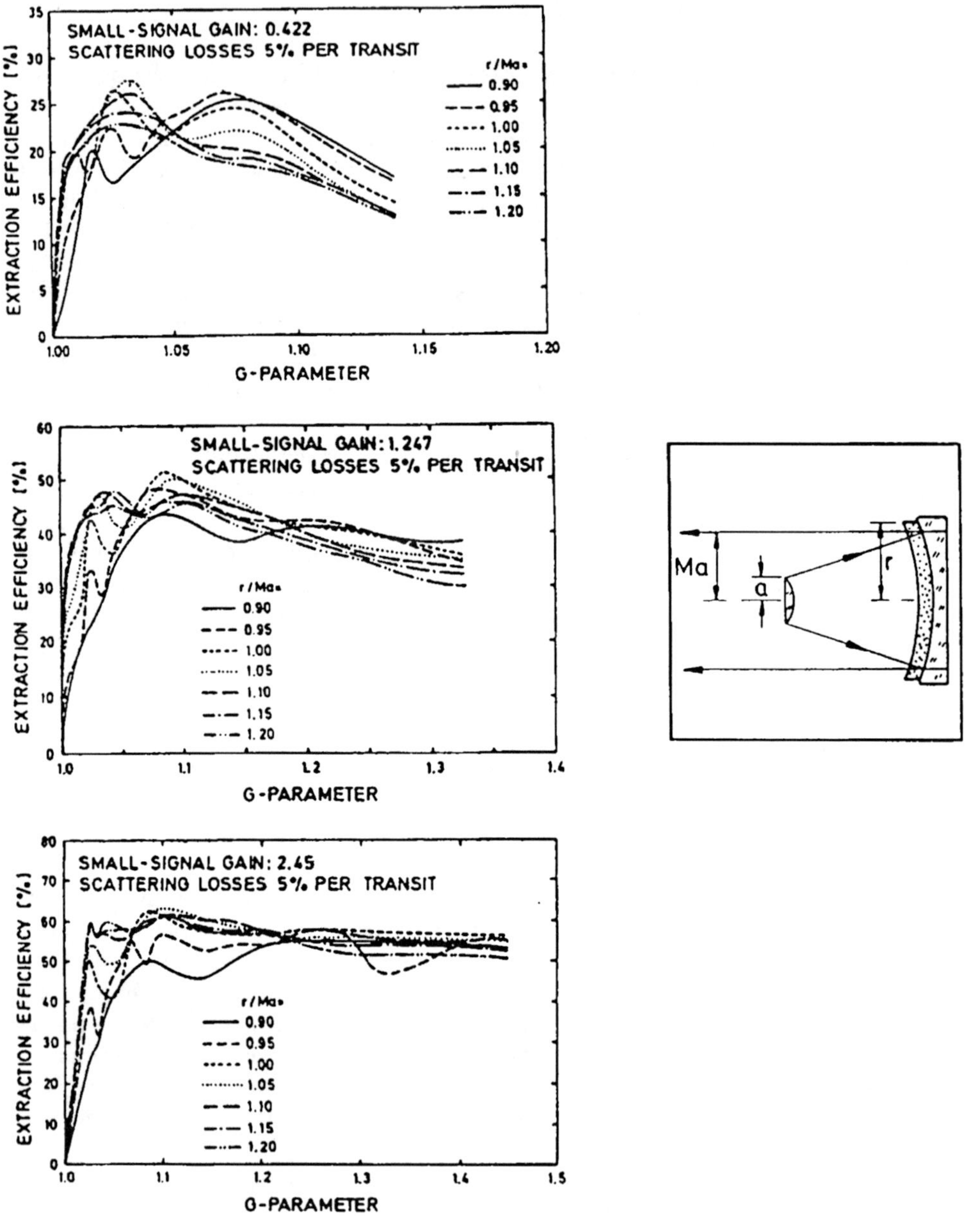

Fig. 11.17 Calculated extraction efficiency as a function of the magnification for three different small-signal gains (confocal resonators in circular symmetry with N_{eq}=0.5). The curve parameter is the ratio of the rod radius r to the beam radius Ma. The loss factor per transit of the medium is V_S=0.95.

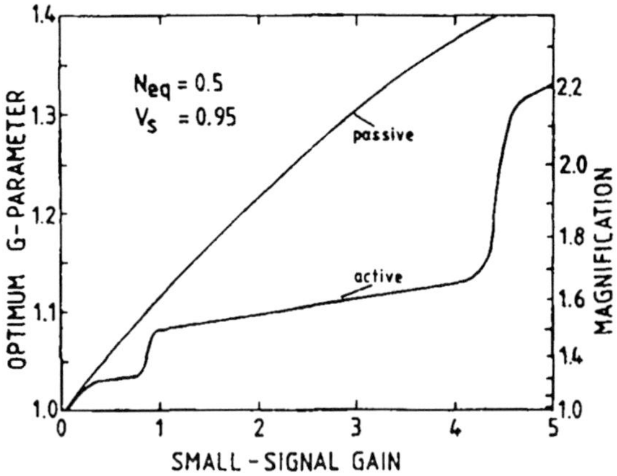

Fig. 11.18 Calculated optimum equivalent g-parameters G and optimum magnifications M as a function of the small-signal gain for unstable resonators in circular symmetry with $N_{eq}=0.5$ and $V_S=0.95$ (curve marked with 'active'). The radius b of the active medium is adapted to the beam radius ($b=1.05Ma$). The curve marked with 'passive' is the optimum G-parameter if no decrease of the loss factor due to gain saturation would occur.

Similar to stable resonators (see Eq. (10.15)), the optimum passive loss factor for homogeneously broadened lasers with medium to high gain ($g_0\ell > 1.5$) can be determined using the expression:

$$\ln V = -2(\alpha_0\ell+0.05)\left[\sqrt{\frac{g_0\ell}{\alpha_0\ell+0.05}}-1\right] \tag{11.7}$$

where the additional term 0.05 reflects the fact that the adaptation of the beam radius to the radius of the active medium generates additional diffraction losses of typically 5% per transit. Compared to stable resonators in multimode operation, the maximum extraction efficiencies attainable with unstable resonators are lower due to the additional diffraction loss and a lower fill factor of about 0.9-0.95 (Fig. 11.20). In practice, the extraction efficiency of unstable resonators is lower than the maximum values shown in Fig. 11.20 because it is more difficult to implement the optimum resonator design (tolerances and alignment). For high gain lasers, unstable resonators providing 70% of the output power attainable in stable multimode operation have been described [4.104,4.107]. Compared to stable resonators in fundamental mode operation, the maximum extraction efficiencies are slightly higher due to a higher fill factor. The higher fill factor reflects the fact that it is easier to adapt the unstable resonator mode to the medium than it is a Gaussian beam because of the steeper slopes of the intensity profile.

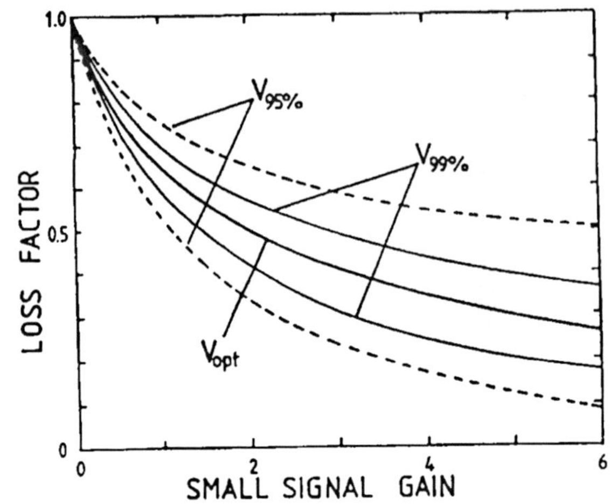

Fig. 11.19 Optimum loss factor per round trip V_{opt} as a function of the small-signal gain. The curves marked $V_{99\%}$ and $V_{95\%}$ define the loss factor range within which 99% and 95% of the maximum output power is attained (Loss factor per transit of the medium: V_S=*0.95*).

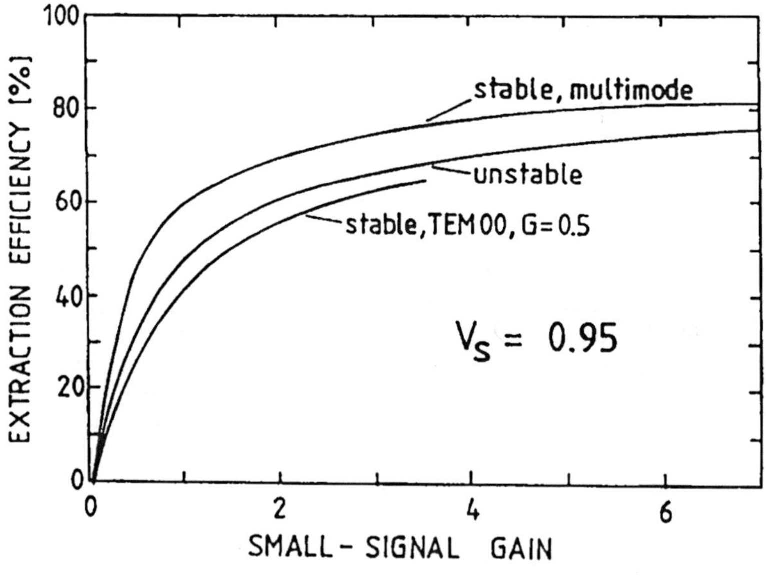

Fig. 11.20 Maximum extraction efficiencies of homogeneously broadened lasers as a function of the small-signal gain with different resonators (Loss factor per transit of the medium V_S=*0.95*). In all three resonators the active medium is located at the high reflecting mirror.

11.4 Mode Structure and Steady State Condition

In steady state operation the gain $g\ell$ will be saturated by the intensity so that the threshold gain $g_{th}\ell$ is obtained. If we assume a homogeneous transverse beam profile and I denotes the sum of intensities of the two counterpropagating fields, the steady state condition reads (see also Sec. 9.3):

$$g\ell = \frac{g_0\ell}{1 + I/I_S} = g_{th}\ell = |\ln(\sqrt{R}V_S)| \tag{11.8}$$

with
R : output coupling mirror reflectance
V_S : loss factor per transit
I_S : saturation intensity of the active material

In general, the intensity profile will not be homogeneous, especially in unstable resonators and stable resonators with a low number of transverse modes. The gain, therefore, will become a function of the transverse coordinate. In this case we cannot directly apply the condition (11.8), but we have to integrate over the cross sectional area of the medium. For a rod with radius b, the steady state condition then reads:

$$\frac{2}{b^2}\int_0^b g(r)\ell \, r dr = \frac{2}{b^2}\int_0^b \frac{g_0\ell}{1 + I(r)/I_S} r dr = \frac{1}{\gamma} g_{th}\ell = \frac{1}{\gamma} |\ln(\sqrt{R}V_S)|$$

with γ : fill factor (11.9)

The left hand side represents the mean gain, averaged over the cross sectional area of the active medium. For a homogeneous beam profile this modified condition becomes equal to (11.8). For a fill factor of less than 1.0, the average gain is higher than the threshold gain. This is clear since a lower fill factor means that the available inversion is not completely used by the mode. However, this does not necessarily mean that a homogeneous intensity profile is required to get a fill factor close to 1.0. As Fig. 11.21 shows for different transverse modes of a stable resonator, the gain is saturated below the threshold gain $g_{th}\ell$ (dotted line) at the intensity peaks, whereas the intensity minima leave the gain at a value higher than the threshold value. The average gain, therefore, can come close to the threshold gain (and this corresponds to a fill factor close to one) even for a highly modulated intensity profile like the one in the lower graph.

How does the resonator mode 'know' how to distribute the intensity over its maxima and minima? There must be some sort of information flow between different parts of the mode structure. This information flow is provided by diffraction. The electric field at each point in the active medium is generated by the whole field distribution that was present one round trip earlier.

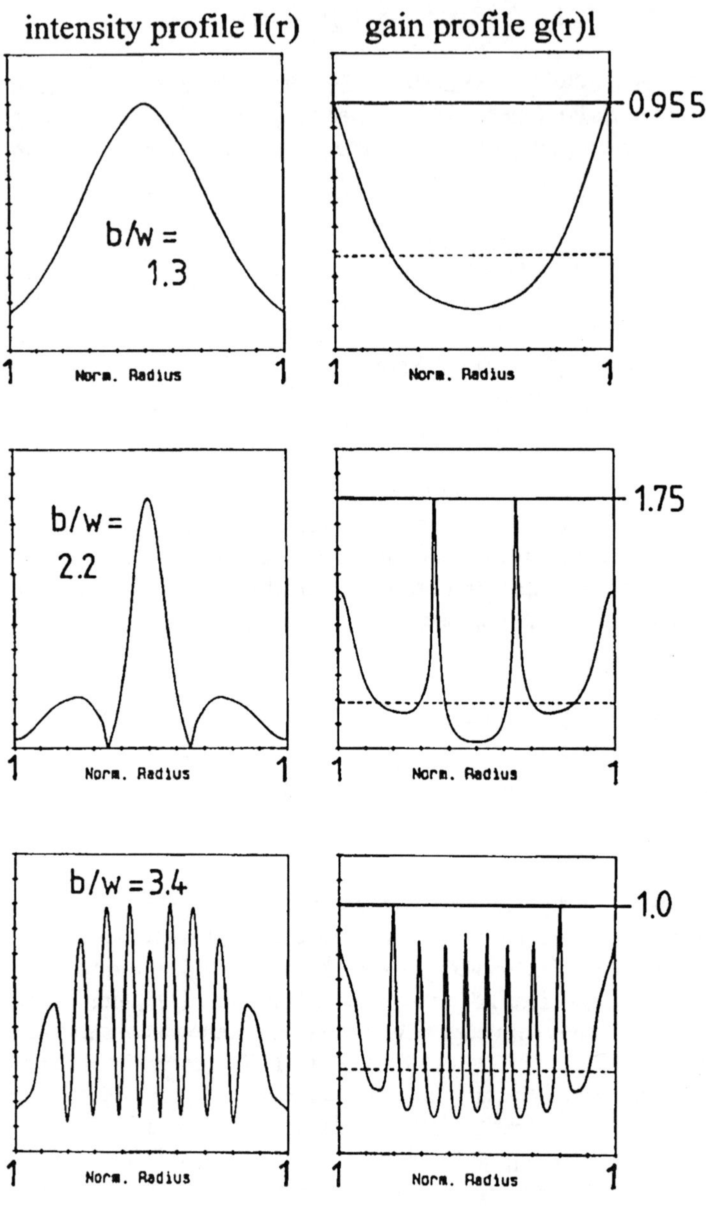

Fig. 11.21 Calculated normalized radial intensity distributions in the active medium and corresponding saturated gain profiles $g(r)\ell$ for a stable resonator with different transverse modes without azimuthal structure (same resonator as in Fig. 11.13, b: radius of the active medium, w: Gaussian beam radius in the active medium). The dotted line is the threshold gain $g\ell_{th}=0.307$ ($R=0.6$, $V_S=0.95$). The small-signal gain is $g_0\ell=2.0$.

Chapter 12

Q-switched Laser Resonators

12.1 General Aspects

In the previous chapters we have discussed the output power of laser resonators emitting in continuous wave operation. For non-military laser systems, the cw output powers are limited to about 10kW for solid state lasers and 20kW for gas lasers. Much higher peak powers can be achieved by utilizing the energy storage capability of the active medium and releasing the energy in short pulses. Peak powers up to several hundred MW with pulse duration on the order of 10ns can be realized using a technique collectively referred to as Q-switching. With this technique, the round trip loss of the resonator is increased to prevent the laser from reaching threshold during a defined pump time interval. The energy stored during this time is then partially released in form of a short pulse after the round trip loss has been decreased to its normal value. In other words, in order to store and release the energy, the quality factor Q of the resonator (see Chapter 4) is switched from a low value to a high value and back again. Figure 12.1 shows the typical time sequence of resonator loss, stored energy and output of a repetitively Q-switched laser.

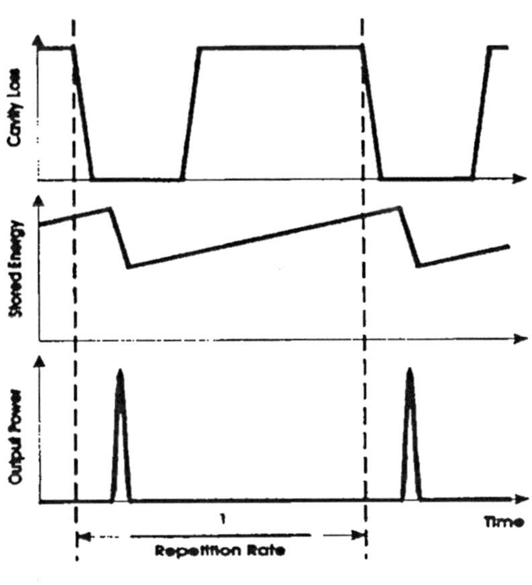

Fig.12.1 Qualitative time sequence of resonator loss, stored energy and output power of a cw pumped Q-switched laser at high repetition rate.

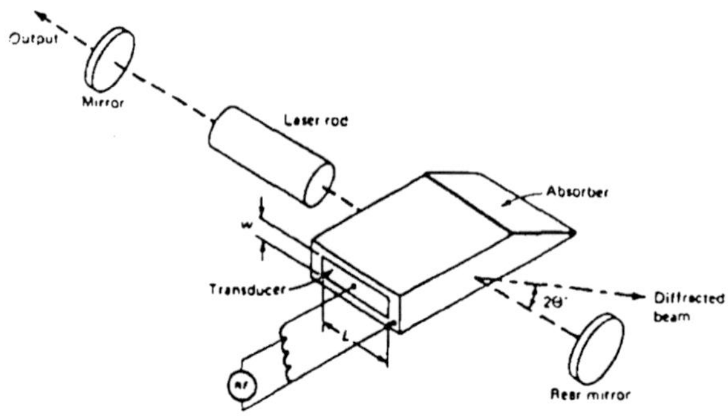

Fig. 12.2 Schematic of a Q-switched solid state laser using an acousto-optic modulator (AOM). The AOM generates a high resonator loss (low Q) by steering the optical axis when RF power is applied [4.2].

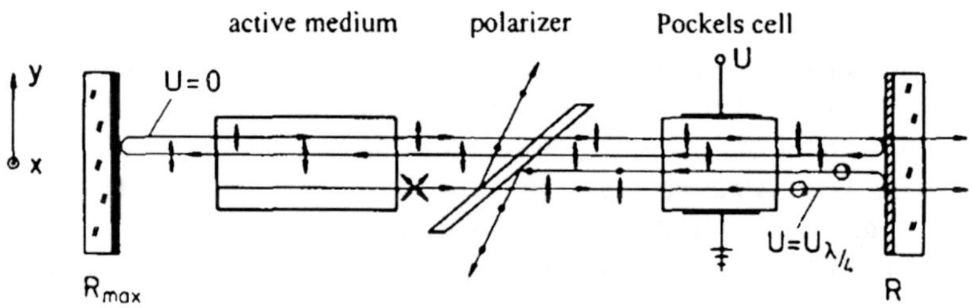

Fig. 12.3 Pockels cell resonator for Q-switch operation. Applying the quarter wave voltage U will rotate the polarization by 90° after a double pass, generating a high loss at the polarizer. Dropping the voltage to zero will lead to the emission of a Q-switch pulse.

A general detailed discussion of Q-switching is given in Siegman's book [4.1]. This description applies to modern practical realization of Q-switched lasers. Commercial laser systems use one of three techniques to temporarily increase the cavity loss: acousto-optic Q-switching using an acousto-optic modulator (Fig.12.2) [4.61,4.76,4.77], electro-optic Q-switching using a nonlinear crystal (Pockels-cell) [4.63,4.65,4.76,4.79], and passive Q-switching with a saturable absorber [4.72-75]. Mechanical Q-switching devices such as rotating mirrors [4.50,4.53,4.54] or piezo-driven etalons [4.56] have not been standardized for commercial applications.

Acousto-optic Q-switching is the method most commonly used because the AOM provides switching capabilities at very high repetition rates (up to tens of MHz, depending on the beam size) and can handle high average beam powers of up to kWs. By applying modulated RF power to a piezo-electric transducer, an ultrasonic wave is launched into the AOM material, typically made of fused silica or crystalline quartz. Due to the photoelastic effect, the acoustic wave generates an index grating that moves across the beam at the speed of sound. This phase grating leads to diffraction of the incoming light, generating a secondary beam at an angle of typically less than 1°. Diffraction efficiencies (deflected beam power/total power) can be as high as 85% for RF powers of about 10W per mm of transducer height *w*, and RF frequencies between 25 and 80MHz. By applying RF power, most of the intracavity power is deflected away from the optical axis for each transit through the AOM, resulting in an additional round trip loss of up to 98%, thereby surpressing laser emission. The fact that there is a residual zero-order transmission of a couple of percent makes it necessary to use several AOMs to prevent laser emission in high gain lasers (small-signal gain > 5). In addition, a certain resonator length (for example > 100 mm) is required to prevent overlap of the deflected and the transmitted beam at the resonator mirrors. Another disadvantage of the AOM is its relatively long turnoff time which is determined by the speed of sound of the AOM material, For typical acoustic speeds of 4-6 km/s it takes the index grating 150-250 ns per mm of beam diameter to clear the beam. In high gain lasers, this time may be too long compared to the build-up time of the Q-switch pulse, resulting in a decrease of laser efficiency due to the residual loss.

Much shorter turnoff times on the order of ns and higher intr acavity losses (>99%) can be realized with electro-optic modulators (Fig.12.3). With this method, voltage is applied to an electro-optic crystal (most commonly KDP, KD*P, $LiNbO_3$, or BBO) which acts as a quarter wave plate with its principal axes rotated by 45° with respect to the polarizer. After a double pass, the polarization vector of the linearly polarized input beam is rotated by 90°, leading to a reflection at the polarizer. Ideally, this should generate a 100% loss inside the resonator, but due to crystal imperfections, stress induced birefringence and polarizer imperfections, a residual transmission of 0.5-1% is typically observed. The high contrast ratio combined with the fact that there is no lower limit on the resonator length, makes the Pockels cell superior to an AOM in Q-switched lasers. However, switching the required voltage of several kVs becomes electronically challenging for switching frequencies beyond 10kHz. Although switching frequencies of up to 50kHz have been demonstrated [4.79], electro-optic Q-switching is generally utilized in low-to-medium repetition rate lasers.

In passive Q-switching a saturable absorber is inserted into the resonator which increases its transmission with increasing intracavity intensity due to the saturation of a spectral transition. The original absorbers were organic dyes dissolved in a solution or incorporated into a thin plastic film. Nowadays, much higher durability and damage thresholds are being provided by crystals doped with absorbing ions, such as Cr^{4+}:YAG [4.72-4.75]. Since the laser radiation itself provides the switching mechanism, no expensive drive electronics are required, making passive Q-switching a low-cost alternative to AO and EO techniques. Unfortunately, the absolute change in transmission is limited to about 30%. A Cr^{4+}:YAG, crystal with an initial transmission of 50% will absorb 20% of the Q-switch pulse energy, resulting in thermal problems in higher power lasers. Another disadvantage is the lack of control of the repetition rate. The Q-switch pulse is generated as soon as the inversion density is high enough to overcome the initial absorber transmission. Therefore, the pulse energy cannot be adjusted and the repetition rate can only be varied by changing the pump power. In addition, passive Q-switching is subject to timing jitter and high rms-noise. Despite these disadvantages, compactness and cost-effectiveness makes passive Q-switching the method of choice for low cost, low power lasers providing pulse durations down to the sub-ns range.

Independent of the Q-switching technique used, the laser performance can be calculated using the rate equations presented in Chapter 9.

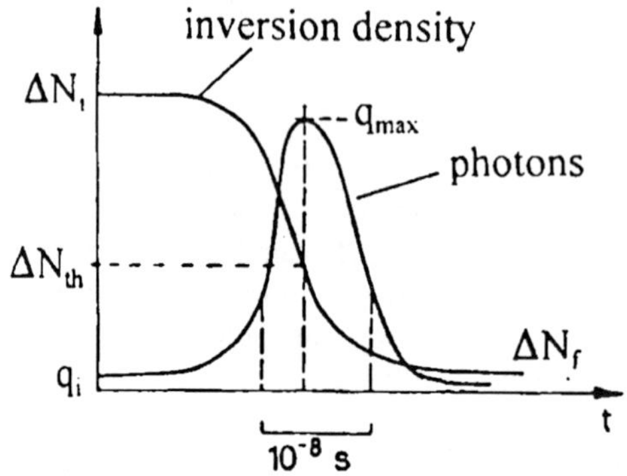

Fig. 12.4 Temporal evolution of inversion density and photon number after the loss induced by the Q-switch has decreased to zero. After a build-up time, the photon number will increase exponentially resulting in a depletion of the inversion. The photon number will decrease again after the inversion density has reached the threshold inversion density ΔN_{th}.

12.2 Rate Equations for Q-switching

12.2.1 Inversion Densities

The analysis of Q-switching performance starts with the rate equations for inversion density ΔN and photon number q inside the resonator, derived in Chapter 9 [4.46,4.49,4.51]:

$$\frac{d\Delta N}{dt} = W(N_0 - \Delta N) - \frac{\Delta N}{\tau} - \frac{\sigma_0 c \Delta N q}{A\,L} \tag{12.1}$$

$$\frac{dq}{dt} = \frac{\sigma_0 c_0 \ell \Delta N}{L}(q+1) - \frac{q}{\tau_C} \tag{12.2}$$

with:

	W	:	pump rate [s^{-1}]
	N_0	:	doping concentration
	σ_0	:	cross section of stimulated emission [cm^2]
	τ	:	upper laser level lifetime
	L	:	length of resonator
	ℓ	:	geometrical length of active medium
	A	:	cross-sectional area of laser beam
	c	:	speed of light in active medium
	τ_C	:	cavity lifetime = $L/[c_0(\alpha_0\ell - \ln\sqrt{R})]$

At the very beginning of the pulse emission process, the inversion density has been built up by the pumping process. This initial inversion density is determined by the time interval Δt during which the Q-switch suppressed laser emission. For an ideal four-level system we have (see Chapter 9):

interval Δt during which the Q-switch suppressed laser emission. For an ideal four-level system we have (see Chapter 9):

$$\Delta N_i = \frac{WN_0}{W+1/\tau}\,(1-\exp[-(W+1/\tau)\Delta t]) \tag{12.3}$$

The emission of the Q-switch pulse happens during a time interval that is much shorter than the upper level lifetime. Any change of the inversion density due to pumping or spontaneous emission can therefore be neglected (first and second term on right hand side of (12.1)). By introducing the intensity $I=qh\nu c_0/(AL)$, the Q-switched rate equations can then be written as:

$$\frac{d\Delta N}{dt} = -\frac{I}{\tau}\frac{\Delta N}{I_S} \tag{12.4}$$

$$\frac{dI}{dt} = \frac{\sigma_0 c_0 \ell(\Delta N-\Delta N_{th})}{L}\,I \tag{12.5}$$

where we used the threshold inversion density:

$$\Delta N_{th} = \frac{\alpha_0\ell-\ln(\sqrt{R})}{\sigma_0\ell} \tag{12.6}$$

Equation (12.5) indicates that the output pulse will have its maximum ($dI/dt=0$) when the inversion density is equal to the threshold inversion density, an important result that will help us later to determine the peak power. It is not possible to solve these equations analytically to obtain inversion and intensity as a function of time. However, since we know the relationship between intensity and inversion density at the beginning and the end of the pulse as well as at the intensity maximum, one possible approach is to express the intensity as a function of the inversion density. With $dI/dt = dI/d\Delta N \cdot d\Delta N/dt$, we get:

$$\frac{dI}{d\Delta N} = \frac{h\nu c_0\ell}{L}\left[\frac{\Delta N_{th}}{\Delta N}-1\right] \tag{12.7}$$

Integration of (12.7) from time 0 to time t yields:

$$I(t) = \frac{h\nu c_0\ell}{L}\left[\Delta N_{th}\ln\left(\frac{\Delta N(t)}{\Delta N_i}\right)+\Delta N_i-\Delta N(t)\right] \tag{12.8}$$

where we used $I(t=0)=0$ and $\Delta N(t=0)$ equals the initial inversion density ΔN_i. At the end of the output pulse, the inversion density has been depleted to the final inversion density

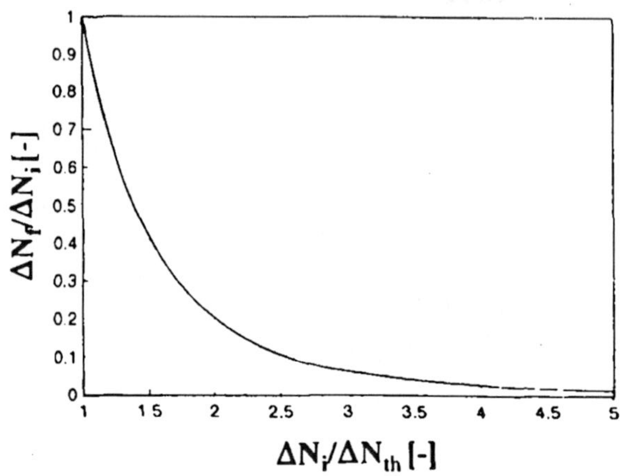

Fig. 12.5 Relationship between final and initial inversion densities according to (12.9).

$$\Delta N_i - \Delta N_f = \Delta N_{th} \ln\left(\frac{\Delta N_i}{\Delta N_f}\right) \tag{12.9}$$

This equation can be solved numerically by successive approximation. Figure 12.5 shows the final inversion density as a function of the initial inversion density. Since the threshold inversion density and the initial inversion density are known, Eq.(12.9) can be used to determine the residual inversion ΔN_f after the pulse has been emitted.

12.2.2 Energy, Pulse Duration and Peak Power

The output energy can be obtained by integ rating (12.7) using the output coupling transmission |lnR|/2:

$$E_{out} = A \frac{h\nu}{\sigma_0} \frac{|\ln R|}{2} \frac{\Delta N_i - \Delta N_f}{\Delta N_{th}} = A \frac{h\nu}{\sigma_0} \frac{|\ln R|}{2} \ln\left(\frac{\Delta N_i}{\Delta N_f}\right) \tag{12.10}$$

The peak power is determined by (12.7) with $\Delta N(t) = \Delta N_{th}$:

$$P_{peak} = A \frac{h\nu c_0 \ell}{L} \frac{|\ln R|}{2} \Delta N_{th} \left[\frac{\Delta N_i}{\Delta N_{th}} - \ln\left(\frac{\Delta N_i}{\Delta N_{th}}\right) - 1\right] \tag{12.11}$$

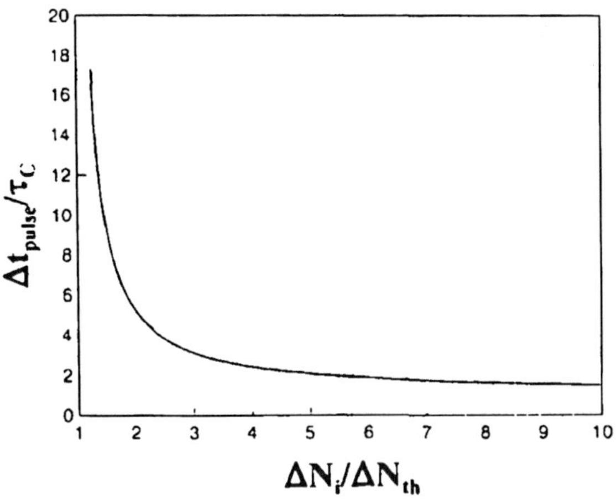

Fig.12.6 Ratio of pulse duration to cavity lifetime as a function of the initial inversion ratio, according to (12.12) and (12.9).

Since we do not know the temporal shape of the intensity, we have to define the pulse duration of the Q-switch pulse by the ratio of energy to peak power:

$$\Delta t_{pulse} = \tau_C \left[\frac{\Delta N_i}{\Delta N_{th}} - \frac{\Delta N_f}{\Delta N_{th}}\right] \Big/ \left[\frac{\Delta N_i}{\Delta N_{th}} - \ln\left(\frac{\Delta N_i}{\Delta N_{th}}\right) - 1\right] \qquad (12.12)$$

where τ_C is the cavity lifetime (see (12.2). By combining (12.12) with (12.9), the ratio of pulse duration to cavity lifetime becomes dependent only on the initial inversion ratio $\Delta N_i/\Delta N_{th}$. Figure 12.6 shows this relationship graphically. The pulse duration is proportional to the resonator length L and roughly inversely proportional to the initial inversion ratio. In other words, for a given hold-off time Δt, the pulse duration will decrease proportionally to the inverse pump power.

Let us now calculate an example to demonstrate how to apply (12.9) through (12.12) to predict the performance of a Q-switched laser.

Example:
A diode-pumped Nd:YAG (τ=230 μs, $\sigma_0=4.1\cdot10^{-19}$cm^2) with a rod diameter of 4mm, a rod length of 100mm, and 0.8% doping concentration ($N_0=1.1\times10^{20}$ cm^{-3}) is pumped with an optical pump power of 1kW at 808nm. The resonator has a length of 0.5m with 50% output coupling (R=0.5) and a loss of 2% per transit ($\alpha_0\ell=0.02$). Assuming 95% pump light absorption, the pump rate can be calculated to be W=27.6 s^{-1}, using $W\cdot N_0\cdot V\cdot h\nu_{pump}=P_{pump,abs}$, where V is the volume of the rod. The Q-switch is holding off the cavity for 2ms. With (12.3) we get the initial inversion density $\Delta N_i= 6.31\cdot10^{-3}\,N_0$. Equations (12.7)-(12.12) then yield:

threshold inversion density ΔN_{th} = $8.13 \cdot 10^{-4} N_0$

initial inversion ratio $\Delta N_i/\Delta N_{th}$ = 7.76

final inversion ratio $\Delta N_f/\Delta N_{th}$ = $3.32\ 10^{-3}$

output energy E_{out} = 156 mJ

cavity lifetime τ_C = 4.55 ns

pulse duration Δt_{pulse} = 7.49 ns

peak power P_{peak} = 20.8 MW

Given a pump time of 2ms, this system can be operated at a maximum repetition rate of nearly 500 Hz, with an average output power of over 70 W.

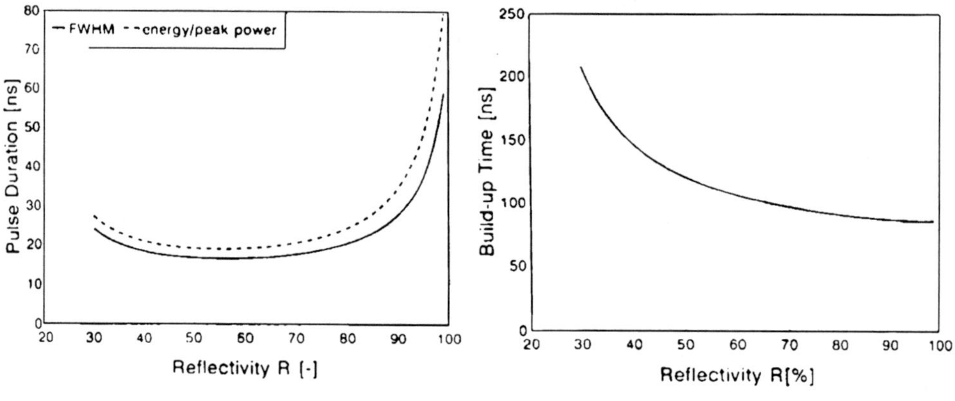

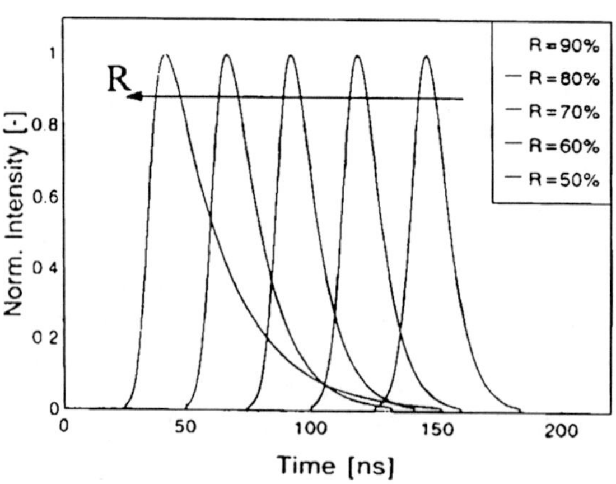

Fig. 12.7 Calculated FWHM pulse duration, build-up time and pulse shapes for a $Nd{:}YVO_4$ laser with $g_0\ell=\Delta N_i\sigma_0\ell=2.0$, $\alpha_0\ell=0.02$, and L= 500mm as a function of the reflectivity of the output coupler. Equations (12.1) and (12.2) were solved numerically. The dotted line shows the pulse duration as defined by (12.12). Note that in the lower graph, the pulses have been equally spaced to improve the presentation. The initial inversion ratio decreases from 79.3 for R=99% to 3.21 for R=30%.

Numerical solutions of the rate equations (12.1) and (12.2) are shown in Fig. 12.7 for a diode-pumped Vanadate laser with 2mm diameter pump spot and 104W absorbed pump power. With increasing output coupling (*1-R*), the pulses become shorter until the decrease in cavity lifetime is overcompensated by the right hand factor of (12.12) containing the inversion densities. This example indicates that increasing the output coupling may not necessarily result in shorter pulse durations. The build-up time is defined as the time interval between switching the Q-switch and the maximum of the output pulse. No analytical expression for this build-up time exists, but similar to the pulse duration (12.12), the build-up time is proportional to the resonator length L and will increase with decreasing ratio $\Delta N/\Delta N_{th}$ As a rule of thumb, for Q-switched lasers with near optimum output coupling, the build-up time is about ten times the pulse duration.

In preparation for the next section, where we will calculate the optimum output coupler reflectivity, let us define the initial and the final small-signal gain:

$$g_0\ell = \Delta N_i\,\sigma_0\,\ell \qquad\qquad (g_0\ell)_f = \Delta N_f\,\sigma_0\,\ell$$

Equations (12.9) through (12.12) can then be expressed in terms of the small-signal gains, the reflectivity R and the loss per transit $\alpha_0\ell$. Equation (12.9) and (12.10) then become:

$$g_0\ell - [g_0\ell]_f = (\alpha_0\ell - \ln\sqrt{R})\,\ln\left(\frac{g_0\ell}{[g_0\ell]_f}\right) \tag{12.13}$$

$$E_{out} = A\,\frac{h\nu}{\sigma_0}\,\frac{|\ln R|}{2}\,\ln\left(\frac{g_0\ell}{[g_0\ell]_f}\right) \tag{12.14}$$

By using these two equations, we can now determine the optimum reflectivity R_{opt} for which the output energy exhibits a maximum.

12.3 Optimum Output Coupling

Similar to the output power of cw laser resonators presented in Chapter 10, maximum output energy and optimum mirror reflectivity are a function of the small-signal gain $g_0\ell$ and the loss per transit $\alpha_0\ell$ [4.71]. The Q-switch extraction efficiency is defined as ratio of the output energy to the stored energy present at the time when the Q-switch opens:

$$\eta_{extr} = \frac{E_{out}}{\Delta N_i\,h\nu\,A\ell} = \frac{|\ln R|}{2g_0\ell}\,\ln\left(\frac{g_0\ell}{[g_0\ell]_f}\right) \tag{12.15}$$

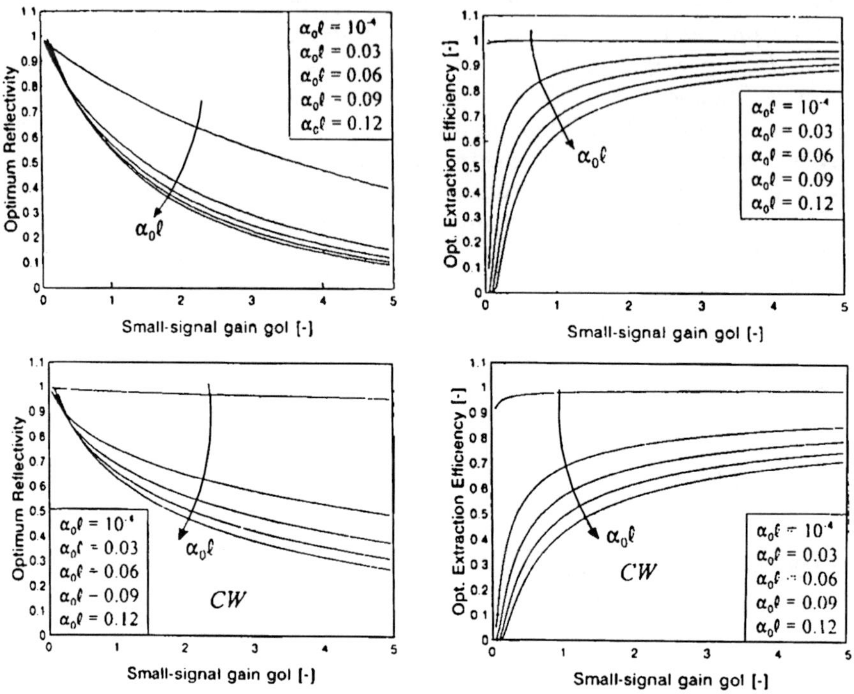

Fig. 12.8 Optimum reflectivity and maximum extraction efficiency for Q-switching (upper graphs) and cw-emission (lower graphs) as a function of the small-signal gain. Curve parameter is the loss per transit $\alpha_0\ell$. For Q-switching, the small-signal gain is determined by the initial inversion density: $g_0\ell=\Delta N_i\sigma_0\ell$.

Optimum reflectivity R_{opt} and maximum extraction efficiency $\eta_{extr,max}$ can be found analytically [4.71]:

$$\ln R_{opt} = -2\alpha_0\ell \left[\frac{x - 1 - \ln x}{\ln x}\right] \tag{12.16}$$

$$\eta_{extr,\max} = \frac{x - 1 - \ln x}{x} \tag{12.17}$$

with $x=g_0\ell/\alpha_0\ell$. These expressions are very similar to the corresponding equations for a cw laser presented in Chapter 10. Figure 12.8 shows optimum extraction efficiencies and optimum reflectivities as a function of small-signal gain and loss for both, Q-switching (Eqs. (12.16) and (12.17)) and cw emission (Eqs. (10.14) and (10.15)). These graphs indicate that for the same ratio x, Q-switched operation requires a higher output coupling. This relative insensitivity to the total cavity loss is also apparent in the extraction efficiency. Compared to a cw laser, Q-switching will provide higher extraction efficiencies for the same loss per transit $\alpha_0\ell$. At the optimum output coupling, the pulse duration (12.12) is given by [4.71]:

$$\Delta t_{pulse} = \frac{L}{c_0\alpha_0\ell}\left[\frac{\ln x}{x[1-y+y\ \ln y)]}\right] \qquad with\ \ y = \frac{x-1}{x\ \ln x} \qquad (12.18)$$

Example:
Application of (12.16) through (12.18) to the diode-pumped Nd:YAG laser discussed on page 439 yields:

small-signal gain $g_0\ell$	=	2.85
ratio $x=g_0\ell/\alpha_0\ell$	=	142.5
optimum reflectivity R_{opt}	=	33.2%
maximum Extraction Efficiency	=	95.8%
pulse duration	=	6.1 ns

For the same laser, decreasing the pump power by a factor 2 will result in:

small-signal gain $g_0\ell$	=	1.425
ratio $x=g_0\ell/\alpha_0\ell$	=	71.25
optimum reflectivity R_{opt}	=	53.9%
maximum extraction efficiency	=	92.6%
pulse duration	=	11.6 ns

12.4 Repetitive Q-switching

In the previous sections, we have assumed that the inversion at the start of the pump process is equal to zero (see Eq. (12.3)). This implies that after a Q-switch pulse, a certain time must elapse in order to let the final inversion density go to zero via spontaneous emission prior to starting the next pump and Q-switch cycle. If, however, the repetition rate f gets close to or is greater than the inverse of the upper laser level lifetime τ, the left over inversion at the start of the pump cycle has to be considered. This condition is referred to as repetitive Q-switching [4.55,4.57,4.59,4.60]. For a cw-pumped laser, the initial inversion density ΔN_i depends on the final inversion density of the previous pulse (we consider an ideal four-level system), adding another term to Eq. (12.3):

$$\Delta N_i = \frac{WN_0}{W+1/\tau}\,(1-\exp[-(W+1/\tau)/f]) - \Delta N_f \exp[-(W+1/\tau)/f] \qquad (12.19)$$

With the assumption $W\tau \ll 1$, this equation can be rewritten in terms of the small-signal gains:

$$g_0\ell = [g_0\ell]_{cw}\,(1-\exp[-1/(\tau f)]) - [g_0\ell]_f \exp[-1/(\tau f)] \qquad (12.20)$$

where $[g_0\ell]_{cw}$ is the small-signal gain of the cw-laser. This equation has to be combined with (12.13) and (12.14) to find a numerical solution through successive approximation. The average output power can then be calculated using P_{out}=f E_{out}. Figs. 12.9 through 12.11 show calculated output power, pulse duration and build-up time for the same diode-pumped solid state laser resonator (L=500mm, 2mm pump spot diameter, 100W pump power, $\alpha_0\ell$=0.02), but for four different commonly used laser materials. As the curves indicate, the output characteristics depend mainly on the upper level lifetime τ and the cross-section for stimulated emission σ_0. The relatively short lifetime of Nd:YVO_4 requires high repetition rates to achieve output powers close to the cw power, but its high emission cross-section results in short pulse duration. Nd:YLF and Yb:YAG, on the other hand, lend themselves well to low repetition rate operation due to their energy storage capacities. Yb:YAG, however, exhibits a low emission cross-section, which leads to relatively long pulses and limits the extraction efficiency due to the low small-signal gain. Note that for all four lasers, the build-up time at the optimum output coupling is about ten times the pulse duration, as stated previously. With increasing repetition rate, the initial small-signal gain becomes lower (see (12.19), resulting in a shift of the optimum output coupling to higher reflectivities.

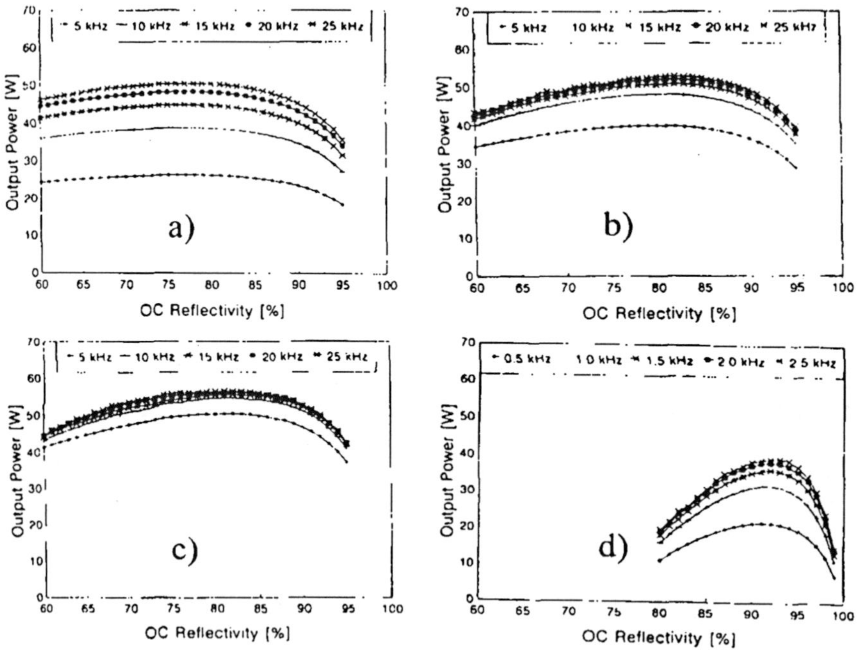

Fig. 12.9 Calculated average output power as a function of the output coupler reflectivity for four different diode-pumped solid state materials and the same resonator (L=500mm, 2mm pump spot diameter, 100W pump power, 2% loss per transit). a) Nd:YVO_4, λ_p=808nm, σ_0=15 x $10^{-19}cm^2$, τ=100μs , b) Nd:YAG, λ_p=808nm, σ_0=4.1x$10^{-19}cm^2$, τ=230μs c) Nd:YLF, λ_p=806nm, σ_0=1.8 x $10^{-19}cm^2$, τ=480μs d) Yb:YAG, λ_p=940nm, σ_0=0.21 x $10^{-19}cm^2$, τ=950μs.

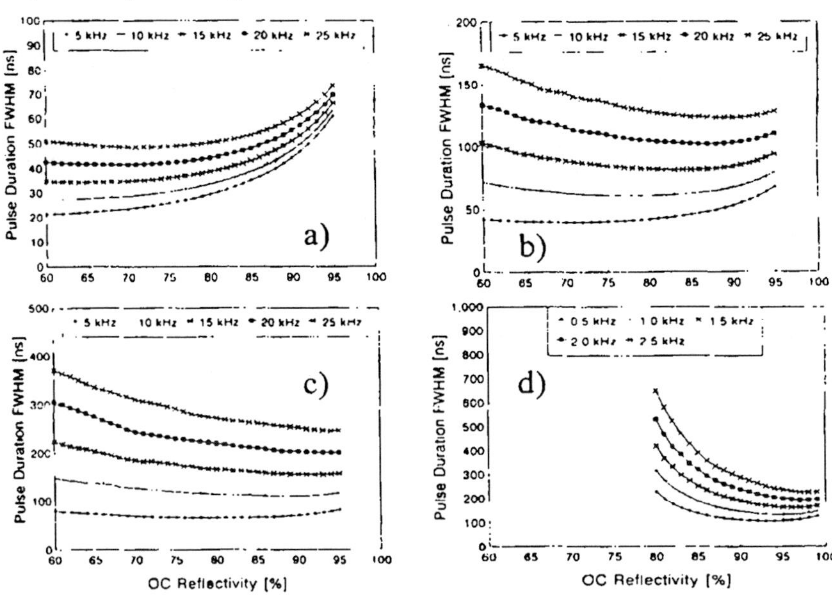

Fig. 12.10 Calculated pulse durations for the resonators of Fig. 12.9 (L=500mm, 2mm pump spot diameter, 100W pump power, 2% loss per transit). a) Nd:YVO_4, λ_p=808nm, b) Nd:YAG, λ_p=808nm, c) Nd:YLF, λ_p=806nm, d) Yb:YAG, λ_p=940nm.

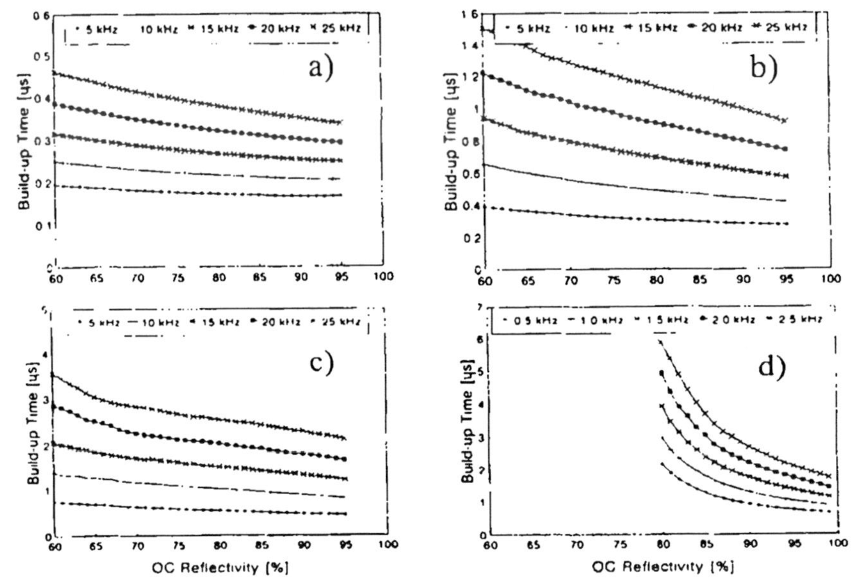

Fig. 12.11 Calculated pulse build-up time for the resonators of Figs. 12.9 and 12.10 (L=500mm, 2mm pump spot diameter, 100W pump power, 2% loss per transit). a) Nd:YVO_4, λ_p=808nm, b) Nd:YAG, λ_p=808nm, c) Nd:YLF, λ_p=806nm, d) Yb:YAG, λ_p=940nm.

Experimental examples of repetitively Q-switched Vanadate lasers are presented in Figs. 12.12 and 12.13. Both systems are designed to provide a diffraction limited output beam. The choice of laser material is usually determined by the application requirements on pulse energy, pulse duration and repetition rate. For repetition rates below 10kHz, Nd:YAG or Nd:YLF are the materials of choice, whereas Nd:YVO_4 finds widespread applicaton at high repetition rates (20-400 kHz).

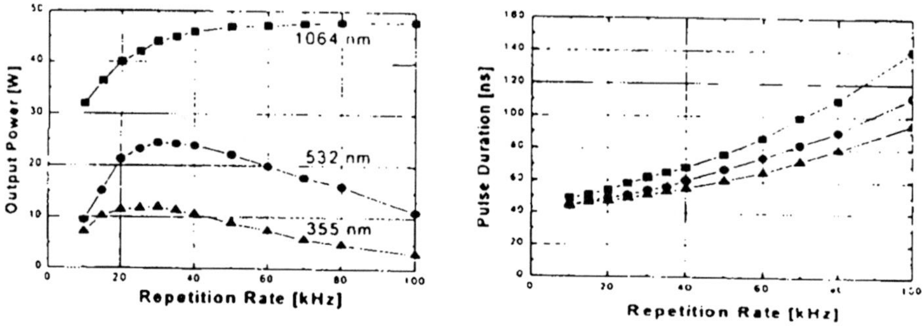

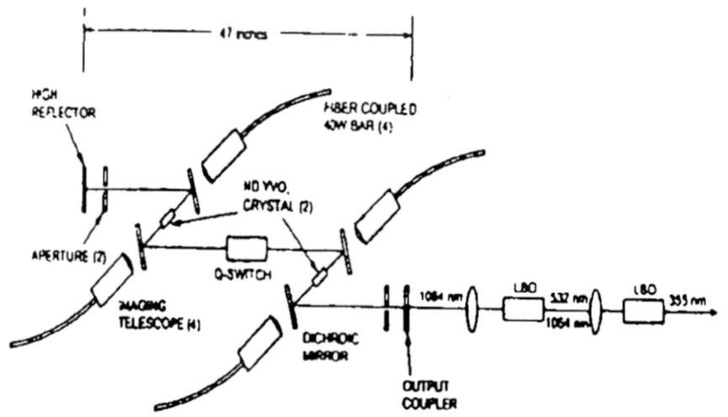

Fig. 12.12 Measured output power and pulse durations of a diode-end-pumped Nd:YVO_4 laser with AOM Q-switch in TEM_{00} mode operation at 1064nm and with external frequency doubling and tripling. The layout for frequency tripling is shown.. Total optical pump power at 808nm incident on the crystals is 104W [4.77] (© OSA 2001).

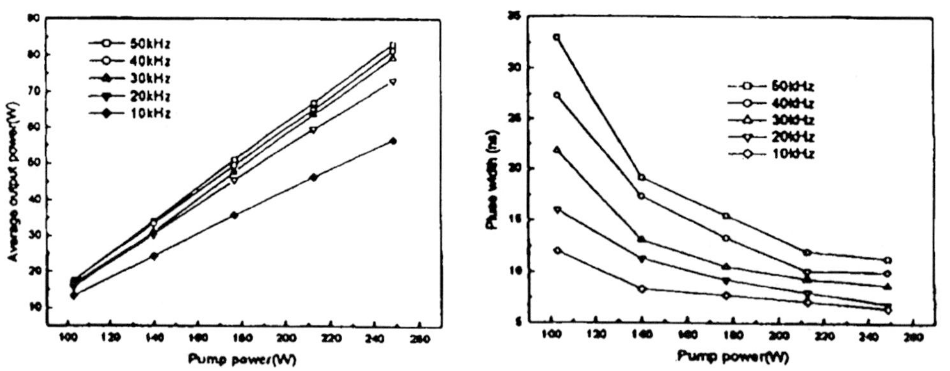

Fig. 12.13 Output power and pulse durations for a diode-pumped Nd:YVO_4 slab laser with an electro-optic BBO Pockels cell. The slab with dimensions 1x12x10mm is end pumped over an area of 0.5mmx12mm. A 67.5mm long flat-unstable hybrid resonator is used to obtain $M^2<1.5$ [4.79] (© OSA 2003).

Optimum Output Coupling

The optimum output coupling now depends on three rather than two parameters: the cw small signal gain $[g_0\ell]_{cw}$, the loss per transit $\alpha_0\ell$, and the product of lifetime and repetition rate τf. This makes it a little difficult to present the results graphically. However, since the loss per transit is usually within a small range of a couple of percent, a good compromise is to present the optimum values as a function of gain and repetition rate. To find the optimum, the three equations (12.13), (12.14), and (12.20) have to be solved numerically via successive approximation. In the limit $\tau f \to 0$, the final inversion density of the previous pulse has been depleted before the next Q-switch cycle starts, which means that the optimum reflectivities for Q-switching shown in Fig. 12.8 apply. For high repetition rates with $\tau f \gg 1$, we expect the optimum reflectivities to be close to those of a cw-laser. In any case, the previously discussed expressions (12.16) and (12.17) are still valid, provided that the initial inversion density is calculated with (12.20) rather than (12.3). Since we are considering a cw-pumped laser, it is reasonable to define the extraction efficiency as the ratio of the average output power to the inversion power:

$$\eta_{extr} = \frac{P_{out}}{A\ I_s\ [g_0\ell]_{cw}} = \frac{E_{out}\,f}{A\ I_s\ [g_0\ell]_{cw}} = \frac{|\ln R|}{2[g_0\ell]_{cw}}\ \ln\left(\frac{g_0\ell}{[g_0\ell]_f}\right)\ \tau f \tag{12.21}$$

Figure 12.14 presents optimum output coupling and maximum extraction efficiencies as a function of the cw small-signal gain $[g_0\ell]_{cw}$ for a loss of $\alpha_0\ell=0.03$. As expected, both parameters approach the cw-values with increasing repetition rates. Typically, for $\tau f>5$, the average output power is within 5% of the cw power. Unfortunately, the repetition rate of repetitively Q-switched lasers can not be increased indefinitely, because the time it takes the inversion to reach the threshold inversion will eventually exceed the inverse frequency $1/f$. The pulse repetition rate will then automatically switch to half the Q-switch repetition rate.

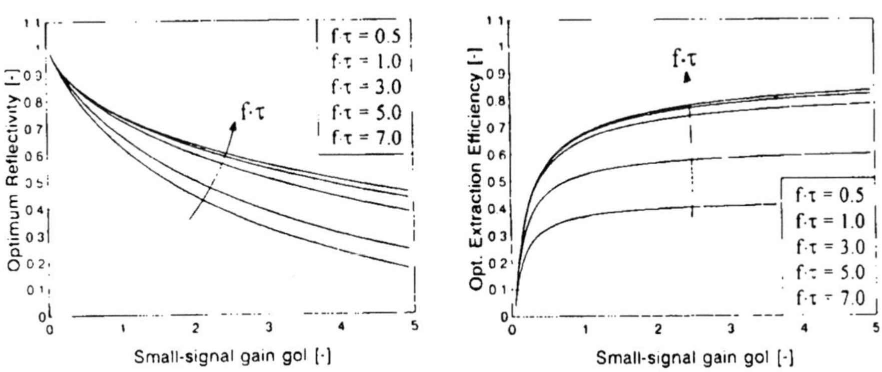

Fig. 12.14 Optimum output coupling reflectivities and maximum extraction efficiencies as a function of the cw small-signal gain $[g_0\ell]_{cw}$ for a loss per transit of $\alpha_0\ell=0.03$. Curve parameter is the product of upper level lifetime and repetition rate.

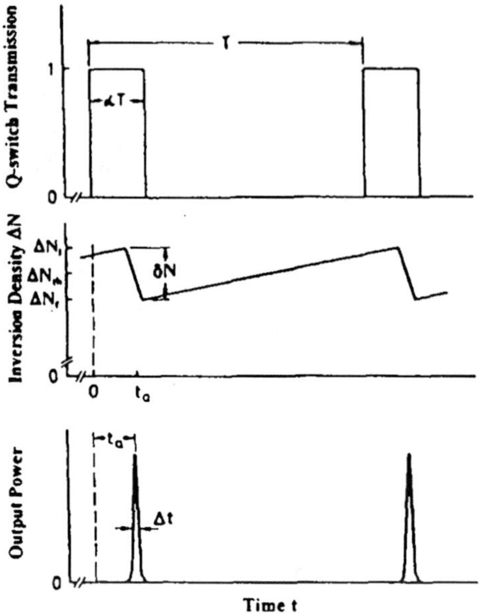

Fig. 12.15 Q-switch transmission, inversion density build-up and pulse emission for repetitive Q-switching with $\tau f>1$. The Q-switch generates no loss during the time αT. The peak power is observed after the build-up time t_a.

At high repetition rates with $\tau f>1$, the inversion increases linearly between pulses and the initial and final inversion densities stay close to the threshold inversion density (Fig. 12.15):

$$\Delta N_i - \Delta N_{th} \approx \Delta N_f - \Delta N_{th} \approx \delta N/2 \tag{12.22}$$

Insertion into (12.10), (12.11) and (12.12) and expansion of the logarithm to the second order results in:

$$E_{out} = \frac{A\,h\nu}{\sigma_0}\,\frac{|lnR|}{2}\,\frac{\delta N}{2\Delta N_{th}} \tag{12.23}$$

$$P_{peak} = A\,\frac{h\nu c_0 \ell}{L}\,\frac{|lnR|}{2}\,\Delta N_{th}\left[\frac{\delta N}{2\Delta N_{th}}\right]^2 \tag{12.24}$$

$$\Delta t_{pulse} = \frac{L}{c_0}\,\frac{2}{\delta N\,\sigma_0 \ell} \tag{12.25}$$

A surprising result is that the product of the pulse energy and the pulse duration is a constant of the laser resonator and does not depend on the pump power:

$$E_{out} \cdot \Delta t_{pulse} = \frac{AL}{c_0} \frac{h\nu}{\sigma_0} \frac{|\ln R|}{|\ln R| + 2\alpha_0 \ell} \tag{12.26}$$

If we introduce the cw-output power P_{cw} of the system:

$$P_{cw} = A\ I_S \frac{1-R}{2\sqrt{R}} \left(\frac{[g_0 \ell]_{cw}}{|\ln(\sqrt{R}V_S)|} - 1 \right)$$

the equations read [4.78]:

$$E_{out} \approx 2\ \tau\ m\ P_{cw} \tag{12.27}$$

$$P_{peak} = \frac{c\tau}{2L} \left| \frac{\ln^2(\sqrt{R}V_S)}{\ln\sqrt{R}} \right| \frac{m^2}{A\ I_S} P_{cw}^2 \tag{12.28}$$

$$\Delta t_{pulse} = A\ I_S \frac{2L}{c} \frac{2}{m\ P_{cw}} \tag{12.29}$$

with the modulation function:

$$m = \frac{1 - \exp[-1/(\tau f)]}{1 + \exp[-1/(\tau f)]} \tag{12.30}$$

Let us now find an expression for the maximum Q-switch repetition rate at which still one pulse is emitted per Q-switch period. It is safe to assume, that this is only possible if the inversion can build up beyond the threshold inversion within the time period $1/f$. By linearizing Eq. (12.19) with $W\tau \ll 1$ and using (12.22), we can find an expression for the modulation depth δN of the inversion density:

$$\frac{\delta N}{2\Delta N_{th}} = \frac{\tau W N_0}{\Delta N_{th}} - 2\tau f + 1 \tag{12.31}$$

This equation can be rewritten in terms of the pump power:

$$\frac{\delta N}{2\Delta N_{th}} = \frac{\eta_{excit}\ P_{pump}}{I_S\ A\ (\alpha_0 \ell - \ln\sqrt{R})} - 2\tau f + 1 \tag{12.32}$$

with the excitation efficiency η_{excit} defined by (9.19)-(9.21). The initial inversion ΔN_i can only be greater than the threshold inversion if the right-hand side of (12.32) is greater than zero. The condition for the allowed Q-switch repetition rates then reads:

$$f < \frac{1}{2\tau}\left[\frac{\eta_{excit}\,P_{pump}}{I_S\,A\,(\alpha_0\ell - \ln\sqrt{R})} + 1\right] \tag{12.33}$$

Example:
Diode pumped Nd:YVO$_4$ laser with emission wavelength λ_L=1064nm, diameter of pumped area: 2mm, pump power at λp=808nm: 100W with 95% absorption, R=0.9, $\alpha_0\ell$=0.02. With I_S=1.257 kW/cm^2 and τ=100 μs, Eq. (12.33) provides the condition:

$$f < 125.7 \text{ kHz}$$

Increasing the repetition rate of the Q-switch beyond the limit given by (12.33) will result in a reduction of the pulse frequency by a factor of 2. It now takes two Q-switch cycles to build up the inversion. Further increase of the repetition rate will then result in further reductions of the pulse repetition rates. An experimental example of an cw flashlamp-pumped Nd:YAG with AOM is shown in Fig. 21.16. It should be mentioned that the time during which the Q-switch generates no loss, which is referred to as the gate-open time, has a strong influence on the laser performance. If the gate-open time is chosen too large, several pulses are emitted, or the laser may even emit in cw-operation after a sequence of spikes. On the other hand, if the gate-open time is too short, it takes two Q-switch cycles before the pulse can be emitted. Typical gate-open times for solid state lasers range from one to ten microseconds, depending on gain, lifetime and repetition rate.

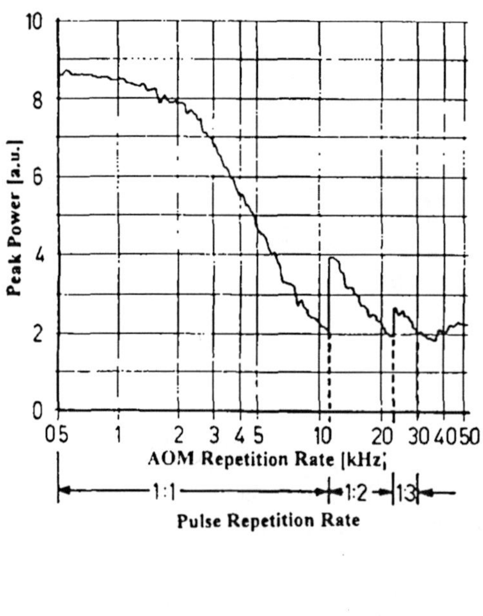

Fig. 12.16 Measured pulse peak power as a function of the AOM repetition rate for a cw flash-lamp pumped Nd:YAG laser in TEM$_{00}$ mode operation with acousto-optic Q-switching. The pulse repetition rate is reduced by a factor 2 and 3 as the AOM repetition is increased and never exceeds 10.5kHz.

Chapter 13

Resonators with Variable Internal Lenses

13.1 General Aspects

13.1.1 Thermal Lensing in Solid State Lasers

In solid state lasers the active medium exhibits the properties of a thermally induced lens when the material is pumped. The refractive power is brought about by a combination of heat generation due to absorption of pump and laser radiation and the flow of heat to the outer periphery due to cooling. For a laser rod this leads to a temperature distribution which is parabolic with respect to the radial position r if the thermal conductivity is constant and the medium is illuminated homogeneously by the excitation source (Fig. 13.1). This results in a similar radial dependence of the index of refraction, neglecting polarization effects:

$$n(r) = n_0 (1 - \gamma r^2) \tag{13.1}$$

where n_0 is the index of refraction at the center. A bundle of parallel rays incident on the laser rod is focused by the thermal lens for $\gamma > 0$ since the optical path length $n(r)\ell$ inside the medium is longer for rays being closer to the center of the rod.

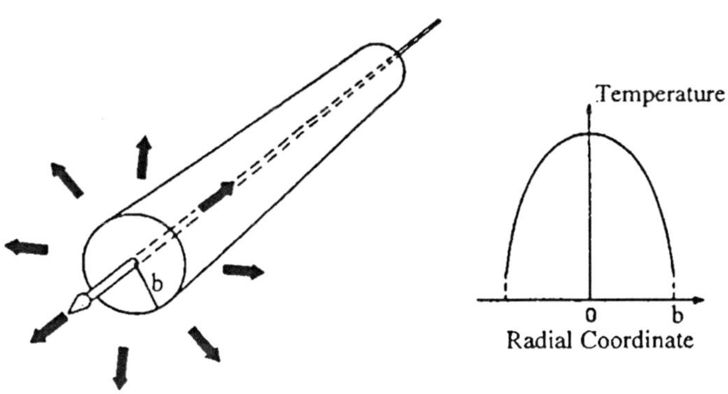

Fig. 13.1 Heat deposition by the pump process and cooling of the outer periphery generates a parabolic temperature profile in solid state laser rods.

This is in contrast to common lenses where the change in the optical path length is generated by an increased thickness at the center of the lens and not by an increase in the index of refraction. To a first approximation, the refractive power D of a rod of length ℓ is given by:

$$D = 2\gamma n_0 \ell \tag{13.2}$$

whereby effects caused by the bending of the rod endfaces are neglected [4.2,4.83, 4.84,4.89]. The proportionality of the parameter γ to the dissipated heat P_H leads to a linear relationship between the pump power and the refractive power. For a rod with radius b, one gets for the averaged value (radial/azimuthal polarization):

$$D = \frac{1}{K\, 2\pi b^2}\left[\frac{dn}{dT} + \epsilon\right] P_H = \frac{1}{K\, 2\pi b^2}\left[\frac{dn}{dT} + \epsilon\right] \chi \eta_{excit} P_{pump} \tag{13.3}$$

with:

K	:	thermal conductivity
dn/dT	:	temperature derivative of the index of refraction
ϵ	:	stress dependent variation of the refractive index
χ	:	thermal load parameter
η_{excit}	:	excitation efficiency (see Sec. 9.3)
P_{pump}	:	electrical pump power for flashlamp pumping and optical pump power for diode pumping

The thermal load parameter χ is the ratio of the heat P_H to the power P_{UL} available in the form of population inversion. For flashlamp pumped Nd:YAG rods in pulsed and cw operation the thermal load parameter typically is between 2-3 and 1-1.5, respectively, depending on the excitation spectrum. For diode pumping at 808nm, the thermal load parameter is about 0.33. The stress dependent parameter ϵ accounts for the variations in refractive power between radial and azimuthal polarization; in Nd:YAG the refractive power for radial polarized light is about 15% higher than the refractive power for azimuthally polarized light. The material constants in (13.3) can be combined in a constant C with:

$$D = \frac{C}{A}\, \chi\, \eta_{excit}\, P_{pump} \tag{13.4}$$

We see that the thermal lens does not depend on the length of the active medium and that it is inversely proportional to the cross sectional area $A = \pi b^2$ of the rod material. The constant C is a characteristic of the laser material. This expression is usually simplified by introducing the thermal lensing coefficient $\alpha = C\chi\eta_{excit}$:

$$D = \frac{\alpha}{A} P_{pump} \tag{13.5}$$

Table 13.1 gives an overview of typical thermal lensing coefficients for flashlamp pumped solid state rods and the corresponding refractive powers for a rod diameter of 10mm. For diode pumping, the thermal lensing coefficient is about 30% lower (P_{pump} is the optical pump power absorbed by the laser material). Thermal lensing coefficients for diode-pumped materials are given in Table 13.2. Measured thermal lensing coefficients for end-pumped Vanadate are depicted in Fig.13.2.

Table 13.1 Thermal lensing coefficients of flashlamp pumped solid state laser rods and the corresponding refractive powers at 1kW of electrical pump power for a rod diameter of 10mm. The range of the thermal lensing coefficient is generated by different pump conditions (efficiency!) and variations in crystal quality.

Material	α [μm/kW]	D[m^{-1}]
Nd:YAG (1064nm)	21 - 30	0.27 - 0.38
Nd:glass (1050nm)	150 -190	1.91 - 2.42
Nd:Cr:GSGG (1064nm)	37 - 57	0.47 - 0.73
Nd:Cr:GGG (1064nm)	75 - 95	0.95 - 1.21
Nd:YAP (1058nm)	37 - 69	0.47 - 0.88
Nd:YLF (1053nm, 1047nm)	-3*; 14**	- 0.048; 0.018
Alexandrite (760nm)	7 - 12	0.09 - 0.15

* σ-polarized at λ=1053nm **π-polarized at λ=1047nm

Table 13.2 Thermal lensing coefficients of optically pumped solid state laser rods and the corresponding refractive powers at 100W of absorbed optical pump power for a rod diameter of 3mm. The range of the thermal lensing coefficient is caused by differences in pumping geometry and doping concentration. In end-pumped lasers, a lower doping concentration results in a lower thermal lensing coefficient due to lower temperatures and less pronounced bulging of the end face. For low doping concentrations (<0.4% for Nd:YAG, and < 0.25% in Nd:YVO_4), the lower value of the thermal lensing coefficient applies. A flat-top transverse pump profile is assumed. For a Gaussian pump profile, the thermal lensing coefficient is two times higher than the values shown.

Material	pump wavelength [nm]	α [μm/W]	D[m^{-1}]
Nd:YAG (1064nm)	808	0.10 - 0.15	1.41- 2.12
Nd:glass (1054nm)	808	0.70 - 1.0	9.9 - 14.1
Nd:YVO_4(1064nm)	808	0.09 - 0.14	1.27 - 1.98
Nd:YVO_4(1064nm)	879	0.06 - 0.10	0.85 - 1.41
Nd:$GdVO_4$ (1064nm)	808	0.11 - 0.15	1.55 - 2.12
Nd:YAP (1079nm)	795	0.18 - 0.26	2.55 - 3.67
Nd:YLF (1053nm, 1047nm)	806	-0.015*; 0.07**	- 0.21 ; 0.99
Yb:YAG	940	0.04 - 0.07	0.56 - 0.99
Ti:sapphire	532	0.07 - 0.12	0.99 - 1.7

* σ-polarized at λ=1053nm **π-polarized at λ=1047nm

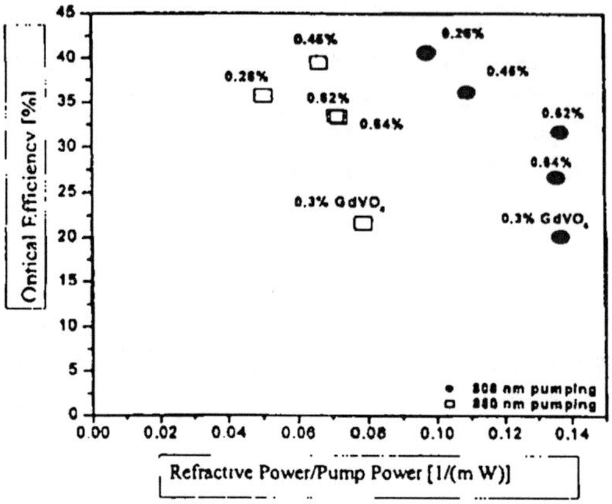

Fig. 13.2 Measured optical efficiencies and refractive powers D for end-pumped Nd:YVO_4 lasers at different pump diode wavelengths and a pump power of 18W (flat-top transverse pump distribution). Parameter is the doping concentration. Measurements for $GdVO_4$ are also shown. Pump spot diameter: 1.1mm [4.122] (©OSA 2002).

Since the lens becomes stronger as the thermal load is increased, the beam propagation inside and outside the resonator will become dependent on the pump power. Therefore, the beam quality and the focusing characteristics of the laser beam change with the laser output power. Thermal lensing is the dominant effect for beam quality and power limitations in solid state lasers. However, thermal lensing also occurs in gas lasers [4.86], but compared to solid state lasers, the focal lengths are two orders of magnitude longer.

13.1.2 Ray Transfer Matrices

Similar to the previous discussion of optical resonators, the theoretical description of resonators with an internal variable lens is based on ray transfer matrices. For a thermal lens with an index of refraction given by (13.1) the ray transfer matrix reads [4.103]:

$$M_{TL} = \begin{pmatrix} \cos(\sqrt{2\gamma}\ell) & \frac{1}{\sqrt{2\gamma}n_0}\sin(\sqrt{2\gamma}\ell) \\ -\sqrt{2\gamma}n_0\sin(\sqrt{2\gamma}\ell) & \cos(\sqrt{2\gamma}\ell) \end{pmatrix} \tag{13.6}$$

In general, the relation $\sqrt{2\gamma}\ell < 1$ holds and the ray transfer matrix can be simplified by expanding the matrix elements into series of linear and quadratic terms:

$$M_{TL} = \begin{pmatrix} 1 - Dh & \ell/n_0 \\ -D & 1 - Dh \end{pmatrix} \tag{13.7}$$

with: $D = 2\gamma n_0 \ell$ and $h = \ell/(2n_0)$

Thus, we get the ray transfer matrix of a thick lens with refractive power D with the principal planes being located inside the rod at a distance h from the endfaces. The ray transfer matrix of a thick lens can be used for $\sqrt{D\ell/n_0}<0.5$. At the upper limit, the difference in beam radii and beam propagation are on the order of one percent compared to the correct ray transfer matrix (13.6). For a 100 mm long Nd:YAG rod ($n_0=1.82$), that means that the refractive power can be as high as 4.5 Diopters before the matrix (13.6) has to be used to calculate the ray propagation. This approximation is usually valid for all solid state lasers so that the ray transfer matrix of the thick lens is used from now on.

In Sec. 8.1, resonators with an internal thin lens were already discussed. All results derived in this section are still valid with the exception that d_1 and d_2 now represent the distances of the principal planes to the mirrors [4.80,4.81,4.93,4.100]. In order to determine the propagation of the Gaussian beam (fundamental mode) inside the resonator, the concept of the equivalent g-parameters is applied. An optical resonator with an internal thermal lens (lens resonator) exhibits the same Gaussian beam radii at the mirrors as the equivalent empty resonator with the equivalent g-parameters g_i^* and the equivalent resonator length L^* with (Fig. 13.3):

$$g_i^* = g_i - Dd_j(1 - d_i/\rho_i) \qquad i,j=1,2;\ i\neq j \tag{13.8}$$

$$g_i = 1 - (d_1 + d_2)/\rho_i \tag{13.9}$$

$$L^* = d_1 + d_2 - Dd_1d_2 \tag{13.10}$$

Note that the term d_1+d_2 is the effective resonator length L_{eff} of the resonator (see Sec. 9.2) which means that we could replace the thick lens by a thin lens with d_1,d_2 being the distances to the two mirrors. The resulting ray transfer matrix for a transit between the mirrors would be identical. By using the equivalent resonator parameters, the Gaussian beam propagation inside the resonator can be calculated by making use of the fact that the mirror surfaces are surfaces of constant phase for the beam.

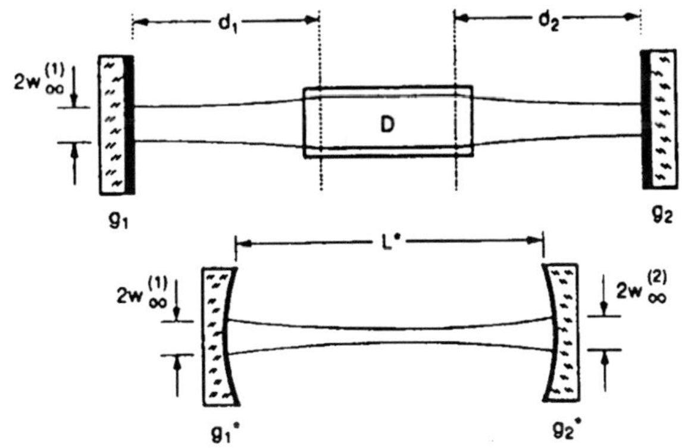

Fig. 13.3 The lens resonator and its equivalent resonator. Both resonators exhibit identical Gaussian beam radii at the mirrors but the beam propagation is different.

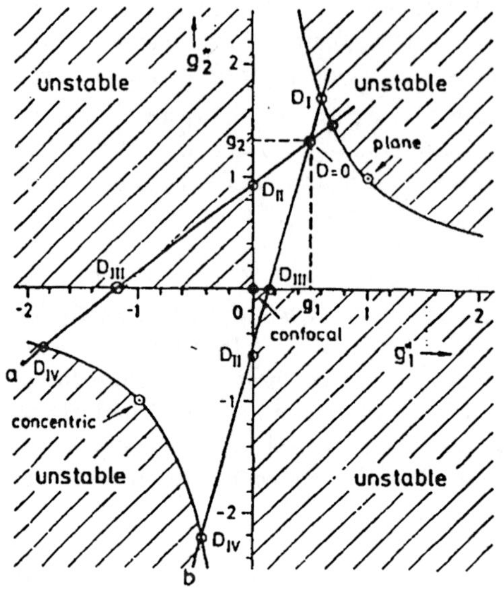

Fig. 13.4 Lens resonators with one internal lens move along straight lines through the equivalent g-diagram as the refractive power D is increased. The slope of the lines depends on the position of the medium and on the mirror curvatures.

The properties of a lens resonator can be visualized in the equivalent g-diagram (Fig. 13.4). The resonator starts at the point (g_1,g_2) and moves along a straight line through the diagram with increasing refractive power. In general, the lens resonator passes through stable zones and unstable zones. A lens resonator is called stable for $0<g_1^*g_2^*<1$ and unstable for $|g_1^*g_2^*|>1$. Four characteristic refractive powers exist at which the resonator intersects stability limits (see Fig. 13.4):

a) $g_1^*g_2^*=1$ and $g_1^*>0$: $$D_I = -\frac{1}{\rho_1 - d_1} - \frac{1}{\rho_2 - d_2} \qquad (13.11)$$

b) $g_1^*=0$: $$D_{II} = -\frac{1}{\rho_1 - d_1} + \frac{1}{d_2} \qquad (13.12)$$

c) $g_2^*=0$: $$D_{III} = \frac{1}{d_1} - \frac{1}{\rho_2 - d_2} \qquad (13.13)$$

d) $g_1^*g_2^*=1$ and $g_1^*<0$: $$D_{IV} = \frac{1}{d_1} + \frac{1}{d_2} \qquad (13.14)$$

If $|D_{II}|<|D_{III}|$ holds, the lens resonator passes through the upper left unstable region in Fig. 13.4. Resonators with the property $D_{II}=D_{III}$ can reach the origin of the diagram (confocal resonator). These resonators are characterized by the condition:

$$\frac{d_2}{d_1} = \sqrt{\frac{g_1}{g_2}} \tag{13.15}$$

which for a resonator with two identical mirrors means that the lens is located in the center.

Example:
Nd:YAG rod, ℓ=150mm, n_0=1.82; resonator : $\rho_1=\infty$, ρ_2=3m, d_1=0.3m, d_2=0.5m
The distance of the mirrors from the rod endfaces are 0.259m and 0.459m which means that the geometrical resonator length is 0.868m. The resonator starts at the point g_1=1, g_2=0.733 in the equivalent g-diagram. The four characteristic refractive powers are:

D_I = -0.4m^{-1}, D_{II} = 2.0m^{-1}, D_{III} = 2.933m^{-1}, D_{IV} = 5.333m^{-1}

13.2 Stable Resonators

13.2.1 Fundamental Mode Operation

By using the equivalent resonator parameters, the beam propagation of the Gaussian beam with wavelength λ inside the lens resonator can be derived from the known beam radii at the resonator mirrors. The Gaussian beam exhibits two waists whose position and radius are a function of the refractive power. The following relations hold for the resonator shown in Fig. 13.5:

Beam radius at mirror i :
$$w_i^2 = \frac{\lambda L^*}{\pi}\sqrt{\frac{g_j^*}{g_i^*(1-g_1^*g_2^*)}} \tag{13.16}$$

Waist radii :
$$w_{0i}^2 = \frac{\lambda L^*}{\pi}\,\frac{\sqrt{g_1^*g_2^*(1-g_1^*g_2^*)}}{g_j^*(L^*/\rho_i)^2 + g_i^*(1-g_1^*g_2^*)} \tag{13.17}$$

Beam radius at the principal planes of the thermal lens:

$$w_L^2 = w_1^2\left[\left(1-\frac{d_1}{\rho_1}\right)^2 + \left(\frac{d_1}{L^*}\right)^2\frac{g_1^*(1-g_1^*g_2^*)}{g_2^*}\right] \tag{13.18}$$

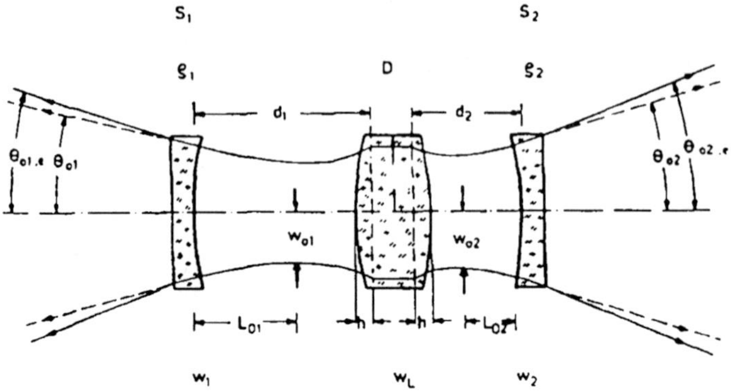

Fig. 13.5 Propagation of the Gaussian beam in a lens resonator.

Location of beam waist i :

$$L_{0i} = L^* \frac{g_j^* L^*/\rho_i}{g_j^*(L^*/\rho_i)^2 + g_i^*(1-g_1^* g_2^*)} \tag{13.19}$$

Half angle of divergence :

$$\theta_{0i} = \frac{\lambda}{\pi\, w_{0i}} \tag{13.20}$$

The beam waist i is located to the left of mirror i for $L_{0i}<0$ and to the right for $L_{0i}>0$. In Fig. 13.5, L_{01} is positive and L_{02} is negative. Due to the thermal lens, the two angles of divergence are different because the two waist radii are not equal. The beam parameter product is the same on both sides of the resonator and is given by:

$$w_{01}\theta_{01} = w_{02}\theta_{02} = \frac{\lambda}{\pi} \tag{13.21}$$

Unfortunately, the above shown expressions are too complicated to discuss in general; therefore, it is necessary to calculate the beam parameters to get a feeling for the beam characteristics. In the following we present the dependence of the beam radii and the angle of divergence on the refractive power for the resonator discussed in the previous example. The Gaussian beam is an eigensolution for this resonator if the refractive power lies within the ranges $0m^{-1}$-$2m^{-1}$ and $2.93m^{-1}$-$5.33m^{-1}$. Outside the stable zones, real values are not provided by (13.16)-(13.20). Figure 13.6 presents the beam radius in the medium w_L, the waist radii w_{0i}, and the half angles of divergence θ_{0i} as a function of the refractive power. The stability limits are characterized by infinite or zero waist radii whereby the divergence exhibits a reciprocal behavior since the beam parameter product has to remain constant.

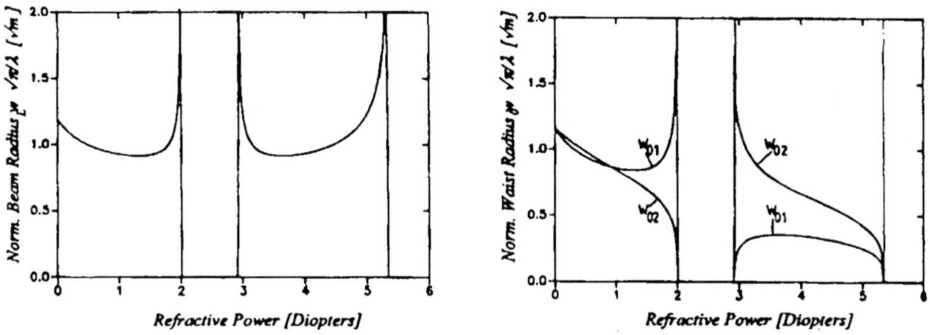

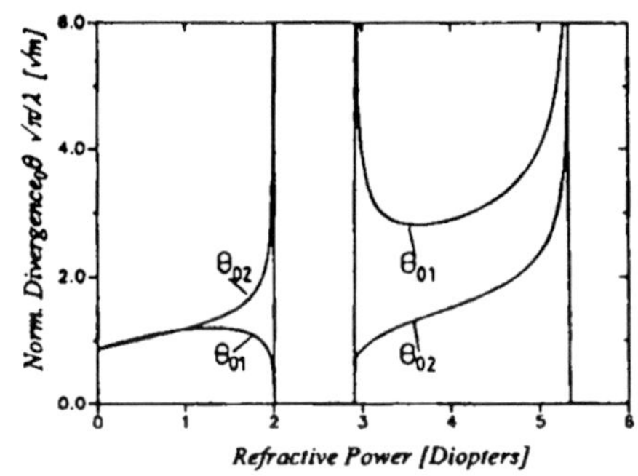

Fig. 13.6 Gaussian beam radius in the active medium w_L, waist radii w_{0i}, and half angles of divergence θ_{0i} as a function of the refractive power of the thermal lens ($\rho_1=\infty$, $\rho_2=3m$, $d_1=0.3m$, $d_2=0.5m$).

Typically, the Gaussian beam radius inside the medium exhibits a minimum which is approximately located in the middle of each stable zone. If ΔD denotes the range of the refractive power within which the resonator is stable, the minimum Gaussian beam radius in the medium reads:

$$w_{L,\min}^2 = \frac{4\lambda}{k\ \pi\ \Delta D} \tag{13.22}$$

with $k = 1$ if the confocal point $g_1^*=g_2^*=0$ is passed and $k = 2$ elsewhere.

For the resonator in Fig. 13.6 and a wavelength of λ=1.064µm, we get a minimum beam radius of 0.531mm in both stable zones. The Gaussian beam radius in the medium stays relatively constant over a range of the refractive power that covers about 90% of the stable zone. Within this range fundamental mode operation can be attained if an aperture is placed in front of the medium whereby the aperture radius a needs to be adapted to the Gaussian beam radius:

$$a = 1.3\ w_{L,\min} \tag{13.23}$$

13.2.2 Transverse Multimode Operation

If no aperture is used inside the resonator, higher order transverse modes will oscillate whose beam radii are smaller than the radius b of the active medium. The number of oscillating modes is equivalent to the beam quality factor M^2 and can, to a good approximation, be calculated with:

$$M^2 = \left(\frac{b}{w_L}\right)^2 \tag{13.24}$$

where w_L is the Gaussian beam radius at the principal planes of the thermal lens. Since the Gaussian beam radius changes with increasing thermal lens power, the beam quality of multimode lasers is a function of the pump power. Insertion of (13.18) into (13.24) yields for the beam propagation factor:

$$M^2 = \frac{\pi b^2}{\lambda L^*}\sqrt{\frac{g_1^*(1-g_1^*g_2^*)}{g_2^*}}\left[\left(1-\frac{d_1}{\rho_1}\right)^2+\left(\frac{d_1}{L^*}\right)^2\frac{g_1^*(1-g_1^*g_2^*)}{g_2^*}\right]^{-1} \tag{13.25}$$

This expression is only valid for $M^2 > 1$ since lower beam quality factors make no physical sense. At the stability limits the beam quality factor approaches $M^2=1$. For rectangular active media, the radius b has to be replaced with half the side length. In general, the beam quality factors in the x- and the y-direction are different.

With the knowledge of the beam quality factor (13.25), the waist radii w_{mi} and the half angles of divergence θ_{mi} can be calculated using (13.17) and (13.20):

$$w_{mi} = \sqrt{M^2}\, w_{0i}\,, \qquad \theta_{mi} = \sqrt{M^2}\,\theta_{0i} \tag{13.26}$$

with the beam parameter product:

$$w\theta = w_{m1}\theta_{m1} = w_{m2}\theta_{m2} = M^2\frac{\lambda}{\pi} \tag{13.27}$$

The locations of the waist radii are the same as for the Gaussian beam. It is interesting to note that the beam parameter product in multimode operation does not depend on the laser wavelength λ, since the beam quality factor scales with $1/\lambda$. We thus cannot improve the focusability by using a smaller wavelength. The beam parameter product of the Gaussian beam becomes lower with a smaller wavelength but the smaller Gaussian beam radius also leads to the oscillation of a higher number of transverse modes. Therefore, the beam parameter product remains constant.

As previously discussed, the Gaussian beam radius in the medium exhibits a minimum near the center of a stable zone. Therefore, the multimode beam parameter product has a maximum at this point and decreases towards both stability limits. We can calculate the maximum beam parameter product by inserting (13.22) into (13.24):

$$(w\theta)_{max} = k\,\frac{b^2 \Delta D}{4} \tag{13.28}$$

with: $k = 1$ if the origin of the equivalent stability diagram is passed
$k = 2$ otherwise
ΔD : refractive power range within which the lens resonator is stable

Figure 13.7 presents the qualitative dependence of the beam parameter product on the refractive power for four different resonators with different refractive power ranges ΔD and their different paths in the equivalent g-diagram. The beam parameter products in the unstable zones are not shown (it has not been determined how to calculate them at this time). This is not a disadvantage because the output power drops drastically as the resonators penetrate into unstable zones due to increasing diffraction losses at the active medium.

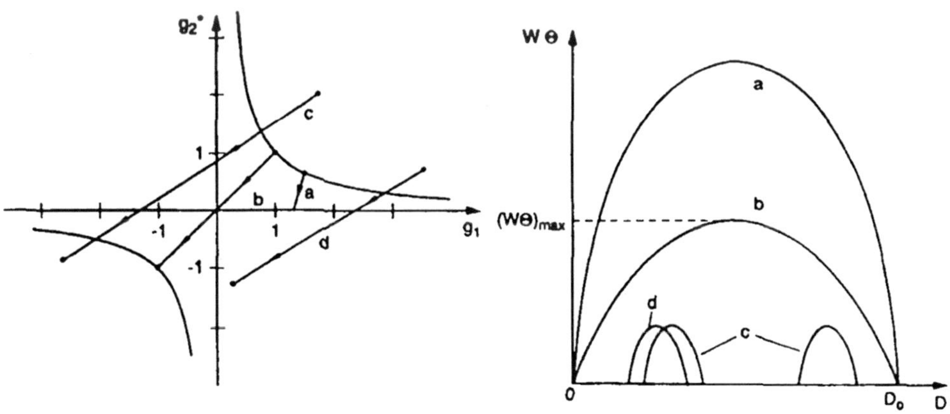

Fig. 13.7 Qualitative dependence of the beam parameter product $w\theta$ on the refractive power D for four resonators. In all resonators the thermal lens power is varied from 0 to D_0. Due to the different resonator geometries, the resonators pass through differently sized stable zones in the equivalent stability diagram.

In order to achieve a good beam quality, the resonator has to exhibit a small refractive range ΔD within which it is stable. This is a problem in most cases since the refractive power of most solid state materials is relatively high, especially for those that provide high output power. In order to attain a certain beam quality it is therefore necessary to restrict the pump power range within which the laser is operated. A practical example may give a better understanding of this design problem.

Example:
A flashlamp-pumped Nd:YAG laser ($\lambda=1.064\mu m$, $n_0=1.82$) with a rod of length $\ell=150mm$ and a radius of $b=5mm$ is required to operate in a stable zone at 12kW of electrical pump power in order to obtain a high output power, but the beam parameter product should not exceed a value of 10 mm mrad at any pump power below 12kW. The refractive power of the rod was determined to be 0.3m^{-1} per kW of pump power.

We choose a resonator that can pass through the origin of the stability diagram since then, according (13.28), the largest stable zone for a given maximum beam parameter product is provided ($k=1$). Equation (13.28) yields for the stable refractive power range $\Delta D=1.6m^{-1}$, a value that is much smaller than the total refractive power range of $\Delta D_L=3.6m^{-1}$ of the laser rod. Since the lens resonator has to be stable at 3.6m^{-1}, we chose a resonator which becomes stable at $D_I=2.2m^{-1}$ and leaves the stable zone at $D_{IV}=3.8m^{-1}$. There is a whole variety of resonator set-ups that would fulfill this requirement; however, we use a symmetric resonator with two equal mirrors and the rod in the middle (Fig. 13.8). Equations (13.11) and (13.14) yield for the mirror curvatures $\rho_1=\rho_2=-0.383m$ and for the distances $d_1=d_2=0.526m$. The geometrical length of the resonator is $L_0=2d+(n_0-1)\ell/n_0=1.12m$. Figure 13.8 presents the expected beam parameter products as a function of the refractive power. The laser can only be operated at pump powers between 7.35kW and 12.67kW.

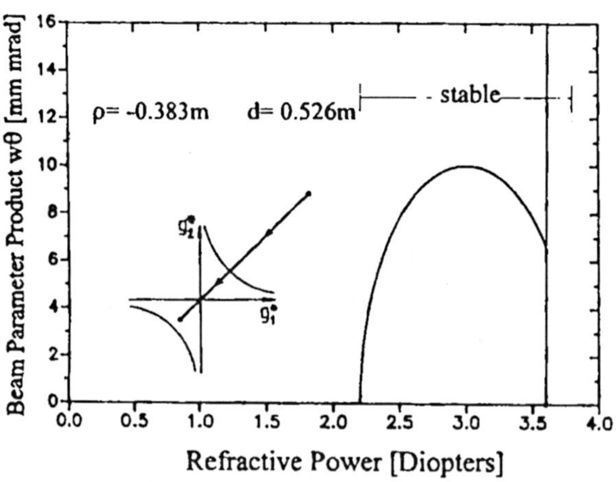

Fig. 13.8 A symmetric resonator with mirror curvatures $\rho_1=\rho_2=-0.383m$ and distances between the mirrors and the principal planes $d_1=d_2=0.526m$ provides a maximum beam parameter product of 10 mm mrad for a Nd:YAG rod with a maximum refractive power of 3.6 Diopters.

As shown in the previous example, the laser resonator may not cover the whole stable refractive power range ΔD, but only the portion ΔD_A (=1.4m^{-1} in the example). How do we calculate the maximum beam parameter product in this case? Two cases have to be distinguished:

a) $\Delta D_A \geq \Delta D/2$

The maximum beam parameter product is given by (13.28)

b) $\Delta D_A < \Delta D/2$

The maximum beam parameter product is lower than the one given by (13.28) since less than half the stable zone is covered by the lens resonator. To a good approximation the maximum beam parameter product can be calculated using:

$$(w\theta)_{max} = k\,\frac{b^2 \Delta D}{4}\left[1 - \left(1 - \frac{2\Delta D_A}{\Delta D}\right)^2\right] \tag{13.29}$$

Figure 13.9 presents a calculated example to clarify this statement. For a symmetric flat-flat resonator the beam parameter product is shown as a function of the refractive power. The curve parameter is the distance *d* from one mirror to the adjacent principal plane of the thermal lens. The operating range of the refractive power is $\Delta D_A = 3m^{-1}$. The refractive power range of the stable zone is given by $\Delta D = 2/d$ and it decreases as the distance *d* is increased. For $d < 1/3m$ the maximum beam parameter product given by (13.28) is reached, for larger distances the maximum beam parameter product is reached at the maximum refractive power, and it can be calculated with (13.29). This example also visualizes the typical dependence of the beam quality on the resonator length; low beam parameter products require long resonators.

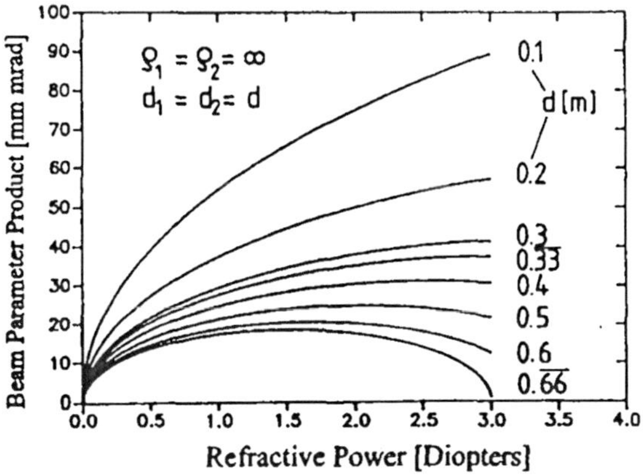

Fig. 13.9 Beam parameter product as a function of the refractive power for symmetric flat-flat resonators calculated with (13.25). The curve parameter is the distance *d* from one mirror to the adjacent principal plane of the laser rod (rod radius *b=5mm*).

Figures 13.10 and 13.11 show measured beam divergences and beam parameter products for Nd:YAG rod lasers with different resonators in comparison with the theoretical curves given by (13.25) and (13.26). If the refractive power of the laser material is known, the theoretical values can be used to determine the beam quality with a good accuracy. On the other hand the measurement of the beam parameter product as a function of the pump power can be used to determine the refractive power (see also Chapter 23 for measurement techniques).

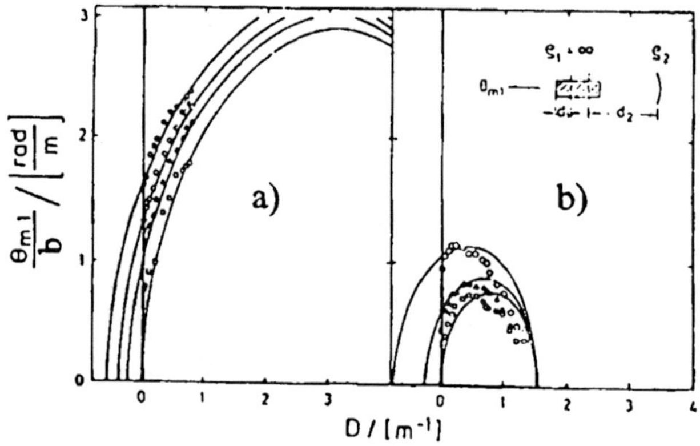

Fig. 13.10 Measured half angles of divergence (normalized to the rod radius b) of a Nd:YAG rod laser (b=3mm, ℓ=90mm) with different plane-spherical resonators as a function of the refractive power of the rod. The curves were calculated using (13.25) and (13.26). a) d_2= 0.165m, b) d_2=0.64m. The curve parameter is the radius of curvature ρ_2 of the high reflecting mirror with: □: $\rho_2=\infty$, Δ: ρ_2=5m, ◇: ρ_2=3m, o:ρ_2=2m. The output coupling mirror is coated onto the left rod endface.

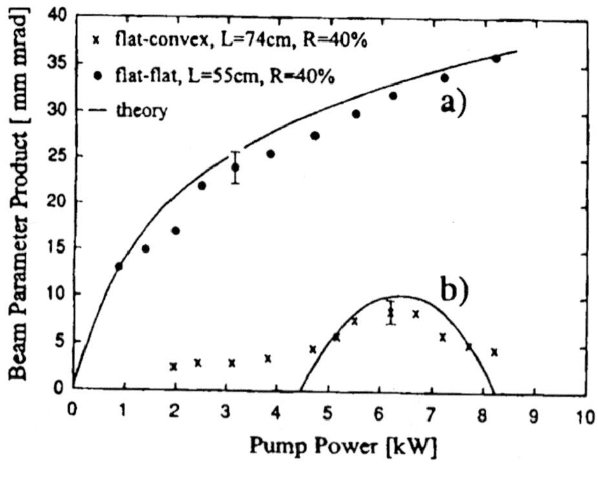

Fig. 13.11 Measured and calculated beam parameter products of a Nd:YAG rod laser (b=10mm, ℓ=150mm) for two resonators as a function of the pump power. a) $d_1=d_2=0.24m$, $\rho_1=\rho_2=\infty$; b) $d_1=0.24m$, $d_2=0.43m$, $\rho_1=\infty$, $\rho_2=-0.34m$.

13.2.3 Beam Radii, Divergences, and Focusing

The thermal lens of the active medium has a considerable impact on the focusing properties of the laser beam. Since the beam waist, the angle of divergence and the position of the beam waist change with the refractive power, the size and the position of the focus spot vary as the pump power is increased (see Sec. 5.2.5). However, over large ranges of the refractive power either the waist radius or the angle of divergence stays fairly constant [4.91,4.93,4.99] (Fig. 13.12). If a flat output coupler is used, the position of the beam waist remains constant. Therefore, by using appropriate focusing optics in combination with a flat output coupler it is possible to maintain the position and the size of the focus over large ranges of the pump power.

If the waist radius is constant (first stable zone in Fig. 13.5), the beam can be focused to a spot of constant size by using a telescope as discussed in Sec.5.2.5. The telescope images the plane of the output coupling mirror to the plane of the focus spot; the focus is the image of the beam waist. The change in the divergence, therefore, will not affect the spot size. In areas of constant divergence, the preferred means of focusing is a focusing lens located at a distance of one focal length from the output coupling mirror. The spot size in the front focal plane only depends on the angle of divergence since the Fourier transform of the field at the output coupling mirror is generated.

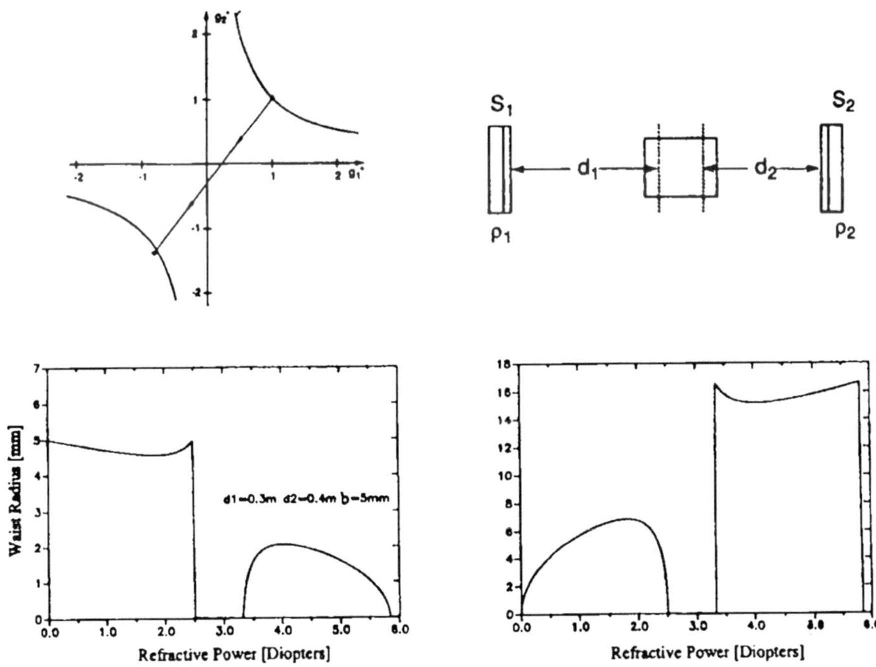

Fig. 13.12 Beam waist radius and half angle of divergence as a function of the refractive power for a flat-flat resonator. The output coupler is mirror 1 (b=5mm, d_1=0.3m, d_2=0.4m, λ=1.064μm).

Which of the two cases, constant waist radius or constant angle of divergence, is observed depends on the resonator set-up. If the output coupler is closer to the active medium than the high reflecting mirror, the waist radius stays constant in the first stable zone and the divergence in the second stable zone. This statement holds for all resonators with one flat mirror. If both mirrors are curved the waist position is a function of the pump power. However, we can still find regions of constant beam radius on one mirror or constant divergence (Fig. 13.13).

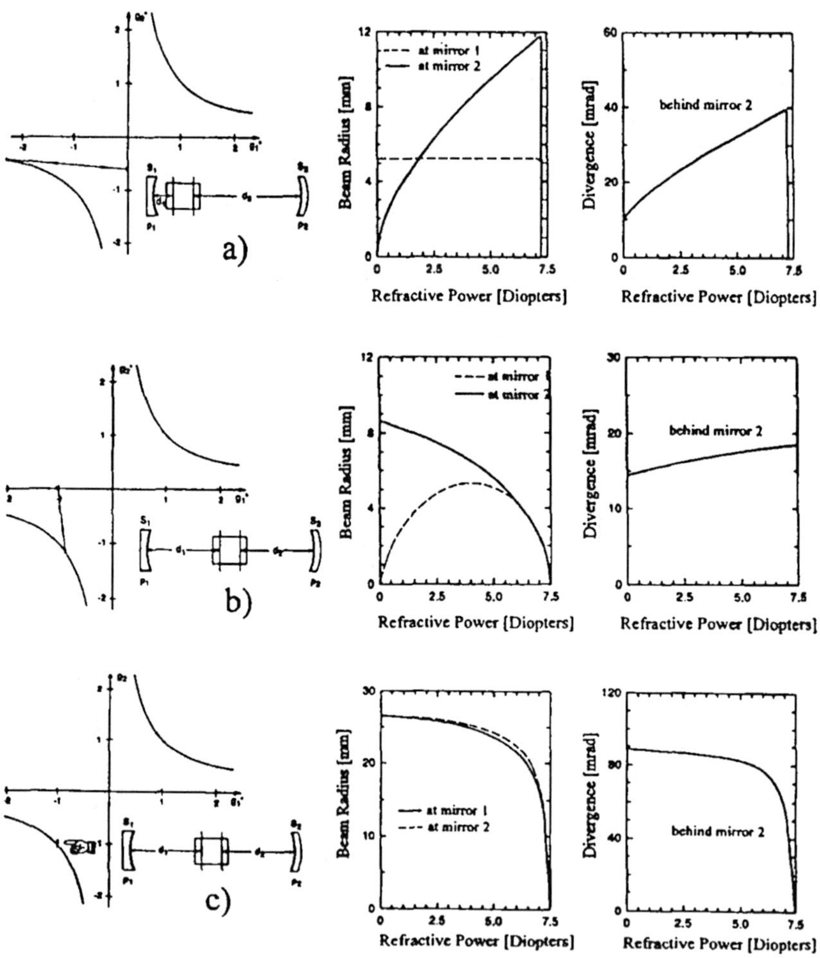

Fig. 13.13 Calculated beam radii at the mirrors and angles of divergence behind mirror 2 as a function of the refractive power for three different resonators in the third quadrant of the stability diagram (rod radius $b=5mm$, $\lambda=1.064\mu m$). a) $d_1=0.065m$, $d_2=0.405m$, $\rho_1=0.5m$, $\rho_2=0.3m$; b) $d_1=0.305m$, $d_2=0.275m$, $\rho_1=0.3m$, $\rho_2=0.6m$; c) $d_1=0.305m$, $d_2=0.275m$, $\rho_1=0.3m$, $\rho_2=0.3m$.

An interesting resonator is the one working close to the concentric point (-1,-1) in the equivalent g-diagram (Fig.13.13c). Both the beam radius and the angle of divergence are relatively constant although refractive power is varied over a large range of 5 Diopters. This constancy is generated by the low sensitivity of the equivalent g-parameters to the refractive power; if we had a symmetric concentric resonator ($d_1=d_2=\rho_1=\rho_2$) the equivalent g-parameters would not change at all. Unfortunately, the proportionality between the maximum beam parameter product and the stable refractive power range ΔD also holds for this near concentric resonator. The price to pay for the beam radius and the divergence being constant simultaneously, therefore, is a high beam parameter product of 260 mm mrad.

12.2.4 Output Power and Beam Quality

Since the refractive power of a solid state laser is proportional to the pump power P_{pump}, we can find a relationship between the maximum beam parameter product and the pump power range ΔP_{pump} within which the lens resonator is stable. Insertion of (13.5) into (13.28) yields [4.100]:

$$\frac{(w\theta)_{max}}{\Delta P_{pump}} = k\,\frac{\alpha}{4\pi} \tag{13.30}$$

where α is the thermal lensing coefficient, ΔP_{pump} is the electrical pump power for flashlamp pumping and the absorbed optical pump power for diode pumping, and $k=1$ if the confocal point is reached and $k=2$ elsewhere. We thus need a small pump power range to realize good beam quality. Unfortunately, a small pump power range is related to a small output power range ΔP_{out} since the resonator provides a useful output only inside the stable zones (Fig. 13.14).:

$$\Delta P_{out} = \eta_{slope}\,\Delta P_{pump} \tag{13.31}$$

where η_{slope} is the slope efficiency (see Sec. 10.1.1). To obtain (13.31), we assumed that laser oscillation ceases once the stability limits are crossed. This assumption is not entirely correct as we shall see. Equation (13.30) can now be written as:

$$\frac{(w\theta)_{max}}{\Delta P_{out}} = \frac{k}{4\pi}\,\frac{\alpha}{\eta_{slope}} \tag{13.32}$$

This expression reveals that a large range of the output power and a good beam quality cannot be achieved simultaneously. For a given excitation source (flashlamp or laser diode), the ratio of the thermal lensing coefficient to the slope efficiency can be considered a characteristic for a solid state laser material (the maximum slope efficiency has to be used). This ratio can serve as a figure of merit for the high power performance of different solid state lasers. In order to combine a large output power range with a reasonable beam quality, materials with a low ratio are preferred. Tables 13.3 and 13.4 present values of α/η_{slope} for common solid state materials.

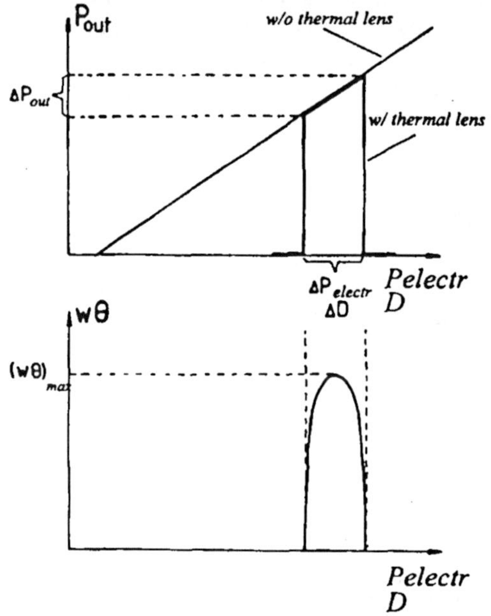

Fig. 13.14 With a thermal lens only the part of the output power curve can be used within which the resonator is stable. A small output power range is linked to a low maximum beam parameter product.

Table 13.3 Values of α/η_{slope} in µm/W for flashlamp pumped solid state laser materials. Variations are due to differences in crystal quality, doping concentration, and slope efficiency.

Nd:YAG	Nd:glass	Nd:Cr:GSGG	Nd:Cr:GGG	Nd:YAP	Nd:YLF	Alexandrite
0.4-0.8	3.0-4.0	0.5-1.0	1.5-2.0	0.7-1.1	-(0.1-0.15)	0.2-0.3

Table 13.4 Values of α/η_{slope} in µm/W for diode-pumped solid state laser materials. Variations are due to differences in doping concentration, absorption efficiency and pump geometry (see Table 13.2 for pump wavelengths).

Nd:YAG	Yb:YAG	Nd:YVO$_4$	Nd:glass	Nd:YLF	Nd:YAP	Ti:sapphire
0.3-0.5	0.1-0.2	0.2-0.3	2.0-3.0	-0.05	0.5-0.75	0.25-0.35

Most commercially available solid state lasers are comprised of resonators that are either flat-flat or already stable at zero pump power. For these lasers the output power range is equivalent to the maximum output power. If we plotted the maximum beam parameter product of commercial Nd:YAG lasers as a function of the maximum output power we would thus expect a linear relationship. This is shown in Fig. 13.15 in comparison with the theoretical curve given by (13.32) with $k=1$. Beam quality and output power for diode-pumped single rod lasers are summarized in Fig. 13.16. Compared to flashlamp pumped systems, the beam parameter products are about 20% lower for the same output power, as expected from the values shown in Tables 13.2 and 13.3. The beam quality is generally worse than predicted by (13.32) due to two reasons. First, many commercial single rod systems do not use resonators that pass through the confocal point $g_1^*=g_2^*=0$. The operation at this point is very sensitive to tolerances in the resonator design and small offsets of the mirror distances may result in a power dip.

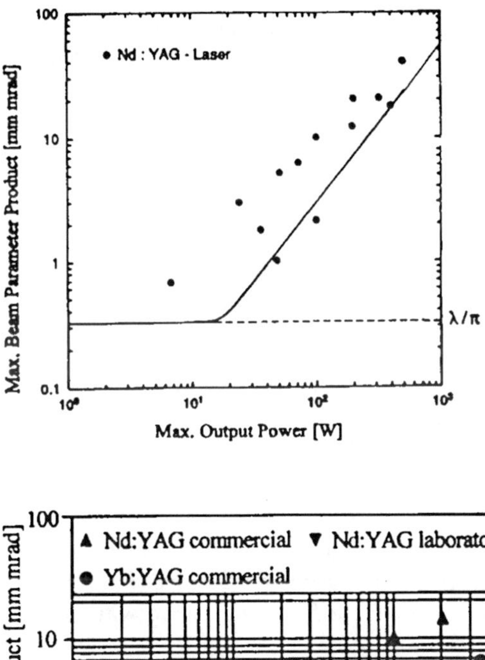

Fig. 13.15 Maximum beam parameter product as a function of the maximum average output power for commercially available flashlamp pumped single rod Nd:YAG lasers. The solid line represents (13.32) with $k=1$.

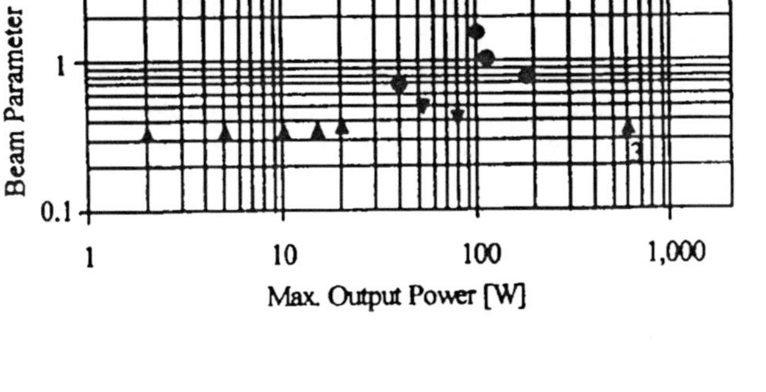

Fig. 13.16 Beam parameter products and output powers for diode-pumped solid state lasers using one Nd:YAG or Yb:YAG rod. For comparison: 1: fiber bundle laser, 2: disk laser with two disks, 3: single fiber laser

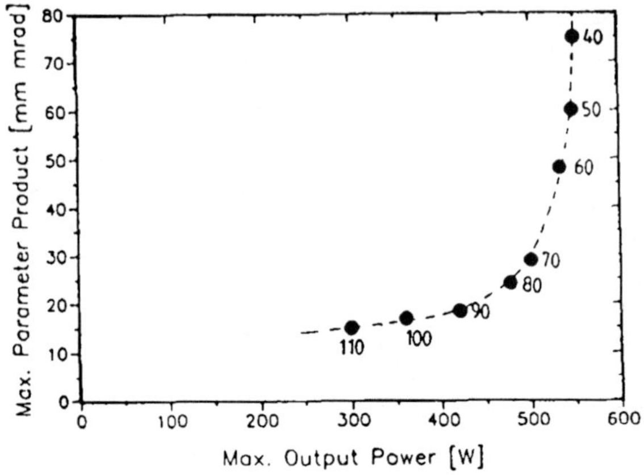

Fig. 13.17 Measured maximum beam parameter products and maximum output powers for a pulsed flashlamp pumped Nd:YAG rod laser (*b=5mm, ℓ=150mm*) with different symmetric flat-flat resonators. The number at each point is the geometrical resonator length in cm. The refractive power of the rod is $0.38m^{-1}$ per kW of pump power and the pump power was varied between 0 and 12kW (pulse width: 2ms, pump energy: 240J, the repetition rate: 0-50Hz). The laser covers the whole stable zone ($\Delta D=4.56m^{-1}$) for an effective resonator length of 93cm.

Secondly, commercial lasers generally do not cover the whole stable zone of the lens resonator. The reason for this is, besides space requirements, the decrease in laser efficiency when the lens resonator gets too close to a stability limit or the stable zone is chosen too small. As we shall see in the next section, this decrease is caused by spherical aberration of the thermal lens leading to a decrease of the fill factor. The modes basically "pull in their skirts" to avoid the strong aberration in the periphery of the medium. An experimental example is presented in Fig. 13.17 in which the maximum beam parameter product is plotted versus the maximum output power for a Nd:YAG laser with symmetric flat-flat resonators of different lengths. The maximum output power is decreased as the resonator length is increased. For a geometrical resonator length of 0.93m, the Nd:YAG laser would cover the whole stable zone with a maximum beam parameter product of 15mm mrad. Unfortunately, the output power is only 400W compared to 550W measured with a short resonator.

The beam quality of lens resonators can be improved by using a graded reflectivity mirror as an output coupler. The reflectivity profile acts as a mode-selecting aperture but in contrast to a conventional aperture, the power hitting the aperture is coupled out of the resonator. By using this technique it is possible to considerably enhance the beam quality of a lens resonator without decreasing the output power too much, as the experimental example in Fig. 13.18 indicates. This resonator scheme is, of course, also applicable to laser systems that do not exhibit thermal lensing.

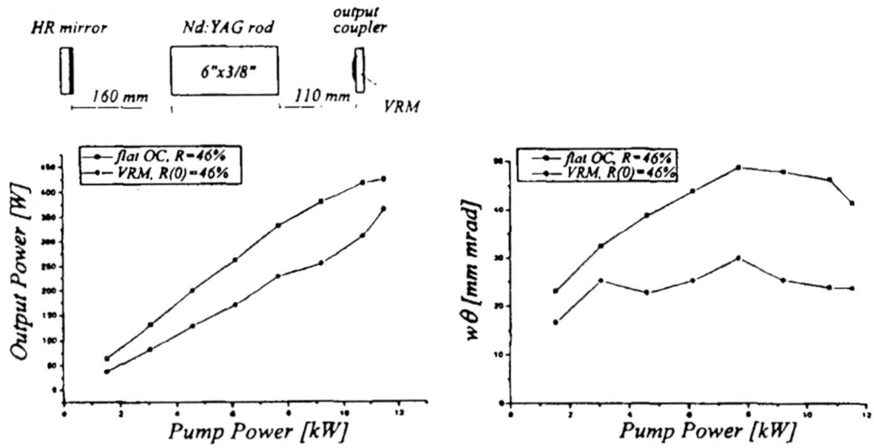

Fig. 13.18 Measured output power (b) and measured beam parameter product (c) as a function of the electrical pump power for a pulsed, flashlamp pumped Nd:YAG rod laser utilizing the resonator set-up depicted in (a) (D=0.3 diopters per kW of pump power). Measurements taken with a conventional flat output coupler ($R=46\%$) and with a VRM with super-Gaussian index $n=12$, center reflectance $R(0)=46\%$, and a profile radius of $w=2.5mm$ are compared. With the VRM, the beam parameter at the maximum output power is decreased by 45% whereas the output power drops only by 15% [S.12].

At the stability limits the lens resonator does not abruptly stop or start laser emission as was assumed in Fig. 13.14. If the resonator approaches a stable zone the output power begins to rise before the stability limit is crossed and a similar smoothing of the power curve is observed when the resonator goes unstable again. A typical example for this behavior is shown in Fig. 13.19 for a flashlamp pump Nd:YAG rod laser. The penetration of the power curve into the unstable zones is more pronounced the higher the small-signal gain of the laser. Since the diffraction losses are increased considerably as the resonator goes unstable, laser oscillation will stop as soon as the gain cannot compensate the losses anymore.

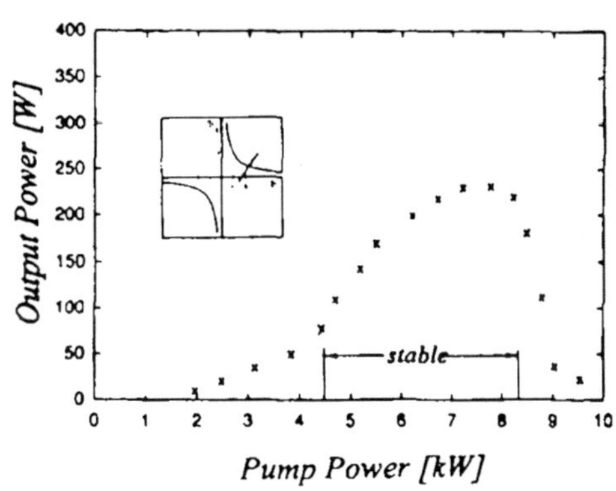

Fig. 13.19 A lens resonator passes through a stable zone. The output power rises and falls outside the stability limits (pulsed Nd:YAG rod laser, $b=5mm$, $\ell=150mm$, small-signal gain $g_0\ell=4.1$, mirror reflectance $R=40\%$).

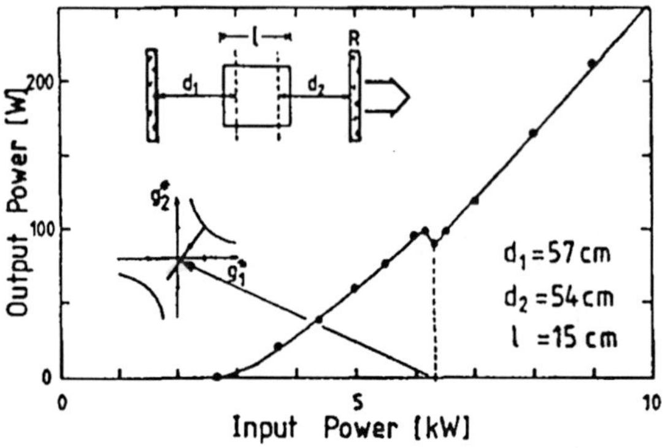

Fig. 13.20 Power dip generated by a small unstable zone. The lens resonator misses the confocal point due to the slight asymmetry of the resonator (pulsed Nd:YAG rod laser, b=5mm, ℓ=150mm, flat-flat resonator).

However, if the resonator only passes through a small unstable zone, as shown in Fig. 13.21, the output power may not go to zero at all. A dip in the output power is observed which can be used to determine the refractive power of the active medium. It should be added that the laser output in the unstable zones is not useful unless an output coupler with a confined high reflecting area (see Sec. 7.1) is used. We will discuss the operation in the unstable zones in Sec. 13.3.

In some cases, two power maxima are observed when the lens resonator penetrates into a wide enough unstable zone (Fig.13.21). The two maxima are generated if the laser material exhibits birefringence. For Nd:YAG the birefringence is caused by thermally induced stress; the refractive power for radially polarized light is about 15-20% higher compared to the refractive power for the azimuthal polarization. The radial polarization reaches the stability limit first and the output power drops as the refractive power is increased further. In the first power maximum the laser beam is radially polarized. Since resonator modes always try to decrease their losses, the polarization of the modes now becomes azimuthal. This polarization switching pushes the lens resonator back into the stable zone and, consequently, the output power rises again until the azimuthal polarization also goes unstable. Except for the operation at the limit of a stable zone, the laser beam is usually unpolarized, and the mean value of the two refractive powers can be used for the effective refractive power of the lens resonator.

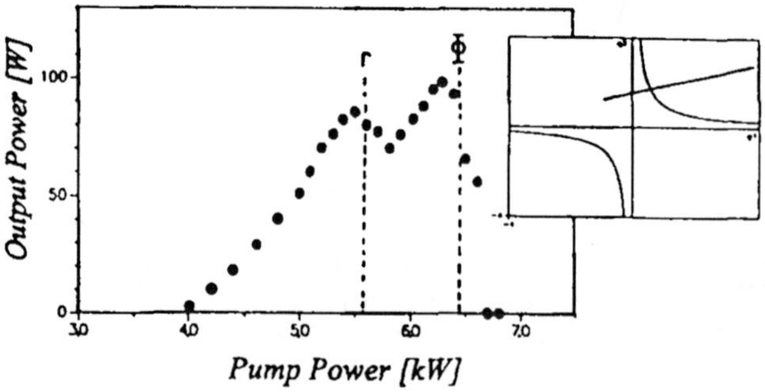

Fig. 13.21 Measured output power for a resonator with a small stability zone. The pump powers at which the second stability limit is reached are indicated for radial and azimuthal polarization (pulsed flashalmp pumped Nd:YAG rod laser, b=5mm, ℓ=150mm, d_1=0.15m, d_2=0.45m, ρ_1=-0.3m, ρ_2=-0.5m).

13.2.5 Output Power in Fundamental Mode Operation

As discussed in Sec. 13.2.1, the Gaussian beam radius at the principal planes w_L exhibits a minimum $w_{L,min}$ near the center of the stable refractive power range ΔD with:

$$w_{L,\min}^2 = \frac{4\lambda}{k\ \pi\ \Delta D} \tag{13.33}$$

with $k=1$ if the confocal point $g_1^*=g_2^*=0$ is reached and $k=2$ else. In order to attain fundamental mode operation an aperture with a radius a that is adapted to the Gaussian beam radius has to be placed in front of the medium:

$$a = x\ w_{L,\min}\ , \qquad x \in [1.3,...,1.4] \tag{13.34}$$

Fundamental mode operation can be maintained over a refractive power range ΔD_{00} which covers about 90% of the stable zone (Fig.13.22):

$$\Delta D_{00} = \frac{3.6\ \lambda}{k\ \pi\ w_{L,\min}^2} \tag{13.35}$$

Close to the stability limits the Gaussian beam radius w_L becomes much larger than the aperture radius and the resulting increase in the diffraction loss pushes the laser below threshold. In order to attain a high output power in fundamental mode operation, the Gaussian beam radius in the medium should be as large as possible. Unfortunately, (13.35) indicates that the refractive power range within which laser oscillation is achieved becomes smaller as the beam size is increased. For a typical Gaussian beam radius in a Nd:YAG laser rod of $w_{L,min}=1mm$, (13.35) yields a refractive power range of 1.15 Diopters ($k=1$).

We can rewrite (13.35) into a more convenient form by using the known expressions for the electrical pump power range and the output power range:

$$\Delta D_{00} = \frac{\alpha}{\pi b^2}\ \Delta P_{pump} \tag{13.36}$$

$$\Delta P_{out} = \frac{a^2}{b^2}\ \eta_{slope}\ \Delta P_{pump} \tag{12.37}$$

with

	α	:	thermal lensing coefficient (see Tables 13.1 and 13.2)
	b	:	rod radius
	η_{slope}	:	slope efficiency without aperture (multimode)

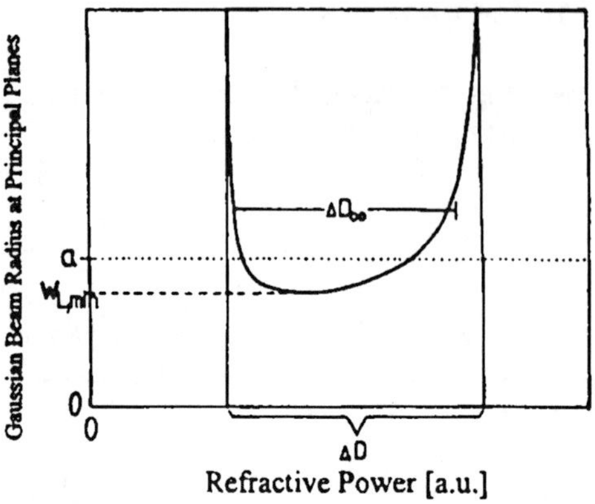

Fig. 13.22 Qualitative dependence of the Gaussian beam radius inside the medium on the refractive power. By limiting the medium with an aperture with radius $a=1.3w_{L,min}$, fundamental mode operation is obtained within the range ΔD_{00}.

Insertion of (13.34), (13.36), and (13.37) into (13.35) results in the following expression for the output power range (for $a=1.4\ w_{L,min}$) [4.97]:

$$\Delta P_{out} = \frac{7.1\ \lambda}{k}\ \frac{\eta_{slope}}{\alpha} \qquad (13.38)$$

Again we find that the laser performance is determined by the ratio of the thermal lensing coefficient to the multimode slope efficiency.

Although the exact value of this ratio depends on the excitation source, the doping level, and the crystal quality, it can be considered a characteristic of a laser material. In order to attain a large power range in fundamental mode operation, this ratio has to be as small as possible. Tables 13.5 and 13.6 give an overview of the output power ranges in TEM_{00} mode operation for different solid state materials for flashlamp pumping and diode pumping, respectively. Note that (13.38) represents the maximum output power if the resonator is already in a stable zone at zero pump power. Since most commercial fundamental mode lasers (mostly Nd:YAG lasers) are designed this way, the maximum fundamental mode power is on the order of 10-15W for flashlamp pumped systems [4.88,4.94,4.95,4.97] and about 20-25W for diode-pumped single rod lasers [4.115] (Fig.13.23). With the restriction that the resonator is stable at all pump powers, higher TEM_{00} mode output power can only be realized if a laser material with better thermal properties, like Nd:YLF, is used. However, by limiting the stable zone of the resonator, it is possible to increase the output power in TEM_{00} mode operation beyond the value given by (13.38). With diode-pumped single Nd:YAG rods, output powers of as high as 80W with $M^2<1.2$ have been reported [4.112,4.113,4.117]. However, limitation of the stability zone does come at the cost of increased misalignment sensitivity and decreased overall efficiency. Figure 13.24 shows experimental results obtained with a side-pumped Nd:YAG using an intracavity 45° Faraday rotator for birefringence compensation [4.123]. The resonator is stable for pump powers between 660W and 725W, resulting in a TEM_{00} mode beam diameter inside the rod of 1.7mm. At a pump power of 720W, the output power was 53W with an M^2 of less than 1.5.

Table 13.5 Output power range ΔP_{out} in TEM_{00} mode operation for different flashlamp pumped solid state laser materials (*k=1*). The shown values correspond to the variations in α/η_{slope}.

Material	wavelength λ [μm]	α/η_{slope} [mm/kW]	ΔP_{out} [W]
Nd:YAG	1.064	0.4 - 0.8	9.5 - 18.9
Nd:glass	1.054	3.0 - 4.0	1.9 - 3.5
Nd:Cr:GSGG	1.064	0.5 - 1.0	7.6 - 15.1
Nd:Cr:GGG	1.064	1.5 - 1.9	4.0 - 5.0
Nd:YAP	1.079	0.7 - 1.1	7.0 - 11.0
Nd:YLF	1.047	-(0.1 - 0.15)	50 - 75

Table 13.6 Output power range in TEM_{00} mode operation for different diode-pumped solid state laser materials (*k=1*). The shown values correspond to the variations in α/η_{slope}.

Material	wavelength λ [μm]	α/η_{slope} [μm/W]	ΔP_{out} [W]
Nd:YAG	1.064	0.3 - 0.5	15.2 - 25.3
Yb:YAG	1.030	0.1 - 0.2	36.6 - 73.1
$Nd:YVO_4$	1.064	0.2 - 0.3	22.8 - 28.4
Nd:glass	1.053	2.0 - 3.0	2.5 - 3.7
Nd:YAP	1.079	0.5 - 0.75	10.2 - 15.3
Nd:YLF	1.047	- (0.05 - 0.07)	103 - 148

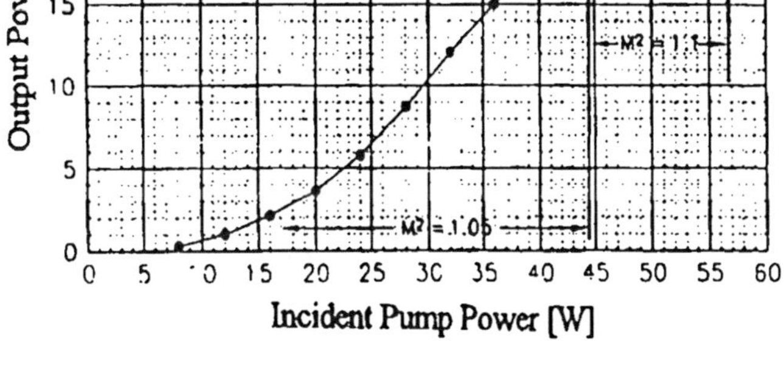

Fig. 13.23 Measured output power in TEM_{00} mode operation for a dual-diode-end-pumped $Nd:YVO_4$ laser with ~45cm long resonator as a function of the incident 808nm pump power. The laser emits a Gaussian mode over the entire output power range [4.115] (© SPIE 1998).

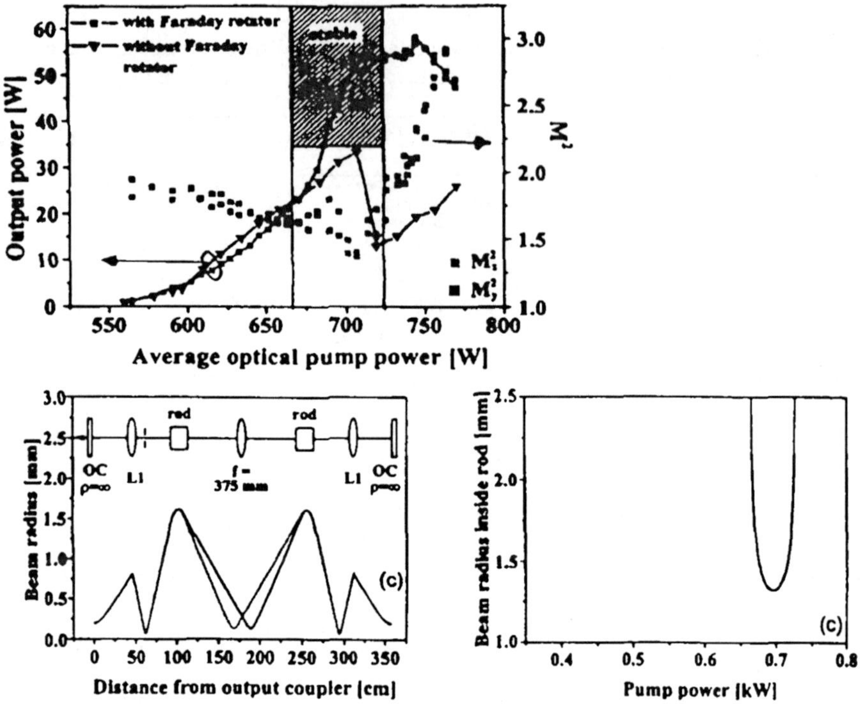

Fig. 13.24 Side-pumped Nd YAG rod laser in TEM_{00} mode operation with restricted stable zone and birefringence compensation using a Faraday rotator. Measured output power and beam quality with and w/o the Faraday rotators are shown. The lower graphs depict the resonator set-up (presented for a round trip) and the calculated Gaussian beam diameter in the rod as a function of the pump power [4.117] (© OSA 2002).

13.2.6 Spherical Aberration

So far we only related the maximum beam parameter product to the output power range. This relationship does not inhibit the simultaneous realization of high output power and good beam quality as was shown in Fig.13.24. Unfortunately, there is experimental evidence that the laser efficiency decreases considerably if the resonator operates close to a stability limit or if the stable zone is chosen small (Fig.13.25). This decrease in efficiency is caused by the pump induced birefringence and by the spherical aberration of the thermal lens. Spherical aberration in solid state lasers is caused by inhomogeneous pump profiles and by the temperature dependence of both the thermal conductivity K and the thermo-optic constant dn/dT, resulting in a dependence of the refractive power on the radial coordinate. The refractive power in high power flashlamp or diode-side-pumped Nd:YAG lasers, for instance, typically is about 20-30% higher in the rod center. As a result, the diffraction

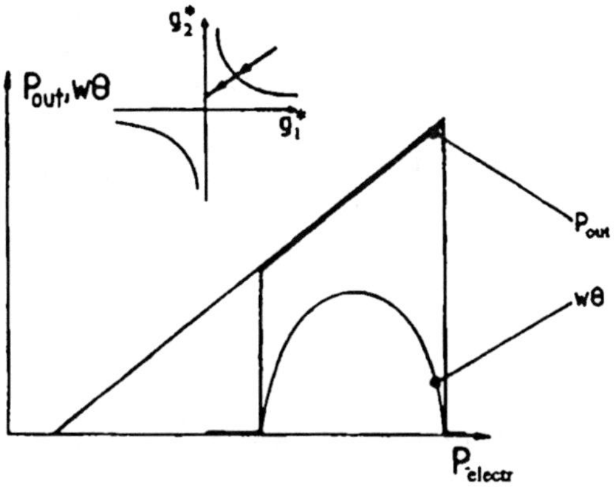

Fig. 13.25 High output power and a low beam parameter product should be attainable near a stability limit of the lens resonator.

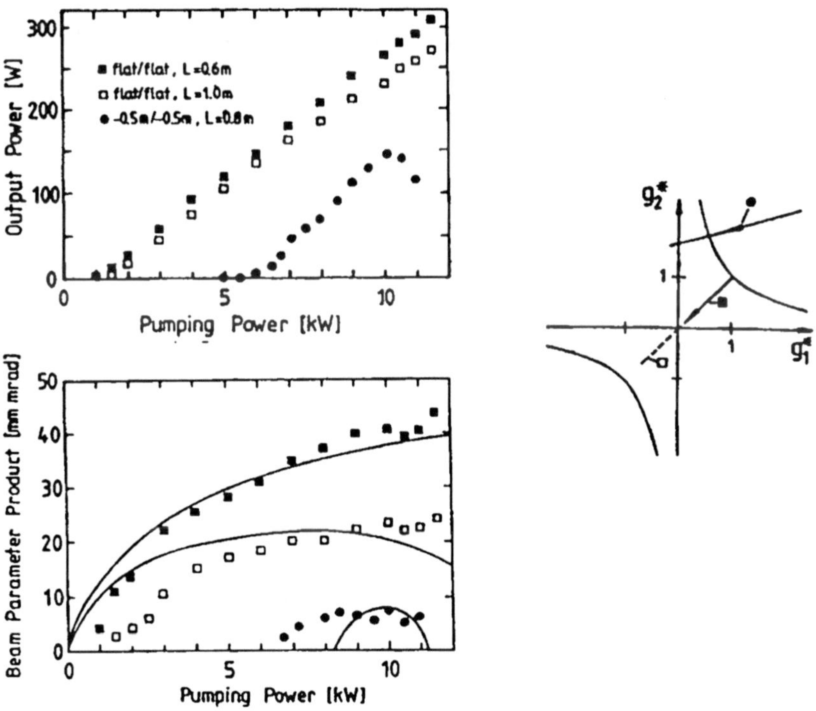

Fig. 13.26 Measured output power of a pulsed Nd:YAG rod laser (*b=5mm, ℓ=150mm, D=0.3m⁻¹ per kW of pump power*) as a function of the average pump power for three different resonators. A smaller stable zone results in a decrease of the output power. The beam parameter product $w\theta$, however, can still be calculated with Eq. (13.25) (solid lines in lower graph). The right graph shows the paths of the three resonators in the equivalent g-diagram [4.108] (© IEEE 1993).

losses become higher and the radius of the beam inside the medium decreases which means that the pump energy stored in the outer periphery of the rod is not being extracted anymore [4.109]. In the following we will derive an analytical expression for the spherically aberrated thermal lens as a function of the radial gain profile in the rod [4.82, 4.107]. Let us consider a rod with radius b and length ℓ in which heat is generated at a rate Q per unit volume with a radial dependence given by (where the total heat is $Q_0 \pi b^2 \ell$):

$$Q(r) = \frac{Q_0}{1 - \beta/2}\left(1 - \frac{\beta}{b^2} r^2\right) \tag{13.39}$$

This heat distribution is generated by a similar pump profile in the rod. For $\beta=0$ the rod is pumped homogeneously, for $\beta=0.5$ the gain at the center is twice as high as at the rod surface. The total heat P_H dissipated is independent of the shape parameter β:

$$P_H = \pi b^2 \ell Q_0 \tag{13.40}$$

The resulting temperature profile $T(r)$ is a solution of the one-dimensional heat conduction equation which for a temperature dependent thermal conductivity $K(T)$ reads:

$$\frac{1}{r}\frac{d}{dr}\left[K(T)\, r \frac{dT}{dr}\right] = -\,Q(r) \tag{13.41}$$

For Nd:YAG, the thermal conductivity $K(T)$ can be approximated by [4.108] (Fig.13.27):

$$K(T) = a/T\,, \qquad \text{with } a = 36\ W/cm \tag{13.42}$$

Solving (13.41) for the radial temperature profile $T(r)$ yields:

$$T(r) = -\,T_0 \exp\left[\alpha\left[1 - \left(\frac{r}{b}\right)^2 - \frac{\beta}{4}\left[1 - \left(\frac{r}{b}\right)^4\right]\right]\right] \tag{13.43}$$

with:

$$\alpha = \frac{P_H}{4\pi a \ell\,(1 - \beta/2)} \tag{13.44}$$

$$T_0 = T_W + P_H/(2\pi b \ell h) \qquad : \text{surface temperature} \tag{13.45}$$

where T_W is the cooling water temperature and h is the heat transfer coefficient. The dissipated heat P_H is related to the electrical pump power P_{electr} (to the flashlamps or the diodes) via:

$$P_H = \chi \, \eta_{excit} \, P_{electr} \tag{13.46}$$

with χ: thermal load parameter (=ratio of heat to power realized as inversion)
~ 2.3-2.7 for pulsed flashlamp pumped Nd:YAG,
~ 1.1-1.5 for cw flashlamp pumped Nd:YAG,
~ 0.33 for diode-pumped Nd:YAG or Nd:YVO_4)

η_{excit}: excitation efficiency (see Sec. 9.3)
~ 0.05-0.07 for flashlamp pumped Nd:YAG,
~ 0.3 for end-pumped Nd:YAG or Nd:YVO_4
~ 0.2 for side pumping

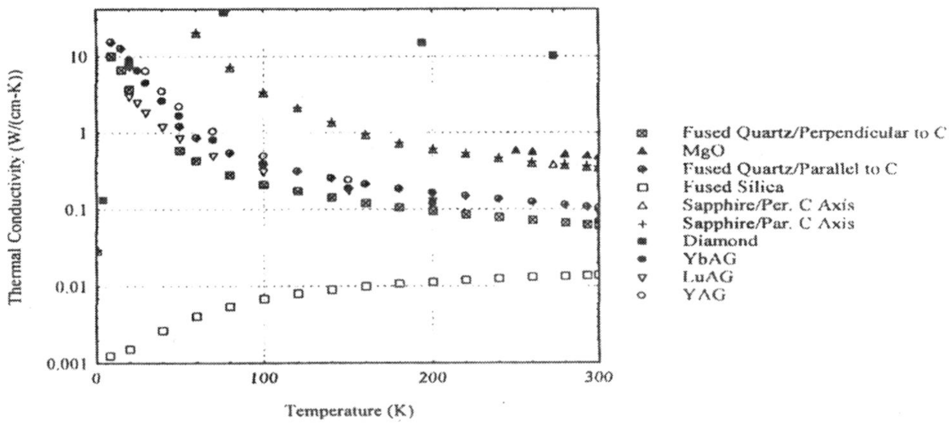

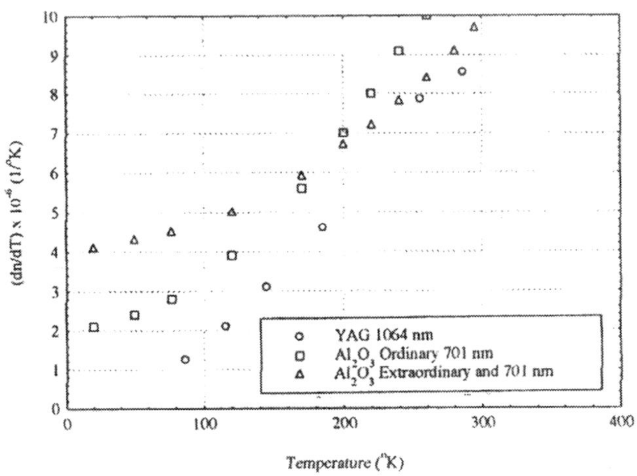

Fig. 13.27 Measured temperature dependence of the thermal conductivity (top) and the thermo-optical constant dn/dT (bottom) for different dielectric materials and solid state laser hosts [S.34, S.35]. Courtesy of Dave Brown of Lynx Lasers, LLC.

Expansion of the exponential, neglecting terms with powers higher than 4, leads to:

$$T(r) = T_C \left[1 - \alpha \left[\frac{r}{b}\right]^2 + \frac{\alpha^2}{2}\left[1 + \frac{\beta}{2\alpha}\right]\left[\frac{r}{b}\right]^4\right] \tag{13.47}$$

with: $$T_C = T_0 \exp\left[\alpha\left(1 - \frac{\beta}{4}\right)\right]$$

The refractive power of the thermal lens is given by:

$$D(r) = -\ell \frac{dn}{dT}\frac{d^2T(r)}{dr^2} \tag{13.48}$$

where dn/dT is the thermo-optical constant (temperature derivative of the index of refraction), which is also a function of the temperature (Fig.13.27):

$$\frac{dn}{dT} = \epsilon + \delta\,(T(r) - T_W) \tag{13.49}$$

where ϵ is dn/dT at the water temperature T_W and δ is the slope of the curve in Fig.13.27 ($=d^2n/dT^2$). Due to the non-parabolic temperature profile and the temperature dependence of dn/dT, the refractive power is a function of the radius. This means that an incident ray parallel to the optical axis at a distance r will intersect the optical axis at a distance $1/D(r)$ from the principal plane of the thermal lens. Inserting (13.49) and (13.47) into (13.48) and neglecting terms depending on a power of the radius higher than 4, the final result for the refractive power reads:

$$D(r) = D_0\left[1 - \gamma\left(\frac{r}{b}\right)^2\right] \tag{13.50}$$

with: $$D_0 = \frac{2\ell\alpha T_C}{b^2}\left[\epsilon + \delta(T_C - T_W)\right] \quad \text{: center refractive power} \tag{13.51}$$

$$\gamma = \frac{3}{2}\beta + \alpha\left(3 + \frac{\delta\, T_C}{\epsilon + \delta(T_C - T_W)}\right) \quad \text{: aberration factor} \tag{13.52}$$

The refractive power decreases parabolically with the radius and since both α and T_C are a non-linear function of the heat load P_H, both the center refractive power D_0 and the aberration factor γ show a non-linear increase with pump power. Figure 13.28 shows calculated center refractive powers and aberration factors for a diode side-pumped Nd:YAG laser rod as a function of the heat load for different rod diameters, assuming a homogeneous heat distribution in the rod (β=0). Measured radial dependence of the refractive power of a flashlamp pumped Nd:YAG laser are depicted in Fig. 13.29. The nonlinear increase of the refractive power with the pump power can be easily observed in diode end-pumped TEM_{00} mode solid state lasers because high pump power of tens of Watts is deposited in small pump spot diameters of typically less than 1 mm (Fig. 13.30). The nonlinear increase of the refractive power in combination with the strong aberration of the thermal lens severely limits the power scalability of these systems. Power scaling therefore requires optimization of the pump geometry by reducing the heat load (e.g. 880nm instead of 808nm pumping for Nd:YVO_4) and reducing the doping concentration while keeping the overall pump light absorption high.

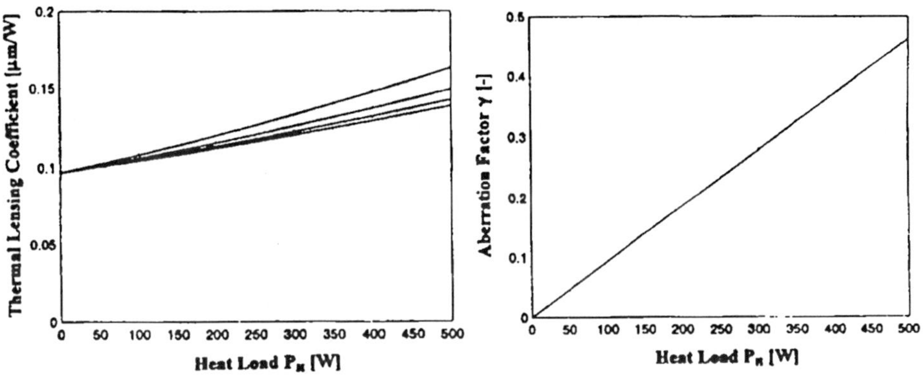

13.28 Calculated thermal lensing coefficient $D_0\pi b^2/P_{abs}$ and aberration factor γ for homogeneously diode-pumped Nd:YAG rods of 10 cm length as a function of the heat load P_H and different rod diameters (using (13.51) and (13.52)). Rod diameters 2b are 4, 6, 8, and 10 mm from bottom to top. (β=0, ϵ=9.5 $\cdot 10^{-6}$ 1/K, δ=0.04 $\cdot 10^{-6}$ 1/K^2, T_W=293K, h=1.0 W/(cm^2 K) P_H=0.25*P_{abs}, P_{abs}: absorbed pump power at 808nm).

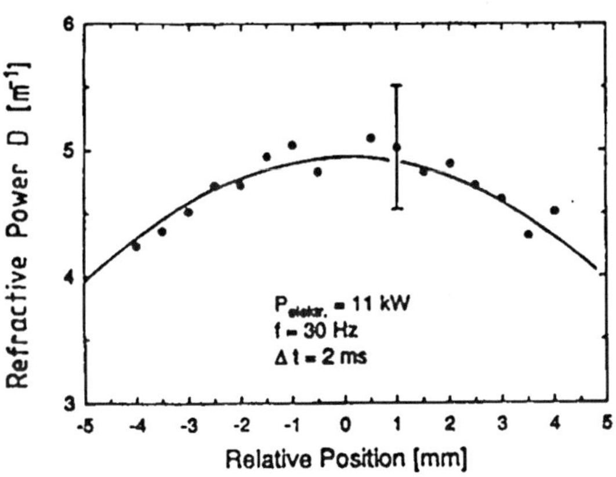

Fig. 13.29 Measured radial profile of the refractive power for a dual flashlamp pumped pulsed Nd:YAG laser rod (10mm diameter) in a ceramic pump cavity for 11kW average electrical pump power. The specular reflection of ceramic generates a nearly homogeneous pump profile in the rod. The solid line represents (13.50) with γ=0.2 [4.108] (© IEEE 1993).

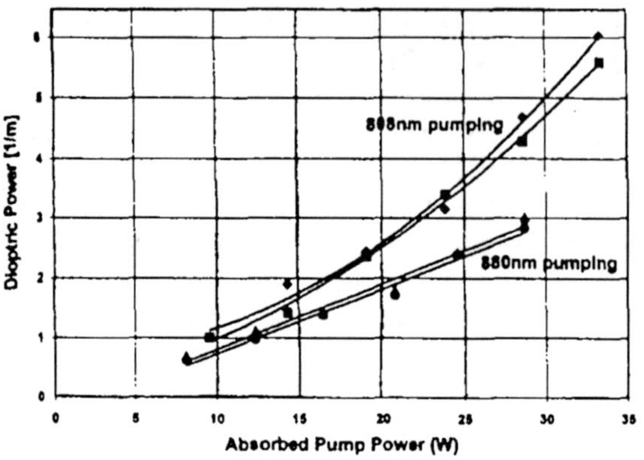

Fig. 13.30 Measured dioptric power at 1064nm of a 15mm long end-pumped Nd:YVO_4 crystal with 0.4at% doping as a function of the absorbed pump power for different diode pump wavelengths. The pump spot diameter is 1.1mm. The 40% higher heat load for 808nm pumping and the shorter absorption lengths leads to a nonlinear increase of the thermal lensing due to the increased temperature [S.32]. For each pump wavelength, the two curves represent measurements in x- and y-direction, respectively.

Equations (13.50-13.52) reveal important properties of spherical aberration in pumped rods:

- the refractive power decreases parabolically with the radius. A constant refractive power can only be achieved if the pump profile exhibits a minimum at the center; for flashlamp pumped systems this is very difficult if not impossible to realize.
- the spherical aberration is also present for homogeneous pumping. For a homogeneously pumped, 150 mm long, Nd:YAG rod, the aberration factor γ is about 0.4 per kW of dissipated heat.
- the aberration factor γ is proportional to the dissipated heat, which means aberration becomes stronger as the pump power is increased.
- the aberration factor γ is inversely proportional to the rod length. Therefore, it is advantageous to use longer rods, which for end-pumped systems means that the doping concentration has to be decreased proportionally.
- the aberration factor γ does not depend on the rod radius ! Although a bigger rod will exhibit a lower center temperature, the relative radial decrease of the refractive power is the same.

In other words, spherical aberration of the thermal lens basically is always noticeable in medium to high power solid state lasers. In multimode lasers, however, this causes problems only near the unstable zones. The typical influence of the spherically aberrated lens on the resonator performance is shown in Fig. 13.31. The calculated output power and beam quality factor of three resonators all having flat rear mirrors but different lengths and output coupler curvatures are compared. The small-signal gain was kept constant and the refractive power was increased up to 2 Diopters. Without aberration (γ=*0*), the output power remains constant if the stable range ΔD is decreased and the beam quality factor shows the expected parabolic dependence on the refractive power. With aberration (γ=*0.45*), the output power is reduced significantly for the resonators with smaller stable zones. The effect on the beam quality is less dramatic, but note that the beam quality is improved by aberration due to a lower fill factor [4.110,4.124]. These theoretical results seem to be in good agreement with the experimental results presented in Figs.13.32 and 13.33.

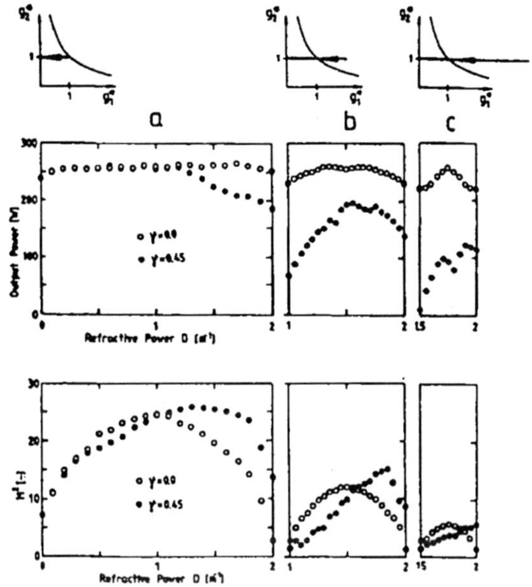

Fig. 13.31 Calculated output power and beam quality factor with (γ=*0.45*) and without aberration (γ=*0.0*) for resonafors covering different large stable zones in the stability diagram. The gain medium with radius $b=3mm$, wavelength $\lambda=1.064\mu m$, and loss factor per transit $V_S=0.95$ is located at the output coupling mirror 1 ($R=80\%$) and exhibits a small-signal gain of $g_0\ell=1.0$ ($I_S=2kW/cm^2$). a) $\rho_1=\rho_2=\infty$, $d_1=0m$, $d_2=0.5m$, a) $\rho_1=-0.5m$, $\rho_2=\infty$, $d_1=0m$, $d_2=1.0m$, a) $\rho_1=-0.66m$, $\rho_2=\infty$, $d_1=0m$, $d_2=2.0m$ [4.108] (© IEEE 1993).

A drastic experimental example of how the output power drops off if the stable refractive power range ΔD is chosen small is presented in Fig. 13.32. This resonator exhibits two stable zones of less than 0.3 Diopters, located at high input power. Compared to a short flat-flat resonator, the output power is decreased by one order of magnitude. The different power maxima correspond to different polarizations of the radiation. The resonator reaches the first stable zone with a mode structure mainly radially polarized. Before the resonator goes unstable, the polarization is changed to azimuthal orientation and a second power peak is observed. In the second stable zone, the stable refractive power range ΔD is larger than the difference of refractive powers of the two polarizations. As a consequence, the two power maxima are not separated anymore (third peak).

Due to spherical aberration and birefringence, high output power is always linked to a deterioration of the beam quality, if common dual-mirror resonators without internal lenses are used. For each output power a lower bound for the beam parameter product exists. This is shown in Fig. 13.33 for a Nd:YAG laser capable of a maximum average output power of 320W. The beam parameter products for a variety of resonators are plotted versus the output power. The minimum beam parameter product observed increases considerably if the maximum output power is approached. Even for low output powers between 50 and 100W it is not possible with this laser to get a beam quality better than 13 times diffraction limited.

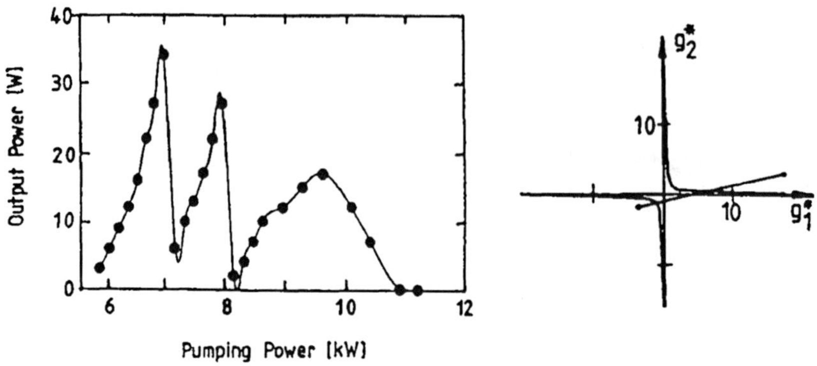

Fig. 13.32 Measured output power of a pulsed Nd:YAG rod laser (b=*5mm*, ℓ=*150mm*) utilizing a resonator with two extremely small stable zones as a function of the electrical pump power. The right graph shows the stability diagram for a pump power range of 12kW. The power maxima are related to different polarizations. Resonator data: ρ_1=*-0.09m*, ρ_2=*-0.65m*, d_1=*76cm*, d_2=*79cm*, R=*80%*, input energy: 80J, pulse duration: 2ms, repetition rate is varied from 0 to 150 Hz, refractive power: 0.3 Diopters per kW of pump power [4.108] (© IEEE 1993).

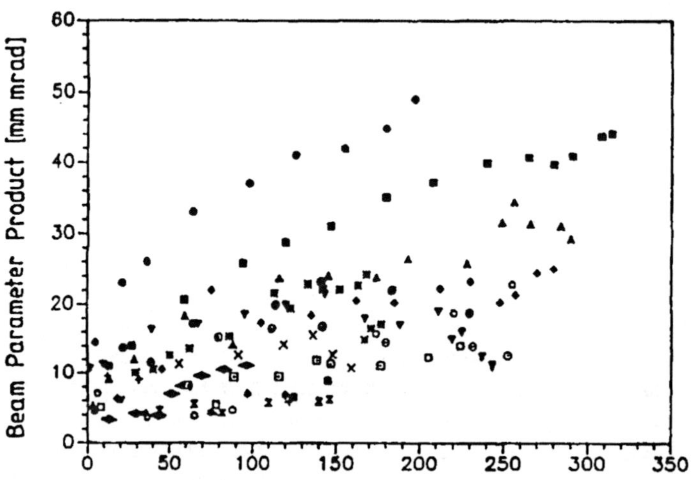

Fig. 13.33 Measured beam parameter product $w\theta$ versus measured output power for a Nd:YAG rod laser (b=*5mm*, ℓ=*150mm*, D=*0.3m*$^{-1}$ per kW of pump power), using 15 resonators with differently large stable zones in the stability diagram. (repetition rate: 30Hz, pulse width: 2ms, input energy was varied from 0 to 400J). The following resonator data $\rho_1/\rho_2/d_1/d_2$ were used: ●: ∞/∞/13cm/13cm; ■: ∞/∞/23cm/23cm; ▲: ∞/∞/33cm/33cm; ◆: ∞/∞/43cm/43cm; ▼: ∞/∞/54cm/54cm; O: ∞/∞/59cm/59cm; O: ∞/∞/63cm/63cm; □: ∞/∞/66cm/66cm; ✱: -0.5m/∞/20cm/45cm; O: -0.3m/∞/13cm/33cm; □: -0.3m/∞/ 43cm/43cm; I: -0.5m/-0.5m/21cm/21cm; ✪: -0.3m/-0.5m/10cm/40cm; ×: -0.3m/∞/10cm/83cm; ♦: -1m/ ∞/90cm/53cm [4.108] (© IEEE 1993).

The temperature induced part of spherical aberration in laser rods is, according to (13.51/13.52), inversely proportional to the rod length ℓ. Therefore, it is possible to attain a higher output power at the same beam quality by distributing the pump power over a longer active medium. Since the available crystal lengths are limited to 200mm, the preferred arrangement is a dual rod resonator that exhibits small refractive power ranges within which the resonator is stable. A quartz rotator (QR) between the rods is used for birefringence compensation so that the resonator stability does not depend on the polarization of the resonator modes. The Nd:YAG laser resonator depicted in Fig. 13.33 utilizes two 150mm long flashlamp pumped Nd:YAG rods to provide a maximum output power of 290W and a beam parameter product of 1.25 mm mrad ($M^2=3.75$) at a total electrical pump power 14.3kW. Although the output power is less than half the maximally possible (a short flat-flat resonator provides 650W at the same pump power), the beam parameter product at this output power could be decreased several times, compared to a single rod resonator providing the same output power. In recent years, similar set-ups have been used to improve the brightness of diode-pumped Nd:YAG lasers: $M^2=4$, P=500W [4.121], $M^2<1.1$, P=208W [4.116], $M^2=1.2$, P=183W [4.123], $M^2<1.2$, P=153W [4.120]. With an Yb:YAG dual rod system, M^2 of 2.2 at an output power of 530W was measured [4.118,4.119].

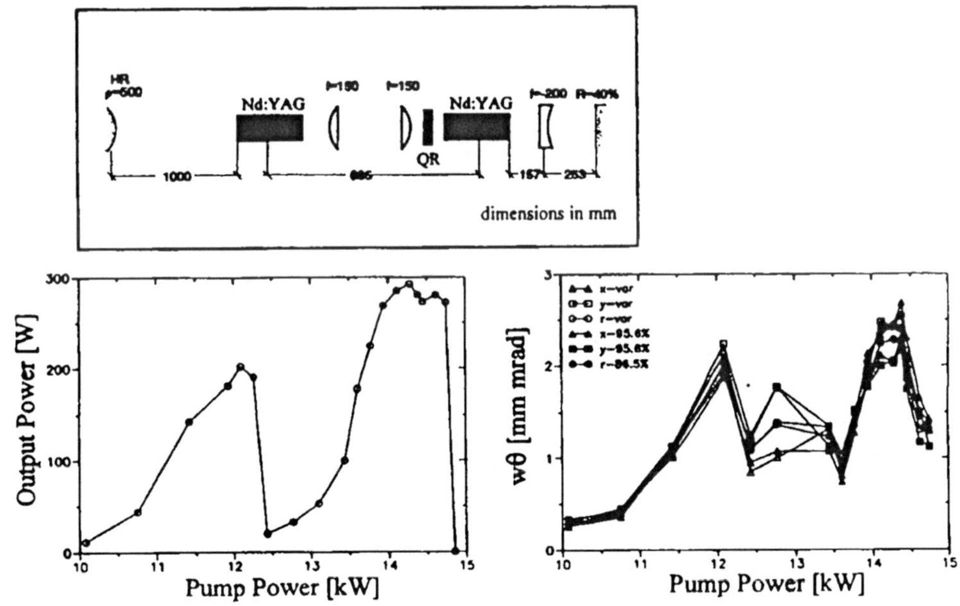

Fig. 13.33 Resonator set-up, measured output power and measured beam parameter product as a function of the total electrical pump power for a pulsed, dual rod Nd:YAG laser ($b=4.75mm$, $\ell=150mm$, refractive power per rod: 0.28m^{-1} per kW of pump power). Pump energy per rod: 168J, pump pulse duration: 4ms, maximum repetition rate: 44 Hz. The beam parameter products (waist radius w times half angle of divergence θ) defined via the power content (full points) and via the second intensity moment (open points) are shown [S.13].

In all of these demonstrations of high power TEM_{00} mode operation, resonators were used that operated only within a limited pump power range. This restriction can be overcome by moving one of the resonator mirrors inwards as the pump power is increased. This way TEM_{00} mode operation can be attained at all pump power levels, and the beam diameter in the active media and on one of the resonator mirrors can be kept constant. An experimental realization of this technique in an industrial diode-pumped Nd:YAG laser is shown in Fig. 13.34.

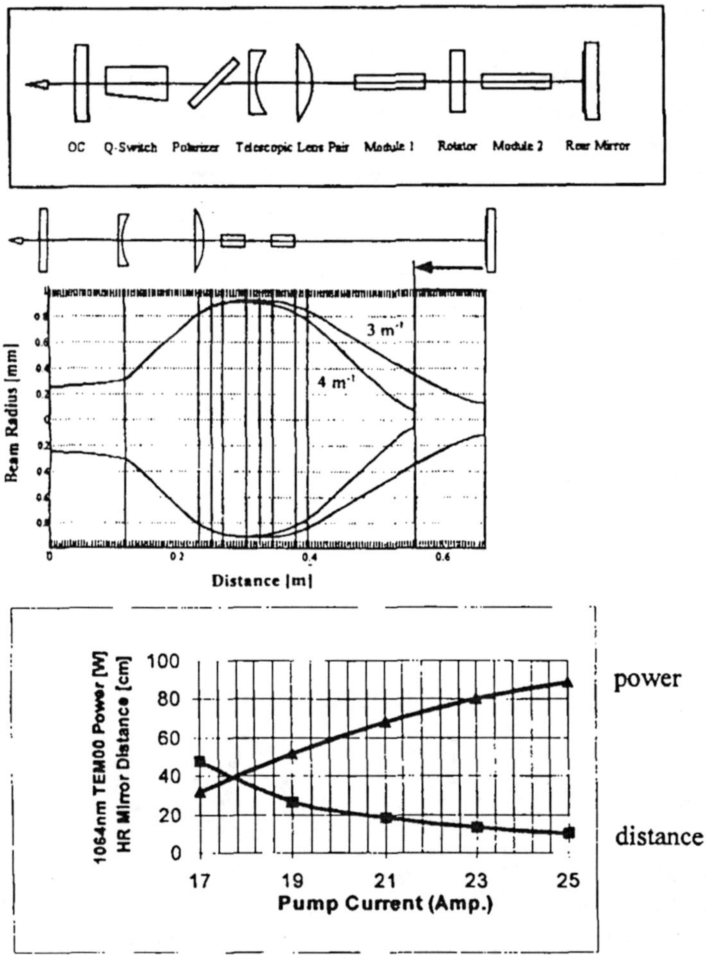

Fig. 13.34 Side-pumped dual rod Nd:YAG laser with moving HR mirror to compensate for a varying refractive power per rod. The middle graph shows the calculated TEM_{00} mode beam diameter for 3 and 4 Diopters of refractive power per rod. The lower graph depicts measured TEM_{00} mode power (upper curve) and the distance of the HR mirror from the adjacent rod end face as a function of the diode current. A current of 25A is equivalent to an optical pump power of 300W per rod. By moving the HR mirror, the TEM_{00} mode diameters at the output coupler and inside the rods remain constant with varying pump power [4.125,S.33].

The aberration of the thermal lens can lead to a significant decrease of the beam quality when near diffraction limited beams are amplified. For a transit in the amplifier, the phase shift generated by the thermal lens is given by:

$$\Delta\Phi(r) = \frac{2\pi}{\lambda}\left[\frac{D(r)\,r^2}{2}\right] = \frac{2\pi}{\lambda}\left[\frac{D_0\,r^2}{2} - \frac{D_0\gamma\,r^4}{2b^2}\right] \tag{13.53}$$

Neglecting gain saturation effects, the beam propagation factor M^2 of the amplified beam can be calculated by using the second intensity moments (see Sec. 2.6.2):

$$M^2 = \frac{\pi}{\lambda}\sqrt{\langle w^2\rangle\langle\theta^2\rangle - \langle w\theta\rangle^2} \tag{13.54}$$

For radially symmetric beam with Gaussian intensity profile and initial beam propagation factor M_i^2, the beam propagation factor M_f^2 after amplification reads [4.115,4.126]:

$$M_f^2 = \sqrt{\left[M_i^2\right]^2 + \left[\frac{\sqrt{8}\,\alpha_T\,P_{abs}\,\gamma}{\lambda}\left(\frac{w}{b}\right)^4\right]^2} \tag{13.55}$$

where α_T is the thermal lensing coefficient ($=D_0\pi b^2/P_{abs}$) and P_{abs} is the absorbed pump power. Figure 13.35 shows a calculated example for a diode-pumped Nd:YAG rod.

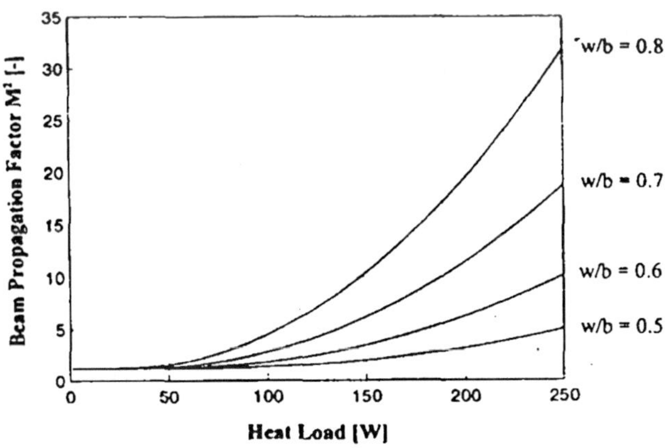

Fig. 13.35 Calculated beam propagation factor M_f^2 of an incident TEM_{00} mode beam ($M_i^2=1.2$) after propagation through a homogeneously diode-pumped Nd:YAG rod amplifier with 4mm diameter and 10 cm length as a function of the heat load P_H, using (13.51,13.52 and 13.55). Curve parameter is the ratio of beam radius w to rod radius b. ($\beta=0$, $\epsilon=9.5\cdot 10^{-6}$ 1/K, $\delta=0.04\cdot 10^{-6}$ 1/K^2, $T_W=293$K, h=1.0 W/(cm^2 K) $P_H=0.25*P_{abs}$, P_{abs}: absorbed pump power at 808nm).

13.3 Unstable Resonators

13.3.1 Beam Propagation

In unstable resonators the refractive power of the thermal lens leads to a decrease of the magnification M and of the equivalent Fresnel number N_{eq}. Furthermore, the radius of curvature of the outcoupled wavefront is a function of the refractive power, which means that the location of the focus spot changes as the pump power is increased [4.39,4.104,4.130]. The concept of the equivalent resonator discussed in the preceding section is also applicable to unstable resonators. The unstable lens resonator with g-parameters g_1,g_2 and distances d_1,d_2 from the mirrors to the principal planes of the lens exhibits the same field distributions at the resonator mirrors and the same output coupling loss as the equivalent, empty resonator with g-parameters g_1^*,g_2^* and the geometrical length L^* (see Eqs. (13.8)-(13.10)).

Figure 13.36 presents the typical behavior of an unstable lens resonator. In this example the thermal lens with a refractive power between 0 and 3.1 Diopters is located at the high reflecting mirror 2. Therefore, the g-parameter of the output coupling mirror stays constant. In order to stay in the unstable zone over the whole refractive power range, the resonator has to start at a high magnification of $M=24.2$. The magnification is decreased down to $M=1.0$ as the resonator travels towards the stability limit with increasing refractive power. The beam propagation shown for three refractive powers demonstrates the sensitivity of the magnification and of the position of the focus spot to the refractive power. It is evident that this resonator can only be operated efficiently over a small refractive power range around the confocal point at $D=3m^{-1}$ at which a magnification of $M=2$ is attained. Fortunately, the sensitivity can be decreased considerably by choosing a proper resonator design.

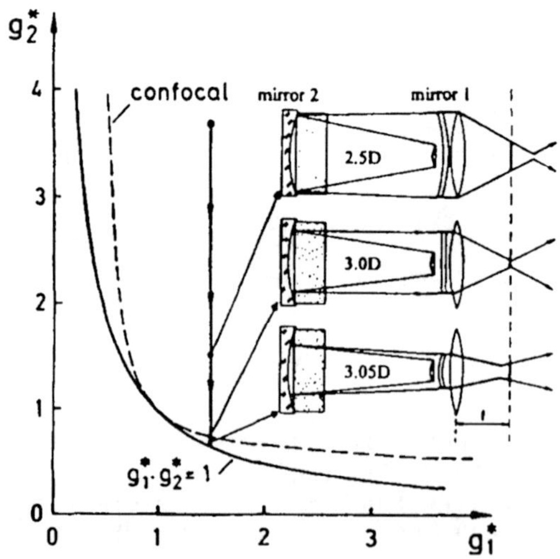

Fig. 13.36 The equivalent g-diagram for a positive branch unstable resonator (ρ_1=-2m, ρ_2=-0.364m, d_1= 1.0m, d_2=0). The refractive power is varied between 0 and 3.1m^{-1}. The beam propagation including focusing is shown for three refractive powers. The resonator is confocal at 3.0m^{-1} with a magnification of $M=2$.

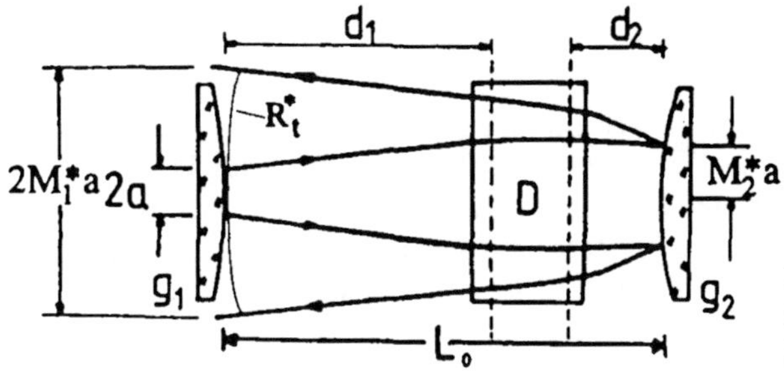

Fig. 13.37 Beam propagation in an unstable resonator with a variable lens.

All relations derived in Sec.7, except for the wavefront curvatures, are still valid for the unstable lens resonators if the equivalent resonator parameters g_1^*, g_2^*, L^* are inserted (Fig.13.37):

Equivalent g-parameters : $g_i^* = g_i - Dd_j(1 - \frac{d_i}{\rho_i})$ $\quad i,j = 1,2;\ i \neq j$ (13.56)

Equivalent length : $L^* = d_1 + d_2 - Dd_1d_2$ (13.57)

Equivalent G-parameter : $G^* = 2g_1^*g_2^* - 1$ (13.58)

Magnification (round trip) : $M_1^* = G^* \pm \sqrt{G^{*2} - 1}$ (13.59)

Magnification (transit) : $M_2^* = g_1^* \pm \sqrt{G^{*2} - 1}/(2g_2^*)$ (13.60)

Equivalent Fresnel number : $N_{eq}^* = a^2\sqrt{G^{*2} - 1}/(2L^*g_2^*\lambda)$ (13.61)

Wavefront curvature : $R_t^* = \dfrac{L^*}{L^*/\rho_1 - g_1^* + M_2^*}$ (13.62)

In (13.59) and (13.60), the positive sign applies to positive branch, the negative sign to negative branch unstable resonators. In order to operate an unstable lens resonator efficiently over a wide refractive power range, it is necessary to minimize the sensitivity of the magnification and the wavefront curvature to the refractive power of the thermal lens. There are three basic resonator schemes to choose from, operating in different quadrants of the stability diagram. Although their sensitivity to the thermal lens is different, all three resonator schemes have in common that the optimum output coupling can only be attained at one fixed refractive power. An optimized unstable resonator design will provide maximum efficiency only at this point. For a positive branch unstable resonator, the resonator should be confocal at the design refractive power.

13.3.2 Positive Branch Confocal Unstable Resonators

We will now derive the design equation for a positive branch unstable lens resonator that becomes confocal at a certain refractive power D_0 and provides maximum extraction efficiency at this point. As depicted in Fig. 13.38, both the magnification M_1^* and the equivalent Fresnel number N_{eq}^* of the lens resonator decrease as the refractive power is increased. When the resonator is confocal, both parameters should assume values M_0 and $N_{eq,0}$, respectively, providing optimum output coupling. In addition, the active medium with radius b should be completely filled by the laser mode as shown in Fig. 13.35. This leads to four conditions:

$$M_1^*(D_0) = M_0 \tag{13.63}$$

$$N_{eq}^*(D_0) = N_{eq,0} \tag{13.64}$$

$$b = 1.05\, M_0 a \tag{13.65}$$

$$M_2^* = g_1^* - L^*/\rho_1 \qquad \textit{(confocality)} \tag{13.66}$$

Since the unstable resonator has five parameters $(a,d_1,d_2,\rho_1,\rho_2)$, the resonator design is not completely defined by the four constraints (13.63)-(13.66). However, in general the distance d_2 should be chosen as small as possible to minimize the influence of the amplifier region on the extraction efficiency. Therefore, it makes sense to express the resonator parameters as a function of d_2.

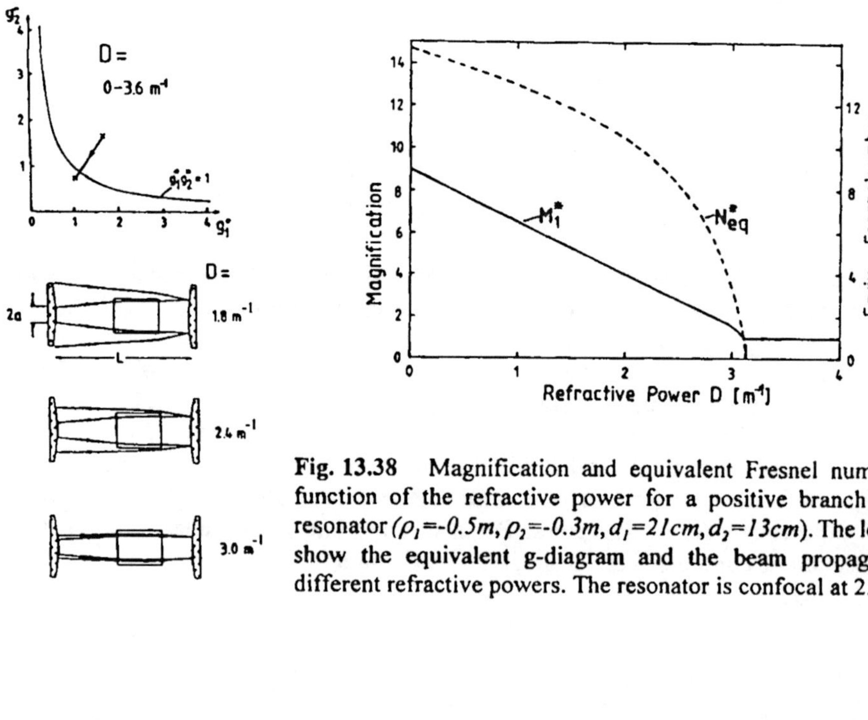

Fig. 13.38 Magnification and equivalent Fresnel number as a function of the refractive power for a positive branch unstable resonator (ρ_1=-0.5m, ρ_2=-0.3m, d_1=21cm, d_2=13cm). The left graphs show the equivalent g-diagram and the beam propagation for different refractive powers. The resonator is confocal at 2.7 m⁻¹.

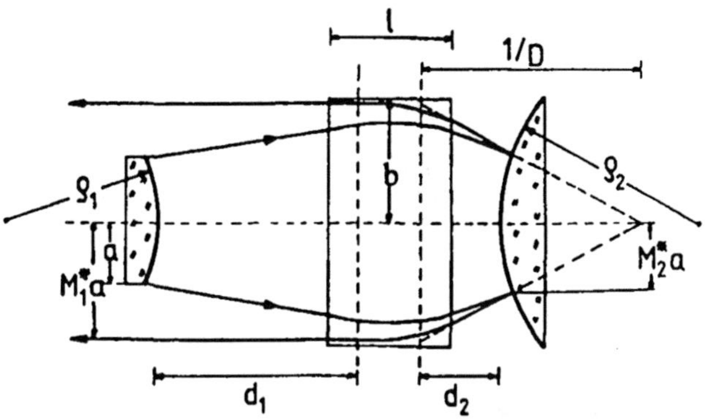

Fig. 13.39 The positive branch unstable resonator with an internal lens operating at the confocal point. The medium is filled by the beam, and the near field exhibits a flat phase distribution.

By using (13.60) and (13.61) in combination with the four constraints, the following system of design equations can be derived [4.104]:

$$a = \frac{b}{1.05\, M_0} \tag{13.67}$$

$$\rho_1 = -\frac{b^2}{\lambda N_{eq,0}(1.05 M_0)^2} \tag{13.68}$$

$$d_1 = \frac{d_2[2 - D_0(M_0-1)\rho_1] + (M_0-1)\rho_1}{2(D_0 d_2 - 1)} \tag{13.69}$$

$$\rho_2 = \frac{L^{*2}}{L^*(1-D_0 d_1) - \dfrac{a^2(M_0^2-1)}{4M_0 N_{eq,0}\lambda}} \tag{13.70}$$

The combination of these four equations provides the resonator parameters for optimum performance at a refractive power D_0. The distance d_2 is a free parameter. Note that the solution of (13.68) has to be inserted into (13.69) to get the distance d_1, and d_1 serves as an input parameter for (13.70). For high power solid state lasers, the optimum magnification is around $M_0=2$ and, for better transverse mode discrimination, the equivalent Fresnel number should be a half integer value. Figure 13.40 presents the solutions of (13.67)-(13.70) for this case as a function of the design refractive power D_0. For these graphs, a distance d_2 between the principal plane of the lens and the unconfined mirror of 10cm was used.

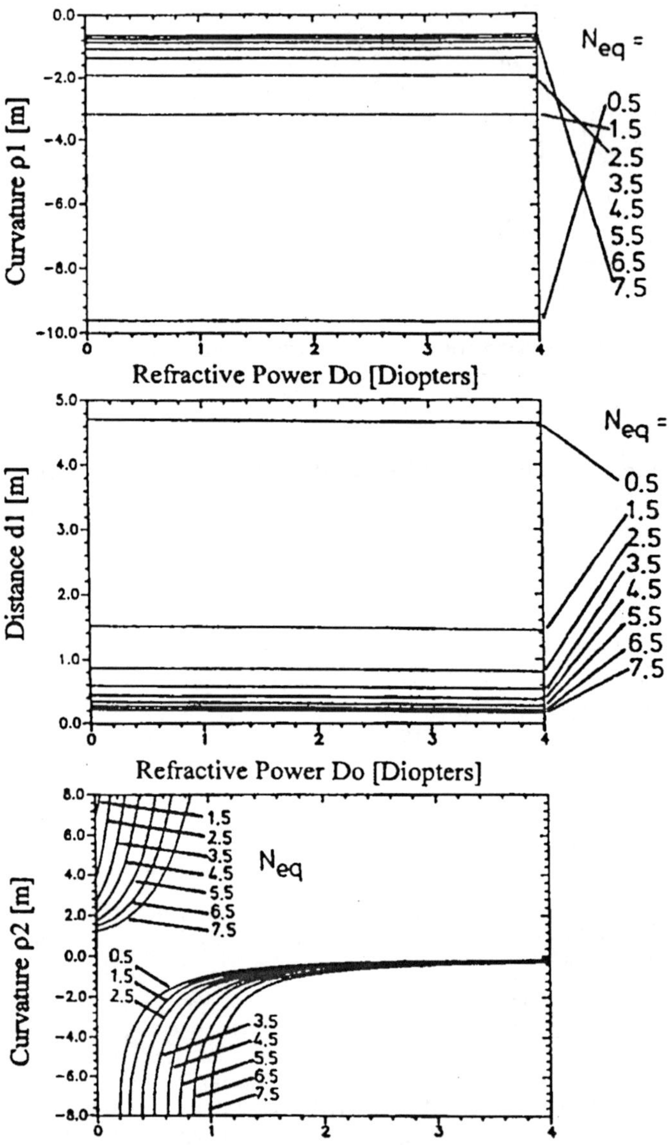

Fig. 13.40 Optimum distance d_1 and optimum mirror curvatures ρ_1,ρ_2 as a function of the design refractive power D_0 for positive branch unstable lens resonators in the first quadrant of the g-diagram. The lens resonator is confocal at the refractive power D_0 with a magnification of M_0=*2.0* and an equivalent Fresnel number $N_{eq,0}$. (*b*=*5mm*, d_2=*10cm*, λ=*1.064μm*). The curve parameter is the equivalent Fresnel number $N_{eq,0}$.

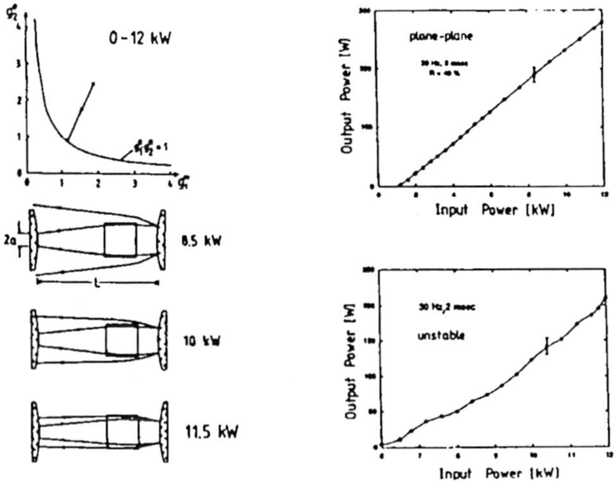

Fig. 13.41 Measured output power for a pulsed Nd:YAG rod laser ($b=5mm$, $\ell=150mm$) with a positive branch unstable resonator (lower right graph). The refractive power of the rod is $0.3m^{-1}$ per kW of input power. The left graph depicts the equivalent g-diagram and the beam propagation for different input powers. The resonator is confocal with a magnification $M_0=2$ at an input power of 11.5kW ($\rho_1=-0.5m$, $\rho_2=-0.3m$, $d_1=27cm$, $d_2=12cm$, $a=2.0mm$. The upper right graph shows the output power measured with a symmetric flat-flat resonator (geometrical length: 1.15m) for comparison (pulse duration: 2ms, repetition rate: 30Hz, maximum pump energy: 720J) [4.104] (© Chapman & Hall 1990).

The experimental data presented in Fig. 13.41 indicate that the positive branch unstable resonator exhibits a high laser threshold. This is due to the high initial magnification of $M_i=14$ that is required to keep the resonator in the unstable zone over the whole pump power range. For lasers that comprise a strong thermal lens, this resonator exhibits a very limited operating range around the confocal point since both the magnification and the wavefront curvature at the output coupler depend strongly on the refractive power. Furthermore, the confocal resonators are very close to the stability limit. A slight increase of the pump power will thus drive the resonator into the stable zone. If a high reflecting output coupler is used, accidental penetration into the stable zone may damage the mirror due to the high intracavity intensity (there is basically no output coupling in the stable zone). Therefore, it is advantageous to use a variable reflectivity mirror with a lower center reflectivity.

Compared to a stable resonator, unstable resonators generally provide only 60-70% of the output power. As was discussed in Sec.11, this decrease is caused by a lower fill factor and diffraction losses generated by the active medium. The sensitivity of the resonator properties to thermal lensing can be decreased considerably by using resonators in other quadrants of the stability diagram. One resonator scheme is the rod-imaging resonator [4.105], a negative branch unstable resonator operating in the fourth quadrant (Fig. 13.42).

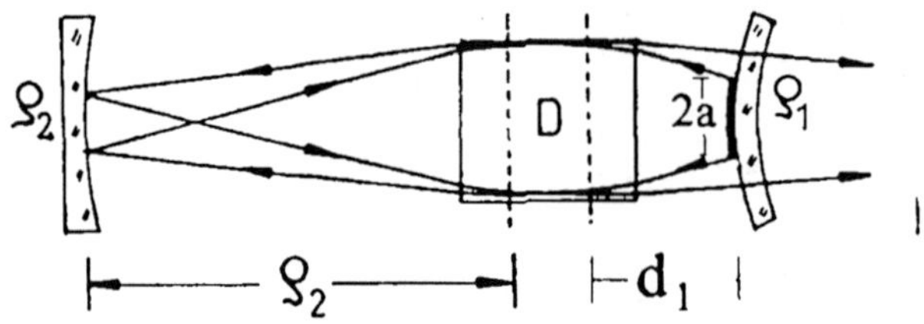

Fig. 13.42 Beam propagation in a rod-imaging unstable resonator (ρ_1=-0.3m, ρ_2=0.4m, d_1=13cm, d_2=ρ_2, a=3mm, b=5mm). Mirror 2 images the left principal plane onto itself. The refractive power is D=4.6m^{-1} [4.130] (© OSA 1993).

12.3.3 Rod-Imaging Unstable Resonators

Negative branch unstable resonators exhibit a much lower sensitivity of the magnification to thermal lensing compared with positive branch resonators. This low sensitivity stems from the positive radius of curvature of one mirror which also produces a focal point inside the cavity. A careful resonator design, however, is required because the focal point must not be located too close to the active medium. A useful design is the rod-imaging resonator [4.105,4.130], as shown in Fig. 13.42. The rear mirror 2 images the principal plane of the lens onto itself which means that the g-parameter of this mirror is constant:

$$g_2^* = g_2 - Dd_1\left(1 - \frac{\rho_2}{\rho_2}\right) = g_2 = -\frac{d_1}{\rho_2} \tag{13.71}$$

and the g-parameter of the output coupling mirror reads:

$$g_1^* = 1 - \frac{d_1 + \rho_2}{\rho_1} - D\rho_2\left(1 - \frac{d_1}{\rho_1}\right) \tag{13.72}$$

The resonator will thus move parallel to the g_1-axis in the equivalent g-diagram. Rod-imaging resonators are confocal if the condition:

$$g_1^* + g_2 = 2g_1^* g_2 \tag{13.73}$$

is met. Note that for rod-imaging resonators, both the magnification M_l^* and the equivalent Fresnel number N_{eq}^* are negative. Besides the lower sensitivity of the magnification (Fig. 13.43), other advantages of rod-imaging resonators are the efficient filling of the active medium and a low misalignment sensitivity of the output coupling mirror.

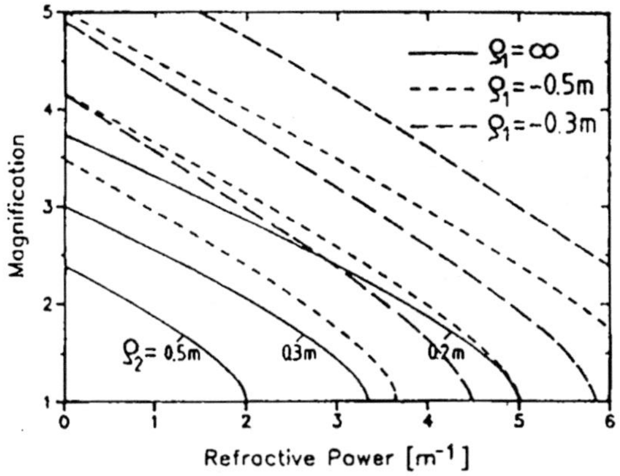

Fig. 13.43 Dependence of the absolute value of the magnification M_l^* of rod-imaging unstable resonators ($d_2=\rho_2$) on the refractive power D for three different mirror curvatures ρ_1. The distance d_1 is always 10cm; curves with the same line type differ in the mirror curvature ρ_2 (higher magnification for smaller curvature) [4.130] (© OSA 1993).

Similar to the positive branch resonators, we can determine an optimized resonator design that provides a certain magnification M_0, a certain equivalent Fresnel number $N_{eq,0}$, and a collimated output beam at a refractive power D_0 by applying the conditions (13.64)-(13.70). Since we have four resonator parameters (ρ_1,ρ_2,d_1,a), there is only one resonator that can meet all four constraints. In order to have more freedom in the resonator design, let us drop the condition on the Fresnel number and express the curvatures as a function of the distance d_1. The following design equations hold for rod-imaging resonators that become confocal at the refractive power D_0 with a magnification of M_0 (b: radius of active medium):

$$a = \frac{-b}{1.05\ M_0} \tag{13.74}$$

$$\rho_1 = d_1 \left[D_0 + \frac{M_0-1}{2d_1}\right] \bigg/ \left[D_0 + \frac{M_0^2-1}{4M_0 d_1}\right] \tag{13.75}$$

$$\rho_2 = d_1 \frac{-2}{M_0 + 1} \tag{13.76}$$

Keep in mind that the magnification M_0 is negative in these equations. The refractive power at which the resonator goes confocal is mainly controlled by the radius of curvature ρ_1 of the output coupler. Figure 13.44 presents the radius of curvature ρ_1 as a function of the design refractive power for different magnifications and different distances d_1.

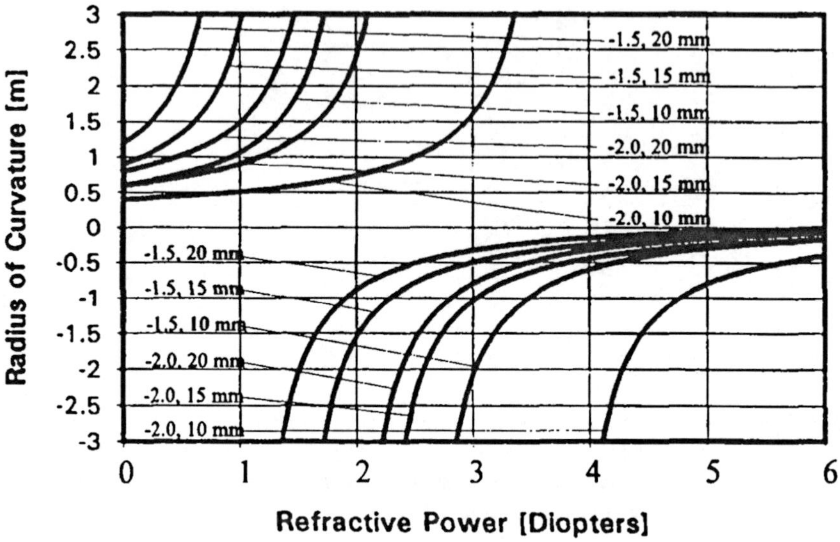

Fig. 13.44 Radius of curvature of the output coupling mirror as a function of the refractive power D_0 for rod-imaging unstable resonators that become confocal at the refractive power D_0 with a magnification of M_0. The curve parameters are the magnification M_0 and the distance d_1.

For high power Nd:YAG lasers that require a magnification of $M_0=2$ at refractive powers around $4m^{-1}$, for instance, mirror radii of $\rho_2=0.3m$ and $\rho_1=-0.5m$ seem to be a reasonable choice if a distance of $d_1=15cm$ is used. An experimental result for a pulsed Nd:YAG laser utilizing a similar resonator design is presented in Fig. 13.45. The resonator provides a magnification of 2.6 at the maximum refractive power of $3.8m^{-1}$. As we can see in the equivalent g-diagram, the rod-imaging condition is not exactly met (the g-parameter g_1 increases slightly since d_2 is 2mm too long). However, this slight mismatch, which is within the typical experimental tolerance, does not have a serious impact on the resonator performance. Since rod-imaging resonators can start at a lower magnification without being driven out of the unstable zone, the laser threshold is considerably lower as compared to confocal positive branch resonators. Again, the output power attained is lower compared to a stable resonator which for this laser head provided 520W at the maximum pump power.

It should be added that the rod-imaging unstable resonator does not necessarily have to be operated in a confocal point. It is true that the confocal resonator provides the highest fill factor, but the increase in extraction efficiency is almost negligible compared to non-confocal configurations. As far as the flat wavefront of the outcoupled beam is concerned, that can be realized in any resonator by using transformation optics right behind the output coupler. The design equations (13.74)-(13.76) should be considered as a suggestion. Any other resonator design that provides the right output coupling at the desired refractive power but is not confocal will probably perform as well.

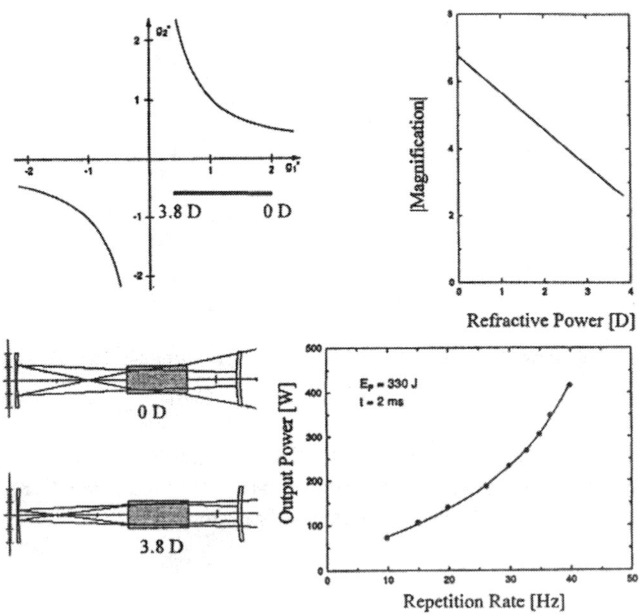

Fig. 13.45 Measured output power of a pulsed Nd:YAG rod laser (b=*5mm*, ℓ=*150mm*) utilizing a rod-imaging resonator as a function of the repetition rate. The refractive power at 40Hz (electrical pump power of 13.2kW, pulse duration: 2 ms) is 3.8 Diopters. Resonator data: ρ_1=*-0.5m*, ρ_2=*0.3m*, d_1=*16cm*, d_2=*32cm*, a=*2.5mm*. The magnification as a function of the refractive power, the equivalent g-diagram, and the beam propagation for minimum and maximum refractive power are also shown.

13.3.4 Near Concentric Unstable Resonators

In the rod-imaging resonator one principal plane was imaged onto itself resulting in one constant g-parameter. If both principal planes are imaged by the mirrors ($d_1=\rho_1$ and $d_2=\rho_2$), both g-parameters become independent of the refractive power and the lens resonator remains concentric with $g_1^*g_2^*=1$. By increasing the distances d_1,d_2 by small amounts Δ_1,Δ_2, respectively, the lens resonator is moved into the unstable zone (Fig. 13.46). In this case, the g-parameters are not constant anymore, but, due to the proximity to the concentric resonators, they show a very low sensitivity to the refractive power. This is the concept of the near-concentric unstable resonator (NCUR) [4.107]. One special configuration is the symmetric NCUR with $\Delta_1=\Delta_2=\Delta$ and equal mirror curvatures $\rho_1=\rho_2=\rho$. In this case, the equivalent g-parameters read:

$$g_1^* = g_2^* = -1 - \frac{\Delta}{\rho}[2 - D(\rho+\Delta)] \tag{13.77}$$

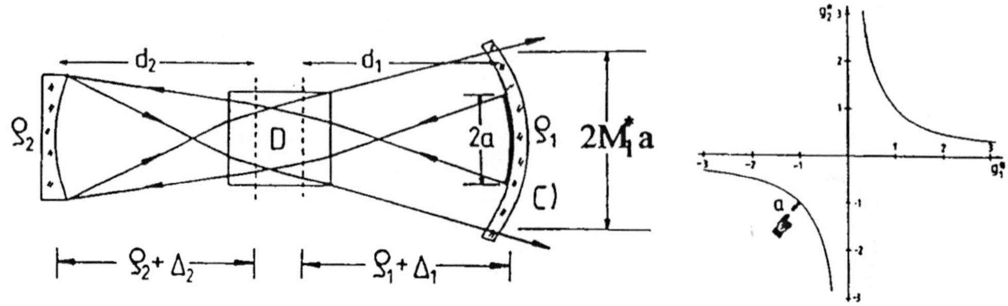

Fig. 13.46 The near concentric unstable resonator (NCUR) and its location in the g-diagram.

The refractive power D_{max} at which the resonator goes concentric again depends only slightly on the additional shift and is mainly determined by the mirror curvature. This refractive power also represents the range within which the NCUR can be operated:

$$D_{\max} = \frac{2}{\rho + \Delta} \tag{13.78}$$

Therefore, it is possible to use this resonator for large ranges of refractive power by choosing the appropriate mirror curvature. The shift Δ typically is on the order of 10-20mm. Figure 13.47 presents the magnification M_t^* as a function of the refractive power for different symmetric NCURs. Compared to the rod-imaging unstable resonator (see Fig. 13.43), the sensitivity of the magnification to thermal lensing is now further reduced. This is particularly true if small shifts Δ and small mirror curvatures are used. However, small shifts Δ are problematic since the two focal spots generated inside the resonator get too close to the active medium and may damage the endfaces. The shift Δ should, therefore, never be smaller than 10mm if mirror curvatures between 0.3m and 0.5m are used. A lower sensitivity to the refractive power is also generated for the radius of curvature of the outcoupled wavefront, resulting in a smaller shift of the focus spot as the refractive power is increased (as we will see in the next section).

As far as the optimized resonator set-up (optimum magnification M_0 at a refractive power D_0) is concerned, the mirror curvatures and the shifts can be determined using a graph like the one in Fig. 13.47. The only parameter that needs a more careful treatment is the radius a of the output coupling mirror. Experimental results indicate that this radius has to be about 50% smaller than predicted by the geometrical beam propagation. If R_t^* denotes the radius of curvature of the wavefront (see Eq. (13.62)) at the refractive power D_0 with magnification M_0, the radius a can, to a good approximation, be determined with:

$$a = 0.5 \frac{R_t^* \, b}{M_0 \, [R_t^* - \rho - \Delta + \ell/(2n)]} \tag{13.79}$$

where ℓ is the length of the active medium, b is its radius, and n its refractive index.

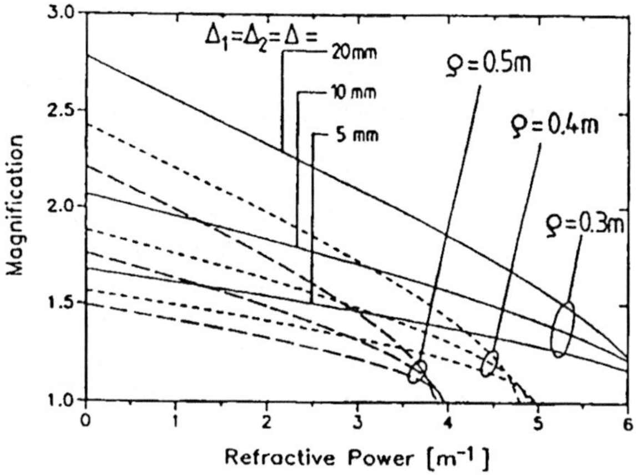

Fig. 13.47 Magnification M_l^* as a function of the refractive power for symmetric NCURs. The curve parameters are the mirror curvature ρ and the shift Δ [4.107] (© OSA 1993).

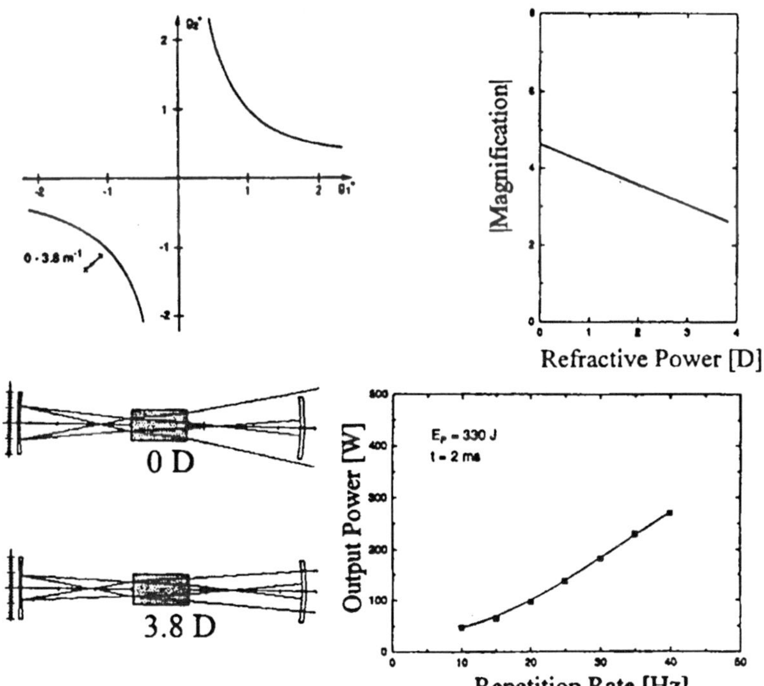

Fig. 13.48 Measured output power of a pulsed Nd:YAG rod laser (b=*5mm*, ℓ=*150mm*) utilizing a NCUR as a function of the repetition rate. The refractive power at 40Hz (electrical pump power of 13.2 kW, pulse duration: 2ms) is 3.8 Diopters. Resonator data: $\rho_1=0.3m$, $\rho_2=0.3m$, $d_1=35cm$, $d_2=36cm$, $a=2.5mm$. The magnification as a function of the refractive power, the equivalent g-diagram, and the beam propagation for minimum and maximum refractive power are also shown.

13.3.5 Beam Quality and Focusing

The beam quality of unstable resonators is strongly affected by the spherical aberration of the thermal lens and the stress-induced birefringence of the active medium. In contrast to the operation of lasers without a thermal lens, where beam propagation factors M^2 between 3 and 6 can be attained, the utilization of unstable resonators in high power solid state lasers provides beam qualities that are one order of magnitude lower. Typically, the focus consists of a bright spot with a diameter that corresponds to a M^2 value of 10-15 surrounded by a wide low-intensity pedestal that contains about 50% of the output power. If a CCD camera or any other detector with spatial resolution is used to determine the beam quality, this pedestal goes undetected due to a low signal-to-noise ratio (typically 100:1). Figure 13.49 shows measured output power and beam parameter products (determined with an 8 bit CCD-imaging system) for a pulsed Nd:YAG rod laser using three different unstable resonator schemes presented earlier. All three resonators provide the same maximum output power and similar beam parameter products. Although these beam parameter products seem very low compared to those of stable resonators in multimode operation (25mm mrad at 400W output power), the power content is only 50% since the low intensity pedestal cannot be resolved by the CCD camera. This is clearly demonstrated in Fig. 13.50 in which two CCD images of the focus spot are shown, one at the nominal intensity used for the measurement and one at a 100 times higher intensity to visualize the side lobes. The corresponding power-in-the-bucket measurement indicates that the center spot contains about 50% of the total power. In order to use the laser beam for any application the pedestal has to be cut off with a spatial filter.

At present, the physical mechanisms leading to the beam quality deterioration are not yet understood. There is evidence that the birefringence in combination with the spherical aberration of the lens has a negative impact on the beam quality [3.209]. On the other hand, similar problems have been reported for low power solid state lasers where thermal lensing should not have much effect on the resonator properties. However, the problem seems to be restricted to solid state lasers since high power gas lasers (in the 10kW range and higher) utilizing unstable resonators provide beam propagation factors in the expected range of M^2=3-6.

One other problem of unstable resonators with a thermal lens is the dependence of the near field phase curvature R_t^* (see Eq. 13.62) on the pump power. The curvature R_t^* will vary between a value R_1^* at the minimum pump power and a value R_2^* at the maximum pump power. If a focusing lens is placed right behind the output coupler, the distance x of the focus spot is given by (Fig. 13.51):

$$x = f \frac{1}{1 - f/R_t^*} \tag{13.80}$$

and the total shift Δx of the focal distance over the whole pump power range reads:

$$\Delta x = f^2 \left| \frac{R_2^* - R_1^*}{(R_2^* - f)\,(R_1^* - f)} \right| \tag{13.81}$$

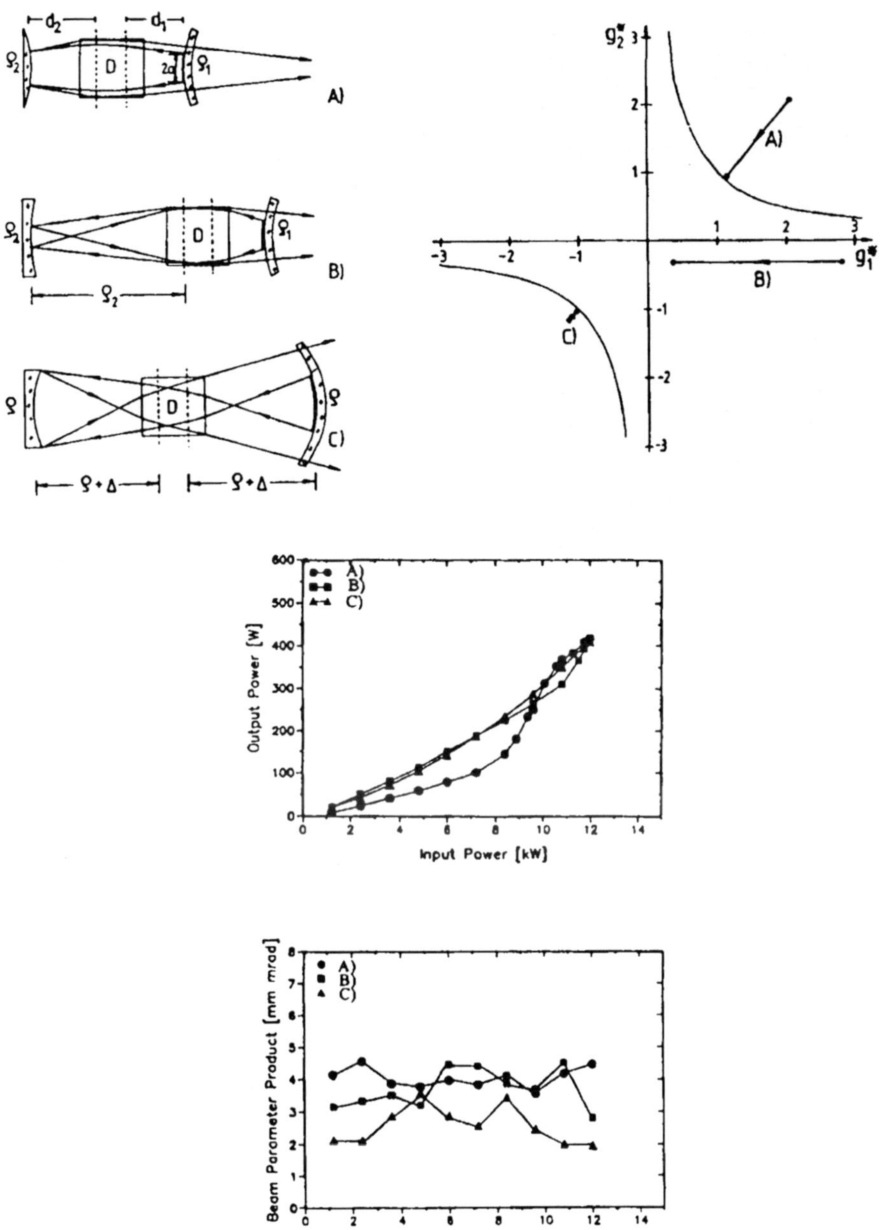

Fig. 13.49 Measured output power and beam parameter products (50% power content) of a pulsed Nd:YAG laser (b=*5mm*, ℓ=*150mm*) with three different unstable resonators. A) positive-branch confocal, $\rho_1=-0.3m$, $\rho_2=-0.3m$, $d_1=15cm$, $d_2=16cm$, $a=2.5mm$, B) rod imaging, $\rho_1=-0.3m$, $\rho_2=0.4m$, $d_1=13cm$, $d_2=40cm$, $a=3mm$, C) near concentric, $\rho_1=0.3m$, $\rho_2=0.3m$, $d_1=32cm$, $d_2=32cm$, $a=4.5mm$. The refractive power of the rod is $0.38m^{-1}$ per kW of pump power (pulse duration: 2ms, pump energy: 240J). The beam propagation inside the resonators at the maximum refractive power of $4.6m^{-1}$ and the paths in the equivalent g-diagram are shown in the upper graphs[4.130] (© OSA 1993).

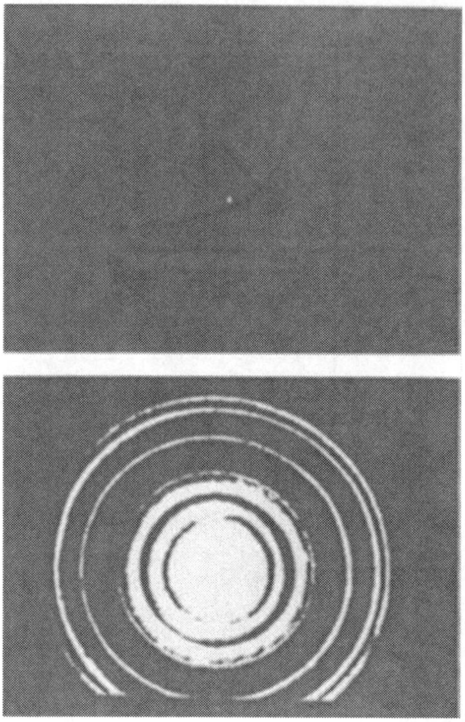

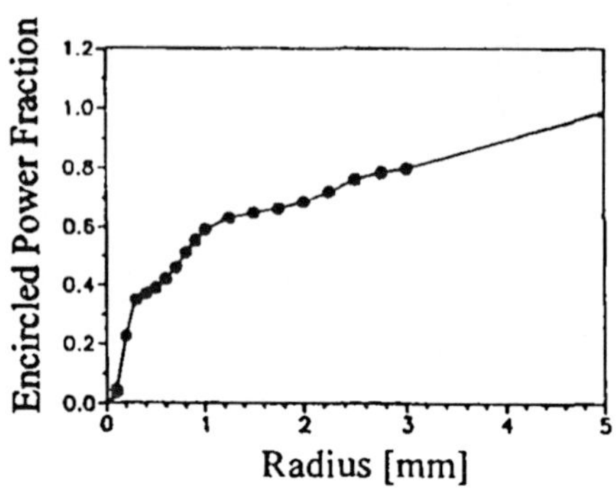

Fig. 13.50 CCD images of the focal spot of an unstable resonator (pulsed Nd:YAG rod laser, output power: 50W) at the nominal intensity and with saturation by a factor of 100 (lower photograph). The graph presents the result of a power-in-the-bucket measurement of this intensity distribution [S.9].

Since the curvatures R_i^* are in general much larger than the focal length f we can approximate this expression by:

$$\Delta x = f^2 \left| \frac{R_2^* - R_1^*}{R_1^* \, R_2^*} \right| = f^2 \left| \frac{1}{R_1^*} - \frac{1}{R_2^*} \right| = f^2 \, S \tag{13.82}$$

The shift sensitivity S is equal to the variation of dioptric power of the output beam. In order to get a small shift of the focus, a resonator set-up providing a low shift sensitivity S must be found. Unfortunately, the determination of the shift sensitivity is not sufficient to characterize the focusing performance of a resonator, since S can be arbitrarily changed by means of transformation optics. An increase of the beam radius will decrease the shift sensitivity. However, the product of the beam radius and the shift sensitivity is a constant of the resonator, which means the shift sensitivities of different resonators can be compared if they exhibit similar beam radii at the output coupler. Figure 13.51 presents average beam radii on the output coupling mirror versus the shift sensitivity S for various positive branch and negative branch unstable resonators. This graph was calculated using (13.82) and the relations (13.56)-(13.62) for a rod radius of b=*5mm* and a maximum refractive power of 4.6m^{-1}. This figure clearly shows that the near concentric resonator (curve *a)* provides the best means to obtain good constancy of the focal distance. The experimental verification of this statement is presented in Fig. 13.52, in which the distance x of the focal spot for the three resonators of Fig. 13.49 is shown as a function of the pump power.

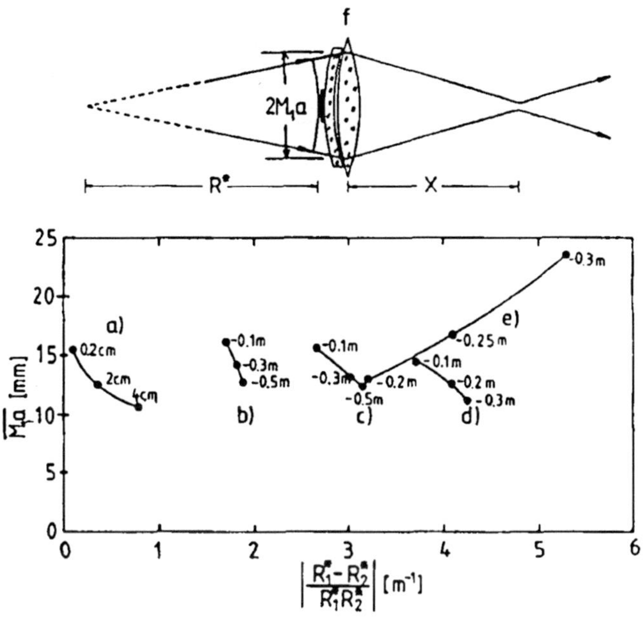

Fig. 13.51 The radius of curvature R^* of the near field wavefront changes as the pump power is increased resulting in a variation of the focal distance x. The lower graph presents the average beam radius on the output coupler versus the shift sensitivity S (13.82) for different unstable resonators for a variation of the refractive power between 0 and 4.6 m^{-1} and a rod radius of 5mm. a) near concentric, $\rho_1=0.3m$, $\rho_2=0.3m$, *parameter:* Δ; b) rod imaging, $\rho_2=0.2m$, $d_1=13cm$, *parameter* : ρ_1; c) rod-imaging, $\rho_2=0.3m$, $d_1=13cm$, *parameter:*ρ_1; d) rod imaging, $\rho_2=0.5m$, $d_1=13cm$, *parameter:*ρ_1; e) positive branch confocal, $\rho_2=0.3m$, $d_1=15cm$, $d_2=16cm$, *parameter:*ρ_1 [4.130] (© OSA 1993).

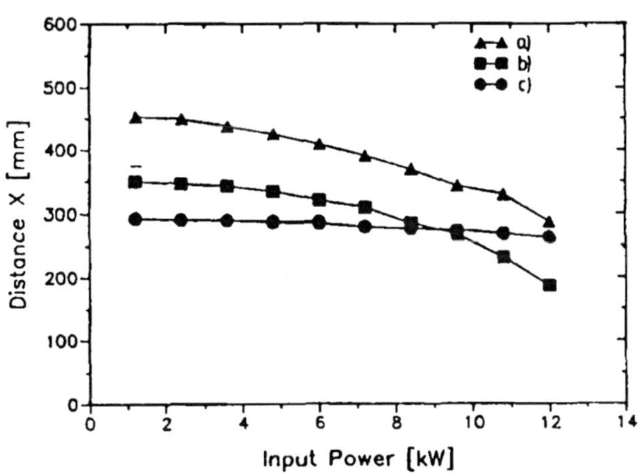

Fig. 13.52 Measured distance x between the beam waist and the focusing lens as a function of the input power for the resonators presented in Fig. 13.46. The lens with focal length $f=200mm$ was positioned 260mm behind the output coupler. a) positive branch confocal; b) rod-imaging, c) near concentric [4.130] (© OSA 1993).

Chapter 14

Resonators with Several Active Media

14.1 General Aspects

In high power lasers, the maximum output power scales proportionally with the length of the active medium. In general, there is an upper bound for the output power that can be attained with a medium of a certain length. For solid state lasers this limit is caused by the pump induced thermal heat P_H which has to stay below the fracture limit to avoid damage to the active material. For a laser rod the maximum output power P_{out} per length is given by:

$$\frac{P_{out,\max}}{\ell} = \frac{\eta_{extr}}{\chi}\frac{P_{H,\max}}{\ell} \tag{14.1}$$

where η_{extr} is the extraction efficiency, χ is the thermal load parameter (ratio of heat to power transferred to inversion, ~0.33 for 808nm diode-pumped Nd:YAG lasers emitting at 1064nm), and $P_{H.\max}$ is the maximally allowed thermal heat (see Table 14.1 for typical values). In gas lasers the limitation of the output power per tube length is also a result of the heat deposition in the gas. For sealed-off CO_2 lasers in which the gas is not constantly removed, the maximum output power is on the order of 100W/m for tube lasers. Better heat removal can be achieved by flowing the gas through the electrical discharge region. In fast flow CO_2 lasers, maximum output powers per length on the order of 2,000W/m can be realized. Unfortunately, using a longer active medium to increase the output power is often not feasible due to space requirements, limitations on the length (solid state laser rods, for instance, are usually not grown longer than 200mm), or technical difficulties in providing efficient heat removal. It is for these reasons that solid state lasers with output powers greater than 600W and CO_2 lasers in the multi-kW power range utilize several active media inside the resonator (Fig. 14.1). The use of additional external amplifiers has also found application, particularly in solid state lasers.

Table 14.1 Maximally allowed heat and maximum output power per unit length of different solid state laser rods (flashlamp pumping)..

Material	Nd:YAG	Nd:glass (LG76)	Nd:Cr:GSGG	Alexandrite
$P_{H,\max}/\ell$ [kW/m]	14	0.7	12.5	64
$P_{out,\max}/\ell$ [kW/m]	4.0	0.2	3.5	13

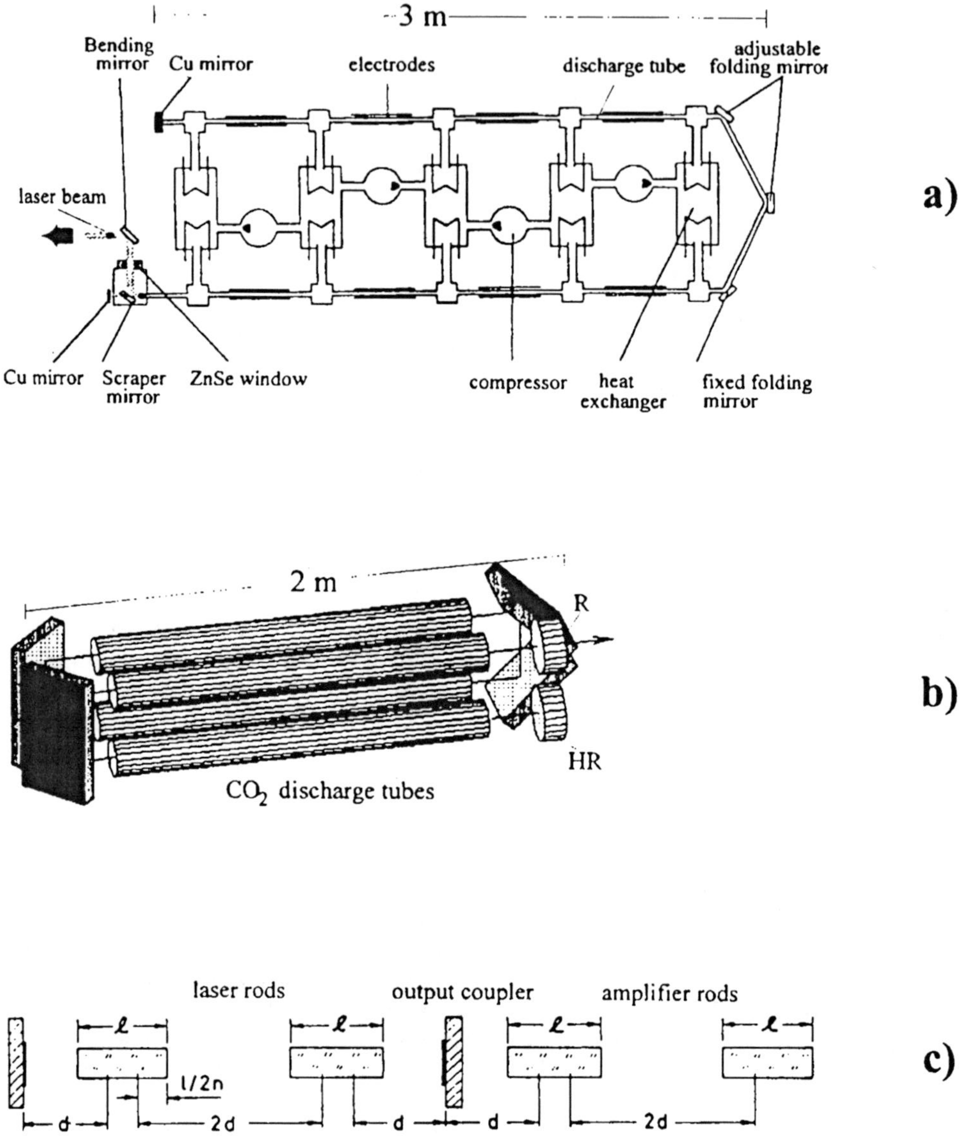

Fig. 14.1 Lasers with several active media. a) 20kW CO_2 laser [4.131] (© SPIE 1994), b) 6kW CO_2 laser [4.132] (© SPIE 1994), c) symmetric flat-flat resonator for a 2kW Nd:YAG rod laser with two amplifiers (*ℓ=15cm, d=25cm*).

14.2 Output Power and Efficiency

14.2.1 Oscillator Arrangement

Let us assume that the laser material is homogeneously broadened. We saw in Chapter 10 that the maximum extraction efficiency $\eta_{extr,max}$ and the corresponding optimum output coupling reflectance R_{opt}, to a good approximation, are given by:

$$\eta_{extr,max} = \frac{\alpha_0 \ell}{g_0 \ell}\left[\sqrt{\frac{g_0 \ell}{\alpha_0 \ell}} - 1\right]^2 \tag{14.2}$$

$$\ln R_{opt} = -2\alpha_0 \ell \left[\sqrt{\frac{g_0 \ell}{\alpha_0 \ell}} - 1\right] \tag{14.3}$$

where g_0 is the small-signal gain coefficient, α_0 is the loss coefficient, and ℓ is the length of the active medium. If N equally pumped active media are used, the small-signal gain $g_0\ell$ and the loss $\alpha_0\ell$ are N times higher. According to (14.2), the maximum extraction efficiency is the same and the optimum output coupling reflectance reads:

$$\ln R_{opt}^{(N)} = -2N\,\alpha_0 \ell \left[\sqrt{\frac{g_0 \ell}{\alpha_0 \ell}} - 1\right] \tag{14.4}$$

Thus we get an N times higher output power if the reflectance of the output coupling mirror is the N-th power of the optimum reflectance for a single medium. An accurate calculation of the output power, however, shows that the extraction efficiency decreases slightly for a high number of active media. Considering the fact that usually no more than ten media are used this roll-off is negligible.

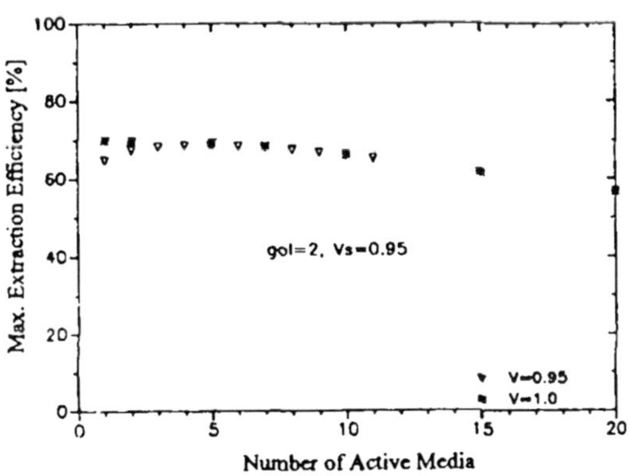

Fig. 14.2 Maximum extraction efficiency versus the number of active media. Each medium exhibits a small-signal gain of $g_0\ell=2$ and a loss of $\alpha_0\ell = -\ln V_S = 0.053$. V is the diffraction loss factor per round trip.

13.2.2 Oscillator-Amplifier Arrangement

If some of the N active media are placed behind the output coupler to amplify the laser beam, the output power behind the amplifiers can approach the maximum output power of the oscillator arrangement with N media. However, the maximum output power can only be achieved if the intensity incident on the first amplifier is greater than the saturation intensity I_S of the active laser material. For Nd:YAG (I_S=2,000W/cm^2), this means that we need an output power from the oscillator of at least 1kW (in cw-operation) to get comparable extraction efficiencies in the amplifier rods. In each amplifier medium the amplification of the intensity I along the optical axis (z-axis) is determined by:

$$\frac{dI}{dz} = \left[\frac{g_0}{1 + I/I_S} - \alpha_0\right] I \tag{14.5}$$

If I_0 denotes the intensity incident on the amplifier and $P_{out,0}$ is the corresponding power, the power $P_{out,1}$ after amplification is given by:

$$P_{out,1} = G\, P_{out,0} \tag{14.6}$$

where the gain factor G is the solution of the equation:

$$\ln G - \frac{g_0}{\alpha_0}\ln\left[\frac{g_0 - \alpha_0(1 + GI_0/I_S)}{g_0 - \alpha_0(1 + I_0/I_S)}\right] = (g_0-\alpha_0)\ell \tag{14.7}$$

The extraction efficiency is related to the gain factor via:

$$\eta_{extr} = \frac{(G-1)\, P_{out,0}}{A\, I_S\, g_0\ell} \tag{14.8}$$

where A is the cross sectional area of the laser beam. For lossy media, the gain factor G as a function of the incident intensity exhibits a maximum and for high intensities the gain factor approaches unity, which means that the amplifier medium does not contribute any power to the beam (zero extraction efficiency) (Fig.14.3). The maximum power that can be attained with amplification can be found by setting the bracket in (14.5) equal to zero:

$$P_{out,max} = A\, I_S\, [g_0/\alpha_0 - 1] \tag{14.9}$$

For Nd:YAG rods with 10mm diameter and 150mm length and each rod being pumped continuously with an electrical pump power to the flashlamps of 10kW ($g_0\ell$=0.4, $\alpha_0\ell$=0.07), Eq. (14.8) yields a maximum output power of 7.4kW. Lower losses will result in higher output power; high quality rods can provide more than 10kW.

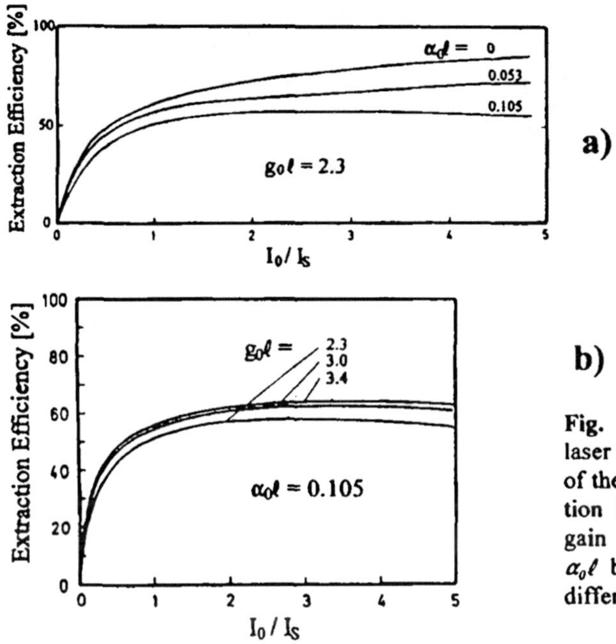

Fig. 14.3 Extraction efficiency of a laser amplifier as a function of the ratio of the incident intensity I_0 to the saturation intensity I_S. a) for a small-signal gain of $g_0\ell=2.3$ and different losses $\alpha_0\ell$ b) for a loss of $\alpha_0\ell=0.105$ and different small-signal gains.

14.3 Multirod Solid State Lasers

14.3.1 The Equivalent g-Diagram

In solid state lasers the active media cannot be placed arbitrarily inside the resonator since the thermal lensing may drive the resonator into unstable regions. Similarly to single rod resonators discussed in the preceding chapter, an equivalent, empty, resonator that exhibits the same beam radii at the mirrors can be defined. If M denotes the ray transfer matrix for the transit from mirror 1 to mirror 2 (remember that each mirror with radius of curvature ρ is replaced by a pair of lenses with focal length $f=\rho$ and the propagation starts and ends at the planes between the lenses of each pair), the equivalent g-parameters and the equivalent length are given by:

$$g_1^* = A \; ; \quad g_2^* = D \; ; \quad L^* = B \tag{14.10}$$

with: $$M = \begin{pmatrix} A & B \\ C & D \end{pmatrix}$$

In general, the equivalent resonator does not drive through the g-diagram on a straight line (Fig. 14.4). If N lenses are inside the resonator, the resonator moves along a curve that exhibits N-1 inflection points and may cross unstable zones up to $2N$ times [4.128,4.129]. In order to prevent the resonator from penetrating into unstable zones, a symmetric resonator set-up with two flat mirrors has to be chosen. The active media are positioned at an equal distance Δ from one another which corresponds to a distance $2d$ between the principal planes of adjacent rods (Fig. 14.5). The equivalent resonator moves along a straight line through the origin of the g-diagram and reverses its direction N-1 times at the stability limits. For N laser rods the refractive power per rod D_{SL} at which the equivalent resonator hits a stability limit is given by:

$$D_{SL} = \frac{2}{d}\sin^2\left[\frac{m\ \pi}{2\ N}\right] \quad ; \quad m = 0,1,2,...,N \tag{14.11}$$

Thus the resonator remains stable for a refractive power per rod of $D < 2/d$.

Example : *d=30 cm, N=5*:

m	1	2	3	4	5
D_{SL} [m^{-1}]	0.64	2.30	4.36	6.03	6.67

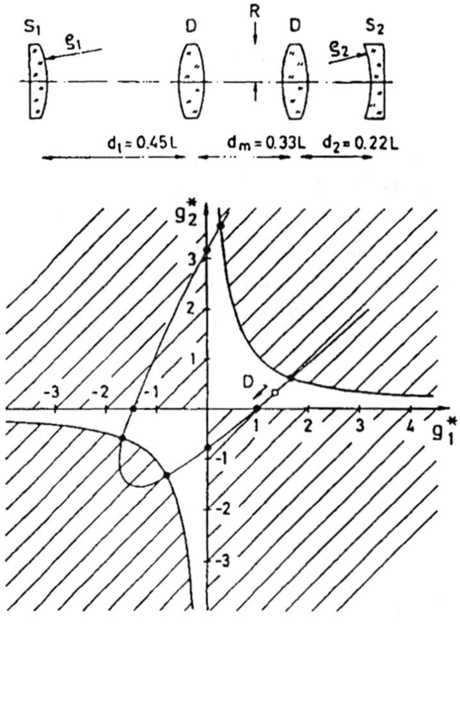

Fig. 14.4 The equivalent g-diagram for a resonator wit h two rods as a function of the refractive power D per rod. Both rods exhibit the same refractive power [4.128] (© IEEE 1988).

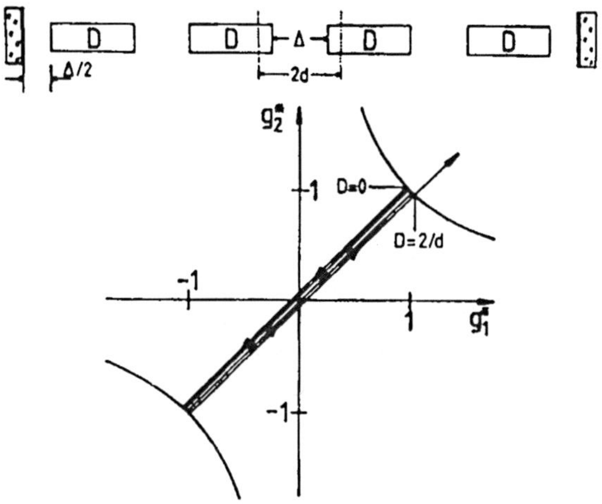

Fig. 14.5 The symmetric flat-flat multirod resonator and its path in the equivalent g-diagram. The distance d is related to the rod spacing Δ via: $d = \Delta/2 + \ell/(2n)$ where ℓ is the rod length.

14.3.2 Beam Quality and Output Power

As was shown in the last chapter, stable resonators with one thermal lens exhibit a fundamental relationship between the output power range ΔP_{out} and the maximum beam parameter product $(w\theta)_{max}$:

$$\frac{(w\theta)_{max}}{\Delta P_{out}} = \frac{k}{4\pi} \frac{\alpha}{\eta_{slope}} \tag{14.12}$$

where α is the thermal lensing coefficient, η_{slope} is the slope efficiency, and the integer k is equal to 1 if the confocal point $g_1^*=g_2^*=0$ is passed, and $k=2$ otherwise. A similar relationship holds for multirod systems. At any refractive power the symmetric flat-flat resonator exhibits planes with mirror symmetry in the middle between two rods. The beam waists are always located at this symmetry plane and the beam caustic is the periodic repetition of the beam caustic in a single rod resonator (Fig. 14.6). Thus, the beam parameter product does not depend on the number N of active elements placed inside or outside the resonator if all media are spaced at an equal distance. Figure 14.7 presents measured beam parameter products of a cw Nd:YAG laser for one and six intracavity rods. Since the output power P_{out} is now N times higher and the lens resonator remains always in the stable zone ($\Delta P_{out}=P_{out}$), expression (14.12) reads for symmetric flat-flat multirod resonators:

$$\frac{(w\theta)_{max}}{P_{out}} = \frac{k}{4\pi} \frac{\alpha}{\eta_{slope}} \frac{1}{N} \tag{14.13}$$

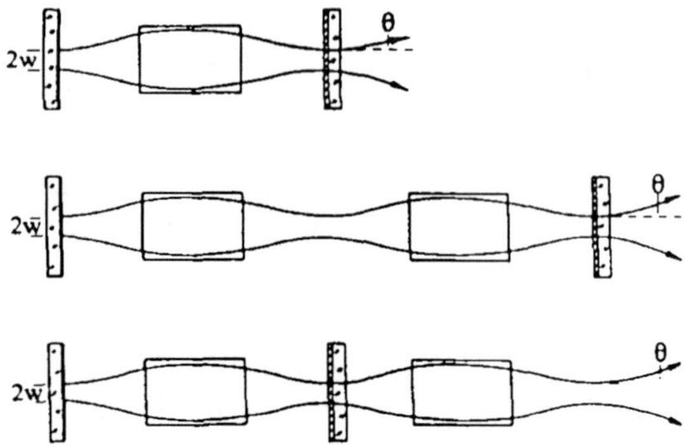

Fig. 14.6 If the rods are equally spaced in a flat-flat resonator the beam parameter product $w\theta$ does not depend on the number N of rods used, inside or outside the resonator.

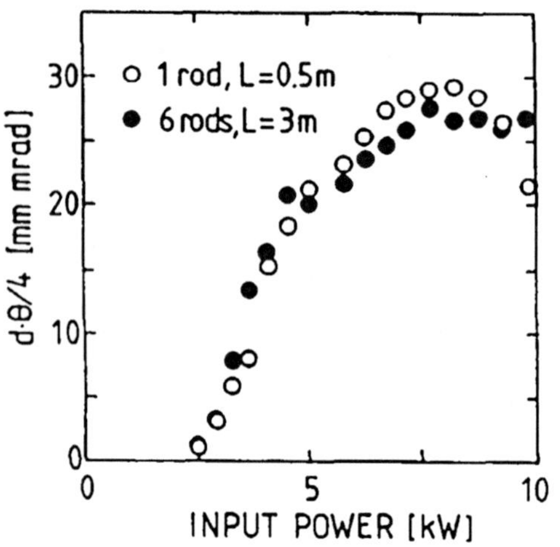

Fig. 14.7 Measured beam parameter product (d: beam waist diameter, θ: full angle of divergence, 86.5% power content) as a function of the electrical input power per rod for a cw flashlamp pumped Nd:YAG laser with one and six rods (diameter: 8mm, length: 150mm). The rods are positioned in a symmetric flat-flat resonator with a distance Δ=20cm between the rod endfaces. The output power is 300W and 1,800W with one and six rods, respectively. L is the geometrical resonator length. The refractive power of each rod is 0.52m^{-1} per kW of input power.

At present multirod solid Nd:YAG lasers providing average output powers of up to 5 kW are commercially available that use 6-8 rods. Figure 14.8 gives an overview of the maximum beam parameter products and the maximum output power of flashlamp-pumped Nd:YAG lasers in comparison with corresponding data of typical high power CO_2 lasers. Similar to single rod systems (Fig. 14.15), the beam quality is worse than predicted by (14.13). For an optimized dual-rod 1kW laser, for instance, we would expect a beam parameter product of about 16 mm mrad (using 0.4 mm/kW for α/η_{slope}, see Table 13.6). One explanation for this discrepancy is the fact that most commercial lasers do not make use of the whole stable refractive power range $\Delta D=2/d$. In general, the resonators are chosen shorter to avoid the decrease in efficiency due to spherical aberration. Furthermore, there are also space requirements to be considered. In order to cover the stable zone with the six rod laser of Fig. 14.7, the resonator length has to be increased from 3m to 4.5m. This is a length that is considered too long, even though the maximum beam parameter product would decrease by a factor of 1.5.

Better beam qualities are generally observed in diode-pumped Nd:YAG lasers (Fig. 14.9) due to the fact that for the same generated inversion, the heat dissipation is about 30% lower compared to lamp pumped systems. This translates into a 30% improvement of the beam quality, which is rather low compared to the expectations people had for diode-pumped solid state lasers a decade ago. The poorer than expected thermal characteristics, together with the high cost and limited lifetime of the pump diodes (~10,000 hours), are still limiting factors for the introduction of DPSSLs into industrial applications that require kW of average output power. At present the vast majority of this market is served by flashlamp pumped YAG lasers. A photograph of the laser head of an industrial flashlamp pumped cw Nd:YAG laser used for cutting and welding is presented in Fig. 14.10. This laser was the first commercially available kW Nd:YAG system and provided a maximum output power of 1.2kW with four rods (length: 6 inch, diameter 5/16 inch) at a total electrical pump power of 40kW. Today, kW lasers use only two rods, and 5kW output power can be easily achieved with 8 rods, at total efficiencies of 4-5% (power out versus electrical power to flashlamps). Multirod Nd:YAG systems are also widely used in laboratories to generate high output powers in TEM_{00} mode operation. By restricting the stability range and compensating for thermally induced birefringence with quartz-rotators, output powers of up to 208W have been achieved with dual-rod systems [4.112,4.116,4.120,4.121,4.123]. Without restricting the stability range, TEM_{00} mode output powers of 48W at 1064nm have been achieved with diode-pumped Nd:YVO_4 [4.77,4.115].

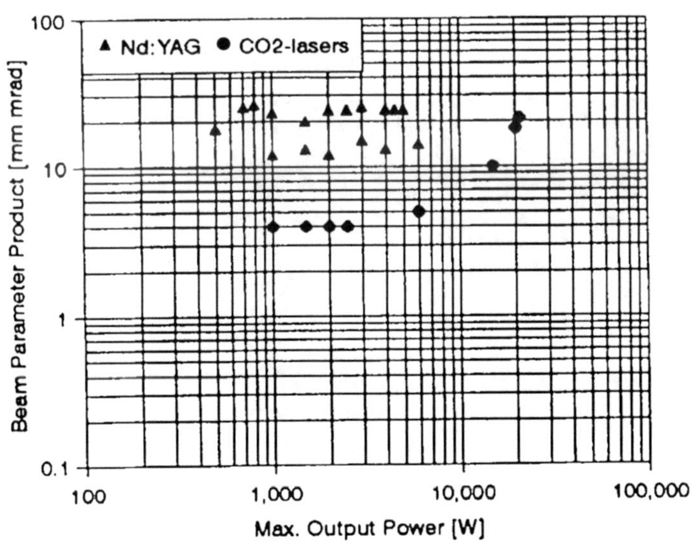

Fig. 14.8 Maximum beam parameter products versus maximum output powers for commercially available flashlamp-pumped Nd:YAG multirod lasers (λ=1.064μm) and high power CO_2 lasers (λ=10.6μm). The YAG data includes systems introduced to the market since 1988. Early systems used more rods to achieve a certain output power, resulting in abetter beam quality (row of points around 15mm mrad). The beam parameter product corresponding to M^2=1 are 0.34mm mrad for Nd;YAG lasers and 3.4 mm mrad for CO_2-lasers.

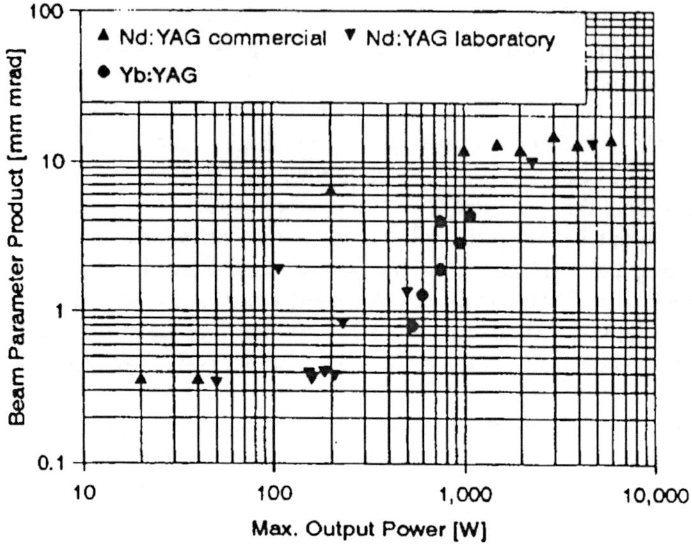

Fig. 14.9 Beam Parameter Product as a function of output power for diode-pumped solid state lasers.

Fig. 14.10 The first industrial kW Nd:YAG laser with four rods in a symmetric flat-flat resonator (NEC 1988). The distance Δ between adjacent rod endfaces is 20cm. The overall length of the laser head is 1.6m (courtesy of Laser- und Medizin-Technologie Berlin gGmbH, Berlin, Germany) [S.14].

14.3.3 Multirod Resonators with Variable Reflectivity Mirrors

It is possible to improve the beam quality of stable solid state multirod resonators by using a variable reflectivity mirror (VRM) as an output coupler. Similar to a hard aperture, the VRM controls the number of transverse modes, but the output power which is absorbed by a conventional aperture is now coupled out of the resonator. As the lens resonator moves through the stability diagram, the multimode beam radius at the output coupler is decreased from a value close to the rod radius at zero pump power, to a radius on the order of 200μm shortly before the resonator goes unstable. Ideally, the resonator should provide the same maximum output power as compared to a conventional output coupling mirror. This can be accomplished by choosing the VRM profile radius about the size of the beam radius at the highest pump power. At lower pump powers, the VRM decreases the number of transverse modes, resulting in a considerable decrease of the maximum beam parameter product. Due to gain saturation, which results in a spreading of the beam profile in the active medium, the decrease of the output power is generally kept low. An experimental example for a dual rod Nd:YAG laser with stable resonator and VRM is presented in Fig. 14.11.

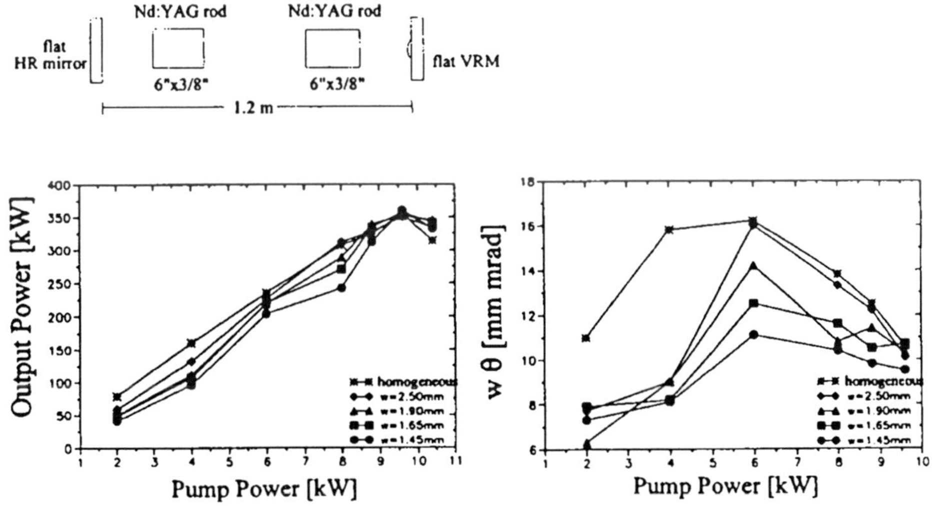

Fig. 14.11 Measured output power and measured beam parameter product (w: beam waist radius, θ: half angle of divergence, 86.5% power content) as a function of the electric pump power per rod for a flashlamp pumped, pulsed, dual rod Nd:YAG laser with different VRM output couplers. The curve parameter is the reflectivity profile radius *w* of the VRM. The center reflectivity of the VRM is 46% and the super-Gaussian index of the VRM is $n=12$. In both graphs, the upper curve corresponds to a conventional output coupling mirror with reflectance $R=46\%$. The refractive power of each rod is 0.38m^{-1} per kW of pump power [S.15].

Chapter 15

Misalignment Sensitivity of the Output Power

15.1 General Properties

The misalignment of the resonator mirrors results in a change of the output power of a laser resonator due to two mechanisms (we intentionally use the term change instead of decrease since in unstable resonators the power can also increase with mirror tilt); the diffraction losses are increased leading to a higher laser threshold and the cross sectional area of the laser beam in the medium is decreased resulting in a lower mode volume (Fig. 15.1). The resonator scheme determines which of these two effects dominates. In stable resonators with multimode operation, the decrease of the power is caused by the decreasing mode volume since additional diffraction losses are only generated if low order transverse modes get clipped by an aperture. In fundamental mode operation, on the other hand, the change in mode volume does not play any role and the output power drops due to the increased losses. We discussed the sensitivity of the losses to misalignment in Sec. 5.4 and Sec. 7.4. The reader who is not familiar with the properties of passive misaligned resonators should first go through Sec. 5.4.

Independent of the resonator type, the output power of a misaligned resonator can be investigated by using the standard model for the output power. The functional dependencies of the resonator parameters on the tilt angle have to be inserted into expression (10.7) for the output power of homogeneously broadened lasers.

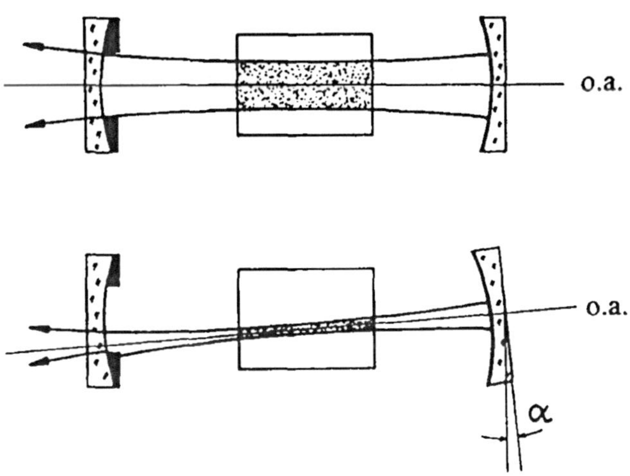

Fig. 15.1 Decrease of the output power of a misaligned stable resonator due to increased diffraction losses and a lower mode volume (o.a.: optical axis).

If we assume that the diffraction losses occur at the high reflecting mirror, the modified expression (10.7) for the output power of a misaligned stable resonator reads:

$$P_{out}(\alpha) = A_b(\alpha)\, I_S \, \frac{1-R}{1-RV(\alpha)+\sqrt{RV(\alpha)}[1/V_S - V_S]} \left[g_0\ell - |\ln\sqrt{RV_S^2 V(\alpha)}| \right] \tag{15.1}$$

where $A_b(\alpha)$ is the cross sectional area of the laser beam, $V(\alpha)$ is the diffraction loss factor per round trip (=1-loss), $g_0\ell$ is the small-signal gain, I_S is the saturation intensity, and V_S is the loss factor per transit of the active medium. This equation can also be applied to unstable resonators if the diffraction loss factor is set to 1.0 and the reflectance R is replaced by the output coupling loss factor V(α). The corresponding expression reads:

$$P_{out}(\alpha) = A_b(\alpha)\, I_S \, \frac{1-V(\alpha)}{1-V(\alpha)+\sqrt{V(\alpha)}[1/V_S - V_S]} \left[g_0\ell - |\ln\sqrt{V_S^2 V(\alpha)}| \right] \tag{15.2}$$

The dependence of the loss factor $V(\alpha)$ on the tilt angle α was discussed in Sec. 5.4 for stable resonators and in Sec. 7.4 for unstable resonators. For single transverse mode operation, the loss factor decreases parabolically with the tilt angle of a resonator mirror:

$$V(\alpha) = V(0) \left[1 - 0.1 \left(\frac{\alpha}{\alpha_{10\%}} \right)^2 \right] \tag{15.3}$$

where $\alpha_{10\%}$ denotes the tilt angle at which the losses have increased by 10%, a parameter used to define the misalignment sensitivity of the resonator. In a similar way, we can define a 10%-angle $\beta_{10\%}$ at which the output power has dropped by 10% and use this quantity as a figure of merit for the tilt sensitivity of the output power. With the results presented in Sec. 5.4 and 7.4 we can already calculate the 10%-angle $\alpha_{10\%}$ for the losses of stable resonators in fundamental mode operation and of unstable resonators. Typically mirrors can be tilted by an angle on the order of 50-100μrad before the diffraction losses increase by 10%. For such small tilt angles, the change of the mode volume is too small to have a noticeable effect on the power. Therefore, we can use a constant cross sectional area A_b in (15.2) and (15.3). This simplification enables us to determine the ratio $\beta_{10\%}/\alpha_{10\%}$ directly from these equations which means we compare the sensitivity of the output power to the known sensitivity of the losses.

For stable resonators in multimode operation, the losses do not increase for small tilt angles. Only if low order transverse modes get limited, do the losses increase. However, tilt angles on the order of 10mrad are necessary to make this happen. For smaller tilt angles the decrease of the output power is only caused by the decreasing mode volume. We can thus investigate the misalignment sensitivity by expressing the mode volume as a function of the tilt angle. The 10%-angle at which the mode volume has decreased by 10% defines the misalignment sensitivity.

15.2 Stable Resonators in Multimode Operation

15.2.1 Without Thermal Lensing

The tilt of one or both resonator mirrors leads to a rotation of the optical axis. The optical axis is defined by the line connecting the center of curvatures of the two resonator mirrors (Fig. 15.2). With both mirrors aligned the modes fill the entire medium since it represents the only mode-selecting aperture in the resonator. As the optical axis is rotated due to misalignment, the mode structure gets cut at one edge of the medium (lower left corner in Fig. 15.2) resulting in a decrease of the mode volume [4.136,4.140]. However, the optical axis remains the center of gravity of the mode structure which means that the beam gets clipped on both sides although only one side is affected by an edge. This effect can be easily observed experimentally by moving a knife edge into a stable multimode resonator, as shown in Fig. 15.3.

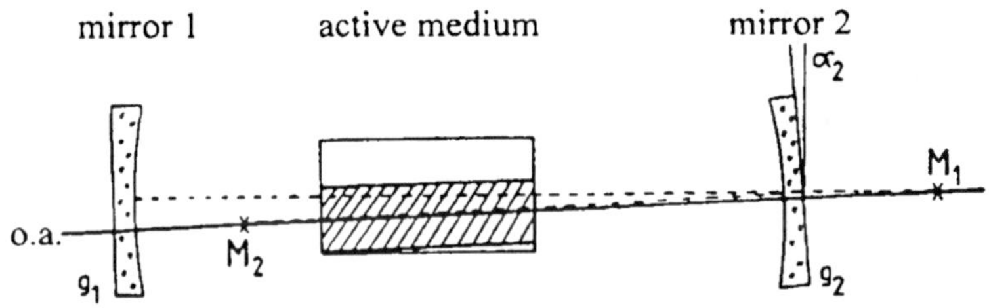

Fig. 15.2 The beam in misaligned stable multimode resonators stays symmetric to the optical axis (a.o.). The optical axis is defined by the line going through both centers of curvature of the mirrors.

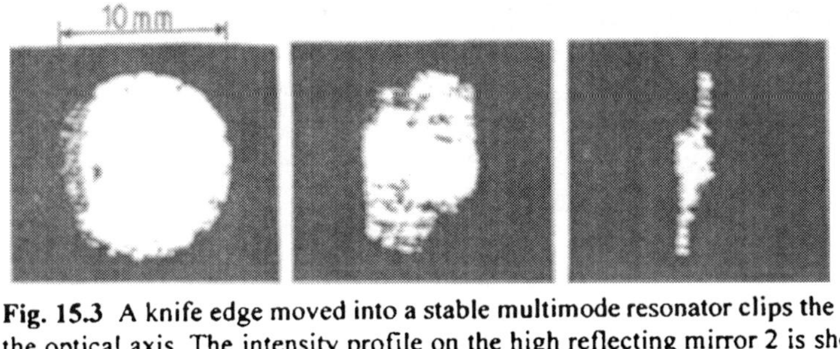

Fig. 15.3 A knife edge moved into a stable multimode resonator clips the beam symmetrically to the optical axis. The intensity profile on the high reflecting mirror 2 is shown (Nd:YAG laser in single shot operation, rod diameter : 9.5mm, $g_1=1$, $g_2=0.75$, $L=0.5m$).

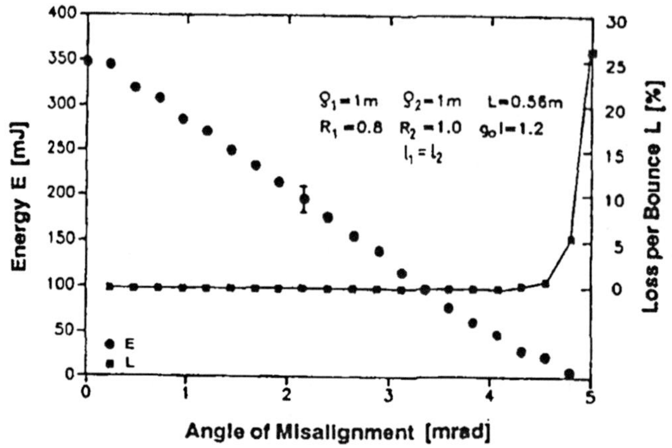

Fig. 15.4 Measured output energy per pulse of a Nd:YAG laser (rod radius: 3.17mm, length: 75mm) with a symmetric stable resonator in multimode operation ($\rho_1=\rho_2=1m$, geometrical resonator length: 0.56m) as a function of the tilt angle of one mirror. The lower curve is the measured diffraction loss per bounce. The diffraction losses increase only at large tilt angles at which the fundamental mode gets clipped. For smaller tilt angles the decrease of the output energy is caused by the decreasing mode volume [4.140] (© Taylor & Francis 1992).

This behavior is to be expected since the knife edge generates losses for the highest order modes and forces these modes to stop oscillating. Since all transverse modes exhibit mirror symmetry with respect to the optical axis, the beam must stay symmetric. The cross sectional area $A_b(\alpha)$ of the beam in the misaligned resonator can now be easily calculated. For a radius b of the active medium and small angles of misalignment one gets (Fig. 15.4):

$$A_b(\alpha) = \pi b^2 \left[1 - \frac{4\,\max(h_1,h_2)}{\pi\, b}\right] \tag{15.4}$$

where the function *max* chooses the maximum of the shifts h_1 and h_2. Since both shifts are proportional to the angle of misalignment, (15.4) predicts that the output power decreases linearly with the mirror tilt. This is in agreement with measurements as shown in Fig. 15.4. Therefore, we can define the 10%-angle $\beta_{10\%}$ at which the output power has decreased by 10% using the following linear expression:

$$A_b(\alpha) = \pi b^2 \left[1 - 0.1\left(\frac{\alpha}{\beta_{10\%}}\right)\right] \tag{15.5}$$

Since both mirrors can be tilted we assign 10%-angles for each mirror, $\beta_{10\%,1}$ if mirror 1 is misaligned and $\beta_{10\%,2}$ for mirror 2.

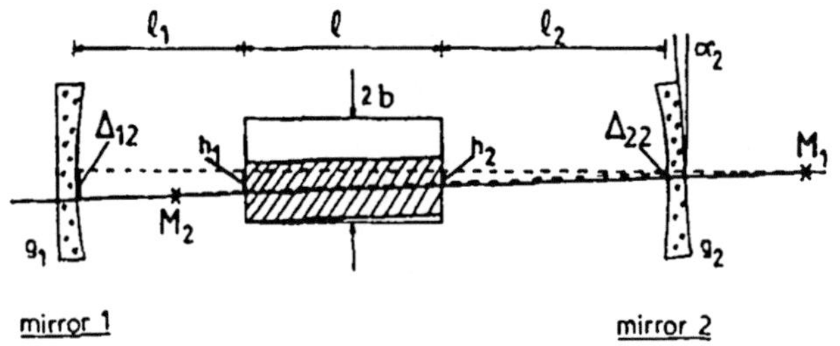

Fig. 15.5 Schematic of a misaligned stable resonator showing all the parameters used in Eq. (15.6) [4.140] (© Taylor & Francis 1992).

The two shifts h_1 and h_2 can be calculated geometrically by using the expressions for the shifts Δ_{ij} of the intersection points of the optical axis on the mirrors (see Sec. 5.4). The combination of (15.4) and (15.5) yields for the 10%-angle $\beta_{10\%,i}$ at which the mode cross section and the output power have decreased by 10% when mirror i is tilted (Fig. 15.5):

$$\beta_{10\%,i} = \frac{0.025\ \pi\ b}{L_{eff}} \frac{|1 - g_1 g_2|}{|1 - x/\rho_i|}\ , \qquad i,j=1,2;\ i \neq j \tag{15.6}$$

with:

$x_j = \ell_j - \ell(n-1)/(2n)$		if $\rho_j > \ell_j + \ell/(2n)$
$x_j = \ell_j + \ell(n+1)/(2n)$		else
$L_{eff} = \ell_1 + \ell_2 + \ell/n$	:	effective resonator length
n	:	index of refraction of the medium

Figure 15.6 shows the experimental verification of this relation for a Nd:YAG laser with different resonators. In these graphs, the 10%-angle is plotted as a function of the effective resonator length. In the case where the resonator goes unstable, (15.6) is not valid anymore and the theoretical curves, therefore, are broken in these areas. The good agreement enables one to use (15.6) to find improved resonator designs providing minimum misalignment sensitivity. For a tilt of mirror i, the maximum 10%-angle is obtained if the center of curvature of mirror j is located in the middle of the medium, which means that $\ell_j=\rho_j-\ell/(2n)$ holds. In this case, (15.6) yields for the 10%-angle:

$$\beta_{10\%,i} = \frac{0.05\ \pi\ b}{L_{eff}\ \ell} |\rho_j\ (1 - g_1 g_2)| \tag{15.7}$$

An example of a stable resonator optimized in this manner is presented in Fig. 15.7. For the rod radius of 3.17mm used, 10%-angles in excess of 5mrad become possible. This is one order of magnitude higher compared to the non-optimized resonators in the previous figure.

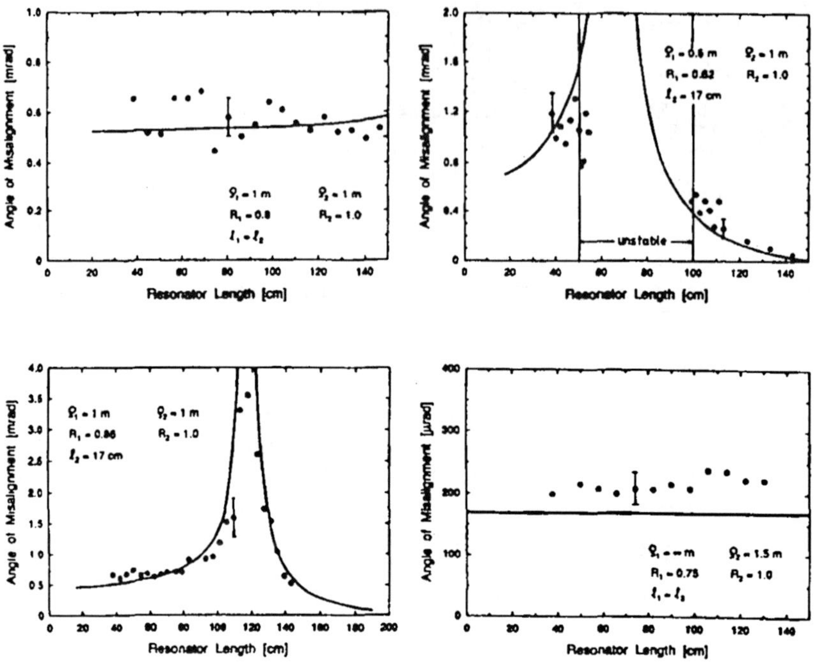

Fig. 15.6 Dependence of the measured 10%-angle $\beta_{10\%,2}$ for four stable resonators on the effective resonator length L_{eff} (pulsed Nd:YAG rod laser in single shot operation, $b=3.17mm$, $\ell=75mm$). The theoretical curves calculated with (15.6) are marked by solid lines. Resonator data are presented in each graph. In the unstable region (top right graph), Eq. (15.6) does not apply (broken line) [4.140] (© Taylor & Francis 1992).

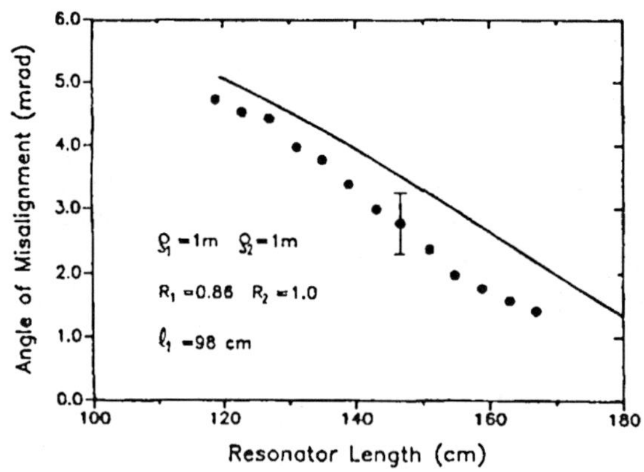

Fig. 15.7 Measured 10%-angles $\beta_{10\%,2}$ for a stable resonator with minimum misalignment sensitivity of mirror 2 according to (14.7) as a function of the effective resonator length L_{eff} (pulsed Nd:YAG rod laser, $b=3.17mm$, $\ell=75mm$). The center of curvature of mirror 1 is in the middle of the rod. The solid line represents (15.7) [4.140] (© Taylor & Francis 1992).

As a rule of thumb, the 10%-angles $\beta_{10\%,i}$ of stable resonators in multimode operation are on the order of 0.2mrad per mm of rod radius. However, by choosing optimized resonator designs (small g-parameter product and centers of mirror curvature inside the active medium) it is possible to increase this value by one order of magnitude.

15.2.2 With Thermal Lensing

In high power solid state lasers the refractive power D of the active medium affects the rotation of the optical axis if the resonator is misaligned. The expression for the mode cross section derived above, however, is still valid. Thus we have only to investigate how the shifts h_1 and h_2 of the optical axis are changed by the refractive power. In order to simplify the discussion, we use a thick lens, the two principal planes of which are located at a distance d_i from mirror i (Fig. 15.8). In this approximation, the optical axis is shifted parallel by a displacement h between the principal planes. The result of any bending of the optical axis inside the medium is considered by different slopes of the axis outside the medium. These slopes are determined by the fact that the centers of curvature of both mirrors act as virtual sources of the optical axis as indicated in Fig. 15.8 [4.136,4.140]. The calculation of the shift h is then simple, and if mirror i is tilted by an angle α_i the final result reads:

$$h = \left| \frac{\rho_i\ \alpha_i}{(d_i-\rho_i)\ [D-1/(d_1-\rho_1)-1/(d_2-\rho_2)]} \right| \tag{15.8}$$

where ρ_i is the radius of curvature of mirror i.

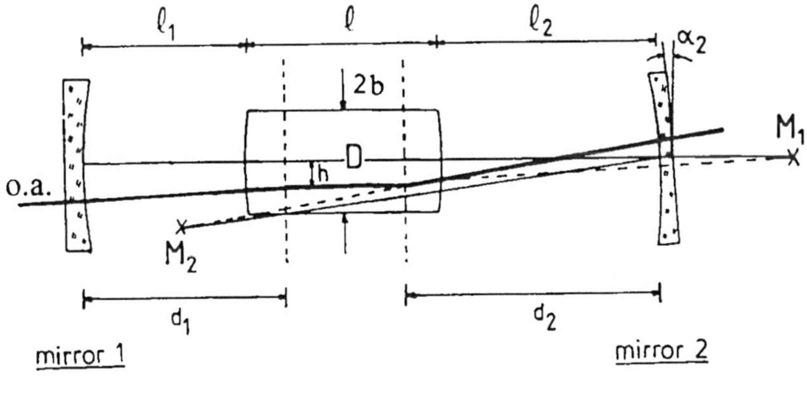

Fig. 15.8 Optical axis in a misaligned resonator with an internal lens. On both sides of the lens the optical axis seems to emerge from a virtual source defined by the centers of curvature M_i of the mirrors [4.140] (© Taylor & Francis 1992).

For a rod with radius b, the 10%-angle $\beta_{10\%,i}$ at which the output power has decreased by 10% when mirror i is tilted becomes:

$$\beta_{10\%,i} = 0.025\ \pi\ b \left| \frac{d_i-\rho_i}{\rho_i} \left[D - \frac{1}{d_1-\rho_1} - \frac{1}{d_2-\rho_2} \right] \right| \tag{15.9}$$

This equation is, of course, only valid if the resonator is in a stable zone in the g-diagram, which means that the refractive power D is within the limits defined by the characteristic refractive powers D_I-D_{IV} (see Sec. 13.1.2). Note that (15.6) is not exactly obtained for D going to zero. This is due to the thick lens approximation used since any difference between the shifts h_1 and h_2 (see Fig. 15.5) has been neglected. Nevertheless, (15.9) can be used to predict the misalignment sensitivity of stable resonators with a thermal lens to a high accuracy as the measurements presented in Fig. 15.9 indicate. No matter what kind of resonator is used, they all have in common the characteristic, that an increasing refractive power D results in a linear increase of the 10%-angle. The thermal lens thus has a stabilizing effect on the resonator.

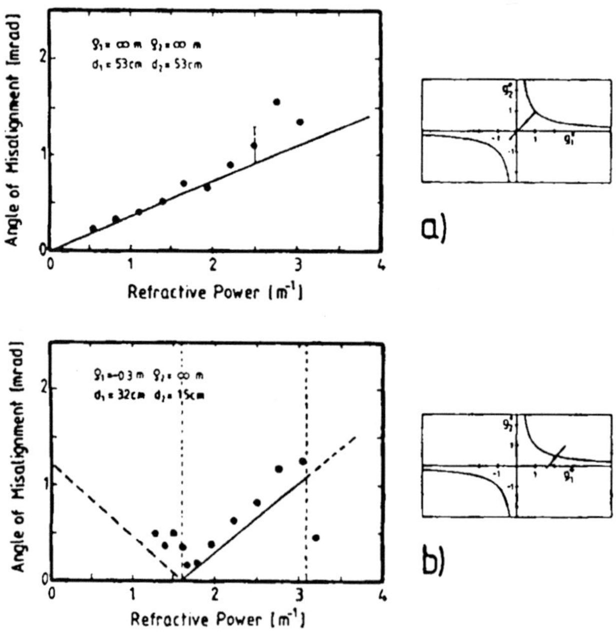

Fig. 15.9 Dependence of the measured 10%-angles $\beta_{10\%,2}$ of a high power Nd:YAG rod laser for two different resonators on the refractive power D of the rod (b=4.75mm, ℓ=150mm). The equivalent g-diagram of these resonators is shown as well. The solid lines represent (15.9) [4.140] (© Taylor & Francis 1992).

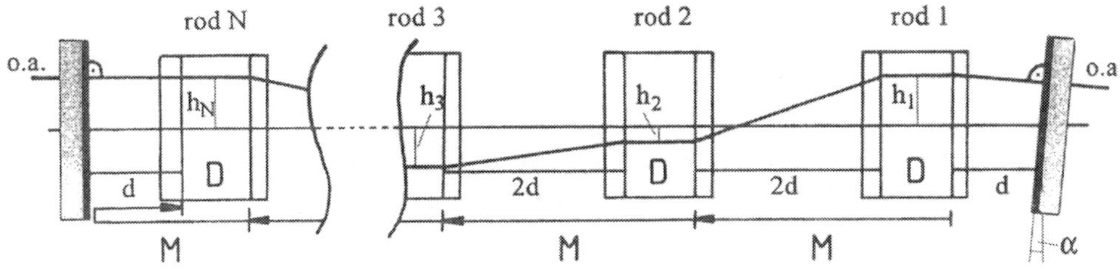

Fig. 15.10 The optical axis in a misaligned symmetric flat-flat resonator with N thermal lenses with refractive power D.

Misalignment in Symmetric Multirod Resonators

The misalignment in symmetric flat-flat resonators with N thermal lenses can be treated in a similar way. The tilt of one mirror by an angle α generates different transverse shifts h_i of the optical axis in the rods (Fig. 15.10). On both mirrors the optical axis is normal to the mirror surface. The shifts can be calculated by repetitively applying the ray transfer matrix M on the ray vector (h_1, α). This ray vector represents the optical axis as it is incident on the first principal plane of rod 1. The ray transformation equations for a total of N rods read:

$$\begin{pmatrix} h_{m+1} \\ \alpha_{m+1} \end{pmatrix} = \boldsymbol{M}^m \begin{pmatrix} h_1 \\ \alpha \end{pmatrix}, \qquad m = 0,\ldots,N-1 \tag{15.10}$$

with: $\boldsymbol{M} = \begin{pmatrix} 1-2dD & 2d \\ -D & 1 \end{pmatrix}$, D: refractive power per rod

d : distance as shown in Fig. 15.10

By using Sylvester's Theorem (1.46) for periodic optical systems one gets:

$$\begin{pmatrix} h_{m+1} \\ \alpha_{m+1} \end{pmatrix} = \frac{1}{\sin\Phi} \begin{pmatrix} (1-2dD)\sin[m\Phi] - \sin[(m-1)\Phi] & 2d\sin[m\Phi] \\ -D\sin[m\Phi] & \sin[m\Phi] - \sin[(m-1)\Phi] \end{pmatrix} \begin{pmatrix} h_1 \\ \alpha \end{pmatrix}$$

with: $\cos\Phi = 1 - dD$ and $m = 0,\ldots,N\text{-}1$ (15.11)

At the left mirror, the angle of the optical axis must be equal to zero. To get there we have to apply another matrix M to the last ray. The second row of (15.11) thus provides the shift h_1 if we set m equal *to N*:

$$h_1 = \frac{\sin[N\Phi] - \sin[(N-1)\Phi]}{D\,\sin[N\Phi]}\,\alpha \tag{15.12}$$

Insertion of (15.12) into (15.11) results in the final expression for the absolute values of the shifts h_m. In contrast to (15.11), the index m is now going from 1 to N:

$$h_m = \left|\frac{(\sin[N\Phi]-\sin[(N-1)\Phi])(\sin[(m-1)\Phi]-\sin[(m-2)\Phi]) + 2dD\sin[(N-1)\Phi]\sin[(m-1)\Phi]}{D\,\sin[N\Phi]\,\sin\Phi}\right|\alpha$$

$$:= \frac{C_{mN}}{D\,|\sin[N\Phi]\,\sin\Phi|}\,\alpha \tag{15.13}$$

The rod that exhibits the largest shift h_m acts as the limiting aperture for the laser beam, whereas the beam can propagate freely in the other rods. The maximum of the shifts h_m, therefore, determines the reduction of the fill factor in all rods. At the angle $\alpha=\beta_{10\%}$, the mode volume and the output power have decreased by 10%, which means that the fill factor γ is equal to 0.9:

$$\gamma = \frac{\pi b^2 - 4b\,\max(h_m)}{\pi b^2} = 0.9 \tag{15.14}$$

where the function *max* determines the maximum of N values ($m=1,..,N$). By inserting (15.13) into (15.14) the 10%-angle is obtained:

$$\beta_{10\%} = \frac{0.025\,\pi\,b\,D\,|\sin[N\Phi]\,\sin\Phi|}{\max(C_{mN})} \tag{15.15}$$

For a single rod ($N=1$) this expression is equivalent to (15.9):

$$\beta_{10\%} = 0.025\,\pi\,b\,D \tag{15.16}$$

For two rods, the shift h_2 in the rod closer to the aligned mirror is larger for $D<1/d$. The 10%-angle reads:

$$\beta_{10\%} = 0.025\,\pi\,b\,2D|(1-dD)|\,, \qquad \text{for } D < 1/d$$

$$\beta_{10\%} = 0.025\,\pi\,b\,2D\frac{|1-dD|}{1-2dD}\,, \qquad \text{for } 1/d \le D \le 2/d \tag{15.17}$$

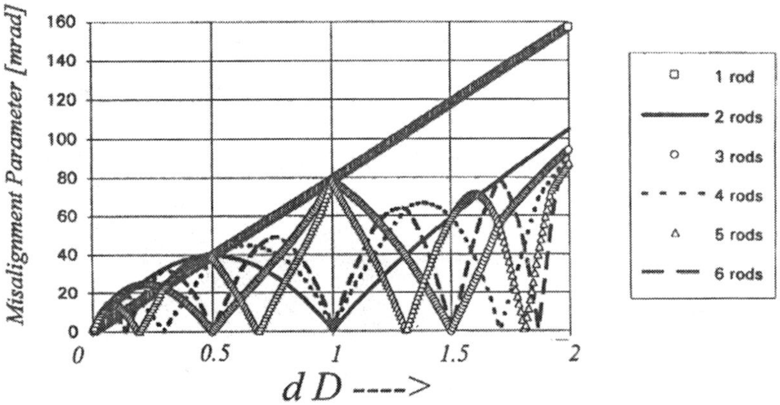

Fig. 15.11 Misalignment parameter $\beta_{10\%}\, d/b$ as a function of dD for symmetric flat-flat resonators with one to six rods (calculated with (15.15)); b: rod radius, d: distance between mirror and the adjacent principal plane of the first rod.

The 10%-angle goes to zero at $D=1/d$ and exhibits a local maximum for $D=1/2d$. The maximum is assumed when the equivalent lens resonator is confocal, a behavior that is to be expected since the confocal resonator generally shows a low misalignment sensitivity. Similar to the single rod, the largest 10%-angle is assumed at the stability limit with $D=2/d$.

For more rods, it is easier to find a numerical solution of (15.15). For the discussion of the misalignment sensitivity, it is more convenient to use the misalignment parameter $\beta_{10\%} d/b$ since, according to (15.13) and (15.15), this parameter is only a function of dD and, therefore, we can use a single graph to read off the 10%-angles of symmetric multirod resonators. This is shown in Fig. 15.11 in which the misalignment parameter is plotted versus dD for one to six rods. The range $0<dD<2$ represents the stable zone of the symmetric flat-flat resonator. The 10%-angle goes to zero if the equivalent resonator hits a stability limit ($g_1^* g_2^*=1$), and minima of the misalignment sensitivity are obtained when the confocal point ($g_1^*=g_2^*=0$) is passed. The refractive powers at which stability limits are reached are given by (see Table 15.1):

$$D = \frac{2}{d} \sin^2\left[\frac{m}{2}\frac{\pi}{N}\right], \qquad m = 0,1,2,...,N \tag{15.18}$$

The equivalent resonator gets confocal at the refractive powers (Table 15.2):

$$D = \frac{2}{d} \sin^2\left[\frac{m}{4}\frac{\pi}{N}\right], \qquad m = 1,3,5,... \textit{ and } m \leq N \tag{15.19}$$

Note that the 10%-angle does not assume a maximum for the last passage through the confocal point. According to Fig. 15.11, the average misalignment parameter is about 40mrad. For typical rod radii b of 3.5-5mm, and distances d between 10-20cm, the corresponding 10%-angles are between 1 and 4mrad (see Fig. 15.12).

Table 15.1 Values of the product dD for which a stability limit is reached (N: number of rods).

m=	1	2	3	4	5	6
N=1	2.0	-	-	-	-	-
N=2	1.0	2.0	-	-	-	-
N=3	0.5	1.5	2.0	-	-	-
N=4	0.293	1.0	1.707	2.0	-	-
N=5	0.191	0.691	1.309	1.809	2.0	-
N=6	0.134	0.5	1.0	1.5	1.866	2.0

Table 15.2 Values of the product dD for which the confocal point is reached (N: number of rods).

m=	1	3	5
N=1	1.0	-	-
N=2	0.293	-	-
N=3	0.134	1.0	-
N=4	0.076	0.617	-
N=5	0.049	0.412	1.0
N=6	0.034	0.293	0.741

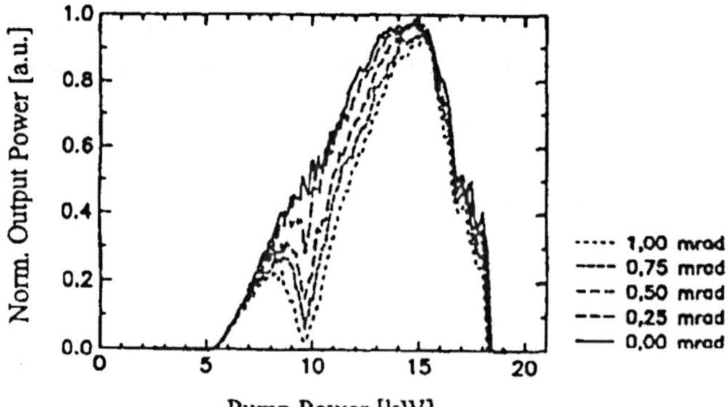

Fig. 15.12 Measured output power of a symmetric flashlamp-pumped dual rod Nd:YAG laser (b=*4mm*, ℓ=*150mm*, d=*0.5m*) with flat mirrors as a function of the total pump power for different tilt angles of one mirror [S.16]. The refractive power of each rod is 0.41 diopters per kW of pump power. The resonator hits the stability limit at a pump power per rod of 4.8kW. Note the high misalignment sensitivity at this point.

15.3 Stable Resonators in Fundamental Mode Operation

In fundamental mode operation the misalignment of the mirrors leads immediately to an increase of the diffraction losses. As mentioned before, the tilt angles at which the losses have increased by 10% typically are between 10 and 50μrad and thus too small to have a noticeable effect on the mode volume. Therefore, we can use the expression (15.1) for the output power with a constant cross sectional area on the beam and insert the tilt dependence of the diffraction loss factor V which is given by:

$$V(\alpha) = V(0)\ [1 - 0.1(\alpha/\alpha_{10\%})^2] \tag{15.20}$$

where $\alpha_{10\%}$ is the angle at which the diffraction losses have increased by 10%. In order to simplify our discussion, we approximate the fraction in (15.1) by $1\text{-}R$. This is a valid approximation for high mirror reflectances ($R>0.9$) and low small-signal gains ($g_0\ell<0.4$). The resulting expression reads:

$$P_{out}(\alpha) = A_b\ I_S\ (1-R)\ [g_0\ell - |\ln\sqrt{RV(\alpha)V_S^2}|] \tag{15.21}$$

Insertion of (15.19) yields the dependence of the output power on the tilt angle:

$$P_{out}(\alpha) = P_{out}(0) \left| 1 - \frac{|\ln\sqrt{1 - 0.1(\alpha/\alpha_{10\%})^2}|}{g_0\ell - |\ln\sqrt{RV(0)V_S^2}|} \right| \tag{15.22}$$

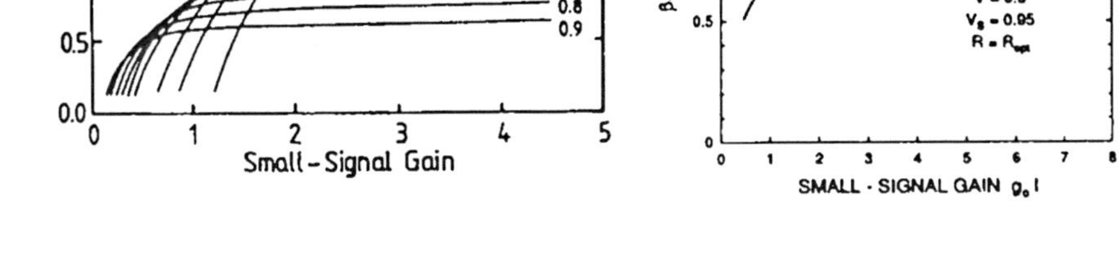

Fig. 15.13 Ratio $\beta_{10\%}/\alpha_{10\%}$ as a function of the small signal gain for $V(0)=0.9$, calculated by combining with (15.1) and (15.20). The angle $\beta_{10\%}$ is the tilt angle at which the output power has dropped by 10%. The left graph shows the angle ratios with the mirror reflectance R being the curve

parameter, the curve in the right graph represents the angle ratios for optimum output coupling.

Thus the output power becomes less sensitive to mirror misalignment as the small- signal gain is increased. The angle $\beta_{10\%}$ at which the power has dropped by 10% is given by:

$$\frac{\beta_{10\%}}{\alpha_{10\%}} = \sqrt{10\ (1\ -\ \exp[-0.2\ (g_0\ell\ +\ \ln\sqrt{RV(0)V_S^2})])} \qquad (15.23)$$

This expression indicates that the power is more sensitive to misalignment than the diffraction losses, which means that the ratio $\beta_{10\%}/\alpha_{10\%}$ is less than 1.0. Unfortunately, due to the approximations made above, this equation is only applicable to low gain lasers with low output coupling. In all other cases the correct expression for the output power (15.1) has to be used to calculate the 10%-angle ratio by inserting (15.23) and numerically varying the tilt angle. However, the statement that the power is more sensitive to the tilt than the losses holds also for high gain lasers. The numerically calculated angle ratios are presented in Fig. 15.13 for stable resonators. These resonators have the mode selecting aperture adapted to the Gaussian beam radius (the adaptation generates diffraction losses of about 10% per round trip, see Sec. 11.2). The higher sensitivity of the output power with respect to the loss can also be confirmed experimentally, as shown in Fig. 15.14. The detailed discussion of the 10%-angles $\alpha_{10\%}$ in Sec. 5.4 enables us now to determine the misalignment sensitivity of any stable resonator in fundamental mode operation. Let us calculate an example to clarify this statement.

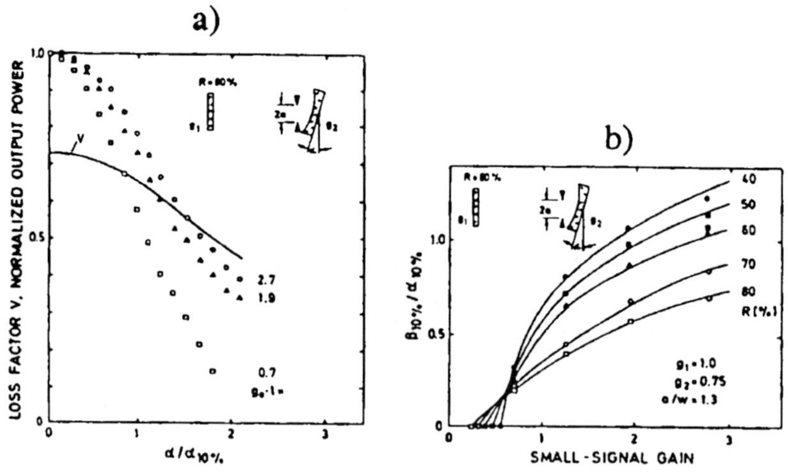

Fig. 15.14 a) Measured output energy per pulse and measured loss factor as a function of the tilt angle for a stable resonator in fundamental mode operation (g_1=*1.0*, g_2=*0.75*, L_{eff}=*0.5m*, a/w_{00}=*1.3*, Nd:YAG laser). The curve parameter is the small-signal gain. b) Measured 10% angle ratios for the resonator of graph a) as a function of the small-signal gain. The curve parameter is the reflectance of the output coupling mirror.

Example:
CO_2-laser (λ=10.6μm), tube diameter $2b = 6mm$, flat mirror 1 (output coupler) and 5m concave high reflecting mirror. The resonator length is 0.5m.

The following resonator properties are obtained:
g-parameters : $g_1 = 1.0,\ g_2 = 0.9,\ G = 2g_1g_2 - 1 = 0.8$
Gaussian beam radii : $w_1 = 2.25mm,\ w_2 = 2.37mm$
The tube is located at mirror 2 and the tube radius a is adapted to the Gaussian beam radius with $a/w_2 = 1.27$. Since the confined mirror is mirror 2, the effective Fresnel number is given by:

$$N_{eff} = \frac{b^2}{2Lg_1\lambda} = 0.849$$

Please do not get confused by the fact that in Sec. 5.4 the aperture is placed at mirror 1. We simply have to swap g_1 and g_2 (as we already did in the expression for the effective Fresnel number). If the unconfined mirror is tilted (output coupler), Fig. 5.49 provides a misalignment parameter $L\alpha_{10\%}/a$ of 45mrad for $G=0.8$ and $N_{eff}=0.849$. This results in a 10%-angle $\alpha_{10\%}$ of 27μrad. If the confined mirror 2 is tilted, the corresponding 10%-angle is $\alpha_{10\%}/g_1$, which means that for this resonator both mirrors exhibit the same 10%-angle. According to Fig. 15.12, the 10%-angle $\beta_{10\%}$ ranges from 13μrad to 25μrad for small-signal gains between 0.5 and 3.0 and optimum output coupling.

15.4 Unstable Resonators

15.4.1 Without Thermal Lensing

In contrast to stable resonators, the mirror misalignment in unstable resonators results in an increase of the output coupling (remember that the loss factor V of unstable resonators corresponds to the mirror reflectance R of a stable resonator). Again, the mode volume is not affected by the misalignment since the tilt angles are too small. In order to investigate the misalignment sensitivity of the output power it is, therefore, sufficient to discuss the dependence of the power on the output coupling (Fig. 15.15). If the aligned resonator is undercoupled, which means that the loss factor is higher than the optimum loss factor V_{opt}, the resonator virtually moves to the left along the power curve and the output power is increased. Accordingly, overcoupled resonators decrease their output power immediately with the mirror tilt. We are, of course, particularly interested in unstable resonators with optimum output coupling. For these resonators, the power decreases as well, but due to the insensitivity of the power to changes in the loss factor, the misalignment sensitivity is relatively low. We get a better understanding of this fact if we expand the output power in a Taylor series around $\alpha=0$:

$$P_{out}(\alpha) = P_{out}(0) + \left.\frac{dP_{out}}{d\alpha}\right|_{\alpha=0}\alpha + \frac{1}{2}\left.\frac{dP^2_{out}}{d^2\alpha}\right|_{\alpha=0}\alpha^2 + ... \quad (15.24)$$

which can be written as:

$$P_{out}(\alpha) = P_{out}(0) + \frac{dP_{out}}{dV}\frac{dV}{d\alpha}\alpha + \frac{1}{2}\left[\frac{dP^2_{out}}{dV^2}\left(\frac{dV}{d\alpha}\right)^2 + \frac{dP_{out}}{dV}\frac{d^2V}{d\alpha^2}\right]\alpha^2 + .. \quad (15.25)$$

Insertion of the loss factor $V(\alpha)$ with:

$$V(\alpha) = V(0)\ [1 - 0.1(\alpha/\alpha_{10\%})^2] \quad (15.26)$$

results in:

$$P_{out} = P_{out}(0) - 0.1\ V(0)\frac{dP_{out}}{dV}\left(\frac{\alpha}{\alpha_{10\%}}\right)^2 + \left[\frac{0.1\ V(0)}{\sqrt{2}}\right]^2\frac{d^2P_{out}}{dV^2}\left(\frac{\alpha}{\alpha_{10\%}}\right)^4 \quad (15.27)$$

The first derivative of the power dP_{out}/dV is negative, positive, or zero, depending whether the output coupling is too low, too high, or optimal [4.83]. For optimum output coupling (15.27) becomes:

$$P_{out}(\alpha) = P_{out}(0)\left[1 - 0.1\left(\frac{\alpha}{\beta_{10\%}}\right)^4\right] \quad (15.28)$$

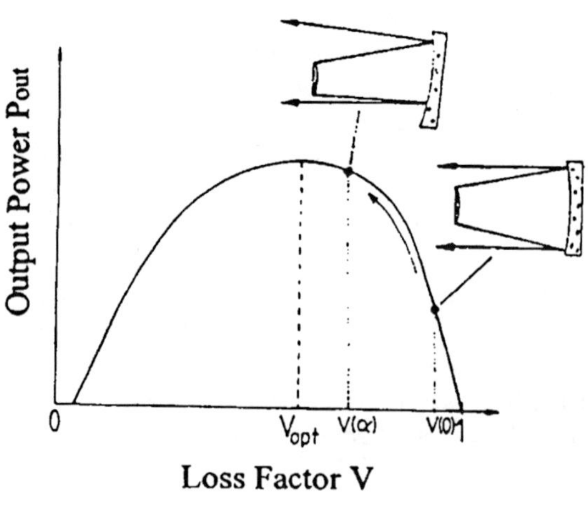

Fig. 15.15 In misaligned unstable resonators the output coupling is increased. The power of under-coupled resonators will thus rise as the mirrors are tilted.

with: $$\frac{\beta_{10\%}}{\alpha_{10\%}} = \left[\frac{200\ P_{out}(0)}{-\ d^2P_{out}/dV^2\ V^2(0)}\right]^{\frac{1}{4}} \tag{15.29}$$

For optimal output coupling, the power decreases only as the fourth power of the angle of misalignment. The 10%-angle $\beta_{10\%}$ at which the power has decreased by 10% is always larger than the 10%-angle of the losses $\alpha_{10\%}$. This can be verified by inserting the expression (15.2) for the output power into (15.29) or (15.27). The resulting analytical expression however, is too complicated so that a numerical solution is more convenient. Numerically calculated angle ratios (15.29) as a function of the small-signal gain are presented in Fig. 15.16. Compared to stable resonators in fundamental mode operation (Fig.15.13), the angle ratio $\beta_{10\%}/\alpha_{10\%}$ is about twice as high. Considering the fact that the 10%-angle of the losses of unstable resonators is already at least twice as high (see Sec. 7.4), the output power of unstable resonators is at least four times less sensitive to misalignment than that of stable resonators in fundamental mode operation with a comparable mode volume. Some experimental results are presented in Fig. 15.17. The right graph shows the increase of the output power for a misaligned, undercoupled, unstable resonator.

Example:
We will use the same CO_2-laser tube as in the last section with a positive branch confocal resonator having a magnification of $M=1.5$ ($\rho_1=-2m, \rho_2=3m, L=0.5m$). The radius a of the high reflecting zone on the output coupling mirror 1 is adapted to the tube radius b with $a=b/M=2mm$. The resonator parameters are:

g-parameters: $g_1 = 1.25\ ,\ g_2 = 0.833, G=1.083$

effective Fresnel number: $N_{eff} = a^2/(2Lg2\lambda) = 0.453$

equivalent Fresnel number: $N_{eq} = N_{eff}\sqrt{G^2-1} = 0.189$

According to Fig. 7.21a, this resonator exhibits a loss factor $V(0)$ of about 0.64, which means that the output coupling is optimum at a small-signal gain of $g_0\ell=3.0$. The misalignment sensitivity of the losses can be calculated from Fig. 7.30. For the tilt of the unconfined mirror 2, the misalignment parameter $L\alpha_{10\%}/a$ is about 80mrad which corresponds to a 10%-angle of $\alpha_{10\%,2}$=320μrad. The 10%-angle for the output coupler is given by $\alpha_{10\%,1} = \alpha_{10\%,2}/g_2$ = 384μrad. With this information we can use Fig. 15.16 to determine the angles $\beta_{10\%}$ at which the power has decreased by 10%:

if the output coupler is tilted : $\beta_{10\%,1} = 0.8\ mrad$

if mirror 2 is tilted : $\beta_{10\%,1} = 0.67mrad$

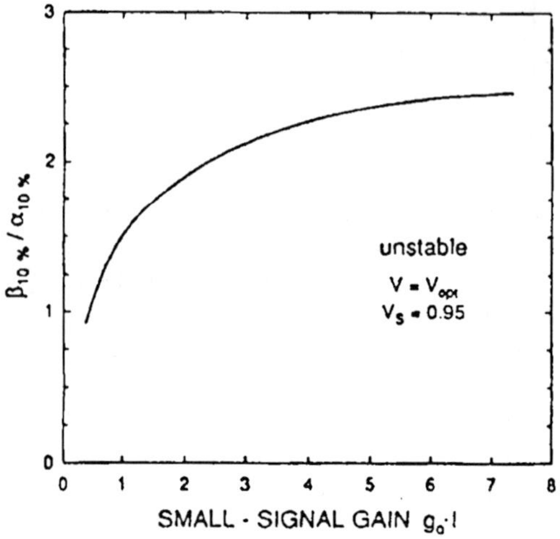

Fig. 15.16 Numerically calculated angle ratios $\beta_{10\%}/\alpha_{10\%}$ for unstable resonators with optimum output coupling as a function of the small-signal gain. Equations (15.2) and (15.26) were used. The 10% angles $\alpha_{10\%}$ and $\beta_{10\%}$ denote the tilt angle at which the losses and the power, respectively, have decreased by 10%.

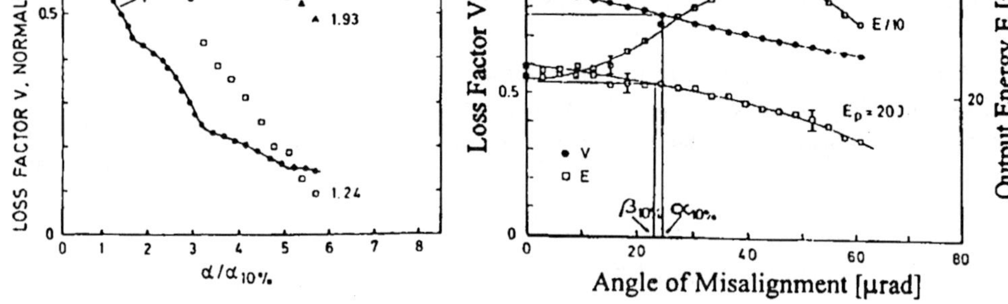

Fig. 15.17 Measured loss factor per round trip and output energy per pulse of a Nd:YAG laser with different unstable resonators as a function of the tilt angle of the unconfined mirror. The curve parameter in both graphs is the small-signal gain. In the right graph, a pump energy of 160J corresponds to a small-signal gain of 3.5.

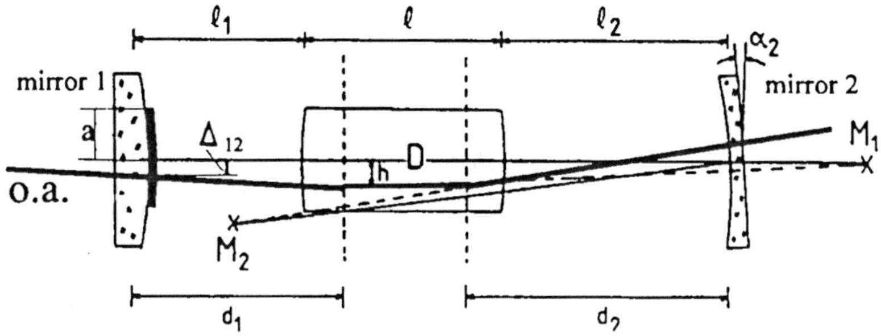

Fig. 15.18 Misaligned unstable resonator with internal variable lens. The tilt of mirror 2 results in a shift Δ_{12} of the intersection point of the optical axis (o.a.) on the output coupler.

15.4.2 With Thermal Lensing

In unstable resonators with an internal variable lens, the additional losses due to misalignment are generated by the shift Δ_1 of the intersection point of the optical axis on the output coupler (Fig. 15.18). For a tilt of mirror i by the angle α_i, the shift is given by:

$$\Delta_{1i} = \left| \frac{\dfrac{(1 - d_1/\rho_1)}{(1 - d_i/\rho_i)} \alpha_i}{\left[D - \dfrac{1}{d_1 - \rho_1} - \dfrac{1}{d_2 - \rho_2} \right]} \right| \tag{15.30}$$

This equation reveals a symmetry relation between the two mirrors. If mirror 2 is tilted by α_2, the same shift on the output coupler is generated for a tilt of mirror 1 by the angle:

$$\alpha_1 = \alpha_2 \frac{1 - d_1/\rho_1}{1 - d_2/\rho_2} \tag{15.31}$$

Having determined the shift, we can now use the equivalent, empty, resonator with g-parameters g_1^*, g_2^*, and length L^* (see Sec. 15.1). The shift of the optical axis at which the losses have increased by 10% was calculated in Sec. 5.4 for the misalignment of mirror 2:

$$\Delta_{1,10\%} = \frac{2L^*}{|G^* - 1|} \alpha_{10\%} = \frac{2a\, D_2}{|G^* - 1|} \tag{15.32}$$

where $G^* = 2g_1^* g_2^* - 1$ is the equivalent g-parameter and D_2 is the misalignment parameter according to Fig. 7.30. A comparison of (15.32) and (15.30) with $\alpha_i = \alpha_{10\%,i}$ yields the following expression for the 10%-angle when mirror i is tilted:

$$\alpha_{10\%,i} = D_2 \frac{2a}{|G^* - 1|} \left| \left(\frac{1 - d_i/\rho_i}{1 - d_1/\rho_1} \right) \left[D - \frac{1}{d_1 - \rho_1} - \frac{1}{d_2 - \rho_2} \right] \right| \tag{15.33}$$

Unfortunately, no analytical expression for the misalignment parameter D_2 exists and we have to get the values from Fig. 7.30. Keep in mind that D_2 is a function G^* and N_{eq}^* and will in general decrease as the refractive power is increased. If we knew the misalignment parameter as a function of the refractive power we could use (15.33) to determine the 10%-angle of the losses and then go into (15.27) or (15.29) to obtain the 10%-angle of the output power $\beta_{10\%}$.

Instead of following this complicated procedure let us make a rough estimate by assuming a constant misalignment parameter of 35mrad (see Fig. 7.30). For a positive branch unstable resonator with ρ_1=-0.5m, ρ_2=-0.3m, d_1=0.3m, d_2=0.13m and a refractive power of 3m^{-1}, Eq. (15.33) yields the 10%-angles $\alpha_{10\%,1}$=280µrad and $\alpha_{10\%,2}$=250µrad (Fig. 15.19). For a small-signal gain of 3.0 we expect, according to Fig. 15.16, an angle of misalignment $\beta_{10\%}$ that is about twice as high as the angle $\alpha_{10\%}$. Our final estimate of 0.56mrad and 0.5mrad for the 10%-angles is in agreement with the measurement presented in Fig. 15.19. Note that for a constant small-signal gain the angle $\beta_{10\%}$ stays at the same value as the refractive power is increased (left graph). This agrees with the above statement that the misalignment parameter decreases. In the right graph the small-signal gain increases linearly with the refractive power resulting in an increase of the 10%-angle.

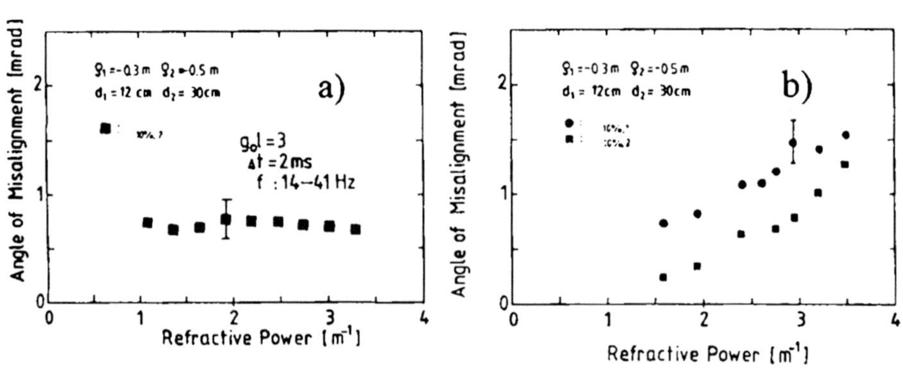

Fig. 15.19 Measured angles of misalignment $\beta_{10\%,i}$ at which the output power has decreased by 10% for a pulsed Nd:YAG laser with a positive branch unstable resonator (rod radius: 5mm). Mirror i is misaligned, mirror 1 is the output coupler with a radius of a=1.75mm. a) for a constant small-signal gain of 3.0 and the repetition rate varied between 14 and 41Hz. b) for a constant repetition rate of 30Hz and the small-signal gain varied between 2.0 and 4.6.

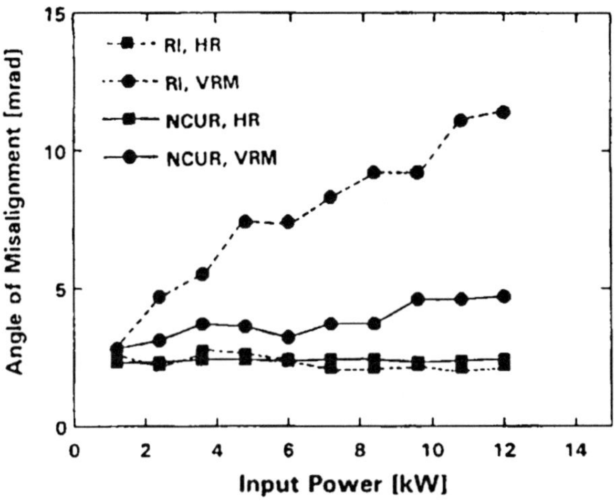

Fig. 15.20 Measured angle of misalignment at which the output power has decreased by 10% of a pulsed Nd:YAG laser (*b=4.75mm, ℓ=150mm*) with a rod-imaging unstable resonator (RI) and a near concentric unstable resonator (NCUR) as a function of the electrical pump power. The output coupler is a variable reflectivity mirror (VRM) with 70% center reflectivity. Please see Fig. 13.49 for resonator geometries and output power. Data marked VRM refer to the tilt of the output coupler, data marked HR refer to the tilt of the high-reflecting mirror [S.9].

According to (15.33), the angles $\alpha_{10\%,i}$ become infinite if the distances d_i are equal to the mirror curvatures ρ_i. Unfortunately, our geometrical model holds only for small angles of misalignment and the infinite solutions of (15.33) thus make no physical sense. However, an infinite or very large 10%-angle indicates that the misalignment sensitivity of the resonator must be very low. Two unstable resonator schemes exhibit large 10%-angles; the rod-imaging unstable resonator with $d_2=\rho_2$, and the near concentric unstable resonator (NCUR) whose distances d_i are slightly larger than the radii of curvature of the mirrors. For both resonator schemes, Fig. 15.20 presents measured angles of misalignment at which the output power decreased by 10%. A comparison with the corresponding 10%-angles of stable lens resonators (Figs. 15.9 and 15.11) indicates that unstable resonators exhibit similar misalignment sensitivities of the output power as stable multimode resonators. By using optimized unstable resonator designs, like the rod-imaging resonator, it is even possible to attain a lower misalignment sensitivity.

Chapter 16

Resonators with Internal Nonlinear Elements

16.1 General Aspects

So far, we assumed that the optical materials inside the resonator exhibit an index of refraction that does not depend on the intensity of the incident light. This is a reasonable assumption as long as the light intensity is low. However, for high intensities the index of refraction n becomes a function of the incident field E (for symplifying the discussion we consider a real, scalar field):

$$n^2(E) = 1 + \chi^{(1)} + \chi^{(2)}E + \chi^{(3)}E^2 + \dots \tag{16.1}$$

where $\chi^{(m)}$ is the m-th order susceptibility of the medium. This expansion of the refractive index holds for all optical materials, but for most materials the higher order susceptibilities are too small to have a noticeable effect on the electric field. However, some crystals exhibit relatively large susceptibilities on the order of 10^{-12} m/V for $\chi^{(2)}$ and 10^{-20} m^2/V^2 for $\chi^{(3)}$. These materials, which are referred to as nonlinear materials, are used to generate a variety of nonlinear effects. Second order effects (generated by $\chi^{(2)}$) include frequency doubling, sum and difference frequency generation, and parametric amplification, to name a few. Among third order effects are frequency tripling, self focusing, and stimulated Brillouin scattering. These nonlinear effects become more understandable if we consider the electric polarization P for an incident field E :

$$P = P^{(1)} + P^{(2)} + P^{(3)} + \dots = \epsilon_0 \left(\chi^{(1)}E + \chi^{(2)}E^2 + \chi^{(3)}E^3 + \dots\right) \tag{16.2}$$

with $\epsilon_0 = 8.854x10^{-12}$ $As/(Vm)$. If two monochromatic fields E_1,E_2 propagating in the z-direction are incident on the nonlinear material, with:

$$E = E_1 + E_2 = E_{10} \sin[k_1 z - \omega_1 t] + E_{20} \sin[k_2 z - \omega_2 t] \tag{16.3}$$

the second order susceptibility $\chi^{(2)}$ generates, besides a constant offset, oscillating components of the dielectric polarization $P^{(2)}$ at the frequencies ω_1, ω_2, $2\omega_1$, $2\omega_2$, $\omega_1+\omega_2$, and $\omega_1-\omega_2$. In the following we restrict our discussion to those two nonlinear effects that are applied in laser resonators and induce a change of the light frequency; intracavity second harmonic generation (ICSHG), intracavity sum frequency generation, and phase conjugation via stimulated Brillouin scattering (SBS).

16.2 Intracavity Second Harmonic Generation

16.2.1 Basic Properties of SHG

We have seen that frequency doubling, also referred to as second harmonic generation, is generated by the second order susceptibility. If we consider a vector field $\boldsymbol{E}=(E_x, E_y, E_z)$, the second order dielectric polarization $\boldsymbol{P}^{(2)}$ can be written as:

$$\boldsymbol{P}^{(2)} = \begin{pmatrix} P_x^{(2)} \\ P_y^{(2)} \\ P_z^{(2)} \end{pmatrix} = \epsilon_0 \begin{pmatrix} d_{11} & d_{12} & d_{13} & d_{14} & d_{15} & d_{16} \\ d_{21} & d_{22} & d_{23} & d_{24} & d_{25} & d_{26} \\ d_{31} & d_{32} & d_{33} & d_{34} & d_{35} & d_{36} \end{pmatrix} \begin{pmatrix} E_x^2 \\ E_y^2 \\ E_z^2 \\ 2E_yE_z \\ 2E_xE_z \\ 2E_xE_y \end{pmatrix} \tag{16.4}$$

with $\epsilon_0=8.85 \cdot 10^{-12}$As/(Vm). For loss-free materials, only 10 of the 18 nonlinearity coefficients d_{ij} are independent (Kleinman symmetry relations). Furthermore, depending on the symmetry of the crystal, the number of independent coefficients is considerably reduced so that for most SHG crystals only two or three independent, nonzero coefficients remain (Table 16.1). In centro-symmetric crystals all coefficients equal zero, which means that these materials cannot generate second order nonlinear effects, except at the surface.

Table 16.1 Nonzero nonlinear coefficients d_{mn} of some materials [4.142].

Material	d_{mn} [10^{-12}m/V]	fundamental wavelength λ [μm]
KDP (KH_2PO_4)	$d_{36} \approx d_{25} = d_{14} = 0.39\pm0.1$	1.06
KD*P (KD_2PO_4)	$d_{36} \approx d_{25} = d_{14} = 0.40\pm0.17$	1.06
AD*P ($NH_2D_2PO_4$)	$d_{36} \approx d_{25} = d_{14} = 0.52\pm0.08$	0.69
KDA (KH_2AsO_4)	$d_{36} \approx d_{25} = d_{14} = 0.52\pm0.03$	0.69
LBO (LiB_3O_5)	$d_{31}=d_{32}=d_{15}=d_{24}=\pm(1.1\pm0.09)$ $d_{33}=0.06\pm0.006$	1.06
$LiNbO_3$	$d_{15}=d_{24}=d_{32}=d_{31}=-5.5\pm0.2$ $d_{16}=d_{21}=-d_{22}=-2.76\pm0.1$, $d_{33}=-34.4\pm2$	1.06
BBO (β-BaB_2O_4)	$d_{15}=d_{24}=d_{32}=d_{31}=\pm(0.12\pm0.06)$ $d_{16}=d_{21}=-d_{22}=\pm(2.0\pm0.25)$	1.06
KTP ($KTiOPO_4$)	$d_{15}=d_{31}=\pm(6.5\pm0.5)$ $d_{24}=d_{32}=\pm(5.0\pm0.5)$, $d_{33}=13.7$	1.06
quartz (SiO_2)	$d_{11}=0.36\pm0.05$	1.06
GaAs	$d_{14} = d_{25} = d_{36} = 148\pm20$	10.6
Banana ($Ba_2NaNb_5O_{15}$)	$d_{31}=d_{32}=d_{15}=d_{24}=-13.2$, $d_{33}=-18.2$	1.06

If a field $E^{(1)}$ at the fundamental frequency ω_1 is incident on the crystal, a field $E^{(2)}$ at the second harmonic frequency $\omega_2=2\omega_1$ is generated at the expense of the fundamental wave. We consider the propagation in the z-direction only and transform the fields and the electric polarizations into the complex notation:

$$E^{(l)} = \frac{1}{2}\left(A^{(l)}\exp[i(\omega_l t - k_l z)] + CC\right) \tag{16.5}$$

$$P^{(l)} = \frac{1}{2}\left(P_c^{(l)} + P_c^{*(l)}\right) \tag{16.6}$$

The electric polarization P acts as a source for both fields $E^{(i)}$, which means that the propagation of each wave is described by the wave equation:

$$\frac{\delta^2 E^{(l)}}{\delta z^2} - \frac{1}{c_0^2}\frac{\delta^2 E^{(l)}}{\delta t^2} = -\frac{1}{\epsilon_0}\frac{\delta^2 P}{\delta t^2} \tag{16.7}$$

The electric polarization P is generated by the sum field $E^{(1)}+E^{(2)}$. After application of the SVE approximation (see Sec. 9.4), the amplitude $A^{(2)}$ of the second harmonic and the amplitude $A^{(1)}$ of the fundamental are given by:

$$2ik_1\frac{\delta A^{(1)}}{\delta z} = \frac{\omega_1^2}{\epsilon_0 c_0^2} P_c^{(2)}(\omega_1)\exp[i(k_1 z - \omega_1 t)] \tag{16.8}$$

$$2ik_2\frac{\delta A^{(2)}}{\delta z} = \frac{\omega_2^2}{\epsilon_0 c_0^2} P_c^{(2)}(\omega_2)\exp[i(k_2 z - \omega_2 t)] \tag{16.9}$$

where $P_c^{(2)}(\omega)$ denotes the component of $P_c^{(2)}$ that oscillates at the frequency ω. The meaning of these equations becomes clear if we insert the nonlinear polarization for a material with the nonzero nonlinear coefficient d_{11} (e.g. quartz). If the incident fundamental wave is linearly polarized in the x-direction, which means $A^{(1)}$ is given by $(A_x^{(1)},0,0)$, the second harmonic is also polarized in the x-direction. According to (16.4) and (16.6), the x-component of the second order polarization $P_c^{(2)}$ then reads:

$$P_{cx}^{(2)} = \frac{\epsilon_0}{2}d_{11}\,(A_x^{(1)}A_x^{*(1)} + A_x^{(2)}A_x^{*(2)} + A_x^{(1)}A_x^{(1)}\exp[2i(\omega_1 t - k_1 z)] +$$

$$+ 2A_x^{*(1)}A_x^{(2)}\exp[i(\omega_1 t - (k_2 - k_1)z)] + \ldots) \tag{16.10}$$

This is a simplified equation, assuming that the coefficient d_{11} is the same for each term. If we insert the terms oscillating at the frequencies ω_1 and $2\omega_1$ into (16.8) and (16.9), respectively, the following set of equations is obtained:

$$\frac{\delta A_x^{(1)}}{\delta z} = -i\frac{\omega_1^2 d_{11}}{2c_0^2 k_1} A_x^{*(1)} A_x^{(2)} \exp[-i\Delta kz] \tag{16.11}$$

$$\frac{\delta A_x^{(2)}}{\delta z} = -i\frac{\omega_2^2 d_{11}}{4c_0^2 k_2} A_x^{(1)} A_x^{(1)} \exp[i\Delta kz] \tag{16.12}$$

where $\Delta k = k_2 - 2k_1$. These equations reveal that the amplitude of the second harmonic can only increase significantly for $\Delta k=0$. This condition is referred to as the *phase-matching condition.* If we consider the SHG as a transformation of two photons with energy $\hbar\omega_1$ into one photon with energy $\hbar\omega_2$, the phase-matching condition is equivalent to the conservation of the momentum:

$$\hbar k_1 + \hbar k_1 = \hbar k_2 \tag{16.13}$$

Since the wave number is related to the frequency and the speed of light via $k=\omega/c$, this relation means that the fundamental wave and the second harmonic must propagate with the same speed to avoid destructive interference of the second harmonic along the propagation direction (Fig. 16.1):

$$c(\omega_1) = \frac{c_0}{n(\omega_1)} = \frac{c_0}{n(2\omega_1)} = c(2\omega_1) \tag{16.14}$$

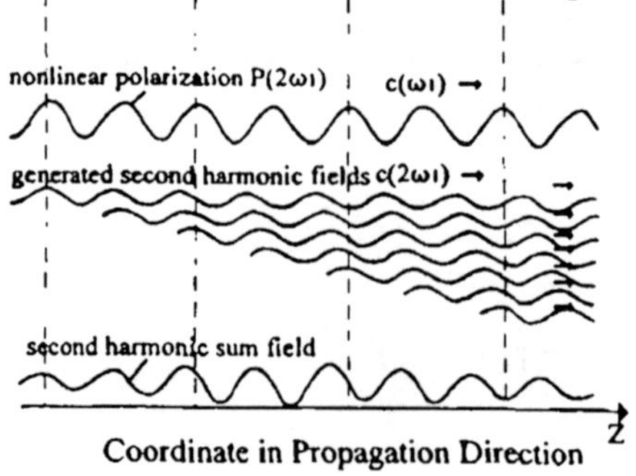

Fig. 16.1 Destructive interference of the second harmonic wave for $c(2\omega_1) < c(\omega_1)$.

For efficient SHG it is, therefore, necessary to match the indices of refraction n at the two frequencies. In isotropic crystals this condition can never be fulfilled because of normal dispersion ($n(2\omega_l)>n(\omega_l)$). In uniaxial crystals, phase matching can be achieved by letting the fundamental be an ordinary wave propagating at an angle θ to the optic axis of the crystal while the harmonic is propagating as an extraordinary wave in this direction, or vice versa (*Type I phase matching*). The refractive index n_e of the extraordinary wave is a function of the angle θ:

$$\frac{1}{n_e^2(\omega,\theta)} = \frac{\cos^2\theta}{n_o^2(\omega)} + \frac{\sin^2\theta}{n_e^2(\omega)} \tag{16.15}$$

where $n_o(\omega)$ is the index of refraction for the ordinary wave and $n_e(\omega)=n_e(\omega, \theta=\pi/2)$. The phase matching angle θ_p is defined by $n_o(\omega_l)=n_e(2\omega_l, \theta_p)$ for an ordinary fundamental wave and by $n_o(2\omega_l)=n_e(\omega_l, \theta_p)$ for an extraordinary fundamental wave. It is also possible to split the fundamental wave into an extraordinary wave and an ordinary wave to attain phase matching. This type of phase matching is referred to as *Type II phase matching.*

In general the fundamental wave and the second harmonic wave exhibit different polarizations and for intracavity SHG a polarizer has to be placed in the resonator to define the polarization of the fundamental wave. The reader may refer to [4.141] or [4.142] to get more insight into the physical aspects of phase matching and how it affects the differential equations (16.8) and (16.9). In the following we will assume the ideal case that the phase matching condition is met. The effect of a slight phase mismatch on the SHG efficiency shall be discussed later. Without lack of generality we use the scalar equations (16.11) and (16.12) with an effective nonlinear coefficient d_{eff} to discuss the conversion efficiency for SHG. The two equations can be solved analytically. For a nonlinear crystal with index of refraction n, the intensities I_1 and I_2 of the fundamental wave and the second harmonic wave, respectively, read as a function of the distance z in the crystal (Fig. 16.2):

$$I_2(z) = I_1(0)\ \tanh^2\left[\sqrt{\frac{d_{eff}^2\,\omega_1^2 I_1(0)}{n^3\ 2\epsilon_0 c_0^3}}\ z\right] = I_1(0)\ \tanh^2\left[\frac{z}{L}\right] \tag{16.16}$$

$$I_1(z) = I_1(0) - I_2(z) \tag{16.17}$$

where n is the index of refraction at both the fundamental and the second harmonic wavelengths. The conversion efficiency is defined as the fraction of the fundamental beam power that is converted into the second harmonic:

$$\eta_{SHG} = \frac{I_2(z)}{I_1(0)} = \tanh^2\left[\frac{z}{L}\right] \tag{16.18}$$

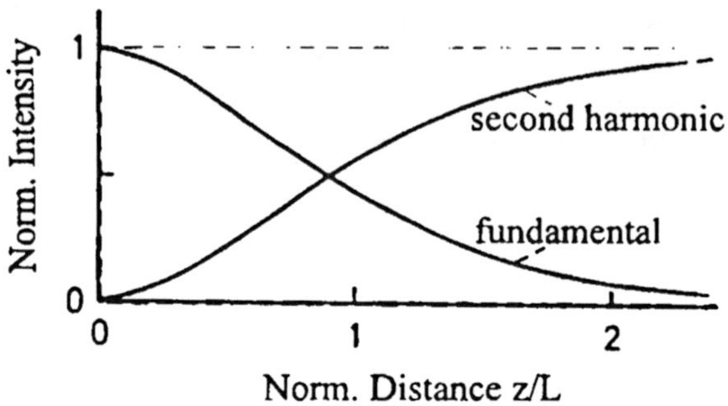

Fig. 16.2 Intensity of the second harmonic wave and of the fundamental wave as a function of the propagation distance z in the nonlinear crystal, according to (16.16) and (16.17).

At the characteristic distance $z=L$, 57% of the fundamental beam power has been converted into the second harmonic beam. For common nonlinear crystals an incident intensity of $I_1(0)=10^8\ W/cm^2$ results in a characteristic distance on the order of 10-20 mm. For low conversion efficiencies (<20%), Eq. (16.16) can be approximated by:

$$I_2(z) = I_1(0)\left[\frac{\omega_1^2}{2\epsilon_0 c_0^3}\, M\, I_1(0)\, z^2\right] \tag{16.19}$$

where M is the figure of merit of the crystal with (Table 16.2):

$$M = \frac{d_{eff}^2}{n^3} \tag{16.20}$$

The experimental values of the conversion efficiency are generally lower than predicted by (16.16) (Fig. 16.3). This is to be expected since we assumed a perfect crystal in which the phase matching condition is exactly met and a fundamental beam that exhibits a flat-top intensity profile. In reality the intensity is nonuniform and the conversion efficiency thus drops at the edges of the beam. Furthermore, due to the small spot size at the crystal the beam will have a small but finite divergence. For a 1.064µm Gaussian beam, a waist diameter of 0.5mm generates a half angle of divergence of 1.3mrad. For most crystals a deviation $\Delta\theta$ on the order of mrads from the phase matching angle θ_P will already decrease the conversion efficiency by 50%. Only the Fourier components of the fundamental beam that lie well within this acceptance angle will experience efficient conversion into the second harmonic. Other crystal properties that affect the conversion efficiency are the optical homogeneity, the absorption losses, and the temperature range within which the phase matching can be realized (typically ±1°C).

Table 16.2 Relative figure of merit M/M_{KDP} ($M_{KDP} = 0.027 x 10^{-24}\ m^2/V^2$), and the fundamental beam intensity $I_1(0)$ required to convert 57% of the fundamental power into the second harmonic for a crystal length of 10mm and a fundamental wavelength of 1.06μm, according to (16.16). The damage threshold represents the incident intensity at which surface damage occurs for a pulse duration of 15ns and a wavelength of 1.06μm (experimental data) [4.141,4.142].

Crystal (Phase Matching Type)	M/M_{KDP}	$I_1(0)$ [GW/cm²]	Damage Threshold [GW/cm²]
KDP (I)	1.0	5.642	14.4
KD*P (I)	1.0	5.642	0.5
ADP (I)	1.3	4.340	6.4
CDA (I)	1.7	4.,340	0.6
CD*A(I)	1.7	3.319	>0.3
LBO (I)	7.0	0.806	---
BBO (I)	25	0.226	23
LAP (I)	42	0.134	<13
$LiNbO_3$ (I)	105	0.054	0.05
KTP (II)	220	0.026	0.8
POM (I)	350	0.016	0.06
Banana (I)	570	0.010	<0.003

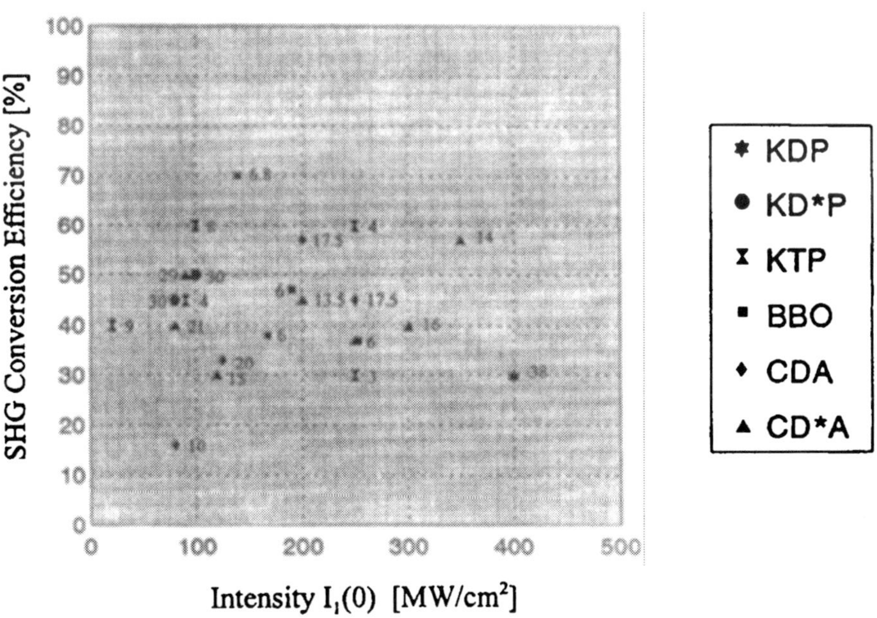

Fig. 16.3 Experimental values of the conversion efficiencies for external frequency doubling of Nd:YAG 1.064μm radiation versus the intensity of the fundamental wave at the nonlinear crystal [4.141,4.142]. The number at each data point is the crystal length in mm.

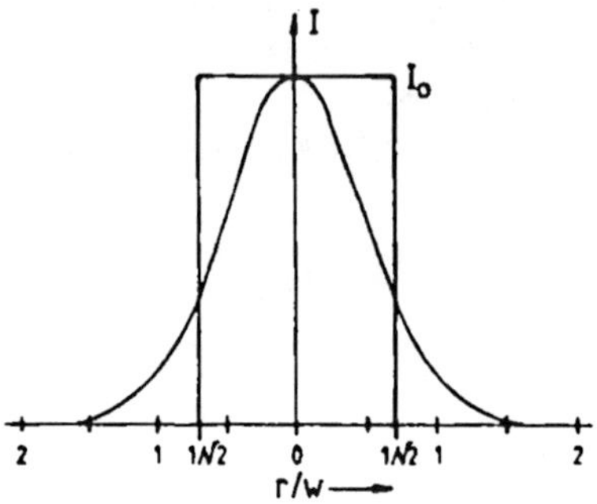

Fig. 16.4 A circular Gaussian beam with beam radius w has the same power content as a flat-top intensity profile with a radius of $w/\sqrt{2}$ and the same peak intensity I_0.

In most applications of SHG, a Gaussian beam at the fundamental wavelength is incident upon the nonlinear crystal. In order to apply the above discussed theoretical description, we have to approximate the Gaussian intensity profile by a flat top profile having the same power content. In circular symmetry, the power P_ω of the Gaussian beam with beam radius w and peak intensity I_0 is given by (Fig. 16.4):

$$P_\omega = I_0 \, 2\pi \int_0^\infty \exp\left[-2\left(\frac{r}{w}\right)^2\right] r dr = I_0 \, \pi \frac{w^2}{2} \tag{16.21}$$

Thus the radius of the flat-top profile is $w/\sqrt{2}$. For low conversion efficiencies, the intensity distribution of the second harmonic is also Gaussian. Neglecting the transverse spreading of the beam inside the crystal, we find with (16.19) that:

$$I_2(r) = \left[\frac{\omega_1^2 M z^2}{2\epsilon_0 c_0^3}\right] I_0^2 \exp\left[-4\left(\frac{r}{w}\right)^2\right] := B I_0^2 \exp\left[-4\left(\frac{r}{w}\right)^2\right] \tag{16.22}$$

The beam radius of the second harmonic wave is $\sqrt{2}$ times smaller than the beam radius of the fundamental wave. Both beams exhibit the same Rayleigh range and are, therefore, always in phase during propagation. The power of the second harmonic wave reads, according to (16.21) and (16.22):

$$P_{2\omega} = B I_0^2 \, \pi \frac{w^2}{4} \tag{16.23}$$

Thus, the SHG conversion efficiency for a Gaussian fundamental beam is a factor of 2 lower compared to that of an incident beam with a flat-top intensity profile:

$$\eta_{SHG} = \frac{1}{2} B \, I_0 \tag{16.24}$$

This decrease is caused by the lower conversion efficiency at the outer areas of the beam. By increasing the fundamental beam intensity this influence of the mode structure becomes less pronounced since the wings of the beam are converted more efficiently. If we use the correct expression (16.16) for the conversion efficiency, the intensity distribution of the second harmonic field is not Gaussian anymore and the power reads:

$$P_{2\omega} = I_0 \, 2\pi \int_0^{\infty} \exp\left[-2\left(\frac{r}{w}\right)^2\right] \tanh^2\left[\frac{z}{L}\exp\left(-\frac{r}{w}\right)\right] r dr \tag{16.25}$$

Compared to a flat-top input beam with intensity I_0, the SHG conversion efficiency is lower by a factor γ with:

$$\gamma = 2 \int_0^{\infty} \exp[-y^2] \tanh^2\left[\frac{z}{L}\exp\left(-\frac{y}{\sqrt{2}}\right)\right] y dy \Big/ \tanh^2\left[\frac{z}{L}\right] \tag{16.26}$$

The SHG conversion efficiency of a Gaussian fundamental beam thus reads:

$$\eta_{SHG} = \gamma \tanh^2\left[\frac{z}{L}\right] \tag{16.27}$$

For low fundamental beam intensities the factor γ is equal to 0.5 and it approaches unity for high conversion efficiencies. Figure 16.5 presents γ as a function of z/L.

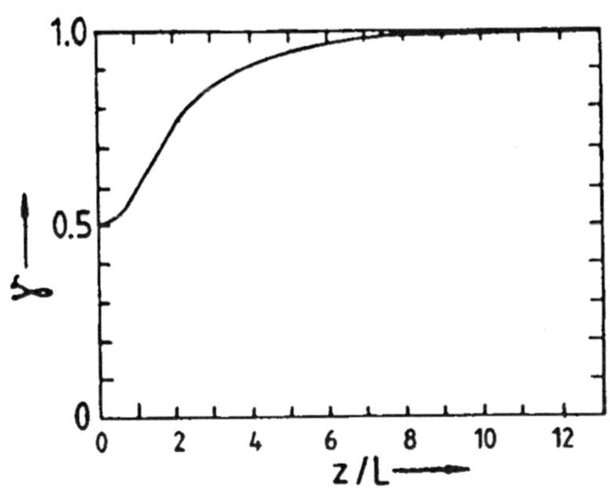

Fig. 15.5 Correction factor for Gaussian beam SHG according to (16.26).

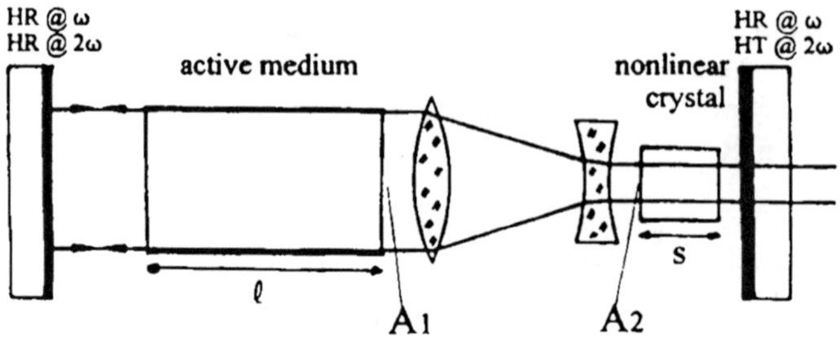

Fig. 16.6 Resonator model for Intracavity Second Harmonic Generation (ICSHG).

16.2.2 Efficiency of Intracavity Second Harmonic Generation

Since the intensity of the beam inside a laser resonator is much higher than the intensity of the laser output beam it is reasonable to place the nonlinear crystal inside the resonator to generate an output at the doubled frequency. Figure 16.6 depicts the resonator model we will use in the following to determine the laser efficiency at the second harmonic wave. The fundamental beam is generated by the laser medium and focused onto the nonlinear crystal to generate a high intensity. Both mirrors are highly reflecting for the fundamental wavelength and the right mirror transmits the second harmonic 100%. This means that the conversion efficiency of the nonlinear crystal acts as the output coupling loss of the resonator. If P_ω is the fundamental beam power incident on the nonlinear crystal and $P_{2\omega}$ is the power of the second harmonic wave generated, the effect of the nonlinear crystal on the fundamental beam can be described by a reflectance R with:

$$R = 1 - \frac{P_{2\omega}}{P_\omega} = 1 - \eta_{SHG} \tag{16.28}$$

where η_{SHG} is the conversion efficiency for the round trip. We can now calculate the average intensity I of the fundamental beam inside the active medium by using the steady-state condition for the round trip:

$$\exp\left[\frac{2g_0\ell}{1 + 2I/I_S} - 2\alpha_0\ell\right] = 1/R \tag{16.29}$$

where I_S is the saturation intensity, $g_0\ell$ is the small-signal gain and $\alpha_0\ell$ is the loss per transit. This expression can be written as:

$$\frac{2g_0\ell}{1 + 2I/I_S} = 2\alpha_0\ell + |\ln(1-\eta_{SHG})| \tag{16.30}$$

Note that the steady state condition (16.29) is only valid for a low small-signal gain and a high reflectance R because we neglect the z-dependence of the fundamental wave intensity inside the medium (see Section 10.1.1). This means that (16.30) is only applicable for low conversion efficiencies and, as a consequence, we cannot only replace the logarithmic term by η_{SHG} but also use (16.19) for its calculation. Equation (16.30) then becomes:

$$\frac{2\,g_0\ell}{1 + 2I/I_S} = 2\alpha_0\ell + 2BI \tag{16.31}$$

$$B = \frac{k\pi^2}{\epsilon_0 c_0}\,\gamma\,\frac{A_1}{A_2}\,\frac{M\,s^2}{\lambda_1^2} \qquad \text{with } k = \begin{bmatrix} 2 \text{ for single pass SHG (see Fig. 16.6)} \\ 4 \text{ for dual pass SHG (see Fig. 16.10b)} \end{bmatrix} \tag{16.32}$$

where A_1, A_2 is the cross-sectional area of the fundamental beam in the active medium and in the nonlinear crystal, respectively, s is the length of the nonlinear crystal, M its figure of merit, λ_1 is the fundamental wavelength, and γ is the correction factor according to (16.26) ($\gamma \approx 0.5$). After solving (16.31) for the intensity I the second harmonic output power can be determined with:

$$P_{2\omega} = A_1\, 2B\, I^2 \tag{16.33}$$

The final result reads:

$$P_{2\omega} = \frac{A_1}{8B}\left[\sqrt{(2\alpha_0\ell + BI_S)^2 + 8BI_S(g_0\ell - \alpha_0\ell)} - (2\alpha_0\ell + BI_S)\right]^2 \tag{16.34}$$

The second harmonic output power can be maximized by adjusting the parameter B of the nonlinear crystal. The optimum parameter B_{opt} is given by:

$$B_{opt} = \frac{2\alpha_0\ell}{I_S} \tag{16.35}$$

and the corresponding maximum output power reads:

$$P_{2\omega} = A_1\, I_S\, \alpha_0\ell \left[\sqrt{\frac{g_0\ell}{\alpha_0\ell}} - 1\right]^2 \tag{16.36}$$

A comparison with (10.13) indicates that this is exactly the output power we would achieve at the fundamental wavelength with optimum output coupling. In other words, if a laser resonator provides a maximum output power at the fundamental wavelength the same output power is provided at the second harmonic wavelength if the nonlinear crystal parameter B is optimized according to (16.35). This is, of course, only true if no additional losses occur due to absorption in the nonlinear crystal (typical loss coefficients are on the order of 0.01 per cm) and the active medium does not absorb the doubled frequency. Keep in mind that although all the fundamental beam power is converted into the second harmonic, the conversion efficiency of the nonlinear crystal may be extremely low. For a cw-laser that operates at an optimum output coupling of 5% for the fundamental wavelength, the conversion efficiency of the crystal in the optimized SHG resonator is also only 5%.

Example: Nd:YAG laser (λ_1=1.064µm, I_S=2kW/cm^2), rod diameter: 3mm (A_1= 7.1mm^2), rod length: 50mm, $g_0\ell$=0.4, $\alpha_0\ell$=0.02. To achieve optimum laser performance at the fundamental wavelength a mirror reflectance of R=*0.87* is required (see (10.15)). According to (10.13), the corresponding maximum output power is 34W. Theoretically, this laser rod is capable of the same output power at the wavelength λ_2=0.532µm, provided that the nonlinear crystal is optimized according to (16.35). If we use LBO (see Table 16.2 for figure of merit M) and focus the intracavity beam to a diameter of 50 µm in the nonlinear crystal (A_1/A_2=3,600), Eqs. (16.32) and (16.34) yield an optimum crystal length of s = 21.2 mm for single pass and s=15 mm for double pass SHG.

For high gain or high output coupling, Eq. (16.29) cannot be used for the calculation of the second harmonic power. Instead, as was shown in Chapter 10, the differential equation (10.1) for the light amplification inside the active medium has to be solved using the boundary conditions at the resonator mirrors. In the case of ICSHG, however, it is not possible to find a simple analytical expression for the intracavity intensity I of the fundamental wave at the active medium since the mirror reflectance R is a function of the intracavity intensity. Furthermore, the correct expression (16.16) has to be used both for the mirror reflectance R and for the output power $P_{2\omega}$ of the second harmonic wave:

$$R = 1 - \tanh^2[\sqrt{2\,B\,I}] \qquad (16.37)$$

$$P_{2\omega} = A_1\, I \tanh^2[\sqrt{2\,B\,I}] \qquad (16.38)$$

Figure 16.7 presents the results of a numerical treatment of this problem. The maximum extraction efficiencies at the second harmonic wavelength are plotted versus the optimum crystal parameter B_{opt} for a saturation intensity of 2kW/cm^2. The maximum extraction efficiencies are the same as those shown in Fig. 10.4. As to be expected, expression (16.35) for the optimum crystal parameter represents a good approximation for low small-signal gains and low losses. The extraction efficiency as a function of the crystal parameter B is shown in Fig. 16.8 for different gains and losses. This graph provides the reader with a feeling for the allowed deviation of the crystal parameter from the optimum value.

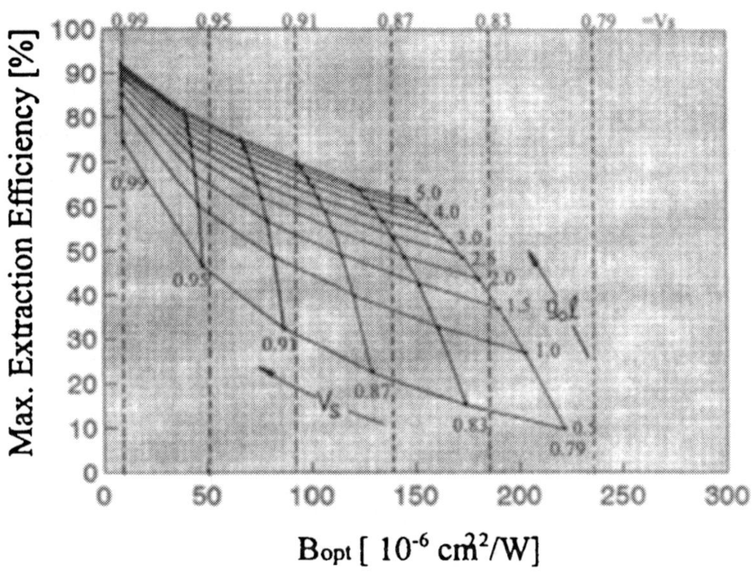

Fig. 16.7 Maximum extraction efficiencies versus the optimum crystal parameter B_{opt} for ICSHG (numerical calculation, I_S=2kW/cm²). The curve parameters are the small-signal gain $g_0\ell$ and the loss factor per transit $V_S=exp[-\alpha_0\ell]$. The vertical lines represent B_{opt} calculated with (16.35). The extraction efficiency is the ratio of the output power at the second harmonic wavelength to the power P_{UL} available in the active medium in form of inversion with $P_{UL}= g_0\ell\ A_1\ I_S$(see also Chapter 10).

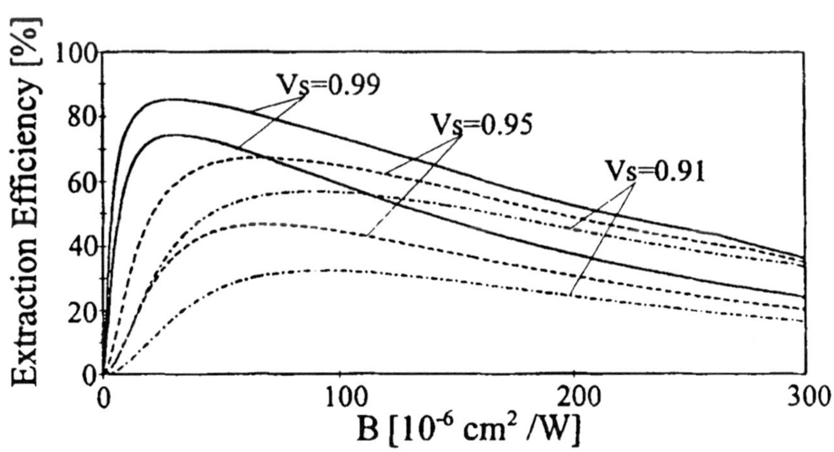

Fig. 16.8 Extraction efficiency as a function of the crystal parameter B. For each loss factor per transit $V_S=exp[-\alpha_0\ell]$, the upper and lower curve holds for a small-signal gain $g_0\ell$ of 1.5 and 0.5, respectively (I_S=2kW/cm²).

16.2.3 Phase Mismatch, LongitudinalModes, and Conversion Efficiency

Efficient second harmonic generation requires the fundamental and the second harmonic fields to be phase-matched, which means that both fields must exhibit the same phase velocity or index of refraction in the nonlinear crystal. The conversion efficiency is very sensitive to a phase mismatch Δkz between the two fields that occurs over a distance z inside the crystal. As long as the depletion of the fundamental field is low, the conversion efficiency for phase mismatched SHG is given by [4.141]:

$$\eta_{SHG} = \left[\frac{\omega_1^2}{2\epsilon_0 c_0^3}\, M\, I_1(0)\, z^2\right]\left[\frac{\sin(\Delta kz/2)}{\Delta kz/2}\right]^2 \tag{16.39}$$

with: $\Delta k = 2[n(\omega_1) - n(2\omega_1)]\omega_1/c_0$

For $\Delta k=0$ this expression is equivalent to (16.16). The mismatch of the wave vector can be caused by an angular deviation from the phase matching direction or by deviations in the temperature or in the fundamental wavelength since the indices of refraction are functions of all three parameters. For a crystal length of 10mm and a fundamental wavelength of 1.064μm, even a small refractive index difference of 0.0001 decreases the conversion efficiency by 26 %. Table 16.3 presents, for different nonlinear crystals, measured deviations in angle, temperature, and wavelength at which the SHG efficiency has dropped to one-half of the maximum value. When choosing a nonlinear crystal for ICSHG, these acceptance ranges are design parameters that are as important as the figure of merit M. It may even be advantageous to trade in the figure of merit for large acceptance ranges since the optimum crystal parameter B_{opt} can still be achieved by increasing the crystal length. Note in Table 16.3 that crystals for which a phase matching angle of 90° can be realized exhibit a high angular tolerance. In addition to slight phase mismatches, the output power at the second harmonic wavelength is also affected by the number m of longitudinal modes oscillating in the laser resonator. Assuming a statistical distribution of the phases of the axial modes, the conversion efficiency (16.39) for second harmonic generation as function of the number m of longitudinal modes reads:

$$\eta_{SHG}(m) = \frac{1}{2}\left[2-\frac{1}{m}\right]\left[\frac{\omega_1^2}{2\epsilon_0 c_0^3}\, M\, I_1(0)\, z^2\right]\left[\frac{\sin(\Delta kz/2)}{\Delta kz/2}\right]^2 = \frac{1}{2}\left[2-\frac{1}{m}\right]\eta_{SHG}(\infty) \tag{16.40}$$

In single mode operation, the conversion efficiency is reduced to half the maximum value. This means that in order to attain the maximum output power (16.36), the area of the fundamental beam in the nonlinear crystal has to be chosen to be half the area of the multimode case. An ICSHG resonator that has been optimized for single longitudinal mode operation, may therefore show a drop in the second harmonic output power if several modes start oscillating (resonator becomes overcoupled).

The major challenge in ICSHG is the realization of stable second harmonic emission with a low rms noise. Simultaneous oscillation of several longitudinal modes generates the so-called "green problem" [4.148,4.154], an amplitude instability due to longitudinal mode coupling caused by sum frequency generation of two different longitudinal modes. This chaotic amplitude fluctuation of the second harmonic can be suppressed by single-mode operation [4.155], by preventing the formation of standing waves in the resonator (unidirectional ring resonator [4.175], twisted mode resonator), by allowing two longitudinal modes that are orthogonally polarized (Fig. 16.9), or by using a high number *m* of longitudinal modes (long resonator) [4.153]. In [4.174] an rms stability of better than 0.1% was reported for a diode pumped Nd:YVO_4/LBO system with on the order of 100 axial modes. In recent years, quiet second harmonic multimode operation has also been reported for longitudinal mode numbers below 30, in some cases as low as 3. At present, there is no satisfactory explanation why quiet ICSHG operation can occur for a low number of longitudinal modes.

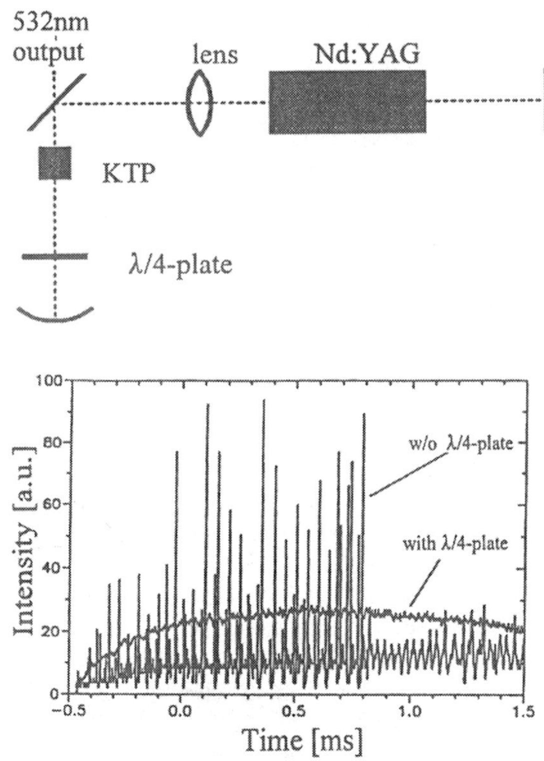

Fig. 16.9 Measured second harmonic intensity of a pulsed, flashlamp pumped Nd:YAG laser with ICSHG in type II KTP. By inserting a quarter-wave plate, whose principal axes are rotated by 45° with respect to the axes of the KTP, sum frequency generation of different longitudinal modes is suppressed, resulting in considerably reduced rms noise [S.31].

Table 16.3 Experimental values of the phase matching angle and the angular, spectral, and temperature deviations at which the SHG conversion efficiency has dropped by 50% (λ_1=1.064μm, crystal length s=10mm) [4.141,4.142].

Crystal (Phase matching type)	θ_P [degree]	$\Delta\theta$ [mrad]	ΔT[°C]	$\Delta\lambda$ [nm]
KDP (I)	41	2.7	11	5.6
KDP(II)	59	5.3	13.2	7.3
KD*P (II)	53.5	5.0	6.7	5.6
CDA (I)	84	12.3	2	-
CDA (I)	90	69	5.8	-
CD*A (I)	79.3	7.2	3.3-6.4	2.45
Banana (II)	76	5.4	0.5	-
Banana (I)	90	43	0.5	-
KTP (II)	25	15-68	25	0.56
KTP (I)	90	15	25	0.56
BBO (I)	23	1.5	55	-
LBO (I)	90	52	4	-

16.2.4 Resonator Configurations

In order to realize the optimum crystal parameter B_{opt} given by (16.32) and (16.35) we can play with three parameters: the crystal length s, the figure of merit M, and the ratio of the cross sectional areas A_1/A_2. However, the ranges of the first two parameters are more or less limited and the conversion efficiency has to be optimized by utilizing a resonator configuration that provides a high ratio A_1/A_2 . Since we have to limit the angle of divergence of the fundamental beam, we are particularly interested in stable resonators that exhibit a large Gaussian beam diameter at one resonator mirror, where the active medium is placed, and have a small beam waist at the location of the nonlinear crystal. If mirror 1 is the resonator mirror at which the active medium is to be placed, there are two regions in the stability diagram where these Gaussian beam properties can be attained; near the stability limit $g_1g_2=1$ with $g_2<-1$ (concentric resonators) and near the axes $g_1=0$ with $g_2<1$.

Resonators in the latter region exhibit an effective length which is equal or slightly shorter than the radius of curvature of mirror 1 (Fig. 16.10a). If a concave mirror 2 with a relatively large radius of curvature is used, the beam waist is slightly shifted inwards and the nonlinear crystal can be placed close to this mirror. In solid state lasers that are end-pumped with laser diodes with output powers in the Watt range, the active medium is also placed at this side since the diode beam has to be tightly focused in order to reach laser threshold, resulting in a high fundamental beam power (Fig. 16.11).

In concentric resonators, the centers of curvatures of both concave mirrors are on top of each other and this point is the location of the beam waist. The active medium is placed close to the mirror with the larger radius of curvature. Fig. 16.10b presents a resonator that is equivalent to a concentric resonator (see also Fig. 16.12).

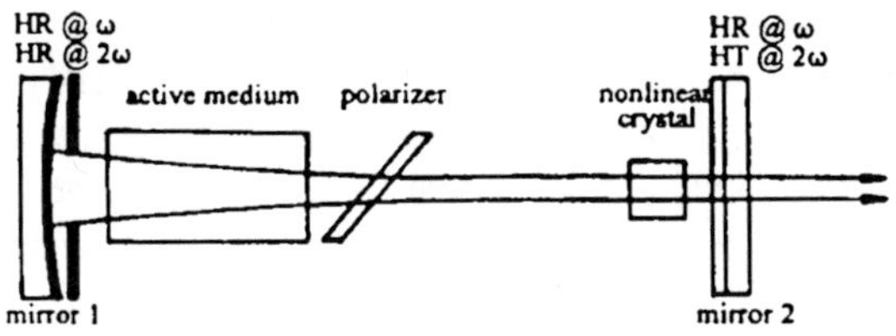

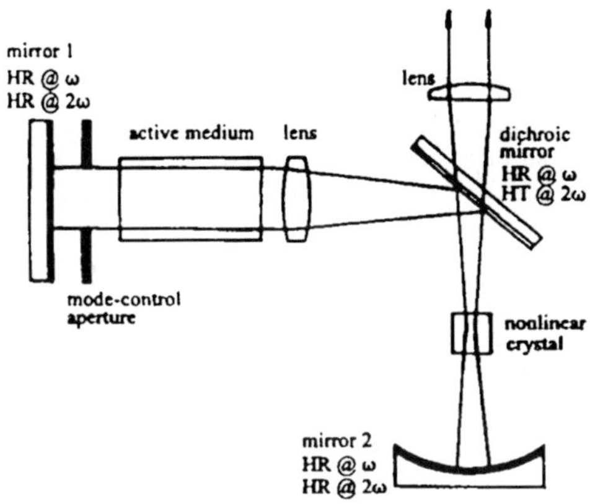

Fig. 16.10 Resonator configurations for ICSHG. a) resonator near the axis $g_1=0$. The radius of mirror 1 is slightly shorter than the effective resonator length. b) concentric resonator with separation of the fundamental field and the second harmonic field by a dichroic mirror.

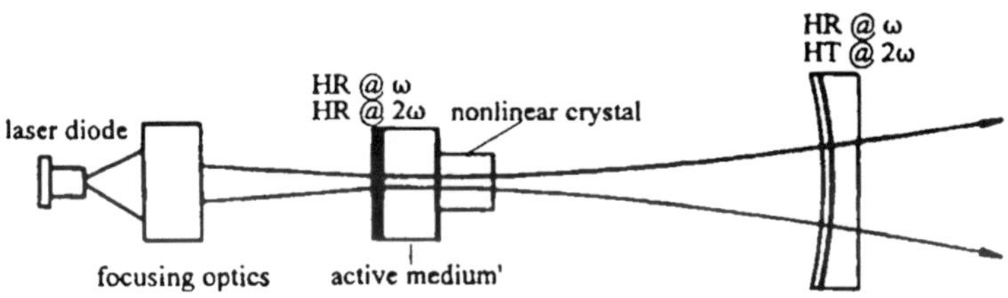

Fig. 16.11 Second harmonic generation in a diode pumped solid state laser.

Note that the combination of the flat mirror and the focusing lens exhibits the same imaging properties as a concave mirror but provides a higher fill factor in the active medium. The use of the dichroic beamsplitter prevents the second harmonic field from entering the active medium and being partially absorbed. Since the second harmonic generation occurs in both propagation directions, the fundamental field and the second harmonic field must be in phase at both endfaces of the nonlinear crystal. Due to the dispersion of air, the propagation from the crystal to mirror 2 and back generates a phase shift between the two fields which, depending on the propagation distance and the phase shift induced by the HR mirror, may lead to destructive interference of the second harmonic. This phase shift of 27.4° per cm of propagation distance has to be compensated by slightly translating mirror 1 until the second harmonic power is maximized [4.146].

Experimental examples of ICSHG in neodymium doped YAG and Vanadate lasers are presented in Figs. 16.12 to 16.14. These three systems represent the state of the art of second harmonic generation for diode pumped and flashlamp pumped Nd:YAG lasers. An overview of the performance of solid state lasers with ICSHG is given in Table 16.4. In general, the second harmonic power can be as high as η=60-70% with respect to the power attained when the laser is operated at the fundamental wavelength (with optimum output coupling). In some instances, conversion efficiencies as high as 90% have been reported.

Table 16.4 Intracavity SHG of different solid state lasers. $P_{2\omega}$ is the power at the second harmonic wavelength, η is the ratio of $P_{2\omega}$ to the power achieved at the fundamental wavelength, P_{electr} is the electrical pump power to the flashlamps, and P_{opt} is the optical power incident on the active medium.

Material	λ_2[nm]	crystal	mode of operation	η[%]	$P_{2\omega}$	Ref.
Nd:YAG	532	KTP	cw-flashlamp, P_{electr}=4.9kW	-	25W	[4.162]
Nd:YAG	532	KTP	cw-flashlamp, Q-switch, P_{electr}=4kW	53	21.5W	[4.152]
Nd:YAG	532	KTP	cw-flashlamp, Q-switch, P_{electr}=8.5kW	21	53W	[S.17]
Nd:YAG	532	KTP	cw-flashlamp, Q-switch, P_{electr}=8kW	-	97W	[4.161]
Nd:YAG	532	KTP	cw-laser diode, Q-switch, P_{opt}=2.2W	50	340mW	[4.157]
Nd:YAG	532	KTP	cw-laser diode, P_{opt}=11W	48	1.05W	[4.163]
Nd:YAG	532	KTP	cw-laser diode, P_{opt}=15.3W	-	3.5W	[4.166]
Nd:YAG	532	KTP	cw-laser diode, P_{opt}=14W	-	3.1W	[4.175]
Nd:YAG	532	KTP	cw-laser diode, modelocked, P_{opt}=55W	56.6	3.0W	[4.158]
Nd:YAG	532	LBO	cw-laser diode, Q-switch, P_{opt}=19W	84	3.2W	[4.188]
Nd:YAG	532	LBO	cw-laser diode, Q-switch, P_{opt}=40W	67	4W	[4.167]
Nd:YAG	532	LBO	cw-laser diode, Q-switch, P_{opt}=800W	90	138W	[4.184]
Nd:YAG	473	$KnbO_3$	cw-laser diode, P_{opt}=279mW	9.5	43mW	[4.173]
Nd:YAG	473	LBO	cw-laser diode, P_{opt}=22.7W	39.4	1.3W	[4.182]
Nd:YAG	473	BiBO	cw-laser diode, P_{opt}=21W	61	2.8W	[4.190]
$Nd:YVO_4$	532	KTP	cw-laser diode, P_{opt}=900mW	27.6	105mW	[4.171]
$Nd:YVO_4$	532	KTP	cw-laser diode, P_{opt}=1.2W	-	250mW	[4.185]
$Nd:YVO_4$	532	KTP	cw-laser diode, P_{opt}=4W	-	1.05W	[4.172]
$Nd:YVO_4$	532	KTP	cw-laser diode, P_{opt}=12.6W	46	3.2W	[4.179]
$Nd:YVO_4$	532	LBO	cw-laser diode, P_{opt}=19.5W	65	6.1W	[4.174]
$Nd:YVO_4$	532	LBO	cw-laser diode, P_{opt}=20W	-	8.5W	[S.18]
$Nd:YVO_4$	532	LBO	cw-laser diode, P_{opt}=30W	-	8.9W	[4.187]
Nd:YLF	523.5	LBO	cw-laser diode, P_{opt}=10W	24	0.9W	[4.164]
Nd:LSB	531	KTP	cw-laser diode, P_{opt}=2.05W	55	522mW	[4.165]
Nd:LiLuF	455	LBO	cw-laser diode, P_{opt}=1.8W	5.7	42mW	[4.186]
Cr:LiSAF	428-444	$KnbO_3$	cw-laser diode, P_{opt}=680mW	6.8	13mW	[4.170]
Cr^{4+}:forsterite	613-655	PPLN	cw-1064nm YAG, P_{opt}=6.8W	18.4	45mW	[4.189]

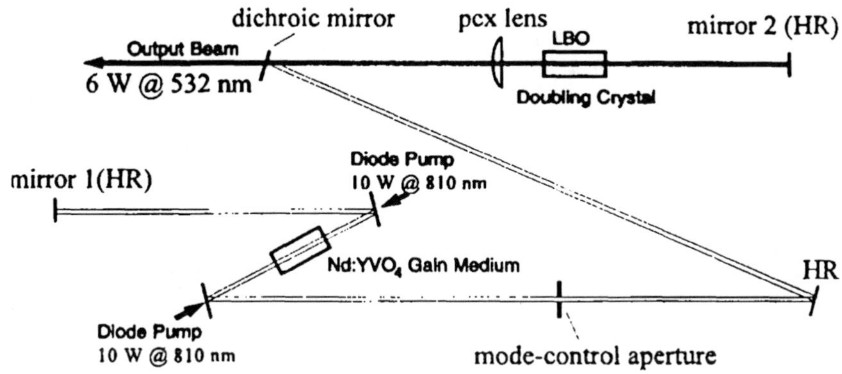

Fig. 16.12 ICSHG in a diode pumped Nd:YVO_4 laser. The LBO crystal is held at a temperature of 150°C to achieve non-critical phase matching (Type I). The maximum second harmonic output power is 6.1 W at an optical pump power from the laser diodes of 20 W. The resonator length is 1m, allowing a large number of axial modes (~100) to oscillate. The axial multimode operation results in a high stability and a low RMS noise (<0.04%). Without frequency doubling, the output power at the fundamental wavelength is 9.4W [4.174] (Millennia V, courtesy of Spectra-Physics, Mountain View, CA, 1996). A shorter version of this resonator, released in 2000, provides green output powers in excess of 12W at about twice the pump power (Millennia X).

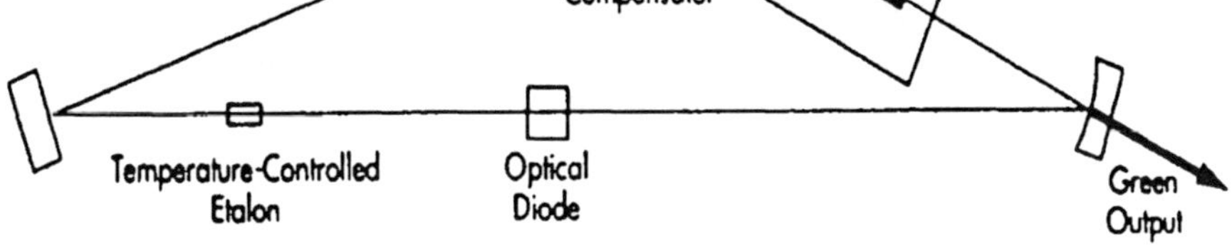

Fig. 16.13 ICSHG in a diode-pumped Nd:YVO_4 ring laser. An output power at 532nm in excess of 18W is obtained in single mode operation with an rms noise of less than 0.03% (18W Verdi, courtesy of Coherent,Santa Clara, CA, 2004).

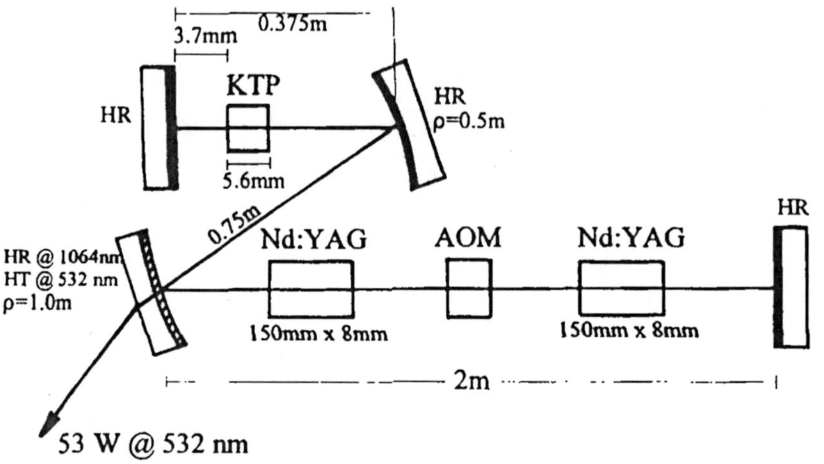

Fig. 16.14 Cw flashlamp pumped Nd:YAG laser with ICSHG providing a maximum average output power of 53 W at the second harmonic wavelength of 532nm for an electrical pump power of 8.5kW (4.25 kW per rod). Pulse duration: 500ns, pulse repetition rate: 5kHz; AOM: acousto-optic modulator [S.17] (courtesy of Laser- und Medizin Technologie Berlin gGmbH, Berlin, Germany).

16.3 Intracavity Third Harmonic Generation

16.3.1 General Properties of Third Harmonic Generation

In general, tripling the frequency of the fundamental wave can be accomplished directly by a one step, third order nonlinear process, in which three fundamental photons are combined into one photon with three times the energy. However, the nonlinear coefficients for this direct process are relatively small, resulting in very low conversion efficiencies in the sub-percent range for typical fundamental beam intensities of hundreds of MW per cm^2. A more efficient way to generated the tripled frequency is the doubling of a portion of the fundamental and subsequent sum frequency generation (SFG) of the two resulting fields (Fig. 16.15). This process is mainly used for frequency tripling of pulsed infrared solid state lasers [4.191-4.193, 4.203-4.205]. In order to achieve 100% conversion efficiency, 67% of the fundamental power has to be converted to the second harmonic since this results in an equal number of fundamental and second harmonic photons in the SFG crystal. Figure 16.16 shows calculated conversion efficiencies as a function of the total input intensity on the SFG crystal (type II KDP, 1064nm fundamental) for different ratios M of second harmonic to fundamental power, assuming a flat-top in intensity profile. A mismatch of green and infrared photon numbers will result in back-conversion and complete suppression of the third harmonic at certain input intensities. However, with matched photon numbers, it is possible to attain 100% conversion efficiency.

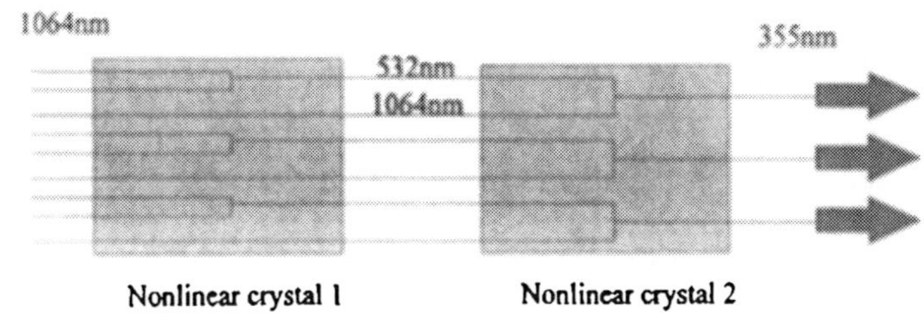

Fig. 16.15 Third harmonic generation via sum frequency generation of the fundamental and the doubled wavelength. 100% conversion occurs, if 67% of the fundamental power is converted into the second harmonic in the doubling crystal.

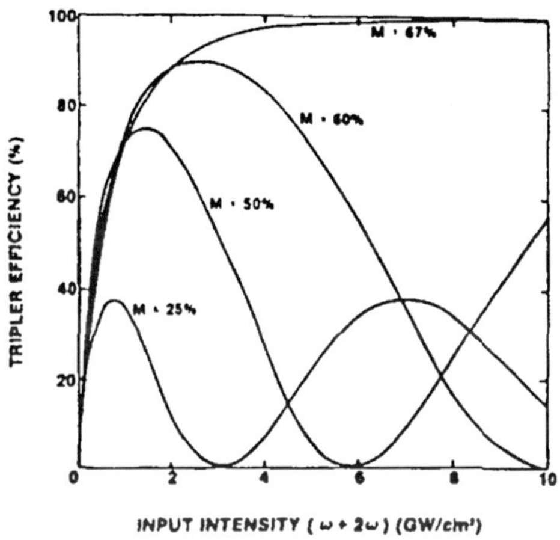

Fig.16.16 Tripling efficiency of a 9mm thick phase-matched KDP type II crystal as a function of the total input intensity for various percentages M of second harmonic power content in the input [4.191] (© Optics Communications, 1980).

The presented calculations are numerical solutions to two sets of differential equations, one describing the doubling process, and the subsequent one for sum frequency generation. In both equations flat-top intensity distributions both in space and time are assumed [4.191]. Assuming perfect phase matching and no losses, the equations for the doubling process read:

$$\frac{\delta A_1}{\delta z} = -i\,\frac{\pi d_2}{n_2\,\lambda}\,A_1^* A_2 \tag{16.41}$$

$$\frac{\delta A_2}{\delta z} = -i\,\frac{\pi d_2}{n_2\,\lambda}\,A_1^* A_1 \tag{16.42}$$

where d_2 is the effective nonlinear coefficient for SHG, n_2 is the refractive index of the crystal and λ is the fundamental wavelength. The fundamental field A_1 and the second harmonic field A_2 at the end of the SHG crystal serve as input fields for the second set of equations describing the sum frequency generation:

$$\frac{\delta A_1}{\delta z} = -\,i\,\frac{\pi d_3}{n_3\,\lambda}\,A_2^{*}A_3 \tag{16.43}$$

$$\frac{\delta A_2}{\delta z} = -2i\,\frac{\pi d_3}{n_3\,\lambda}\,A_1^{*}A_3 \tag{16.44}$$

$$\frac{\delta A_3}{\delta z} = -3i\,\frac{\pi d_3}{n_3\,\lambda}\,A_1 A_2 \tag{16.45}$$

where d_3 is the effective nonlinear coefficient for SFG (see Table 16.5), n_3 is the refractive index of the crystal and A_3 is the field of the third harmonic. The differential equations (16.41-16.45) can also be used to generate a numerical model that simulate third harmonic generation for more realistic beams that include focusing as well as a transverse and temporal intensity distribution. In practice, most frequency-tripled solid state lasers are Q-switched or mode-locked Nd:YAG or Nd:YVO_4 lasers in fundamental mode operation, exhibiting output pulses with near-Gaussian temporal pulse shapes. In this case, matching the fundamental and the second harmonic photon numbers can only be partially realized, resulting in tripling conversion efficiencies that remain well below 100% .

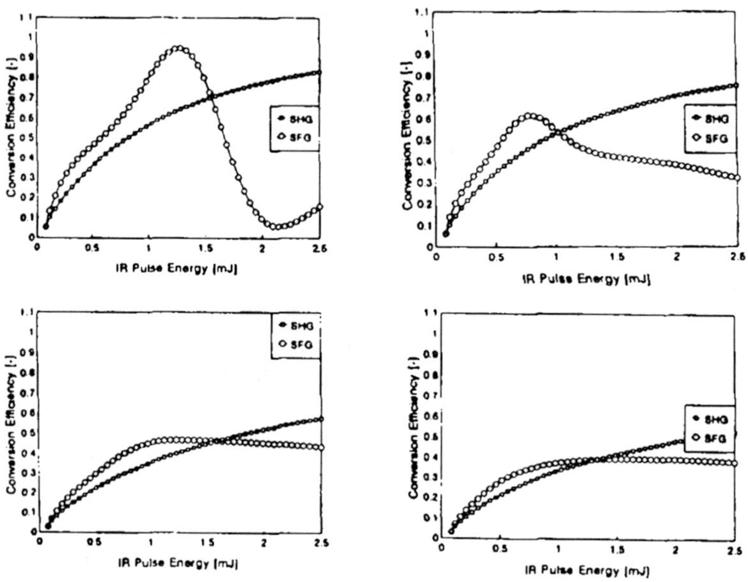

Fig. 16.17 Calculated conversion efficiencies in the SHG crystal (20mm long LBO type I, d_2=0.83pm/V) and the SFG crystal (20mm long LBO type II, d_3=0.66 pm/V) as a function of the total input energy at 1064nm for different spatial and temporal intensity distributions. a) flat-top in spce in time, b) flat-top in space, Gaussian in time, c) Gaussian in space, flat-top in time, d) Gaussian in space and time. Pulse duration: 50ns, IR beam diameter in SHG crystal: 140 μm, IR beam diameter in SFG crystal: 70 μm. Equations (16.41-45) were solved numerically.

While the beam center may exhibit the perfect balance of IR and green photons, the wings of the pulse will experience a much lower conversion efficiency. Increasing the input intensity does not necessarily lead to an enhancement of the over efficiency, because the increased conversion in the wings of the pulse is off-set by a decreased conversion in the beam center. This may result in an intensity dip in the center of the UV beam. An example is shown in Fig. 16.17, in which calculated IR to UV conversion efficiencies are shown for 50ns long input pulses at 1064nm using type I LBO and type II LBO, the crystals most commonly used for third harmonic generation [4.195,4.198,4.203], as the SHG and SFG crystal, respectively. For an intensity profile that is flat-top in time and space, a maximum UV conversion of close to 100% is obtained at a pulse energy at which the doubling efficiency is 67%. Introducing a Gaussian intensity distribution in time or space leads to a substantial decrease of the maximum conversion efficiency. For a Gaussian beam with Gaussian-shaped temporal pulses, the maximum tripling efficiency that can be attained is about 43%, with a similar value for the doubling efficiency.

However, this maximum possible conversion efficiency can only be obtained for a perfect spatial and temporal overlap between the fundamental and the second harmonic intensities. Inside commonly used SFG crystals, the two beams will propagate at a slight angle, referred to as the walk-off angle. This results in a decreased effective interaction length inside the crystal. If the walk-off angle is not too large, walk-off compensation schemes like the one shown in Fig.16.18 can be applied to improve the tripling efficiency [4.196,4.199]. Measured IR to UV conversion efficiencies of up to 40% have been reported for extra-cavity frequency tripled Nd:YVO_4 lasers with walk-off compensation [4.125, 4.203].

The choice of the SFG crystal is a balance of finding a high nonlinear coefficient and a large acceptance angle, combined with a small walk-off angle (Table 16.5). At present, the majority of ultraviolet solid state lasers use LBO as the type II SFG crystal, because the relatively large acceptance angle allows spot sizes of the fundamental beam of 100 microns or less inside the crystal. In addition, the relatively small walk-off angle makes compensation schemes easier to implement.

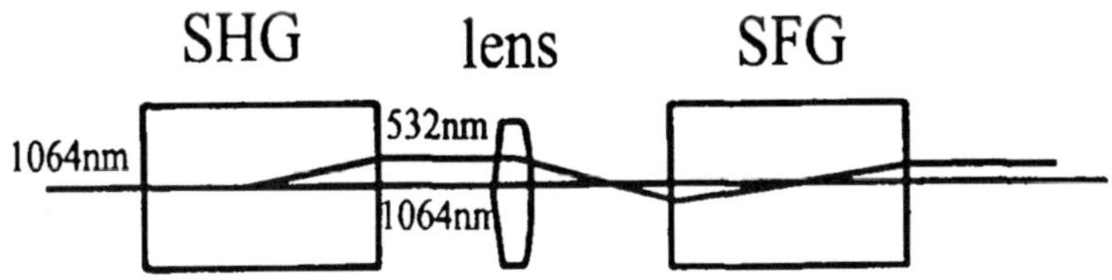

Fig. 16.18 Walk-off compensation scheme, using a focusing lens to cross the IR and the green beam near the center of the SFG crystal. In order for this scheme to work, the two beams need to be slightly separated in front of the lens. In this case, the walk-off in the SHG crystal is used to generate the separation.

Table 16.5 Nonlinear coefficients, acceptance angles and walk-off angles of different nonlinear crystals for 355nm generation via sum frequency generation (1064nm + 532nm) [4.208]

crystal		nonlinear coefficient d_3 [pm/V]	acceptance angle x length [mrad cm]	walk-off angle [mrad]
LBO,	type II	0.66	5.0	9.3
	type I	0.90	---	18.3
CLBO,	type II	0.93	1.2	37.3
	type I	0.62	0.8	37.1
CBO,	type II	1.15	----	16.2
	type I	0.91	1.9	17.5
BBO,	type II	1.2	0.6	77.7
	type I	1.72	0.4	72.1

Figure 16.19 depicts calculated IR to UV conversion efficiencies (1064nm to 355nm) for extra-cavity third harmonic generation as a function of the beam diameter in the type II LBO SFG crystal for a Gaussian beam with 20kW peak power. Parameter of the curves is the beam diameter in the doubling crystal (non-critically phase matched type I LBO). A plane-wave approximation is used (no beam spreading due to diffraction) and walk-off is not included in the numerical model. This graph indicates that for Q-switched or modelocked IR lasers, which typically exhibit peak powers between 5 and 50 kW, spot diameters on the order of 100-200 μm are required in both crystals to attain the maximum conversion efficiency of about 43%.

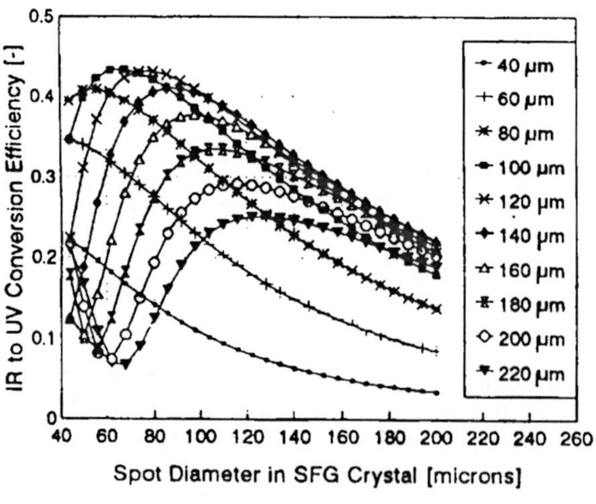

Fig. 16.19 Calculated 1064nm to 355nm conversion efficiency for a Gaussian beam with 1mJ pulse energy and 50ns pulse duration (Gaussian pulse, $1/e^2$ full width) as a function of the 1064nm beam diameter in the SFG crystal (20mm long type II LBO, d_3=0.66 pm/V). Curve parameter is the 1064nm beam diameter in the doubler cyrstal (20mm long LBO, d_2=0.83 pm/V). Equations (16.41)-(16.45) were applied to each of 100 x 100 points in time and space, using a Gaussian input field distribution. No walk-off included in the model.

16.3.2 Properties of Intracavity Third Harmonic Generation

Similar to intracavity second harmonic generation, the two nonlinear crystals can be placed inside the resonator to make use of the high intra-cavity intensity of the fundamental. The major advantage of this technique is that, compared to the extra-cavity arrangement, the beam diameters in the nonlinear crystals can be increased considerably, typically by a factor of five for Q-switched systems. As a result, it is not necessary to use walk-off compensation schemes to attain high conversion efficiencies since the walk off (typically 50-100 microns) is one order of magnitude smaller than the beam size. Lifetime of the SFG crystal is also enhanced due to the lower UV power density. In addition, the pulse-to-pulse stability is improved since resonator output coupling is a function of the fundamental beam intensity. Similar to a standard resonator, a change in output coupling generates only small changes in output power if the resonator is operated near the optimum output coupling. Thus, a variation of the fundamental beam intensity, due to pump power variations or Q-switch jitter, has a much lower effect on the UV output as compared to extra-cavity systems. A typical Q-switched infrared solid state laser resonator with intracavity third harmonic generation is shown in Fig. 16.20. An intra-cavity beam splitter and a prism are used to extract and separate the residual green and the UV beam. Alternative output coupling techniques include transmission through one of the infrared HR mirrors [4.202,4.205], and wavelength separation using intracavity prisms [4.202] or a Brewster-cut SFG crystal [4.200].

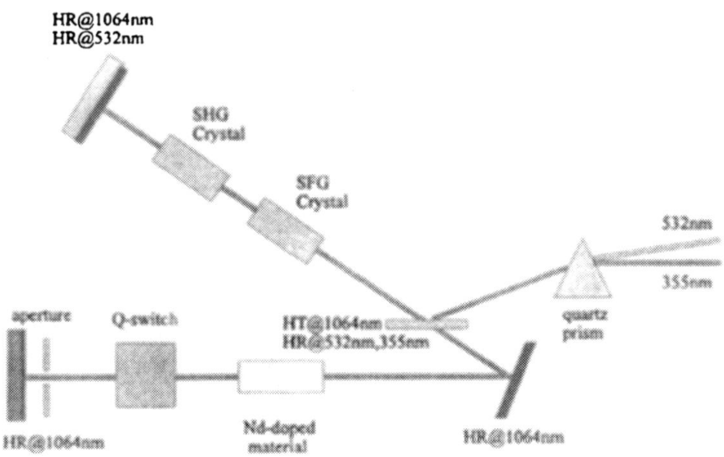

Fig. 16.20 Schematic of a Q-switched solid state laser with intracavity harmonic generation (after [4.210]). The resonator for the fundamental wavelength is formed by the three 1064nm HR mirrors (marked in black). After a double pass through the SHG crystal, the generated green power is mixed with the fundamental in the SFG crystal. Output coupling of the resonator is provided by the nonlinear conversion to the green and the UV.

In an ideal arrangement, the green power is completely depleted in the SFG crystal and the third harmonic power is equal to the power the resonator would provide at the fundamental wavelength with optimum output coupling (see Sec.16.2.2). For flat-intensity profiles, this can be accomplished by properly choosing the fundamental beam diameters in the nonlinear crystals as well as optimizing the crystal lengths. Let us use a simplified model to understand the key issues of this optimization. After the dual pass in the SHG crystal, the green power is given by:

$$P_G = \eta_{SHG}\, P_{IR} \tag{16.46}$$

where P_{IR} is the fundamental power entering the SHG crystal and η_{SHG} is the dual pass conversion efficiency. If we assume that the green power is completely depleted in the SFG crystal, the third harmonic power must be given by (remember that each green photon is added to an infrared photon with half the energy!):

$$P_{UV} = \frac{3}{2}\,\eta_{SHG}\, P_{IR} \tag{16.47}$$

The combination of the two nonlinear crystals acts as an output coupler for the fundamental wavelength resonator with a reflectivity:

$$R = \frac{P_{IR} - P_{UV}}{P_{IR}} = 1 - \frac{3}{2}\,\eta_{SHG} \tag{16.48}$$

Maximum UV power is obtained if the reflectivity R is equal to the optimum reflectivity of the resonator when operated at the fundamental wavelength (see Eq. (10.15)). The optimum SHG efficiency is thus given by:

$$\eta_{SHG,opt} = \frac{2}{3}\left[1 - R_{opt}\right] \tag{16.49}$$

Typically, solid state lasers in Q-switched operation exhibit optimum reflectivities between 50% and 90%, which translates into optimum SHG efficiencies between 33% and 7%. Therefore, optimized intracavity third harmonic generation requires considerably lower green conversion than the 67% needed in the extra-cavity scheme. Once the optimum conversion is realized by choosing the correct beam size and SHG crystal length, the SFG conversion efficiency can be maximized separately by changing the focusing into the SFG crystal and the crystal length. Theoretically, this results in a third harmonic power that is equal to the maximum output power at the fundamental wavelength. A calculated example for a Q-switched, diode pumped Nd:YVO_4 laser is shown in Fig.16.21. Graph a) shows the performance for beam diameters in the nonlinear crystals that are close to optimal. At a repetition rate of 23kHz, 30W of 355nm power is obtained with 22% IR to green conversion and 1W of residual green power.

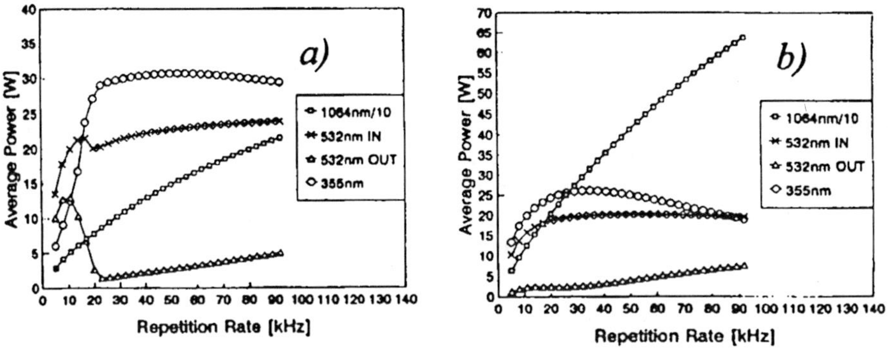

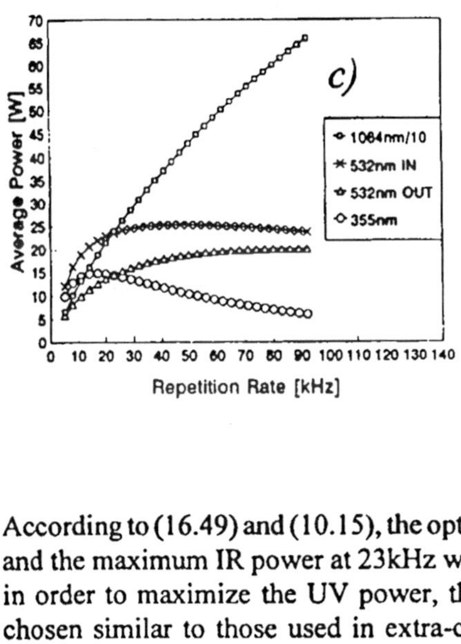

Fig. 16.21 Calculated third harmonic power, green power in front and after the SFG crystal and intracavity IR power in front of the SHG crystal for a Q-switched diode-pumped Nd:YVO_4 laser, with a resonator similar to the one shown in the previous figure. Pump power at 808nm: 50W, small-signal gain: g_0l=3.85, loss per transit $\alpha_0 l$=0.01, SHG crystal: LBO, d_2=0.83pm/V, length: 15mm, SFG crystal: LBO, d_3=0.66pm/V. IR beam diameter in SHG/SFG crystals: a)125µm/60µm, b)500µm/100µm, c)500µm/ 500µm. Flat-top spatial intensity profile, no walk-off. The IR resonator has a length of 0.5m, an optimum reflectivity of 68.9% and a maximum IR output power of 32W at 50kHz (34.2W in cw).

According to (16.49) and (10.15), the optimum green conversion for this laser is about 21%, and the maximum IR power at 23kHz would be 31W. This example clearly indicates, that in order to maximize the UV power, the beam diameters inside the crystals have to be chosen similar to those used in extra-cavity sum frequency generation. This is easy to understand because for an optimum output coupling of 50%, the intracavity IR power is only twice as high as the extra-cavity power. For common nonlinear crystals, the major advantages of intracavity third harmonic generation, i.e. increased crystal lifetime and insensitivity to walk-off, cannot be realized when the resonator is designed for maximum third harmonic power. For this reason, commercial intracavity third harmonic lasers are under-coupled with beam diameters inside the nonlinear crystals that are several times larger than the ones providing optimum coupling (see Fig. 16.21c). This results in IR to UV efficiencies that are close to the extra-cavity case (up to 50%), but with improved pulse-to-pulse stability, especially at high repetition rates (>100kHz). However, despite this superior performance, intracavity third harmonic generation is less common in solid state UV lasers, because of technical problems that are generated by the coupling of the fundamental and the harmonic intensities and the degradation of optical components due to UV irradiation.

16.4 Resonators with Phase-Conjugate Mirrors

16.4.1 General Properties of a Phase-Conjugate Mirror

Compared to a conventional mirror, the imaging properties of a phase conjugating mirror are quite different [4.232,4.242]. Whereas a conventional mirror reflects light rays according to the reflection law of geometrical optics, rays hitting a phase-conjugate mirror (PCM) are simply reversed, independent of the angle of incidence (Fig. 16.22a). If a spherical wave is incident on the PCM, with:

$$E(r,t) = \frac{E_0}{r} \exp[-i\omega t] \exp[i(kr+\phi)] \tag{16.50}$$

where r is the distance from the origin and ϕ is an arbitrary phase shift, the reflected wave converges back into its origin:

$$E_{ref}(r,t) = C \frac{E_0}{r} \exp[-i\omega t] \exp[-i(kr+\phi)] \tag{16.51}$$

where C is the amplitude reflectance of the PCM. The term phase conjugation now becomes obvious since the reflected wave is obtained by replacing the spatial phase term with its complex conjugate. If a plane wave experiences a phase distortion ϕ by passing through a medium, the PCM exactly reverses the phase and after the second transit through the medium the plane wave front is restored (Fig. 16.22b). Thus we can use a PCM to compensate the phase distortions induced by the active medium. The main application of PCMs is the compensation of thermal lensing in solid state laser materials (Fig. 16.23).

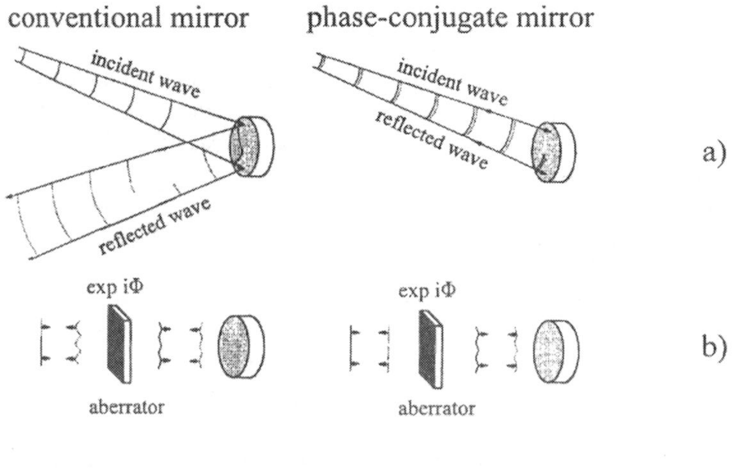

Fig. 16.22 Properties of a conventional mirror and a phase-conjugate mirror [S.19].

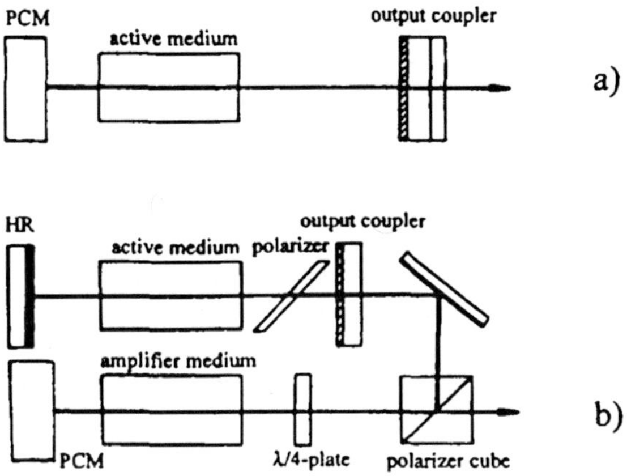

Fig. 16.23 Utilization of PCMs in laser systems. a) oscillator with phase-conjugate resonator mirror; b) PCM for compensating the phase distortions induced by the amplifiers.

Phase conjugation is present in all nonlinear optical processes since an intensity dependence of the index of refraction is sufficient to conjugate the phase. Phase conjugation has been extensively studied for three wave mixing, four wave mixing and stimulated scattering [4.232]. The most commonly used physical mechanism to realize PCMs is stimulated Brillouin scattering in gases and fluids. Before we discuss the physics of stimulated Brillouin scattering and the properties of PCMs based on this effect, let us first investigate how the characteristics of an optical resonator are affected by a phase-conjugate mirror.

16.4.2 Optical Resonators with a Phase-Conjugate Mirror

The basic properties of phase conjugate resonators can be evaluated by applying the ray transfer matrix formalism [4.213,4.216,4.222,4.223, 4.232]. If a Gaussian beam with a radius of curvature R and a beam radius w is incident on the PCM, its q-parameter q_1 reads:

$$\frac{1}{q_1} = \frac{1}{R} - i\frac{\lambda}{\pi w^2} \tag{16.52}$$

After reflection by the PCM the Gaussian beam must exhibit a conjugate phase, which means that the q-parameter q_2 of the reversed beam is given by:

$$\frac{1}{q_2} = -\frac{1}{R} - i\frac{\lambda}{\pi w^2} \tag{16.53}$$

Since the ABCD law $q_2=(Aq_1+B)/(Cq_1+D)$ must hold, the ray transfer matrix of the PCM is found to be:

$$M_{PCM} = \begin{pmatrix} A & B \\ C & D \end{pmatrix} = \begin{pmatrix} 1 & 0 \\ -2/R & 1 \end{pmatrix} \tag{16.54}$$

This is the well-known ray transfer matrix of a mirror with radius of curvature R. Therefore, a conventional mirror will act as a PCM, but only for incident fields whose phase curvatures match the curvature of the mirror. For a true PCM, phase conjugation takes place for arbitrary phase curvatures since the radius of curvature R of the PCM is induced by the field itself.

Using the Gaussian ABCD law for a resonator round trip, starting either on the PCM or the second, conventional mirror, one can derive the self-consistent Gaussian beams. For a resonator round trip, the Gaussian beams that are self-reproducing exhibit, like in conventional resonators, a wavefront that matches the radius of curvature ρ_1 of mirror 1. However, there is no constraint on the beam radius w_1 and an infinite number of Gaussian beams thus exist that represent eigensolutions of the phase-conjugate resonator (Fig. 16.24). For a Gaussian beam with radius w_1 on mirror 1, the following relations hold:

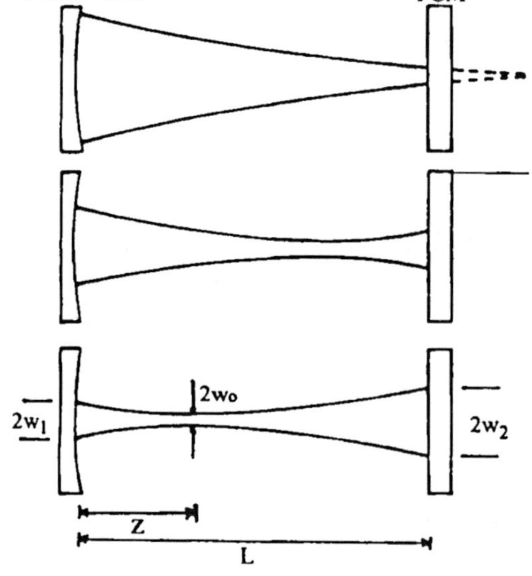

Fig. 16.24 Gaussian beams that exhibit a constant phase on mirror 1 are eigensolutions for the round trip in the resonator with an ideal phase-conjugate mirror 2.

Rayleigh range : $$z_0 = \frac{\pi w_1^2/\lambda}{1 + [\pi w_1^2/(\lambda \rho_1)]^2} \qquad (16.55)$$

beam waist radius : $$w_0 = \sqrt{z_0 \lambda / \pi} \qquad (16.56)$$

distance of waist
from mirror 1 : $$z = z_0 \sqrt{\pi w_1^2/(\lambda z_0) - 1} \qquad (16.57)$$

Using these relations, the radius of curvature R on the PCM and the beam radius w_2 can be easily obtained. The phase conjugation generates a freedom of stability constraints. Regardless of the mirror spacing and the curvature of mirror 1 we can always find self consistent Gaussian beams and the resonator will never become unstable. Another class of eigensolutions for the phase conjugate resonator can be found by applying the self-consistent field analysis to two round trips. It follows that electric fields exhibiting arbitrary amplitude and phase distributions on mirror 1 are exactly reproduced after two round trips, a result that can be easily verified by tracking a spherical wave through the resonator.

Which field distribution establishes itself in a real phase conjugate resonator depends on the physical process used for phase conjugation (e.g. the reflectance depends on the beam quality) and the size of apertures that are generally placed inside the PCR to control the transverse mode spectrum. It was shown both theoretically and experimentally that by using an apodized aperture with a Gaussian transmission (i.e. imposing a Gaussian reflectivity profile on the PCM, or generating a radial gain profile) only one self-consistent Gaussian beam is found [4.232]. Furthermore, the nonlinear nature of phase conjugation leads to transverse mode discrimination.

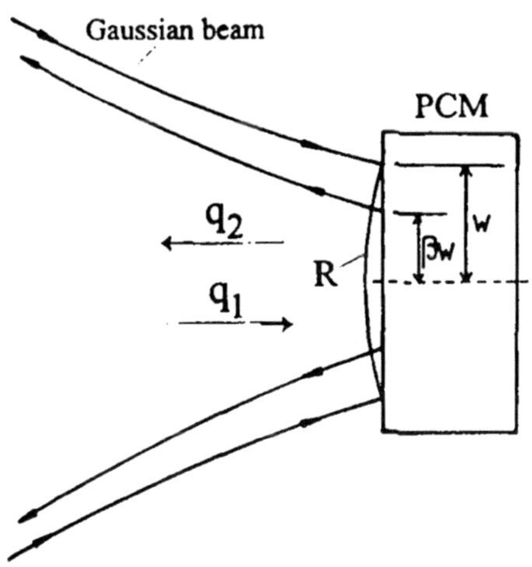

Fig. 16.25 A Gaussian beam is reflected by a real PCM. The beam radius w is reduced by a factor β.

Since the slopes of a beam incident on the PCM contain less power per unit area than its center, the reflectance of a PCM based on a nonlinear optical process is lower in the periphery of the beam. In addition to the reversal of the wavefront, a second characteristic of a real PCM, therefore, is the reduction of the beam diameter. For a Gaussian beam with beam radius w at the PCM (Fig. 16.25), it is reasonable to assume that the reflected beam is still Gaussian, with a beam radius that is reduced by a factor β (with $0<\beta<1$) [4.212,4.223]. The factor β is an empirical parameter that depends on the physical properties of the PCM and the intensity of the incident beam. The additional constraint on the beam size selects one well-defined Gaussian beam as the fundamental eigenmode of the phase conjugate resonator. Although the combination of wavefront reversal and beam size reduction is definitely a model too idealized to describe the properties of a real phase conjugate mirror, this description was found to be useful to theoretically predict the beam propagation inside phase-conjugate resonators (see Fig. 16.25).

In the following we will discuss this model in more detail [4.223]. According to Fig. 16.25, the q-parameters of the Gaussian beam with wavelength λ before and after reflection by the PCM are given by:

$$\frac{1}{q_1} = \frac{1}{R} - \frac{i\lambda}{\pi w^2} \tag{16.58}$$

$$\frac{1}{q_2} = -\frac{1}{R} - \frac{i\lambda}{\pi \beta^2 w^2} \tag{16.59}$$

If the PCM is used as a resonator mirror, the q-parameter q_2 must transfer into q_1 after one round trip. For a general resonator, the ray transfer matrix M for the round trip, starting at the PCM (but not incorporating it), is given by (Fig. 16.25):

$$M = \begin{pmatrix} A & B \\ C & A \end{pmatrix} \tag{16.60}$$

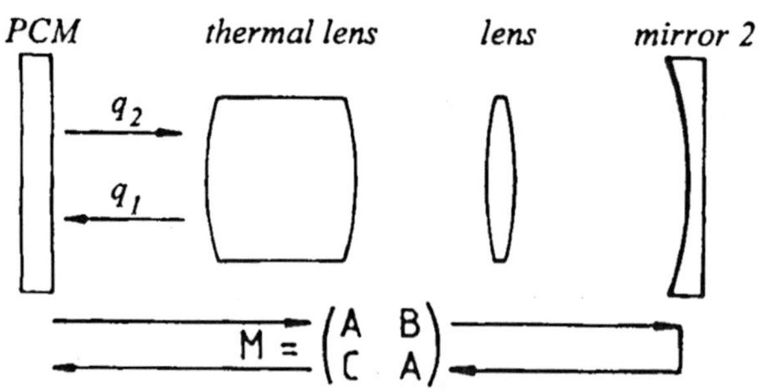

Fig. 16.26 Phase-conjugate resonator with round trip ray transfer matrix M, starting at the PCM. Arbitrary ABCD-type optical elements can be placed inside the resonator.

Application of the ABCD-law:

$$q_1 = \frac{Aq_2 + B}{Cq_2 + A} \tag{16.61}$$

yields for the wavefront curvature R and the beam radius w of the self-consistent Gaussian beam at the PCM (going in):

$$R = \frac{B}{A} \tag{16.62}$$

$$w = \sqrt{\frac{\lambda B}{\pi \beta}} \tag{16.63}$$

With the knowledge of the Gaussian beam parameter at the PCM, the beam diameter at any plane inside the resonator can be calculated by using the well-known propagation law of Gaussian beams. Let us assume that the ray transfer matrix for the transit from the PCM to the considered plane is given by:

$$M^* = \begin{pmatrix} A^* & B^* \\ C^* & D^* \end{pmatrix}$$

The beam radius w^* at this plane then reads:

$$w^{*2} = (A^*\beta w)^2 - 2A^*B^*\frac{\beta^2 w^2}{R} + \left(\frac{B^*\beta w}{R}\right)^2 + \left(\frac{B^*\lambda}{\pi\beta w}\right)^2 \tag{16.64}$$

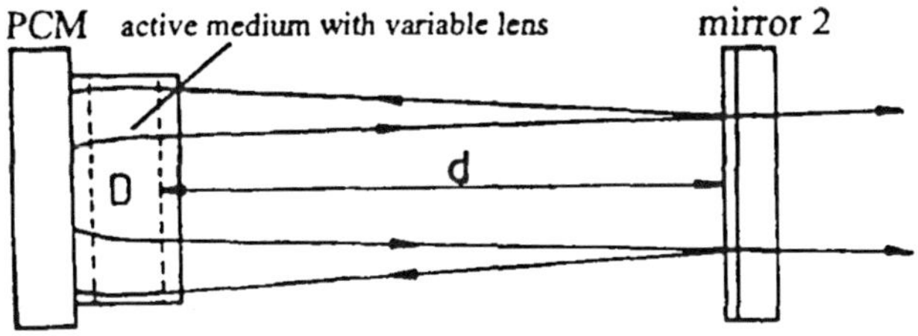

Fig. 16.27 Phase-conjugate resonator with an active medium exhibiting a variable refractive power D. The PCM is attached to the active medium. The broken, vertical lines indicate the principal planes of the lens. Mirror 2 is a flat output coupler.

In a preferred arrangement, the PCM is placed as close as possible to the lens medium as depicted in Fig. 16.27. For convenience we treat the active medium as a thin lens attached to the PCM. Starting at the PCM, the ray transfer matrix for the round trip reads:

$$M = \begin{pmatrix} 1-2dD & 2d \\ -2D(1-dD) & 1-2dD \end{pmatrix} \tag{16.65}$$

Equations (16.62) and (16.63) yield for the parameters of the Gaussian beam:

$$R = \frac{2d}{1-2dD} \quad , \quad w = \sqrt{\frac{2d\lambda}{\pi\beta}} \qquad (16.66)\ (16.67)$$

Application of (16.64) with $A^*=1-dD$ and $B^*=d$, yields for the beam radius w^* at the output coupling mirror:

$$w^* = \sqrt{\frac{d\lambda}{2\pi}\,\frac{1+\beta^2}{\beta}} \tag{16.68}$$

The beam radii w and w^* at the mirrors remain constant as the refractive power D of the lens is varied. In the thin lens approximation used, this means that the mode volume of the Gaussian beam in the active medium is constant as well. Regardless of the refractive power, the PCR is always stable. This can also be demonstrated by calculating the g-parameters of the equivalent lens resonator. The combination of the PCM and the variable lens is equivalent to a conventional mirror with radius of curvature:

$$\rho = \left(\frac{1}{R} + D\right)^{-1} \tag{16.69}$$

Thus the equivalent g-parameters of the two mirrors (mirror 1 is the PCM-lens combination) are given by:

$$g_1^* = 1 - d(1/R + D) = \frac{1}{2} \tag{16.70}$$

$$g_2^* = 1 \tag{16.71}$$

The equivalent resonator is semi-confocal and remains at the same location in the stability diagram. Unfortunately, this resonator scheme only provides a relatively small Gaussian beam radii in the active medium (for $d=1m$ and $\lambda=1\mu m$, the beam radii are on the order of 1mm). Larger Gaussian beam radii can be attained by separating the active medium from the PCM and incorporating a negative lens into the resonator (see Fig. 16.35). The separation may lead to a dependence of the Gaussian beam radii on the refractive power. However, the resonator always remains stable since the equivalent resonator remains in a stable zone. This can be shown for a general resonator set-up by using the ray transfer matrix (16.60) for a round trip. As was shown in Sec. 1.3, the ray transfer matrix for a resonator round trip starting at mirror 1 reads:

$$M_{RT} = \begin{pmatrix} 2g_1^* g_2^* - 1 & 2L^* g_2^* \\ \dfrac{(2g_1^* g_2^* - 1)^2 - 1}{2L^* g_2^*} & 2g_1^* g_2^* - 1 \end{pmatrix} \tag{16.72}$$

where g_1^*, g_2^* are the equivalent g-parameters, and L^* is the equivalent length. The round trip matrix for the phase-conjugate resonator depicted in Fig. 16.26 is given by (using (16.54),(16.60), and (16.62):

$$M_{RT} = \begin{pmatrix} 1 & 0 \\ -1/R & 1 \end{pmatrix} \begin{pmatrix} A & B \\ C & A \end{pmatrix} \begin{pmatrix} 1 & 0 \\ -1/R & 1 \end{pmatrix} = \begin{pmatrix} 0 & B \\ -1/B & 0 \end{pmatrix} \tag{16.73}$$

A comparison with (16.72) leads to the generalized stability condition for phase conjugate resonators:

$$g_1^* g_2^* = \frac{1}{2} \tag{16.74}$$

The first operation of a phase-conjugate resonator was reported in 1979 using a pulsed ruby laser and four-wave-mixing in a cell containing CS_2[4.213]. The CS_2 cell, which acts as the PCM, was pumped by two counterpropagating waves generated by a second ruby laser. Phase-conjugate resonators were also implemented in dye lasers and argon lasers using four wave-mixing or self-pumped photorefractive crystals [4.234]. In recent years, solid state lasers, excimer lasers, and dye lasers with intracavity stimulated Brillouin scattering (SBS) as the phase-conjugate process have been successfully operated [4.234-4.237,4.244-4.273] (Fig. 16.28).

Average output powers of tens of Watts have been demonstrated using Nd:YAG lasers utilizing SBS in liquids (CCl_4, Acetone, CS_2) and gases (CH_4, SF_6). An average output power of 27 W and 50W have been reported for Nd :YAG and Nd:YALO laser oscillators, respectively, both with a phase-conjugate mirror using SBS in SF_6 [4.262,5.268, 4.272].

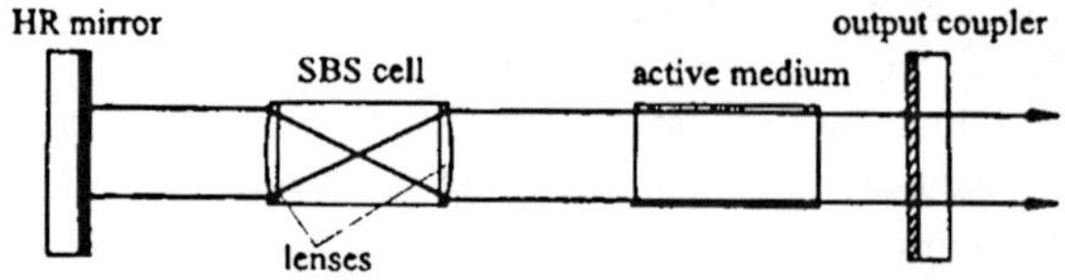

Fig. 16.28 A phase-conjugate resonator employing a SBS cell. Laser oscillation first occurs between the two conventional mirrors until the SBS threshold is exceeded.

The advantage of SBS is its self-pumping nature; no additional pump waves are required to initiate phase-conjugation. Unfortunately, the PCMs based on SBS exhibit a high threshold power on the order of 10 kW, which means that the phase-conjugate resonators can only be operated in a Q-switch mode with typical pulse lengths on the order of 10-100ns. Lower threshold powers are possible by using multiple SBS cells or tapered fibers [4.274].

16.4.3 Phase-Conjugate Resonators using SBS

Brillouin scattering is caused by the interaction of the incident field with the periodic density variations generated by a sound wave (Fig.16.29). For a monochromatic field with wavelength λ_0 incident at an angle θ with respect to the propagation direction of the sound wave, maximum scattering is observed if the fields scattered at the different density maxima interfere constructively (Bragg condition):

$$2\ \Lambda\ \sin\theta = \lambda_0 \tag{16.75}$$

where Λ is the wavelength of the sound wave.

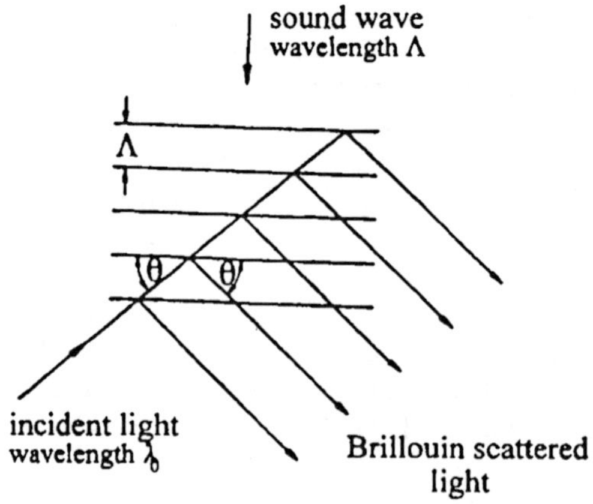

Fig. 16.29 Brillouin scattering of light by a sound wave.

Since the sound wave propagates with a speed V, the frequency of the scattered light is shifted by the Brillouin shift $\Delta\nu_B$ with respect to the incident frequency ν_0 due to the Doppler effect:

$$\Delta\nu_B = \pm\, \nu_0\, 2\, \frac{V}{c}\, \sin\theta \tag{16.76}$$

where c is the speed of light in the medium and the upper sign holds for sound waves propagating towards the incoming wave vector. Since the speed of sound V is on the order of 100 m/s in gases and 1,000 m/s in liquids, the relative frequency shift ν/ν_0 is very small (10^{-6}-10^{-5}). The sound waves are a manifestation of the thermal energy in the medium which means that all propagation directions and wavelengths down to a cut-off wavelength are present. Therefore, the incident electric field is scattered under all angles and the frequency is broadened and shifted according to (16.67). This scattering process is referred to as *spontaneous Brillouin scattering.*

If the intensity of the incident light is sufficiently high, the sound wave that scatters light is produced by the light itself (*stimulated Brillouin scattering (SBS)*). The physical process involved is electrostriction, the variation in volume due to an applied electric field. For SBS, the frequency of the scattered light is always reduced so that only the negative sign in (16.76) applies. If I_{in} denotes the intensity incident upon the SBS medium and I_{out} is the scattered intensity, the following coupled equations hold for the change of the intensities along the propagation distance z in the medium (assuming plane waves, a stationary sound wave, and pulse durations greater than the decay time of the sound wave):

$$\frac{dI_{in}}{dz} = -g_B\, I_{in}\, I_{out} - \alpha\, I_{in} \tag{16.77}$$

$$\frac{dI_{out}}{dz} = -g_B\, I_{in}\, I_{out} + \alpha\, I_{out} \tag{16.78}$$

with g_B : gain coefficient [cm/GW]
α : absorption loss coefficient [1/cm]

The gain coefficient, a characteristic of the medium (Table 16.6), is proportional to the frequency shift $\Delta\nu_B$ and thus maximum for backscattering ($\theta=\pi$). If we assume that the reflectivity generated by Brillouin scattering is low and the medium is loss-free, the incident intensity can be considered as being constant inside the medium. Equation (16.78) then yields for the reflected intensity:

$$I_{out} \sim \exp[g_B\, I_{in}\, s] \sim G \tag{16.79}$$

where s is the interaction length within which stimulated Brillouin scattering occurs.

Since the reflected wave is amplified with the gain factor G and the gain coefficient is maximum for backscattering, the incident wave with frequency ν_0 is reversed and exhibits the frequency:

$$\nu_B = \nu_0 \left(1 - 2\frac{V}{c}\right) \tag{16.80}$$

Although the gain factor, according to (16.79), seems to be a function of the incident intensity, the reflectance of a real SBS cell depends on the power P_{in} of the input beam since the beam is focused into the cell. This becomes understandable if we consider an incident beam with a radius w and a beam propagation factor M^2 focused into the SBS medium to a spot radius w_0 (Fig. 16.30). If the incident power is not too high, efficient scattering can only occur within an interaction length s that is x times larger than the Rayleigh range z_0 (Fig. 16.33) [4.255]:

$$s = x\,\frac{\pi w_0^2}{\lambda_0 M^2} \tag{16.81}$$

where the factor x depends slightly on the intensity. By using the relation $I_{in} = P_{in}/(\pi w_0^2)$ and inserting (16.81) into (16.79), the reflectance R of the SBS cell reads:

$$R = \frac{I_{out}}{I_{in}} \sim \exp\left|\frac{x\, g_B\, P_{in}}{\lambda_0 M^2}\right| \tag{16.82}$$

The pump power required to attain a reflectivity of the SBS cell in the percent range is defined as the threshold pump power, with:

$$P_{th} = \frac{C}{x}\,\frac{\lambda_0 M^2}{g_B} \tag{16.83}$$

where the ratio C/x was determined experimentally to be on the order of 10 [S.19].

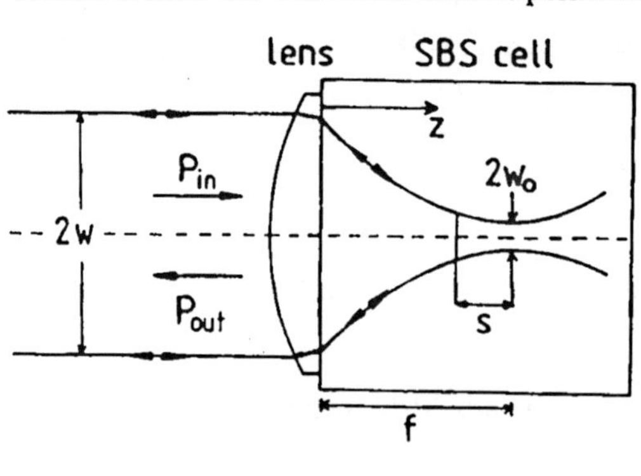

Fig. 16.30 Focusing of a laser beam into a SBS cell.

Keep in mind that the term "threshold" is physically not correct since SBS is also present at low pump powers, but with negligible efficiency. Typical threshold powers are in the range of 10-100kW which means that pulsed lasers are needed to reach noticeable reflectances. If a pulse duration of 30ns is used, the threshold pump power of 100kW corresponds to a beam energy of 3mJ. Note that the threshold becomes higher for higher order transverse modes. If the SBS cell is used as a resonator mirror, it can provide transverse mode discrimination such that the fundamental mode exhibits the lowest loss. The threshold defined by (16.83) holds only for cw-irradiation. If the pulse duration of the incident beam is on the same order of magnitude as the phonon lifetime τ_B, the threshold pump power can be several times higher. Table 16.6 lists gain coefficients, phonon lifetimes and threshold powers (16.83) for different SBS media. Measured reflectances for different SBS media are presented in Fig. 16.31.

Table 16.6 Gain coefficient g_B, phonon lifetime τ_B, and threshold power P_{th}, according to (16.83), of different SBS media ($\lambda_0=1\mu m$, $M^2=1$, $C/x=10$) [4.232,4.242,4.266,S.19,S.20].

Medium	g_B[cm/GW]	τ_B [ns]	P_{th} [kW]
H_2O	5	---	200
CCl_4	6	0.6	167
$C_2Cl_3F_3$	6	0.84	167
Methanol	13	0.13	58
Benzene	13	---	58
SF_6	14	17.3	71
C_2F_6	18	6.6	55
Acetone	20	2.1	50
CO_2	20	10	50
CS_2	130	4.9	8

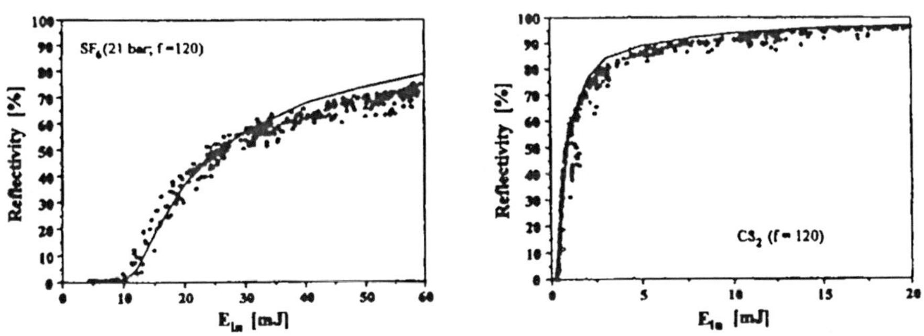

Fig. 16.31 Measured energy reflectance of two different SBS media [4.266]. The beam of a single longitudinal mode Q-switched Nd:YAG laser in fundamental mode operation was focused into the cells using a focal length of f=120mm. The minimum spot diameter in the cells was 100μm, the duration of the laser pulses was 25ns (FWHM), and the linewidth was 100MHz. The threshold input powers are: SF_6: 440kW; CS_2: 14kW. The solid lines are theoretical curves calculated with the model described in [4.255].

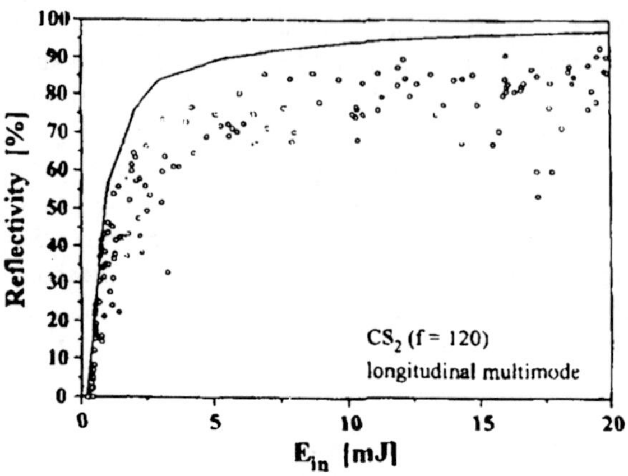

Fig. 16.32 Measured energy reflectance of CS_2 for a small coherence length of the incident Nd:YAG laser beam [4.266]. The same laser as in Fig. 16.31 was used, but with operation in longitudinal multimode. Compare this graph with the lower left graph of the previous picture to see the influence of the coherence length on the SBS reflectance. The solid line is the theoretical reflectance for single longitudinal mode operation, calculated with the model described in [4.255].

An important requirement for efficient SBS is a coherence length of the incident beam that is much greater than the interaction length s in the SBS cell (Fig. 16.32). This is necessary since the sound wave grating is generated by the interference of the incident and the reflected light wave. To accomplish this, the beam can either be focused tighter into the SBS cell or the coherence length can be increased with interference filters.

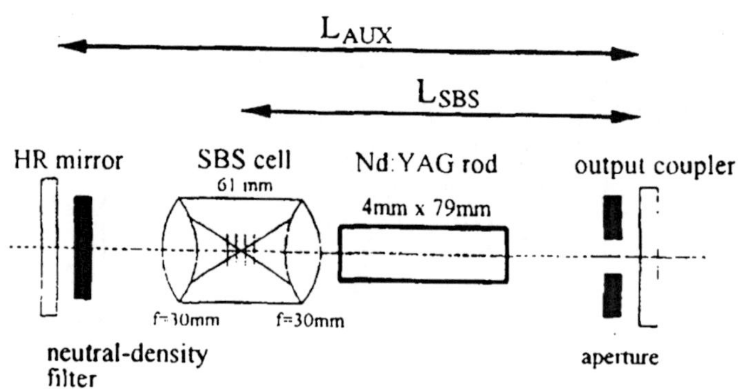

Fig. 16.33 A pulsed Nd:YAG laser with a SBS cell containing SF_6 at a pressure of 20 bar. The system uses an auxiliary resonator to reach the SBS threshold. An average output power of 10W in a near diffraction limited beam was obtained (pump pulse duration: 0.2ms, pump energy per pulse: 42J, L_{AUX}=1.8m, L_{SBS}=1.2m) [4.252] (© OSA 1992).

If the SBS cell is to replace a conventional mirror in a laser resonator it is necessary to use an auxiliary resonator to reach and exceed the threshold of the SBS cell. The necessary intracavity powers on the order of 10kW can be attained by either placing a Q-switch into the auxiliary resonator [4.211,4.218,4.244], or by using the Q-switching properties of the SBS cell itself [4.215,4.237,4.250,4.268,4.272]. An SBS Nd:YAG laser using the latter technique is depicted in Fig.16.33 [4.252]. The laser rod is pumped in a pulsed mode with a pump pulse duration of 0.2ms and a pump energy of 42J. At a repetition rate of 45Hz this phase-conjugate laser resonator provides a near diffraction limited output beam with an average output power of 10W. Between two and three pulses with a FWHM duration of 30ns were observed per pump pulse. Compared to a conventional resonator in multimode operation with the same pumping conditions, the efficiency was 65%. The neutral-density filter is needed to prevent stable laser oscillation in the auxiliary resonator (the filter can be removed if a partially reflecting rear mirror is used). By using a pump cavity that generated a peaked radial gain profile in the Nd:YAG rod, fundamental mode operation was also observed without the mode-control aperture.

In order to achieve efficient operation of a SBS laser resonator it is crucial to meet two design criteria. The optical length L_{SBS} of the SBS resonator and the optical length L_{AUX} of the auxiliary resonator have to be matched so that both resonators share axial mode frequencies. This ensures a smooth transition when the laser oscillation switches from the auxiliary resonator to the SBS resonator. Furthermore, the Brillouin shift should be multiples of the frequency spacing between the axial modes of the SBS resonator. This leads to the following two design equations:

$$\Delta\nu_B = n\,\frac{c_0}{2L_{SBS}} \quad , \quad n = 1,2,3,... \tag{16.84}$$

$$\Delta\nu_B = m\,\frac{c_0}{2L_{AUX}} \quad , \quad m = 2,3,4,... \tag{16.85}$$

where c_0 is the speed of light in vacuum. For SF_6 at 20 bar, the Brillouin shift is 250MHz. The shortest resonator (n=1,m=2) exhibits the effective lengths L_{AUX}=1.2m and L_{SBS}=0.6m. In addition, there is experimental evidence that the transverse mode structure of the two resonators should be matched too. This can be accomplished by an appropriate choice of the curvature of the high reflecting mirror. In addition, the auxiliary resonator must be stable over the whole refractive power range of the active medium, since the thermal lens is not compensated until the SBS cell becomes reflective.

At present, Nd:YAG lasers employing SBS resonator mirrors provide average output powers in excess of 50W in a near diffraction-limited output beam [4.272]. In order to achieve a high efficiency in fundamental mode operation it is necessary to adapt the Gaussian beam diameter to the diameter of the active medium. Although SBS oscillators have to be chosen long to meet the design criteria (16.84) and (16.85), the Gaussian beam fills only a fraction of the active volume. By inserting a negative lens into the resonator, Gaussian beam diameters in excess of 5mm can be realized in the active medium as the measured and the calculated beam diameters in Fig. 16.34 illustrate [4.267]. The resonator set-up depicted meets (16.84) and (16.85) with n=2 and m=3 and a Brillouin shift of 250MHz (SF_6 at 20 bar). Note that the measured beam diameters in the medium are larger than predicted by the Gaussian beam propagation model described in (16.58)-(16.64), a discrepancy that cannot be explained satisfactorily by gain saturation.

Figure 16.35 presents results reported for a flashlamp-pumped phase-conjugate Nd:YAG laser oscillator using SF_6 at 20 bar as the SBS medium. This laser provides a maximum average output power of 27W in TEM_{00} mode operation [4.262,4.268]. Figure 16.35a shows one of the resonator designs used including the calculated fundamental mode beam diameter using the Gaussian beam propagation model (Eqs. (16.58)-(16.64)) in the limit β-->1. By using a slightly modified resonator scheme, a maximum output power of 27W was achieved at an electrical pump power of 3.9kW and a repetition rate of 50Hz, corresponding to a total efficiency of 0.7% (Fig. 16.35b).

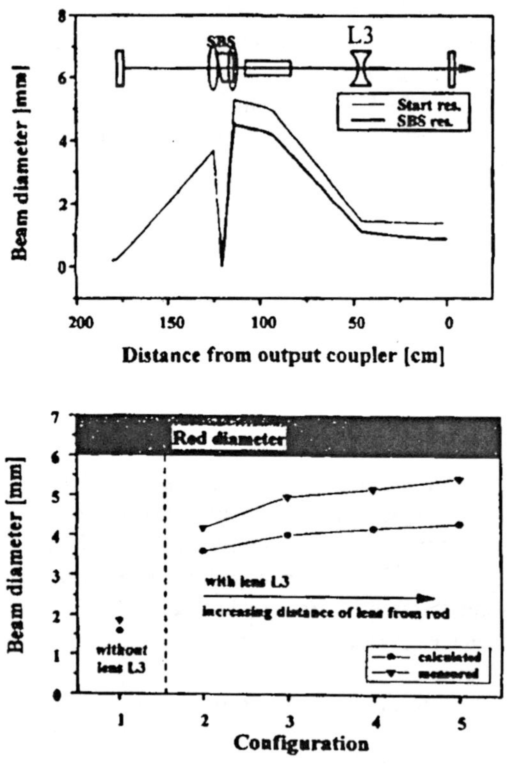

Fig. 16.34 Measured and calculated beam diameters for different phase-conjugate resonators using SF_6 at 20 bar as the SBS medium. The different configurations differ in the position of the -200mm lens L3. The upper graph depicts configuration 5. The beam diameters of the SBS resonator were calculated with (16.64) in the limit β-->1 (pump power: 390W, focal length of the rod: 84cm) [4.267].

The modulation of the temporal intensity profile of one laser pulse is due to the Brillouin shift of 250 MHz resulting in periodic spikes with a spacing of 4ns (Fig. 16.35c). A measurement of the beam quality using a moving knife edge (see Sec. 24.1) yielded a beam propagation factor of M^2=1.2 over the whole output power range.

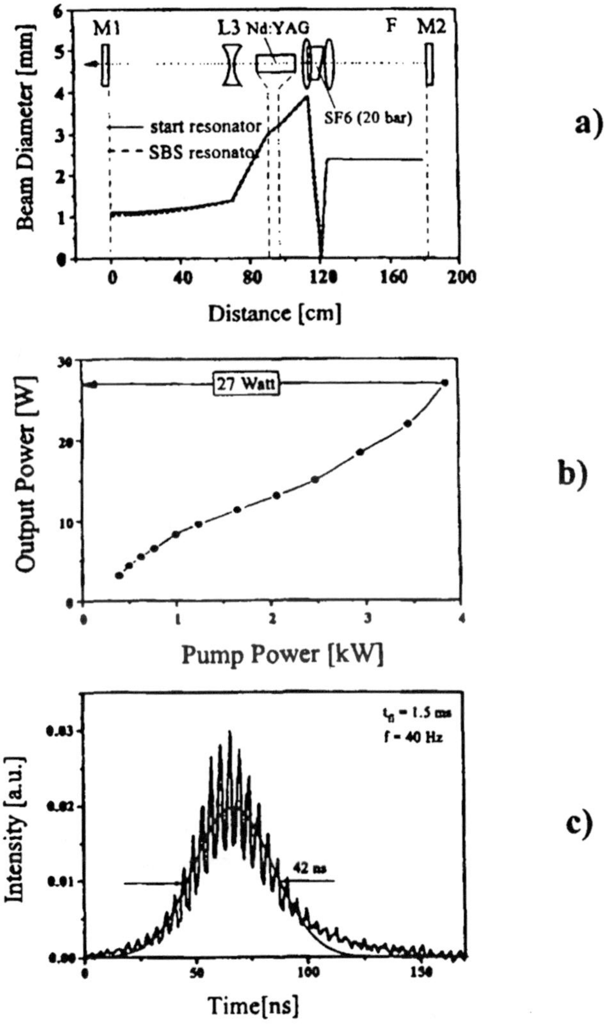

Fig. 16.35 A pulsed, phase-conjugate Nd:YAG rod laser using SBS in SF_6. (rod diameter: 4mm, rod length: 79mm, refractive power of rod: 1.1 diopter per kW of pump power, pump pulse duration: 1.5ms, pump energy: 77J, repetition rates: 10-50Hz). a) one of the resonator set-ups used showing calculated beam diameters, b) measured average output power as a function of the pump power, c) temporal emission profile of one laser pulse at a pump power of 2.8kW [4.262,4.268] (© IEEE 1998).

Higher output powers in the multi-100W range have already been realized by using the SBS cell in amplifier chains [4.259,4.263-4.265,4.273]. The Gaussian beam of a low power Q-switch oscillator with average output powers in the Watt range is used to generate high powers via amplification in one, two or more amplifier rods. In this arrangement the SBS cell is used to compensate for phase distortions induced by the amplifiers. The set-up depicted in Fig. 16.36 provides average output powers between 20 W and 215 W with beam propagation factors M^2 between 1.7 and 2.8 [4.264].

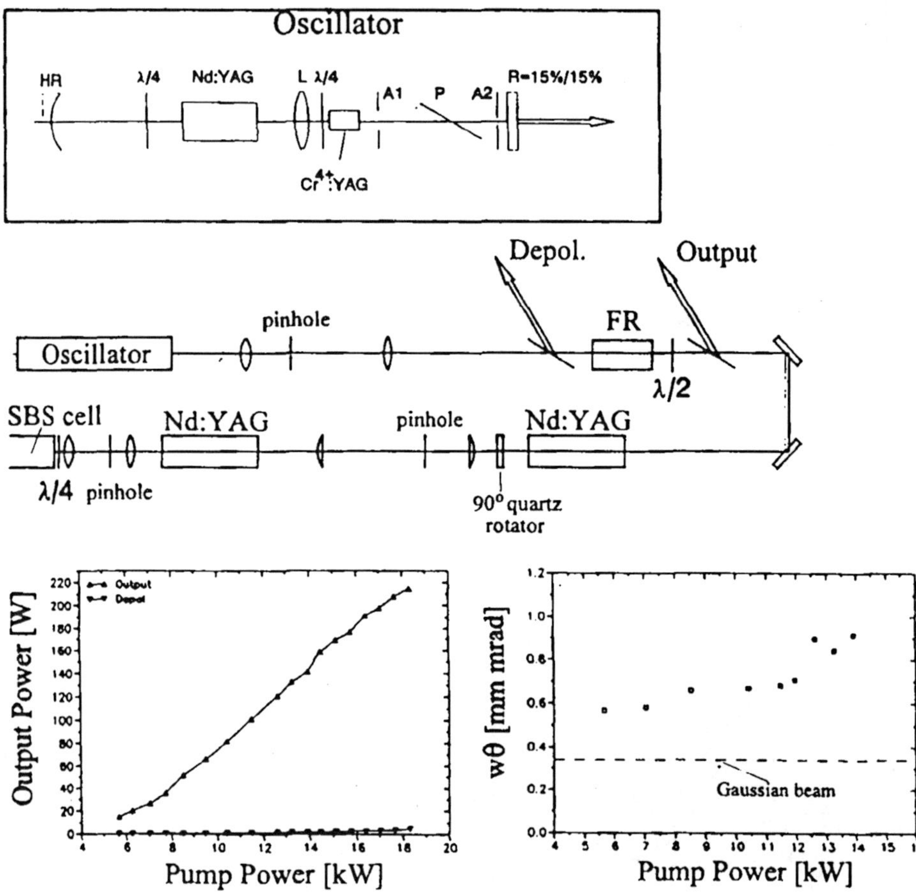

Fig. 16.36 Pulsed Nd:YAG laser with two amplifiers and an SBS cell containing CS_2. The upper graphs show the laser set-up including a schematic of the Q-switch oscillator (FR: Faraday rotator, P: polarizer, A: aperture, L: lens, λ/4: quarter wave plate, λ/2: half-wave plate, rod dimensions: 6"x3/8"). The oscillator, with Cr^{4+}:YAG as a passive Q-switch, provides 20 Q-switch pulses per pump pulse with an average output power of 3W (pump pulse duration: 0.5ms, pump pulse repetition rate: 100Hz, laser pulse duration: 200ns). The lower graphs present the average output power and the beam parameter product (waist radius w x half angle of divergence θ, 86.5% power content) as a function of the total electrical pump power [4.264] (© SPIE 1996).

Part V

Special Resonator Concepts

Chapter 17

Prism Resonators

17.1 Porro Prism Resonators

Stable resonators as well as unstable resonators exhibit a relatively high sensitivity to mirror misalignment. Typical tilt angles at which the output power has decreased by 10% are in the range between 0.1mrad and 1mrad. In some applications these angles may be too low to ensure a reliable performance over a long period of time. The misalignment sensitivity can be decreased considerably by using prisms as the resonator mirrors. For a right-angle prism (Porro prism), an incident ray is reflected back parallel to its propagation direction, independent of the angle of incidence (Fig.17.1). A planar wave front remains planar after the reflection by the prism. Except for the mirroring at the prism edge and the change in polarization, a Porro prism exhibits the same imaging properties as a flat, high-reflecting mirror. It is for this reason that high-reflecting resonator mirrors can be replaced by Porro prisms [5.5,5.7,5.8,5.13-5.15]. Mirror curvatures can be simulated by using a curved prism interface. If one prism is used in the resonator set-up, the laser beam is coupled out conventionally through the output coupling mirror (Fig.17.2a). With both resonator mirrors replaced by roof prisms, the output coupling is realized by using a retardation plate in combination with a polarizer (Fig. 17.2b). The rotation of the retardation plate leads to a variation of the output coupling (see also Chapter 8), which means that the output coupling can be adjusted to its optimum value. The decrease in misalignment sensitivity, however, is only attained for a tilt around the roof edge. A tilt of the prism around the perpendicular axis results in the same sensitivity as that of a flat mirror. In order to attain a low misalignment sensitivity with respect to arbitrary tilt axes, a corner cube prism has to be used.

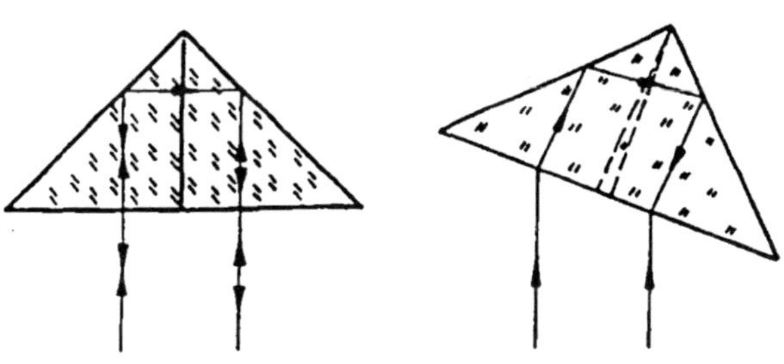

Fig. 17.1 Reflection of a light ray by a Porro prism.

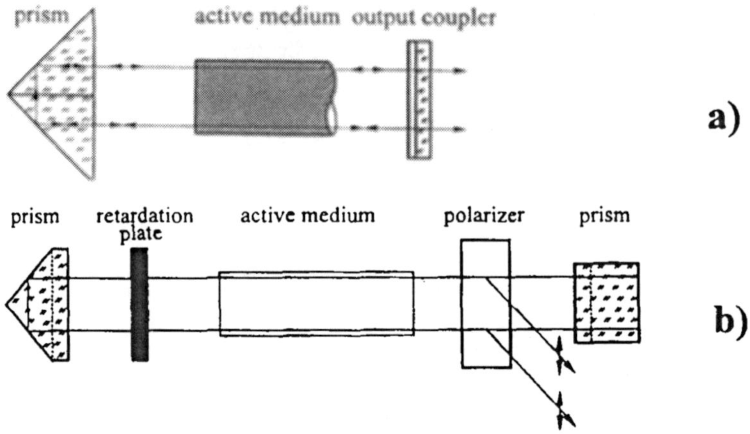

Fig. 17.2 Porro prism resonators with one (a) and two prisms (b). If two prisms are used, the radiation is coupled out of the resonator by means of a polarizer and a rotatable retardation plate.

The beam quality of the laser resonator is not affected by the prism, which means that the same beam parameter product is attained if the prism is replaced by a conventional resonator mirror (Fig. 17.3a). However, depending on the optical quality of the prism, the diffraction losses may be slightly increased due to output coupling and phase distortions induced at the prism edge (Fig. 17.3b). For high quality prisms, which exhibit edge widths of less than 5μm, the additional loss per round trip is less than 0.5%. In this case, the extraction efficiency is as high as for a conventional resonator provided that the small-signal gain is chosen high enough.

The misalignment of the Porro prism around an axis parallel to the prism edge results in a decrease of the mode volume in the active medium (Fig. 17.4). For a resonator in multimode operation, the smaller mode volume is the only cause for the decrease of the output power, since additional diffraction losses are only generated if the fundamental mode gets clipped by the active medium. A simple geometrical analysis yields for the angle $\beta_{10\%}$, at which the mode volume, and consequently the output power, have decreased by 10%:

$$\beta_{10\%} = \frac{0.025\ \pi\ b}{d - h(1-1/n)} \tag{17.1}$$

with:

- b : radius of the active medium
- d : distance of the tilt axis to the prism edge
- h : prism height
- n : refractive index of prism material

For a radius b of the active medium of 3.15mm and a glass prism (n=*1.5*) with a height of h=*20mm*, the 10%-angle is 37mrad if the prism is rotated around the roof edge (d=*0*).

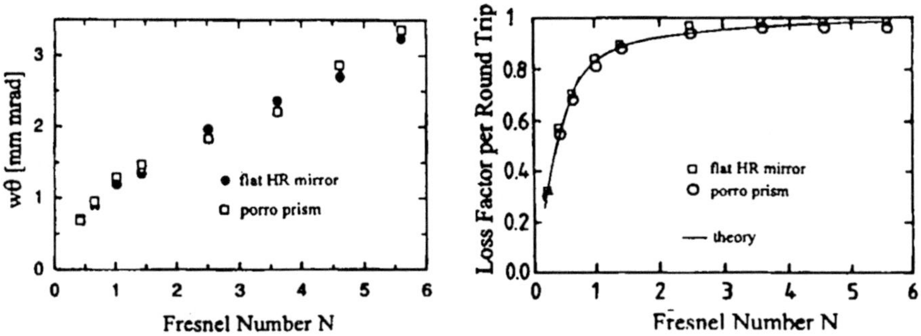

Fig. 17.3 Measured beam parameter product and measured loss factor per round trip for a Nd:YAG rod laser with a flat-flat resonator as a function of the effective Fresnel number N. Measurements with a flat HR mirror and a Porro prism are compared. The output coupling mirror is a conventional flat mirror (see Fig. 17.2a). The effective Fresnel number is given by $N = b^2/(2\lambda L)$ with b: rod radius, L: effective resonator length (b=3.15mm, λ=1.064μm). The theoretical loss factor was calculated using diffraction theory.

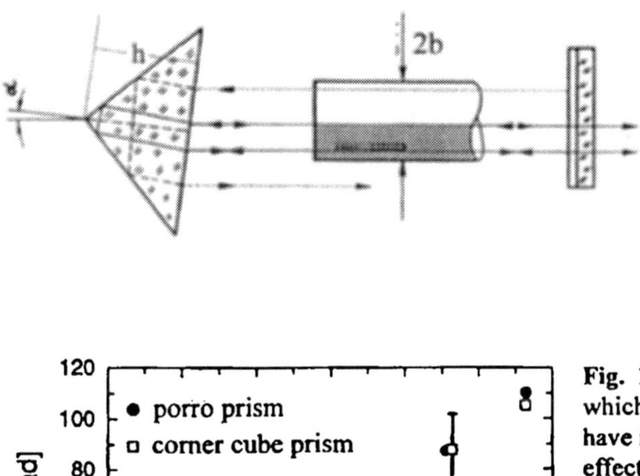

Fig. 17.4 Beam propagation in a misaligned Porro prism resonator. The hatched area indicates the mode volume.

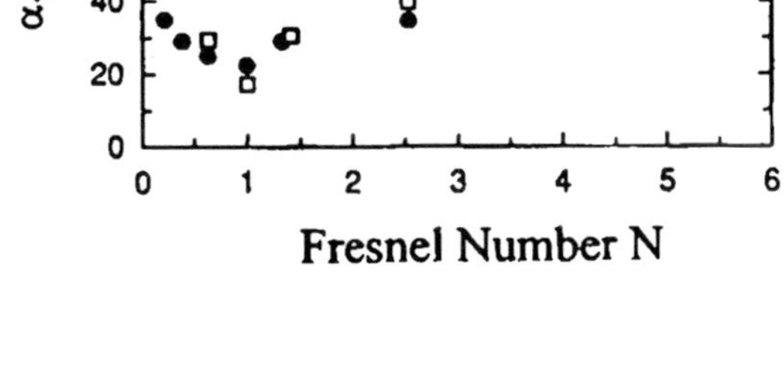

Fig. 17.5 Measured prism tilt angles at which the diffraction losses per round trip have increased by 10% as a function of the effective Fresnel number for a flat-flat resonator with one prism (Porro prism and corner cube prism are compared). The resonator set-up is the one depicted in Fig. 16.2a (Nd:YAG rod laser, rod radius b=3.15mm, λ=1.064μm). The effective resonator length is 0.75m, the Fresnel number was varied by using an intracavity aperture. The prisms are rotated around the edge (d=0).

The sensitivity to prism misalignment is only low if the prism or the output coupling mirror is rotated around an axis parallel to the roof edge. In order to extend the low sensitivity to both axes, two crossed Porro prisms may be used, as depicted in Fig. 17.6. The output coupling is accomplished by a polarizer in combination with a retardation plate (see also Sec. 8.2.2). The output coupling loss can be calculated by determining the eigenvalues of the Jones matrix for the resonator round trip (see Sec. 1.3). The Jones matrix of a Porro prism is identical to that of a retardation plate since the p-polarized and the s-polarized light exhibit different phase shifts δ_P and δ_S, respectively, due to total internal reflection. According to the Fresnel equations, the net phase shift $\delta=\delta_P-\delta_S$ after two reflections in a Porro prism with index of refraction n is given by:

$$\tan\frac{\delta}{4} = \sqrt{1 - \frac{2}{n^2}} \tag{17.2}$$

Note that the index of refraction must be greater than $\sqrt{2}$ in order to attain total internal reflection at the angle of incidence of 45°. For a BK7 glass prism (n=1.5) the net phase shift amounts to δ=73.7°. Furthermore, an additional phase shift of π has to be taken into account due to the reversal of the beam propagation. The Jones matrix for a Porro prism thus reads:

$$M_{PP}^{P} = \begin{pmatrix} 1 & 0 \\ 0 & \exp[i\Phi] \end{pmatrix} \tag{17.3}$$

with: $\Phi = \delta - \pi$ if the prism edge is parallel to the y-direction

$\Phi = \pi - \delta$ if the prism edge is parallel to the x-direction

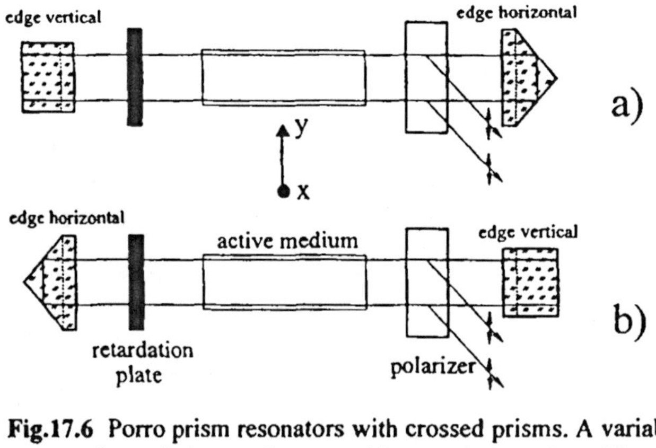

Fig.17.6 Porro prism resonators with crossed prisms. A variable output coupling is achieved by means of a polarizer and a rotatable retardation plate. The p-polarization passes through the polarizer and becomes elliptically polarized after one round trip. The s-polarized portion of the radiation is then reflected by the polarizer.

The resulting Jones matrix yields p-polarized and s-polarized light as the eigenvectors for a round trip. The loss factor V per round trip (1-loss) is given by the square of the eigenvalue for p-polarized light. Provided that the active medium does not affect the polarization, the loss factor reads:

$$V = \left| \left(\cos^2\alpha + \sin^2\alpha \exp[-i\gamma]\right)\exp[i\Phi] + \cos^2\alpha\sin^2\alpha(1-\exp[-i\gamma])^2 \right|^2 \tag{17.4}$$

where α and γ are the angle of rotation and the phase shift of the retardation plate, respectively, and Φ is the phase shift of the left Porro prism (see Fig. 17.6), according to (17.3). Figure 17.7 presents the loss factor per round trip of a Porro prism resonator as a function of the angle of rotation of the retardation plate, calculated with (17.4) using different phase shifts of the retardation plate. The output coupling can be optimized by adjusting the angle of rotation of the retardation plate such that the maximum output power is extracted for a given small-signal gain $g_0\ell$. This adjustment is the second advantage of this resonator scheme, besides the low misalignment sensitivity. Conventional resonators provide optimum output coupling for one fixed value of the small-signal gain only. The loss factor V can be considered as the output coupling reflectance in a conventional resonator. We can therefore use (10.8) in combination with (17.4) to calculate the output power of a Porro prism resonator as a function of the parameters of the retardation plate for a homogeneously broadened active medium:

$$P_{out} = A_b\, I_S \frac{1-V}{1 - V + \sqrt{V}\,(1/V_S - V_S)} \left[g_0\ell - \left|\ln\sqrt{VV_S^2}\right| \right] \tag{17.5}$$

where A_b is the cross sectional area of the mode in the active medium, I_S is the saturation intensity of the active medium, and V_S is the loss factor per transit due to scattering, absorption, and diffraction. The angles of rotation which provide maximum output power are determined by the equation:

$$\ln V = 2\ \ln V_S \left[\sqrt{\frac{g_0\ell}{|\ln V_S|}} - 1 \right] \tag{17.6}$$

in combination with (17.4). In general, the prism resonator will provide optimum performance at several angles of rotation. An experimental verification of (17.4) is presented in Fig. 17.8 with the corresponding output energy characteristics shown in Fig. 17.9. Note that the difference between a horizontal and a vertical orientation of the left prism is a 90° shift of both the loss factor curve and the output power curve with respect to the angle of rotation of the retardation plate. The right prism does not affect the output coupling at all and may be replaced by a conventional high reflecting mirror.

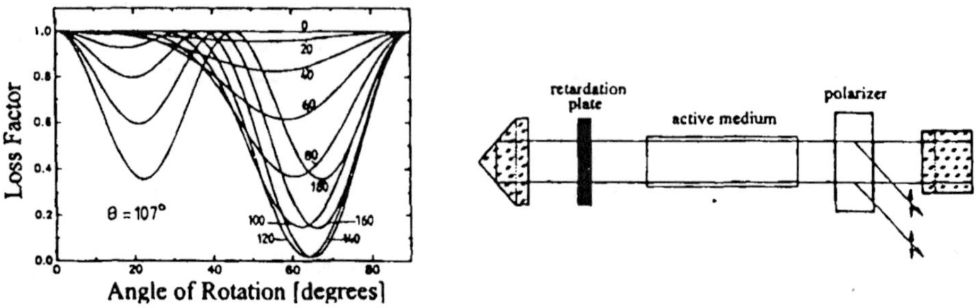

Fig. 17.7 Loss factor per round trip of the depicted Porro prism resonator as a function of the angle of rotation of the retardation plate, calculated with (17.4). The curve parameter is the phase shift γ of the retardation plate. The phase shift of the left Porro prism is Φ=107° (BK7).

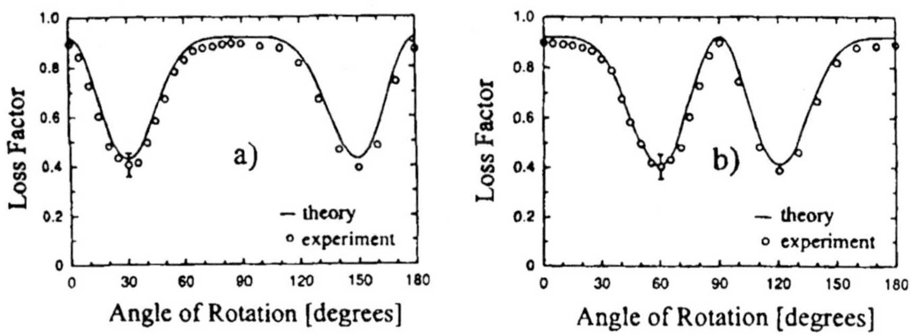

Fig. 17.8 Calculated and measured loss factor per round trip of the prism resonators shown in Fig. 17.6 as a function of the angle of rotation of the retardation plate. The phase shift of the retardation plate is γ=73° and the phase shift of each Porro prism is 73.7°. (Nd:YAG laser with 3"x1/4" rod, single shot operation, effective resonator length $L=0.75m$, 10% diffraction loss per round trip) [5.21].

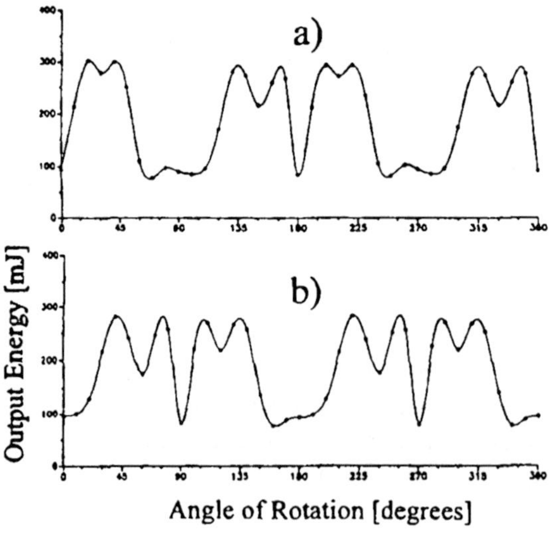

Fig. 17.9 Measured output energy per pulse for the two Porro prism resonators shown in Figs. 17.6 and 17.8 as a function of the angle of rotation of the retardation plate (Nd:YAG laser with 3"x1/4" rod, single shot operation, electrical pump energy : 100J, small-signal gain $g_0\ell=1.0$, effective resonator length $L=0.75m$) [5.21].

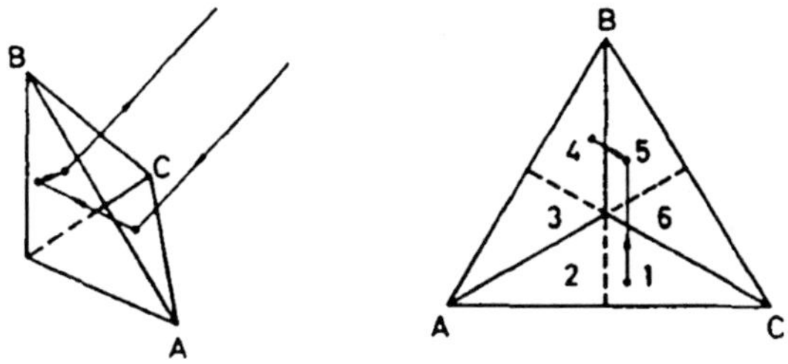

Fig. 17.10 Reflection of a ray by a corner cube prism (*side view and front view*).

17.2 Corner Cube Prism Resonators

If the high-reflecting mirror is replaced by a corner cube prism, the beam quality and the misalignment sensitivity are the same as for a resonator with one Porro prism (Fig. 17.11 and Fig.17.4). However, the corner cube can be tilted around any axis perpendicular to the optical axis whereas the Porro prism exhibits a low misalignment sensitivity only when rotated around an axis parallel to the prism edge. Furthermore, the corner cube prism resonator exhibits a higher diffraction loss per round trip compared to a Porro prism resonator, as shown in Fig. 17.11. This increase in the loss is a result of the geometry of the corner cube (Fig. 17.10) and is not generated by scattering or output coupling losses at the edges. For a high quality corner cube with typical edge widths on the order of µm, these losses are negligible. Three different modes oscillate independently in a corner cube prism resonator. Each of these modes experience reflections in different areas of the corner cube prism. If we look towards the corner cube prism in direction of the beam propagation, as shown in Fig. 17.10, the three different beams can be related to reflections at the areas 1-5-4, 2-4-5, and 3-1-6, respectively. Each of the three modes, therefore, is controlled by an effective aperture given by one of the six segments of the corner cube. Thus, the effective Fresnel number is lower than the Fresnel number defined by the size of the active medium, resulting in a higher diffraction loss.

The phase shift experienced due to the total internal reflection in the corner cube is different for each of the three modes. This difference in phase shift in combination with the transverse separation is the reason that the three modes establish themselves. The phase shift experienced at each surface can be evaluated by a simple geometrical analysis. Regardless of which surface of the corner cube prism is considered, the angle of incidence of the beam is always 54.4°. Each reflection generates the same phase shift δ between the p-polarization and the s-polarization, given by:

$$\tan\frac{\delta}{2} = \frac{\cos\theta\,\sqrt{\sin^2\theta - 1/n^2}}{\sin^2\theta} \tag{17.7}$$

where θ is the angle of incidence on the surface and n is the index of refraction of the

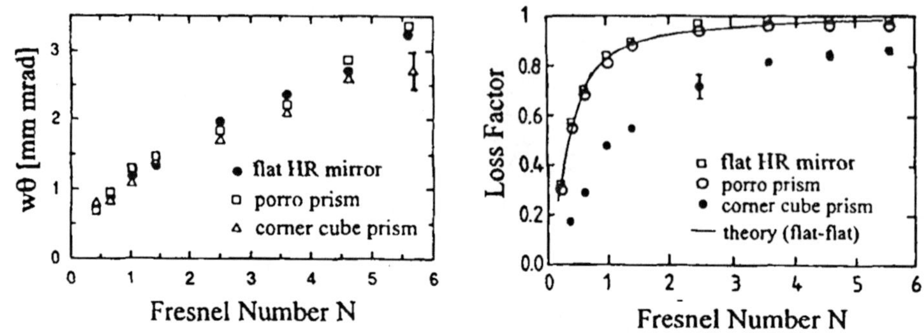

Fig. 17.11 Measured beam parameter product (w: waist radius, θ: half angle of divergence) and measured loss factor per round trip for a Nd:YAG rod laser with a flat flat resonator as a function of the effective Fresnel number N. Measurements with a flat HR mirror, a Porro prism and a corner cube prism are compared. The output coupling mirror is a conventional flat mirror (see Fig. 17.3 for comparison). The effective Fresnel number is given by $N = b^2/(2\lambda L)$ with b: aperture radius, L: effective resonator length (L=0.75m, λ=1.064μm) [5.21].

corner cube prism (δ=44.6° for n=1.5). The phase difference between the three modes is a result of the different orientations of the surface normals with respect to the wave vector incident on each surface. A round trip analysis of the polarization properties of a corner cube prism resonator, therefore, requires the evaluation of a Jones matrix for each of the three possible reflection paths inside the corner cube prism.

Let us consider the reflection at the areas 1-5-4 and determine the Jones matrix for beams incident upon section 1 or 4 of the corner cube prism. The Jones matrices for the other two reflection geometries can be obtained by performing a rotation of the corner cube prism around the optical axis by ±120°. In order to match the orientation of the next reflecting surface, the coordinate system has to be rotated by +60° after the first reflection (at surface 1) and by -60° after the reflection at surface 5. After the final reflection at surface 4, the coordinate system has to be transformed back into the coordinate system of the resonator. The combination of the rotation matrices with the three Jones matrices for the reflections results in the final Jones matrix:

$$M^P_{1\text{-}4} = \frac{1}{8}\begin{pmatrix} 1 & -\sqrt{3} \\ -\sqrt{3} & -1 \end{pmatrix}\begin{pmatrix} 1 & 0 \\ 0 & z \end{pmatrix}\begin{pmatrix} 1 & \sqrt{3} \\ -\sqrt{3} & 1 \end{pmatrix}\begin{pmatrix} 1 & 0 \\ 0 & z \end{pmatrix}\begin{pmatrix} 1 & -\sqrt{3} \\ \sqrt{3} & 1 \end{pmatrix}\begin{pmatrix} 1 & 0 \\ 0 & z \end{pmatrix}$$

$$= \frac{1}{8}\begin{pmatrix} 1+6z-3z^2 & -\sqrt{3}(z+2z^2+z^3) \\ -\sqrt{3}(1+2z+z^2) & z(3-6z-z^2) \end{pmatrix} \; ; \quad z=\exp[i\delta] \tag{17.8}$$

The Jones matrix $M_{4\text{-}1}$ for the reversed beam path is the same. The other two Jones matrices are found by rotating the corner cube by ±120°:

$$M_{2\text{-}5} = M_{5\text{-}2} = \frac{1}{4}\begin{pmatrix} -1 & -\sqrt{3} \\ \sqrt{3} & -1 \end{pmatrix} M_{1\text{-}4} \begin{pmatrix} -1 & \sqrt{3} \\ -\sqrt{3} & -1 \end{pmatrix} \tag{17.9}$$

$$M_{3\text{-}6} = M_{6\text{-}3} = \frac{1}{4}\begin{pmatrix} -1 & \sqrt{3} \\ -\sqrt{3} & -1 \end{pmatrix} M_{1\text{-}4} \begin{pmatrix} -1 & -\sqrt{3} \\ \sqrt{3} & -1 \end{pmatrix} \tag{17.10}$$

In order to determine the output coupling loss of a corner cube prism resonator with variable output coupling using a rotatable retardation plate and a polarizer (see resonator scheme in Fig. 17.12), the eigenvalues of the Jones matrix for a resonator round trip have to be calculated for each of the three modes. For the mode which is incident on segment 1 or segment 4 of the corner cube prism, for instance, the round trip Jones matrix, starting at the polarizer, reads:

$$M_1^P = \begin{pmatrix} 1 & 0 \\ 0 & 0 \end{pmatrix} M^P_{\lambda/4}(\alpha)\; M^P_{1\text{-}4}\; M^P_{\lambda/4}(\alpha) \begin{pmatrix} 1 & 0 \\ 0 & 0 \end{pmatrix} \tag{17.11}$$

where $M^P_{\lambda/4}(\alpha)$ is the Jones matrix of the quarter wave plate as a function of the angle of rotation α (see (3.13)). The two other Jones matrices are obtained by replacing $M^P_{1\text{-}4}(\alpha)$ with the corresponding expressions (17.9) and (17.10).

Figure 17.12 presents the calculated loss factors per round trip for the three modes reflection paths as a function of the angle of rotation of a quarter wave plate (the loss factor represents the output coupling reflectance of the polarizer). The average loss factor shown in the right graph indicates that a rotation of the retardation plate will not have much effect on the output power, provided that all three modes are well above the laser threshold. An experimental verification of this statement is given in Fig. 17.13 for a Nd:YAG rod laser with a small-signal gain of 2. Considering this property, output coupling via polarization does not seem to be advantageous at all. However, since the different modes exhibit different output coupling reflectances, it is possible to control their relative power content by rotating the retardation plate. This may be interesting if a laser system operating at three different wavelengths is considered (i.e. operation of Nd:YAG at 946nm, 1064nm, and 1320nm). By subdividing the flat mirror into six segments using wavelength selective coatings, the emission of the laser can be distributed over the three wavelengths. If the small-signal gain of the medium is not too high, emission at one, two, or three wavelengths can be attained simply by rotating the retardation plate. An experimental example of this mode-switching behavior is presented in Fig. 17.14. The recorded near field intensity distributions of a Nd:YAG laser are shown for different angles of rotation of the retardation plate in this figure.

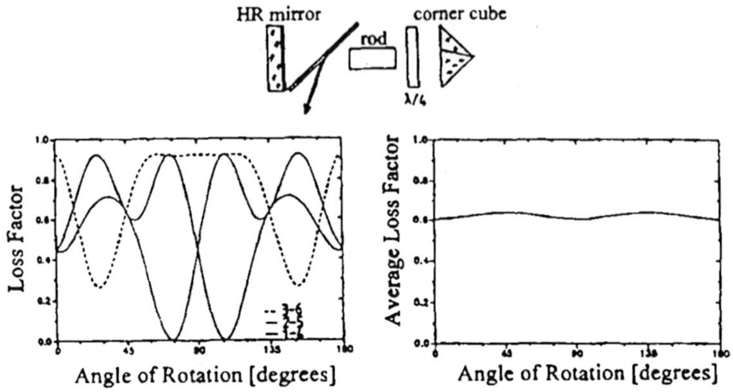

Fig. 17.12 Calculated loss factor per round trip for a corner cube prism resonator with an internal polarizer as a function of the angle of rotation of the quarter wave plate. The loss factor corresponds to the output coupling reflectance of the resonator. The left graph shows the loss factor for the three different modes, the right graph is the mean loss factor, averaged over the three modes.

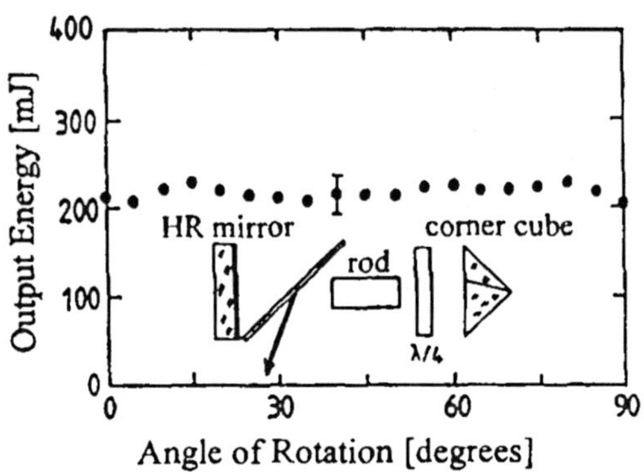

Fig. 17.13 Measured output energy per pulse of a Nd:YAG laser with a corner cube prism resonator as a function of the angle of rotation of the quarter wave plate. (3"1/4" rod, small-signal gain $g_0\ell=1.0$, single shot operation, pulse duration: 500 μs).

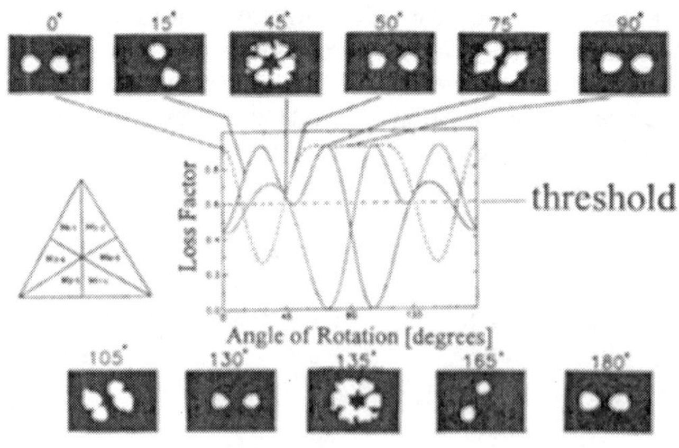

Fig. 17.14 Photographs of the near field intensity of the Nd:YAG laser used in Fig. 17.13 for different angles of rotation of the quarter wave plate. The small-signal gain is $g_0\ell=0.3$, which means that the loss factor must be higher than 0.6 for a mode to exceed the laser threshold (5% loss per transit) [S.10].

Chapter 18

Fourier Transform Resonators

18.1 Self-Filtering Unstable Resonators

As was discussed in Sec.7.2.2, unstable resonators exhibit a considerable amount of power in the side lobes of the far field, especially if the magnification is low. One way to decrease the power fraction in the side lobes is the utilization of variable reflectivity mirrors as output couplers, as seen in Sec. 7.7. A different approach to enhance the shape of the far field intensity distribution is the spatial filtering of the electric field inside the resonator. If the resonator exhibits an intracavity focus point at a certain plane, this plane is conjugate to the focal plane generated by the focusing optics outside the resonator. The insertion of a spatial filter (e.g. an aperture or a filter with a defined lateral transmission profile) at the intracavity focal plane will, therefore, generate the same intensity distribution at the extracavity focal plane. In general, the field distribution at the intracavity focal plane is the Fourier transform of the field at the resonator mirrors. This type of resonator, therefore, is referred to as the Fourier transform resonator or self-filtering resonator [5.22-5.26,5.31,5.32]. The schematic of a self-filtering unstable resonator (SFUR) is shown in Fig. 18.1. The resonator is confocal which means that both mirrors generate a focal spot at the same plane. The basic principle of this resonator is the transformation of a rectangular intensity profile into a Gaussian one by cutting off the side lobes of the Fourier transform with an aperture.

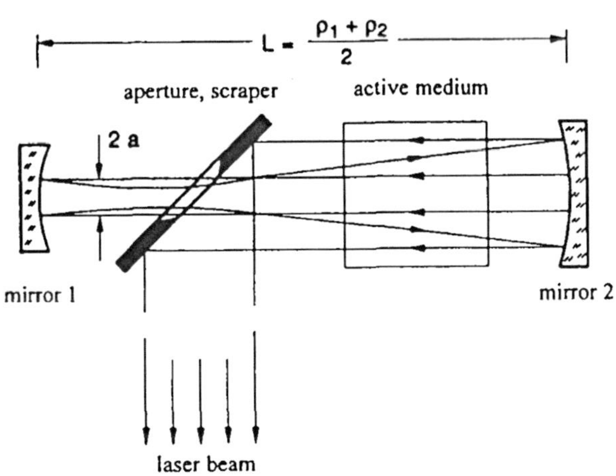

Fig. 18.1 Negative branch confocal unstable resonator with self-filtering. The aperture shapes the intensity profile by filtering the Fourier transform and serves as an output coupler at the same time (ρ_i: radius of curvature of mirror i).

The electric field incident on the aperture from the right mirror generates its Fourier transform at the aperture after being reflected by mirror 1 (Fig. 18.2). If we assume that the intensity distribution of the field traveling towards mirror 1 exhibits a flat-top profile behind the aperture, the intensity distribution at the aperture is determined by the Fourier transform of a circular aperture with radius a. The first minimum of this intensity distribution is located at the radius:

$$r_0 = 0.61 \frac{\lambda}{a} \frac{\rho_1}{2} \tag{18.1}$$

where λ is the wavelength of the electric field and ρ_l is the radius of curvature of mirror 1. The aperture radius a is chosen equal to r_0 to cut off the side lobes of the Fourier transform:

$$a = \sqrt{0.305 \ \lambda \ \rho_1} \tag{18.2}$$

The truncation of the Fourier transform generates an intensity profile that can be approximated by a Gaussian beam with a beam waist radius of:

$$w_{Bl} = 0.43 \frac{\lambda}{a} \frac{\rho_1}{2} \tag{18.3}$$

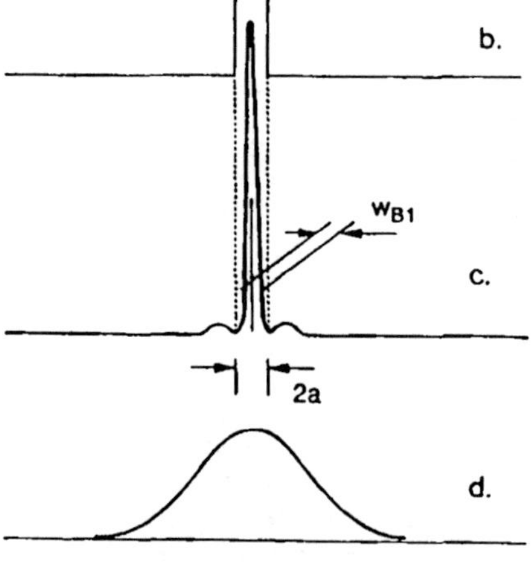

Fig. 18.2 Radial intensity distributions inside a self-filtering unstable resonator (SFUR). a) at the aperture, incident from mirror 2, b) behind the aperture, incident from mirror 2, c) at the aperture, incident from mirror 1, and d) at mirror 2.

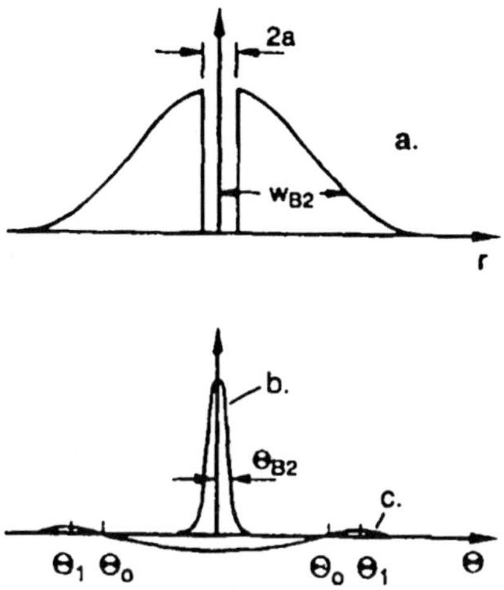

Fig. 18.3 Radial field distributions in the near field and the far field of a self-filtering unstable resonator (SFUR). a) at the aperture, b) far field of the Gaussian profile, c) far field of the circular aperture.

After reflection from mirror 2, the beam radius of the Gaussian intensity distribution at the aperture is given by:

$$w_{B2} = \frac{a}{0.43\ \pi}\ \frac{\rho_2}{\rho_1} = 0.74\ M\ a \tag{18.4}$$

where $M = \rho_2/\rho_1$ is the magnification of the confocal unstable resonator. Unfortunately, the approximation of the truncated Airy pattern by a Gaussian profile is too rough. Since the Airy pattern exhibits steeper slopes, it is to be expected that the beam radius at the aperture is larger than predicted by (18.4). A more accurate treatment of the beam propagation, in fact, shows that the magnified intensity distribution at the scraper can be described by a Gaussian profile with a beam radius of [5.28]:

$$w_{B2} = 0.974\ M\ a \tag{18.5}$$

The near field intensity distribution is thus given by a Gaussian beam profile with the center core of radius a missing. The diffraction at the aperture generates small side lobes in the far field. The far field amplitude distribution $E(\theta)$ can be calculated by subtracting the far field of a circular aperture with radius a from the far field of the Gaussian beam (Fig. 18.3). With the assumption that $w_{B2} \gg a$, the final result reads:

$$E(\theta) = C\left(\exp\left[-\frac{\theta\pi w_{B2}}{\lambda}\right] - \left[\frac{a}{w_{B2}}\right]^2 \frac{J_1(2\pi\theta a/\lambda)}{\pi a\theta/\lambda}\right) \tag{18.6}$$

where J_1 is the Bessel function of order 1 and θ is the angle of divergence. Compared to a conventional unstable resonator with the same magnification M, the self-filtering decreases the side lobes considerably (Fig. 18.4). However, this improvement of the beam quality is a result of an increase in the diffraction losses. The loss factor V (=1-loss) of the SFUR due to output coupling can be calculated assuming a Gaussian intensity profile with the beam radius w_{B2} given by (18.5). For $a \ll w_{B2}$, we get:

$$V = \frac{\pi a^2}{0.5 \pi w_{B2}^2} = \frac{2.108}{M^2} \tag{18.7}$$

In addition to the output coupling loss, the truncation of the Airy pattern results in an additional loss of 16% per round trip. Thus, the total loss factor per round trip of the SFUR is given by:

$$V_{tot} = 0.84 \ V = \frac{1.771}{M^2} \tag{18.8}$$

A comparison of the far field properties of conventional and self-filtering unstable resonators indicates that at the same round trip loss, both resonators provide a similar power content in the side lobes (Fig. 18.4). For conventional unstable resonators with equivalent Fresnel numbers N_{eq} around 0.5, the round trip loss factor is higher than the one for the SFUR given by (18.8). Therefore, the conventional unstable resonators can be operated at a higher magnification $M^* > M$, resulting in near field and far field distributions similar to those of a SFUR with magnification M. This leads us to the conclusion that the SFUR is not superior to a conventional unstable resonator as far as the beam quality is concerned. Furthermore, the loss generated by the truncation of the Fourier transform may considerably decrease the output power if a low gain medium is used. However, for lasers that emit at small wavelengths (like in excimer lasers), the design of conventional unstable resonators with equivalent Fresnel numbers on the order of 0.5 would lead to a small diameter of the high reflecting spot on the output coupling mirror (on the order of 0.5mm). High peak powers might damage the output coupler and it is, therefore, safer to use a scraper at an angle of 45° to couple out the beam. Placing the scraper at the focal plane is just a convenient way to realize the output coupling. SFURs have been successfully implemented in excimer, CO_2, and Nd:YAG lasers [5.27-5.34]. In addition to the Gaussian intensity distribution in the near field, a second advantageous property of the SFUR is the high discrimination against higher order transverse modes. This results in a fast establishment of the steady-state fundamental mode, a feature that is of crucial importance for active media with a very short population inversion lifetime. However, the disadvantage of the SFUR is the dependence of the mode volume on the wavelength. Combination of (18.2) and (18.5) yields for the cross sectional area of the beam in the active medium:

$$A_b = \pi w_{B2}^2 = 0.91 \ M^2 \ \lambda \ \rho_1 \tag{18.9}$$

At short wavelengths, where the SFUR is of particular interest, it is not possible to fill a large diameter active medium without choosing a long resonator (Fig. 18.5). However, shorter resonator set-ups can be realized with nonconfocal SFURs [5.31].

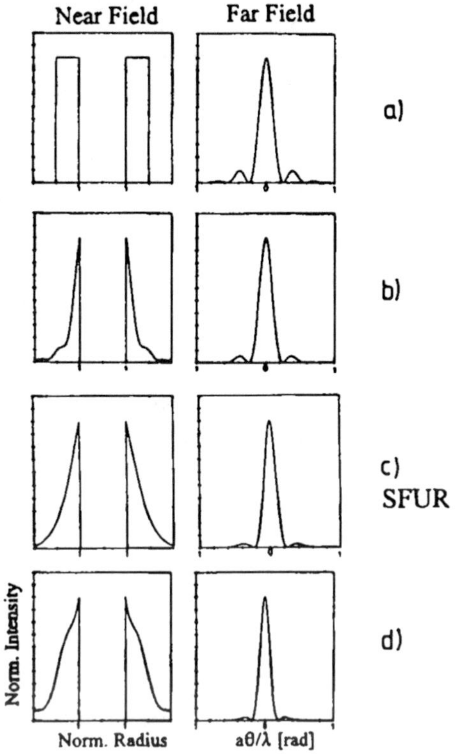

Fig. 18.4 Comparison of the radial intensity distribution in the near field and the far field of unstable resonators with magnification $M=2$. a) conventional unstable, geometric, loss factor =0.25, b) conventional unstable, with diffraction and $N_{eq}=0.5$, loss factor =0.55, c) SFUR, loss factor=0.44. Figure d) presents the far field for a conventional unstable resonator with $M=2.5$ and $Neq=0.5$, which means that this resonator exhibits a total loss similar to that of the SFUR in c).

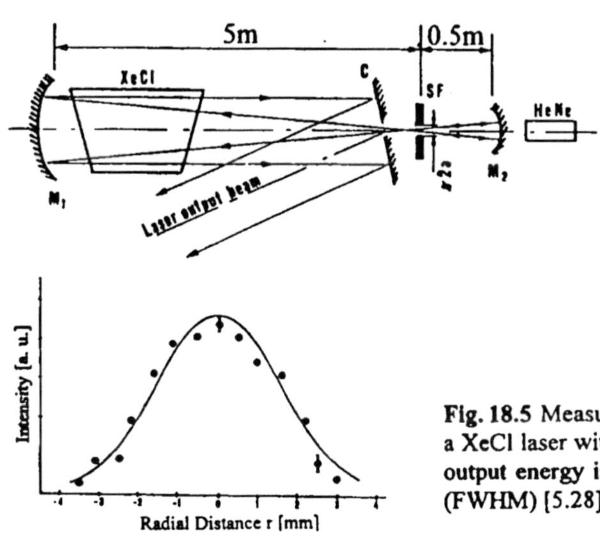

Fig. 18.5 Measured near field intensity distribution of a XeCl laser with SFUR ($\lambda=308nm$, $a=0.3mm$). The output energy is 120mJ at a pulse duration of 90ns (FWHM) [5.28] (© IEEE 1987).

18.2 Stable Fourier Transform Resonators

The concept of self-filtering can also be applied to stable resonators. In contrast to the SFUR, the spatial filter is only used to put a constraint on the Fourier transform and output coupling is accomplished by means of a conventional partially reflecting mirror. The design of a stable Fourier transform resonator is determined by the fact that the field at the active medium or at the output coupling mirror is the Fourier transform of the field at the spatial filter. Therefore, we need appropriate intracavity transformation optics that provide a Fourier transformation between two planes. In general, such transformation optics are described by a ray transfer matrix of the following form:

$$M_{FT} = \begin{pmatrix} 0 & B \\ C & 0 \end{pmatrix} \tag{18.10}$$

The two simplest optical systems that provide an optical Fourier transformation between the input and the output plane are depicted in Fig. 18.6. Based on these two set-ups we can design Fourier transform resonators by placing the spatial filter at the input plane and the active medium at the output plane. The two resonator mirrors have to be added in such a manner that the fields at both planes represent self-consistent solutions for a resonator round trip. With design b), the use of flat mirrors at both the input and output plane results in a resonator that is equivalent to a confocal resonator ($g_1=g_2=0$) (Fig.18.7a). A true confocal resonator is obtained with design a) by replacing the two lenses with resonator mirrors that provide the same imaging properties ($\rho=f$) which means that the mirror distance is equal to the curvature of both mirrors (Fig. 18.7b). We already discussed the Fourier transform properties of the confocal resonator in Sec. 6.3 using circular apertures as spatial filters.

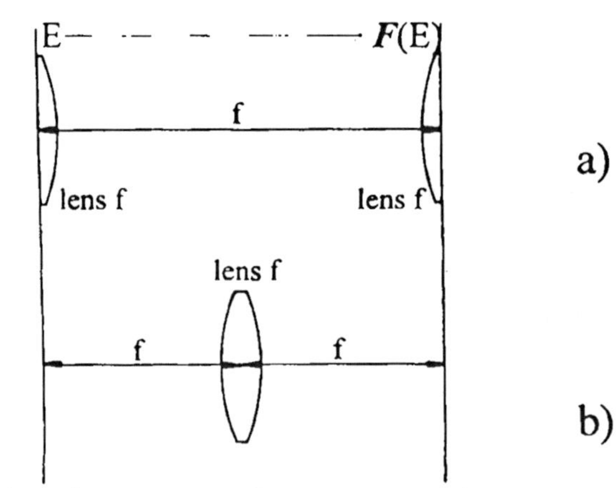

Fig. 18.6 Optical systems that provide the Fourier transform $F(E)$ of the input field E at the output plane.

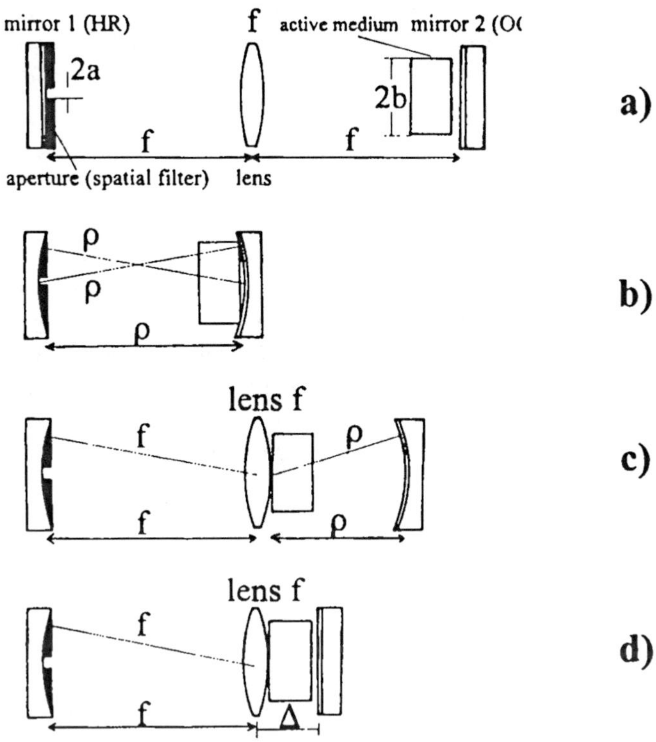

Fig. 18.7 Stable Fourier transform resonators. a) equivalent confocal ($g_1=g_2=0$), b) confocal ($g_1=g_2=0$), c) concentric ($g_1g_2=1$ and $g_1<0$), d) vanishing g-parameter of mirror 1 ($g_1=0$).

Whereas in Fig. 18.7a) and b) the fields on the two resonator mirrors are related to each other via a Fourier transform, the resonator in Fig. 18.7c) generates the Fourier transform of the field at the aperture at the left face of the active medium. The curvature of the right mirror is chosen so that this plane is imaged onto itself. The equivalent g-parameters of this resonator are given by:

$$g_1 = -\frac{\rho}{f} \quad , \quad g_2 = -\frac{f}{\rho} \tag{18.11}$$

Thus the resonator is concentric. The resonator in Fig. 18.7d) does not exactly reproduce the field distribution at the left face of the active medium. However, this Fourier transform resonator will also work provided that the Fresnel number of the right resonator segment $b^2/(2\Delta\lambda)$ is high enough (greater than 30). In this case the propagation of the field to the right mirror and back can be described by geometric optics since the spread of the field due to diffraction is negligible.

All four resonators are equivalent as far as the Fourier transformation and the optimization of the resonator set-up are concerned. If we assume that the spatial filter supports a Gaussian beam with a beam radius w_1 at mirror 1, the beam radius in the active medium is given by:

$$w_2 = f\,\theta = \frac{f\,\lambda}{\pi\,w_1} \tag{18.12}$$

where θ is the half angle of divergence of the Gaussian beam. For the confocal resonator in Fig. 18.7b), the focal length f has to be replaced by the mirror curvature ρ in (18.12) and all equations derived below. Let us first make the assumption that we can use a circular aperture to generate the Gaussian beam (which means that diffraction losses at the aperture are negligible). We saw in Chapter 11.2 that the maximum output power in fundamental mode operation is attained if the radii of intracavity apertures are about 1.3 times larger than the Gaussian beam radii. Equation. (18.12), therefore, can be written as:

$$\frac{a\,b}{f\,\lambda} = \frac{(1.3)^2}{\pi} \approx 0.54 \tag{18.13}$$

where a is the aperture radius and b is the radius of the active medium. This gives us a first idea of how to choose the aperture dimensions to attain a near diffraction limited beam. An exact treatment of the field propagation using diffraction integrals reveals that the intensity distribution of the fundamental mode is in fact almost Gaussian, and optimum performance in fundamental mode operation is obtained for [5.36]:

$$\frac{a\,b}{f\,\lambda} \approx 0.6 \tag{18.14}$$

Unfortunately, the Gaussian beam only provides a fill factor in the active medium of less than 0.9, which means that at least 10% of the available power cannot be extracted. In order to realize a fill factor close to 1.0 it is necessary to generate a flat-top intensity profile with radius $\boldsymbol{b}$ at the active medium (with a planar wave front). In order to generate this homogeneous filling of the active medium, the radial field distribution at mirror 1 must be the Airy pattern (see Sec. 2.2.2):

$$E(r) = \frac{2\pi b^2}{\lambda f}\,E_0\,\frac{J_1[2\pi r b/(\lambda f)]}{2\pi r b/(\lambda f)} \tag{18.15}$$

where E_0 is the electric field amplitude at the active medium and r is the radial coordinate at mirror 1.

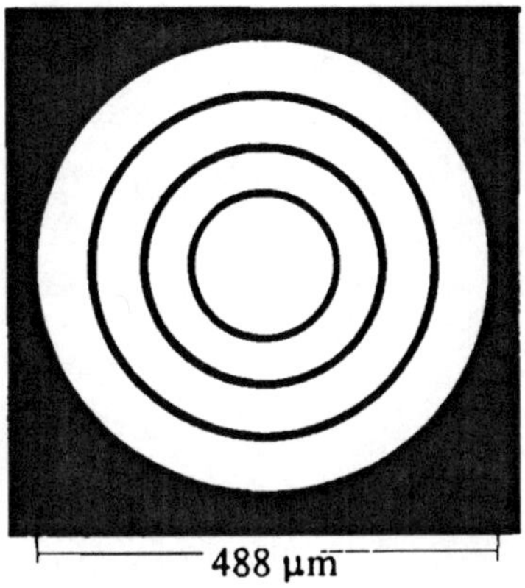

Fig. 18.8 A spatial filter which supports the Airy pattern in a Fourier transform resonator (*λ=1μm, f=300mm, b=3mm*). The black areas absorb the electric field, the white areas exhibit 100% transmission. This filter is placed in front of the high reflecting mirror 1.

To a good approximation, the Airy pattern exhibits zero intensity at the radii:

$$r_m = m\ 0.61\ \frac{f\lambda}{b}\ , \qquad m = 1,2,3,... \tag{18.16}$$

An appropriate spatial filter to generate a flat-top beam profile in the active medium, therefore, should absorb or deflect the electric field in the vicinity of the zero intensity lines of the Airy pattern. In practice, it is sufficient to take only the first couple of rings into consideration (Fig. 18.8). The widths of the rings have to be optimized to efficiently discriminate against unwanted transverse modes without generating losses that are to high for the fundamental mode. Such a spatial filter can be produced via photoedging using a focused laser beam with a flat-top near field beam profile. This technique was successfully applied in [5.37] to generate a diffraction limited output beam in a Q-switch Nd:YAG laser using a confocal resonator (Fig. 18.9). Other examples of stable Fourier transform resonators can be found in [5.38-5.40].

The reader should keep in mind that stable Fourier transform resonators are extremely sensitive to variations in the distances between the optical elements. This is the result of their location on the stability limits in the g-diagram. As was presented in Sec. 6.3 for the confocal resonator, a tight length tolerance of less than ±0.5% must be maintained to prevent a decrease of output power by more than 10%. If the active medium exhibits thermal lensing, the corresponding refractive power has to be compensated by the resonator design. Therefore, a similarly tight tolerance has to be kept on the pump power going into the active medium which means that the output power cannot be varied. Furthermore, the gain saturation inside the active medium will modify the mode structure. For the confocal Fourier transform resonators (Fig. 18.7a,b), the field distribution at mirror 2 is not the exact Fourier transform of the field distribution at the spatial filter. The losses generated by the filter, therefore, will slightly rise as the small-signal gain of the active medium is increased.

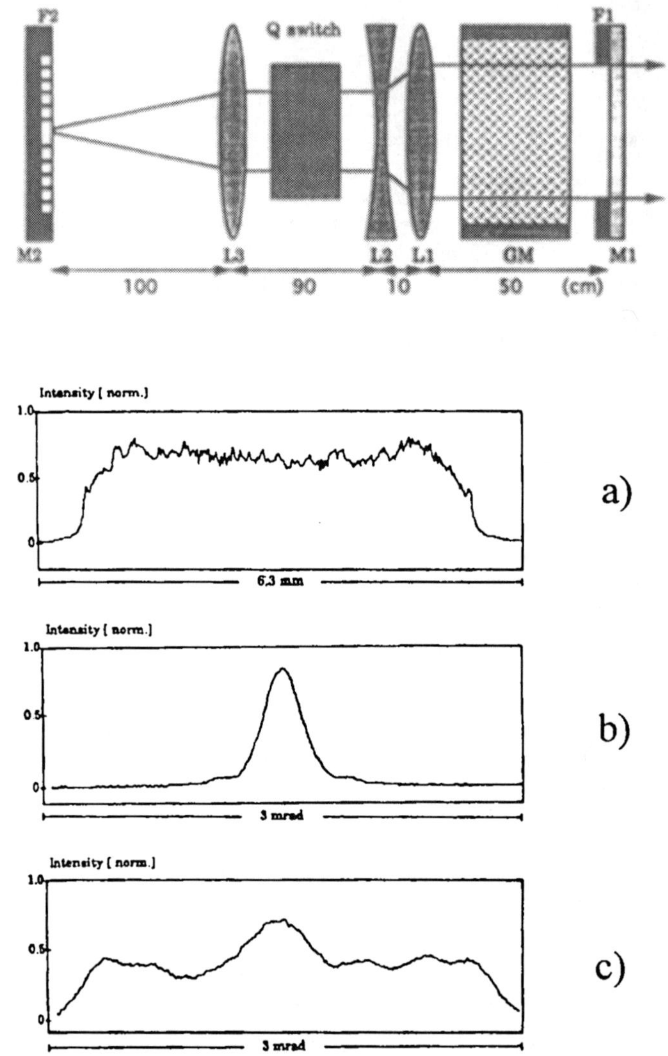

Fig. 18.9. Measured radial intensity profiles in the near field a) and in the far field b) of a Nd:YAG Q-switch laser (λ=1.064μm) with a confocal Fourier transform resonator. (rod diameter: 0.25", focal lengths: L_1=*12.5cm*, L_2=*-5cm*, L_3=*100cm*). The spatial filter consists of eleven rings of 80μm width that correspond to the nodal lines of the Airy pattern. The output energy per pulse is 200mJ at a pulse duration of 60ns and a repetition rate of 10Hz (electrical input energy per pulse: 25J, pump pulse duration: 200μs). The repetition rate could be varied between 5Hz and 10Hz with minor variations in the distance between the lenses L_1 and L_2. Without the spatial filter an output energy of 210mJ and a full angle of divergence of about 3mrad was measured (Fig. c) [5.37] (© OSA 1993).

Chapter 19

Hybrid Resonators

19.1 General Aspects

If one or both resonator mirrors exhibit different radii of curvature in the x- and the y-directions (cylinder mirrors or toric mirrors), the g-parameters of the resonator, and consequently the resonator properties, become different for the two directions. Different mode properties in the x and the y directions are also obtained for spherical resonator mirrors if the corresponding effective resonator lengths are different. This happens if the surface normals of intracavity optical elements exhibit an angle with respect to the optical axis. Examples are solid state slab lasers and ring resonators. In most of these cases, the resonator works stable or unstable in both directions and the output beam can be described by astigmatic stable or unstable resonator modes.

For rectangular active media that exhibit large aspect ratios (ratio of height to width), it is advantageous to design resonators that work stable in one direction and unstable in the perpendicular direction (Fig. 19.1). These resonators are referred to as hybrid resonators. As was discussed in Sec. 7.3, unstable resonators exhibit the unique property that near diffraction limited beam quality can be attained regardless of the transverse dimensions of the active medium. Therefore, the beam quality can be improved considerably by choosing an unstable resonator design along the wide dimension of the medium. In the orthogonal direction a stable or flat-flat resonator is usually sufficient to achieve good beam quality provided that the width of the medium generates a low Fresnel number.

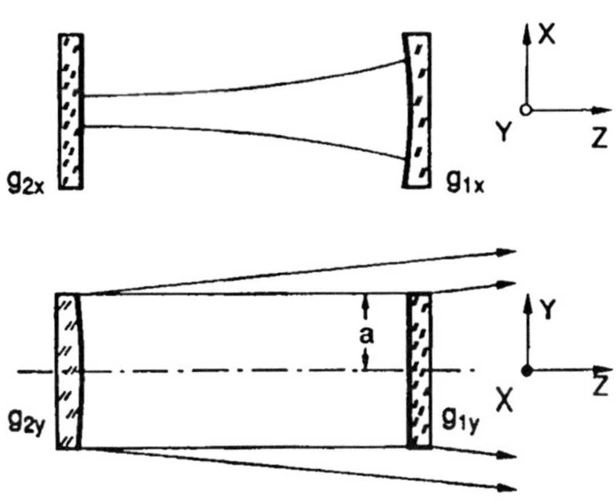

Fig. 19.1 Hybrid resonator with cylinder mirrors arranged perpendicularly to each other. The resonator is stable in the x-direction and unstable in the y-direction.

Another common resonator design is the combination of an unstable resonator with a waveguide resonator along the smaller dimension of the active medium. This resonator scheme is used commercially in sealed-off CO_2 slab lasers with average output powers of up to 2.5kW [5.81,5.86,5.87,5.92,5.96-5.106].

19.2 Unstable-Stable Resonators

The resonator depicted in Fig. 19.1 represents the general unstable-stable resonator. In the x direction, the resonator exhibits the g-parameters $g_{2x}=1$ and $0<g_{1x}<1$ and the transverse eigenmodes are Gauss-Hermite modes. In the perpendicular direction, the resonator is unstable with $g_{1y}=1$ and $g_{2y}>1$. In each direction the mode structure can be calculated separately using the methods discussed in Part III. Note that the output coupling is realized only in the y direction since both resonator mirrors are highly reflective. The near field of this resonator consists of two bars which are modulated in the x-direction with the Gauss-Hermite polynomials and exhibit the typical near field structure of an unstable resonator in the y direction.

If the axes of both cylinder mirrors are parallel, a flat-flat resonator is attained in the direction of the cylinder axes. In the perpendicular direction, the unstable resonator can be designed either in on-axis or off-axis geometry (Fig. 19.2). In the off-axis unstable resonator, the optical axis is located at the bottom side of the active medium. The off-axis design has the potential advantage that the power content in the side lobes of the far field is lower [5.46-5.51]. The safety glass burn pattern shown in Fig. 19.3 represents the near field intensity distribution of a commercial 3kW CO_2 laser utilizing such an off-axis plane-unstable resonator. The edge of the output coupling mirror is located on the right side of the structure. Note the intensity modulation in the direction of the mirror edge, generated by the mode structure of the flat-flat resonator.

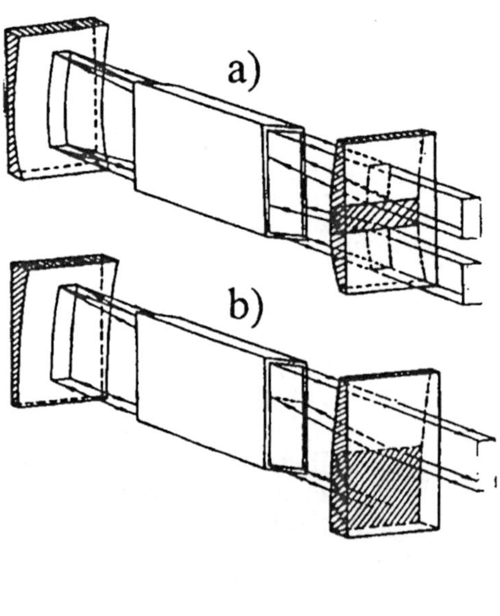

Fig. 19.2 Plane-unstable resonators in on-axis (a) and off-axis geometry (b).

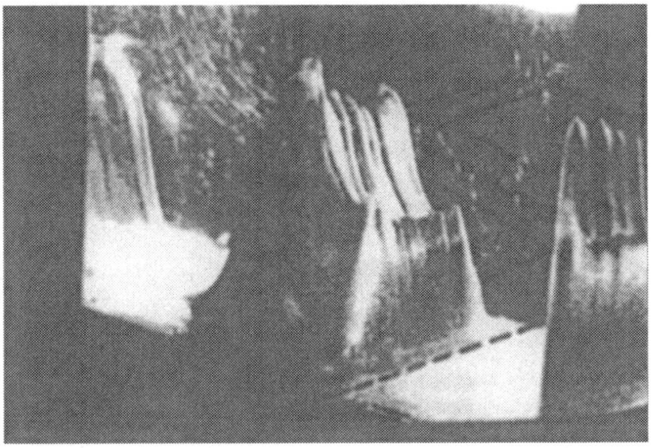

Fig. 19.3 Near field intensity distribution of a 3kW CO_2 laser employing a plane-unstable resonator (burn pattern in safety glass). The transverse beam dimensions are 40mm along the mirror edge (black broken line) and 22mm in the perpendicular direction [S.21].

19.3 Waveguide Resonators

19.3.1 Motivation

In all high power laser systems, the removal of the heat generated by the pump process is crucial for efficient operation. In sealed-off gas lasers, the gas is mainly cooled through heat conduction to the walls of the tube, a process which is not very efficient if cylindrical discharge tubes are used. Regardless of the tube diameter, the maximum output power attainable per discharge length is limited meaning that very long tubes are required to reach output powers in the kW range. In fact sealed-off CO_2 lasers using circular tubes can hardly extract more than 50W of output power per meter of tube length. One solution to this problem is to flow the gas longitudinally or transversally through the discharge region with subsonic speed, a technique that is nowadays used in all industrial high power CO_2 lasers. Although these industrial CO_2 lasers provide output powers on the order of 1kW per meter of discharge length, they have relatively large dimensions due to the incorporation of several pumps and heat exchangers.

A more elegant solution is the utilization of discharge geometries that exhibit a larger ratio of the cooling surface to the discharge volume than cylindrical tubes. Besides the annular geometry (see next chapter), the most promising concept pursued in the last decade is the slab waveguide laser, as depicted in Fig. 19.4. Heat removal in a slab laser scales inversely with the electrode separation *d*. This is quite understandable since the molecules need less time to reach the electrodes when the electrode distance is small. Typical electrode distances for CO_2 slab waveguide lasers are between 1.5mm and 2.5mm. The discharge volume can be controlled via the cross sectional area of the electrodes. This means that the output power of slab lasers is scaled by area rather than length as in conventional geometries.

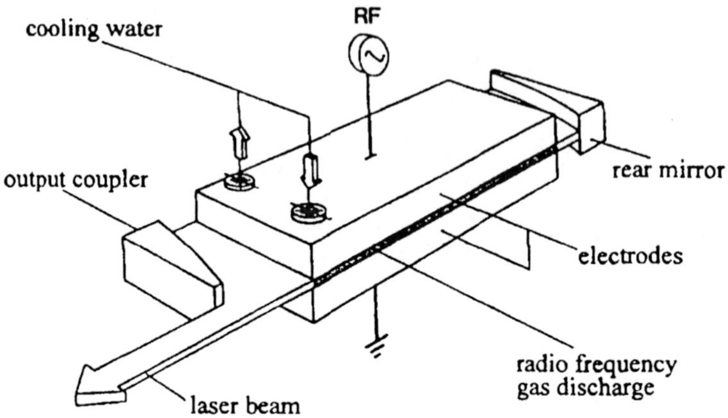

Fig. 19.4 CO_2 slab waveguide laser with unstable waveguide resonator and a RF excited gas discharge (RF frequencies between 50 and 100MHz) [5.92] (© AT Fachverlag 1991). Electrode separations between 1.5 and 2.5mm are used. The width and the length of the discharge regions typically are 4-7cm and 60-100cm, respectively. Electrode materials commonly used are aluminum and copper.

Assuming a constant laser efficiency η_{tot} of 10%, the scaling law for the maximum achievable output power per area of one electrode can be approximated by [5.92]:

$$P_{out,\max} = \frac{3}{d}\,\frac{W}{cm^2} \tag{19.1}$$

where the electrode distance d is given in mm. This expression holds only for distances d greater than 1.5mm since smaller electrode gaps will decrease the laser efficiency due to an increase in diffraction losses and a decrease of the excitation efficiency. The scaling law (19.1) indicates that for a typical electrode gap of 2mm, it should be possible to realize a 1kW sealed-off CO_2 laser with reasonable dimensions of the electrodes (e.g. 7cm wide and 100cm long). In [5.87,5.92], a CO_2 slab laser with an electrode area of 10cm x 60cm and an electrode separation of 1.5mm providing a maximum output power of 1,060W was reported.

These dimensions of the active medium leave us with some real challenges regarding the resonator design. As done previously, we can take care of the larger dimension by using an unstable resonator to achieve near diffraction limited beam quality. Unfortunately, in the perpendicular direction the small gap width in combination with the length of the electrodes makes it virtually impossible to propagate a Gaussian beam through the discharge region without touching the electrodes. If we position the Gaussian beam waist with a diameter of 1mm in the middle of a CO_2 slab discharge area, the Rayleigh range of the Gaussian beam of 74mm is much too small to prevent the beam from being reflected off the electrodes. A stable resonator used in this direction, therefore, will exhibit modes that are guided by the electrodes rather than modes that follow the laws of free space propagation as in conventional open resonators. This resonator concept using guided mode propagation is referred to as a waveguide resonator [5.53,5.62,5.74]. Guided mode propagation occurs if the Fresnel number of the waveguide $a^2/(\lambda L)$ is much lower than 1, where L is the length of the waveguide and $2a$ its diameter. The mode properties of waveguide resonators are quite different from those of open resonators discussed so far. Therefore, it is necessary to get a basic understanding of waveguide physics before we can derive design rules for the waveguide unstable resonator shown in Fig. 19.4.

19.3.2 Eigenmodes of Hollow Waveguides

We consider a waveguide resonator comprising a hollow, rectangular waveguide as depicted in Fig. 19.5. The walls of the waveguide are characterized by a complex dielectric constant $\epsilon = \epsilon_1 + i\,\epsilon_2$ which is related to the index of refraction $n = n_1 + i\,n_2$ via:

$$n_1^2 - n_2^2 = \epsilon_1 \quad (19.2) \qquad , \qquad 2n_1 n_2 = \epsilon_2 \quad (19.3)$$

The imaginary part of the index of refraction is proportional to the absorption coefficient of the waveguide material. The dielectric constant of the interior is set to $\epsilon = 1$ (free space) and the permeability of all media is μ_0. Let us first remove the resonator mirrors and determine the properties of the waveguide alone.

The modes of such a waveguide can be found by solving the wave equation using the appropriate boundary conditions for the electric and the magnetic fields at the surfaces. The natural modes of any hollow dielectric waveguide exhibit all three components of both the electric and the magnetic fields, and as a consequence rather complicated expressions are obtained for the fields in the waveguide and in the walls [5.62].

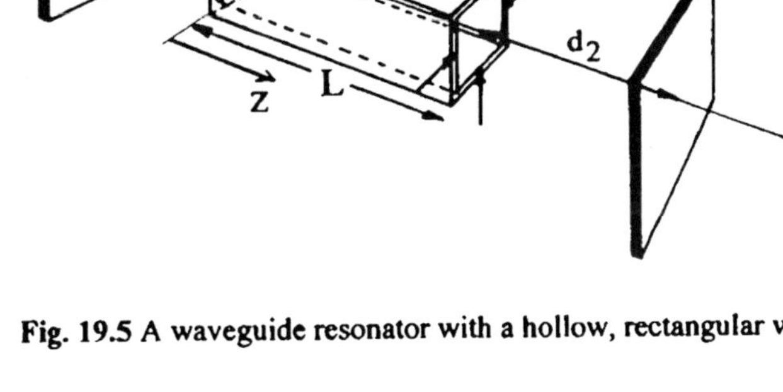

Fig. 19.5 A waveguide resonator with a hollow, rectangular waveguide.

Fortunately, in waveguide lasers, the cross section of the waveguide is very large compared to the wavelength λ. It is thus justified to neglect terms of first order or higher order in λ/a and λ/b. With this approximation, the fields in the walls and at the wall surface vanish and the internal fields become transverse. The following electric fields $E(x,y,z,t)=(E_x,E_y,E_z)$ and magnetic fields $H(x,y,z,t)=(H_x,H_y,H_z)$ are obtained for the rectangular waveguide shown in Fig. 19.5, provided that the waveguide length is large compared to its width and its height [5.62]:

a) y-polarized hybrid modes $E^yH^x_{mn}$:

$$E_x = E_z = 0 \quad , \quad H_y = H_z = 0$$

$$E_y = \frac{1}{\sqrt{ab}} \exp[i(\beta_{mn}z-\omega t)] \exp[-\alpha_{mn}z] \begin{bmatrix} \cos\left[\frac{m\pi x}{2a}\right]\cos\left[\frac{n\pi y}{2b}\right] & m,n \text{ odd} \\ \cos\left[\frac{m\pi x}{2a}\right]\sin\left[\frac{n\pi y}{2b}\right] & m \text{ odd}, n \text{ even} \\ \sin\left[\frac{m\pi x}{2a}\right]\cos\left[\frac{n\pi y}{2b}\right] & m \text{ even}, n \text{ odd} \\ \sin\left[\frac{m\pi x}{2a}\right]\sin\left[\frac{n\pi y}{2b}\right] & m,n \text{ even} \end{bmatrix}$$

$$H_x = -\sqrt{\frac{\epsilon_0}{\mu_0}}\, E_y \tag{19.4}$$

where $\omega=2\pi c_0/\lambda$ is the angular light frequency, ϵ_0 is the permittivity of free space, μ_0 is the permeability of free space, and m,n are integer numbers with $m,n\geq 1$. For $\lambda \ll a$ and $\lambda \ll b$, the propagation constants are given by:

$$\alpha_{mn} = \frac{\pi^2}{4k^2}\left[\frac{m^2}{a^3}Re\left(\frac{1}{\sqrt{\epsilon-1}}\right) + \frac{n^2}{b^3}Re\left(\frac{\epsilon}{\sqrt{\epsilon-1}}\right)\right] \tag{19.5}$$

$$\beta_{mn} = k\left[1 - \frac{\pi^2}{8k^2}\frac{m^2}{a^2}\left(1-\frac{2}{ak}Im\left(\frac{1}{\sqrt{\epsilon-1}}\right)\right) - \frac{\pi^2}{8k^2}\frac{n^2}{b^2}\left(1-\frac{2}{bk}Im\left(\frac{\epsilon}{\sqrt{\epsilon-1}}\right)\right)\right] \tag{19.6}$$

Where $k=2\pi/\lambda$ is the wave number and Re,Im denote the real part and the imaginary part of the expression in the bracket, respectively. These field distributions represent the eigenmodes if the following relations hold:

$$\frac{m\lambda}{4a} < 1 \quad , \quad \frac{n\lambda}{4b} < 1 \tag{19.7}$$

$$\frac{m\lambda}{4a} < |\sqrt{\epsilon-1}| \quad , \quad \frac{n\lambda}{4b} < \left|\frac{\sqrt{\epsilon-1}}{\epsilon}\right| \tag{19.8}$$

b) x-polarized hybrid modes $E^xH^y_{mn}$:

$$E_y = E_z = 0 \quad , \quad H_x = H_z = 0$$

$$E_x = \frac{1}{\sqrt{ab}} \exp[i(\beta_{mn}z-\omega t)] \exp[-\alpha_{mn}z] \begin{bmatrix} \cos\left[\frac{m\pi x}{2a}\right]\cos\left[\frac{n\pi y}{2b}\right] & m,n \text{ odd} \\ \cos\left[\frac{m\pi x}{2a}\right]\sin\left[\frac{n\pi y}{2b}\right] & m \text{ odd, } n \text{ even} \\ \sin\left[\frac{m\pi x}{2a}\right]\cos\left[\frac{n\pi y}{2b}\right] & m \text{ even, } n \text{ odd} \\ \sin\left[\frac{m\pi x}{2a}\right]\sin\left[\frac{n\pi y}{2b}\right] & m,n \text{ even} \end{bmatrix}$$

$$H_y = -\sqrt{\frac{\epsilon_0}{\mu_0}}\, E_x \tag{19.9}$$

$$\alpha_{mn} = \frac{\pi^2}{4k^2}\left[\frac{m^2}{a^3}Re\left(\frac{\epsilon}{\sqrt{\epsilon-1}}\right) + \frac{n^2}{b^3}Re\left(\frac{1}{\sqrt{\epsilon-1}}\right)\right] \tag{19.10}$$

$$\beta_{mn} = k\left[1 - \frac{\pi^2}{8k^2}\frac{m^2}{a^2}\left(1-\frac{2}{ak}Im\left(\frac{\epsilon}{\sqrt{\epsilon-1}}\right)\right) - \frac{\pi^2}{8k^2}\frac{n^2}{b^2}\left(1-\frac{2}{bk}Im\left(\frac{1}{\sqrt{\epsilon-1}}\right)\right)\right] \tag{19.11}$$

with $n \geq 1, m \geq 1$, and the following restrictions:

$$\frac{m\lambda}{4a} < 1 \quad , \quad \frac{n\lambda}{4b} < 1 \tag{19.12}$$

$$\frac{m\lambda}{4a} < \left|\frac{\sqrt{\epsilon-1}}{\epsilon}\right| \quad , \quad \frac{n\lambda}{4b} < |\sqrt{\epsilon-1}| \tag{19.13}$$

Both sets of hybrid modes (x- and y-polarized) are orthonormal:

$$\int_{-b}^{+b}\int_{-a}^{+a} E_{mn}E^{*}_{m'n'}\,dxdy = \delta_{mm'}\delta_{nn'}\exp[-2\alpha_{mn}z] \tag{19.14}$$

where E_{mn} stands for E_y in a) and for E_x in b). The eigenmodes form a complete set and can thus be used as a basis set for a mode expansion. Any continuous field distribution $E(x,y)$ launched into the waveguide can be expanded in a linear combination of the waveguide eigenmodes. Let us first get a feeling for the properties of these waveguide eigenmodes. For the sake of simplicity we restrict the discussion to one dimension (we let b go to infinity) and consider the y-polarized eigenmodes exhibiting the electric field:

$$E_y = \frac{1}{\sqrt{a}}\exp[-\alpha_m z]\exp[i(\beta_m z-\omega t)]\begin{bmatrix}\cos\left[\dfrac{m\pi x}{2a}\right] & \text{m odd}\\ \sin\left[\dfrac{m\pi x}{2a}\right] & \text{m even}\end{bmatrix} \tag{19.15}$$

with:

$$\alpha_m = \frac{m^2\lambda^2}{16\,a^3}Re\left(\frac{1}{\sqrt{\epsilon-1}}\right) \tag{19.16}$$

$$\beta_m = \frac{2\pi}{\lambda}\left[1-\frac{m^2\lambda^2}{32\,a^2}\left(1-\frac{\lambda}{a\pi}Im\left(\frac{1}{\sqrt{\epsilon-1}}\right)\right)\right] \tag{19.17}$$

(19.13) can also be written as:

$$E_y = \frac{1}{2\sqrt{a}}\exp[-\alpha_m z]\left[\exp[i(\frac{m\pi}{2a}x+\beta_m z-\omega t)] + \exp[-i(\frac{m\pi}{2a}x+\beta_m z-\omega t)]\right]$$

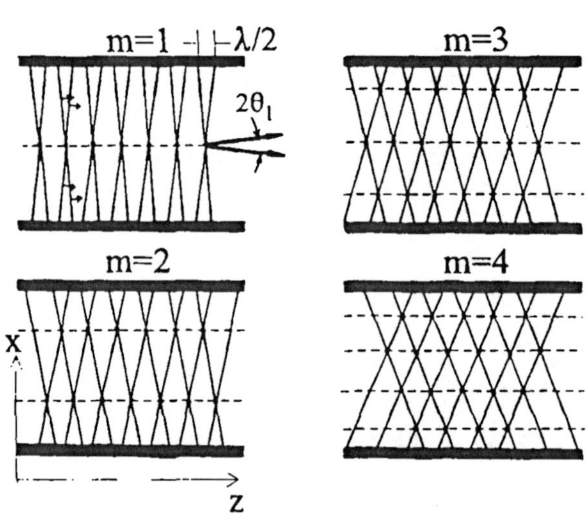

Fig. 19.6 The eigenmodes of a rectangular waveguide can be considered a superposition of two plane waves propagating at angles $\pm\theta_m$ to the z-axis. The horizontal, broken lines indicate the location of the intensity maxima. Note the phase difference of $\lambda/2$ between the two waves at the wall.

The symmetric fields in This expression represents a decomposition into two plane waves propagating at angles $\pm\theta_m$ to the z-axis. Using the above conditions that $\lambda \ll a$ and $m\lambda/(4a) \ll 1$, we can approximate β_{mn} by $2\pi/\lambda$. The angle θ_m is then given by:

$$\theta_m = \frac{m\,\lambda}{4\,a} \tag{19.18}$$

Thus, each eigenmode can be described as a superposition of two plane waves being reflected off the waveguide walls at an angle θ_m. A similar decomposition into two plane waves can also be performed for the asymmetric modes (m even), resulting in the same expression (19.18) for the propagation angles. For all waveguide modes, the intensity maxima can now be understood as a result of constructive interference of the two plane waves (Fig. 19.6). Waves propagating at any other angles than the ones given by (19.18) do not generate a constant intensity distribution in the z direction due to a varying phase shift between the waves. The waveguide eigenmodes are determined by the fact that each of the plane waves has to interfere constructively with the other wave after being reflected by the wall. The propagation angles for the fundamental waveguide mode ($m=1$) of CO_2 lasers typically are between 1 and 3mrad. We are thus dealing with reflection at grazing incidence. An accurate treatment of the reflection at the interface of an absorbing medium reveals that at grazing incidence both p- and s-polarized waves experience a phase shift of π. Taking this phase shift into account, a zero phase shift between incoming wave 1 and the reflected wave 2 can only be realized at the angles θ_m (note in Fig. 19.6 that the two waves indeed exhibit a phase shift of π at the wall to compensate the phase shift due to the reflection). The absorption loss of waveguide modes can now be simply explained by applying the Fresnel equations for the reflection and absorption of light at the interface of an absorbing medium. For grazing incidence, the power loss due to absorption increases proportionally to the angle between the wave vector and the surface (Fig.19.7).

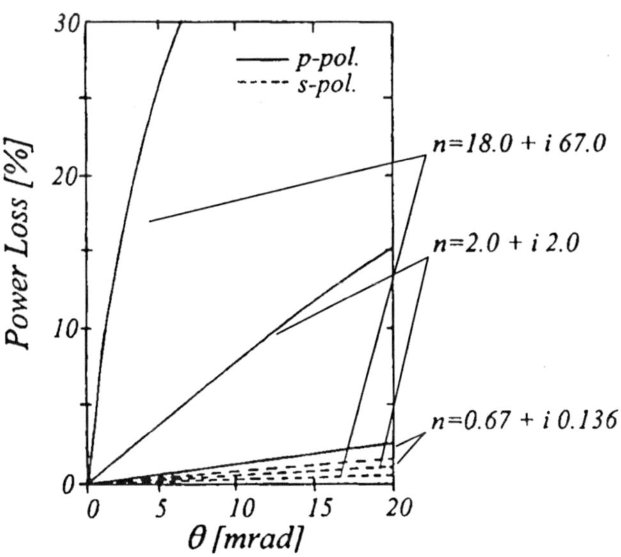

Fig. 19.7 Reflection of a plane wave at the interface of an absorbing medium. The power loss of p-polarized and s-polarized waves as a function of the angle θ between the wave vector and the surface is shown for different refractive indices n of the medium (calculation using Fresnel equations).

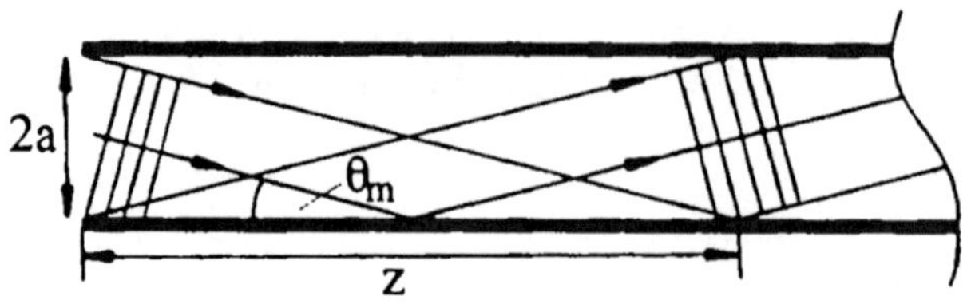

Fig. 19.8 Reflection of a plane wave propagating at the angle θ_m to the waveguide wall.

In general, p-polarized light experiences higher losses than s-polarized light. For metals, the difference in loss is on the order of 10^4 which means that only the y-polarized modes should be observed in the one-dimensional waveguide. This linearity of the loss versus the angle enables us to write the absorption loss ΔV experienced by one of the plane waves due to the reflection as:

$$\Delta V = \gamma\, \theta_m = 1 - \exp[-2\alpha_m z] \approx 2\alpha_m z \tag{19.19}$$

where α_m is the absorption coefficient of the electric field, z is the interaction length as shown in Fig. 19.8, and γ is a proportionality factor. The interaction length z depends on the propagation angle θ_m according to:

$$z = \frac{2a}{\tan\theta_m} \approx \frac{2a}{\theta_m} \tag{19.20}$$

Insertion of (19.20) into (19.19) in combination with (19.18) yields for the absorption coefficient:

$$\alpha_m = \gamma\, \theta_m \frac{\theta_m}{4a} = \gamma \frac{\lambda^2}{64} \frac{m^2}{a^3} \tag{19.21}$$

Thus, we obtain an expression identical to (19.16), a result that is very satisfying. In a next step, the proportionality constant γ could be determined by inserting the correct expression for the loss given by the Fresnel equations. The more or less phenomenologic approach to the waveguide modes by assuming that the propagation occurs by repeated glancing-incidence Fresnel reflection has definitely helped to understand the mode properties. Actually, we could have used this approach right from the start and we would have gotten almost the exact mode properties. This is, of course, not very surprising since we only separated the physical problem into two: interference of plane waves and grazing incidence Fresnel reflection at the waveguide wall.

Table 19.1 Refractive indices and calculated absorption coefficients of the x-polarized (α_l^x) and the y-polarized (α_l^y) fundamental mode for different waveguide materials (one-dimensional waveguide in the x direction with a=1mm and λ=10.6μm, using Eqs. (19.5) and (19.10)) The refractive indices $n=n_1+in_2$ of the metals are from: Handbook of Optics, vol.II. Washington: Optical Society of America 1995, Chapter 35 (extrapolated data). The refractive indices of the dielectrics are from [5.93].

Material	n_1	n_2	α_l^x[1/mm]	α_l^y[1/mm]
Cu	15	63	1.054×10^{-4}	2.51×10^{-8}
Au	18	67	1.264×10^{-4}	2.63×10^{-8}
Al	28	95	1.966×10^{-4}	2.00×10^{-8}
Al_2O_3	0.67	0.136	2.238×10^{-6}	1.40×10^{-6}
BeO	0.64	0.59	4.688×10^{-6}	2.13×10^{-6}
SiO_2	1.9	0.73	1.506×10^{-5}	3.37×10^{-6}

Waveguide modes exhibit some interesting features, some of which are quite contrary to those of the modes of open, stable resonators:

- The beam diameters of the transverse modes scale with the waveguide cross section.
- Due to the reflection losses, the active medium (waveguide) provides a transverse mode discrimination proportional to m^2/a^3. Since higher order modes exhibit higher absorption losses, only a limited number of transverse waveguide modes (let us say not more than 20) will oscillate in a waveguide laser.

Table 19.1 presents the calculated amplitude absorption coefficients for different waveguide materials in a one-dimensional waveguide with a=1mm and b-->∞. For a waveguide length of L=1m, the power loss per transit for the y-polarized fundamental mode is on the order of 0.005% for metals and 0.4% for dielectric materials. We see that high order modes will be strongly attenuated in the waveguide. However, as we shall discuss in detail later, the properties of waveguide resonators can only be adequately described if several transverse waveguide modes are taken into account. This is due to the fact that waveguide modes are not eigenmodes of free space propagation. If the mirrors are not attached to the waveguide, the transverse modes of a waveguide resonator, therefore, must be linear combinations of several waveguide modes.
At the end of the waveguide, the transverse waveguide mode of order m propagates into free space at a full divergence angle of $2\theta_m$. Assuming that the waist diameter is about $1.5a$, we can make a rough estimate for the beam propagation factor M^2 of each mode:

$$M^2 = 0.75\ a\ \theta_m \frac{\pi}{\lambda} \approx m \qquad (19.22)$$

In order to attain a diffraction limited laser beam, the lowest loss mode of the waveguide resonator should closely resemble the fundamental waveguide mode.

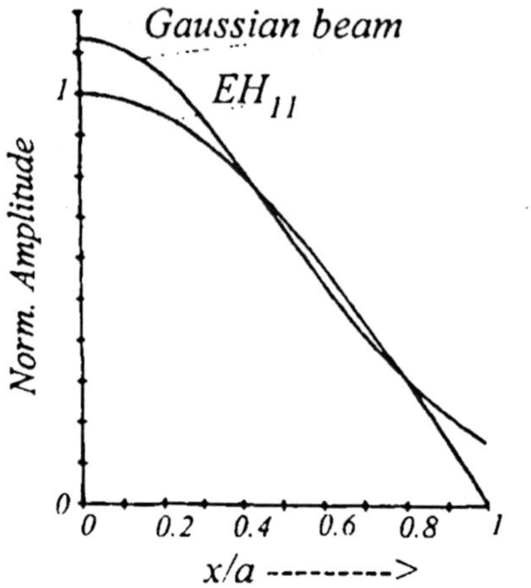

Fig. 19.9 Approximation of the field of the fundamental waveguide mode EH_{11} by a Gaussian beam with beam radius $w_0=0.7032a$.

The goal of waveguide resonator design, of course, is to attain a near diffraction limited output beam, preferably with properties similar to those of a Gaussian beam. Since the field distributions of the waveguide modes form a complete orthogonal set, we can write the one-dimensional field distribution $E_m(x)$ of the m-th resonator mode at the end of the waveguide as a linear combination of waveguide mode field distributions $\Phi_n(x)$:

$$E_m(x) = \sum_{n=1}^{\infty} c_{mn}\ \Phi_n(x) \tag{19.23}$$

where the c_{mn} are complex constants. The propagation of this field distribution to the mirror and back to the waveguide will generate higher losses for the high order waveguide modes since they propagate at higher angles of divergence. The free space propagation thus provides an additional waveguide mode discrimination. By using appropriate mirror curvatures and mirror distances, it should, therefore, be possible to generate a lowest loss resonator mode exhibiting most of its power in the fundamental waveguide mode. As shown in Fig. 19.9, the fundamental waveguide mode EH_{11} can be approximated by a Gaussian beam with beam radius $w_0=0.703a$ resulting in a power overlap of about 98% [5.64]. If we added EH_{13} and EH_{15} modes into the expansion, the truncation error would be in the sub-percent range. This Gaussian beam could be similar to the fundamental resonator mode, for instance, if we imaged the waveguide end onto itself by means of the resonator mirror. Higher order waveguide modes would be suppressed because they are coupled back into the waveguide less efficiently. Before discussing waveguide resonator design in detail, let us first list the waveguide modes of hollow, circular waveguides. We do this without much explanation since the mode properties are basically the same. Those readers who need to get more, detailed information on circular waveguide modes are referred to [5.52,5.53,5.61,5.63,5.67]. The eigenmodes of coaxial waveguides can be found in [5.166].

Transverse Eigenmodes of Hollow, Circular Waveguides

We consider a waveguide with inner radius a having the same properties of the wall material as the rectangular waveguide discussed above. Circular waveguides support three different sets of eigenmodes. Using the assumption that the waveguide radius a is very large compared to the wavelength, the transverse field components become negligibly small and the modes can be classified as circularly polarized transverse electric (TE modes), transverse magnetic modes (TM modes), and hybrid modes (EH modes). The modes in each class form a complete orthogonal set. A fourth class of modes exhibiting linear polarization can be obtained by a combination of two hybrid modes with identical propagation constants. These waveguide modes are referred to as LP modes. The modes of the waveguide resonator can be expanded as a linear combination of waveguide eigenmodes of a single class. However, linear combinations of TM or TE modes yield field distributions without azimuthal structure and with zero intensity on the optical axis. Therefore, they are usually not considered in waveguide resonator theory. In most publications, the resonator properties are analyzed by using the linearly polarized, circularly symmetric hybrid modes EH_{1n} for the resonator mode expansion since they form a complete set for circularly symmetric field distributions. In the following we list the field distributions of the four classes of waveguide eigenmodes in polar coordinates [5.63]:

a) Transverse Electric Modes (TE_{0m},m>0):

$$E_\phi \propto \exp[i(\beta_{0m}z-\omega t)]\ \exp[-\alpha_{0m}z]\ J_1\left[\frac{u_{0m}r}{a}\right] \tag{19.24}$$

$$H_r = -\sqrt{\frac{\epsilon_0}{\mu_0}}\ E_\phi \tag{19.25}$$

$$H_\phi = H_z = 0 \quad , \quad E_r = E_z = 0$$

b) Transverse Magnetic Modes (TM_{0m}, m>0):

$$E_r \propto \exp[i(\beta_{0m}z-\omega t)]\ \exp[-\alpha_{0m}z]\ J_1\left[\frac{u_{0m}r}{a}\right] \tag{19.26}$$

$$H_\phi = -\sqrt{\frac{\epsilon_0}{\mu_0}}\ E_r \tag{19.27}$$

$$H_r = 0 \quad , \quad E_\phi = 0$$

c) Hybrid Modes (EH_{nm}, n<0 or n>0, m>0):

$$E_r = A_{nm} \exp[i(\beta_{nm} z - \omega t)] \exp[-\alpha_{nm} z] J_{n-1}\left[\frac{u_{nm} r}{a}\right] \sin[n\phi] \tag{19.28}$$

$$E_\phi = A_{nm} \exp[i(\beta_{nm} z - \omega t)] \exp[-\alpha_{nm} z] J_{n-1}\left[\frac{u_{nm} r}{a}\right] \cos[n\phi] \tag{19.29}$$

$$H_r = -\sqrt{\frac{\epsilon_0}{\mu_0}}\, E_\phi \quad , \quad H_\phi = -\sqrt{\frac{\epsilon_0}{\mu_0}}\, E_r \tag{19.30}$$

where $k=2\pi/\lambda$ is the wave number and u_{nm} is the m-th root of the equation $J_{n-1}(u_{nm})=0$. This equation sets the fields to zero at the waveguide wall (infinite conductivity of the wall). The constants A_{nm} normalize the power at $z=0$ with:

$$A_{1m} = \frac{1}{a\sqrt{\pi}J_1(u_{1m})} \quad , \quad and \quad A_{nm} = \frac{\sqrt{2}}{a\sqrt{\pi}J_{n-1}(u_{1m})} \quad for\ n>1$$

The propagation constants for all three sets of modes are given by:

$$\alpha_{nm} = \left[\frac{u_{nm}}{k}\right]^2 \frac{1}{a^3} Re(X) \tag{19.31}$$

$$\beta_{nm} = k\left(1 - \frac{1}{2}\left[\frac{u_{nm}}{ka}\right]^2 \left[1 + \frac{2}{ka} Im(X)\right]\right) \tag{19.32}$$

The function X is different for the three modes:

$$X = \begin{cases} 1/\sqrt{\epsilon - 1} & for\ TE\ modes \\ \epsilon/\sqrt{\epsilon - 1} & for\ TM\ modes \\ (\epsilon + 1)/(2\sqrt{\epsilon - 1}) & for\ EH\ modes \end{cases} \tag{19.33}$$

In a waveguide resonator, each mode propagates to the resonator mirror and is, to a certain degree, coupled back into the waveguide. In order to calculate the free space propagation of these modes using diffraction integrals, the electric field distributions first have to be separated into linearly polarized fields in the x and the y direction. The cartesian field amplitudes are given by:

$$E_x = E_r \cos\phi - E_\phi \sin\phi \tag{19.34}$$

$$E_y = E_r \sin\phi + E_\phi \cos\phi \tag{19.35}$$

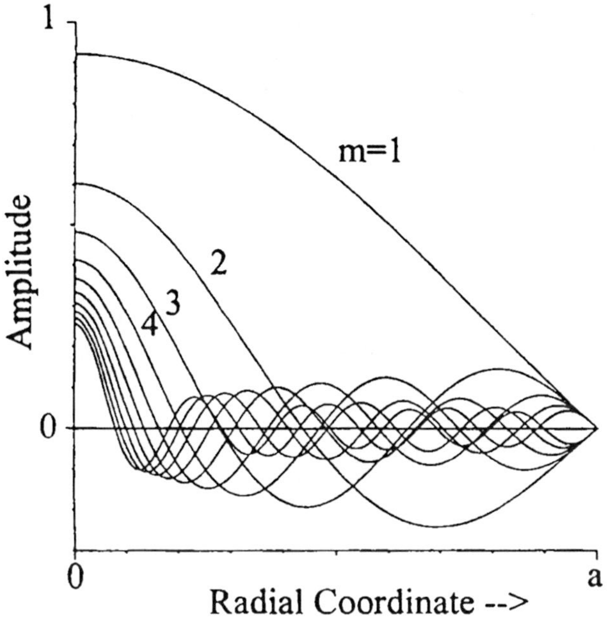

Fig. 19.10 Radial electric field distribution of EH_{1m} modes for m=1-10. The field distributions are not power normalized [5.83].

By using this transformation it can be easily shown that the hybrid modes EH_{1m} are linearly polarized in the y direction. Owing to the linear polarization and the circular symmetry, these waveguide modes are commonly used to analyze waveguide resonators by expanding each resonator mode into a linear combination of free space modes (Gauss-Laguerre modes) outside the waveguide and into a linear combination of waveguide modes inside the waveguide. Figure 19.10 presents the radial amplitude distribution of EH_{1m} modes for m=1 to 10.

Since the field of the EH_{1m} mode is p-polarized with respect to the waveguide surface, the absorption coefficients α_{1m} should be similar to those of x-polarized hybrid modes in one-dimensional rectangular waveguides (with the waveguide gap along the x direction). Therefore, metallic waveguides exhibit a relatively high power loss per transit for linearly polarized hybrid modes (we ignore the fact that it would be technically impossible to create a gas discharge in a circular, metallic structure unless a cathode is inserted in the center). Table 19.2 presents the absorption coefficients α_{1m}, calculated with (19.31), and the corresponding power loss per transit for a 1m long waveguide (see Table 19.1 for comparison). In metallic, circular waveguides, the TE modes should dominate due to the azimuthal polarization of the electric field.

Table 19.2 Refractive indices and calculated absorption coefficients of the fundamental hybrid mode EH_{11} for di fferent waveguide materials (circular waveguide with radius $a=1mm$ and wavelength $\lambda=10.6\mu m$, using Eqs. (19.31) and (19.33)). For a waveguide length of $L=1m$, the corresponding power loss ΔV per transit is shown. The refractive indices $n=n_1+in_2$ of the metals are from: Handbook of Optics, vol.II. Washington: Optical Society of America 1995, Chapter 35 (extrapolated data). The refractive indices of the dielectrics are from [5.93].

Material	n_1	n_2	α_1^*[1/mm]	ΔV [%]
Cu	15	63	1.235×10^{-4}	21.88
Au	18	67	1.482×10^{-4}	25.65
Al	28	95	2.305×10^{-4}	36.93
Al_2O_3	0.67	0.136	4.264×10^{-6}	0.85
BeO	0.64	0.59	7.986×10^{-6}	1.58
SiO_2	1.9	0.73	2.161×10^{-5}	4.23

d) Linearly Polarized Modes (LP_{nm}, $n \geq 0, m \geq 1$):

By combining equal amounts of hybrid modes $EH_{n-1,m}$ and $EH_{n+1,m}$ for $n \geq 2$, linearly polarized field distributions are generated exhibiting intensity profiles that closely resemble those of Gauss-Laguerre modes of free space propagation. The hybrid modes have, to the order discussed here, identical propagation constants and the combination, therefore, is also an eigenmode of the circular waveguide. The modes LP_{0m} are identical to the hybrid modes EH_{1m}, and the modes LP_{1m} are generated by a superposition of an EH_{2m} mode and a TE_{0m} mode. Unfortunately, the latter two modes have only similar propagation constants in weakly guiding waveguides ($\epsilon \approx 1$). Only under this condition do the linearly polarized LP_{1m} modes represent true eigenmodes. The propagation constants of the modes LP_{nm} with $n=0$ or $n \geq 2$ and of the LP_{1m} modes with $\epsilon \approx 1$ are equal to those of the hybrid modes EH_{nm}. The electric field of the LP modes is given by:

$$E_{nm} = B_{nm} \exp[i(\beta_{nm} z - \omega t)] \exp[-\alpha_{nm} z] \, J_n\left[\frac{u_{nm} r}{a}\right] \cos[n\phi] \tag{19.36}$$

with $n \geq 0$, $m \geq 1$, and the normalization constants:

$$B_{0m} = \frac{1}{a\sqrt{\pi} J_1(u_{0m})} , \quad and \quad B_{nm} = \frac{\sqrt{2}}{a\sqrt{\pi} J_{n+1}(u_{nm})} \quad for\ n \geq 1$$

Provided that the weak guidance condition is met, the LP modes may be used as a basis set for expanding linearly polarized field distributions inside the waveguide, preferably those exhibiting noncircular symmetry.

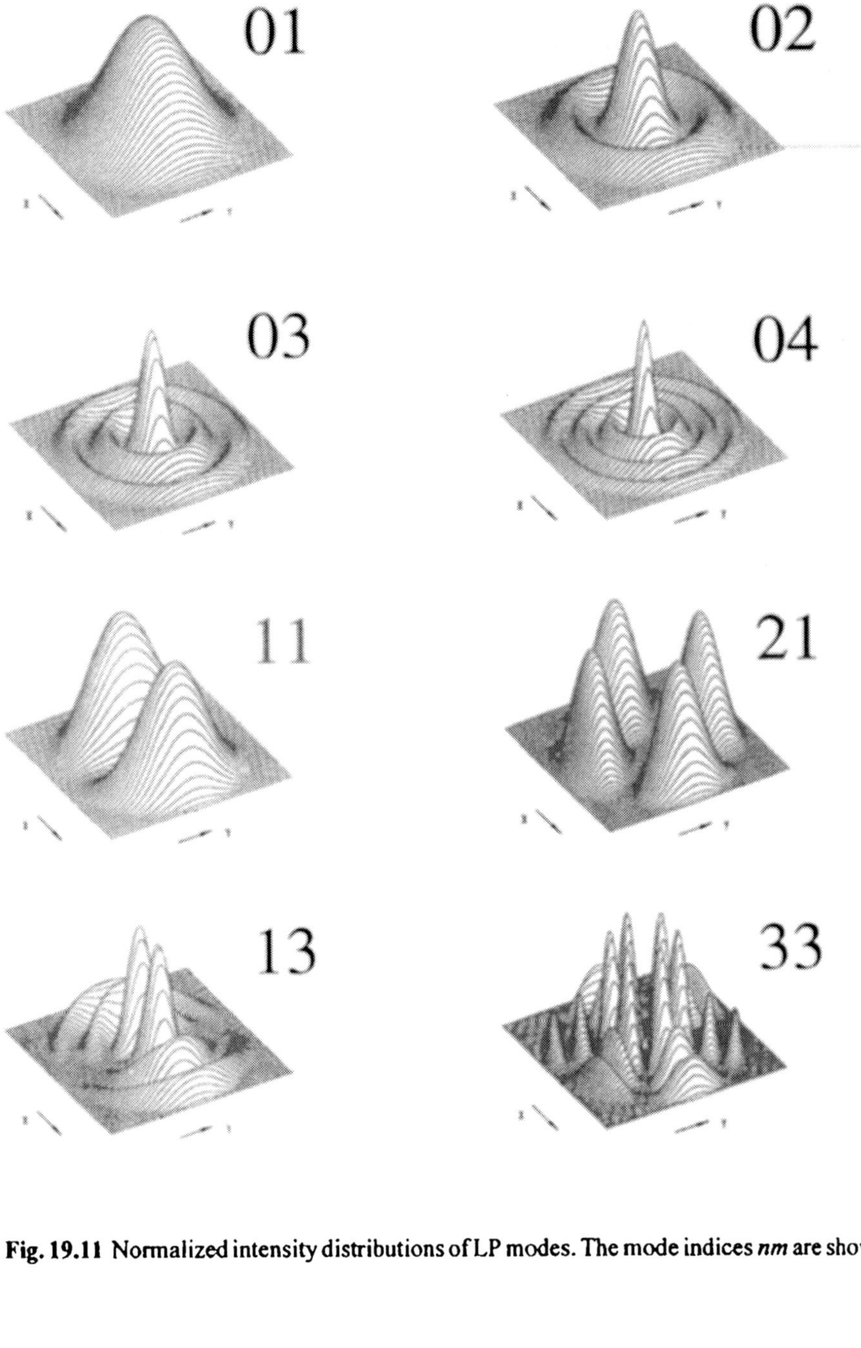

Fig. 19.11 Normalized intensity distributions of LP modes. The mode indices *nm* are shown [5.83].

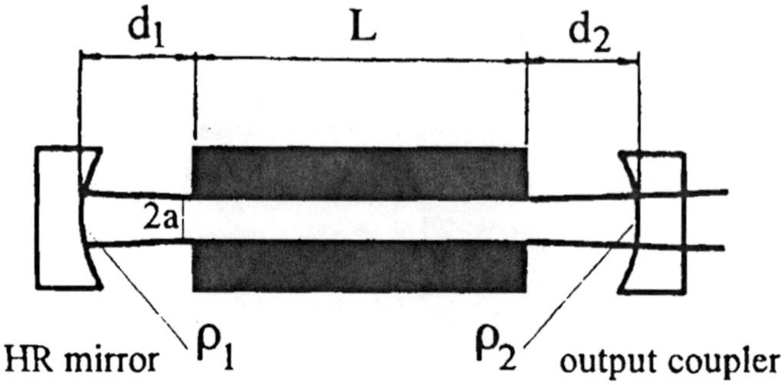

Fig. 19.12 Schematic of a waveguide resonator.

19.3.3 Properties of Waveguide Resonators

With the knowledge of the waveguide eigenmodes and their propagation constants, it is now possible to evaluate the mode properties of the waveguide resonator itself. The general waveguide resonator consists of two mirrors with radii of curvature ρ_1 and ρ_2 positioned at a distance d_1 and d_2 from the waveguide ends (Fig. 19.12). The free space propagation from the waveguide to the mirrors and back generates coupling losses for the waveguide modes. We are, of course, interested in resonator configurations that provide a fundamental resonator mode with low coupling losses and a good mode discrimination against higher order modes. Before determining the resonator properties in a mathematical way, let us first try to find low loss configurations by using our intuition.

If we place two flat mirrors right against the waveguide, the transverse modes of the resonator are identical with the waveguide eigenmodes. The round trip loss of each mode is due to its waveguide absorption loss $2\alpha L$. Thus, the mode discrimination is rather low and, depending on the small-signal gain, the output coupling, the type of line broadening, and the dimensions of the waveguide, the output beam will be a combination of several transverse waveguide modes. Assuming a refractive index of $n=1$ of the gas, the resonance frequencies of the modes are determined by:

$$2\,\beta_{nm}\,L = q\,2\pi \tag{19.37}$$

where q is the axial mode order, L is the waveguide length, and β_{nm} is the propagation constant according to (19.6) or (19.32). To a good approximation, (19.37) yields for the mode frequencies of rectangular waveguides:

$$\nu_{nmq} = \frac{c_0}{2L}q + \frac{c_0\lambda}{32}\left[\frac{m^2}{a^2} + \frac{n^2}{b^2}\right] \tag{19.38}$$

where λ is the center wavelength of the gain profile and c_0 is the speed of light in a vacuum. In circular symmetry one obtains from (19.37) and (19.32):

$$\nu_{nmq} = \frac{c_0}{2L}q + \frac{c_0 \lambda}{8\pi^2}\left[\frac{u_{nm}^2}{a^2}\right] \tag{19.39}$$

In order to achieve fundamental mode operation, it is necessary to choose the waveguide cross section a small so that higher order modes cannot reach the laser threshold. Unfortunately, the coupling losses of the fundamental waveguide mode will increase too, resulting in a decrease in laser efficiency. A better mode discrimination is attained if the mirrors are slightly separated from the waveguide. The distance is typically on the order of ten times the waveguide cross section. Waveguide resonators with mirrors close to the waveguide are referred to as *Case I Resonators.*

Similar mode properties are to be expected in *Case II Resonators* in which the waveguide ends are imaged onto themselves by the resonator mirrors. Imaging requires a distance d_i of each mirror equal to its radius of curvature ρ_i. Each waveguide mode is imaged back onto itself and no coupling losses occur for infinite mirrors. However, the transverse mode discrimination is as poor as in true Case I resonators. Note that for the calculation of the mode frequencies, the free space propagation has to be taken into account in (19.37).

As was shown in Fig.19.9, the fundamental waveguide mode in rectangular symmetry EH_{11} can be approximated by a Gaussian beam with beam radius $w_0=0.7032a$ with a power overlap of 98% [5.66]. In circular waveguides a similar overlap between the lowest order hybrid mode and the Gaussian beam can be achieved with $w_0=0.6345a$ [5.56] (Fig. 19.13). Thus, it is reasonable to describe the EH_{11} mode propagation into free space like a Gaussian beam propagation. Given this assumption, the wavefront curvature of the mode at a distance z from the waveguide reads:

$$R(z) = 2z_0\left[\frac{z}{z_0} + \frac{z_0}{z}\right] \tag{19.40}$$

where $z_0=\pi w_0^2/\lambda$ is the Rayleigh range. If we choose the mirror curvature ρ equal to $R(z)$ at the mirror position $d=z$, we simply reverse the beam propagation and the EH_{11} mode will be coupled back into the waveguide with low losses. This configuration combines low fundamental mode loss and excellent transverse mode discrimination since higher order modes will not couple back efficiently. It is for this reason that most waveguide resonators investigated in the past were based on this priniciple. Numerical calculation of the coupling losses of the EH_{11} mode indicated that positioning the mirrors at $d=z_0$ with $\rho=2z_0$ yields the lowest loss. Waveguide resonators with $d\approx\rho/2$ are referred to as *Case III Resonators.*

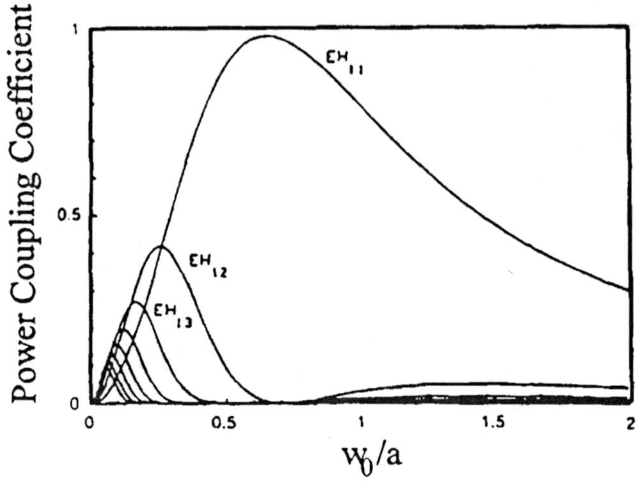

Fig. 19.13 Power coupling coefficients for the EH_{lm} modes of circular waveguides as a function of the ratio of the beam waist radius w_0 of the Gaussian beam to the guide radius a [5.94] (© OSA 1992).

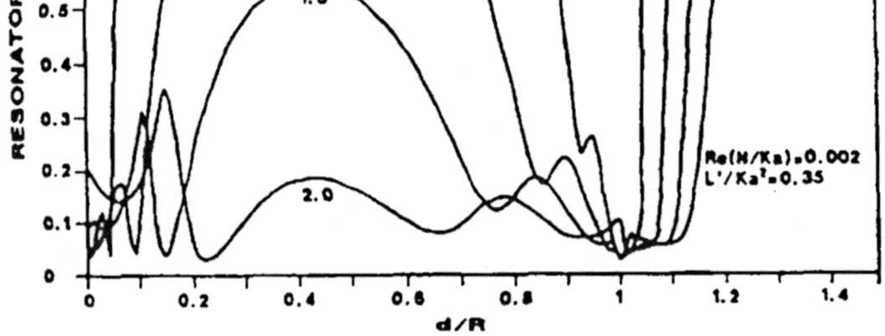

Fig. 19.14 Calculated round trip loss (lowest loss mode) for symmetric waveguide resonators with a circular waveguide of radius a and effective length L' as a function of the mirror distance to radius of curvature ratio d/R. The curve parameter is the inverse of the normalized mirror curvature $R/(ka^2)$; K is the wave number and N is the function X in our notation (see (19.32). Both mirrors have a radius of curvature R and are located at a distance d from the adjacent waveguide end. Calculation using multiple waveguide modes[5.89] (© IEEE 1990).

Calculated round trip losses of the lowest loss resonator mode as a function of d/ρ are presented in Fig. 19.14 (ρ=R) for symmetric resonators and circular waveguides. The parameter of the curves is $\alpha = 2\pi a^2/(\lambda\rho)$. The meaning of this parameter and others shown in the graph shall be explained in the next paragraph. Case I and Case II resonators are given by $d/\rho=0$ and $d/\rho=1$, respectively. Case III coupling with $d=z_0$ is equivalent to $\alpha=2.41$ and $d/\rho=0.5$. Although the curve with $\alpha=2.41$ is not shown, it is obvious that Case III resonators provide similar low losses as the other two resonators. However, this graph indicates that low losses can also be realized with resonators that do not belong to one of the three resonator classes. These additional loss minima are a result of the interference between different waveguide modes. A realistic description of waveguide resonator properties, therefore, requires the incorporation of multiple waveguide modes.

Waveguide Resonator Mode and Loss Calculations

Since the waveguide modes and the free space modes (Gauss-Laguerre polynomials in circular symmetry and Gauss-Hermite polynomials in rectangular symmetry) both form complete sets, they can be used as basis sets for the resonator modes. In general, a limited number of functions (on the order of 10 waveguide modes and 50 free space modes) from both sets are taken into account. This is mathematically not correct, but the truncation errors can be kept in the sub-percent range if only the first three lowest loss resonator modes are considered. For simplicity we restrict the discussion to one-dimensional rectangular waveguides [5.76,5.84]. A similar treatment for circular waveguides can be found in [5.71]. The field distribution $E_p(x)$ of the p-th resonator mode at the end of the waveguide is written as a linear combination of N waveguide eigenmode field distributions $\Phi_n(x)$ and as a linear combination of M free space modes $\Psi_m(x)$. For the free space modes we use Gauss-Hermite polynomials with the beam waist with radius w_0 located at the waveguide end. The Gaussian beam waist radius w_0 is preferably chosen in the range $0.6a<w_0<0.8a$. The two power normalized function sets are given by:

$$\Phi_n(x) = \frac{1}{\sqrt{a}} \begin{bmatrix} \cos\left[\frac{n\pi x}{2a}\right], & n \text{ odd} \\ \sin\left[\frac{n\pi x}{2a}\right], & n \text{ even} \end{bmatrix} \qquad n=1,\ldots,N \qquad (19.41)$$

$$\Psi_m(x) = \left(\frac{2}{\pi}\right)^{1/4} \frac{1}{\sqrt{w_0 2^m m!}} \exp\left[-\frac{x^2}{w_0^2}\right] H_m\left[\sqrt{2}\frac{x}{w_0}\right] \qquad m=1,\ldots,M \qquad (19.42)$$

The mode expansions of the field distribution of the p-th resonator mode read:

$$E_p(x) = \sum_{n=1}^{N} c_{pn}\Phi_n(x) \tag{19.43}$$

$$E_p(x) = \sum_{m=1}^{M} d_{pm}\Psi_m(x) \tag{19.44}$$

with $$c_{pn} = \int_{-a}^{a} E_p(x)\Phi_n^*(x)dx \quad and \quad d_{pm} = \int_{-a}^{a} E_p(x)\Psi_m^*(x)dx \tag{19.45a,b}$$

The expansion coefficients are related to one another via:

$$c_{pn} = \sum_{m=1}^{M} d_{pm}\int_{-a}^{a}\Psi_m(x)\Phi_n^*(x)dx \tag{19.46}$$

$$d_{pm} = \sum_{n=1}^{N} c_{pn}\int_{-a}^{a}\Psi_m^*(x)\Phi_n(x)dx \tag{19.47}$$

The free space modes can now be propagated to the mirror and back to the waveguide. The new field distributions $\Psi_m'(x)$ of the propagated modes can be calculated analytically using the Gaussian ABCD law. The new field distribution of the resonator mode *p* is now given by:

$$E_p'(x) = \sum_{m=1}^{M} d_{pm}\Psi_m'(x) \tag{19.48}$$

This field can again be expanded into waveguide modes $\Phi_q(x)$ with expansion coefficients:

$$c_{pn}' = \sum_{m=1}^{M} d_{pm}\int_{-a}^{a}\Psi_m'(x)\Phi_n^*(x)dx \tag{19.49}$$

Insertion of (18.45b) into (18.49) yields a matrix equation relating the wave guide mode expansion coefficients before to the ones after the free space propagation:

$$\boldsymbol{C}_p' = \boldsymbol{M}_1\ \boldsymbol{C}_p \tag{19.50}$$

where the components of the vectors $\boldsymbol{C}_p$ and $\boldsymbol{C}_p'$ are the *N* expansion coefficients c_{pq} (q=1,...,N) and c_{pn}' (n=1,...,N), respectively. The components of the NxN matrix read:

$$M_{nq} = \sum_{m=1}^{M} \int_{-a}^{a} \Psi_m^*(x)\Phi_q(x)dx \int_{-a}^{a} \Psi_m'(x)\Phi_n^*(x)dx \ , \quad n,q=1,2,...,N \tag{19.51}$$

Due to the symmetry of the eigenfunctions, all integrals in (19.51) with $m+q$ or $m+n$ even are zero. This reduces the number of integrals to be calculated to M times N.

Without gain, the propagation through the waveguide induces a phase shift and absorption losses to each waveguide mode, but the modes will not mix since they represent eigenmodes of the waveguide. Therefore, each expansion coefficient is simply multiplied by the corresponding propagation factor. The new mode expansion coefficients are thus determined by:

$$C_p'' = M_{WG}\, C_p' \tag{19.52}$$

where the waveguide matrix M_{WG} has only diagonal elements given by the propagation factor $exp[i(\beta+i\alpha)L]$ for the corresponding mode. For y-polarized modes, the diagonal elements of the waveguide matrix read (see (19.16) and (19.17)):

$$M_{nn} = \exp\left[-i\left(\frac{n\pi}{\sqrt{8}}\right)^2 \frac{L'}{ka^2}\right] \exp\left[-\left(\frac{n\pi}{2}\right)^2 \frac{L}{ka^2}\frac{1}{ka} Re\left(\frac{1}{\sqrt{\epsilon-1}}\right)\right] \tag{19.53}$$

where we dropped the phase factor $exp[ikL]$ for it does not affect the loss or the mode structure of the resonator modes (it has to be taken into account for the resonance frequencies though). The effective waveguide length L' is defined as:

$$L' = L\left(1 - \frac{\lambda}{ka} Im\left(\frac{1}{\sqrt{\epsilon-1}}\right)\right) \tag{19.54}$$

If we continue this procedure and complete the round trip in the waveguide resonator, we find a resulting vector C_p''''. The field distribution $E_p(x)$ can only represent a resonator eigenmode if this last vector is proportional to the initial vector C_p. Thus, the resonator modes are eigenvectors of the eigenvalue equation:

$$\Lambda_p\, C_p = M_1 M_{WG} M_2 M_{WG}\, C_p \tag{19.55}$$

where M_2 is the free space propagation matrix at mirror 2 and Λ_p is the eigenvalue associated to the resonator mode p. The round trip loss ΔV of the p-th resonator mode is given directly by the eigenvalue:

$$\Delta V = 1 - |\Lambda_p|^2 \tag{19.56}$$

The round trip loss includes the absorption losses inside the waveguide and the coupling losses at both waveguide apertures. This matrix method enables us to determine the losses and the mode structure of passive waveguide resonators. However, if higher order resonator modes ($p>3$) are to be calculated, at least 5 waveguide modes and 20 free space modes have to be used to reduce truncation errors. This will not only greatly increase the computation time, but may also result in ill-conditioned matrices (the eigenvalues become sensitive to small variations in the matrix elements). It is recommended to use the presented matrix method only for the first three lowest loss resonator modes.

Before we discuss some results of waveguide resonator mode calculations, let us first define the dimension of the problem with which we are dealing. How many independent parameters are needed to entirely describe a waveguide resonator ? For the free space propagation, four independent parameters can be found in the phase term of the two Collins integrals (see Section 2.3). Those parameters are: ka^2/d_1, ka^2/d_2, d_1/ρ_1, and d_2/ρ_2. The independent parameters governing the waveguide propagation are contained in the propagation constant (19.50): L'/ka^2, L/ka^2, and $Re(1/\sqrt{\epsilon-1})/ka$. A generalized presentation of waveguide resonator properties, therefore, is a seven-dimensional problem. However, in most cases the effective length L' differs from the geometrical length L by no more than a few percent and we may skip L as a parameter. In both circular waveguides and one-dimensional rectangular waveguides, we are dealing with the following six independent, dimensionless parameters:

- *the effective length parameter*: $\dfrac{L'}{ka^2}$
- *the distance parameters*: $\dfrac{d_1}{ka^2}, \dfrac{d_2}{ka^2}$
- *the distance-curvature ratios*: $\dfrac{d_1}{\rho_1}, \dfrac{d_2}{\rho_2}$
- *the waveguide loss parameter*: $\dfrac{Re(X)}{ka}$ with: $X = \begin{bmatrix} \epsilon/\sqrt{\epsilon-1} \\ 1/\sqrt{\epsilon-1} \\ (\epsilon+1)/\sqrt{2(\epsilon-1)} \end{bmatrix}$ *rect., x-pol.* / *rect., y-pol..* / *circular*

Looking back at Fig. 19.14, the different parameters indicated in the graph are now understandable. In order to simplify the discussion, we focus on resonators with two flat mirrors. This restriction leaves us with four independent parameters. The properties of waveguide resonators with curved mirrors can be found in [5.56,5.58,5.72].

Resonator Losses for Pure EH_{11} Oscillation

The historical approach to waveguide resonator theory was based on the assumption that the fundamental resonator mode must closely resemble the fundamental waveguide mode EH_{11} [5.56,5.60,5.64,5.66,5.67]. Considering the excellent power overlap of 98% with a Gaussian beam with beam radius $w_0=0.7032a$ in rectangular geometry, and $w_0=0.6435a$ in circular geometry, this assumption is certainly justified. In fact, later mode calculations using multiple waveguide modes proved that over wide ranges of the resonator parameters, the fundamental resonator mode is predominantly EH_{11} with only very little fractions of next order waveguide modes. Coupling losses in both circular and rectangular geometries were calculated utilizing the discussed matrix method (note that the matrix becomes a scalar) with between 6 and 30 free space modes and the fundamental waveguide mode EH_{11} [5.58,5.64,5.66,5.72]. Different analytical approximations for the coupling losses were derived for flat mirrors located at relatively small distances z from the waveguide (less than 20 times the guide diameter). The numerically calculated coupling losses Γ are best approximated by the following expressions providing the loss in percent [5.72]. To avoid possible confusion we show the functions in their original notation (labeled a) and after conversion into the notation using dimensionless resonator parameters (labeled b):

a) rectangular geometry, square guide with side length 2a [5.70,5.72]:

$$\Gamma = 64.5\left(\frac{z}{z_0}\right)^{\frac{3}{2}} , \quad for \ \frac{z}{z_0} < 0.2 \tag{19.57a}$$

with: $z_0 = \pi w_0^2/\lambda \ , \quad w_0 = 0.7032a.$

$$\Gamma = 524\left(\frac{d}{ka^2}\right)^{\frac{3}{2}} , \quad for \ \frac{d}{ka^2} < 0.05 . \tag{19.57b}$$

b) circular geometry, waveguide diameter 2a [5.58,5.72]:

$$\Gamma = 158\left(\frac{z}{z_0}\right)^{2} , \quad for \ \frac{z}{z_0} < 0.13 \tag{19.58a}$$

$$\Gamma = 57\left(\frac{z}{z_0}\right)^{\frac{3}{2}} , \quad for \ 0.13 \le \frac{z}{z_0} < 0.25 \tag{19.59a}$$

with: $z_0 = \pi w_0^2/\lambda \ , \quad w_0 = 0.6435\ a.$

$$\Gamma = 3{,}686 \left(\frac{d}{ka^2} \right)^2, \qquad for \ \frac{d}{ka^2} < 0.027 \tag{19.58b}$$

$$\Gamma = 605 \left(\frac{d}{ka^2} \right)^{\frac{3}{2}}, \qquad for \ 0.027 \le \frac{d}{ka^2} < 0.052 \tag{19.59b}$$

As we shall see in Fig. 19.16, the latter two expressions deviate sightly from the true coupling losses around the crossing point $d/(ka^2)=0.027$. By using these analytical expressions in combination with the known absorption losses of the EH_{11} mode in the waveguide, we can derive equations for the round trip power loss of the fundamental resonator mode. The resonator model used in this derivation is depicted in Fig. 19.15. The round trip loss ΔV is given by:

$$\Delta V = 1 - V^2(1-\Gamma_1)(1-\Gamma_2) \tag{19.60}$$

where V is the absorption loss factor per transit, and Γ_i is the coupling loss at mirror i (note that $\Gamma_i = \Gamma/100$ since (19.57) through (19.59) give the loss in percent). For low absorption losses ($V>0.9$), insertion of all functions and parameters into (19.60) results in the following final expressions:

a) rectangular symmetry, square waveguide with side length 2a:

$$\Delta V = 1 - \left(1 - \frac{\pi^2 L}{ka^2}\frac{Re(X)}{ka}\right)\left(1 - 5.24\left(\frac{d_1}{ka^2}\right)^{\frac{3}{2}}\right)\left(1 - 5.24\left(\frac{d_2}{ka^2}\right)^{\frac{3}{2}}\right)$$

with: $$X = \frac{\epsilon+1}{\sqrt{\epsilon-1}} \tag{19.61}$$

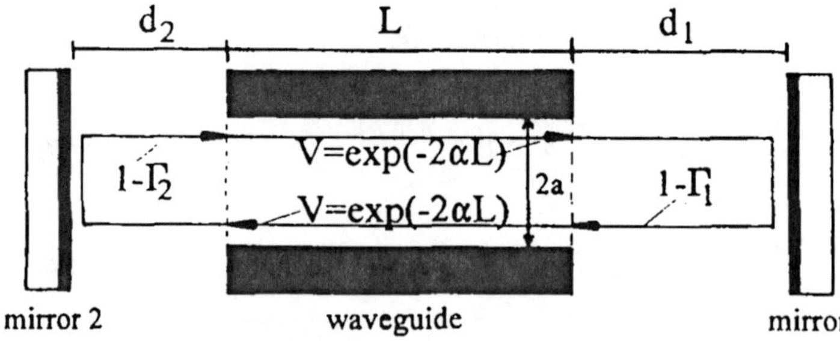

Fig. 19.15 The round trip loss factor (=1- loss) in an EH_{11} waveguide resonator can be calculated by multiplying the loss factors $1-\Gamma_i$ due to coupling with the loss factor V^2 due to absorption in the waveguide.

b) rectangular symmetry, one-dimensional waveguide with side length 2a (slab waveguide with s-polarized hybrid mode):

$$\Delta V = 1 - \left(1 - \frac{\pi^2 L}{ka^2}\frac{Re(X)}{ka}\right)\sqrt{\left(1-5.24\left(\frac{d_1}{ka^2}\right)^{\frac{3}{2}}\right)\left(1-5.24\left(\frac{d_2}{ka^2}\right)^{\frac{3}{2}}\right)}$$

with: $$X = \frac{1}{\sqrt{\epsilon - 1}} \qquad (19.62)$$

c) circular geometry, waveguide diameter 2a:

Four combinations, corresponding to the definition ranges of (19.58) and (19.59), are possible. For simplicity we consider only symmetric resonators with $d_1=d_2=d$:

$$\Delta V_1 = 1 - \left(1 - 4u_{11}^2\frac{L}{ka^2}\frac{Re(X)}{ka}\right)\left(1-\gamma\left(\frac{d}{ka^2}\right)^{\delta}\right) \qquad (19.63)$$

with: $$X = \frac{\epsilon+1}{2\sqrt{\epsilon-1}} ,$$

and $\gamma=36.86$ *and* $\delta=2$ for $d/(ka^2)<0.027$, $\gamma=6.05$ *and* $\delta=1.5$ for $0.027 \le d/(ka^2) < 0.052$, and $u_{11}=2.4048$. A graphical presentation of (19.63) is shown in Fig. 19.16 for alumina. The comparison with the numerically calculated loss indicates that the analytical expressions represent a good approximation for the round trip loss. Note that the loss difference between curve a) and curve b) is due to the higher absorption loss in the waveguide (the waveguide length is 5 times greater for curve b).

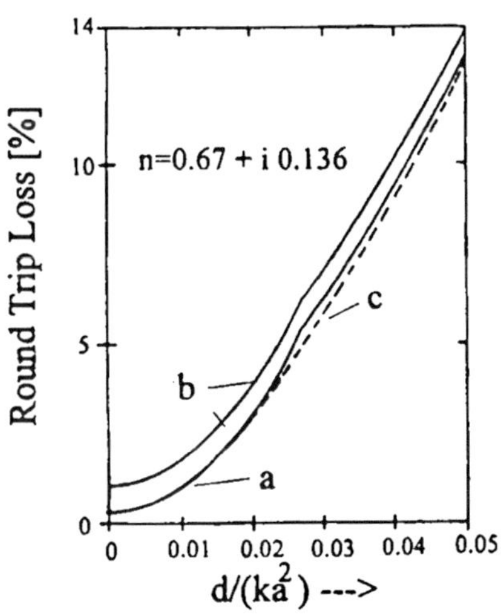

Fig. 19.16 Resonator round trip loss calculated with (19.63) for a circular alumina waveguide with refractive index $n=0.67 + i\ 0.136$. Curve (a): $L/(ka^2)=0.2$; curve (b): $L/(ka^2)=1.0$. Curve (c) is the correct, numerically determined loss for $L/(ka^2)=0.2$ (using data presented in [5.72]). For a=1mm and λ=10.6μm, an effective length parameter $L/(ka^2)$ of 1.0 corresponds to a guide length of $L=592.7mm$.

We could now use the expressions for the round trip loss for the waveguide resonator design since the standard equations for the laser threshold and the output power of open resonators (see Chapter 10) also apply. Unfortunately, the assumption that the resonator is always operating in an EH_{11} mode is in contradiction with experimentally observable changes in mode structure as the resonator parameters are varied. This is a clear indication that the lowest loss mode is a mixture of low order waveguide modes whose relative amplitudes and phases change with the resonator parameters. The necessary incorporation of multiple waveguide eigenmodes into the mathematical description will lead to deviations from the round trip losses presented above due to interference between the modes.

Multimode Calculation of Resonator Losses

Calculated round trip losses of a symmetric, circular waveguide resonator with flat mirrors using the matrix method are presented in Fig. 19.17 for λ=10.6μm. The length of the 2mm cross section aluminum waveguide is varied, but the distance from the waveguide ends to the mirrors remains fixed at $d=20mm$. Over several relatively wide ranges of the waveguide length, low losses of the fundamental mode and an excellent discrimination against the next order mode can be simultaneously realized. In these length ranges the lowest loss mode is predominantly EH_{11} with small amounts of higher order waveguide modes. Operation near points of degeneracy, like at $L/ka^2=0.25$ or 0.5, has to be avoided since the output beam may switch between EH_{11} and EH_{12} like intensity distributions. The most surprising behavior in this graph is the periodic oscillation of the loss curves. This structure can be attributed to interference among two or more waveguide eigenmodes.

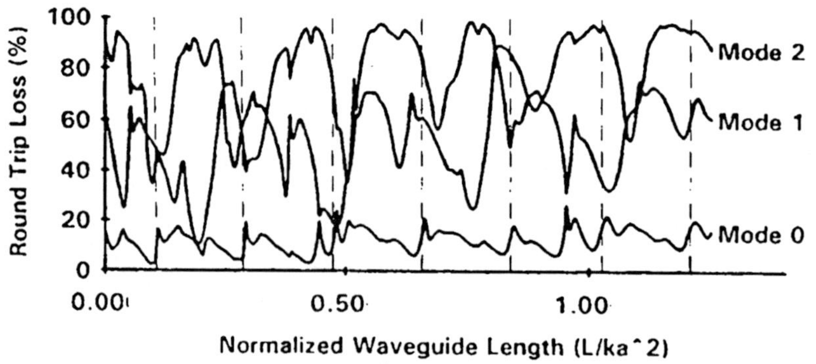

Fig. 19.17 Calculated resonator losses versus the effective length parameter for the three lowest loss modes of a symmetric waveguide resonator with flat mirrors (circular waveguide, $a=1mm$, $\lambda=10.6\mu m$, refractive index of the wall material: $n= 0.67 + i\,0.136$ (alumina), mirror distances: $d_1=d_2=20mm$, $d/(ka^2)=0.0337$. Six hybrid modes EH_{1m} and 12 free space modes TEM_{0m} were used as basis sets. An effective length parameter L/ka^2 of 1.0 corresponds to a waveguide length of L=593mm. See text for an explanation of the vertical lines [5.103] (© SPIE 1994).

At the loss minima, two or more waveguide modes emerge from the guide with relative amplitudes and phases such that the spot size of the beam coupled back into the waveguide assumes a minimum. The coupling loss at that point is lower than the coupling loss of the EH_{11} mode discussed earlier. These resonances were investigated in detail in [5.71] (circular guide) and [5.76] (rectangular guide) using two waveguide modes. The periodicity of the resonances in $L/(ka^2)$ can be attributed to the periodic regeneration of the relative phase between the modes at the waveguide ends (Fig. 19.18). For the two waveguide modes EH_{1n} and EH_{1m} we thus get the condition:

$$\Delta\phi = (\beta_{1m}-\beta_{1n})L = \frac{L}{2ka^2}(u_{1n}^2 - u_{1m}^2) = 2\pi\, p\ , \quad p=1,2,... \tag{19.64}$$

where we used the propagation constant (19.31). For the three lowest loss waveguide modes EH_{11}, EH_{12}, and EH_{13}, (19.64) yields the following distances ΔL between the resonances:

$$EH_{11} - EH_{13} \quad : \quad \frac{\Delta L}{ka^2} = 0.183 \tag{19.65}$$

$$EH_{12} - EH_{13} \quad : \quad \frac{\Delta L}{ka^2} = 0.283 \tag{19.66}$$

$$EH_{11} - EH_{12} \quad : \quad \frac{\Delta L}{ka^2} = 0.509 \tag{19.67}$$

Although the resonator modes in Fig. 19.17 represent combinations of more than two waveguide modes, most loss minima of the lowest loss mode can be assigned to a resonance between two of the three waveguide modes EH_{11}, EH_{12}, and EH_{13}. The broken vertical lines indicate resonances that can be attributed to a mixing of EH_{11} and EH_{13}.

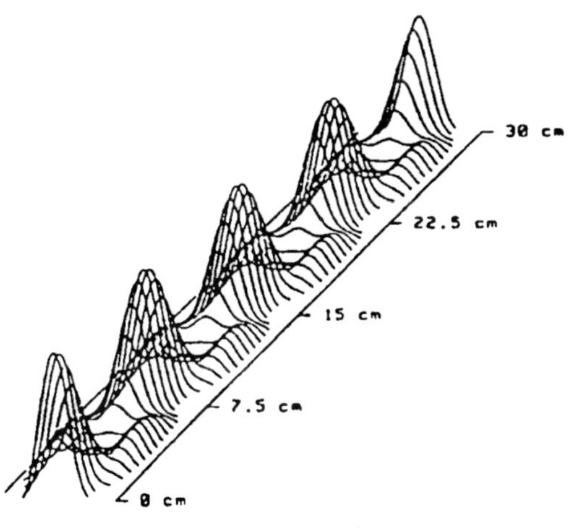

Fig. 19.18 Propagation of a Gaussian beam through a circular waveguide with radius a=1mm (λ=10.6μm). The Gaussian beam with beam waist $w_0=0.4a$ at the entrance plane of the waveguide is expanded into EH_{11} and EH_{12} modes. The initial amplitude and phase distribution is regenerated at the resonant propagation distances given by (19.64) [5.94] (© OSA 1992).

In rectangular, square waveguides with simultaneous oscillation of the two waveguide eigenmodes EH_{mn} and $EH_{m'n'}$, the spacing of the resonances in the effective waveguide length is determined by [5.76]:

$$\frac{\Delta L}{ka^2} = \frac{16}{\pi\,|m^2-m'^2+n^2-n'^2|} \tag{19.68}$$

We can easily show that the two mixing modes share the same frequency at the resonances. For small distance parameters $d/(ka^2)$, the resonance frequencies ν_{mnq} can be found by replacing L with $L+2d$ in (19.37). By using the same approximations as in (19.38), we get:

a) circular symmetry:

$$\nu_{nmq} = q\,\frac{c_0}{2(L+2d)} + \frac{c_0 u_{nm}^2 L}{4\pi k a^2 (L+2d)} \tag{19.69}$$

b) rectangular symmetry, square waveguide:

$$\nu_{nmq} = q\,\frac{c_0}{2(L+2d)} + \frac{c_0 \pi L (m^2+n^2)}{16 k a^2 (L+2d)} \tag{19.70}$$

where k is the wave number at the center wavelength of the gain profile. If the two waveguide modes exhibit axial mode orders q and q', respectively, with $|q-q'|=2p$, Eqs. (19.64) and (19.68) imply that they oscillate at the same frequency. Note that the axial mode orders must differ in multiples of 2. Thus, the frequency degeneracy of two modes does not generate a resonance if the difference in the axial mode order is an odd number. This is due to the fact that (19.64) and (19.68) regenerate the relative phase in a transit, whereas the mode frequencies are determined by a phase regeneration in a round trip.

For small distance parameters, the absolute locations of the mode mixing points, to a good approximation, are given by [5.71]:

a) circular symmetry:

$$\frac{L+2d}{ka^2} = \frac{4\pi p}{|u_{1n}^2 - u_{1m}^2|}, \quad p=1,2,... \tag{19.71}$$

b) rectangular symmetry, square waveguide:

$$\frac{L+2d}{ka^2} = \frac{16\,p}{\pi\,|m^2-m'^2+n^2-n'^2|}, \quad p=1,2,... \tag{19.72}$$

For the waveguide resonator of Fig. 19.17, Eq. (19.71) yields the following values of the effective length parameter $L/(ka^2)$ at which a resonance between two modes occurs:

$$EH_{11} - EH_{13} \quad : \quad \frac{L}{ka^2} = 0.116 \;,\; 0.299 \;,\; 0.482 \;,\; 0.665$$

$$EH_{12} - EH_{13} \quad : \quad \frac{L}{ka^2} = 0.216 \;,\; 0.499 \;,\; 0.782 \;,\; 1.065$$

$$EH_{11} - EH_{12} \quad : \quad \frac{L}{ka^2} = 0.442 \;,\; 0.951 \;,\; 1.459 \;,\; 1.968$$

By using these values we can identify most of the kinks in Fig. 19.17. The pronounced loss minima of the second lowest loss mode (mode 1), for instance, can be attributed to a mixing of the EH_{12} mode with the EH_{13} mode.

In order to attain a near diffraction limited output beam, the fundamental waveguide resonator mode should be predominantly EH_{11} with only small power coupling into higher order waveguide modes. Therefore, it is advantageous to choose the resonator parameters such that operation near mixing points is avoided. The calculated loss and output power of the fundamental resonator mode shown in Fig. 19.19 clearly illustrates how the power coupling coefficients for the higher order waveguide modes become large around the resonance.

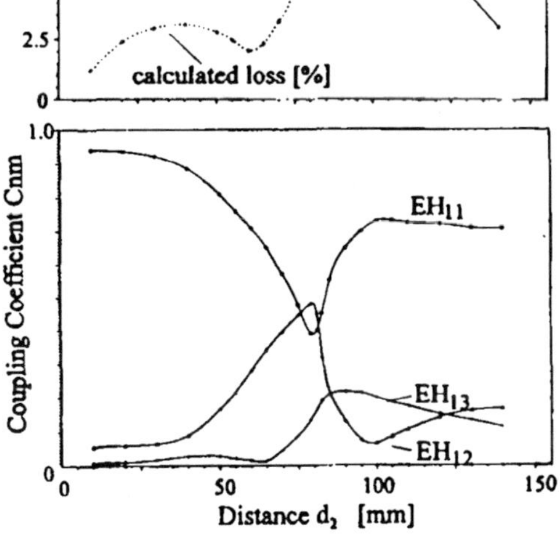

Fig. 19.19 Calculated round trip loss and measured output power of the fundamental mode for a circular waveguide resonator ($L=400mm$, $a=1.75mm$, $\lambda=10.6\mu m$). The distance d_2 to the output coupler is varied, and the HR mirror is attached to the waveguide ($d_1=0$). The waveguide material is alumina with a refractive index of $n=0.7+i0.1$. [5.83].

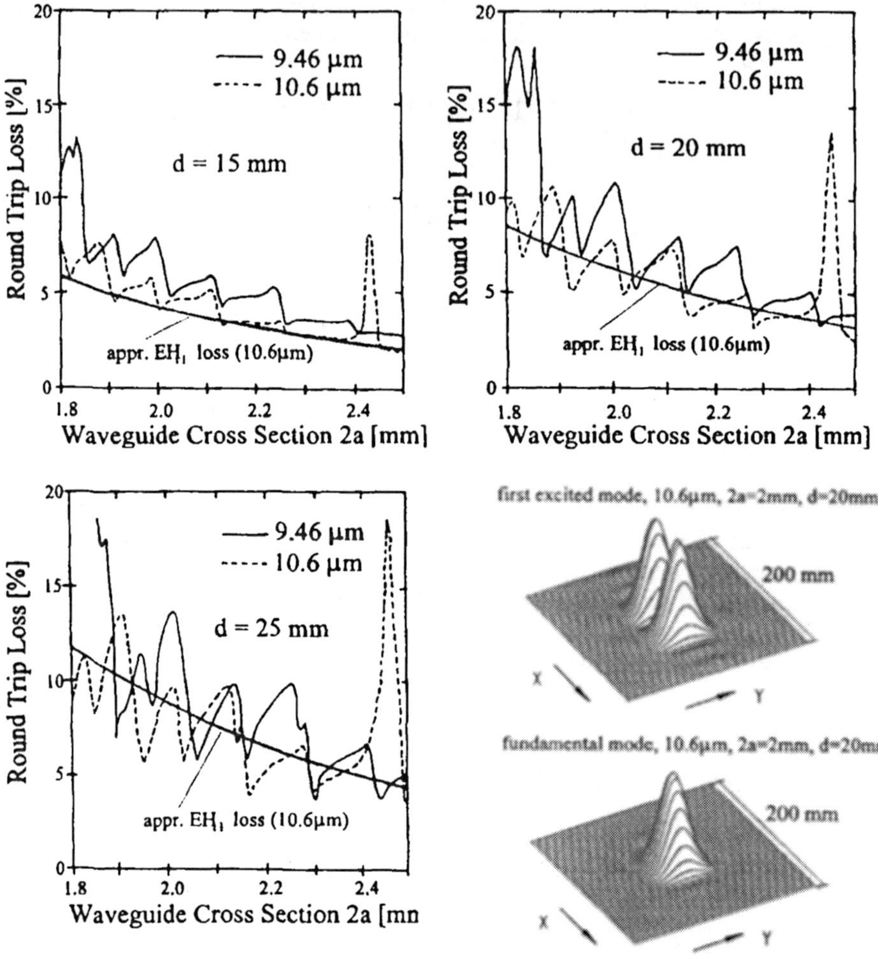

Fig. 19.20 Calculated round trip loss for the fundamental mode in symmetric waveguide resonators with two flat mirrors as a function of the side length *2a* of the square guide (guide material: aluminum, matrix calculation using 3 waveguide modes and 20 free space modes). The losses for the two main emission lines of CO_2 at 10.6µm and 9.46µm are shown for three different mirror distances *d* from the waveguide. The effective length of the waveguide is *L*= 600mm. The absorption losses for s-polarization were used in the x and in the y directions ($b^3=a^3\epsilon$ in (19.5)) with he refractive index $\sqrt{\epsilon}=22.7+i63.6$ at 9.46µm and $\sqrt{\epsilon}=27.9+i70.7$ at 10.6µm. This means that the osses $\Delta V(1D)$ of a slab waveguide can be deduced from the two-dimensional losses $\Delta V(2D)$ shown n the graphs by using the relation $\Delta V(1D)=1-\sqrt{1-\Delta V(2D)}$. The EH_{11} coupling losses were calculated with (19.61) using $X=1/\sqrt{\epsilon-1}$. The lower right graphs present the intensity distributions of the fundamental mode and the first excited mode at a distance of 5m from the output coupler (≈far field), both for $\lambda=10.6\mu m$, $d=20mm$, and $2a=2mm$. The round trip loss of the first excited mode is not shown in the graphs[5.83].

Operation far away from resonances between low order waveguide modes given by (19.71) and (19.72), however, may not automatically ensure that the output beam exhibits the desired Gaussian beam like propagation properties. The mode calculations rely on the exact knowledge of the refractive index of the guide material, a requirement that can be hardly met, especially for metals. Furthermore, the finish of the waveguide walls and heat induced variations in the waveguide dimensions also have a strong influence on the mode properties. In addition, the mode calculations hold only for the passive resonator. The gain saturation and inhomogeneous gain profiles may considerably change the mode properties. Therefore, the numerical evaluation of the resonator properties can only be considered a starting point of the laser design, and the optimization of the laser properties have to be accomplished experimentally. Furthermore, in gas lasers with relatively large gain bandwidths (like CO_2 lasers), wavelengths other than the desired one may exhibit lower round trip losses resulting in the oscillation at the unwanted wavelength or a switching between wavelengths. This stems from the dependence of the refractive index on the wavelength, which is particularly strong for metallic waveguides. The calculated fundamental mode round trip losses presented in Fig. 19.20 for the two main emission lines of a CO_2 waveguide laser with a square aluminum guide clearly indicate that the resonator parameters have to be carefully chosen to attain pure 10.6μm oscillation. Multi-wavelength emission generally poses no problem for industrial laser systems, but in medical applications operation at a well-defined wavelength is crucial to control the penetration depth in tissue [5.103]. The EH_{11} losses calculated with (19.61) are also shown in the graphs of Fig. 19.20 to give the reader a feeling for the applicability of the approximate expressions for the fundamental mode loss.

Considering all these difficulties, the question arises as how to proceed with the design of a waveguide laser? The dimensions of the waveguide are usually determined by constraints on output power and heat removal which means that the waveguide cross section can only be varied within a small range. With the dimensions of the gain medium being more or less fixed, a first step in the resonator design would be the calculation of the EH_{11} losses using the approximate expressions (19.61)-(19.63) and the evaluation of the extraction efficiency and the laser threshold with the standard models discussed in Chapter 10. In the course of these calculations, the location of the resonances should already be taken into account. Preferably, the resonator parameters should be chosen such that the resonator is working between two resonances of the EH_{12} and EH_{13} modes since the second order resonator mode is most likely to exhibit a loss minimum near these resonances. The experimental set-up can then be optimized experimentally by varying the mirror distances, the mirror curvatures, and, to some extent, the waveguide cross section. A resonator mode calculation using the matrix method may be required if the empirical optimization process does not yield satisfactory laser properties.

In general, waveguide lasers provide excellent beam properties, combining a high fill factor, good mode control, and beam qualities near the diffraction limit. In the first two decades since the first reported implementation of a waveguide laser in 1971 [5.54], mostly circular waveguides were used. In recent years, the interest shifted to slab waveguides due to the area scaling of the output power and the realization of near diffraction limited beam quality using off-axis unstable resonator schemes [5.81,5.86,5.87,5.92,5.101,5.103-5.105].

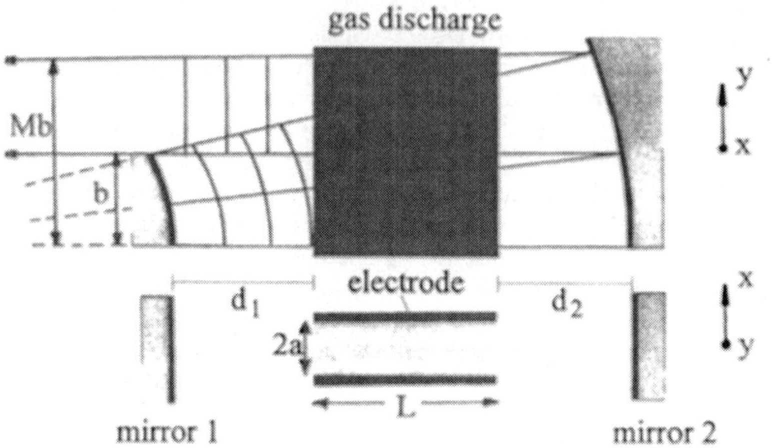

Fig. 19.21 Slab waveguide laser utilizing a positive branch, confocal, off-axis unstable resonator with magnification *M* in the y-direction.

Several commercial CO_2 slab waveguide lasers are currently used in medical applications (100W) and in material processing with output powers in the kW range [5.78,5.86,5.96-5.100,5.103,5.104,5.106]. In order to achieve high output power with a compact laser head, folded waveguide designs are often used. Popular folding schemes are the U-fold, the V-fold, and the Z-fold, all three using plane or curved folding mirrors. The theoretical modeling of folded waveguide resonators is more complicated especially if curved mirrors are used [5.3,5.76,5.84,5.85].

19.3.4 Properties of Slab Waveguide Lasers

One possible set-up of a slab waveguide resonator is depicted in Fig. 19.21. The positive branch confocal unstable resonator in off-axis geometry with magnification M provides near diffraction limited beam quality along the wide dimension of the waveguide. The high-reflecting cylindrical mirrors form a one-dimensional Case I waveguide resonator with flat mirrors in the perpendicular direction. Typical dimensions for CO_2 lasers are b=*30-50mm*; L=*60-120cm*; d_1,d_2 = *15-30mm*; $2a$=*1.5-2.5mm*. Mirror distances of less than 15mm are not feasible due to possible damage by the discharge plasma. In commercial systems, negative branch confocal unstable resonators are preferred due to their lower misalignment sensitivity even though the efficiency is about 20% lower. In the negative branch, with both mirrors being concave in the unstable direction, spherical mirrors can also be used resulting in lower coupling losses for the waveguide modes.

The design of the unstable resonator can be performed geometrically since the equivalent Fresnel numbers are relatively high. As discussed in Chapter 7, the output coupling loss is almost identical to the geometrical loss of $1-1/M$ for equivalent Fresnel numbers higher than 10. Typical equivalent Fresnel numbers are between 20 and 60, and magnifications in the range from 1.2 to 1.4 are used. The mirror curvatures needed to set up a confocal unstable resonator with magnification M are given by:

$$\rho_1 = \frac{2(L+d_1+d_2)}{1-M} \tag{19.73}$$

$$\rho_2 = \frac{-2M(L+d_1+d_2)}{1-M} \tag{19.74}$$

where $M=-\rho_2/\rho_1$ is positive for positive branch resonators and negative for negative branch resonators. The equivalent Fresnel number reads:

$$N_{eq} = \frac{b^2}{2L\lambda}(M-1) \tag{19.75}$$

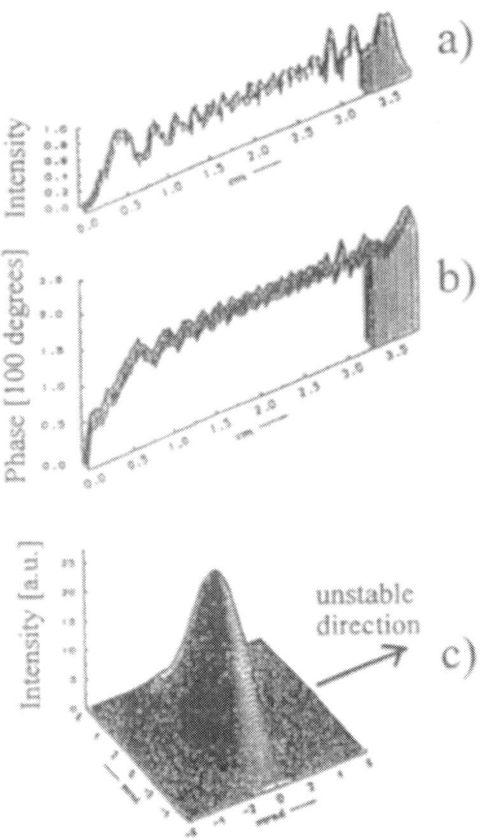

Fig. 19.22 Calculated intensity and phase distributions in the near field and in the far field for an off-axis confocal unstable resonator with equivalent Fresnel number $N_{eq}=20.8$ and magnification $M=1.15$ ($\lambda=10.6\mu m$). The intracavity near field intensity distribution at the output coupling mirror is shown in a) and the corresponding phase distribution is given in b). The shaded area indicates the portion being coupled out of the resonator. In the waveguide direction a homogeneous phase and amplitude profile of width 1.5mm is assumed. Graph c) shows the intensity distribution in the far fiel d. The beam is astigmatic due to a higher divergence in the waveguide direction [5.92] (© AT Fach-verlag 1991).

The electric field coupled out of the confocal unstable resonator exhibits a phase variation of less than $\lambda/5$ over the entire cross section of the beam (Fig.19.22) resulting in a near diffraction limited beam quality. However, due to the different beam dimensions in the x and the y directions, the beam is highly astigmatic. The angle of divergence in the waveguide direction is about $(|M|-1)b/(2a)$ higher than in the perpendicular direction. In order to generates a near circular focus spot, the astigmatism has to be compensated by means of cylindrical lenses. Figure 19.23 presents a compensation scheme using a telescope in the waveguide direction. The measured far field intensity profiles of the transformed, stigmatic beam of a 1kW CO_2 slab waveguide laser is shown in Fig. 19.24. Typically, a Gaussian intensity profile is observed in the waveguide direction with a beam propagation factor of $M^2=1.1$. In the unstable direction, small side lobes are generated and the beam propagation factor is about 10% higher.

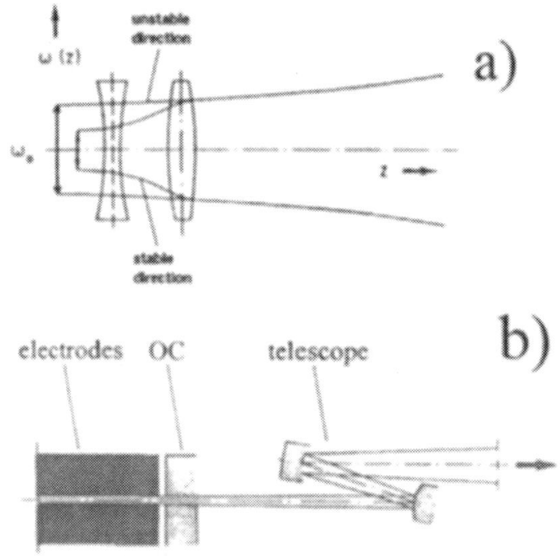

Fig. 19.23 Compensation of the astigmatism by using a telescope in the waveguide direction. a) The virtual beam waist in the waveguide direction has to coincide with the beam waist in the unstable direction (location and diameter). b) Technical realization using an off-axis telescope with cylindrical mirrors [5.92] (© AT Fach-verlag 1991).

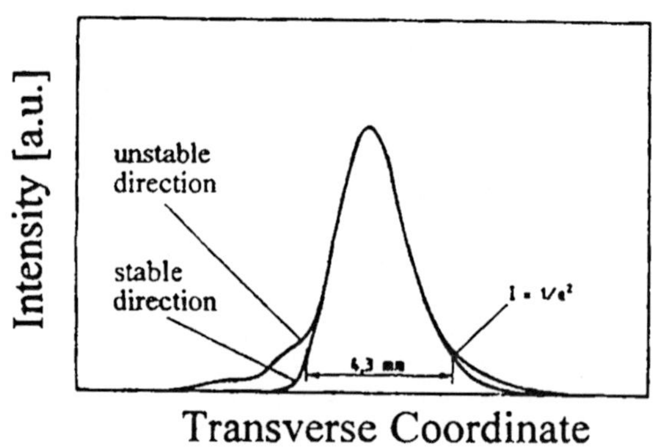

Fig. 19.24 Measured far field intensity distribution of a 1kW CO_2 slab waveguide laser with a positive branch confocal resonator in off-axis geometry. The size of the beam at the output coupler is 1.5mm x 6.3mm. The far field was generated in the focal plane of a concave mirror with focal length f=5m. The beam propagation constants are M^2=1.07 in the waveguide direction and M^2=1.2 in the unstable direction [5.92] (© AT Fach-verlag 1991).

The design of a medical CO_2 slab waveguide laser and its power characteristics are presented in Figs. 19.25 and 19.26 [S.22]. The aluminum waveguide has an active area of 240cm^2 and the electrodes are separated by 2mm. A negative branch confocal unstable resonator with spherical mirrors and a magnification of $M=1.3$ is used to extract a maximum output power of 150W at an RF input power of 1.5kW (RF frequency: 70 MHz). In pulsed operation, with a pulse length of 0.2ms and a repetition rate of 1kHz, the peak power of 700W is close to the theoretical limit predicted by (19.1) for 10% efficiency. Positive branch unstable resonators were also investigated yielding up to 20% higher output power but these designs exhibited an alignment sensitivity that was too high for the intended use. Measured output powers as a function of the RF input power using stable and unstable resonators are shown in Fig. 19.26. The measured power for different output coupling reflectances of the stable resonator can be used to determine the optimum output coupling reflectance R_{opt} at the design input power (this is the maximum input power, which, in a commercial version of this laser, is about 1kW). The nonlinear dependence on the input power is a result of the changing impedance of the gas discharge (fixed RF matching network). The optimum magnification M of the unstable resonator is roughly determined by $M=1/R_{opt}$. The measured output power for different magnifications of the unstable resonator is shown in the lower graph of Fig. 19.26 for pulsed operation at a repetition rate of 1kHz. The optimum magnification of $M=1.3$ yields a total efficiency of 11% at 1kW RF input power. The saturation of the output power at higher input powers is due to insufficient cooling of the gas.

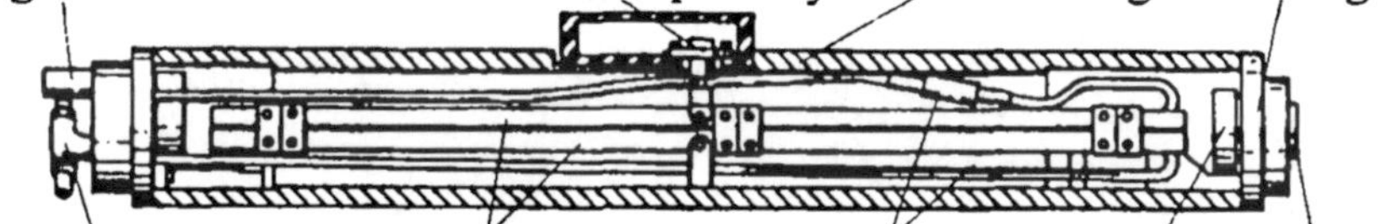

Fig. 19.25 A 200W CO_2 slab waveguide laser with 10% total efficiency (top: cross sectional view, left: photograph of the electrode structure). The aluminum electrodes are 44mm wide and 616mm long. The electrode spacing is 2mm. A negative branch confocal resonator with magnification M=1.3 is used in the wide direction. The mirrors are both spherical and are placed at a distance of 20mm from the waveguide. A standard CO_2 gas mix (3.6% Xe, 10.7% CO_2, 21.4% N_2, 64.3% He) at a pressure of 100mbar is used [S.22].

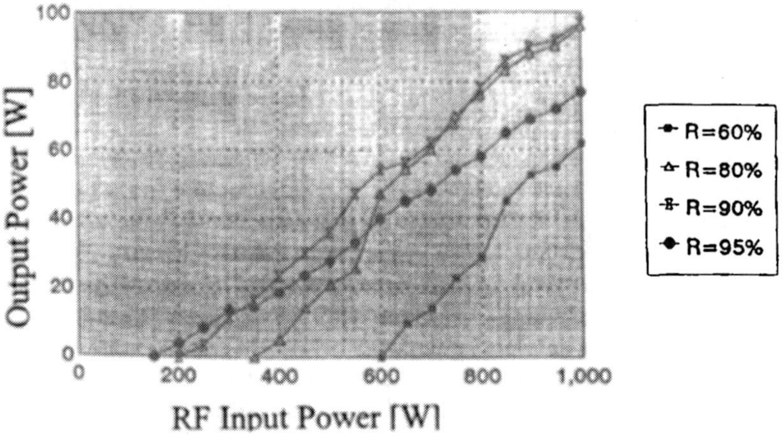

a)

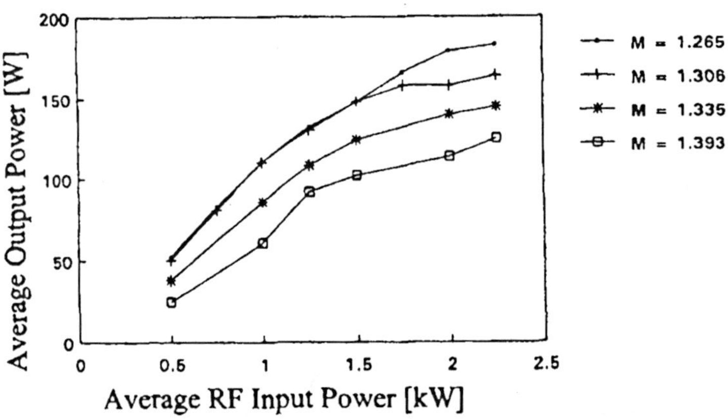

b)

Fig. 19.26 Measured output power versus RF input power for the CO_2 slab waveguide laser described in Fig. 19.25. a) Stable resonator with different mirror reflectances, cw operation, and a fixed RF matching network. b) Unstable resonator with different magnifications *M*, pulsed operation, repetition rate: 1kHz, peak power: 5kW, variable pulse width, and variable impedance matching [S.22].

19.3.5 Solid State Waveguide Lasers

In recent years, slab waveguide resonators have also been demonstrated in diode-pumped Nd:YAG and Yb:YAG laser systems. In contrast to metallic and ceramic waveguides where higher order mode discrimination is generated by reflection losses at the waveguide walls, single transverse mode operation in solid state waveguide relies on the optical confinement of only the lowest order waveguide mode by surrounding the waveguide with layers having a lower refractive index. In general, this is accomplished by sandwiching the doped YAG layer between non-doped YAG claddings using diffusion bonding techniques (Fig. 19.27). If n_W and n_C denote the refractive indexes of the waveguide and the cladding layers, respectively, single transverse mode operation is attained if the following relation holds for the waveguide thickness d_W [S.36]:

$$d_W \frac{2\pi}{\lambda_0} \sqrt{n_W^2 - n_C^2} = V < \pi \tag{19.76}$$

where λ_0 is the free-space wavelength of the laser light and the parameter V characterizes the waveguide. The intensity distribution of the lowest order waveguide mode can be approximated by a Gaussian profile. The beam waist radius w_0 and half angle of divergence of the emerging beam are given by [S.36]:

$$w_0 = d_W \left[0.31 + \frac{3.15}{V^{1.5}} + \frac{2}{V^6} \right] \quad ; \qquad \theta_0 = \frac{\lambda_0}{\pi\, w_0} \tag{19.77}$$

In general, the mode is spread out beyond the waveguide layer with the power fraction of approximately $V^2/(V^2 + 2)$ residing inside the cladding. For 1 at% Nd:YAG (n_W=1.8216) and 10 at% Yb:YAG (n_W=1.8166), the index step between undoped and doped layer is about 4×10^{-4} and 12×10^{-4}, respectively, resulting in single mode waveguide thicknesses of less than 13.9 μm for Nd:YAG and less than 7.8 μm for Yb:YAG [5.113,5.116,5.120].

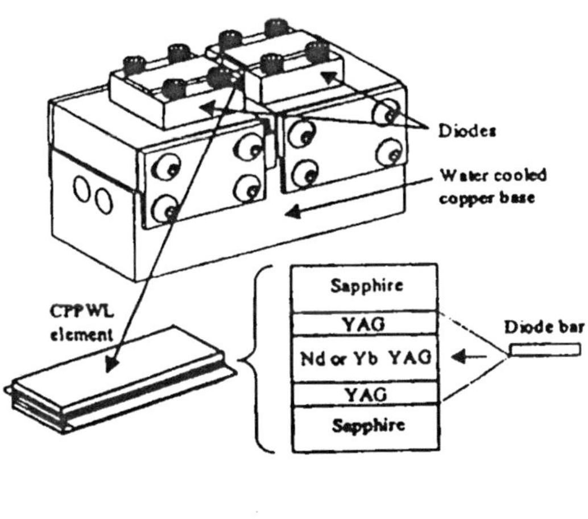

Fig. 19.27 Diode-pumped Nd or Yb doped YAG waveguide laser with waveguide thickness of 20 μm and 8 μm, respectively. The undoped YAG provides the guiding of the laser modes, and the sapphire layers confine the pump light through total internal reflection [5.120] (© OSA 2001).

By adding additional sapphire layers, total internal reflection can be realized for the pump light as well (numerical aperture of 0.46), allowing end- or side-pumping by proximity coupling of diode bars (see Fig. 19.27). Similar to CO_2 slab waveguides, the advantage of a solid state waveguide is that the maximum attainable output power scales with the slab's cross-sectional area and inversely to the waveguide thickness due to the increased ratio of cooled surface to the volume of the gain material. This implies that high single transverse mode output power also requires the implementation of unstable resonator schemes perpendicular to the waveguide direction.

The first solid state waveguide lasers were demonstrated with Tm:YAG as the active waveguide layer. The Tm laser transition at a wavelength of 2.012 μm is sensitive to temperature, making a thin waveguide structure a preferred geometry to implement compact eye-safe lasers. Output powers of up to 15W have been reported [5.111,5.121b, 5.125a] using multimode stable resonators perpendicular to the waveguide direction. The application of side-pumped or end-pumped double-clad single-mode waveguide structures to Nd:YAG resulted in output powers of up to 12W [5.113, 5.120, 5.122]. Similar cw output powers were achieved with Yb:YAG single mode waveguides [5.113, 5.120, 5.124]. Passive Q-switching with average output powers of up to 9W were also demonstrated both for Nd:YAG [5.120] and Yb:YAG [5.124] by incorporating a Cr^{4+}:YAG section into the waveguide layer. The implementation of an unstable resonator in the direction perpendicular to the waveguide can be found in [5.120] for Nd:YAG and in [5.124] for Yb:YAG.

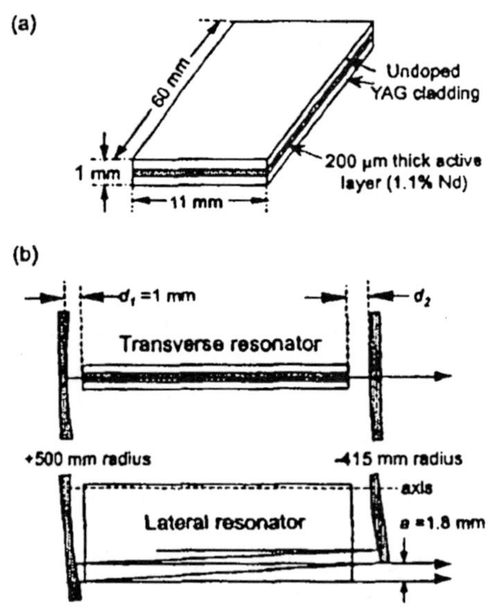

Fig. 19.28 Diffusion bonded planar Nd:YAG waveguide laser with off-axis unstable resonator perpendicular to the waveguide. The central 50mm of the waveguide are side-pumped by ten quasi-cw laser diode bars. M^2 values of 1.5 and 4 along the lateral direction and the waveguide direction, respectively, were obtained with an average output power of 4 W [5.119] (© Elsevier Science B.V. 2001)

In spite of the obvious advantages of a high fracture limit and improved thermal management, single mode waveguide structures have not gained widespread application. This is due to the small thickness of the active layer which does not provide enough pump light absorption to efficiently pump the waveguide through the large side-faces. Face pumping would allow power scaling by increasing the waveguide area. With the demonstrated side-pumping arrangements, the output power can only be scaled by increasing the waveguide length. In addition, side pumping imposes tight tolerances on the position of the diode bars. For power scaling reasons waveguide designs therefore moved to using multimode Nd:YAG and Yb:YAG waveguides with waveguide thicknesses between 100 μm and 200 μm [5.109,5.119]. An Nd:YAG multimode waveguide laser using a stable-unstable hybrid resonator is depicted in Fig. 19.29 [5.119]. The central 50 mm of the waveguide are face-pumped using multipass absorption in a high reflection pump chamber with ten quasi-cw diode bars. This waveguide structure was scaled to cw output powers of 160W in multimode operation and 140W in a high brightness configuration [5.123a]. Similar to CO_2 waveguide lasers, near diffraction-limited beam quality in the waveguide direction requires discrimination of higher order waveguide modes in an external stable resonator. The three different types of waveguide resonators discussed in Sec. 19.3.3 (Case I - III resonators) can also be applied to solid state multimode waveguides. In [5.119], transverse mode discrimination was accomplished by adjusting the distances of the resonator mirrors from the waveguide ends (Case I). A Case II resonator (imaging of the waveguide ends) was used in [5.123b] to generate single mode operation in a 100 μm thick Nd:YAG waveguide, end-pumped by a 3.5W single emitter 808 nm diode (Fig.19.29).

A further increase of the waveguide thickness leads to the transition from a multimode waveguide laser to a slab laser with non-guided modes. In general, a 200 μm thick multimode waveguide solid state structure provides only little advantage over a face-pumped slab that has a thickness between 0.5 mm and 1.0 mm. The possible disadvantage of the lower gain of the slab is more than outweighed by a simpler laser head design (no cladding layers required) that lends itself better to power scaling to the kW level.

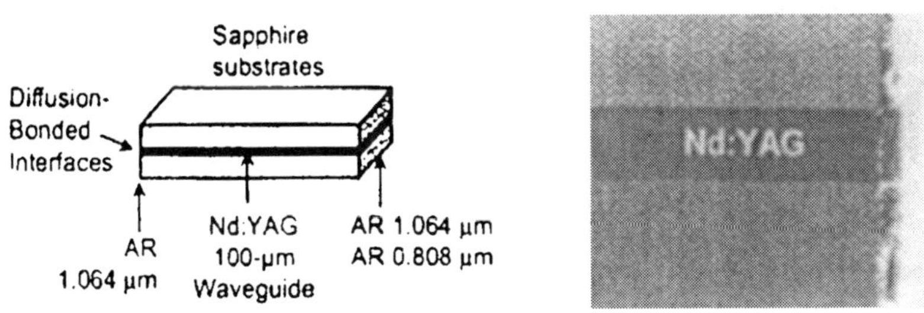

Fig. 19.29 Nd:YAG waveguide with a 100 μm Nd:YAG layer diffusion-bonded between two sapphire substrates. A cross-sectional view is shown in the photograph on the right. The waveguide was end pumped in a self-imaging resonator and provided an average output power of 1.5W with $M^2 < 1.5$ in the index-guided direction and a slope efficiency of 39% [5.123] (© OSA 2002).

Since TEM_{00} mode diameters of 1 mm can be easily realized in stable solid state laser resonators, waveguiding is not required to attain diffraction-limited beam quality in the thin direction of the slab. For this reason, the highest powers with near diffraction-limited beam quality have been achieved with non-guiding thin slab structures like the ones depicted in Figs 19.30 and 19.31. Similar to other solid state geometries, the output power in fundamental transverse mode operation is limited by the aberration of the thermal lens. However, even for refractive powers as high as 15 Diopters along the direction of the slab thickness, stable resonator schemes can be found that provide high power TEM_{00} mode operation with M^2 of less than 1.5 [5.125b].

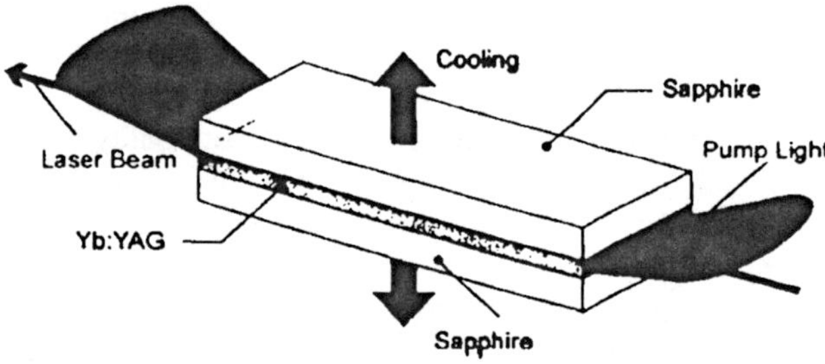

Fig. 19.30 End-pumped Yb:YAG waveguide with a 500 μm thick, 0.5 at % doped, Yb:YAG layer diffusion-bonded between two sapphire substrates. A maximum cw output power of 257W was attained at 600W pump power (pump wavelength: 940 nm) [5.125c] (© OSA 2004).

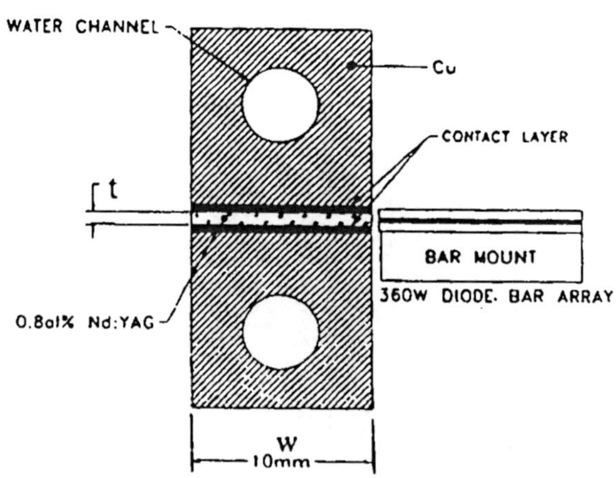

Fig. 19.31 Dual side-pumped Nd:YAG slab laser with a 100 mm long slab with width w=10 mm and thickness *t* between 0.8mm and 1.0mm (only one diode bar array is shown). At a total pump power of 720W an output power of 220W was obtained with $M^2 < 1.5$ using a stable-unstable hybrid resonator with a variable reflectivity mirror [5.125b] (© OSA 2002).

Chapter 20

Resonators for Annular Gain Media

20.1 Characteristics of Annular Gain Lasers

In high power gas and solid state lasers the maximum output power is limited by the efficiency of heat removal. The temperature rise associated with the heat generated by the pump process leads to a decrease of the stimulated emission cross section and an increase of the lower laser level population. For sealed-off CO_2 lasers in cylindrical geometry, these effects reduce the maximum output power per length to about 80W/m, independent of the tube diameter. Furthermore, in solid state laser materials, the pump induced stress causes irreversible damage to the medium if the pump power is increased beyond the fracture limit. The experimentally established fracture limit of flashlamp pumped and diode-pumped Nd:YAG rod allows up to 40W and 60Wof output power per cm of rod length, respectively.

Lower temperatures and higher fracture limits can be attained by choosing a geometry of the gain medium that exhibits a larger surface to volume ratio than the cylindrical geometry. Besides the slab geometry (see Sec. 19.2), an annular active medium provides a much larger cooling surface than a rod of equal volume. If a and b denote the inner and outer radii of the annulus, respectively, the maximum output power P_A that can be extracted from an annular gain medium reads [5.162,5.167]:

$$P_A = P_R \, 2 \, \frac{b+a}{b-a} \qquad \textit{for gas lasers} \tag{20.1}$$

$$P_A = P_R \, \frac{3}{2} \, \frac{b+a}{b-a} \qquad \textit{for solid state lasers} \tag{20.2}$$

where P_R is the maximum output power attainable for a rod (or gas tube) of equal length, and b-a is much smaller than a. The power scaling is different for gas and solid state lasers due to the physical mechanisms involved. In gas lasers, the output power is limited by the gas temperature, whereas in solid state materials the surface stress has to remain below the fracture limit. For an inner radius of 30mm and a wall thickness of 5mm, an annular gas medium provides a 26 times higher output power per length compared to a conventional tube. Annular gain media have been realized in dye lasers and He-Xe lasers [5.131,5.137], HF lasers [5.134], CO_2 lasers [5.133,5.135,5.141,5.151,5.163-5.170], Nd:glass lasers [5.130], and Nd:YAG lasers [5.157,5.158,5.161,5.162] (Figs.20.1, 20.2). A maximum output power of 1.86kW was reported for an inside pumped Nd:YAG tube laser (tube length: 130mm, R_i=*17.5mm*, R_o=*26.5mm*)[5.162]. An output power of 1kW was achieved with a sealed-off CO_2 laser (electrode length: 1.15m, R_i=*48mm*, R_o=*55mm*) [5.170]. Both systems operated at a total efficiency of 10%.

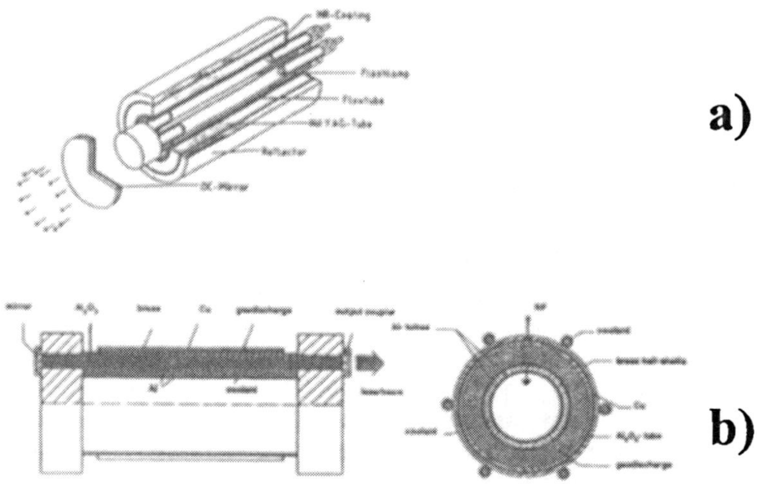

Fig. 20.1 a) Nd:YAG laser [5.157] (© OSA 1991) and b) CO_2 laser with annular gain medium [S.23].

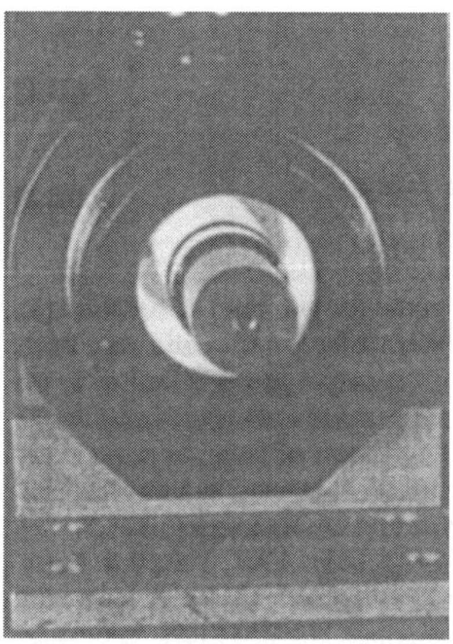

Fig. 20.2 Inside pumped Nd:YAG tube laser providing an average output power of 1.8kW at a total efficiency of 10%. The left photograph is a front view of the tube with the four flashlamps in simmer operation. The upper photograph presents a side view of the laser head including the output coupling mirror . The rear endface of the tube is HR coated [5.157,5.158] (courtesy of Laser- und Medizin-Technologie Berlin gGmbH, Germany).

Although the tube geometry is very well suited to provide high output powers, this concept has not yet found commercial application. This is not only due to the more sophisticated, and therefore more expensive, laser head design. The main problem of the annular geometry is the realization of good beam quality without decreasing the output power too much. A variety of optical resonators suitable for annular gain media have been studied in the last two decades. Nevertheless, none of these resonator schemes are capable of providing beam properties that would justify a replacement of commercial rod or slab lasers. In the following, an overview of the performance of different annular resonator schemes is given.

20.2 Stable Resonators with Toric Mirrors

20.2.1 Transverse Mode Structure

Resonators with toric mirrors exhibit mode structures similar to those of conventional spherical mirror resonators. Both mirrors exhibit a radius of curvature ρ_i in the radial direction and are flat in the azimuthal direction (Fig. 20.3). Both stable and unstable resonators can be realized [5.133,5.146]. The resonator properties are determined by the g-parameters $g_i=1-L/\rho_i$, the effective resonator length L, the inner radius a, and the wall thickness of the tube $d=b-a$. If the toric mirror resonator is unwound, a one dimensional strip resonator is obtained (Fig.20.4). The circumference of the annular mirrors is now represented by the height of the cylindrical mirrors. For a large radius r_0 of the mirror vertex, the radial modes of the unconfined stable toric resonator can therefore be approximated by one-dimensional Gauss-Hermite modes [5.153]. In this so-called slab approximation, the mode diameter and the number of oscillating radial modes can be calculated by using the well-known relationships for conventional resonators in rectangular symmetry. However, in general, the vertex diameter is not large enough to neglect the influence of the bending on the radial mode structure. In this case, the transformation from a cylindrical to a toric mirror shifts the radial intensity maxima slightly inward.

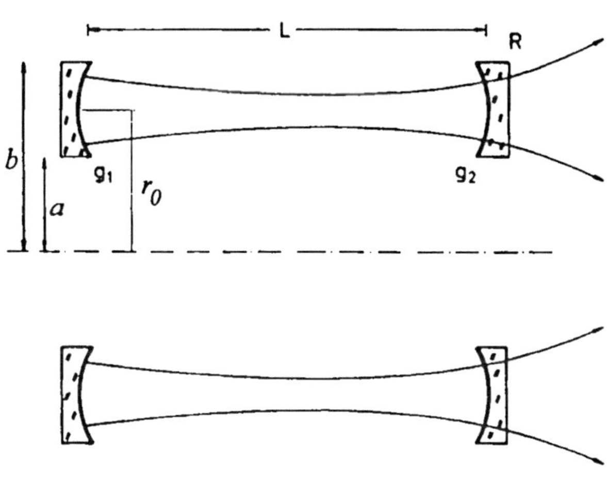

Fig. 20.3 Stable resonator with toric mirrors (cross sectional view).

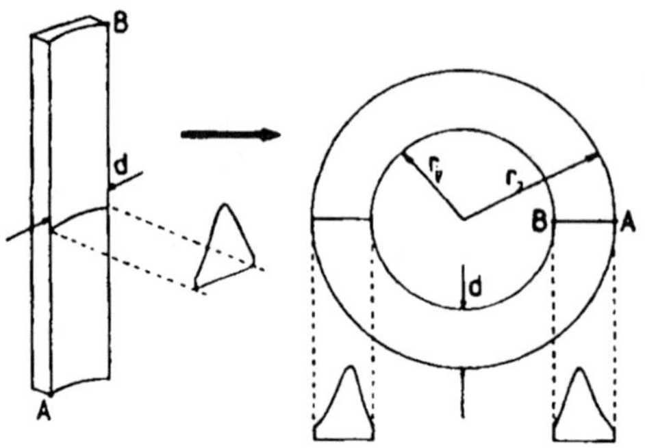

Fig. 20.4 A toric mirror is generated by bending a cylindrical mirror. For vertex diameters r_1+r_2 that are much larger than the mirror width d, the radial mode structures of toric resonators are similar to those of resonators in rectangular geometry.

If both mirrors are unconfined, the field distributions of the transverse eigenmodes of passive toric mirror resonators can be calculated analytically [5.153,5.160]. In polar coordinates, the field distributions read:

$$E_{p\ell}(r,\phi) \propto \frac{1}{\sqrt{r}} \exp\left[-\frac{(r-r_0)^2}{w_0^2}\right] H_p\left[\frac{\sqrt{2}(r-r_0)}{w_0}\right] \exp[\pm i\ell\phi] \tag{20.3}$$

where p,ℓ are the radial and the azimuthal mode index ($p, \ell \geq 0$), respectively, r_0 is the radius of the mirror vertex, w_0 is the Gaussian beam radius, and $H_p[x]$ is the Hermite polynomial of order p. Note that for large vertex radii r_0 and small beam diameter, the influence of the factor $1/\sqrt{r}$ on the mode structure becomes negligible. This is the mathematical representation of the slab approximation. The radial intensity distributions can then be approximated by Gauss-Hermite polynomials.

If both mirrors are unconfined, the beam radius of the transverse modes depends only on the radial mode order p. Thus, the transverse modes are highly degenerate and the simultaneous oscillation of on the order of a hundred azimuthal modes is to be expected. This also holds true if the mirrors are limited by annular apertures. The numerical investigation of the diffraction integral for a resonator round trip reveals that the azimuthal mode discrimination remains weak, although the radial mode structure now depends on the azimuthal index [5.163]. The radial mode profile is shifted outward as the azimuthal mode index is increased, resulting in an increase in the diffraction losses. Figure 20.5 presents numerically calculated round trip losses of the modes with $p=0$ as a function of the azimuthal order for symmetric toric stable resonators. The azimuthal mode discrimination increases with the mirror curvature and the mode with $\ell=0$ experiences the lowest loss if the mirror vertex is in the center of the annular gap. However, the losses are nearly degenerate for low order modes.

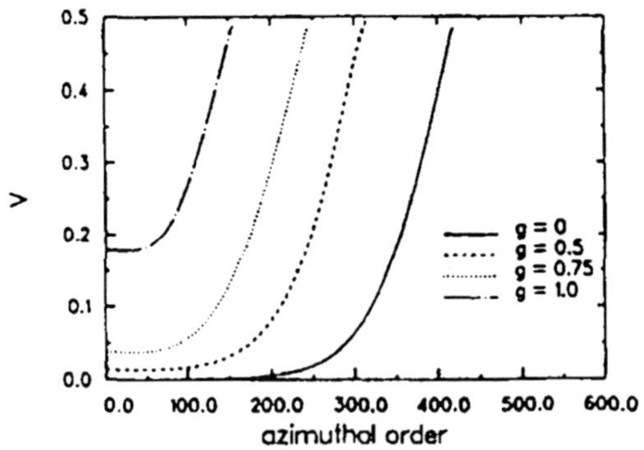

Fig. 20.5 Calculated diffraction losses per round trip for the fundamental radial mode (p=0) of symmetric toric resonators as a function of the azimuthal mode order. The curve parameter is the g-parameter of the mirrors. The mirror vertex is centered with the annular gap. Vertex radius r_0=50mm, resonator length L=1m, wavelength λ=10.6µm, gap width b-a=6mm [5.163] (© OSA 1993).

The loss minimum can be shifted towards higher azimuthal orders if the radius r_0 of the mirror vertex is chosen smaller than the radius $(a+b)/2$ of the gap center. This will improve the azimuthal mode discrimination, but, unfortunately, will also prevent circularly symmetric modes ($\ell=0$) from oscillating. This is a major drawback because only the modes with $\ell=0$ exhibit a centered intensity profile in the far field. The azimuthal index ℓ of the lowest loss mode for symmetric resonators is given by [5.163]:

$$\ell = \frac{\pi(a+b)^2}{2\lambda L}\sqrt{2(1-g)\left(1 - \frac{2r_0}{a+b}\right)} \tag{20.4}$$

where g is the g-parameter of the resonator mirrors and λ is the wavelength.

19.2.2 Beam Quality

The beam quality of toric stable resonators is determined by the mode with the highest radial mode order p and only slightly depends on the azimuthal structure. Similar to conventional resonators, the highest mode order can be found by adapting the beam diameter to the gap width of the annular aperture:

$$w_{00}\sqrt{p+1} \approx \frac{b-a}{2} \tag{20.5}$$

where w_{00} is the Gaussian beam radius at the aperture.

The beam waist w_p of this radial mode and the corresponding half angle of divergence θ_p can then be directly calculated using the expressions derived in Chapter 5. However, the waist radius w of the laser beam is given by the outer radius of the annular intensity distribution and not by the radius w_p of the mode profile in the annular gap. For a given mode order, the beam parameter product, therefore, is higher than for conventional resonators. In a geometrical approximation, the laser beam waist radius is larger than the waist radius of the radial mode in the annular gap by the factor $2b/(b-a)$, where a,b are the inner and the outer radii of the annular aperture, respectively. Thus, the beam parameter product can be approximated by:

$$w\theta = \frac{2b}{b-a} w_p \theta_p = (p+1) \frac{2b}{b-a} \frac{\lambda}{\pi} \tag{20.6}$$

where θ is the half angle of divergence. Figure 20.6 presents calculated radial intensity distributions in the near field and the far field of the fundamental mode for a toric resonator with different inner diameters of the annulus. The annular shape of the near field generates a high power content in the side lobes of the far field. An on- axis intensity maximum in the far field is only found for modes without azimuthal structure (ℓ=0). For higher azimuthal orders, the far field intensity distribution remains annular since all waves originating from the annular medium interfere destructively on the optical axis. Considering the lack of azimuthal mode discrimination, it is no surprise that measured far field intensity distributions of toric resonators generally contain most of the power in an annular ring (Fig. 20.6).

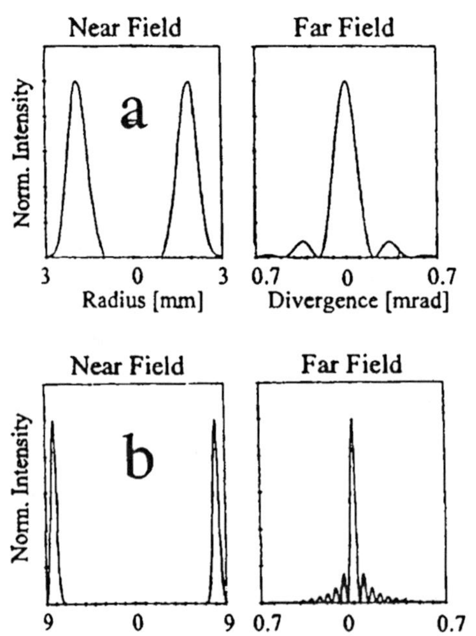

Fig. 20.6 Calculated intensity distributions of the TEM_{00} mode for a semi-confocal toric resonator (g_1=1, g_2=*0.5*, L=*0.5m*, λ=*1.06μm*). The normalized intensity distributions at the flat mirror and in the far field are shown for two different tube dimensions. a) a=*1mm*, b=*3mm*, b) a=*7mm*, b=*9mm*.

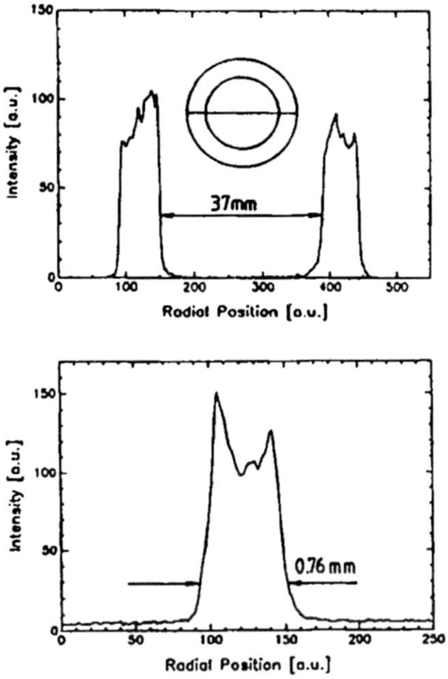

Fig. 20.7 Recorded radial intensity profiles in the near field and the far field of a Nd:YAG tube laser with a flat-flat resonator (tube dimensions: a=17.5mm, b=26.5mm, tube length: 100mm, resonator length: 1.2m). The equivalent resonator is toric due to the thermal lensing of the tube. The far field was recorded in the focal plane of a 200mm lens. The beam parameter product is 50 mm mrad (pump power: 3kW, output power: 200W) [5.158] (© AT Fachverlag 1991).

The beam quality of annular laser beams can be improved by reducing the diameter of the central hole without increasing the beam divergence. This is not in contradiction with Liouville's Theorem because the area in phase space is not decreased. The removal of the central hole can be accomplished by an axicon telescope, as shown in Fig. 20.8. The beam parameter product is reduced by the same factor as the outer beam diameter. However, the axicon telescope only preserves the beam divergence if the azimuthal mode index is low. Calculated beam parameter products with and without the axicon telescope are presented in Fig. 20.8. This figure indicates that the beam quality will only be significantly improved if the azimuthal mode order is lower than about twice the radial mode index. Note that without the axicon telescope, the beam parameter product is almost independent of the azimuthal mode order.

In summary, toric resonators are not very suitable for annular gain lasers due to the low azimuthal mode discrimination. Even if fundamental mode operation in the radial direction can be attained (like in CO_2 lasers), the oscillation of higher order azimuthal modes results in an annular far field. Furthermore, due to the annular shape of the near field, the beam parameter product is an order of magnitude higher than for a conventional Gaussian beam. The goal of annular resonator design, therefore, is to break the circular symmetry of the near field by generating a compact output beam at one side of the annular medium. This not only eliminates the deterioration of the beam quality due to the annular geometry, but also prevents the formation of ring patterns in the far field. One such resonator scheme is the Multipass Herriot Cell.

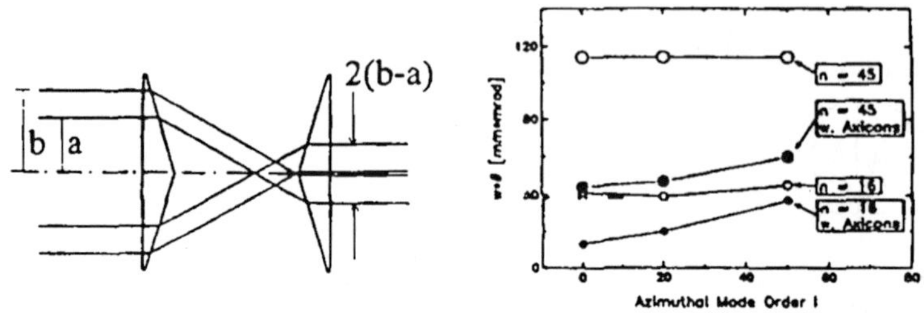

Fig. 20.8 Calculated reduction of the beam parameter product for an annular beam of a stable toric resonator with inner radius $a=17.5mm$ and outer radius $b=26.5mm$ as a function of the azimuthal mode order ℓ. The results for two different radial mode orders n, both with and without the axicon telescope, are shown. The beam parameter product is defined via the 86.5% power content (numerical calculation using diffraction integrals) [S.24].

20.3 Herriot Cell Resonators

A common resonator used in annular CO_2 lasers to extract the power in a compact, low order transverse mode is the Herriot cell resonator [5.127,5.149,5.151,5.156]. The Herriot cell can be viewed as a folded, linear resonator that generates multiple passes through the medium (Fig.20.9). The two high reflecting spherical mirrors with off-axis apertures form an optical delay line. The number of reflections and the orientation of the beams can be controlled by the mirror curvatures, the mirror distances, the location of the apertures, and the inclination of the incident beam [5.126,5.140,5.154]. In general, the reflected beams trace an elliptical curve on each mirror, but for special geometries the beams lie on a surface of a hyperboloid of revolution and the intersecting points move around the mirror in a circular pattern. The number of reflections can be increased by increasing the radius of curvature of the mirrors. There are several constraints in the design of Herriot cell resonators. The beam diameter has to be adapted to the wall thickness of the tube and with each round trip in the Herriot cell, and the beam has to be shifted in the azimuthal direction by slightly more than its diameter. Furthermore, laser oscillation between the two high reflecting folding mirrors has to be prevented by inserting a multi-segment aperture that generates losses for fields that do not propagate along the intended beam path. The output power of optimized Herriot cell resonators typically is 40% lower as compared to resonators with toric or spherical mirrors. This is due to diffraction losses generated at the apertures and the incomplete filling of the gain medium. An experimental example is shown in Fig. 20.10 for a linearly folded Herriot cell resonator. Furthermore, wavefront aberrations induced by the active medium sum up due to the high number of transits, resulting in increased diffraction losses. The Herriot cell resonator, therefore, is not suitable for annular solid state lasers.

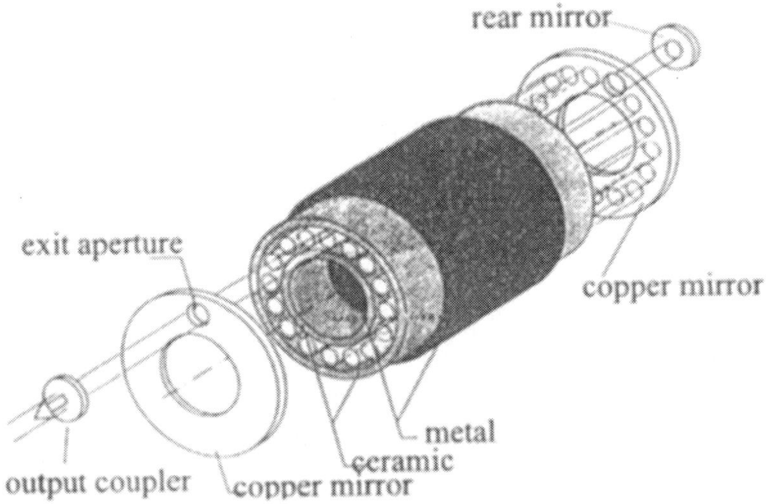

Fig. 20.9 Herriot cell resonator for an annular CO_2 laser [S.25].

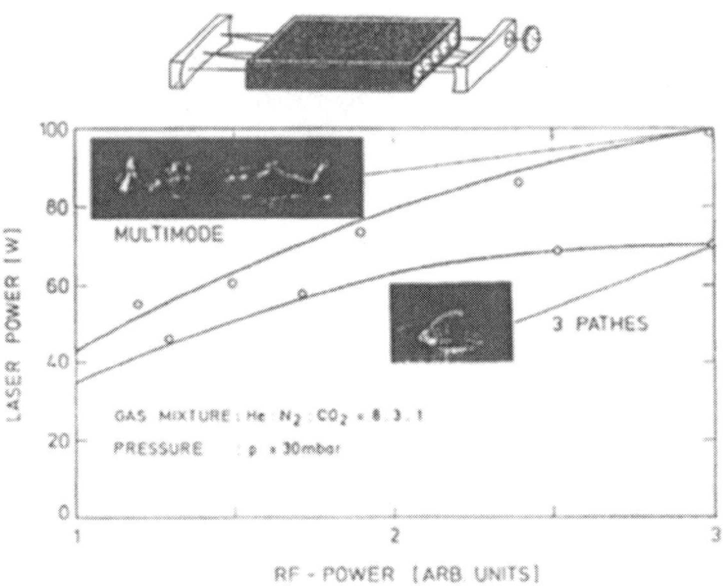

Fig. 20.10 Measured output power for a linear Herriot cell resonator with three passes in comparison to the output power attained with a conventional resonator. The upper graph presents the resonator scheme (with 5 passes) [S.25].

The beam propagation in Herriot cells can be mathematically described with ray transfer matrices [5.126,5.127,5.140,5.154]. Let us assume that a ray is launched into the Herriot cell at the point (x_0,y_0) with inclination angles (α_0,β_0) (Fig. 20.11). For a symmetric Herriot cell comprised of two spherical mirrors with radius of curvature ρ, the ray transfer matrix for a round trip reads:

$$M = \begin{pmatrix} 1 & 0 \\ -2/\rho & 1 \end{pmatrix}\begin{pmatrix} 1 & L \\ 0 & 1 \end{pmatrix}\begin{pmatrix} 1 & 0 \\ -2/\rho & 1 \end{pmatrix}\begin{pmatrix} 1 & L \\ 0 & 1 \end{pmatrix}$$

$$= \begin{pmatrix} 2g-1 & 2Lg \\ 4g(g-1)/L & 4g^2-2g-1 \end{pmatrix} \qquad (20.7)$$

where $g=1-L/\rho$ is the g-parameter of the mirrors and L is the mirror distance. After m round trips the ray vectors are given by:

$$\begin{pmatrix} x_m \\ \alpha_m \end{pmatrix} = M^m \begin{pmatrix} x_0 \\ \alpha_0 \end{pmatrix} , \quad \begin{pmatrix} y_m \\ \beta_m \end{pmatrix} = M^m \begin{pmatrix} y_0 \\ \beta_0 \end{pmatrix} \qquad (20.8)\ (20.9)$$

Application of Sylvester's Theorem (1.46) yields for the ray transfer matrix:

$$M^m = \frac{1}{\sin\Phi}\begin{pmatrix} (2g-1)\sin[m\Phi]-\sin[(m-1)\Phi] & 2Lg\sin[m\Phi] \\ 4g(g-1)\sin[m\Phi]/L & (4g^2-2g-1)\sin[m\Phi]-\sin[(m-1)\Phi] \end{pmatrix}$$

$$= \begin{pmatrix} A_m & B_m \\ C_m & D_m \end{pmatrix} , \quad \sin\Phi = 2g\sqrt{1-g^2} \qquad (20.10)$$

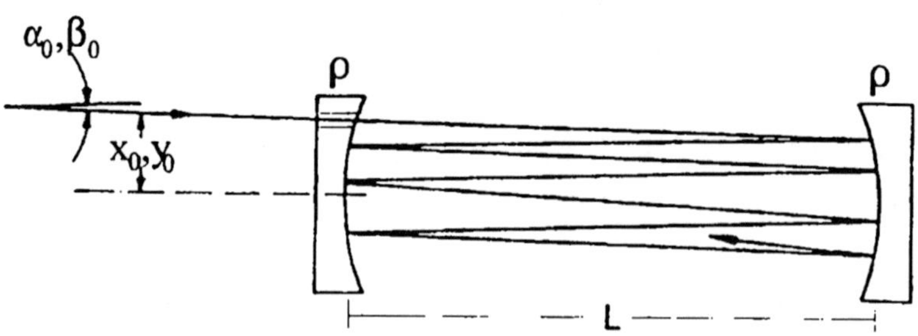

Fig. 20.11 Ray propagation in a symmetric Herriot cell.

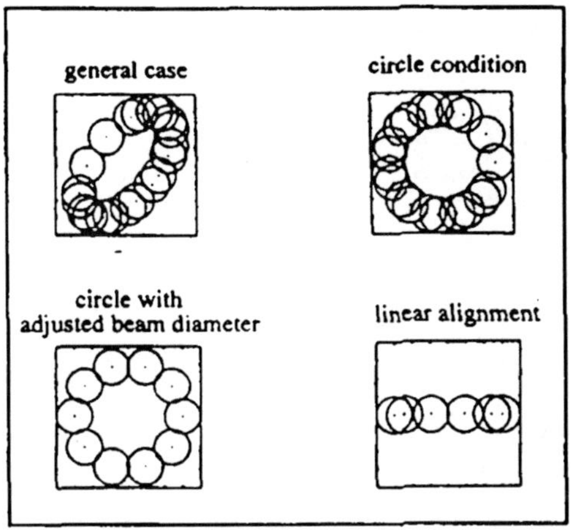

Fig. 20.12 Spot pattern on one mirror of the Herriot cell [S.26].

The coordinates of the ray after m round trips are given by:

$$x_m = A_m x_0 + B_m \alpha_0 \qquad (20.11)$$

$$y_m = A_m y_0 + B_m \beta_0 \qquad (20.12)$$

These two equations can be rewritten as:

$$x_m = R_x \sin(m\Phi + \delta_x) \qquad (20.13)$$

$$y_m = R_y \sin(m\Phi + \delta_y) \qquad (20.14)$$

with:

$$R_x^2 = x_0^2 + \frac{(\alpha_0 L + x_0(1-g))^2}{1-g^2} \qquad (20.15)$$

$$R_y^2 = y_0^2 + \frac{(\beta_0 L + y_0(1-g))^2}{1-g^2} \qquad (20.16)$$

$$\tan\delta_x = \frac{x_0\sqrt{1-g^2}}{\alpha_0 L + x_0(1-g)} \qquad (20.17)$$

$$\tan\delta_y = \frac{y_0\sqrt{1-g^2}}{\beta_0 L + y_0(1-g)} \qquad (20.18)$$

In general, the points where the rays intersect the mirror lie on an ellipse whose shape is determined by the launching conditions and the resonator set-up. For $\delta_y - \delta_x = k\pi$ the rays trace a circle, and for $\delta_y - \delta_x = (k+1/2)\pi$ a linear alignment is obtained (Fig. 20.12).

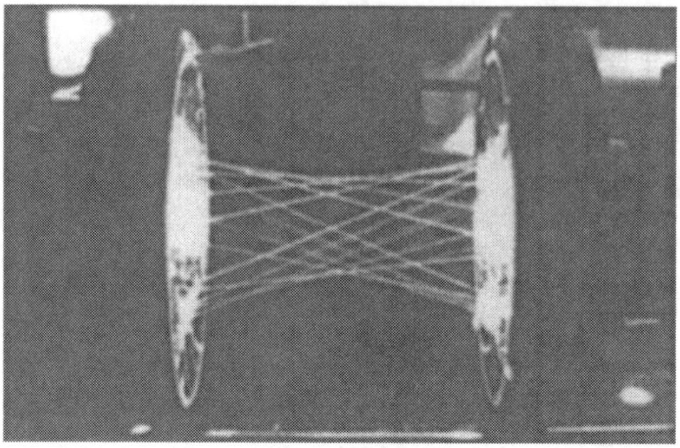

Fig. 20.13 The beam of an argon laser in a multipass Herriot cell used for stimulated Raman scattering (courtesy of Laser- und Medizin-Technologie Berlin gGmbH, Berlin, Germany).

In addition to their application in annular gain lasers, multipass Herriot cells have also been used as optical delay lines to increase the cavity round trip time. Effective resonator lengths of several kilometers have been achieved [S.25]. A further application of the Herriot cell is stimulated Raman scattering (SRS). The long interaction length with the gas significantly reduces the pump power required to achieve SRS threshold and high conversion efficiency [5.140] (Fig. 20.13).

20.4 Unstable Resonators

20.4.1 Toric Unstable Resonators

Similar to stable resonators with toric mirrors, the transverse modes of toric unstable resonators are closely related to those of unstable strip resonators with cylindrical mirrors [5.146]. The toric mirrors are obtained by rotating an arc of a circle about the optical axis (Fig. 20.14). The common parameters which describe conventional unstable resonators, namely the magnification M and the equivalent Fresnel number, also apply to toric unstable resonators. A further similarity to conventional unstable resonators is the applicability of geometric optics to predict the mode structure and the output coupling losses in the first order. By using geometrical beam propagation, a set of self-reproducing radial intensity distributions $I_p(r)$ at the output coupler can be found, given by [5.146]:

$$I_p(r) = const. \frac{(r-r_0)^{2p}}{r} \tag{20.19}$$

where r_0 is the radius of the mirror vertex and p is the radial mode order.

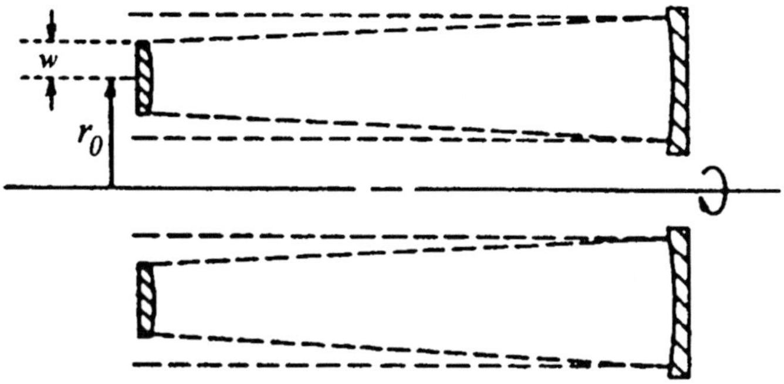

Fig. 20.14 Toric Unstable Resonator [5.146] (© OSA 1984).

These geometrical eigenmodes exhibit a round trip loss factor (=1-loss) of:

$$V_p = \frac{1}{M^{2p+1}} \tag{20.20}$$

where M is the magnification. For high equivalent Fresnel numbers, the actual intensity profile of the lowest loss mode will converge towards the $1/r$ distribution given in (20.19) for $p=0$, and the round trip loss is that of a strip resonator. Higher order modes will generally not be observed due to the high radial mode discrimination. For lower equivalent Fresnel numbers, the geometric approach is not suitable, and the mode structure has to be calculated using diffraction theory. If $u_p(x)$ denotes the transverse field distribution of the p-th eigenmode in the equivalent unstable strip resonator, the transverse field distributions of the eigenmodes of toric unstable resonators can be approximated by:

$$E_{p\ell}(r,\phi) = \frac{1}{\sqrt{r}}\, u_p(r-r_0)\, \exp[\pm i\ell\phi] \tag{20.21}$$

where r_0 is the radius of the mirror vertex. Since the output coupling loss is generated only in the radial direction, all azimuthal modes experience the same loss and no azimuthal mode discrimination is provided. The correct solution of the diffraction integral shows that there is a dependence of the radial mode profile on the azimuthal mode order, but this diffraction effect is too small to discriminate low order azimuthal modes. Therefore, the simultaneous oscillation of multiple azimuthal modes is to be expected unless mode selecting apertures are inserted into the resonator. As discussed in 20.2, the presence of azimuthally structured modes poses a serious problem as far as the focusability is concerned. On-axis intensity in the focal plane can only be generated with $\ell=0$ modes.

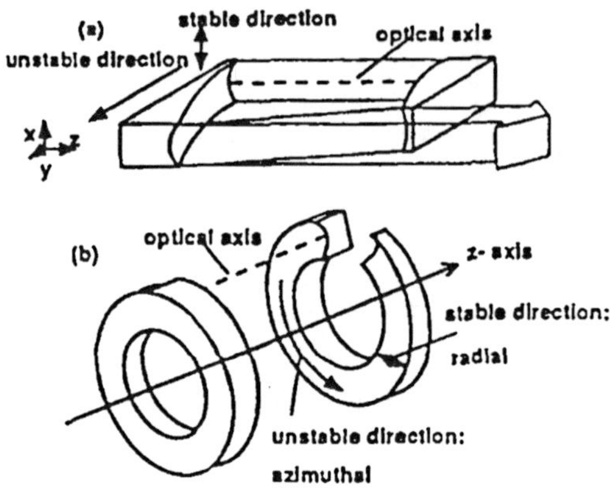

Fig. 20.15 Azimuthally unstable resonator (b). If the resonator is unwound, a one-dimensional off-axis unstable resonator is obtained (a) [5.167] (© SPIE 1994).

20.4.2 Azimuthally Unstable Resonators

The lack of azimuthal mode discrimination in toric unstable resonators is basically a result of the resonator being unstable in the radial direction. Improved mode properties should thus be expected if the azimuthal direction is chosen as the unstable direction [5.167,5.170]. Such an azimuthally unstable resonator can be obtained by winding an off-axis unstable resonator around the optical axis, as shown in Fig. 20.15. In the radial direction either a flat-flat or a stable resonator can be chosen since the radial dimension of the annulus is usually small enough to limit the radial mode order to low values.

Instead of using a parabolic mirror surface in the azimuthal direction, near diffraction limited beam quality can also be achieved with a helical output coupling mirror [5.167,5.170]. The set-up of a helical resonator is similar to that shown in Fig. 20.15b, except that the output coupling mirror has a constant slope in the azimuthal direction. In a geometrical model, the rays in the resonator are driven towards the output coupling aperture by the helical surface. However, the resonator is not unstable in the usual sense. Since the helical mirror is the annular version of a tilted flat mirror, the output coupling losses are given by the diffraction losses of a misaligned stable resonator and no geometric optics approximations for the mode structure exist like for unstable resonators. The output coupling can be increased with the slope of the helical mirror. For CO_2 lasers, typical slopes are on the order of $10\mu m/2\pi$. Measured and calculated intensity profiles in the near and the far field of a helical mirror CO_2 laser resonator are presented in Fig. 20.16. This resonator provided a maximum output power of 1,100W (the maximum total efficiency of 9% was attained at 450W) and beam parameter products close to the diffraction limit [5.170]. Note that for both resonator concepts, the output beam is highly astigmatic and beam shaping is necessary to generate round focus spots.

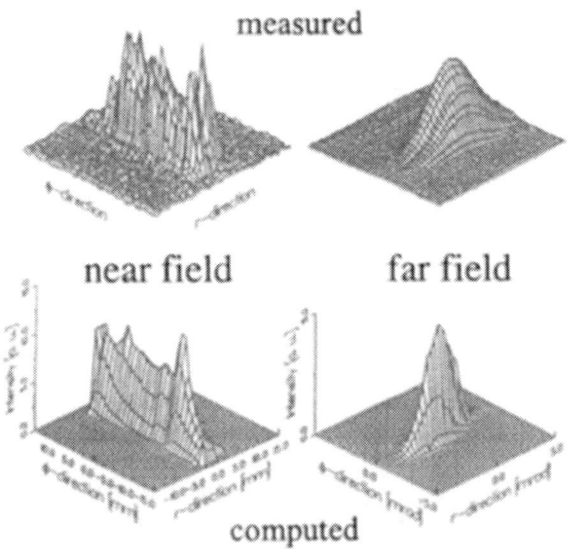

Fig. 20.16 Measured and calculated intensity distributions in the near field and the far field of an annular resonator with a helical output coupling mirror. Resonator length: 1.36m, discharge length: 1.15m, inner radius $a=48mm$, outer radius $b=55mm$ [5.170] (© IEEE 1994).

The application of azimuthally unstable resonators is currently limited by the fabrication of the resonator mirrors. For mid- to far-infrared lasers, the mirrors can be made of metal and flexible diamond turning techniques are available that are capable of generating non-rotationally symmetric surfaces. However, it is difficult to achieve optical surface quality and a commercial availability of these mirrors is not yet in sight. In the visible and the near-infrared spectral region, where glass is the preferred substrate material and a tighter tolerance on the surface roughness must be met, no successful manufacturing of toric or helical mirrors with optical quality has been reported yet.

An alternate annular resonator scheme that uses mirrors which are already available and that provides near diffraction limited beam quality is the tilted annular resonator as depicted in Fig. 20.17 [5.166]. Instead of a helically shaped mirror, a tilted toric or flat output coupling mirror is used. By using two flat mirrors, this resonator concept is also applicable to solid state lasers [5.161]. Similar to the helical resonator, the intracavity rays are guided towards the output coupling aperture. Unfortunately, no discrimination between the positive and the negative azimuthal direction is provided. Consequently, two nonparallel beams emerge from the aperture (Fig. 20.18). In order to achieve unidirectional operation, one of the beams can be reflected back into the resonator with an external mirror. Two times diffraction limited beam quality at a maximum output power of 700W (7% total efficiency) was reported for a diffusion-cooled CO_2 laser utilizing a tilted annular resonator [5.166]. A similar resonator provided a maximum output power of 2kW [5.165]. Output coupling in this resonator scheme is also present without tilting the resonator mirrors [5.169]. For the aligned resonator, the azimuthal radiation flux is solely induced by the diffraction at the output coupling aperture. Unidirectional operation can be achieved through the mirror alignment (Fig. 20.19), but random jumping between one and two output beams is observed for slight mirror tilts. In [5.169], the unidirectional energy flux could be stabilized in an annular CO_2 laser by inserting diffuse-reflecting copper strips at different locations on the circumference of the mirrors. An output power of 250W at a total efficiency of 7.8% and beam propagation factors M^2 between 2 and 4 were reported.

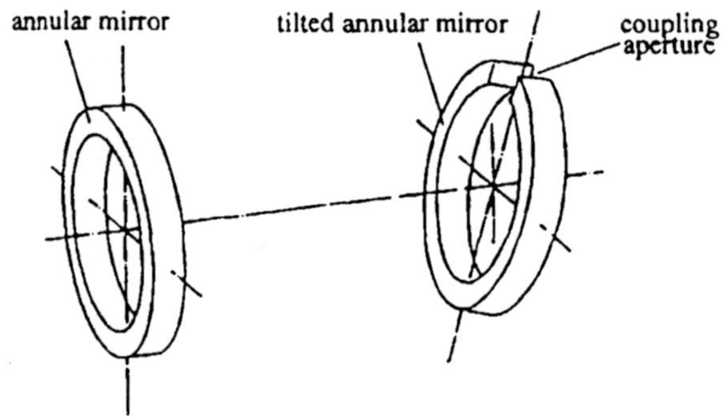

Fig. 20.17 Annular resonator with toric mirrors and an off-axis output coupling aperture. The tilt of the output coupling mirror induces a bidirectional azimuthal energy flux towards the aperture [5.166] (© IEEE 1993).

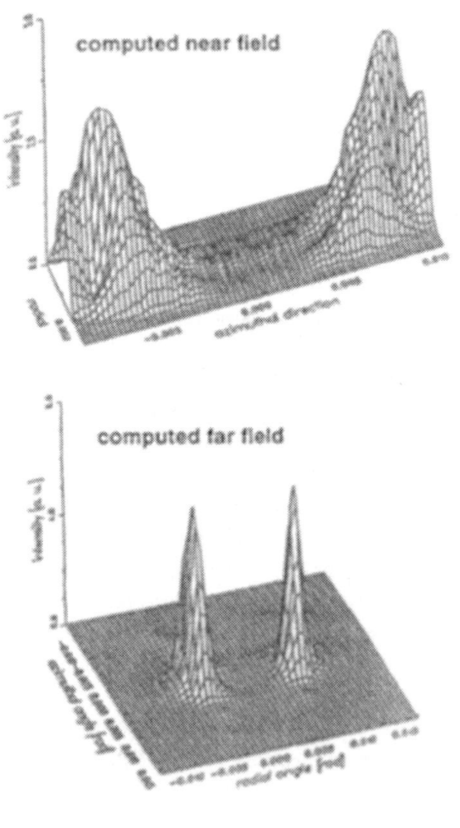

Fig. 20.18 Calculated near and far field intensities of a tilted annular resonator. The wave vectors of the two beams are not parallel, resulting in two peaks in the far field [5.166] (© IEEE 1993).

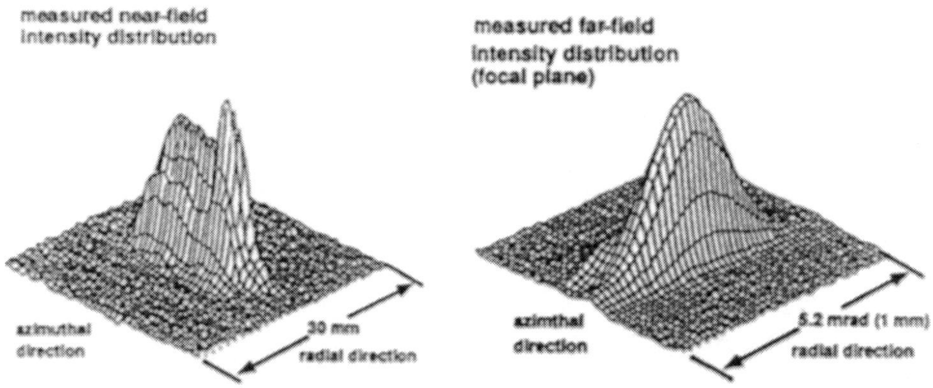

Fig. 20.19 Measured near field and far field intensity distributions for an annular resonator like the one depicted in Fig. 20.17, but with both mirrors aligned. An annular CO_2 discharge with an outer diameter of 110mm, a gap width of 7mm, and a length of 1.15m was used. The output coupling mirror was flat with an angular width of the aperture of 24°. The far field was recorded in the focal plane of a 7.5" focusing lens. Output power: 250W, pump power: 3.2kW [5.169] (© OSA 1994).

20.4.3 Spherical Unstable Resonators

If the unstable resonator comprises spherical mirrors, the intracavity beam has to be adapted to the annular shape of the active medium by using axicons, waxicons, or scrapers. The first experimental investigations of annular resonator geometries were focused on spherical unstable resonators because toric mirrors were not available at the time [5.131-5.137,5.139-5.144]. Some of these resonator schemes are presented in Fig. 20.20. The converging wave resonator uses a standard confocal unstable resonator placed outside the active region to form the mode. In the dual leg set-up shown in Fig. 20.20a, the feedback through the active medium is provided by a high reflecting mirror and a flat feedback mirror. The feedback arm excites the converging wave of the unstable resonator. After a couple of round trips the converging wave reaches the diffraction limited spot size and transforms into a diverging wave. After the diverging wave is subsequently expanded, a fraction of the power is coupled out and a fraction is reflected into the feedback arm. Diffraction limited beam quality has been attained with the converging wave resonator [5.134]. However, this resonator scheme suffers from parasitic oscillation between the flat mirror and the concave spherical mirror as well as from a high misalignment sensitivity. The linear version of the converging wave resonator shown in Fig. 20.20c exhibits similar problems.

A high alignment sensitivity is also a drawback of the HSURIA scheme depicted in Fig. 20.20b [5.141]. Furthermore, the complicated polarization properties of axicons and waxicons lead to a deterioration of the beam quality [5.142,5.164].

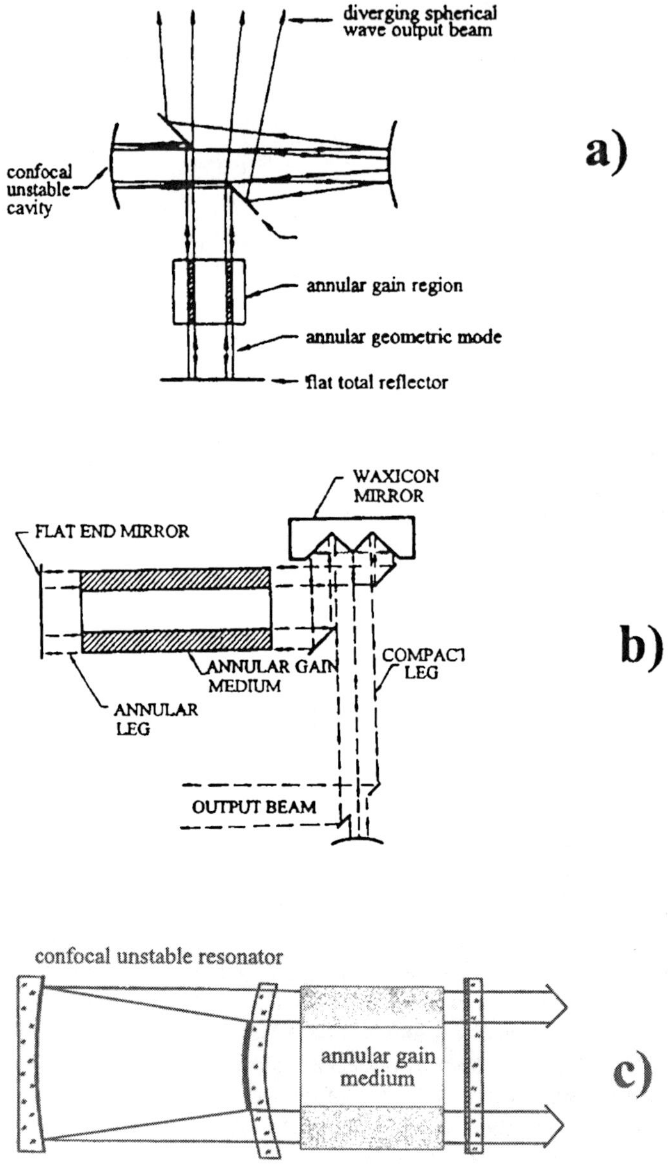

Fig. 20.20 Spherical unstable resonators for annular gain media. a) converging wave unstable resonator [5.134] (© OSA 1976), b) half-symmetric unstable resonator with intracavity axicon (HSURIA) [5.141] (© OSA 1980), c) linear converging wave unstable resonator.

Chapter 21

Ring Resonators

21.1 General Properties of Ring Resonators

In linear resonators, the interference of the two counterpropagating waves generates a modulation of the intensity along the optical axis. As discussed in Sec. 9.5.2, the intracavity intensity distribution of an axial mode of order q outside the active medium reads:

$$I(z) = I_0 \left[1 - \frac{2R}{1+R^2} \cos(2\pi q z/L) \right] \tag{21.1}$$

where L is the resonator length, R is the reflectance of the output coupling mirror, and z is the coordinate along the optical axis, starting at a mirror surface. The intensity minima exhibit a periodicity of half the wavelength of the axial mode $\lambda_q = 2L/q$. Due to gain saturation this results in a modulated axial gain distribution, referred to as *spatial holeburning*. The modulation frequency depends on the axial mode order.

In homogeneously broadened lasers, only one axial mode should theoretically be observed since the axial mode with the highest small signal gain has access to the whole gain and will suppress the oscillation of adjacent modes (see Sec. 9.5.1). However, the spatial hole burning induces axial mode competition because different modes are amplified in different areas of the gain medium. This may lead to multimode oscillation with random jumping between axial modes. Furthermore, the lower extraction efficiency at the intensity minima results in a decreased output power unless a high number of axial modes can coexist. In order to attain single axial mode operation with a high output power it is necessary to prevent spatial holeburning. In the twisted mode resonator (see Sec. 8.2.1), this is accomplished by using orthogonal polarization states for the back and the forth traveling wave [5.220]. Another way to generate a homogeneous intensity profile along the optical axis is unidirectional beam propagation in a ring resonator [5.173-5.178] (Fig. 21.1). By using an optical diode inside the resonator the axial modes are forced to propagate in one direction only. The optical diode can be realized with a retardation plate, a Faraday rotator, or a combination of both such that the losses generated at polarizers are higher for one propagation direction (Fig. 21.2).

In asymmetric unstable ring resonators, the different propagation characteristics associated with the oppositely traveling waves can be used to generate unidirectional beam propagation [5.180-5.182] (e.g. by inserting apertures).

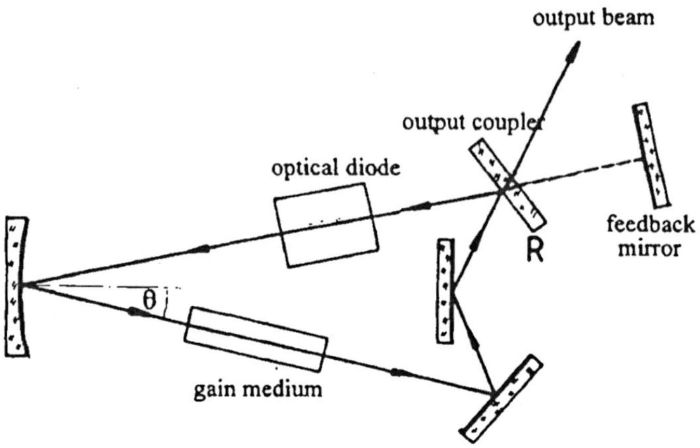

Fig. 21.1 Concept of a stable ring resonator. Unidirectionality can be achieved with an intracavity optical diode or an external mirror that feeds back the wave having the wrong propagation direction.

Unidirectionality can also be attained by means of an external feedback mirror which reflects the reverse oscillation [5.174]. However, with an external mirror, the ratio of the forward to the reverse power is only on the order of 1000:1 because a fraction of the power always has to propagate in the reverse direction to make this scheme work [5.182]. A twisted mode technique has to be applied in addition to suppress the residual hole burning (Fig. 21.3).

In general, ring resonators are set-up with more than three mirrors to decrease the astigmatism induced by the mirror tilt. For an angle of incidence of θ, a curved mirror approximately exhibits a focal length of $f\cos\theta$ in the tangential plane (paper plane in Fig. 21.1) and a focal length of $f/\cos\theta$ in the sagittal plane where f is the focal length at normal incidence (see Sec. 1.2.5). By using more mirrors or a z-fold geometry (Fig.21.2) the angles of incidence can be kept small. It is also possible to compensate the astigmatism by using special resonator designs [5.178].

Similar to linear stable resonators, the Gaussian beam propagation in stable ring resonators can be evaluated by means of the Gaussian ABCD law. In order to include the astigmatism, four-dimensional ray transfer matrices have to be used. Starting at an arbitrary plane, the ray transfer matrix for a round trip is calculated. The $\mathbf{Q}^{-1}$-matrix of the elliptical Gaussian beam (Sec. 2.5.2) is a solution of the generalized ABCD law:

$$Q^{-1} = (C + DQ^{-1})(A + BQ^{-1})^{-1} \tag{21.2}$$

where $A,B,C,\mathbf{D}$ are the 2x2 submatrices of the 4x4 ray transfer matrix (see Sec. 1.2.4),and the superscript -1 denotes the inverse matrix.

The matrix Q^{-1} contains the q-parameters of the Gaussian beam at the reference plane:

$$Q^{-1} = \begin{pmatrix} 1/q_{xx} & 1/q_{xy} \\ 1/q_{yx} & 1/q_{yy} \end{pmatrix} \tag{21.3}$$

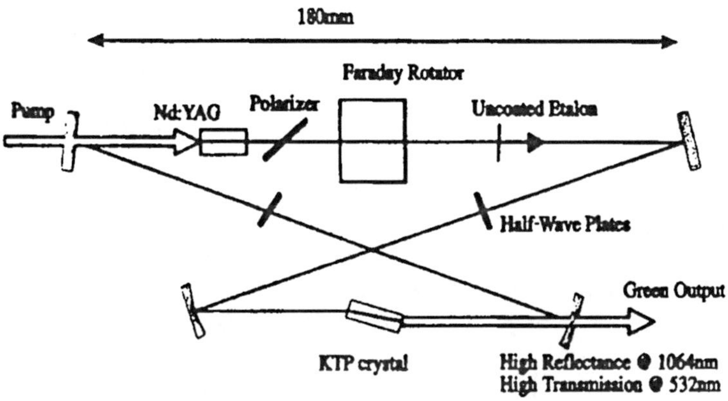

Fig. 21.2 A cw diode laser pumped Nd:YAG laser with intracavity second harmonic generation. The Faraday rotator produces a polarization rotation by 7.5° at 1.064μm. The same rotation is provided by the two half-wave plates. The two rotations add in one direction generating losses at the polarizer. In the intended propagation direction, the two rotations cancel. The laser provides an output power of 3.1W at 532nm in single mode operation for a pump power of 14W at 808nm [5.218] (© OSA 1996).

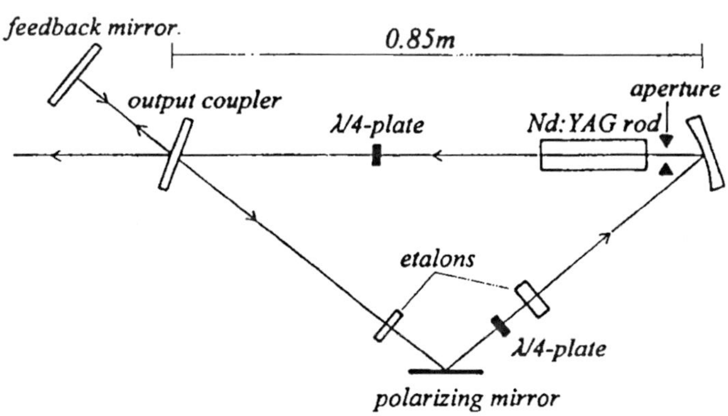

Fig. 21.3 A pulsed, flashlamp pumped, Nd:YAG rod laser using a unidirectional ring resonator. With the feedback mirror, the finite extinction ratio between the two propagation directions cannot completely prevent hole burning. For this reason, a twisted mode technique is applied to generate circular polarization in the active medium. The aperture controls the transverse mode and the two etalons generate a single lowest loss axial mode [S.27].

For planar ring resonators, the Gaussian beam is simple astigmatic, which means that the nondiagonal elements of $Q^{\prime}$ are zero. In this case, the beam propagation can be calculated separately for the sagittal plane and the tangential plane by using the common 2x2 round trip matrices. The ring resonator is stable in one plane if the following relation holds for the elements of the corresponding ray transfer matrix:

$$|A + D| < 2 \tag{21.4}$$

In contrast to a linear resonator, the mirror surfaces of a ring resonator generally do not match the phasefronts of the Gaussian beam. The Gaussian beam waist is always located at that plane at which the round trip matrix exhibits equal diagonal elements ($A=D$).

Example:
We determine the stability condition and the location of the Gaussian beam waist for the resonator depicted in Fig. 21.1. The reference plane is chosen at a distance x from the curved mirror (in the propagation direction). If L is the perimeter length of the resonator, and ρ is the radius of curvature of the mirror, the ray transfer matrix for the sagittal plane reads:

$$M_S = \begin{pmatrix} 1 & x \\ 0 & 1 \end{pmatrix} \begin{pmatrix} 1 & 0 \\ -2\cos\theta/\rho & 1 \end{pmatrix} \begin{pmatrix} 1 & L-x \\ 0 & 1 \end{pmatrix} = \begin{pmatrix} 1-2x\cos\theta/\rho & L-2x(L-x)\cos\theta/\rho \\ -2\cos\theta/\rho & 1-2(L-x)\cos\theta/\rho \end{pmatrix}$$

In the tangential plane, one gets:

$$M_T = \begin{pmatrix} 1 & x \\ 0 & 1 \end{pmatrix} \begin{pmatrix} 1 & 0 \\ -2/(\rho\cos\theta) & 1 \end{pmatrix} \begin{pmatrix} 1 & L-x \\ 0 & 1 \end{pmatrix} = \begin{pmatrix} 1-2x/(\rho\cos\theta) & L-2x(L-x)/(\rho\cos\theta) \\ -2/(\rho\cos\theta) & 1-2(L-x)/(\rho\cos\theta) \end{pmatrix}$$

In both planes, the condition $A=D$ leads to $x=L/2$ for the location of the beam waist. By using (21.4), the stability condition for the ring resonator is obtained:

$$-1 < (1 - L\cos\theta/\rho) < 1 \qquad \textit{in the sagittal plane}$$

$$-1 < (1 - L/(\rho\cos\theta) < 1 \qquad \textit{in the tangential plane}$$

The resonator first goes unstable in the tangential plane as the perimeter length L is increased.

The unidirectionality in ring resonators generates different beam properties as compared to those of linear resonators. First, the electric field distribution is reproduced after one transit. Therefore, the frequency separation of the axial modes is twice that of linear resonators with length L:

$$\Delta\nu = \frac{c}{L} \tag{21.5}$$

Secondly, light amplification takes place only once per round trip. The steady-state condition for ring oscillators, therefore, reads:

$$G\,R\,V_S = 1 \tag{21.6}$$

where G is the gain factor, R is the reflectance of the output coupler, and V_S is the loss factor (=1-loss) per transit. For a linear resonator, the corresponding steady state condition is given by $G\sqrt{R}V_S = 1$. In order to reach laser threshold for the same output coupling, the electrical pump power in a ring resonator has to be higher than in a linear resonator by a factor γ, with:

$$\gamma = \frac{\ln(RV_S)}{\ln(\sqrt{R}V_S)} \tag{21.7}$$

This implies that at identical pump powers, the output power of a ring oscillator is lower. However, this is only due to the fact that the ring resonator is overcoupled when the linear resonator exhibits optimum output coupling. If the output coupling is adjusted, the ring resonator provides the same maximum output power as the linear resonator. Maximum extraction efficiencies of ring resonators and the corresponding optimum output coupling transmission are visualized in Fig. 21.4 for different values of the small-signal gain and the loss per transit (see also Secs. 10.1.1 and 10.1.4). As was already shown in Fig. 10.17, the maximum extraction efficiencies of ring resonators are identical to those attainable with linear resonators.

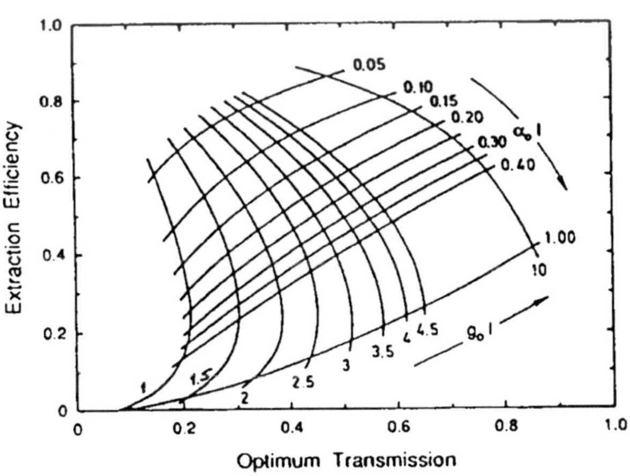

Fig. 21.4 Calculated maximum extraction efficiency of ring resonators with a homogeneously broadened active medium as a function of the optimum transmission of the output coupling mirror. The curve parameters are the small-signal gain $g_0\ell$ and the loss per transit $\alpha_0\ell$ (see also Secs. 10.1.1 and 10.1.4).

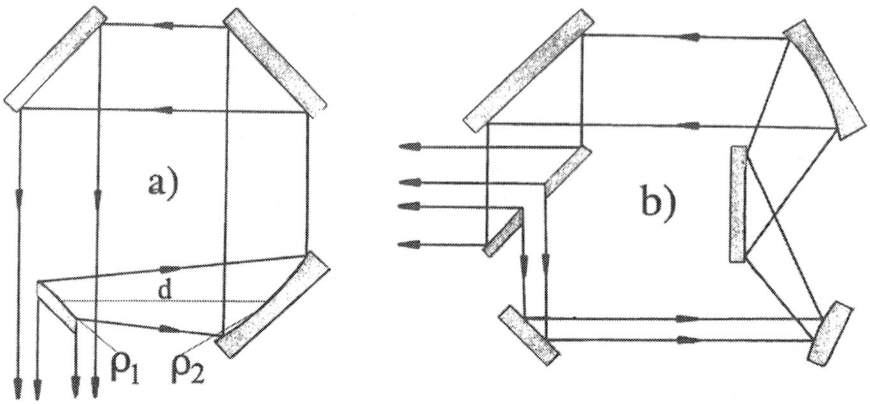

Fig. 21.5 Asymmetric confocal unstable ring resonators. a) confined output coupler, b) scraper.

21.2 Unstable Ring Resonators

The ring geometry can also be applied to unstable resonator schemes as shown in Fig. 21.5 [5.175,5.180-5.182]. Similar to stable ring resonators the resonator properties can be evaluated by unfolding the ring into a linear lens waveguide. Preferably, an asymmetric confocal resonator set-up is chosen which means that the transfer matrix for one round trip, starting at the output coupler, is different for the two propagation directions. The main advantage associated with the asymmetric design is a greater design flexibility. The flat mirrors in the collimated beam path can be placed arbitrarily without changing the magnification of the resonator. This enables one to incorporate several active media or to increase the round trip time. Unfortunately, there is no discrimination against one propagation direction since the forward and the reverse wave exhibit the same output coupling losses. However, the different propagation characteristics of the two waves can be used to suppress one of the wave. Let us clarify this by evaluating the round trip ray transfer matrices for the resonator of Fig. 21.5a. The round trips in the clo ckwise and the counterclockwise directions are equivalent to the transits in the lens waveguides depicted in Fig. 21.6. Starting at the output coupling mirror, the ray transfer matrix for the clockwise direction reads:

$$M_{CW} = \begin{pmatrix} 1 & d \\ 0 & 1 \end{pmatrix} \begin{pmatrix} 1 & 0 \\ -2/\rho_2 & 1 \end{pmatrix} \begin{pmatrix} 1 & 3d \\ 0 & 1 \end{pmatrix} \begin{pmatrix} 1 & 0 \\ -2/\rho_1 & 1 \end{pmatrix}$$

$$= \begin{pmatrix} 2+2\rho_1/\rho_2-\rho_2/\rho_1 & -\rho_1+\rho_2/2-3\rho_1^2/(2\rho_2) \\ 4/\rho_1+4/\rho_2 & -2-3\rho_1/\rho_2 \end{pmatrix} \quad (21.7)$$

where we used the confocality condition $\rho_1+\rho_2=2d$. In the counterclockwise direction we get:

$$M_{CCW} = \begin{pmatrix} 1 & 3d \\ 0 & 1 \end{pmatrix}\begin{pmatrix} 1 & 0 \\ -2/\rho_2 & 1 \end{pmatrix}\begin{pmatrix} 1 & d \\ 0 & 1 \end{pmatrix}\begin{pmatrix} 1 & 0 \\ -2/\rho_1 & 1 \end{pmatrix}$$

$$= \begin{pmatrix} -\rho_2/\rho_1 & -\rho_1+\rho_2/2-3\rho_1^2/(2\rho_2) \\ 0 & -\rho_1/\rho_2 \end{pmatrix} \qquad (21.8)$$

In both directions, we can find two self-reproducing spherical wavefronts, referred to as the diverging wave and the converging wave (see Sec. 7.2.1). The corresponding radii of curvature R and the magnifications M can be calculated using the relations given in Sec. 1.2.7. For the resonator parameters $\rho_1=-0.5m$, $\rho_2=1m$, and $d=0.25m$, we obtain:

ccw-direction, diverging wave:	$M=2.0$,	$R=\infty$
ccw-direction, converging wave:	$M=0.5$,	$R=-0.417m$
cw-direction, diverging wave:	$M=2.0$,	$R=-0.625m$
cw-direction, converging wave:	$M=0.5$,	$R=-0.25m$

As was discussed in Sec. 7.2, the converging wave in unstable resonators is generally not observed since it will transform itself into a diverging wave after a few round trips. Thus, the asymmetric unstable ring resonator supports two diverging waves having opposite propagation directions. Although the asymmetry generates different propagation characteristics for these counterpropagating waves they exhibit the same magnification.

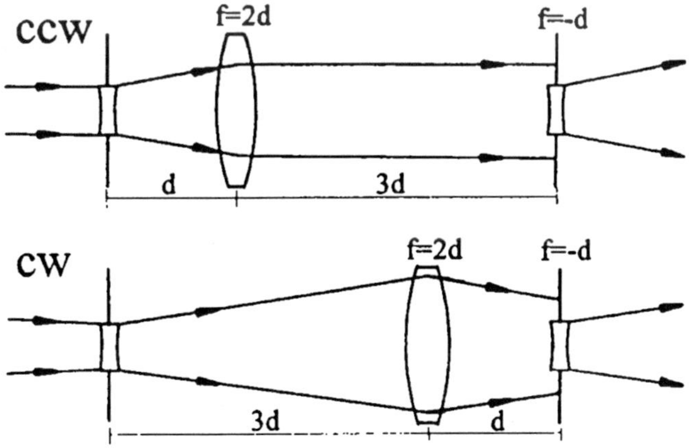

Fig. 21.6 Beam propagation of the diverging waves in a confocal unstable ring resonator with magnification $M=2$. The resonator shown in Fig. 20.5a is represented as a lens waveguide. A mirror with radius of curvature ρ is replaced by a lens with focal length $f=\rho/2$. In the counterclockwise direction (ccw), the output beam (beam portion missing the biconcave lens) is collimated.

The different propagation characteristics can be used to discriminate against one direction. Figure 21.6 presents the beam propagation of the two diverging waves for the resonator discussed above. Note that in the ccw direction a collimated output beam is obtained. The oscillation in the clockwise direction can be suppressed by limiting the focusing mirror with an aperture.

Unstable ring resonators have been realized in CO_2 lasers [5.175,5.180-5.182,5.205], dye lasers [5.184,5.191], and solid state lasers [5.198]. The effect of thermal lensing on the performance of unstable solid state ring lasers is discussed in [5.200]. A discrimination against the reverse traveling wave can be achieved by locating the active medium where the forward traveling mode exhibits the larger mode volume [5.181,5.182]. By using this technique, power ratios of 20:1 have been observed. Unidirectional operation, however, requires intracavity apertures to discriminate against the reverse oscillation. Alternatively, an optical feedback of the reverse traveling wave can be applied such that the radius of curvature and the propagation direction of the forward traveling wave is matched. In [5.182], the use of a feedback mirror increased the ratio of the forward to the reverse power of a CO_2 laser from 16:1 to 1,600:1.

The suppression of the reverse traveling wave by means of intracavity apertures is very effective in negative branch unstable ring resonators [5.185,5.192,5.193, 5.199]. Both the forward and the reverse wave exhibit an intracavity focus at different locations in the resonator (Fig. 21.7a). By placing an aperture in the focus plane of the forward traveling wave, a high loss is generated for the reverse traveling wave. The aperture can also serve as a spatial filter to filter out phase distortions and smooth out the spatial intensity distribution (see Sec. 18.1). A special negative branch ring resonator is the self-imaging unstable resonator as depicted in Fig. 21.7b [5.185, 5.193]. The scraper is in the common focal plane of the two lenses.

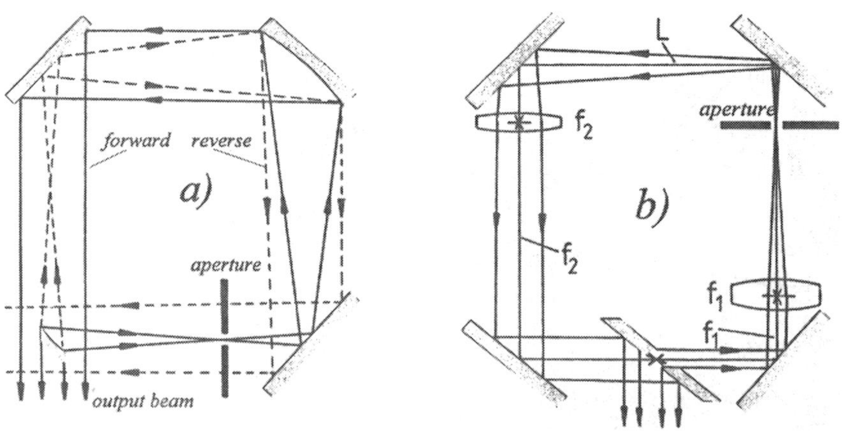

Fig. 21.7 Beam propagation in negative branch unstable ring resonators. a) Fourier transform resonator with the forward wave (solid line) and the reverse wave (broken line), b) Forward traveling wave in a self-imaging unstable resonator with $L=f_1+f_2$. The reverse traveling wave (not shown) is collimated in the upper section and it exhibits a focus at the scraper.

Starting at the scraper, the ray matrix for a round trip in the ccw direction reads:

$$M_{ccw} = \begin{pmatrix} -f_2/f_1 & 0 \\ -1/f_1 - 1/f_2 + L/(f_1 f_2) & -f_1/f_2 \end{pmatrix} \tag{21.9}$$

Since the B-element of the ray transfer matrix is equal to zero, the round trip generates an image of the starting field with a magnification of $-f_2/f_1$. This imaging condition holds for arbitrary lengths L of the upper section. Again, diverging waves will exist in both directions and a pinhole is required to discriminate against the reverse traveling wave. Furthermore, the self-imaging implies that the resonator modes are degenerate. Thus, the pinhole is also needed to select the transverse mode structure. If the length of the upper section L equals f_1+f_2, the output beam becomes collimated. In this case, the scraper can be moved arbitrarily and the two lenses do not need to have a common focal plane in the lower section.

21.3 Nonplanar Ring Resonators

In planar ring resonators with the mirrors tilted around the same axis, s-polarized and p-polarized light represent the polarization eigenstates for a resonator round trip. The phase shifts induced at the mirror generally are different for the two polarizations, but no mixing between s- and p-polarization can occur. This is different in the nonplanar geometry in which the mirrors are tilted around different axes. The polarization properties of nonplanar rings can be evaluated by using the Jones matrix formalism [5.1,5.209,5.211]. It can be shown that nonplanar ring resonators act as polarization rotators having left circularly and right circularly polarized light as the eigenstates. When combined with a Faraday rotator and a polarizing element, the rotation of the polarization plane at the mirrors can be used to discriminate against one propagation direction.

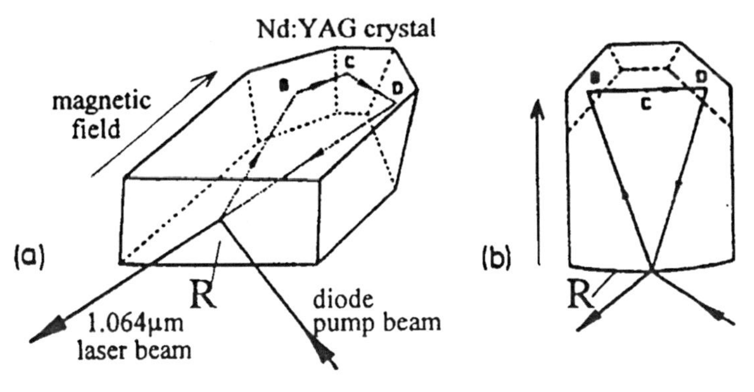

Fig. 21.8 Beam propagation in a diode-pumped monolithic non-planar Nd:YAG ring oscillator. The size of the crystal is 11x7x3 mm³. a) side view, b) top view [5.206] (© AT Fachverlag 1988).

A widely used laser employing this technique is the diode-laser-pumped monolithic nonplanar Nd:YAG ring oscillator [5.196,5.202,5.206,5.207,5.209,5.215] (Fig. 21.8). A dc magnetic field is used to generate a polarization rotation via the Faraday effect (Verdet constant of YAG: 103°/(Tm)).In the clockwise direction, the rotation of the polarization plane caused by the internal reflections at the surfaces B, C, and D and the Faraday rotation cancel, giving a s- polarized eigenmode at the front surface. In the opposite direction, the two rotations add, resulting in an elliptically polarized eigenmode. The reflectance R of the front surface (Fresnel-reflection) is slightly higher for the s-polarization than for the p-polarization. The wave in the ccw direction, therefore, has a higher laser threshold and is suppressed. The output coupling reflectance and the polarization rotation caused by the internal reflections can be controlled by the geometry of the monolithic structure. The suppression also works if the two polarization rotations do not cancel in the clockwise direction. In [5.206], the 17° rotation caused by the internal reflections was superimposed by a 0.6° Faraday rotation. At the coated front surface of the crystal, the resulting difference in polarization for the two propagation directions generated a loss difference of 0.6 percent. This small difference was sufficient to realize unidirectional single mode operation with an output power of 100mW (400mW optical pump power).

Due to the mechanical stability and the highly stable pump source, the monolithic Nd:YAG ring resonator has the capability of providing single mode operation with narrow linewidths and high output stability. Commercial lasers provide output powers in the 100mW range with linewidths of less than 3kHz and an rms stability of better than 0.05%. With active frequency stabilization, very narrow linewidths down to the sub-Hertz range have been achieved [5.207,5.208,5.216].

Chapter 22

Single Mode Resonators

22.1 Axial Mode Spectrum of Lasers

Without additional frequency selecting elements, most lasers utilizing linear resonators emit at several axial modes. A resonator operating in a single transverse mode, exhibits a frequency spectrum which consists of multiple lines. If we neglect the frequency pulling induced by the gain profile, these emission lines are spectrally separated by $c_0/(2L)$, where L is the optical resonator length and c_0 is the speed of light in a vacuum. In inhomogeneously broadened lasers, all axial modes are observed that experience a gain high enough to overcome the resonator losses (Fig. 22.1a). Since no interaction between different axial modes occurs, the number of oscillating axial modes increases as the gain is increased. Well above laser threshold, the number of observed axial modes n can be approximated by:

$$n = \frac{\Delta\nu \, 2L}{c_0} \tag{22.1}$$

where $\Delta\nu$ is the gain bandwidth of the active medium (see Table 4.2 for bandwidths of common lasers). Inhomogeneous line broadening is found in low pressure gas lasers (HeNe, Ar, CO_2) and in semiconductor lasers at high injection currents. In both laser types, typically on the order of 10 axial modes oscillate simultaneously.

The majority of lasers utilize active media with homogeneous line broadening. In a homogeneously broadened laser, each axial mode has access to the whole gain spectrum. The axial mode that first reaches the laser threshold saturates the gain at all frequencies. This automatically ensures single mode operation because the other axial modes are kept below threshold (Fig. 22.1b). However, in a standing wave resonator, the gain saturation leads to a longitudinal modulation of the gain referred to as spatial hole burning. This may result in the simultaneous operation of several axial modes (Fig. 22.1c) because different modes are amplified in different gain regions. Furthermore, small-scale variations of the optical resonator length due to mechanical vibrations or varying pumping conditions lead to random jumping between modes. In general, the bandwidth of the emission is smaller than the gain bandwidth of the active medium. In free-running Ti:sapphire lasers, for instance, the bandwidth of the laser emission is only a few nm, although the gain profile would allow laser oscillation over a range of 300nm. Spatial holeburning is particularly strong in solid state and liquid lasers because the gain modulation cannot be smoothed out by a fast movement of the active atoms.

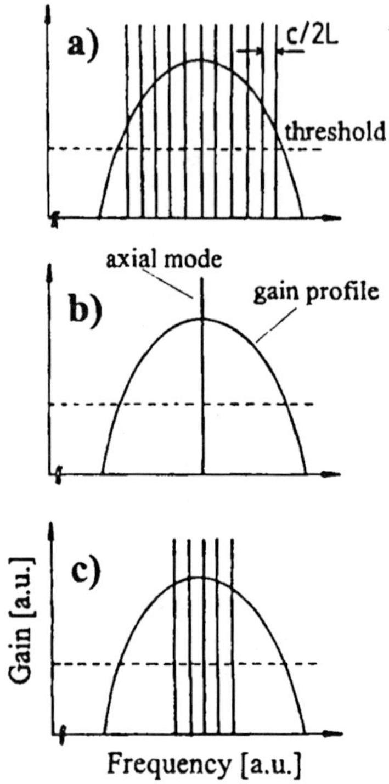

Fig. 22.1 Unsaturated gain profile and axial mode spectrum for a) inhomogeneous line broadening, b) homogeneous line broadening without spatial holeburning, c) homogeneous line broadening with spatial holeburning.

Due to the difference in gain saturation, different mode selecting techniques have to be applied to generate single axial mode operation in homogeneously and inhomogeneously broadened lasers. In the latter systems, losses have to be generated for the other axial modes, whereas in homogeneously broadened lasers, single axial mode operation can also be attained by preventing the spatial holeburning.

Single axial mode operation is needed in applications where a low bandwidth, a high temporal coherence, or a low noise is required. Examples are the measurement of atomic/molecular absorption spectra, the determination of small-scale shifts using Michelson interferometers (e.g. gravitational wave detector), and laser radar. Narrow linewidth lasers are also of interest for the development of length and wavelength standards. For all these applications, a relative frequency stability of better than 10^{-12} is required. In order to apply active frequency stabilization techniques it is necessary that the laser resonator oscillates in a single axial mode. Without frequency stabilization, the linewidth of a laser operating at one axial mode is on the order of MHz due to frequency variations caused by environmental fluctuations. The reduction of the mechanical instabilities by using rigid structures and the control of the pumping conditions can reduce the linewidths to values of several kHz. With active frequency stabilization techniques which lock the laser frequency to a reference frequency using a feedback loop, linewidths down to the mHz level have been attained [5.244,5.246,5.265-5.268,5.272].

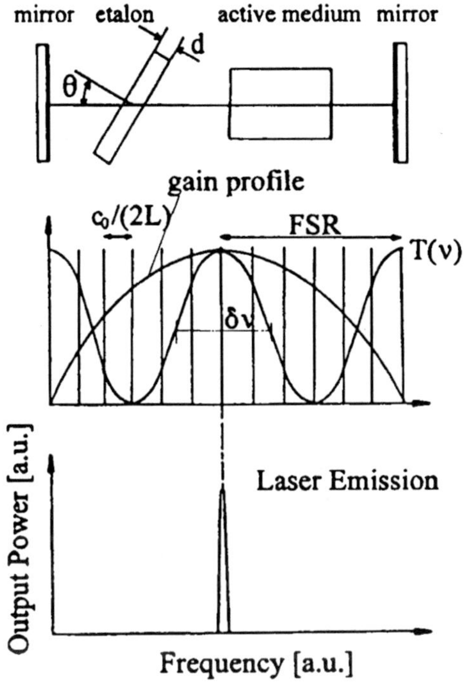

Fig. 22.2 Realization of single axial mode operation with an intracavity etalon with spectral transmission $T(\nu)$. The free spectral range (FSR) of the etalon is adapted to the gain bandwidth and the bandwidth $\delta\nu$ of the etalon is on the same order of magnitude as the axial mode separation. $c_0/(2L)$.

22.2 Axial Mode Selection with Intracavity Elements

In both homogeneously and inhomogeneously broadened lasers, single axial mode operation can be attained by generating losses for the other axial modes. This can be accomplished by using intracavity elements that exhibit a frequency dependent transmission. The most common technique is the insertion of an etalon into the resonator as illustrated in Fig. 22.2. An etalon of thickness d with an angle of θ between the surface normal and the optical axis and a reflectance R of both surfaces exhibits a spectral transmission of (we neglect any absorption losses and assume that the tilt angle θ is small):

$$T(\nu) = \frac{1}{1 + F\sin^2\delta} \tag{22.2}$$

with: $$F = 4R/(1-R)^2$$

$$\delta = 2\pi\frac{d}{\lambda_0}\sqrt{n^2-\sin^2\theta}$$

where n is the index of refraction and λ_0 is the vacuum wavelength. The maximum transmission of $T=1.0$ is obtained at the frequencies:

$$\nu_q = q \frac{c_0}{2d\sqrt{n^2-\sin\theta^2}} = q\,\Delta\nu \tag{22.3}$$

The frequency separation $\Delta\nu$ is called the free spectral range of the etalon. The spectral width (FWHM) of the transmission maxima is given by:

$$\delta\nu = \Delta\nu \frac{1-R}{\pi\sqrt{R}} \tag{22.4}$$

The etalon has to be designed such that the spectral transmission $T(\nu)$ forces all axial modes except one below the laser threshold. If $G(\nu)$ is the gain profile and V is the loss factor (=1-loss) per transit of the resonator (without etalon, but including output coupling), single mode oscillation occurs if $TGV>1$ for one axial mode and $TGV<1$ for all other modes. This is certainly the case, if the free spectral range $\Delta\nu$ of the etalon is larger than the gain bandwidth and the bandwidth $\delta\nu$ is smaller than twice the axial mode spacing. However, depending on the gain factor, the gain profile, and the homogeneous linewidth, etalons having a smaller FSR and a larger bandwidth may also work.

Example:
An Argon laser has a gain bandwidth of 5GHz (see Table 9.2). For a resonator length of 0.5m, the axial mode separation is 300MHz. In order to realize single mode operation, we use an etalon with a FSR $\Delta\nu$ of 5GHz and a bandwidth $\delta\nu$ of 600MHz. For a glass etalon (n=1.5) inserted at an angle of 5°, Eq. (22.3) yields a thickness of 2cm. According to (22.4), the bandwidth of 600MHz is attained for a reflectance of $R=69\%$.

According to (22.4), it is difficult to realize etalons that combine a large free spectral range and a small spectral width. For lasers with a wide gain profile (>30GHz), it is common to use two etalons having different spectral ranges (Fig. 22.3). The etalon with the smaller FSR suppresses the axial modes adjacent to the center mode and the second etalon discriminates against the axial modes located in the outer area of the gain profile. For a Nd:YAG laser with a resonator length of 1m (gain bandwidth: 120GHz, axial mode spacing: 150 MHz), single axial mode operation can be achieved with the following glass etalons [S.26]:

	d[mm]	R[%]	Δν[GHz]	δν[GHz]
etalon 1	20	60	7.5	1.2
etalon 2	0.5	60	300	49

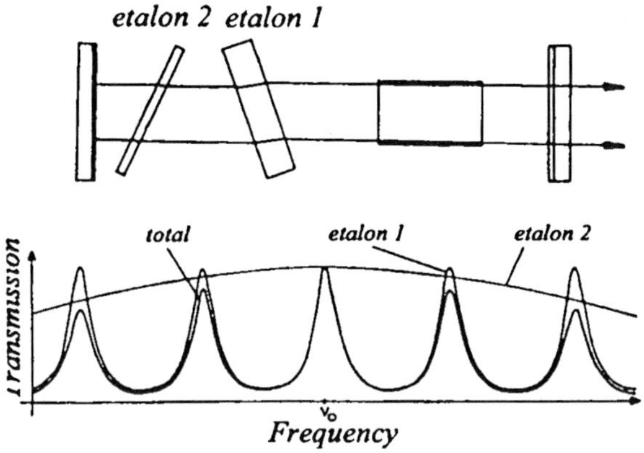

Fig. 22.3 Resonators with single axial mode selection using two etalons with different free spectral ranges. The lower graph shows the spectral transmission of each etalon and the resulting transmission when the etalons are combined [S.27].

Single frequency TEM_{00} mode operation operation with output powers of up to 35W have been attained with Yb:YAG disk lasers with two etalons inserted [5.274,5.275].

Axial mode selection can also be realized with intracavity Fabry Perot interferometers or Lyot filters [5.219,5.223,5.228]. The Lyot filter consists of a birefringent crystal placed between parallel polarizers. For a crystal length L and an index difference Δn between the ordinary and the extraordinary wave, the spectral transmission reads:

$$T(\nu) = T_0 \cos^2\left[\frac{\pi \Delta n L}{c_0}\nu\right] \tag{22.5}$$

where T_0 is the transmission of the crystal. The spectral width of the transmission maxima is equal to half the free spectral range of $\Delta\nu = c_0/(\Delta n L)$. The spectral width can be decreased by combining several Lyot filter with different lengths. For lasers with a wide gain profile (e.g. dye lasers, tunable solid state lasers), several etalons or Lyot filters have to be used in combination with prisms or gratings to achieve single mode operation.

22.3 Axial Mode Selection in Coupled Resonators

An alternate way to select one axial mode, used in gas and semiconductor lasers, is the coupling of two resonators. The simplest coupling scheme is the three mirror cavity depicted in Fig. 22.4. The axial mode spectrum of such a coupled resonator can be evaluated by using the matrix method presented in Sec. 4.3. After calculating the reflectance coefficient for light entering the system from the left, the resonance frequencies are determined by an infinite reflectance of the resonator [5.240]. Instead of going through these calculations, let us try to gain a basic understanding of the axial mode discrimination by considering the coupled resonator as a single resonator with length L_2 having an FPI with length L_1 as the output coupling mirror.

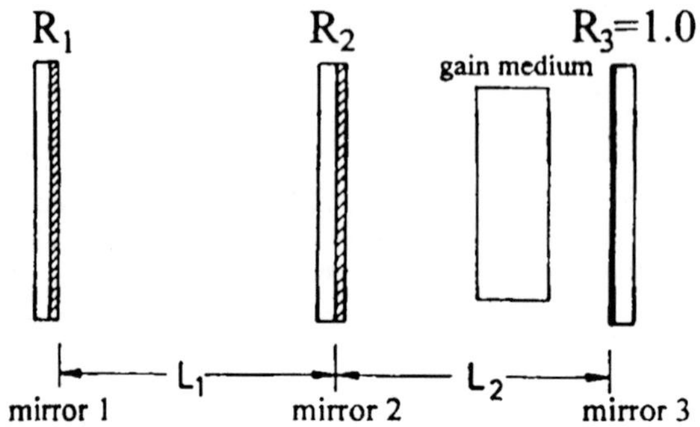

Fig. 22.4 Three mirror resonator.

As discussed in Sec. 4.2.1, the reflectance R of the FPI formed by mirror 1 and mirror 2 assumes a maximum when odd multiples of a quarter wavelength fit into the mirror spacing L_l (anti-resonance). The maximum reflectance reads:

$$R_{max} = \left| \frac{\sqrt{R_1} + \sqrt{R_2}}{1 + \sqrt{R_1 R_2}} \right|^2 \tag{22.6}$$

and the frequency spacing of the reflectance maxima is given by the free spectral range:

$$\Delta\nu = \frac{c_0}{2L_1} \tag{22.7}$$

We can now tune the length L_l until one axial mode of the resonator (mirror 2 and 3) is matched to the anti-resonance condition of the FPI. The wavelength λ of the matched axial mode fulfills the two conditions:

$$q\frac{\lambda}{2} = L_2 \;, \qquad (2p-1)\frac{\lambda}{4} = L_1 \qquad p,q > 0 \qquad \begin{matrix}(22.8)\\(22.9)\end{matrix}$$

This axial mode experiences an effective output coupling reflectance given by (22.6). All other axial modes within the free spectral range of the FPI have a lower reflectance and, consequently, a higher laser threshold. Depending on the gain, the resonator geometry and the mirror reflectances, this loss difference may be sufficient to suppress the adjacent axial modes (see also Fig. 22.5.c). Due to the coupling between the two segments, the mode frequency will shift if one of the two lengths L_l, L_2 is varied.

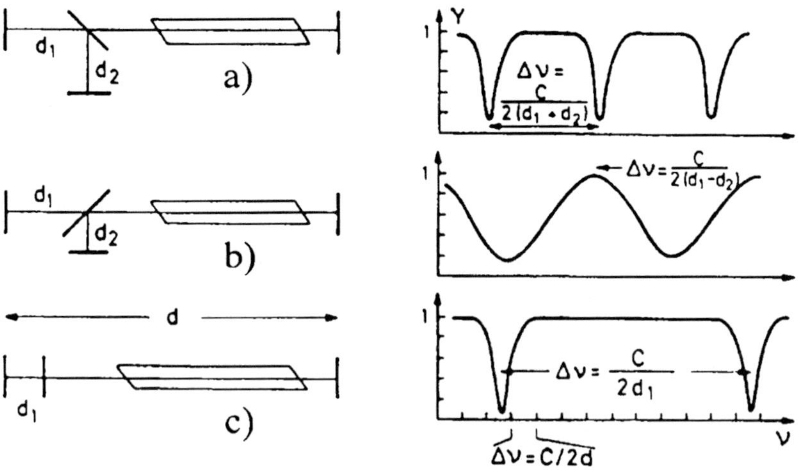

Fig. 22.5 Three mirror resonator schemes used for axial mode selection. The right curves are the round trip losses as a function of the light frequency [5.260]. a) Fox-Smith interferometer, b) Michelson interferometer, c) etalon mirror resonator. d_1,d_2,d are the geometric distances, c is the effective speed of light taking into account the refractive index of the medium.

As the length of the FPI section is scanned, the axial mode of the resonator will try to stay in the reflectance maximum by changing its wavelength. However, if the length L_2 of the resonator is not adjusted according to (22.8), this wavelength shift will generate losses and a jumping to the next order axial mode will occur. Tunability over a frequency range wider than the axial mode spacing $c_0/(2L_2)$, therefore, can only be realized if both resonator lengths are controlled.

Three mirror resonators are widely used in semiconductor lasers to generate single mode operation. In the cleaved coupled cavity approach, two diode lasers are attached with their cleaved end faces forming the resonator mirrors [5.235,5.238,5.262]. Axial mode selection and tunability is provided by controlling the length and the gain of each laser. In the external cavity technique, an external mirror and one endface of the semiconductor laser form the FPI [5.232-5.234,5.242,5.259,5.263,5.270]. The external mirror is moved by means of a piezoelectric transducer to select the axial mode. Some multi-mirror resonator designs used for axial mode control and frequency tuning in gas lasers are shown in Fig. 22.5 [5.224,5.226,5.240].

22.4 Resonators for Homogeneously Broadened Lasers

The basic principle of the resonator concepts discussed in the previous two sections is the generation of frequency dependent losses to control the axial mode spectrum. These resonator schemes can be applied to both inhomogeneously and homogeneously broadened lasers. Additionally, axial mode control in homogeneously broadened lasers can be achieved by preventing spatial holeburning or reducing its effect on the mode spectrum. At low to medium output powers, this second class of resonators is capable of providing single axial mode output without additional dispersive elements. At higher gain levels, however, etalons may have to be inserted to suppress the oscillation of adjacent axial modes. In the following, the different resonator schemes are briefly discussed:

a) Microchip Laser

For some solid state materials, the gain bandwidth is small enough to realize single axial mode operation with a short resonator. For a Nd:YAG laser with an optical resonator length of 3mm, the spectral mode separation $c_0/(2L)$ of 50GHz is comparable to the gain bandwidth of 120GHz. This means that no more than two axial modes can be supported. This concept of axial mode control is inherent to laser diode pumped microchip lasers. The resonator is formed by the HR coated endface of the crystal and an external output coupling mirror. The external mirror can be moved by piezoelectric transducers to maintain single mode operation. Nd:YAG and Nd:YVO_4 microchip lasers having crystal lengths of less than 1mm and optical resonator lengths between 3mm and 5mm provide single axial mode operation with output powers in the 100mW range and optical-to-optical efficiencies of better than 30% [5.247,5.256,5.257,5.272]. For materials with a larger bandwidth (e.g. Nd:glass, Er:Yb:glass), uncoated etalons are inserted to provide axial mode control and frequency tuning [5.269]. In 1996, an actively frequency stabilized Nd:YVO_4 microchip laser was used to attain a linewidth of less than 1mHz [5.272].

b) Resonator with a Thin Gain Medium

In standing wave resonators, all axial modes have a common node at the resonator mirrors. Over a short distance from the mirrors (on the order of $L/100$ where L is the optical resonator length), the peaks of the standing wave intensity patterns of different axial modes will therefore nearly coincide. If one mirror is attached to a thin homogeneously broadened gain medium, the nearly identical spatial holeburning of different modes will lead to single axial mode operation. The axial mode with the highest small-signal gain (in the center of the gain profile) will oscillate first and saturate the gain to its threshold value. The other axial modes are suppressed since they are amplified in the same gain areas and have a higher threshold. This scheme has been used in dye lasers and diode pumped solid state lasers. In [5.252], a 2mm thick Nd:YVO_4 crystal was end-pumped by a 200mW laser diode array to generate a single frequency output power of 35mW in a 20mm long resonator. The scheme works in this case because Nd:YVO_4 exhibits a very small absorption length of 90μm. At higher pump powers, however, the gain deposited farther inside the crystal became high enough to support several axial modes. This is due to the increasing spatial dephasing of the standing wave patterns at larger distances from the mirror. To sustain single mode emission, etalons have to be inserted into the cavity. This is also true for thin disk lasers, where the disk thickness of several 100 microns is too large to sustain single mode lasing. Single mode output powers of up to 35W have been achieved with Nd:YAG disk lasers using two etalons (e.g. 0.1mm and 1.0mm thick fused silica) [5.274,5.275]

In some commercial diode pumped solid state lasers, the thin gain medium is placed in the middle of a symmetric resonator. In cw operation, these lasers oscillate on two axial modes. Inside the medium, the standing wave pattern of one mode is shifted by $\lambda/4$ with respect to the second pattern, leading to a homogeneous axial gain saturation.

c) Twisted Mode Resonator

In a standing wave resonator, spatial holeburning can be prevented by using orthogonal polarization states for the back and the forth traveling wave. In the twisted mode resonator [5.220, 5.285], this is accomplished by placing the active medium between two quarter wave plates. The fast axes of the plates are rotated by +45° and -45° with respect to the pass direction of a polarizer placed in front of one mirror. Inside the active medium, the counterpropagating waves are circularly polarized (see Sec. 8.2). As a result the average intensity is homogeneous along the optical axis. This resonator concept, however, has not found widespread application.

d) Unidirectional Ring Resonators

In solid state and dye lasers, ring resonators with unidirectional beam propagation are usually used to achieve single frequency output [5.176,5.186,5.225,5.227,5.231, 5.239,5.260,5.271] (see Chapter 21). The unidirectionality prevents spatial holeburning, thereby enabling one axial mode to saturate the entire gain. The conventional ring geometry is more versa tile than the resonator schemes presented above because there are no constraints on the size of the active medium or the resonator length. Furthermore, additional tuning elements like prisms, gratings, Lyot filters, or etalons can be easily incorporated into the resonator. This is of particular importance for lasers having a large gain bandwidth like dye lasers and tunable solid state lasers. Typical output powers of single mode dye ring lasers are in the Watt range. Five times higher single frequency output power has been achieved with Ar laser pumped Ti:Al_2O_3 ring oscillators with a tuning range of 650-1200nm in pulsed operation. The single frequency output powers of diode pumped solid state ring lasers are as high as 20W with optical-to-optical efficiencies of 50% and more. With internal frequency doubling, a single mode output powers of greater than 8.5W at 532nm have been achieved in [5.273] using a diode pumped Nd:YVO_4 ring oscillator with LBO (20W optical pump power). Recently the output power of this single mode ring resonator has been increased to 18W at 532nm (Verdi V18). Very high stability and narrow linewidths can be obtained with diode pumped monolithic nonplanar solid state laser oscillators (Nd:YAG, Nd:GGG) (see Sec. 21.3) [5.239,5.248,5.255]. Due to the rigid structure, linewidths of less than 3kHz can be attained in free-running mode. With active frequency stabilization, linewidths in the sub-Hertz level have been measured [5.261,5.265].

22.5 Micro-Optical Resonators

Micro-optical resonators have dimensions comparable to the wavelength of light. The eigenmodes have, in principle, the same structure as those of macroscopic resonators, as long as real eigenfrequencies exist. In most cases closed cavities are used as shown in Figs. 22.6 and 22.7. The eigenmodes depend on the geometry. For a cubic cavity (Fig. 22.8) with length a in the three directions and refractive index n, the normalized amplitude reads:

$$f(x) = \sin(2\pi x/\lambda_x)\ \sin(2\pi y/\lambda_y)\ \sin(2\pi z/\lambda_z) \tag{22.10}$$

with the resonance condition for metallic walls:

$$a = q\ \lambda_{x,y,z}/(2n) \tag{22.11}$$

The active media of micro-lasers are quantum dots (semiconductor emitters), as indicated in Fig. 22.7. The wavelength of these InAs-dots is about 1 μm and they excite the gallery modes of the cylindrical resonator modes travelling around the cylinder envelope. The dots are optically pumped by a Ti-sapphire laser.

The interesting feature of micro-resonators is the possibility to control the spontaneous emission, which means that the spontaneous emission can be enhanced or suppressed. A review of these effects was given by Haroche and Kleppner [5.276].

Fig. 22.6 A cylindrical micro-resonator. The height is approximately 4μm (courtesy of M. Bayer, Universität Würzburg, Germany) [5.283].

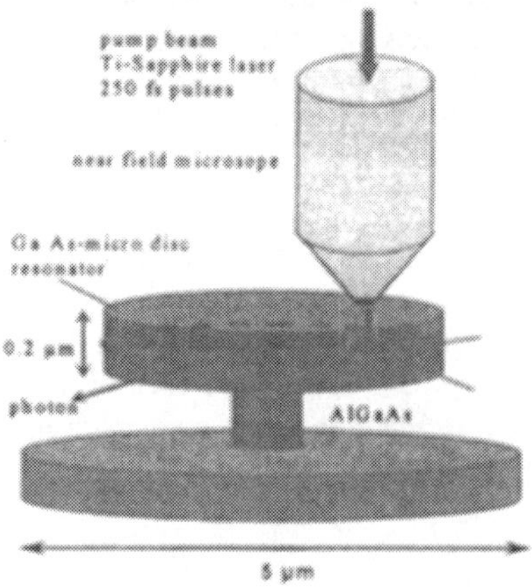

Fig.22.7 A GaAs micro-disk resonator (Michler and Becher). Universität Würzburg, Germany) [5.282]. The quantum dots represent the active medium.

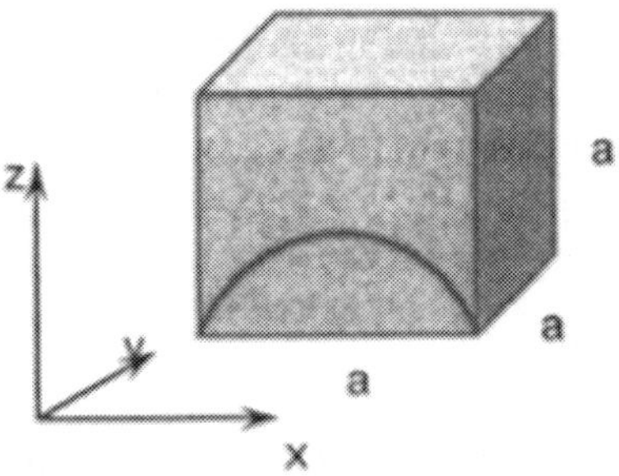

Fig. 22.8 The cubic micro-cavity with the fundamental mode in the x-direction.

Spontaneous emission as well as the photon character of light are a result of the quantization of the electromagnetic field. This theory had already been developed in 1930 by Wigner and Weisskopf [5.277], and Purcell was the first to discuss the influence of a cavity on spontaneous emission in 1946 [5.280]. With the advent of micro-resonators, this theory can now finally be proven. One theoretical prediction is the exponential decay of excited atoms, even if they do not interact with a field. The number n_2 of atoms in the upper level decays according to:

$$n_2(t) = n_2(0)\exp(-\Gamma t) \tag{22.12}$$

Fermi's golden rule can be applied to the decay rate Γ: the decay rate of an excited system is proportional to the density $D(\omega_E)$ of the lower states. As shown in many textbooks [5.278, 5.279], the decay rate is given by:

$$\Gamma = \frac{2\pi}{3\epsilon_0}\,\frac{|\mu_{12}|^2\omega_E}{\hbar V}\,D(\omega_E)\,|f(r_E)|^2 \tag{22.13}$$

with:

μ_{12}	:	dipole moment of the transition
V	:	volume of the cavity
$f(r_E)$	:	amplitude of the normalized eigenmode (22.10)
r_E, ω_E	:	emitter position and emitter frequency

For free space, the density of modes is given by Rayleigh's formula:

$$D(\omega_E) = \frac{\omega_E^2\,V}{\pi^2\,c^3} \tag{22.14}$$

Insertion of (22.14) into (22.13) yields the spontaneous emission rate of free atoms:

$$\Gamma_0 = \frac{1}{\tau} = \frac{|\mu_{12}|^2\,\omega_E^2}{2\pi\epsilon_0\,\hbar c^3} \tag{22.15}$$

where τ is the upper state lifetime due to spontaneous emission. For a small resonator with a single resonance in the range of the gain profile, the density of states is given by the resonator response. It can be approximated by a Lorentzian line profile (see Fig. 22.9):

$$D(\omega_E)_{resonator} = \frac{2}{\pi\Delta\omega_C}\frac{1}{1+(\omega_E-\omega_C)^2/(\Delta\omega_C/2)^2} \tag{22.16}$$

where $\Delta\omega_C$ is the bandwidth of the resonator and ω_C is the resonance frequency. Insertion into (22.13) delivers the decay rate [5.279]:

$$\Gamma_C = \frac{1}{\tau_C} = \Gamma_0\frac{4c^3}{\Delta\omega_C\omega_E^2 V}\frac{1}{1+(\omega_E-\omega_C)^2/(\Delta\omega_C/2)^2} \tag{22.17}$$

The decay time $\tau_C = 1/\Gamma_C$ can be longer or shorter than the natural decay time $\tau = 1/\Gamma_0$ depending on the frequency and position of the emitters and the bandwidth of the cavity. An example is shown in Fig. 22.10. The ratio of cavity decay time τ_C to free space decay time τ is plotted versus detuning . In resonance the decay time is reduced by a factor of three. For the detuned resonator the decay time is much longer than the natural decay time. The effect strongly depends on the resonator bandwidth. A high-Q (coated resonator) produces a much stronger effect than a low-Q resonator.

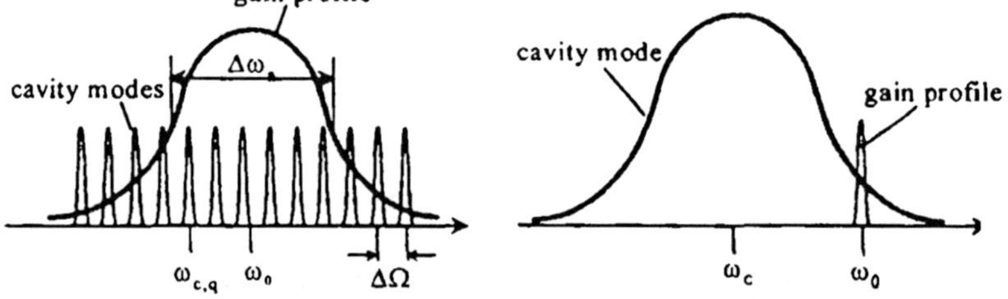

Fig.22.9 Gain profile and the cavity modes for a large (left) and a small (right) resonator.

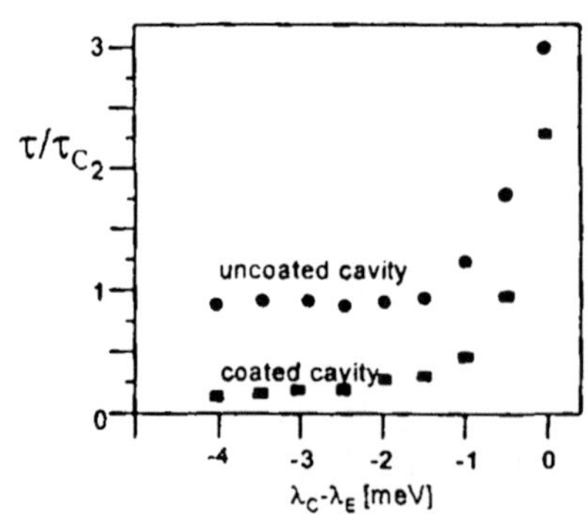

Fig.22.10 Ratio of spontaneous decay time τ to spontaneous emission time τ_C of a coated and an uncoated cavity as a function of the wavelength mismatch between emission and cavity resonance [5.284].

Part VI

Measurement Techniques

Chapter 23

Measurement of Laser Head Parameters

23.1 Measurement of Losses, Gain, and Efficiency

The performance of a laser system depends on a variety of physical parameters that characterize the properties of the radiation. A successful optimization of the laser performance can only be achieved if these parameters are known. Considering the number of independent parameters, an experimental enhancement based on trial and error would be too time-consuming and would hardly result in an optimized laser design. A better procedure, therefore, is to use the measured values of these parameters to compare the experimental laser performance with the one predicted by theory. A possible disagreement with the theoretically determined laser properties provides the researcher with useful information regarding the possible cause of the problem. If a good agreement between theory and experiment is found, the theoretical model can be a powerful tool to find laser designs with an improved performance. In this section we will present common measurement techniques for laser parameters that control the output power and the beam properties. These parameters either classify the active medium (i.e. the small-signal gain, the saturation intensity, the absorption loss, and the phase front distortion induced in a transit) or the resonator (diffraction losses of the passive and the active resonator).

23.1.1 Findlay-Clay Analysis

The excitation efficiency and the round trip losses determine the pump power necessary to reach the laser threshold. The lower the excitation efficiency and the higher the losses due to diffraction and output coupling are, the higher the pump power has to be to achieve laser oscillation. The Findlay-Clay analysis makes use of this relationship between loss and threshold gain [6.1,6.2]. For different output couplings, the pump power is decreased until laser oscillation is barely noticeable. For a resonator with output coupling reflectance R, the small-signal gain $g_0\ell_{th}$ generated at this pump power exactly compensates all losses induced by output coupling, diffraction, scattering, and absorption. The threshold condition reads:

$$\exp[g_0\ell_{th}]\ \sqrt{R}\ V_D\ V_S\ =\ 1 \tag{23.1}$$

where V_D is the loss factor per transit due to diffraction (=1-loss) and V_S is the loss factor per transit due to absorption and scattering.

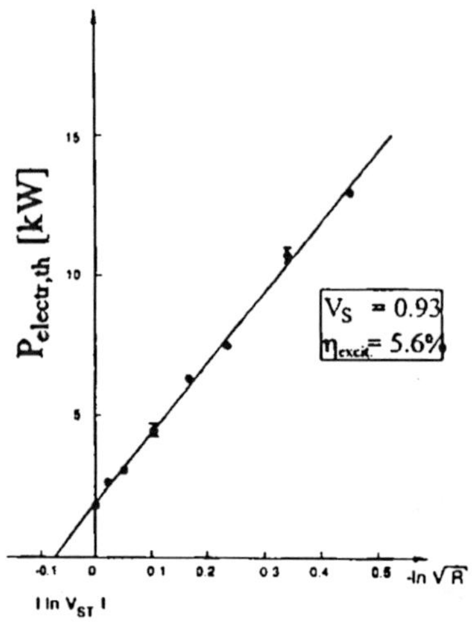

Fig. 23.1 Findlay-Clay diagram for a pulsed flashlamp-pumped Nd:YAG laser with a stable resonator in multimode operation (rod dimensions: 3/8"x 6"). The pump power per pulse was measured. The slope of the interpolated, linear curve yields an excitation efficiency of 5.6% (I_S=$2kW/cm^2$), and the x-axis shift represents a loss of 7% per transit.

Let us assume that the active medium exhibits a constant small-signal gain over the entire cross section and that the laser transition occurs in an idealized four level system. The relation between the electrical pump power and the small-signal gain is then given by (see Chapter 9):

$$P_{electr} = \frac{A\ I_S}{\eta_{excit}}\ g_0\ell \tag{23.2}$$

where A is the cross-sectional area of the active medium, I_S is the saturation intensity, and η_{excit} is the excitation efficiency. The excitation efficiency is the ratio of the available power in the form of inversion to the electrical pump power. Insertion of (23.2) into (23.1) yields for the electrical pump power at the laser threshold:

$$P_{electr,th} = \frac{A\ I_S}{\eta_{excit}}\left[|\ln\sqrt{R}| + |\ln(V_D V_S)|\right] \tag{23.3}$$

If we plot the threshold pump power $P_{electr,th}$ as a function of $|\ln\sqrt{R}|$ we get a straight line having the slope:

$$m = \frac{A\ I_S}{\eta_{excit}} \tag{23.4}$$

and the x-axis shift $|ln(V_D V_S)|$. This plot is referred to as the Findlay-Clay diagram. By measuring the threshold pump power $P_{electr,th}$ as a function of the output coupling reflectance R, the total loss factor $V_D V_S$ is obtained. Provided that the saturation intensity is known, the excitation efficiency can be determined through the slope (see Table 9.1 for saturation intensities of some laser media). With the knowledge of the excitation efficiency, we can now determine the small-signal gain at any pump power by using (23.2).

The Findlay-Clay analysis provides only the product of the individual loss factors. If the losses due to scattering and absorption have to be determined separately, a resonator has to be used that exhibits negligible diffraction losses. This is the case for a stable resonator oscillating in multimode operation. At the laser threshold, the laser will oscillate in a mode that exhibits the lowest possible diffraction loss. If the resonator is designed such that the fundamental mode is not truncated by any aperture (the radius of the apertures should be at least two times the Gaussian beam radius), the loss factor provided by the Findlay-Clay plot is only due to absorption and scattering inside the medium and at the resonator mirrors. Once the intrinsic losses are determined, the Findlay-Clay analysis can be applied to arbitrary resonator configurations to yield the diffraction losses for the fundamental mode. The statement that the diffraction losses at the laser threshold are attributed to a low loss mode is illustrated in Fig. 23.2.

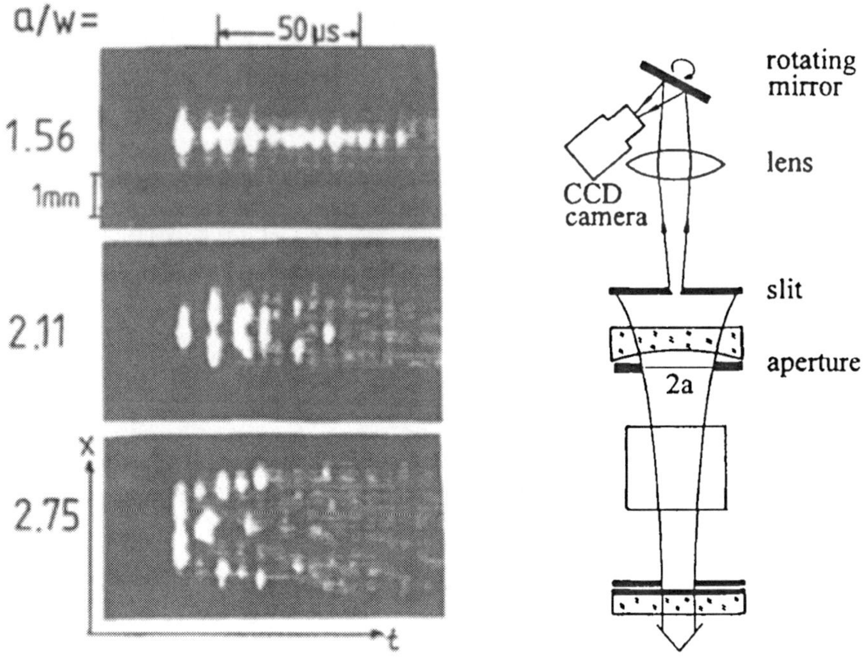

Fig. 23.2 Measured, temporally resolved mode structure of a stable resonator with different radii a of the mode selecting aperture (pulsed Nd:YAG rod laser, w is the Gaussian beam radius at the aperture). The right drawing shows the experimental set-up.

The temporally resolved mode structure at one resonator mirror of a pulsed Nd:YAG laser with a stable resonator is shown for different radii of the mode selecting aperture. Each vertical bar can be assigned to a spike in the laser emission. In the first spike a mode is observed that is not truncated by the aperture. The Findlay-Clay diagram thus cannot provide the loss that occurs when the laser is operated above threshold. We only get the diffraction losses of the passive resonator in fundamental mode oscillation.

However, the Findlay-Clay analysis is a powerful and easy method to characterize the gain and the losses of the active medium. It can be utilized in both cw and pulsed lasers. For the latter systems, Eq. (23.3) applies to the pump power per pulse. The accuracy strongly depends on the technique used to determine the threshold. The measurement of the output power is not the best method since the power is very low near threshold and a cease of the laser oscillation is difficult to detect. A better way is to record the mode structure by means of a camera. The Findlay-Clay analysis in its presented form cannot be used if:

a) the pump pulse cannot be approximated by a rectangular profile
b) the pump pulse duration is less than three times the lifetime of the upper laser level
c) the excitation efficiency is a function of the pump power
d) the transverse gain profile is not homogeneous

Except for case a), in which wrong values for the loss are obtained (Fig.23.3), the Findlay-Clay analysis can be modified to work for the other three cases. These modifications are discussed in the following.

Case b) Short Pump Pulses

If the temporal distribution of the pump pulse can be approximated by a rectangular profile, relation (23.3) can be modified to yield correct results for pump pulse durations Δt that are comparable or shorter than the upper level lifetime τ. For a four level system with pump rates W much lower than $1/\tau$, the temporal build-up of the population inversion density ΔN is given by:

$$\Delta N(t) = W\tau N_0 \; (1 - \exp[-t/\tau]) \tag{23.5}$$

where t is the time since the onset of the pump pulse and N_0 is the density of active atoms. In the Findlay-Clay analysis, the laser threshold is reached at the end of the pump pulse. At this point, the power P_{UL} available in the form of inversion is given by:

$$P_{UL} = \Delta N(\Delta t) \; A\ell \; h\nu/\tau = P_{electr,th}\eta_{excit} \; (1 - \exp[-\Delta t/\tau]) \tag{23.6}$$

where $A\ell$ is the volume of the active medium and $h\nu$ is the photon energy. The excitation efficiency seems to be reduced by the factor $(1\text{-}exp[-\Delta t/\tau])$. Thus, the modified expression (23.3) reads:

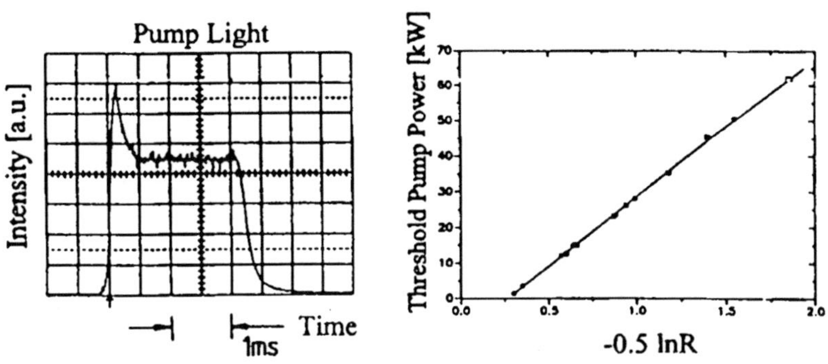

Fig. 23.3 Findlay-Clay plot for a pulsed Nd:YAG laser with a pump profile that exhibits a strong overshoot. The plot yields a loss factor of greater than one. For the excitation efficiency the correct value is obtained.

$$P_{electr,th} = \frac{A\ I_S}{\eta_{excit}(1 - \exp[-\Delta t/\tau])} \left[|\ln\sqrt{R}| + |\ln(V_D V_S)| \right] \tag{23.7}$$

For long pulses, this equation becomes identical to (23.3). If the pump pulse cannot be approximated well enough by a rectangular profile, the Findlay-Clay analysis cannot be applied.

Case c) Dependence of the Excitation Efficiency on the Pump Power

If the excitation efficiency is a function of the pump power, the Findlay-Clay plot will show a nonlinear relationship between the threshold pump power and the logarithm of the output coupling reflectance. The slope of the curve will increase or decrease with the output coupling, depending on whether the excitation efficiency decreases or increases with the pump power, respectively. However, we can determine the relative dependence of the excitation efficiency on the pump power by measuring the power of the spontaneously emitted light and using this information to correct the Findlay-Clay analysis. A possible set-up for this measurement is depicted in Fig. 23.4a. A portion of the spontaneous emission is collected with a lens and the intensity is measured with a photodiode. The geometry has to be chosen such that the influence of amplified spontaneous emission is negligible. Furthermore, for flashlamp-pumped media, the influence of residual signals generated by the pump light has to be eliminated. Provided that these conditions are met, the measured intensity I_{SE} is proportional to the small-signal gain of the active medium. It should be mentioned that for gas lasers and flashlamp-pumped lasers a threshold pump power may be observed. This offset has to be taken into account. Figure 23.4b presents the Findlay-Clay plot for three different dependencies of the excitation efficiency on the pump power and the corresponding intensity of the spontaneously-emitted light.

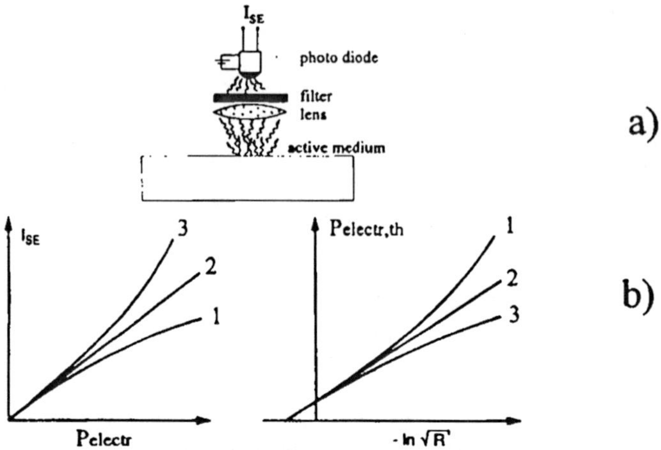

Fig. 23.4 a) Relative measurement of the excitation efficiency by collecting spontaneously-emitted light at the laser wavelength. b) The intensity of the spontaneous emission (left) and the Findlay-Clay plot for three dependencies of the excitation efficiency on the pump power: decreasing (1), constant (2), and increasing (3).

The modified Findlay-Clay plot can be generated in the following way. If $I_{SE}(P_{electr})$ is the intensity of the spontaneous emission as a function of the pump power, we first calculate the normalized signal:

$$S(P_{electr}) = \frac{I_{SE}(P_{electr})}{P_{electr}} \tag{23.8}$$

Then we choose one threshold pump power in the original Findlay-Clay plot. Let us denote this threshold pump power with $P^0_{electr,th}$. We now multiply the measured threshold pump powers $P_{electr,th}$ with the function:

$$f(P_{electr,th}) = \frac{S(P_{electr,th})}{S(P^0_{electr,th})} \tag{23.9}$$

If the corrected threshold pump powers $f(P_{electr,th})$ are plotted versus $|ln\sqrt{R}|$, a linear curve is obtained (Fig. 23.5). The slope provides the excitation efficiency η^0_{excit} at the pump power $P^0_{electr,th}$. The excitation efficiency as a function of the pump power can be calculated by using the relation:

$$\eta_{excit} = \eta^0_{excit}\, f(P_{electr}) \tag{23.10}$$

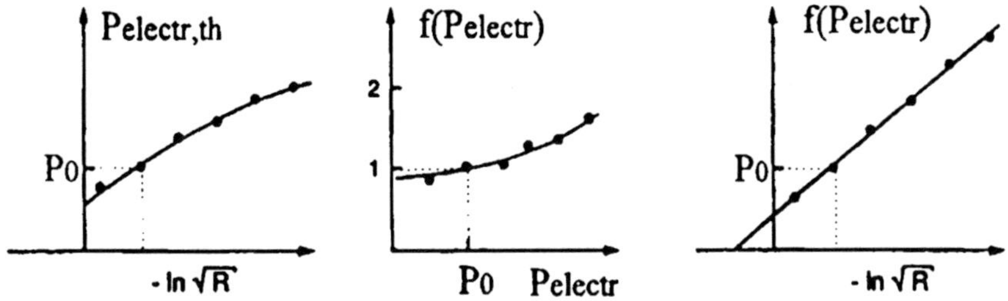

Fig. 23.5 Correction of the Findlay-Clay plot. a) Initial Findlay-Clay plot, indicating an increase in efficiency; b) The measured correction function $f(P^0_{electr,th})$; c) the corrected Findlay-Clay plot which provides the excitation efficiency at $P^0_{electr,th}$.

Case d) Nonhomogeneous Transverse Gain Profile

In most cases the transverse gain profile in the active medium is not homogeneous. For longitudinally-diode-pumped solid state lasers, for instance, the gain along the optical axis is considerably higher than in the periphery of the active medium. Centered gain profiles are also found in flashlamp-pumped lasers in which the flashlamp axis is imaged onto the optical axis inside the active medium by means of elliptical pump reflectors. When the Findlay-Clay analysis is applied to lasers with nonhomogeneous gain profiles the correct values for the losses are obtained, but the excitation efficiency turns out to be incorrect. This is quite understandable since the laser threshold is determined by the maximum small-signal gain. The observed mode pattern is confined to areas around the maxima of the gain profile (Fig. 23.6). The Findlay-Clay analysis yields the correct small-signal gain within the area of the mode, but in Eq. (23.3) it is assumed that this small-signal gain is present over the whole cross sectional area A of the active medium. Therefore, the excitation efficiency is found to be too high. Fortunately, this problem can be corrected by measuring the transverse gain profile.

A very effective method to determine the lateral gain distribution is to move the active medium transversally in a stable resonator that exhibits a small Gaussian beam radius in the active medium (Fig. 23.7). A suitable resonator for this purpose is the confocal resonator (resonator length L, mirror curvatures $\rho=L$) with both mirrors confined by apertures that are adapted to the Gaussian beam radius. The active medium is placed in the middle of the resonator and, for a fixed output coupling reflectance R, the threshold pump power is measured as a function of the transverse shift of the medium. Note that the gain measured is the average over the mode cross section. The outer area of the active medium, with a width approximately given by the Gaussian beam diameter, cannot be probed this way, because losses are generated for the fundamental mode due to beam truncation. The normalized gain profile $g(x,y)$ is obtained by taking the inverse of the threshold pump power and normalizing it to 1.0 at the maximum.

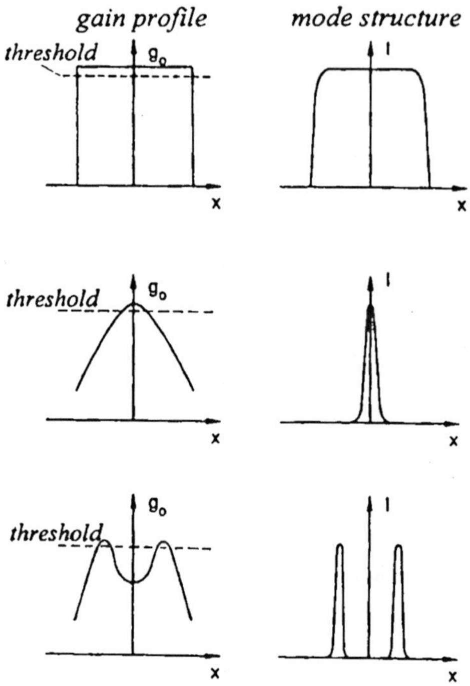

Fig. 23.6 The transverse mode structure at the laser threshold is determined by the maxima of the transverse gain profile.

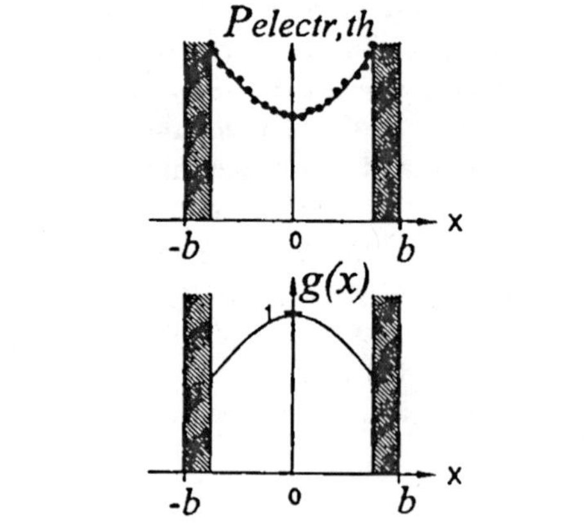

Fig. 23.7 Measurement of the threshold pump power as a function of the transverse shift of the active medium. The normalized gain profile is inversely proportional to the threshold pump power. The hatched areas cannot be probed due to diffraction losses generated by the rim of the active medium.

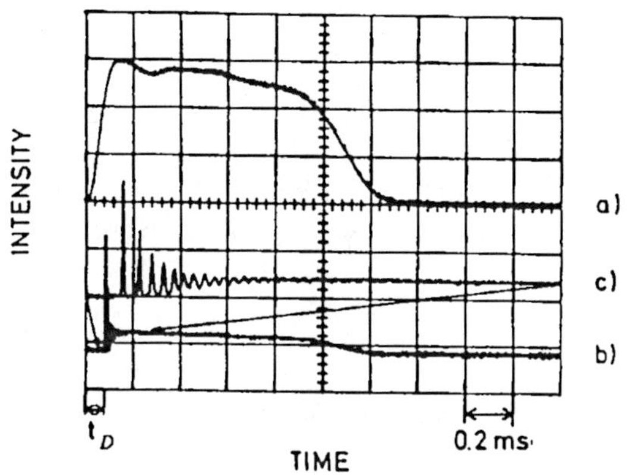

Fig. 23.8 The delay time t_D between the start of the pump pulse and the onset of laser oscillation can be used to determine gain and losses in pulsed lasers. Shown is the pump intensity a) and the laser emission b) for a Nd:YAG laser. Curve c) is a magnified view of b).

In the outer area of the medium, the normalized gain profile has to be determined via extrapolation. The correct value for the excitation efficiency can now be calculated by using the expression:

$$\eta_{excit} = \frac{\eta^0_{excit}}{A} \int_A g(x,y)\; dx dy \tag{23.11}$$

where η^0_{excit} is the excitation efficiency yielded by the Findlay-Clay analysis and A is the cross sectional area of the medium.

23.1.2 Delay-Time Analysis

This method can be applied to pulsed laser systems to determine the excitation efficiency and the losses [6.5]. Similar to the Findlay-Clay analysis, the threshold condition (23.1) is used. However, there are no constraints on the pump pulse shape, the pump pulse duration, and the dependence of the excitation efficiency on the pump power since the rate equation for the population inversion is solved to evaluate the gain in the active medium. Furthermore, this method can be used for four-level and three-level systems. In the following, we derive the delay-time method for ideal four-level lasers. Three level systems and four level systems with a non-negligible population of the lower laser level can be treated similarly. Let us first consider a rectangular pump pulse of duration Δt and pump rate W. Arbitrary pulse shapes will be discussed later.

Below the laser threshold, the rate equation for the inversion density ΔN may be written as:

$$\frac{d\Delta N}{dt} = W(N_0 - \Delta N) - \frac{\Delta N}{\tau} \tag{23.12}$$

where N_0 is the density of active atoms and τ is the upper level lifetime due to spontaneous emission. For $W << 1/\tau$, which holds for almost all four-level laser materials, the solution of (23.12) is given by:

$$\Delta N(t) = W\tau N_0 \left(1 - \exp[-t/\tau]\right) \tag{23.13}$$

At the delay time $t = t_D$, the inversion density has reached the threshold value:

$$\Delta N_{th} = \frac{1}{\sigma \ell} \left|\ln(\sqrt{R} V_S V_D)\right| \tag{23.14}$$

By setting (23.14) equal to (23.13) one gets:

$$W\tau N_0 \sigma \ell \left(1 - \exp[-t_D/\tau]\right) = \left|\ln\sqrt{R}\right| + \left|\ln(V_S V_D)\right| \tag{23.15}$$

The relationship between the electrical pump power and the pump rate is given by:

$$P_{electr} = \frac{W\, N_0\, h\nu\, A\ell}{\eta_{excit}} = \frac{W\tau N_0 \sigma \ell\, A\, I_S}{\eta_{excit}} \tag{23.16}$$

where $I_S = h\nu/(\sigma\tau)$ is the saturation intensity. Insertion into (23.15) results in:

$$\left(1 - \exp[-t_D/\tau]\right) = \frac{A\, I_S}{\eta_{excit} P_{electr}} \left[\left|\ln\sqrt{R}\right| + \left|\ln(V_S V_D)\right|\right] \tag{23.17}$$

By plotting the parameter $(1-\exp[t/\tau])$ as a function of $|\ln\sqrt{R}|$ a straight line is obtained from which the excitation efficiency and the loss can be determined (Fig. 23.9). Note that the measurements are performed at one fixed value of the pump power. This eliminates the problem of a varying excitation efficiency as the pump power is changed. The excitation efficiency as a function of the pump power can be obtained in a second step by using (23.17) at different values of the pump power for a fixed output coupling reflectance R. The time t_D at which threshold inversion is reached is, to a good approximation, given by the time at which the first spike maximum appears minus half the separation between the two first spikes.

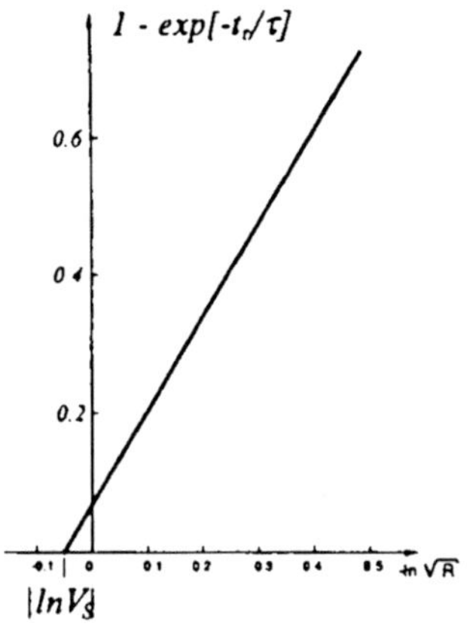

Fig. 23.9 Determination of the loss and the excitation efficiency of a laser with the delay-time method.

If the pump profile is not rectangular, the differential equation (23.12) has to be solved numerically. The input parameter is a recorded pump signal $I(t)$ that is proportional to the pump rate $W(t)$:

$$I(t) = a\ W(t) \tag{23.18}$$

with an unknown proportionality constant a. For optically pumped lasers, the pump signal $I(t)$ can be obtained by recording the pump light intensity at the main absorption bands of the active medium. For gas lasers, the pump signal is given by the electrical power absorbed by the gas and off-sets and nonlinearities have to be considered. The time is measured starting at an arbitrarily set trigger point. At the laser threshold ($t=t_D$), Eqs. (23.12) and (23.14) yield with $W<<1/\tau$:

$$B(t_D) := \frac{1}{\tau} \exp[-t_D/\tau] \int_0^{t_D} I(x)\exp[x/\tau]dx = b\ [|\ln\sqrt{R}| + |\ln(V_S V_D)|]$$

with: $$b = \frac{a}{\tau N_0 \sigma \ell} \tag{23.19}$$

If we plot the term $B(t_D)$ as a function of $|\ln\sqrt{R}|$, a linear curve is obtained from which the total loss factor $V_S V_D$ and the constant b can be determined (Fig. 23.10). The excitation efficiency can then be obtained with:

$$\eta_{excit} = \frac{A\ I_S}{b\ E_{electr}} \int_0^{\infty} I(t)\ dt \tag{23.20}$$

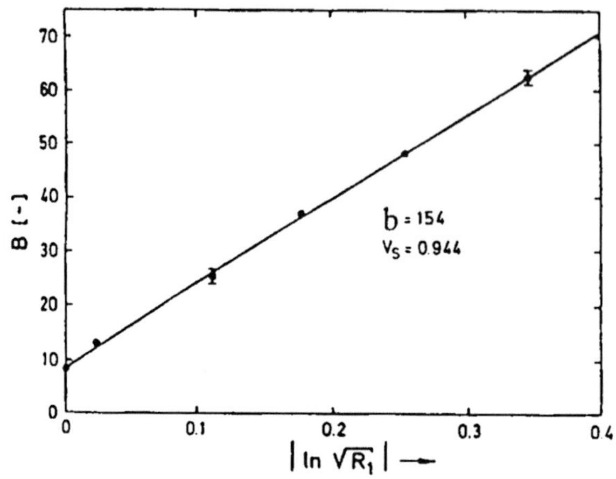

Fig. 23.10 Numerically calculated values $B(t_D)$ as a function of the mirror reflectance for a pulsed, flashlamp-pumped Nd:YAG laser with a stable resonator in multimode operation (the pump profile is shown in Fig. 23.8, electrical pump energy: 112J, pulse duration: 1ms, rod diameter: 1/4", rod length: 3").

where E_{electr} is the electrical pump energy per pulse and A is the cross section of the active medium. For laser materials with a high threshold, like Alexandrite, $W << 1/\tau$ does not hold. In this case the exact solution of (23.12) has to be used. An equation similar to (23.19) can be derived, which yields the factor b by successive approximation. Although the delay time analysis is more time-consuming than the Findlay-Clay analysis, this method is more accurate to determine the losses and the gain of a pulsed laser. The numerical evaluation of (23.19) is worth the effort to attain more reliable values for the laser parameters. Once the constant b is determined, the laser head basically is calibrated, and we can determine loss and gain of any resonator configuration simply by measuring the delay time and applying (23.19). Similar to the Findlay-Clay analysis, a stable resonator in multimode operation has to be used to measure the intrinsic losses separately. Furthermore, any diffraction losses measured represent the losses of the fundamental mode for the passive resonator. Above threshold, the diffraction losses generally are higher due to the oscillation of higher order modes and due to changes in the mode structure generated by gain saturation.

If one is only interested in the measurement of passive diffraction losses, it is not necessary to calibrate the pump profile numerically as discussed above. It is sufficient to measure the delay time as a function of the mirror reflectance R for a stable resonator in multimode operation at a fixed pump power. A graph in which the reflectance R is plotted versus the delay time serves as a calibration curve from which the diffraction loss factor per transit V_D of other resonator set-ups can be determined (Fig.23.11). If the reflectance of the resonator is R' and the measured delay time is linked to a reflectance R in the graph, the diffraction loss factor per transit is given by $V_D = \sqrt{R/R'}$.

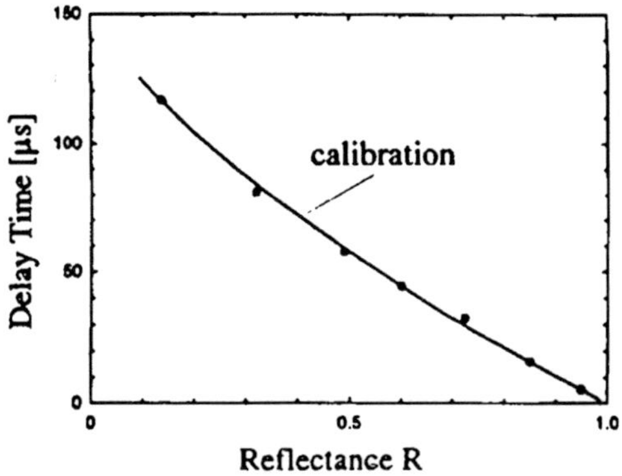

Fig. 23.11 Measured delay times as a function of the output coupling reflectance for a flashlamp-pumped Nd:YAG laser with a stable resonator in multimode operation (electrical pump energy: 75J, pulse duration: 1ms, rod diameter: 3/8", rod length: 6").

23.1.3 Measurement of Diffraction Losses

Both methods, the Findlay-Clay analysis and the delay-time analysis, can be used to measure the diffraction losses of a laser resonator. Once the intrinsic losses are determined by using a stable resonator in multimode oscillation, the diffraction loss factor V_D per transit can be separated from the total loss factor V_SV_D measured. As the measurements shown in Fig. 23.12 illustrate, both methods are very useful to evaluate the diffraction losses of the fundamental mode in stable and unstable resonators. Unfortunately, the measured diffraction losses represent the losses at laser threshold, which means that no gain saturation is present and the mode structure is not the one observed at higher pump powers or at later times during the pump pulse. In order to measure the true losses, the power absorbed by the intracavity apertures has to be measured directly. Figure 23.13 presents two techniques to perform such a measurement. By using an intracavity beamsplitter with low reflectance (e.g. an AR coated glass plate), the power P_1 in the left-traveling beam and the power P_2 in the right-traveling beam are measured. The relative signal $P_2/(RP_1)$ is proportional to the diffraction loss factor per round trip (it is only proportional since the two power meters may exhibit different sensitivities). The proportionality constant can be determined at the laser threshold because the diffraction loss factor must be identical to the loss provided by the Findlay-Clay analysis (see Figs. 11.14 and 11.15 for experimental examples). A better technique, which does not affect the intracavity beam propagation, is the direct measurement of the power hitting the aperture. The aperture is simulated by the elliptical hole of a high reflecting scraper (Fig. 23.13b). The measured relative signals $P_2/(P_2+TP_1)$ and $(P_2-TP_1/R)/P_2$ are proportional to the loss factor due to beam truncation on the right side and the left side of the aperture, respectively ($T=1-R$). Again, a calibration at the laser threshold with the Findlay-Clay analysis or the delay-time analysis is required.

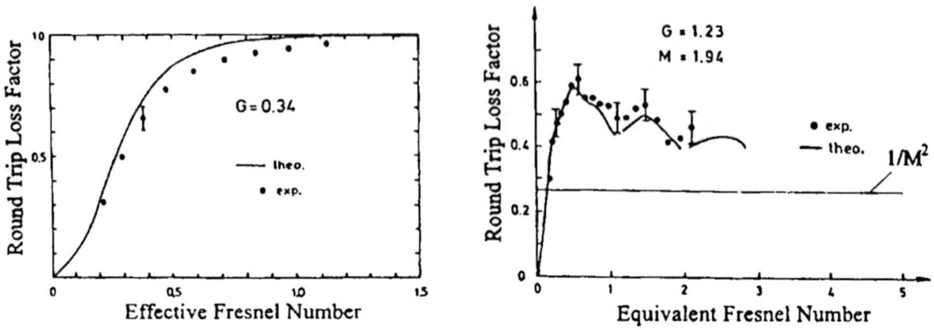

Fig. 23.12 Measured diffraction loss factor per round trip of stable and unstable resonators using the delay time method (pulsed Nd:YAG laser). The solid lines represent the theoretical diffraction losses of the fundamental mode for the passive resonator ($G=2g_1g_2-1$; $N_{eff}=a^2/(2Lg_2\lambda)$; $N_{eq}=N_{eff}\sqrt{G^2-1}$).

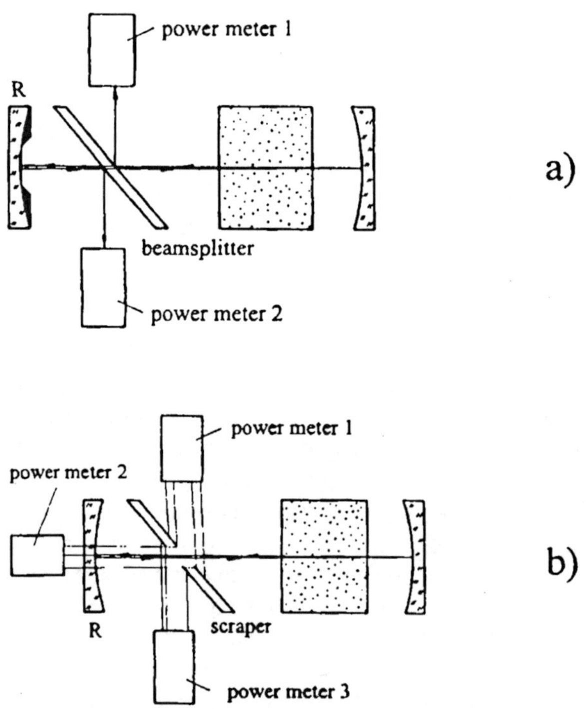

Fig. 23.13 Experimental set-ups for the measurement of the diffraction losses of the active resonator. a) intracavity beamsplitter with low reflectance, b) high reflecting scraper to measure the power hitting the aperture on both sides. The reflectance of the left mirror is *R*.

23.1.4 Measurement of the Saturation Intensity

The saturation intensity I_S is probably the most important parameter of a laser material. It is required to determine the excitation efficiency and calculate the output power of the laser system. Unfortunately, it is also the parameter that is most difficult to determine experimentally since a spectroscopic measurement of the lifetime τ and of the cross sections for stimulated emission σ_i of all transitions contributing to the laser transition is required. For Nd:YAG, for instance, transitions from both Stark levels of the upper laser level $^4F_{3/2}$ contribute to the laser emission at 1.0641 μm. The spectroscopically determined individual cross sections are then combined into an effective cross section σ_{eff} of the laser transition by taking the relative population of the sublevels and the lineshapes of the transitions into account. The saturation intensity then reads:

$$I_S = \frac{h\nu}{\tau\sigma_{eff}} = \frac{h\nu}{\tau\sum_i b_i\sigma_i} \tag{23.21}$$

where the sum of all factors b_i is normalized. Furthermore, the cross sections are a function of the laser wavelength, with the highest value at the center of the gain profile. If only one value for the saturation intensity is used to calculate the output power, a higher saturation intensity is needed to describe lasers operating at a single axial mode than for axial multimode lasers. Considering all thes e difficulties and the limited accuracy of the spectroscopic measurement, it is no surprise that published values of the saturation intensities may differ by a factor of 2.

However, since the output power is proportional to the saturation intensity, a power measurement can be used to determine the effective saturation intensity with a reasonable accuracy. In order to control as many parameters as possible, a stable resonator should be used with an aperture of cross sectional area A_b placed in front of the medium such that high order multimode oscillation is obtained and the gain profile is homogeneous across the aperture. This ensures that the diffraction losses can be neglected when calculating the output power and a well-defined volume of the active medium is used by a homogeneous beam. Furthermore, the mirror reflectance should be chosen relatively high to minimize power losses due to amplified spontaneous emission. For a homogeneously broadened laser, the output power is then given by (see Chapter 10):

$$P_{out} = A_b\, I_s\, \frac{1 - R}{1 - R + \sqrt{R}(1/V_S - V_S)}\left[g_0\ell - |\ln(\sqrt{R}V_S)|\right] \tag{23.22}$$

where A_b is the cross section of the beam inside the active medium, V_S is the loss factor per transit, and R is the reflectance of the output coupling mirror. If we consider only the inversion within the area defined by the aperture, the small-signal gain is related to the electrical pump power via:

$$g_0 \ell = \frac{\eta_{excit}}{A_b\, I_S} P_{electr} = \frac{1}{m} P_{electr} \tag{23.23}$$

The combination of (23.22) and (23.23) can be written as:

$$I_S = \frac{P_{out}}{A_b} \frac{1 - R + \sqrt{R}(1/V_S - V_S)}{(1-R)(P_{electr}/m - |\ln(\sqrt{R}V_S)|)} \tag{23.24}$$

First, a Findlay-Clay analysis is performed with the described resonator to obtain the intrinsic loss factor V_S and the factor m, which is the slope in the Findlay-Clay plot. In a second step, the output power is measured for the high reflectance at an electrical pump power well above threshold. The measured data are then inserted into (23.24) to yield the saturation intensity. Owing to the many parameters that have to be measured separately (P_{electr}, R, V_S, m, A_b), this method provides the saturation intensity with an accuracy of not better than 20%. The accuracy can be improved by calculating the output power numerically with diffraction theory using the saturation intensity as a fit parameter [6.6]. Note that the measured saturation intensity represents the mean value over the laser bandwidth. The determination of the saturation intensity at the center wavelength requires the utilization of a single axial mode resonator. An alternative measurement technique for diode end pumped solid state lasers is described in [6.8]. This method is also based on laser rate equations, and uses the fact that the increasing intra-cavity intensity stimulates a faster decay of excited atoms and reduces the saturation of the pump absorption [6.8].

23.2 Measurement of Thermal Lensing

The determination of the refractive power of the active medium is very important for the optimization of the brightness of solid state lasers. As discussed in Chapter 13, thermal lensing in solid state laser materials depends on a variety of material parameters (thermal conductivity, temperature dependence of the refractive index, geometry of the medium, doping concentration) as well as on the mode of excitation (spectral distribution of the pump light). Furthermore, the refractive power also depends on the round trip losses of the resonator. In general, the refractive power is lower above laser threshold, which means that less heat is dissipated in the material when laser oscillation takes place. This observation can be explained if nonradiative transitions from the upper laser level are assumed [6.7]. The physical mechanism behind these transitions is not fully understood. One explanation is that the stored energy is transferred from the active ion to impurities or to transient color centers generated by the UV portion of the flashlamp spectrum. The measured refractive power of a flashlamp pumped Nd:YAG rod as a function of the output coupling reflectance is shown in Fig. 23.14.

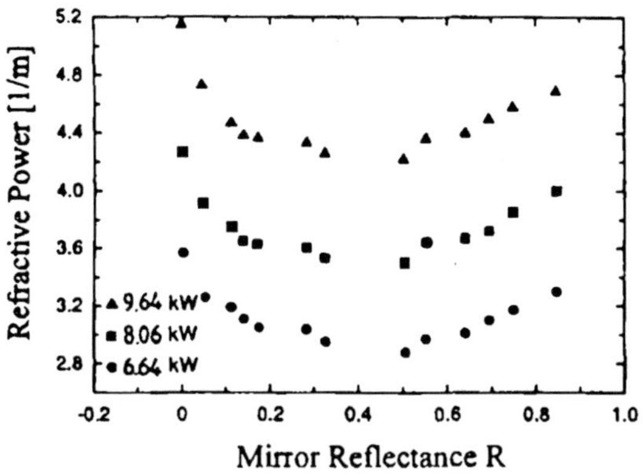

Fig. 23.14 Measured refractive power for the radial polarization of a flashlamp pumped Nd:YAG rod as a function of the reflectance R of a flat-flat resonator. The curve parameter is the average pump power. Rod dimensions: 3/8" x 6", pump pulse duration: 2ms, repetition rate: 30 Hz, ceramic cavity with two Xenon flashlamps [6.7] (© Elsevier Advanced Technology 1993).

As the reflectance is increased, the population density N of the upper laser level is decreased ($N \propto |lnR|$), resulting in a decrease of the power loss due to nonradiative transitions. The rise at high reflectances is a result of the reabsorption of the laser light. Since we are interested in the resonator performance, the measurement of the refractive power should be performed with laser oscillation present, preferably at an output coupling close to the one at which the optimized resonator will be operated.

23.2.1 Focusing of an Expanded Probe Beam

The transverse refractive index profile of a pumped solid state material depends only slightly on the wavelength of the transmitted light. It is therefore not necessary to probe the refractive power with a beam that exhibits an emission spectrum that matches the spectrum of the laser transition. If a probe beam in the visible or near infrared is used, the dispersion is generally too low to have a noticeable effect on the result. Figure 23.15 presents an experimental set-up for the measurement of the refractive power using a collimated probe beam. The beam is expanded with a telescope, and aperture 1 with diameter d_1 defines the area of the active medium probed by the beam. The intensity transmitted through a second aperture with diameter d_2 is measured with a photodiode. This method assumes that the refractive power is constant over the cross section of the probe beam. The measured intensity I increases quadratically with the refractive power D, provided that the thermal lens focuses the probe beam. The refractive power can be calculated by using the expression:

$$D = \frac{1}{L + \ell/(2n)} \left[1 - \sqrt{\frac{I_0}{I}} \right] \tag{23.25}$$

where I_0 is the intensity measured without thermal lensing ($D=0$), ℓ is the length of the medium, and n is its average refractive index.

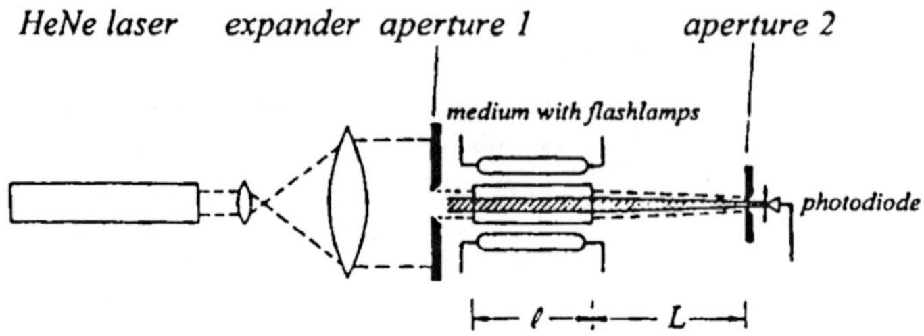

Fig. 23.15 Measurement of the refractive power of a pumped solid state laser medium using an expanded probe beam [S.28].

This equation can only be applied to refractive powers D that generate truncation losses at the second aperture:

$$D < \frac{1 - d_2/d_1}{L + \ell/(2n)} \tag{23.26}$$

This measurement technique is very useful for investigating the temporal behavior of the refractive power. The measured intensity I for a Nd:YAG rod pumped in pulsed operation with a repetition rate of 3Hz and a pump pulse duration of 0.5ms is presented in Fig. 23.16. In general, it takes several seconds for the average refractive power to reach a steady-state value. As far as resonator performance is concerned, only the refractive power during the pump pulse is of interest. In steady-state each pump pulse induces a change in the refractive power that has vanished when the next pump pulse arrives. This transient refractive power can be both positive or negative, depending on the laser material and the repetition rate (Fig. 23.17).

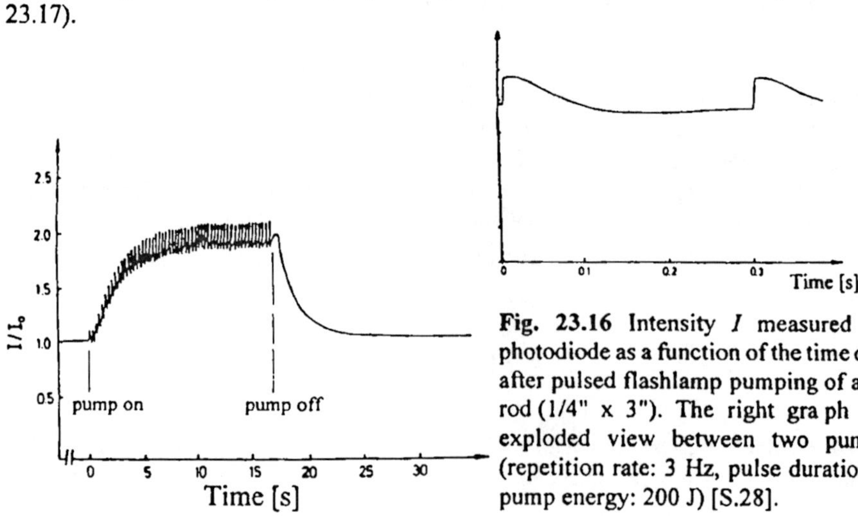

Fig. 23.16 Intensity I measured with the photodiode as a function of the time during and after pulsed flashlamp pumping of a Nd:YAG rod (1/4" x 3"). The right graph shows an exploded view between two pump pulses (repetition rate: 3 Hz, pulse duration: 0.5 ms, pump energy: 200 J) [S.28].

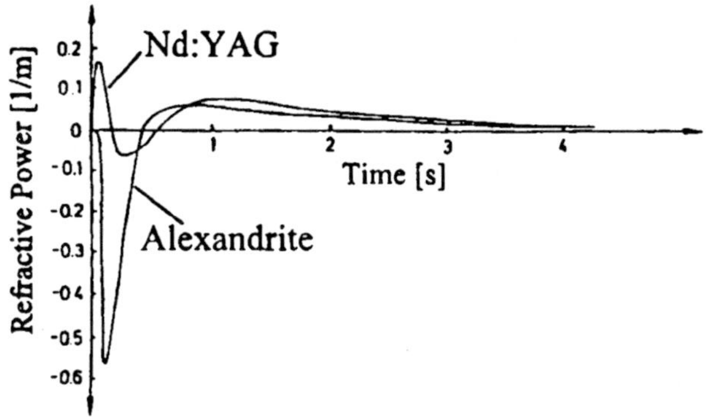

Fig. 23.17 Measured transient refractive powers of Nd:YAG and Alexandrite after a single flashlamp pulse of 0.5 ms duration and 200 J pulse energy (at t=0) (rod dimensions 1/4"x 3") [S.28].

23.2.2 Deviation of a Collimated Probe Beam

Instead of expanding the probe beam, a thin collimated beam with a beam diameter on the order of 0.5mm can be used that passes through the active medium in a distance a from the optical axis (Fig. 23.18). The refractive power leads to a deviation of the probe beam. This results in a shift x of the intersecting point of the beam with a measurement screen located at a distance L from the end face of the active medium. The refractive power D is given by:

$$D = \frac{x}{a\,(L + h)} \tag{23.27}$$

where $h=\ell/(2n)$ is the distance of the principal plane of the thermal lens from the end face of the medium. For a set-up with $L=1m$ and $a=3mm$, the shift of the probe beam is 3mm per diopter of refractive power. For laser materials that exhibit pump-induced birefringence (such as Nd:YAG), an unpolarized probe beam is split into two beams that are perpendicularly polarized. The shifts of the two beams are related to the refractive powers for radially and azimuthally polarized light. In addition to the capability of measuring the refractive power for different polarizations, this method can provide a spatially resolved measurement of the refractive power. The accuracy depends on the exact determination of the distance a of the probe beam. It is highly recommended to first find the optical axis of the medium by transversally scanning the medium with the probe beam and by aligning the beam direction if necessary. The probe beam coincides with the optical axis if no shift on the observation screen is observed when the pump power is switched on and off. This defines the zero position of the probe beam. The distance a can now be controlled with a translation stage. The main difficulty of this method is the determination of the shift x. Since the beam diameter on the observation screen is relatively large (on the order of a cm), a CCD camera with image processing is required to determine the center of gravity accurately. Similar to the previous probe beam technique, this method can be applied both with and without a laser resonator.

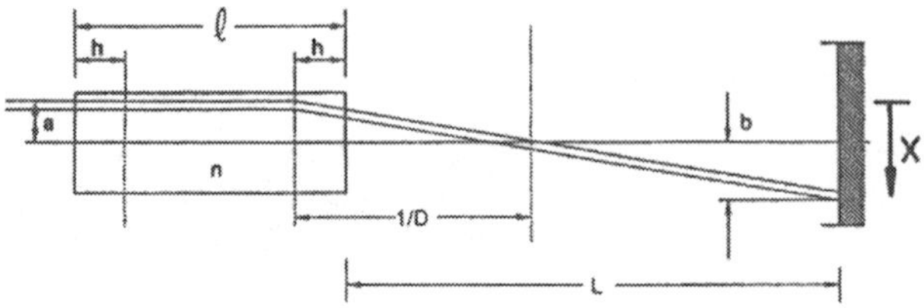

Fig. 23.18 Measurement of the refractive power by deviating a probe beam.

23.2.3 Change in Laser Properties

As discussed in Chapter 13, the refractive power of the active medium influences the properties of the laser resonator. The equivalent lens resonator moves along a straight line through stable and unstable regions of the stability diagram. In the unstable regions, the diffraction losses are considerably increased resulting in a decrease of output power. The laser oscillation may even stop if the resonator penetrates deeply into an unstable region. The decrease in output power can be used to determine the refractive power because the refractive power required to reach a stability limit is known.

a) Power Minimum of Near Symmetric Flat-Flat Resonators

If the active medium is positioned almost in the middle of a flat-flat resonator, the equivalent lens resonator shortly penetrates the unstable region near the origin of the stability diagram as the pump power is increased (Fig. 23.19). For a resonator length of 1m, the distances from the mirrors to the active medium should differ by 3-5cm. When the resonator goes unstable, the output power decreases and exhibits a minimum near the middle of the unstable region. This power minimum is reached for a refractive power D given by:

$$D = \frac{1}{2}\left[\frac{1}{x_1 + \ell/(2n)} + \frac{1}{x_2 + \ell/(2n)}\right] \tag{23.28}$$

where x_i is the distance of mirror i to the next end face of the active medium, ℓ is the length of the active medium, and n is its refractive index. By varying the resonator length, the refractive power at different pump powers can be determined. The advantage of this method lies in the fact that the refractive power as it is 'experienced' by the resonator modes is measured. The influence of the polarization of the laser radiation on the thermal lensing as well as the cooling of the medium due to stimulated emission are automatically taken into account.

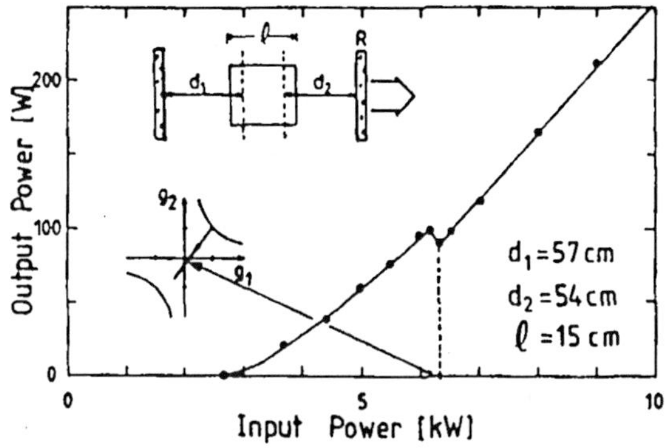

Fig. 23.19 Determination of the refractive power of a Nd:YAG rod by using the power minimum generated with a near-symmetric flat-flat resonator. The refractive power is $1.8m^{-1}$ at a pump power of 6.3kW (rod dimensions: 3/8" x 6", pulse duration: 2ms, repetition rate: 30Hz).

b) Power Maxima of Asymmetric Flat-Flat Resonators

If the active medium is located close to one resonator mirror, the laser oscillation will stop when the resonator penetrates into the unstable region (Fig. 23.20). After passing the stability limit, however, the output power does not decrease continuously. At a higher pump power a second power maximum is observed. The two maxima are related to different polarizations of the laser radiation. For Nd:YAG the refractive power is 15-20% higher for radially polarized light than for azimuthally polarized light. The first power maximum is found at a pump power at which the resonator reaches the stability limit for the radial polarization. Further penetration into the unstable region now leads to a change of the polarization since the resonator modes with azimuthal polarization exhibit lower diffraction losses. The second power maximum is generated when the azimuthal polarization reaches the stability limit. For both polarizations, the refractive power at the power maximum is given by:

$$D = \frac{1}{x_1 + \ell/(2n)} \tag{23.29}$$

where x_1 denotes the larger of the two mirror distances. Unfortunately, this technique is not as accurate as the previous one because it is not ensured that the power maximum can exactly be related to the intersection of a stability boundary. Depending on the resonator set-up and the pumping conditions, the output power may already drop before the unstable region is reached. The exact location of stability limit can be determined if the mode structure on the distant mirror is observed. At the stability limit the mode diameter exhibits a minimum, and the penetration into the unstable region leads to a sudden increase of the beam diameter.

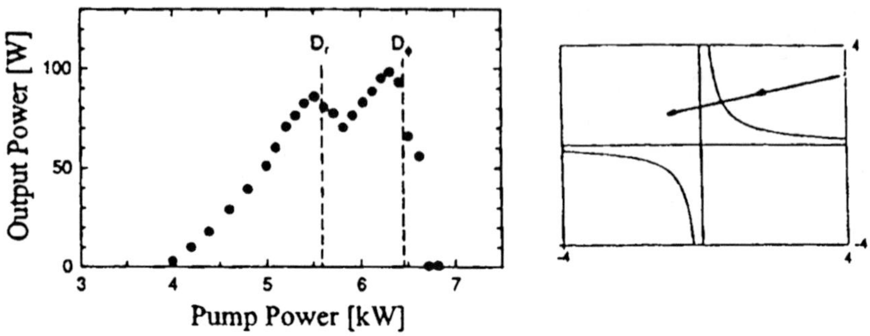

Fig. 23.20 Determination of the refractive power for radially (D_r) and for azimuthally polarized light (D_ϕ) for a Nd:YAG laser with an asymmetric flat-flat resonator. The right graph shows the equivalent stability diagram of the resonator.

c) Mode Structure and Beam Quality

The refractive power can also be determined by comparing the measured beam quality or the measured angle of divergence as a function of the pump power with the theoretical predictions (see Chapter 13). Unfortunately, these measurements usually are not very accurate, and it is therefore difficult to determine the refractive power with an accuracy of better than 10%.

A better technique is provided by the observation of the intensity profile of the laser beam. At the stability limits the beam profile becomes inhomogeneous and a strong change in beam size is observed as the pump power is slightly varied. Figure 23.21 presents measured beam radii at the output coupling mirror of an asymmetric flat-flat resonator illustrating the sudden increase in beam diameter when the unstable region is penetrated.

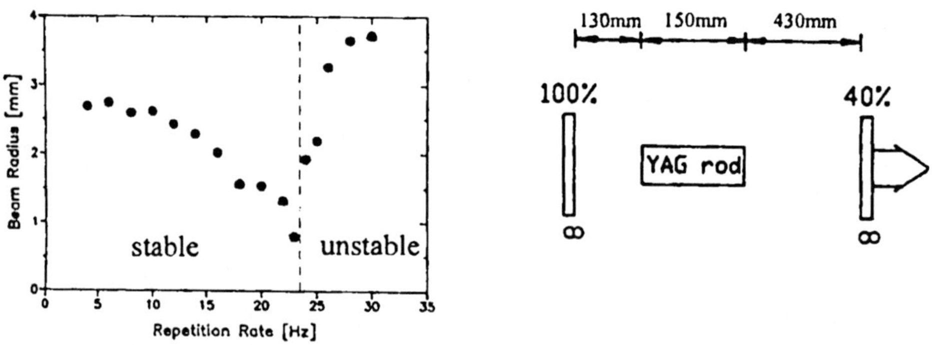

Fig. 23.21 Measured beam radii (86.5% power content) at the output coupling mirror (R=40%) of a pulsed, multimode Nd:YAG laser with an asymmetric flat-flat resonator (3/8" x 6" Nd:YAG rod, pump energy: 308 J, pump pulse duration: 2 ms). The resonator goes unstable at a repetition rate between 23 and 24 Hz. According to (23.29), the refractive power at this point is 2.12 m^{-1}.

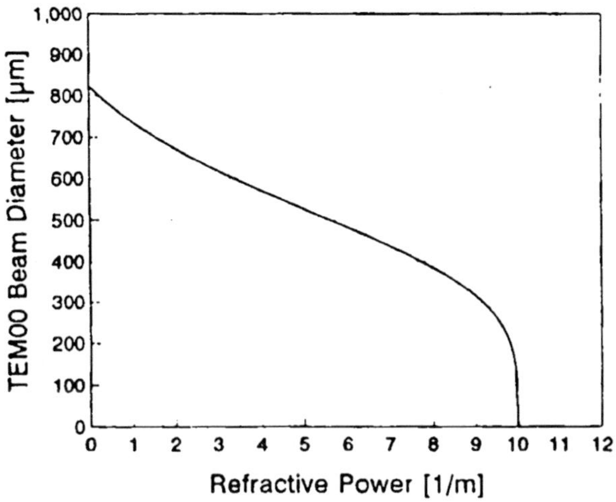

Fig. 23.22 Calculated TEM_{00} mode beam diameters at the resonator mirrors for a symmetric resonator with length 0.4 m and mirror curvatures of 1 m. The thermal lens is located in the middle of the resonator, the laser wavelength is 1064 nm.

A more quantitative approach to determining the the thermal lens via the mode structure is to compare the calculated TEM_{00} mode diameters at the output coupling mirror with the measured beam diameters at this location. In general, the TEM_{00} mode beam diameter exhibits a distinctive decrease with increasing refractive power (Fig. 23.22). When plotted versus the pump power, both the shape of the resulting curve and the absolute diameter values can be used to determine the refractive power with high accuracy. In general, this technique requires the measurement of the beam propagation factor M^2 at each pump power level (see Chapter 24). The beam diameter at the mirror can be determined by imaging the resonator mirror onto a CCD camera and calculating the second order intensity moments. The TEM_{00} mode diameter is obtained by dividing the measured multimode beam diameter by $\sqrt{M^2}$. This method is particularly useful for laser resonators that are operating close to the diffraction limit, like in diode end-pumped solid state lasers.

Chapter 24

Measurement of Laser Beam Parameters

24.1 Measurement of Beam Quality

24.1.1 The Beam Propagation Factor

As was shown in Sec. 2.6, the propagation of arbitrary field distributions can be described by the generalized ABCD law if the beam radii are defined via the second order intensity moments. If a focus spot with waist radii in x- and y-direction w_{0x} and w_{0y}, respectively, is generated, the beam radii as a function of the distance z from the focal plane read:

$$<w_{x,y}(z)^2> = <w_{0x,0y}^2> \left[1 + \frac{z^2}{<z_{0x,0y}^2>}\right] \tag{24.1}$$

where the brackets denote the second order intensity moments. Equation (24.1) applies in both, the x- and the y-directions. The beam radii at a distance z for the beam intensity distribution *I(x,y,z)* are defined by:

$$<w_x(z)^2> = 4\,\frac{\iint (x-x_c)^2 I(x,y,z)dxdy}{\iint I(x,y,z)dxdy} \tag{24.2}$$

$$<w_y(z)^2> = 4\,\frac{\iint (y-y_c)^2 I(x,y,z)dxdy}{\iint I(x,y,z)dxdy} \tag{24.3}$$

where x_c,y_c are the centers of gravity of the intensity distribution at the plane z:

$$\begin{pmatrix} x_c \\ y_c \end{pmatrix} = \frac{\iint \begin{pmatrix} x \\ y \end{pmatrix} I(x,y,z)dxdy}{\iint I(x,y,z)dxdy} \tag{24.4}$$

Similarly, the half angles of divergence are defined as:

$$\begin{pmatrix} <\theta_x^2> \\ <\theta_y^2> \end{pmatrix} = 4\,\frac{\iint \begin{pmatrix} (\theta_x-\theta_{xc})^2 \\ (\theta_y-\theta_{yc})^2 \end{pmatrix} I_F(\theta_x,\theta_y)d\theta_x d\theta_y}{\iint I_F(\theta_x,\theta_y)d\theta_x d\theta_y} \tag{24.5}$$

where $I_F(\theta_x,\theta_y)$ is the far field intensity distribution and (θ_x,θ_y) its center of gravity. In both, the x- and the y-direction, the Rayleigh range is related to the beam waist radius via:

$$<z_0^2> = \frac{<w_0^2>}{<\theta^2>} = \left[\frac{\pi<w_0^2>}{\lambda\ M^2}\right]^2 \tag{24.6}$$

where M^2 is the beam propagation factor, λ is the wavelength of the light, and θ is the half angle of divergence. The beam propagation factor characterizes the beam quality, since it relates the focus area to the Rayleigh range:

$$\frac{\pi<w_0^2>}{\sqrt{<z_0^2>}} = \lambda\ M^2 \tag{24.7}$$

Thus, a good beam quality corresponds to a low beam propagation factor M^2. The lower limit for the beam propagation factor is assumed for a Gaussian beam ($M^2=1$); multimode beams exhibit propagation factors of greater than 1. The measurement of the beam quality of a laser beam, therefore, requires the determination of the beam propagation factor M^2. Equivalent parameters that can be used to characterize the beam quality are the beam quality factor $K=1/M^2$ and the beam parameter product:

$$\sqrt{<w_0^2><\theta^2>} = M^2\,\frac{\lambda}{\pi} \tag{24.8}$$

Unfortunately, the beam radius defined by the second intensity moment does not provide a constant value for the encircled power. Depending on the field distribution, the encircled power fraction may range from 86.5% (Gaussian beam) to 93% (flat top intensity profile). In addition, the second order intensity moments are not defined for intensity distributions that decrease laterally at a rate of $1/x^2$ or less (e.g. far field of a flat top beam. For this reason, attempts were made in the early 1990s to standardize beam quality measurements by defining the beam radii through the 86.5% power content [6.25]. However, in this case the generalized propagation law (24.1) holds only for Gaussian beams. Therefore, the current ISO standard EN ISO 11146 [6.33] requires the application of the second order intensity moments to define the beam diameters, but also allows for alternate measurement methods based on 86.5% power content if the particular laser source does not lend itself to the measurement of the second moments.

If $d(z)$ denotes the beam diameter at a distance z from the beam waist with diameter d_0 (both defined via the second order intensity moment), the basic equations used for the determination of the beam propagation factor M^2 read:

$$d(z)^2 \approx d_0^2 \left[1 + \left(\frac{4\lambda M^2 z}{\pi d_0^2} \right)^2 \right] \tag{24.9}$$

$$\frac{d_0 \theta}{2} = \frac{\lambda}{\pi} M^2 \tag{24.10}$$

where θ is the half angle of divergence as defined by (24.5). Thus, the beam propagation factor can either be determined by measuring the beam diameter at different distances around the waist (24.9), or by measuring the waist diameter and the beam divergence (24.10). The procedures standardized by ISO [6.33] make use of both methods. The measurement of the second order intensity moments is generally performed by recording the two-dimensional intensity distribution with a CCD-camera and calculating the integrals (24.2)-(24.5). Since noise in the wings of the intensity distribution may dominate the second moment, it is necessary to substract a background map from the signal map. In addition, the x-y reference frame of the measurement system should be chosen such that its axes are located along the principal axes of the intensity distribution. If this is not possible, the angle of rotation of the reference frame to the beam axes needs to be recorded.

24.1.2. ISO Standardized Methods

1) Combined Determination of Laser Beam Parameters [6.33]

If the beam waist is accessible for direct measurement, the beam diameters $d(z)$ are measured in at least ten locations around the waist position $z=s_2$ with approximately half the measurement points being distributed within one Rayleigh range on either side of the beam waist and the other half distributed beyond two Rayleigh ranges from the beam waist.

The measured beam diameters are fitted with a hyperbola using the independent fit parameters A,B,C:

$$d^2(z) = A + B\cdot z + C\cdot z^2 \tag{24.11}$$

The laser beam parameters are then given by:

waist diameter $$d_0 = \sqrt{A - \frac{B^2}{4C}} \tag{24.12}$$

waist location $$s_2 = -\frac{B}{2C} \tag{24.13}$$

Rayleigh range $$z_0 = \frac{1}{C}\sqrt{A\cdot C - \frac{B^2}{4}} \tag{24.14}$$

half angle of divergence $$\theta_0 = \frac{1}{2}\sqrt{C} \tag{24.15}$$

beam propagation factor $$M^2 = \frac{\pi}{4\lambda}\sqrt{A\cdot C - \frac{B^2}{4}} \tag{24.16}$$

If the beam waist is not accessible, as is the case for most laser sources, an artificial waist has to be created by using an aberration-free focusing element, like the lens shown in Fig.24.1. The laser beam parameters are determined for this part of the beam (artificial beam) and Eqs.(24.12)-(24.16) apply. By using the parameters of the artificial beam (denoted by superscript *a*), the beam parameters of the original laser beam can be calculated (see Figs. 24.1 and 24.2):

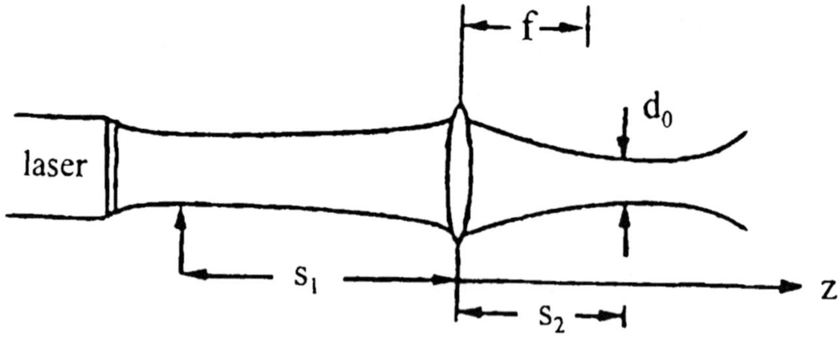

Fig. 24.1 Measurement of the beam parameters by generating a beam waist with a lens [6.22]. (© Chapman and Hall 1992).

waist location in front of lens

$$s_1 = \frac{f \cdot s_2 \cdot (s_2 - f) + f [z_0^a]^2}{s_2^2 - 2 \cdot f \cdot s_2 + f^2 \cdot [z_0^a]^2} \tag{24.17}$$

waist diameter

$$d_0 = d_0^a \sqrt{\frac{2(s_1 - f)^2}{f^2 + \sqrt{(f^4 - 4[z_0^a]^2 \, (s_1 - f)^2}}} \tag{24.18}$$

half angle of divergence

$$\theta_0 = \frac{2 \cdot M^2 \cdot \lambda}{\pi \cdot d_0} \tag{24.19}$$

Rayleigh range

$$z_0 = \frac{d_0}{2\theta_0} \tag{24.20}$$

The beam propagation factor M^2 is assumed to be the same for both the artificial and the original laser beam.

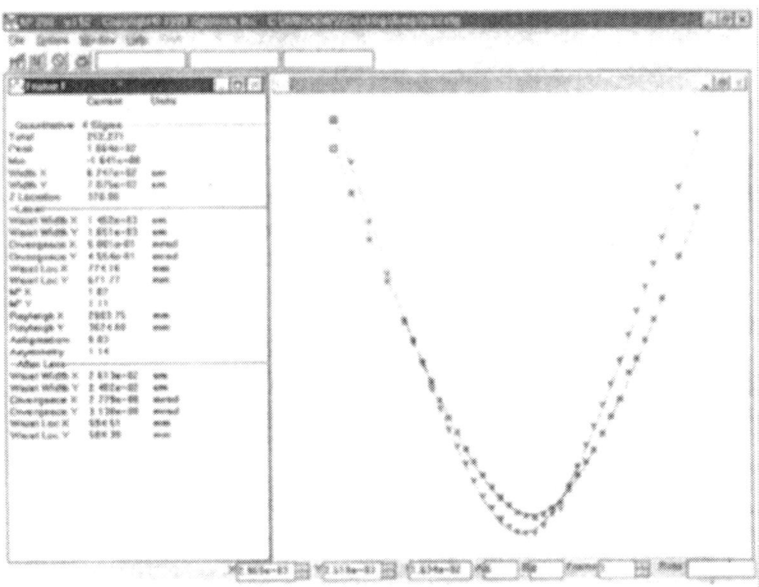

Fig. 24.2 Measured beam diameters in x- and y-direction of the artificial beam of a diode-pumped Vanadate laser emitting at 532nm (courtesy of Spectra-Physics, Inc., Mountain View, CA). The artificial beam is created inside the measurement device, a M^2-200 system by Spiricon, Inc., which uses second order intensity moments to determine beam diameters. Note that the y-axis does not start at zero. The calculated parameters of the original laser beam are: $M_x^2 = 1.07$, $M_y^2 = 1.11$, $d_{0x} = 1.45$ mm, $d_{0y} = 1.65$ mm, $\theta_{0x} = 0.5$ mrad, $\theta_{0y} = 0.46$mrad.

2) Separate Measurement of Beam Waist and Far Field Divergence [6.33]

A beam waist is generated with a lens and the minimum beam diameter d_0 is measured (Fig. 24.3). By using a second lens with focal length f_2, placed behind the beam waist, the angle of divergence of the focused beam can be determined. The full angle of divergence is given by d_l/f_2, where d_l is the beam diameter in the back focal plane of the second lens. In order to reduce measurement errors, the front focal plane of the second lens should coincide with the location of the beam waist. The beam will then be collimated resulting in a higher tolerance in regards to the determination of the exact location of the back focal plane. Furthermore, the focal length of the first lens should be relatively large to reduce errors in the waist diameter measurement. The beam propagation factor is given by:

$$M^2 = \frac{\pi}{\lambda}\,\frac{d_0 d_1}{4 f_2} \tag{24.16}$$

The position of the beam waist behind lens 1 is a function of the Rayleigh range, a parameter that may change as the pump power is varied. If the beam propagation factor is to be measured at several pump power levels, a fixed waist location would be quite helpful. This can be accomplished by using a telescope, provided that a flat output coupling mirror is used (Fig.24.4). The output coupler is located in the front focal plane of the lens with focal length f_l and the image of the beam waist is generated in the back focal plane of the second lens with focal length f_2, magnified by f_2/f_l. Again, the measurement of the divergence is performed in the back focal plane of an additional lens.

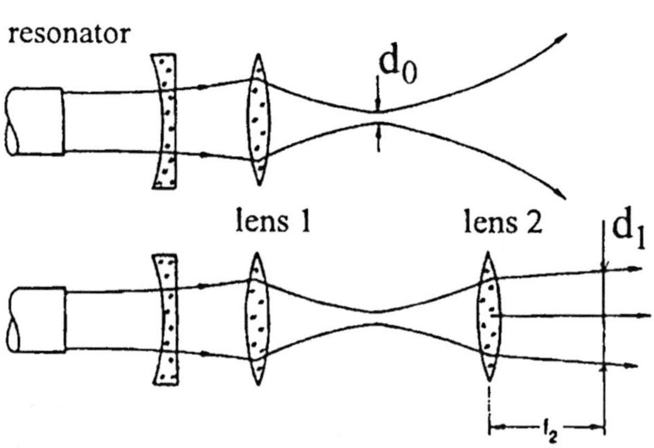

Fig. 24.3 Determination of the beam propagation factor by measuring the beam waist diameter and the corresponding angle of divergence.

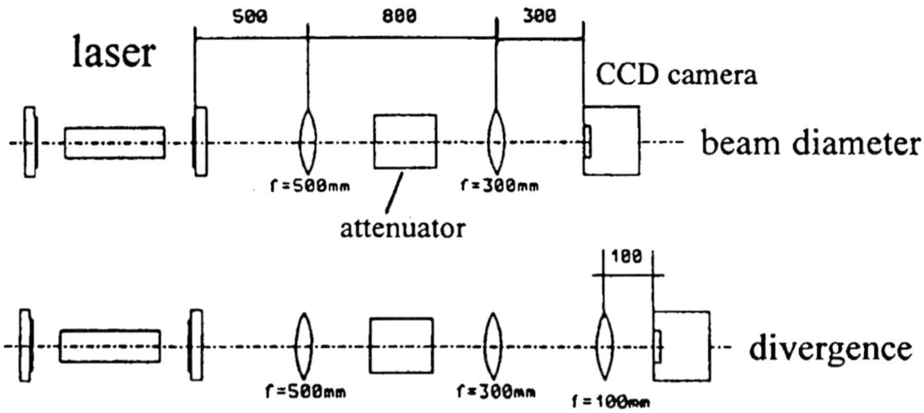

Fig. 24.4 Measurement of the beam quality by using a telescope to generate a beam waist [S.29].

3) Alternative Methods for Beam Width Measurements [6.33]

The two standardized methods of beam quality measurement presented in 1) and 2) can only be applied if the second order intensity moments can be determined with reasonable accuracy. If measurement equipment with high enough signal-to-noise ratio and spatial resolution is not available, or if the intensity profile of the laser beam does not allow the calculation of the intensity moments, the ISO standard provides three alternate methods for measuring the beam diameter, based on the power content. In order to get a correlation with the beam diameter based on the second order intensity moments, correction factors have to be introduced. These correction factors (see Table 24.1) have been verified experimentally for stable resonators with beam quality factors of up to 4. For higher M^2 values, different correction factors have to be used. Figure 24.5 presents the experimental set-up of the three methods. The power is measured with a detector whose sensitive area is large enough to capture at least 99% of the total beam power. Furthermore, a linear response of the detector and a homogeneous sensitivity across the beam is required. For high power lasers, the beam also has to be attenuated to prevent damage to the apertures.

a) Variable Aperture

The diameter of a variable iris is decreased until the power is 86.5% of the power measured without the iris. Alternatively, round apertures with different diameters can be used to extrapolate to the diameter that generates 86.5% power transmission. The step size between apertures should be such that the power is reduced by less than 10% from one diameter to the next.

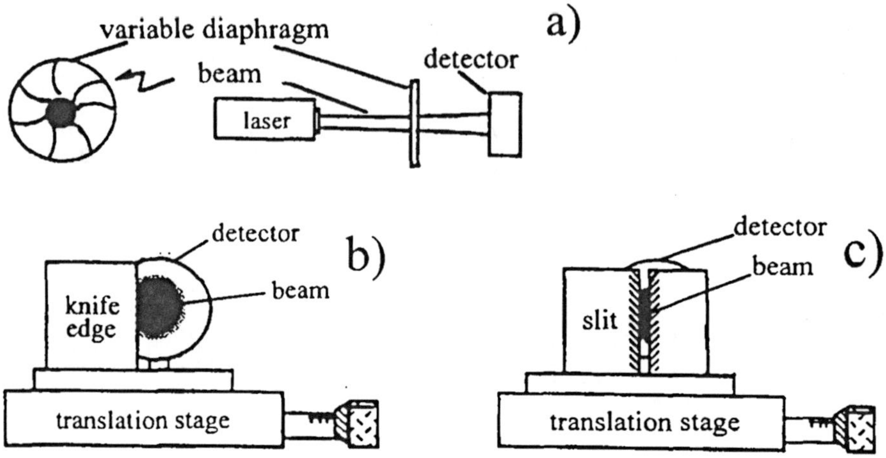

Fig. 24.5 Three methods applicable to the measurement of beam diameters. a) variable diaphragm, b) moving knife edge, c) moving slit [6.23] (© Chapman and Hall 1992).

b) Moving knife edge

The knife edge is moved through the laser beam by means of a translation stage. The positions x_1 and x_2 of the translation stage at which the power is 84% and 16% of the total power, respectively, are recorded. The beam diameter d_{uc} (the subscript *uc* stands for uncorrected) is given by:

$$d_{uc} = 2\,|x_1 - x_2| \tag{24.17}$$

c) Moving slit

A slit with a width b smaller than 1/20 of the approximate beam diameter is moved laterally across the beam by means of a translation stage. First, the maximum power transmitted through the slit is determined. The slit is then moved into the beam at the two opposite edges of the beam so that 13.5% of the maximum power is transmitted (positions x_1 and x_2). The uncorrected beam diameter is given by:

$$d_{uc} = |x_1 - x_2| \tag{24.18}$$

These measured beam diameters are used to calculate the uncorrected beam propagation factor M^2_{uc} according to the methods described in 1) or 2) of this Section. A correction factor c_i is then introduced to determine the corrected beam diameter d, and the corrected beam propagation factor M^2:

$$d = d_{uc} \frac{c_i \cdot (M_{uc} - 1) + 1}{M_{uc}} \tag{24.19}$$

$$M^2 = [c_i \cdot (M_{uc} - 1) + 1]^2 \tag{24.20}$$

where M_{uc} denotes the square-root of M^2_{uc}. The correction factor is different for the three methods, as shown in Table 24.1.

Table 24.1 Correction factor c_i for the three alternative methods for beam width measurements [6.33].

measurement method	correction factor c_i
variable aperture	1.14
moving knife-edge	0.81
moving slit	0.95

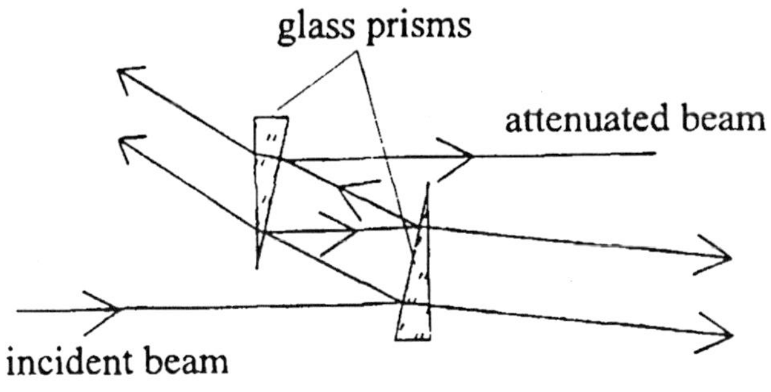

Fig. 24.6 Beam attenuation by using the Fresnel reflection at glass surfaces [6.19]. (© Chapman and Hall 1992)

24.1.3 Beam Attenuation

In general, the laser beam has to be attenuated to record the intensity structure or determine the beam parameters due to a high detector sensitivity and a relatively low damage threshold of the components. Even at low output powers in the Watt range, a simple attenuation using neutral density filters is not feasible. The absorbed power is still sufficient to change the beam properties due to thermally induced lensing and may lead to a melting of the filters. Therefore, it is necessary to decrease the power to a sub-Watt level by a suitable beam attenuation device before any excess power is absorbed by neutral density filters. Beam attenuation devices must preserve both the intensity profile and the polarization of the beam. If a reflecting mirror is used to reflect a portion of the power into a beam dump or a power meter, the laser beam should be incident on the mirror at small angles to the surface normal. At larger angles of incidence the reflectance of the mirror becomes different for s- and for p-polarized light. Any variation in the polarization state of the laser beam will then be mistaken for a change of the output power. In order to avoid this problem, it is recommended to use a glass plate that is AR coated on one side to pick up 4% of the beam power. If an angle of incidence of less than 10° is chosen, the difference in the reflectance for the two polarizations becomes negligible. A similar technique using multiple reflection at parallel glass surfaces is shown in Fig. 24.6. Each reflection attenuates the beam by a factor of 0.04. The number of reflections can be controlled by the relative position and the tilt of the two prisms. Another possibility is the use of a grating (etched quartz) that reflects a very small amount into the first order.

24.1.4 Beam Quality Analyzers

The methods of beam quality measurement discussed thus far all require the separate measurement of at least two beam diameters. In order to apply these techniques it is assumed that the properties of the laser beam do not change during the measurement. Furthermore, the laser has to be running during the time required to perform the measurements (on the order of ten minutes). This is not a problem for cw lasers and pulsed lasers operating at high repetition rates. High power lasers using amplifier stages, however, are sometimes operated only in single shot operation with very low repetition rates. For these systems it is necessary to determine the beam quality for a single laser pulse. Figure 24.7 presents two set-ups that are capable of analyzing the beam quality using a single measurement. These beam quality analyzers rely on the processing of the recorded beam intensity profile.

The phase space beam analyzer (Fig. 24.7a) provides the beam diameter in the x direction of the source and the corresponding angle of divergence. At the observation screen, the intensity distribution in the x direction represents the image of the entrance slit (see Sec. 1.2.5). In the y direction, the optics generate the Fourier transform of the entrance slit. By recording the intensity pattern with a CCD camera and processing the image, it is thus possible to calculate the beam parameter product provided that the beam waist is located at the entrance slit. Unfortunately, the phase space beam analyzer needs to be calibrated to compensate for the divergence generated by the diffraction at the entrance slit [6.15,6.17].

The analyzer presented in Fig. 24.7b images the intensity distributions at different distances from a focusing lens onto one plane. The reflections between the parallel surfaces of two prisms are used to generate a two-dimensional array of beams on a photographic plate. The set-up can be designed such that an image of the focusing region of the lens is generated. By determining the beam diameters on the plate, the beam propagation factor and the waist diameter can be calculated with (24.11). This method is similar to the standardized method using a focusing lens as discussed in Sec. 24.1.1, but less accurate due to the nonlinearity of photographic films.

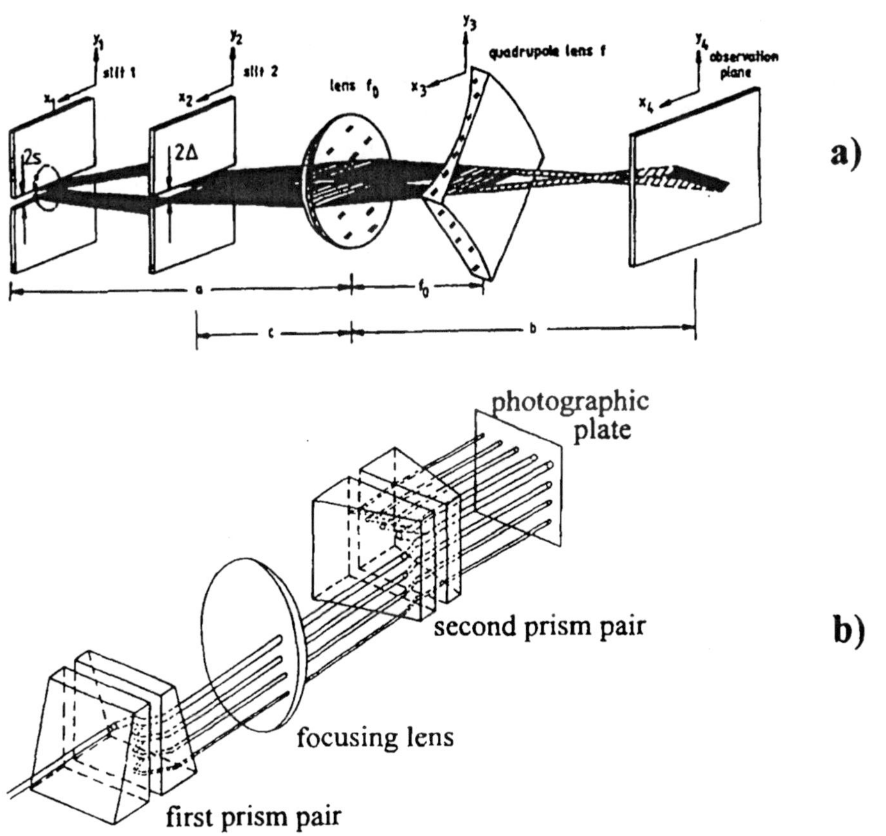

Fig. 24.7 Beam quality analyzers capable of determining the beam propagation factor for one laser pulse: a) phase space beam analyzer [6.17] (© Chapman and Hall 1992), b) analyzer generating an image of the focus region on a photographic plate [S.30].

24.2 Measurement of Polarization

As was discussed in Sec. 1.3, the general polarization state of light is characterized by the electric field amplitudes in two orthogonal directions and the relative phase between them. In a cartesian coordinate system, the polarization is described by the field vector:

$$\boldsymbol{E} = \begin{pmatrix} E_x \\ E_y \exp[i\Phi] \end{pmatrix} \tag{24.21}$$

where we assume a completely polarized beam. Partial polarization has to be treated by using four Stokes parameters [6.12]. A measurement of the polarization of a laser beam, therefore, requires the determination of three parameters using three separate measurements. The easiest way to accomplish this is to measure the beam intensity transmitted through a polarizer for three different pass directions (Fig. 24.8). This measurement has to be repeated for different locations across the beam. If the transmitted intensity is recorded with a CCD camera, the equations (24.21)-(24.25) apply to each pixel.

Measurement 1: E_x
The polarizer transmits along the x direction. The transmitted intensity is given by:

$$I_1 = \frac{1}{2} c_0 \epsilon_0 \, |E_x|^2 \tag{24.22}$$

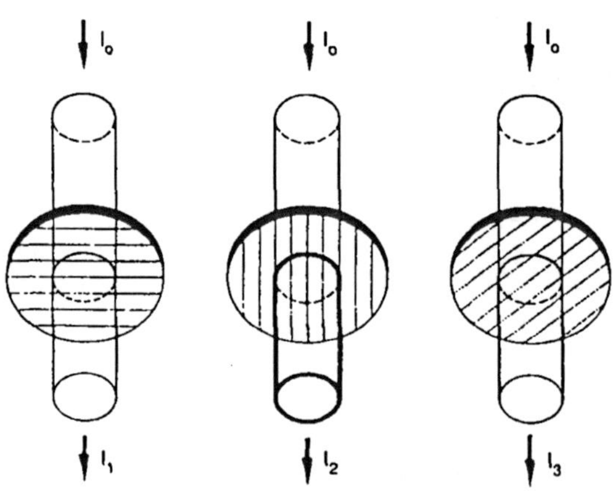

Fig. 24.8 Measurement of the polarization by recording the intensity transmittance at three pass directions of a polarizer.

Measurement 2: E_y

The polarizer is turned by 90° to transmit the field components along the y direction. The transmitted intensity is:

$$I_2 = \frac{1}{2} c_0 \epsilon_0 \, |E_y|^2 \tag{24.23}$$

Measurement 3: Φ

If the pass direction of the polarizer is rotated by 45° with respect to the x-axis, the measured intensity is given by:

$$I_3 = \frac{1}{2}(I_1 + I_2 + 2\sqrt{I_1 I_2}\cos\Phi) \tag{24.24}$$

The phase can be calculated using:

$$\Phi = acos\left[\frac{2I_3 - I_2 - I_1}{2\sqrt{I_1 I_2}}\right] \tag{24.25}$$

The semimajor axis of the polarization ellipse is rotated with respect to the x-axis by the angle (Fig. 24.9):

$$\begin{aligned} \alpha &= atan\sqrt{I_2/I_1} && for \ -\pi/2 \le \Phi < \pi/2 \\ \alpha &= \pi - atan\sqrt{I_2/I_1} && for \ \ \pi/2 \le \Phi < 3\pi/2 \end{aligned} \tag{24.26}$$

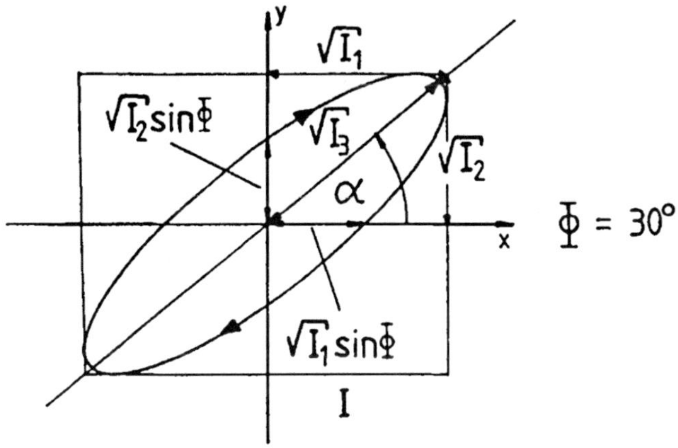

Fig. 24.9 Dependence of the shape and the orientation of the polarization ellipse on the three measured intensities.

This measurement technique can characterize completely polarized light in arbitrary polarization states. Unfortunately, it cannot distinguish between circularly polarized light and unpolarized light. In both cases, the measured intensity is constant as the pass direction of the polarizer is rotated. According to (24.25), the phase is then determined to be 90°. A fourth measurement is necessary to differentiate unpolarized light from circularly polarized light. An optical element that induces a known phase shift between the field components is inserted into the beam path in front of the polarizer. Suitable optical elements are retardation plates or prisms that use total internal reflection to deviate the beam (Fig. 24.10). If the light incident on the optical element is circularly polarized, the rotation of the polarizer will lead to a variation of the transmitted intensity. For unpolarized light, the transmittance still does not depend on the orientation of the polarizer.

The measured polarization of a Nd:YAG laser beam is shown in Fig. 24.11. This measurement was taken at the second power maximum of the laser used in Fig. 23.20. The intensity distribution behind the polarizer was recorded with a CCD camera and the above shown equations were solved for an array of 25x25 pixels. The lines in the plot indicate the orientation of the semimajor axis of the polarization ellipse. The beam is more or less azimuthally polarized because the phase Φ in each point is close to zero (the phase is not shown).

It is important to note that the polarization state of the beam may also change temporally. The measurement method discussed above requires a constancy of the polarization over a certain time period (let us say 1-2 minutes). If the polarization varies within this time period, the measurement results, of course, are falsified. For fast variations of the polarization state, the beam will appear unpolarized. By recording the transmitted intensity with a fast photo diode, temporal changes in the polarization can be detected.

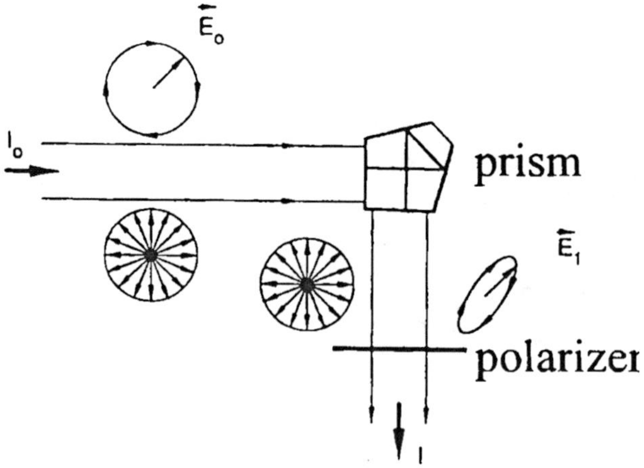

Fig. 24.10 The total internal reflection in a prism can be used to distinguish between unpolarized and circularly polarized light. For circular polarization of the incident beam, the transmitted intensity varies as the polarizer is rotated.

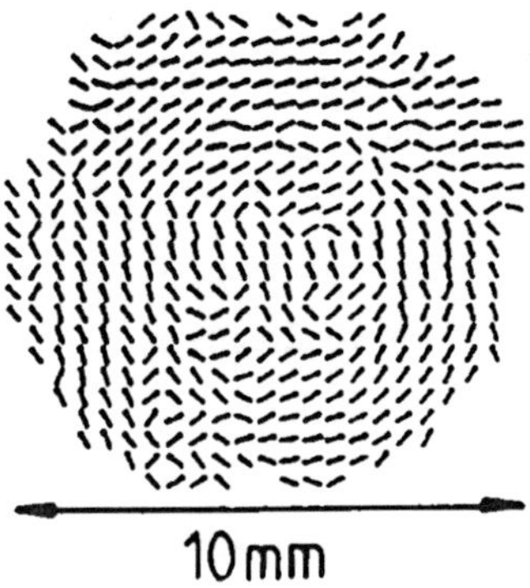

Fig. 24.11 Measured, spatially resolved polarization of a Nd:YAG laser beam. The lines indicate the orientation of the electric field vector. The beam is predominantly azimuthally polarized [S.10].

24.3 Measurement of the Phase

The quasi-monochromatic and coherent electric field, propagating in z-direction, is characterized by:

$$E(x,y,z,t) = E_0(x,y,z)\ \exp[i(\omega t - \phi(x,y,z,t))]$$

with the phase ϕ, which generally depends on the spatial coordinates and time. The wave front is defined as the surface of constant phase ϕ. Some special fields are compiled in Table 24.2. For partially coherent fields, the phase difference between two points of the field may become time dependent and a more sophisticated definition is required [6.34].

Table 24.2 Special beams and their wave fronts.

type of wave	phase	Wave front
plane	$\Phi = k_x x + k_y y + k_z z$	planes normal to (k_x,k_y,k_z)
spherical	$\Phi = kr$	spheres with origin at point-like emitter
Gaussian TEM_{00}	$\Phi = k\left[\frac{r^2}{2R(z)} - z\right] - atan(z/z_0)$	paraboloid around z-axis

The emission of multi-mode lasers consists of many modes with a complicated, time dependent phase structure. Beam propagation (M^2-factor) and focusability (Strehl Definition [6.31]) depend strongly on the phase. Especially for interferometry, the phase is of interest; in most cases the time average value $\langle\phi\rangle$ is measured.

Interferometric measurement
The interferometric measurements, using a shear-interferometer, were already discussed in Sec. 2.8. They are very precise, but also time consuming.

Phase retrieval from intensity measurements
The phase can be retrieved numerically from the measured near- and far-field intensities or from the z-dependent intensity distribution $I(x,y,z)$ [6.11, 6.34]. However, the accuracy is low and quite often the result is not unambiguous. A more precise method uses the Wigner-function (see Sec. 2.8), a kind of tomography, which requires sophisticated numerical procedures [6.36]

Wave front sensor
A quick and simple instrument to determine the phase surface is the wave front- or Hartmann-Shack-sensor [6.35], which in principle allows the measurement of all 10 second intensity moments [6.37, 6.38]. The set-up is shown in Fig. 24.12. Using a lens array, the beam under investigation is divided into a large number of sub-rays. Each sub-ray is focused onto a CCD-camera with 1280 x 1024 pixels (diameter 6.8μm).

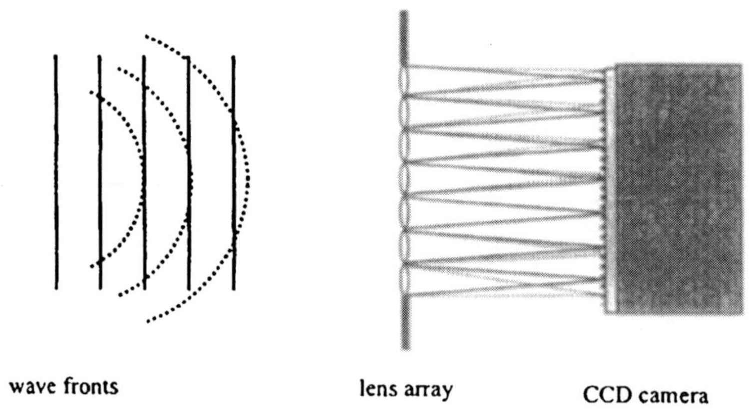

Fig. 24.12 Schematic set-up of a wave front sensor. Typical parameters: micro lenses with 7mm focal length and 200μm pitch, 1280 x 1024 pixels, pixel diameter of 6.8μm. Lens array dimension is approximately 30mm x 40mm.

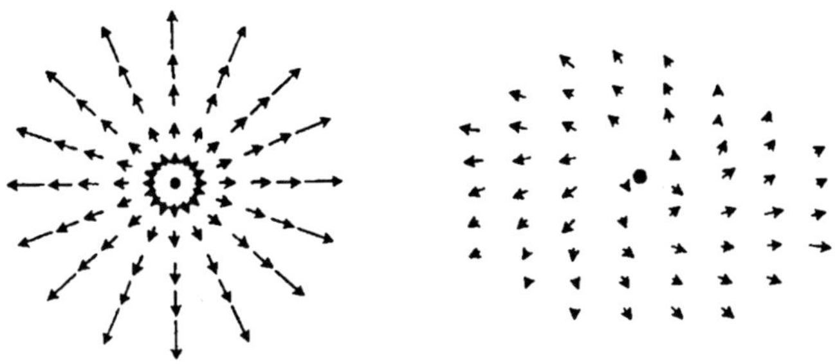

Fig, 24.13 Measured phase gradients of an ideal spherical wave (left) and of a real, non-symmetric beam with twist (right) [6.39].

After calibrating the CCD-array with a plane reference wave, the intensity and the position of each sub-ray is measured. The position of the sub-rays delivers the local phase gradients of the beam as shown in Fig. 24.13. The most important tool of the wave front sensor is the software, which calculates

- local phase gradients
- contour lines of constant phase
- Zernike polynomials
- the 10 second order intensity moments (see Sec. 2.6).

These parameters are delivered by a single measurement, but only if the beam is coherent. For partially coherent beams an additional measurement in the far field is required [6.38]. A set-up to measure all beam parameters for partially coherent beams is depicted in Fig. 24.14. Compared with the interferometric measurements or the phase retrieval using the Wigner-function, the accuracy of the wave front sensor is lower, mainly due to cross talk between adjacent pixels.

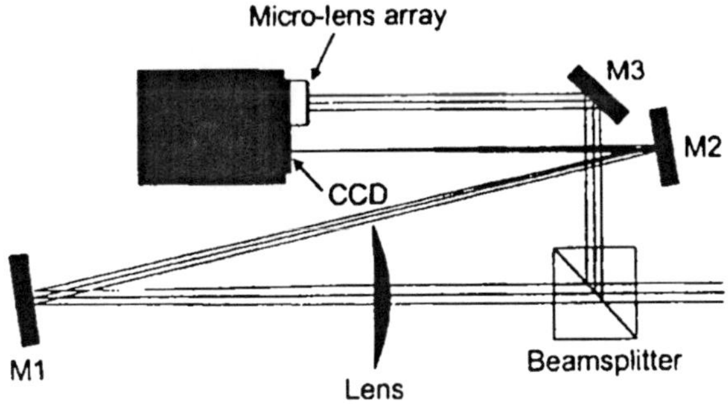

Fig. 24.14 Extended Hartmann-Shack system for simultaneous near-field (via mirror M3), far field (via mirror M2) and wavefront measurement. This set-up allows the determination of all ten second order intensity moments for partially coherent beams [6.40].

References

Part I The Electromagnetic Field

Overview Literature

1.1 M.V. Klein, T.E. Furtak: Optics, 2nd edition. New York, Chichester, Brisbane, Toronto, Singapore: Wiley & Sons 1986, pp. 15-75, 129-191, 337-443, 585-624, 647-653

1.2 M. Born, H. Wolf: Principles of optics. Cambrige, New York, Melbourne, Singapore: Cambridge University Press 1999, Chapters I, III, VIII, X, XV

1.3 E. Hecht: Optik. Munich, New York, Sydney, Tokyo: Addison-Wesley 1989, pp. 135-235, 386-346

1.4 A. Sommerfeld, Vorlesungen ueber theoretische Physik, Bd. IV, Optik. Leipzig: Akademische Verlagsgesellschaft 1959

1.5 F.A. Jenkins, H.E. White: Fundamentals of optics. London, New York, Sydney: McGraw Hill 1957, pp. 3-148, 259-354, 497-518

1.6 S.G. Lipson, H. Lipson, D.S. Tannhauser: Optical physics. Cambridge, New York, Melbourne: Cambridge University Press 1995, pp. 48-98, 152-190

Chapter 1 Geometrical Optics

1.7 J. Liouville, Sur la theorie de la variation des constantes arbitraires, Journales de Mathématiques Pures et Appliquées, 342, 1838

1.8 G.G. Stokes, On a mode of measuring the astigmatism of a defective eye, Report of the British Association for 1849, vol. II, 10, 1849

1.9 M. Bertolotti, Matrix representation of geometrical properties of laser cavities, Il Nuovo Cimento **32**, 1242, 1964

1.10 A. Gerrard, J. Burch: Introduction to matrix methods in optics. London, New York, Sydney, Toronto: John Wiley & Sons 1975

1.11 M. Nemes, G. Nemes, Optical quadrupole, Romanian patent no. 72789, 1977

1.12 J.J. Sylvester, Sur les puissances et les racines de substitutions linéares, Compt. Rendu. Sci. **XCIV**, 55, 1982

1.13 M. Nazarathy, A. Hardy, I. Shamir, Misaligned first-order canonical operator theory, J. Opt. Soc. Am. **A 3**, 1360, 1986
1.14 N. Hodgson, T. Haase, R. Kostka, H. Weber, Determination of laser beam parameters with the phase space beam analyser, Opt. Quantum Electron. **24**, 927, 1992
1.15 G. Nemes, Measuring and Handling General Astigmatic Beams, Proc. of Workshop on Laser Beam Characterization, Sociedad Espanola de Optica, Madrid, 1993
1.16 G. Nemes, A.E. Siegman, Measurement of all ten second-order moments of an astigmatic beam by the use of rotating simple astigmatic (anamorphic) optics, J. Opt. Soc. Am. **A 11**, 2257, 1994
1.17 A.A. Tovar, L.W. Casperson, Generalized Sylvester theorem for periodic applications, J. Opt. Soc. Am. **A 12**, 578, 1995
1.18 T. Graf, J.E. Balmer, Laser beam quality, entropy and the limits of beam shaping, Opt. Commun. **131**, 77, 1996
1.19 S. Wang, D. Zhao, Matrix Optics. Berlin, Heidelberg: Springer-Verlag 2000
1.20 G. Machavariani, N. Davidson, E. Hasman, S. Blit, A.A. Ishaaya, A.A. Friesen, Efficient formation of a high-quality beam from a pure high order Hermite-Gaussian mode, Opt. Commun. **209**, 266, 2002
1.21 J. Serna, G. Nemes, The phase space analyzer with Gaussian slits, Proceedings of the Society of Photo-Optical Instrumentation Engineers vol. **4933**, 2003

Chapter 2 Wave Optics

1.22 E. Waetzmann, Interferenzmethode zur Untersuchung der Abbildungsfehler optischer Systeme, Ann. Phys. **39**, 1043, 1912
1.23 A. Rubinowicz, Die Beugungswelle in der Kirchhoffschen Theorie der Beugungserscheinungen, Annalen der Physik **53**(12), 257, 1917
1.24 E. Wigner, On the quantum correction for thermodynamic equilibrium, Phys. Rev. **40**, 749, 1932
1.25 P.M. Woodward, Probability and Information Theory. London: Pergamon Press 1953, Chapter 7.
1.26 A.G. Fox, T. Li, Modes in a maser interferometer with curved and tilted mirrors, Proc. IEEE **51**, 80, 1963
1.27 E. Wolf, E.W. Marchand, Comparison of the Kirchhoff and the Rayleigh-Sommerfeld theories of diffraction at an aperture, J. Opt. Soc. Am. **54**(5), 587, 1964
1.28 M. Abramowitz, A. Stegun: Handbook of mathematical functions. New York: Dover Publ. 1964
1.29 H. Kogelnik, T. Li, Laser beams and resonators, Proc. IEEE **54**, 1312, 1966
1.30 A. Walther, Radiometry and coherence, J. Opt. Soc. Am. **58**, 1256, 1968
1.31 P. Baues, Huygens' principle in inhomogeneous isotropic media and a general integral equation applicable to optical resonators, Opto-Electr.**1**, 37, 1969
1.32 J.A. Arnaud, H. Kogelnik, Gaussian light beams with general astigmatism, Appl. Opt. **8**(8), 1686, 1969

1.33 M.V.R.K. Mruthy, A compact lateral shear interferometer based on the Michelson interferometer, Appl. Opt. **9**, 1146, 1970
1.34 S.A. Collins, Diffraction-integral written in terms of matrix-optics, J. Opt. Soc. Am. **60**(9), 1168, 1970
1.35 J.A. Arnaud, Nonorthogonal optical waveguides and resonators, Bell. Syst. Tech. J. **49**, 2311, 1970
1.36 J.A. Arnaud, Hamiltonian theory of beam mode propagation, in: Progress in Optics XI, editor: E. Wolf, Amsterdam: North Holland 1973, pp. 247
1.37 A. Papoulis, Ambiguity function in Fourier optics, J. Opt. Soc. Am. **64**, 779, 1974
1.38 M.Lax, W.H. Louisell, W.B. McKnight, From Maxwell to paraxial wave optics, Phys.Rev. **A 11**, 1365, 1975
1.39 M.J. Bastiaans, Wigner distribution function and its application to first order optics, J. Opt. Soc. Am. **69**, 1710, 1979
1.40 M.J. Bastiaans, Wigner distribution function and its application to first order optics, J. Opt. Soc. Am. A **69**, 1710, 1979
1.41 G.P. Argawal, D.N. Pattanayak, Gaussian beam propagation beyond the paraxial approximation, JOSA **69**, 575, 1979
1.42 F. Gori, Collett-Wolf sources and multimode lasers, Opt. Commun. **34**, 301, 1980
1.43 H.O. Bartelt, K.-H. Brenner, A.W. Lohmann, The Wigner distribution function and its optical production, Opt. Commun. **32**, 32, 1980
1.44 M.J. Bastiaans, The Wigner distribution function of partially coherent light, Optica Acta **28**, 1215, 1981
1.45 E. Wolf, New theory of partial coherence in the space-frequency domain, J. Opt. Soc. Am. **72**, 343, 1982
1.46 M.V. Klein, T.E. Furtak: Optics. 2nd edition, New York, Chichester, Brisbane, Toronto, Singapore: Wiley & Sons 1986, pp. 647-653
1.47 R. Simon, E.C.G. Sudarshan, N. Mukunda, Anisotropic Gaussian Schell-model beams; passage through optical systems and associated invariants, Phys. Rev. **A 31**, 2419, 1985
1.48 M.J. Bastiaans, Application of the Wigner distribution function to partially coherent light, J. Opt. Soc. Am. **A 3**, 1227, 1986
1.49 V.V. Korobkin, L.Y. Polonskii, V.P. Poponin, L.N. Pyanitskii, Sov. J. Quant. Electron. **16**, 178, 1986
1.50 J. Turunen, A.T. Friberg, Matrix representation of Gaussian Schell model beams in optical systems, Opt. Las. Tech. **18**, 259, 1986
1.51 J. Durnin, Exact solutions for nondiffracting beams, J. Opt. Soc. Am. A **4**(4), 651, 1987
1.52 J. Durnin, J.J. Miceli, J.H. Eberly, Diffraction-free beams, Phys. Rev. Lett. **58**(15), 1499, 1987
1.53 R. Simon, N. Mukunda, E.C.G. Sudarshan, Partially coherent beams and a generalized ABCD-law, Opt. Commun. **65**(5), 322, 1988
1.54 R. Simon, E.C.G. Sudarshan, N. Mukunda, Opt. Commun. **65**, 322, 1988
1.55 E. Tervonen, J. Turunen, A.T. Friberg, Transverse laser-mode structure determination from spatial coherence measurements, Appl. Phys. **B 49**, 409, 1989

1.56 G. Indebetouw, Nondiffracting optical fields, some remarks on their analysis and synthesis, J. Opt. Soc. Am. **A 6**, 150, 1989

1.57 L. Quigang, W. Shaomin, J. Alda, J. Bernabeu, Transformation of nonsymnmetric Gaussian beams into symmetric ones by means of tensor ABCD law, Optic **85**, 67, 1990

1.58 A.E. Siegman, New developments in laser resonators, Proceedings of the Society of Photo-Optical Instrumentation Engineers vol. **1224**, 2, 1990

1.59 H.S. Lee, B.W. Stewart, D. Will, H. Fenichel, Holographic Bessel beam amplification, Appl. Phys. Lett. **59(54)**, 3096, 1991

1.60 P.A. Belanger, Beam propagation and the ABCD ray matrices, Opt. Lett. **16**, 196, 1991

1.61 M.J. Bastiaans, Second-order moments of the Wigner distribution function in first-order optical systems, Optik **88**, 163, 1991

1.62 J. Serna, R. Martinez-Herrero, P.M. Mejias, Parametric characterization of generally partially coherent beams propagating through ABCD optical systems, J. Opt. Soc. Am. **A 8**, 1094, 1991

1.63 K. E. Oughstun, editor, Selected papers on scalar wave diffraction, SPIE Milestone Series Volume MS **51**. Bellingham, WA: The Society of Photo-Optical Instrumentation Engineers 1992

1.64 H. Weber, Propagation of higher-order intensity moments in quadratic-index media, Opt. Quantum Electron. **24**, 1027, 1992

1.65 G. Scott, N. McArdle, Efficient generation of nearly diffraction-free beams using an axicon, Opt. Eng. **31**, 2640, 1992

1.66 N. Reng, B. Eppich, Definition and measurements of high-power laser beam parameters, Opt. Quantum Electron. **24**, 973, 1992

1.67 G. Nemes, Measuring and handling general astigmatic beams, Proceedings of the workshop on laser beam characterization, eds.: P.M. Mejias, H. Weber, R. Martinez-Herrero, A. Gonzales-Urena, Sociedad Espanola de Optica, Madrid 1993, pp. 325-358.

1.68 R. Simon, K. Sundar, N. Mukunda, Twisted Gaussian Shell-model beams, J. Opt. Soc. Am. **A 10**, 2008, 1993

1.69 Document ISO 11145:1994, Optics and optical instruments - Lasers and laser-related equipment - Vocabulary and symbols, 1994

1.70 B. Eppich, R. Kostka, N. Reng, H. Weber, High power laser beam characterisation, Proceedings of the Society of Photo-Optical Instrumentation Engineers vol. **2206**, High power gas and solid state lasers, 469, 1994

1.71 G. Nemes, A.E. Siegman, Measurement of all ten second-order moments of an astigmatic beam by use of rotating simple astigmatic (anamorphic) optics, J. Opt. Soc. Am. A **11**(8), 2257, 1994

1.72 A. Yariv, Imaging of coherent fields through lenslike systems, Opt. Lett. **19**(20), 1607, 1994

1.73 R. Martinez-Herrero, P.M. Mejias, C. Martinez, Parametric characterization of the phase at the far field, Opt. Lett. **20**(7), 651, 1995

1.74 L. Mandel, E. Wolf, Optical coherence and quantum optics, Cambridge, New York, Madrid, Melbourne, Singapore: Cambridge University Press, 1995

1.75 T. Graf, J.E. Balmer, Laser beam quality, entropy and the limits of beam shaping, Opt. Commun. **131**, 77, 1996

1.76 H. Laabs, B. Ozygus, Excitation of Hermite Gaussian modes in end-pumped solid-state lasers via off-axis pumping, Optics & Laser Technology **28**, 213, 1996

1.77 J.W. Goodman: Introduction to Fourier optics, Physical and Quantum Series. London, New York, San Francisco: McGraw Hill 1996

1.78 B. Eppich, Complete characterization of partially coherent 2D-beams, Proc. of the 4th International Workshop on Laser Beams and Optics Characterization, Munich 1997, eds.: A. Giesen, M. Morin, Düsseldorf: VDI-Technologiezentrum, 1997

1.79 D.W. Coutts, A versatile angular shearing interferometer for measurement of spatial coherence, Conference on Lasers and Electro-Optics Europe 1998, Technical Digest, paper CWI4, 217, 1998

1.80 B. Eppich, Die Charakterisierung von Strahlungsfeldern mit der Wigner-Verteilung und deren Messung, Dissertation, Optical Institute, Technical University Berlin, Germany, 1998

1.81 F. Gori, M. Santarsiero, R. Borghi, G. Guattari, Intensity-based modal analysis of partially coherent beams with Hermite-Gauss modes, Appl. Opt. **23**(13) 989, 1998

1.82 P. Paakonen, J. Turunen, Resonators with Bessel-Gauss modes, Opt. Commun. **156**, 359, 1998

1.83 H. Laabs, Einfluss nichtparaxialer Strahlenausbreitung und räumlicher Strukturen auf Lasermoden in stabilen Resonatoren, Dissertation D 83, Technical University Berlin, Institute of Optics, 1998

1.84 M. Born, E. Wolf, Principles of Optics, Cambridge University Press, 1999

1.85 D. Lorenz, Verstärkungsgeschaltete Cr:YAG und Ti:Saphir Laser mit groβer spektraler Breite und guter Strahlqualität, PhD Dissertation, University of Potsdam, Germany, 1999

1.86 Document EN ISO 11146, Lasers and laser related equipment - Test methods for laser beam parameters: Beam widths, divergence angle and beam propagation factor (ISO 11146:1999), June 1999

1.87 S.M. Jackisch, Shearing Interferometrie zur Laserstrahlcharakterisierung, Diploma Thesis, Technical University Berlin, Germany, Optical Institute, 1999

1.88 C.M. Warnsky, B.L. Anderson, C.A. Klein, Determining spatial modes of lasers by coherence measurements, Appl. Opt. **39**, 6109, 2000

1.89 J.C. Gutiérrez, M.D. Iturbe-Castillo, S. Chávez-Cerda, Alternative formulation for invariant optical fields, Opt. Lett. **25**, 1493, 2000

1.90 E. Trebino, Frequency resolved optical gratings - the measurement of ultrashort laser pulses, Norwell,MA: Kluver Academic Press, 2000

1.91 R. Oron, N. Davidson, A.A. Freisen, E. Hasman, Continuous-phase elements can improve laser beam quality, Opt. Lett. **25**, 939, 2000

1.92 V. Vaicaitis, Cherenkov-type phase-matched third harmonic generation in air, Opt. Commun. **185**, 197, 2001

1.93 B. Eppich, Definition, meaning and measurement of coherence parameters, Proceedings of the Society of Photo-Optical Instrumentation Engineers vol. **4270**, 71, 2001

1.94 G. Cincotti, A. Ciattoni, C. Palma, Hermite Gauss beams in uniaxially anisotropic crystals, IEEE J. Quant. Electron. **37**(12), 1517, 2001

1.95 R. Grunwald, U. Griebner, E.T.J. Nibbering, A. Kummrow, M. Rini, T. Elsaesser, V. Kebbel, H.J. Hartmann, W. Jüptner, Spatially resolved small-angle noncollinear interferometric auto-correlation of ultrashort pulses with microaxicon arrays, J. Opt. Soc. Am. **A 18**, 2923, 2001

1.96 V.E.R. Peet, R.V. Tsubin, Generation of resonance-enhanced fifth harmonic in Xenon with Bessel-Gauss laser beams, Opt. Commun. **189**, 267, 2001

1.97 H. Laabs, B. Eppich, H. Weber, Modal decomposition of partially coherent beams using the ambiguity function, J. Opt. Soc. Am. **A 19**, 497, 2002

1.98 A. Ciattoni, G. Cincotti, C. Palma, H. Weber, Energy exchange between cartesian components of a paraxial beam in an uniaxial crystal, J. Opt. Soc. Am., Sept. 2002

1.99 A. Ciattoni, G. Cincotti, C. Palma, Propagation of cylindrical symmetric fields in uniaxial crystals, J. Opt. Soc. Am. **A 19**, 792, 2002

1.101 G. Machavariani, N. Davidson, E. Hasman, S. Blit, A.A. Ishaaya, A.A. Friesen, Efficient formation of a high-quality beam from a pure high order Hermite-Gaussian mode, Opt. Commun. **209**, 266, 2002

1.102 B. Eppich, G. Mann, H. Weber, Spatial coherence-definition and measurements, Proceedings of 35. School on Quantum Electronics, Erice, Italy, 2002

1.103 G. Cincotti, A. Ciattone, C. Palma, Hermite Gauss beams in uniaxially anisotropic crystals, IEEE J. Quant. Electron. **37**, 12, 2001

1.104 A. Ciattoni, G. Cincotti, C. Palma, H. Weber, Energy Exchange between cartesian components of a paraxial beam in a uniaxial crystal, J. Opt. Soc. Am. **A 19**, 1894, 2002.

1.105 A. Ciattoni, G. Cincotti, C. Palma, Propagation of cylindrical symmetric fields in uniaxial crystals, J. Opt. Soc. Am. **A 19**, 792, 2002

1.106 P. Muys, E. Vandamme, Direct generation of Bessel beams, Appl. Opt. **41**(30), 6375, 2002

1.107 J. Serna, F. Encinas-Sanz, G. Nemes, Complete Characterization of a pulsed doughnut-type beam by use of spherical optics and a cylindrical lens, J. Opt. Soc. Am. **A 18**, 1726, 2001

1.108 B. Eppich, G. Mann, H. Weber, Photonics West, San Jose 2003, Coherence: comparison of interferometric and non-interferometric measurements, Proceedings of the Society of Photo-Optical Instrumentation Engineers vol. **4969**, 137, 2003

1.109 R. Martinez-Herrero, P.M. Mejias, H. Weber, On the different definitions of laser beam moments, Opt. Quantum Electron. **25**, 423, 1993

1.110 S. Lavi, R. Pronchaska, E. Keren, Generalized beam parameters and transformation laws for partially coherent light, Appl. Opt. **27**, 3696, 1988

Chapter 3 Polarization

1.111 R.C. Jones, A new calculus for the treatment of optical systems, J. Opt. Soc. Am. **32**, 486, 1942

1.112 J. Junghans, M. Keller, H. Weber, Laser resonators with polarizing elements-eigenstates and eigenvalues of polarization, Appl. Opt. **13**, 2793, 1974

1.113 E. Collett: Polarized light. New York, Basel, Hong Kong: Marcel Decker, Inc., 1993, pp. 187- 218

1.114 P.M. Mejias, R. Martinez-Herrero, G. Piquero, J.M. Movilla, Parametric characterization of the spatial structure of non-uniformly polarized laser beams, Progress in Quantum Electronic **26**, 65, 2002

1.115 M. Mansuripur, Classical optics and its applications. Cambridge University Press 2002

1.116 J.L. Jackson, Classical Electrodynamics, J. Wiley & Sons, New York 1975

1.117 E. Hecht, A. Zajac, Optics, Addison-Wesley Publ. Comp., Reading MA, 1974

1.118 R.O. Frisch, Experimenteller Nachweis des Einsteinschen Srahlungsrückstoßes, Zs f. Physik, **86**, 42, 1933

1.119 A. Ashkin, The Pressure of Light, Scient. American **226**, 63(2), 1972

1.120 A. Ashkin, Optical Levitation by Radiation Pressure, Appl. Phys. Lett.**19**, 283, 1971

1.121 Y.R. Shen, Laser Manipulation of Particles, in Principles of Nonlinear Optics, J.Wiley & Sons, NY 1984

1.122 M.J. Padgett, L. Allen, The angular momentum of light: optical spanners and the rotational frequency shift, Optical and Quantum Electr. **31**, 1, 1999

1.123 B. Eppich, Die Charakterisierung von Strahlungsfeldern mit der Wigner-Verteilung und deren Messung, Dissertation, D 83, Optisches Institut, Technische Universität Berlin, 1998

1.124 A. Friberg, Optical Twist, International Trends in Optics and Photonics ICO IV, ed. T. Asakura, Springer Berlin 1999, p.3

1.125 C. Gao, Characterization and transformation of astigmatic laser beams, Dissertation, D 83, Optisches Institut, Technische Universität Berlin, Wissenchaft und Technik Verlag, Berlin 1999

1.126 E. Abramochkin, V. Volostnikov, Beam transformations and nontransformed beams, Opt. Comm. **83**,123, 1991

1.127 A. Friberg, C. Gao, B. Eppich, H. Weber, Generation of Partially Coherent fields with Twist, SPIE **3110**, 317, 1997

1.128 C. Tamm, C.O.Weiss, Bistability and optical switching of spatial patterns in a laser, JOSA **B7**, 1034, 1990

1.129 M.W. Beijersbergen, L.Allen, H.E.L.O. van der Veen, J.P. Woerdman, Astigmatic laser mode conversion and transfer of orbital angular momentum, Opt. Comm. **96**, 123, 1993

1.130 M.W. Padgett, L. Allen, Optical tweezers and spanners, Physics World, **Sept.**, 35, 1997

1.131 C. Gao, H. Laabs, H. Weber, T. Brand, N. Kugler, Symmetrization of astigmatic high power diode laser stacks, Optical and Quantum Electr. **31**, 1207, 1999

1.132 N.B. Simpson, K. Pholakia, L. Allen, M.J. Padgett, Mechanical equivalence of spin and orbital angular momentum of light: an optical spanner, Opt. Lett.**22**, 52, 1997
1.133 M.E. Friese, T.A. Nieminen, N.R. Heckenberg, H. Rubinztein-Dunlop, Optical alignment and spinning of laser-trapped microscopic particles, Nature **394**, 348, 1998
1.134 R.A. Beth, Mechanical detection and measurement of the angular momentum of light, Phys. Rev. **50**, 115, 1936
1.135 J. Serna, G. Nemes, Decoupling of coherent Gaussian beams with general astigmatism, Opt. Lett. **18**, 1774, 1993
1.136 J. Serna, Angular momentum and vortices in optical beams, Proceedings 35th International School of Quantum Electronics, Erice 2002, to be published,. American Institute of Physics, 2004

Part II Basic Properties of Optical Resonators

Overview Literature:

2.1 A.E. Siegman: Lasers. Mill Valley: University Sciences Books 1986, pp. 399-456
2.2 F.A. Jenkins, H.E. White, Fundamentals of optics. New York, London, Sydney: McGraw Hill 1957, pp. 286-314
2.3 E. Hecht: Optics. Munich New York Sydney Tokyo: Addison-Wesley 1989, pp. 383-413
2.4 H.A. Macleod: Thin film optical filters. Bristol: Adam Hilger 1986
2.5 M. Bass, E.W. Van Stryland, D.R. Williams, W.L. Wolfe (editors): Handbook of optics, vol. I. New York, London, Sydney: McGraw-Hill 1995, part 11

Chapter 4 The Fabry Perot Resonator

2.6 K. Kotik, M.C. Newstein, Theory of laser oscillation in Fabry-Perot interferometer, J. Appl. Phys. **32**, 178, 1961
2.7 G. Koppelmann, K. Krebs, Zur Technologie des Fabry-Perot-Interferometers, Optik **18**, 358, 1961
2.8 P. Connes, High Resolution Spectroscopy, in C. Townes: Quantum Electronics and Coherent Light, Academic Press 1964
2.9 G. Herziger, H. Lindner, H. Weber, Messung geringer Absorptions- und Brechungs-indexaenderungen mit dem Laserverstaerker, Zeitschrift für angewandte Physik **17**, 67, 1964
2.10 H. Ogura, Y. Yoshida, J. Ikenone, Theory of deformed fabry perot resonator, J. Phys. Soc. Jap. **20**, 598, 1965
2.11 M. Francon: Optical interferometry. New York London Sydney Tokyo: Academic Press 1966, 368
2.12 T. Li, H. Zucker, Modes of a Fabry-Perot laser resonator with output coupling apertures, J. Opt. Soc. Am. **57**, 984, 1967

2.13 M. Herscher, The spherical mirror Fabry-Perot-Interferometer, Appl. Opt. **7**, 951, 1968
2.14 H.K.V. Lotsch, The FPI-Resonator, Part I,II,III, Optik**28**, 65, 328, 555, 1968
2.15 H.K.V. Lotsch, The FPI-Resonator, Part IV,V, Optik **29**, 130, 622, 1969
2.16 J.D. Rancourt: Optical Thin Films Users' Handbook. New York: Macmillan 1987, pp. 289
2.17 A. Thelen: Design of optical interference coatings. New York, London, Sydney: McGraw-Hill 1988, pp. 256

Part III Passive Open Resonators

Overview Literature:

3.1 A.E. Siegman: Lasers. Mill Valley: University Science Books 1986, pp. 559-913
3.2 W. Koechner: Solid state laser engineering. Berlin Heidelberg New York London Paris Tokyo: Springer 1976, pp. 168-244
3.3 D.R. Hall, P.E. Jackson (eds.): The physics and technology of laser resonators. Bristol, Philadephia: Institute of Physics Publishing Ltd. 1989
3.4 Yu. A. Anan'ev: Laser resonators and the beam divergence problem. Bristol, Philadelphia, New York: Adam Hilger 1992
3.5 P.W. Miloni, J.H. Eberly: Lasers. New York, Chichester, Brisbane, Toronto, Singapore: John Wiley & Sons 1988

Chapter 5 Stable Resonators

3.6 G.D. Boyd, J.P. Gordon, Confocal multimode resonator for millimeter through optical wavelength masers, Bell Syst. Tech. J. **40**, 489, 1961
3.7 A.G. Fox, T. Li, Resonant modes in a maser interferometer, Bell. Syst. Tech. J. **40**, 453, 1961
3.8 G.D. Boyd, H. Kogelnik, Generalized confocal resonator theory, Bell. Syst. Tech. J. **41**, 1347, 1962
3.9 L.A. Vainshtein, Open resonators for lasers, Sov. Phys. JETP **17**,709, 1963
3.10 A.G. Fox, T. Li, Modes in a maser interferometer with curved and tilted mirrors, Proc. IEEE **51**, 80, 1963
3.11 J.P. Gordon, A circle diagram for optical resonators, Bell Syst. Tech. J. **43**, 1826, 1964
3.12 J.P. Gordon, H. Kogelnik, Equivalence relations among spherical mirror optical resonators, Bell Syst. Tech. J. **43**, 2873, 1964
3.13 M. Abramowitz, A. Stegun: Handbook of mathematical functions. New York: Dover Publ. 1964
3.14 W. Magnus, F. Oberhettinger, R.P. Soni, Formulas and theories for special functions of mathematical physics, Berlin Heidelberg New York London Paris Tokyo: Springer 1966

3.15 J.P. Gordon, A circle diagram for optical resonators, Bell Syst. Tech. J. **43**, 1826, 1964
3.16 T. Li, Diffraction loss and selection of modes in maser resonators with circular mirrors, Bell Sys. Tech. J. **44**, 917, 1965
3.17 J.C. Heurtley, W. Streifer, Resonator modes: spherical reflectors, J. Opt. Soc. Am. **55**, 1472, 1965
3.18 W. Streifer, Optical resonator modes - rectangular reflectors of spherical curvature, J. Opt. Soc. Am. **55**, 10, 1965
3.19 H. Kogelnik, T. Li, Laser beams and resonators, Appl. Opt. **5**, 1550, 1966
3.20 H. Kogelnik, T.Li, Laser beams and resonators, Proc. IEEE **54**, 1312, 1966
3.21 H. Laig-Hörstebrock, H. Weber, Regelmaessiges und unregelmaessiges Spiken eines Rubinlasers, Z. f. angewandte Physik **23**, 1, 1967
3.22 H.K.V. Lotsch, The FPI-Resonator, Part I,II & III, Optik **28**, 65, 328, 555, 1968
3.23 H.K.V. Lotsch, The FPI-Resonator, Part IV & V, Optik **29**, 130, 622, 1969
3.24 P. Baues, Huygens' principle in inhomogeneous isotropic media and a general integral equation applicable to optical resonators, Opto-Electr. **1**, 37, 1969
3.25 R.L. Sanderson, W. Streifer, Comparison of laser mode calculations, Appl. Opt. **8**, 131, 1969.
3.26 A.E. Siegman, Hermite-gaussian functions of complex arguments as optical beam eigenfunctions, J. Opt. Soc. Am. **63**, 1093, 1973
3.27 N.K. Berger, N.A. Deryugin, Y.N. Lukyanov, Y.E. Studenikin, Open misaligned spherical mirror resonators, Opt. Spectrosc. (USSR) **43**, 176, 1977
3.28 R. Patresi, L. Ronchi, Generalized Gaussian beams in free space, J. Opt. Soc. Am. **67**, 1274, 1977
3.29 A.N. Gromov, S.I. Trashkeev, Opt. Spectr. (USSR) **62**, 369, 1987
3.30 A.E. Siegman, Orthogonality properties of optical resonator eigenmodes, Opt. Commun. **31**, 369, 1979
3.31 S. Nemoto, T. Makimoto, Generalized spot size for a higher order beam mode, J. Opt. Soc. Am. **69**, 578, 1979
3.32 W.H. Carter, Spot-size and divergence for Hermite Gaussian beams of any order, Appl. Opt. **19**, 1027,1980
3.33 R. Hauck, H.P. Kortz, H. Weber, Misalignment sensitivity of optical resonators, Appl. Opt. **19**, 598, 1980
3.34 J.L. Remo, Diffraction losses for symmetrically tilted plane reflectors in open resonators, Appl. Opt. **19**, 774, 1980
3.35 M. Piché, P. Lavigne, F. Martin, P.A. Belanger, Modes of resonators with internal apertures, Appl. Opt. **22**, 1999, 1983
3.36 W.W. Rigrod, Diffraction loss of stable optical resonators with internal limiting aperture, IEEE J. Quantum Electron. **19**, 1679, 1983
3.37 G. Herziger, H. Weber, Equivalent optical resonators, Appl. Opt. **23**, 1450, 1984
3.38 J.P. Taché, Diffraction losses of an asymmetric stable laser resonator using an equivalent resonator, Opt. Commun. **55**, 419, 1985
3.39 O.O. Silichev, Analytical calculation of the lowest mode of a stable resonator, Sov. J. Quantum Electron. **17**, 530, 1987

3.40 P. Ru, L.M. Narducci, J.R. Tredicce, D.K. Bandy, L.A. Lugiato, The Gauss-Laguerre modes of a ring resonator, Opt. Commun. **63**, 310, 1987
3.41 A.N. Gromov, S.I. Trashkeev, Simple loss formulas for symmetric spherical-mirror resonators, Opt. Spectrosc.(USSR) **62**, 369, 1987
3.42 E.A.J. Marcatili, C.G. Someda, Gaussian beams are fundamentally different from free space modes, IEEE J. Quantum Electron. **23**, 164, 1987
3.43 S.D. Brorsen, What is the confocal parameter?, IEEE J. Quantum Electron. **23**, 512, 1988

Chapter 6 Resonators on the Stability Limits

3.44 G.D. Boyd, H. Kogelnik, Generalized confocal resonator theory, Bell Syst. Tech. J. **41**, 1347, 1962
3.45 G.D. Boyd, J.P. Gordon, Confocal multimode resonator for millimeter through optical wavelength masers, Bell Syst. Tech. J, **40**, 489, 1961
3.46 G.D. Boyd, J.P. Gordon, Bell. Syst. Tech. J. **43**, 3009, 1964
3.47 J.C. Heurtley, W. Streifer, Optical resonator modes - circular reflectors of spherical curvature, J. Opt. Soc. Am. **55**, 1472, 1965
3.48 D.E. McCumber, Eigenmodes of a symmetric cylindrical confocal laser resonator and their perturbation by output coupling apertures, Bell Syst. Tech. J. **44**, 333, 1965
3.49 H. Kogelnik, T. Li, Proc. IEEE **54**, 1312, 1966
3.50 G.T. McNice, V.E. Derr, Analysis of the cylindrical confocal laser resonator having a single circular coupling aperture, IEEE J. Quantum Electron. **5**, 569, 1969
3.51 J.M. Moran, Coupling of power from a circular confocal laser with an output aperture, IEEE J. Quantum Electron. **6**, 93, 1970
3.52 H.W. Mueller, W. Rudolph, H. Weber, Optics Commun. **24**, 143, 1976
3.53 N. Hodgson, B. Ozygus, F. Schabert, H. Weber, Degenerated confocal resonator, Appl. Opt. **32**, 3190, 1993

Chapter 7 Unstable Resonators

3.54 A.E. Siegman, Unstable optical resonators for laser applications, Proc. IEEE **53**, 277, 1965
3.55 W.K. Kahn, Unstable optical resonators, Appl. Opt. **5**, 407, 1966
3.56 A.E. Siegman, R. Arrathoon, Modes in unstable optical resonators and lens waveguides, IEEE J Quantum Electron. **3**, 156, 1967
3.57 S.R. Barone, Optical resonators in the unstable region, Appl. Opt. **6**, 861, 1967
3.58 W. Streifer, Unstable optical resonators and waveguides, IEEE J. Quantum Electron. **4**, 156, 1968
3.59 L. Bergstein, Modes of stable and unstable resonators, Appl. Opt. **7**, 495, 1968
3.60 W.F. Krupke, W.R. Sooy, Properties of an unstable confocal resonator CO_2 laser system, IEEE J. Quantum Electr. **QE-5**, 575, 1969
3.61 R.L. Sanderson, W. Streifer, Unstable laser resonator modes, Appl. Opt. **8**, 2129, 1969

3.62 R.L. Sanderson, W. Streifer, Laser resonators with tilted reflectors, Appl. Opt. **8**, 2241, 1969

3.63 A.E. Siegman, H.Y. Miller, Unstable optical resonator loss calculations using Prony method, Appl. Opt. **9**, 2729, 1970

3.64 E.V. Locke, R. Hella, L. Westra, Performance of an unstable oscillator on a 30-kW cw gas dynamic laser, IEEE J. Quantum Electron. **7**, 581, 1971

3.65 A.N. Chester, Mode selectivity and mirror misalignment effects in unstable laser resonators, Appl. Opt. **11**, 2584, 1972

3.66 R.J. Freiberg, P.P: Chenausky, C.J. Buczek, An experimental study of unstable confocal CO_2 lasers, IEEE J. Quantum Electron. **8**, 882, 372, 1972

3.67 Y.A. Anane'ev, Unstable resonators and their applications (review), Sov. J. Quant. Electron. **1**, 565, 1972

3.68 P. Horwitz, Asymptotic theory of unstable resonator modes, J. Opt. Soc. Am. **63**, 1528, 1973

3.69 G.R. Wisner, M.C. Foster, P.R. Blaszuk, Unstable resonators for CO_2 electric discharge convection lasers, Appl. Phys. Lett. **22**, 14, 1973

3.70 R.J. Freiberg, P.P. Chenausky, C.J. Buczek, Unidirectional unstable ring lasers, Appl. Opt. **12**, 1140, 1973

3.71 P.D. Pozzo, R. Polloni, O. Svelto, F. Zaraga, An unstable ring resonator, IEEE J. Quantum Electron. **9**, 1061, 1973

3.72 A.E. Siegman, Unstable optical resonators, Appl. Opt. **13**, 353, 1974

3.73 R.J. Freiberg, P.P. Chenausky, C.J. Buczek, Asymmetric unstable traveling-wave resonators, IEEE J. Quantum Electron. **10**, 279, 1974

3.74 H. Granek, A.J. Morency, Large effective Fresnel number confocal resonator: an experimental study, Appl. Opt. **13**, 368, 1974

3.75 K.I. Zemskóv, A.A. Isaev, M.A. Kazaryan, G.G. Petrash, S. G. Rautian, Use of unstable resonators in achieving the diffraction divergence of the radiation emitted from high-gain pulsed gas laser, Sov. J. Quantum Electron. **4**, 474, 1974

3.76 I.A. Isaev, M.A. Kazaryan, G.G. Petrash, S.G. Rautian, Converging beams in unstable telescopic resonators, Sov. J. Quantum Electron. **4**, 474, 1974

3.77 E.A. Sziklas, A.E. Siegman, Mode calculations in unstable resonators with flowing saturable gain, Appl. Opt. 14, 1874, 1975

3.78 Y.A. Anan'ev, Establishment of oscillations in unstable resonators, Sov. J. Quantum Electron. **5**, 615, 1975

3.79 P. Horwitz, Modes in misaligned unstable resonators, Appl. Opt. **15**, 167, 1976

3.80 A.E. Siegman, A canonical formulation for analyzing multielement unstable resonators, IEEE J. Quantum Electron. **12**, 35, 1976

3.81 C. Santana, L.B. Felsen, Unstable open resonators: two dimensional and three-dimensional losses by a waveguide analysis, Appl Opt. **15**, 470, 1976

3.82 T.F. Ewanitzky, J.M. Craig, Negative-branch unstable Nd:YAG laser, Appl. Opt. **15**, 1465, 1976

3.83 A.A. Isaev, M.A. Kazaryan, G.G. Petrash, S.G. Rautian, A.M. Shalagin, Shaping of the output beam in a pulsed gas laser with an unstable resonator, Sov. J. Quantum Electron. **7**, 746, 1977

3.84 A.A. Isaev, M.A. Kazaryan, G.G. Petrash, S.G. Rautian, A.M. Shalagin, Evolution of Gaussian beams and pulse stimulated emission from lasers with unstable resonators, Sov. J. Quantum Electron. **7**, 746, 1977

3.85 R.L. Herbst, H. Komine, R.L. Byer, A 200 mJ unstable resonator Nd:YAG oscillator, Opt. Commun. **21**,5, 1977

3.86 J.F. Perkins, C. Cason, Effects of small misalignments in empty unstable resonators, Appl. Phys. Lett. **31**, 198, 1977

3.87 C. Santana, L.B. Felsen, Effects of medium and gain inhomogeneities in unstable optical resonators, Appl. Opt. **16**, 1058, 1977

3.88 R.P. Butts, R.V. Avizonis, Asymptotic analysis of unstable laser resonators with circular mirrors, J. Opt. Soc. Am. **68**, 1072, 1978

3.89 C. Santana, L.B. Felsen, Unstable strip resonators with misaligned circular mirrors, Appl. Opt. **17**, 2352, 1978

3.90 T.F. Ewanizky, Ray transfer matrix approach to unstable resonator analysis, Appl. Opt. **18**, 724, 1979

3.91 T.C. Salvi, A.H. Paxton, Calculation of equivalent fresnel numbers for unstable resonators with scraper mirrors, Appl. Opt. **18**, 2098, 1979

3.92 S. Reading, R.C. Sze, C. Tallman, Unstable resonator studies for a 1 joule per pulse KrF avalanche discharge laser, Proceedings of the Society of Photo-Optical Instrumentation Engineers vol. **190**, 311, 1979

3.93 O.L. Bourne, P.E. Dyer, A novel stable-unstable resonator for beam control of rare-gas hailde lasers, Opt. Commun. 31, 193, 1979

3.94 O. Teschke, S.R. Teixeira, Unstable ring resonator N_2 pumped dye laser, Opt. Commun. **32**, 287, 1980

3.95 W.H. Southwell, Virtual source theory of unstable resonator modes, Opt. Lett. **6**, 487, 1981

3.96 W.P.Latham, M.E. Smithers, Diffractive effect of a scraper mirror in an unstable resonator, J. Opt. Soc. Am. **72**, 1321, 1982

3.97 M.E. Smithers, Unstable resonators with aspherical mirrors, J. Opt. Soc. Am. **72**, 1183, 1982

3.98 J.F. Perkins, R.W. Jones, Effects of unstable resonator misalignment in the cusping domain, Appl. Opt. **23**, 358, 1984

3.99 T.R. Ferguson, M.E. Smithers, Optical resonators with nonuniform magnification, J. Opt. Soc. Am. A **1**, 653, 1984

3.100 G.T. Moore, Unstable resonators for the free-electron laser,Proceedings of the Society of Photo-Optical Instrumentation Engineers vol. **453**, 255, 1984

3.101 M.E. Smithers, T.R. Ferguson, Unstable optical resonators with linear magnification, Appl. Opt. **23**, 3718, 1984

3.102 S. Izawa, A. Suda, M. Obara, Experimental observation of unstable resonator mode evolution in a high-power KrF laser, J. Appl. Phys. **58**, 3987, 1985

3.103 W.H. Southwell, Unstable-resonator-mode derivation using virtual-source theory, J. Opt. Soc. Am. A **3**, 1885, 1985

3.104 A.H. Paxton, Unstable resonators with negative fresnel numbers, Opt. Lett. **11**, 76, 1985

3.105 A.H. Paxton, W.P. Latham, Unstable resonator with 90° beam-rotation, Appl. Opt. **25**, 2939, 1986
3.106 J.S. Uppal, J.C. Monga, D.D. Bhawalkar, Performance of a general asymmetric unstable Nd:glass ring laser, Appl. Opt. **25**, 97, 1986
3.107 E. Sklar, The advantages of negative branch unstable resonators for use with free-electron lasers, IEEE J. Quantum Electron. **22**, 1088, 1986
3.108 K.E. Oughstun, Aberration sensitivity of unstable-cavity geometries, J. Opt. Soc. Am. A **3**, 1113, 1986
3.109 K.E. Oughstun, Second-order theory of the aberration sensitivity of a positive-branch, confocal unstable cavity, J. Opt. Soc. Am. A **3**, 1876, 1986
3.110 K.E. Oughstun, Unstable resonator modes, in E. Wolf: Progress in Optics XXIV, Elsevier Science Publishers B.V., 165-387, 1987
3.111 P.A. Apanasevich, V.V. Kvach, V.G. Koptev, V.A. Orlo vich, High power system based on a pulse-periodic YAG:Nd^{3+} laser with an unstable resonator and a two stage amplifier, Sov. J. Quantum Electron. **17**, 160, 1987
3.112 N.G. Vakhitov, M.P. Isaev, V.R. Kushnir, G.A. Sharif, Comparative analysis of single-mode laser resonators, Sov. J. Quantum Electron. **17**, 1037, 1987
3.113 N.D. Cherepenin, Y.Y. Usanov, Simulation of Fresnel diffraction of output beams of unstable resonators, Sov. J. Quantum Electron. **17**, 1404, 1987
3.114 E. Sklar, The tilt sensitivity of a grazing incidence confocal unstable resonator with applications to free-electron lasers, IEEE J. Quantum Electron. **23**, 229, 1987
3.115 A.H. Paxton, Unstable ring resonator with an intracavity prism beam expander, IEEE J. Quantum Electron. **23**, 241, 1987
3.116 R. Hauck, N. Hodgson, H. Weber, Losses and mode structure of unstable resonators with spherical mirrors, J. Appl. Phys. **63**, 628, 1988
3.117 M.A. Malloy, C.M. Clayton, Experimental properties of an unstable resonator with nonuniform magnification using aspheric mirrors, Appl. Opt. **27**, 4407, 1988
3.118 R. Hauck, N. Hodgson, H. Weber, Misalignment sensitivity of unstable resonators with spherical mirrors, J. Mod. Opt. **35**, 165, 1988
3.119 L.N. Litzenberger, M.J. Smith, Direct bandwidth and polarization control of an XeF unstable resonator laser, IEEE J. Quantum Electron. **24**, 2270, 1988
3.120 N. Hodgson, Optical resonators for high power lasers, Proceedings of the Society of Photo-Optical Instrumentation Engineers vol. **1021**, High power solid state lasers, 89, 1988
3.121 J.M. Eggleston, Theory of output beam divergence in pulsed unstable resonators, IEEE J. Quantum Electron. **24**, 1302, 1988
3.122 D. Cooper, L.L. Tankersley, J. Reintjes, Narrow-linewidth unstable resonator, Opt. Lett. **13**, 568, 1988
3.123 K.R. Calahan, C.M. Clayton, A.H. Paxton, Unstable ring resonator with a compact output beam: description and experimental evaluation, Appl. Opt. **27**, 2694, 1988
3.124 J.A. Ruff, A.E. Siegman, Mode characteristics of broad area high power diode lasers in an external stable unstable cavity, Technical Digest CLEO 89, 296, Optical Society of America, Washington DC, 1989

3.125 N. Hodgson, H. Weber, Unstable resonators with excited converging wave, IEEE J. Quantum Electron. **26**, 731, 1990

3.126 N. Hodgson, H. Weber, High power solid state lasers with unstable resonators, Opt. Quantum Electron. **22**, special issue on solid-state lasers, 39, 1990

3.127 K. Yasui, S. Yagi, M. Tanaka, Negative-branch unstable resonator with a phase unifying output coupler for high power Nd:YAG lasers, Appl. Opt. **29**, 1277, 1990

3.128 N. Hodgson, T. Haase, H. Weber, Improved resonator design for rod lasers and slab lasers, Proceedings of the Society of Photo-Optical Instrumentation Engineers vol. **1277**, High power solid state lasers and applications, 70, 1990

3.129 N. Hodgson, T. Haase, Beam parameters, mode structure and diffraction losses of slab lasers with unstable resonators, Opt. Quantum Electron. **24**, 903, 1992

3.130 D. Ehrlichmann, U. Habich, H.D. Plum, P. Loosen, Stable-unstable resonators for annular gain media,Proceedings of the Society of Photo-Optical Instrumentation Engineers vol. **2206**, High power gas and solid state lasers, 54, 1994

3.131 N.G. Vakhimov, Open resonators with mirrors having variable reflection coefficients, Radio Eng. Electron. Phys **10**, 1439, 1965

3.132 H. Zucker, Optical resonators with variable reflectivity mirrors, Bell Syst. Tech. J. **49**, 2349, 1970

3.133 Y.A. Anane'ev, V.E. Sherstobitov, Influence of the edge effects on the properties of unstable resonators, Sov. J. Quantum Electron. **1**, 263, 1971

3.134 G.L. McAllister, W.H. Steier, W.B. Lacina, Improved mode properties of unstable resonators with tapered reflectivity mirrors and shaped apertures, IEEE J. Quantum Electron. **10**, 346, 1974

3.135 A. Yariv, R. Yeh, Confinement and stability in optical resonators employing mirrors with Gaussian reflectivity, Opt. Comm. **13**, 370, 1975

3.136 U. Ganiel, A. Hardy, Eigenmodes of optical resonators with mirrors having Gaussian reflectivity profiles, Appl. Opt. **15**, 2145, 1976

3.137 G. Giuliani, Y.K. Park, R.L. Byer, Radial birefringent element and its application to laser design, Opt. Lett. **5**, 491, 1980

3.138 N. McCarthy, P. Lavigne, Optical Resonators with gaussian reflectivity mirrors, Appl. Opt. **23**, 3845, 1984

3.139 N. McCarthy, P. Lavigne, Large-size gaussian mode in unstable resonators using Gaussian mirrors, Opt. Lett. **10**, 553, 1985

3.140 P. Lavigne, N. McCarthy, J.G. Demers, Design and characterization of complementary gaussian reflectivity mirrors, Appl. Opt. **24**, 2581, 1985

3.141 D.J. Harter, J.C. Walling, Low-magnification unstable resonator used with ruby and alexandrite lasers, Opt. Lett. **11**, 706, 1986

3.142 P. Lavigne, N. McCarthy, A. Parent, D. Pascale, Improved optical resonators for laser radar, Proceedings of the Society of Photo-Optical Instrumentation Engineers vol. **663**, 124, 1986

3.143 A.E. Siegman: Lasers. Mill Valley: University Sciences Books 1986, p. 913-921

3.144 A. Parent, N. McCarthy, P. Lavigne, Effects of hard apertures on mode properties of resonators with gaussian reflectivity mirrors, IEEE J. Quantum Electron. **23**, 222, 1987

3.145 S. deSilvestri, P. Laporta, V. Magni, Laser output coupler based on a radially variable interferometer, J. Opt. Soc. Am. **A4**, 1413, 1987
3.146 K.H. Snell, N. McCarthy, M. Piché, Single transverse mode oscillation from an unstable resonator Nd:YAG- laser using a variable reflectivity mirror, Opt. Commun. **65**, 377, 1988
3.147 P. Lavigne, N. McCarthy, A. Parent, K. Snell, Laser mode control with variable reflectivity mirrors, Canad. J. Phys. **66**, 888, 1988
3.148 S. deSilvestri, P. Laporta, V. Magni, O. Svelto, B. Majocchi, Unstable laser resonators with super-gaussian mirrors, Opt. Lett. **13**, 201, 1988
3.149 C. Zizzo, C. Arnone, C. Cali, S. Sciortine, Fabrication and characterization of tuned Gaussian mirrors for the visible and the near infrared, Opt. Lett. **13**, 342, 1988
3.150 S. deSilvestri, P. Laporta, V. Magni, O. Svelto, Nd:YAG laser with multi-dielectric variable reflectivity output coupler, Opt. Commun. **67**, 229, 1988
3.151 G. Emiliani, A. Piegori, S. DeSilvestri, P. Laporta, V. Magni, Optical coatings with variable reflectance for laser mirrors, Appl. Opt. **28**, 2832, 1989
3.152 A. Parent, P. Lavigne, Increased frequency conversion of Nd:YAG laser radiation with a variable-reflectivity mirror, Opt. Lett. **14**, 399, 1989
3.153 A. Chandonnet, M. Piché, N. McCarthy, Beam narrowing by a saturable absorber in a Nd:YAG laser, Opt. Commun. **75**, 123, 1990
3.154 A. Piegari, A. Tirabassi, G. Emiliani, Thin films for special laser mirrors with radially variable reflectance: production techniques and laser testing, Proceedings of the Society of Photo-Optical Instrumentation Engineers vol. **1125**, 68, 1990
3.155 S. DeSilvestri, V. Magni, O. Svelto, G. Valentini, Laser with super-Gaussian mirrors, IEEE J. Quantum Electron. **26**, 1500, 1990
3.156 V. Magni, S. deSilvestri, O. Svelto, G.L. Valentini, Hundred-watts diffraction limited Nd:YAG laser with super-Gaussian resonator, Technical Digest CLEO 91, Optical Society of America, Washington DC, 1991
3.157 V. Magni, S. deSilvestri, L.-J. Qian, O. Svelto, Rod-imaging supergaussian unstable resonator for high power solid state lasers, Opt. Commun. **94**, 87, 1992
3.158 M.S. Bowers, Diffractive analysis of unstable optical resonators with super-Gaussian mirrors, Opt. Lett. **17**, 1319, 1992
3.159 A. Parent, M. Morin, P. Lavigne, Propagation of super-Gaussian field distributions, Opt. Quantum Electron. **24**, 1071, 1992
3.160 G. Duplain, P.G. Verlyu, J.A. Dobrowolski, A. Waldorf, S. Buissiere, Graded-reflectance mirrors for beam quality control in laser resonators, Appl. Opt. **32**, 1145, 1993
3.161 M. Montecchi, E. Nichelatti, A. Colucci, E. Masetti, Reflectance and phase profile analyser for CO_2 laser graded reflectance mirrors, Technical Abstracts, ENEA-SIOF workshop on laser resonators with graded reflectance mirrors, Florence, Italy, 1993
3.162 A. Piegari, G. Emiliani, Laser mirors with variable reflected intensity and uniform phase shift, Appl. Opt. **32**, 5454, 1993
3.163 N. Hodgson, G. Bostanjoglo, H. Weber, Multirod unstable resonators for high power solid state lasers, Appl. Opt. **32**, 5902, 1993

3.164 N. Hodgson, G. Bostanjoglo, H. Weber, The near-concentric unstable resonator (NCUR) - an improved resonator design for high power solid state lasers, Opt. Commun. 99, 75, 1993
3.165 N. Hodgson, S. Dong, Q. Lü, Performance of a 2.3-kW Nd:YAG slab laser system, Opt. Lett. **18**, 1727, 1993
3.166 G. Bostanjoglo, A. Bernhardt, Variable reflectivity mirrors for Nd:YAG (1.06 μm) and Er:YAG lasers (2.94 μm), Proceedings of the Society of Photo-Optical Instrumentation Engineers vol. **2253**, 1994
3.167 R. Burnham, G. Witt, D. DiBiase, K. Le, W. Koecher, Diode-pumped solid state lasers with kilowatt average power, Proceedings of the Society of Photo-Optical Instrumentation Engineers vol. **2206**, High power gas and solid state lasers, 489, 1994
3.168 G. Bostanjoglo, N. Hodgson, H. Weber, Design of variable reflectivity mirrors and unstable resonators for Nd:YAG lasers with high average power, Pure Appl. Opt. **3**, 497, 1994
3.169 G. Bostanjoglo, N. Hodgson, Unstable multirod Nd:YAG lasers with variable reflectivity mirrors, Proceedings of the Society of Photo-Optical Instrumentation Engineers vol. **2206**, High power gas and solid state lasers, 459, 1994
3.170 A. Dinca, V. Lupei, A.P. Dinca, Design of graded reflectivity mirrors for YAG:Nd lasers, Proceedings of the Society of Photo-Optical Instrumentation Engineers vol. **2206**, High power gas and solid state lasers, 554, 1994
3.171 S. Chandra, T.H. Allik, J.A. Hutchinson, Nonconfocal unstable resonator for solid-state dye lasers based on a gradient-reflectivity mirror, Opt. Lett. **20**(23), 2387, 1995
3.172 K. Du, N. Wu, J. Xu, J. Giesekus, P. Loosen, R. Poprawe, Partially end-pumped Nd:YAG slab laser with a hybrid resonator, Opt. Lett. **23**, 370, 1998
3.173 C. Schnitzler, G. Schmidt, K.M. Du, P. Loosen, R. Poprawe, Improving the brightness of a diode end pumped slab laser by a new pumping scheme, OSA TOPS vol. **50**, Advanced Solid State Lasers, ed.: C.. Marshall, p. 5, Washington: Optical Society of America 2001
3.174 C. Schnitzler, G. Schmidt, M. Hofer, D. Hoffmann, R. Poprawe, A 500W high brightness diode end pumped Nd:YAG slab laser, OSA TOPS vol. **68**, Advanced Solid State Lasers, ed.: C.. Marshall, Washington: Optical Society of America 2001
3.175 C. Schnitzler, M. Höfer, J. Luttmann, D. Hoffmann, R. Poprawe, A cw kW-class diode end pumped Nd:YAG slab laser, Conference on Lasers and Electro-Optics CLEO 2002, Technical Digest, Optical Society of America 2002
3.176 N. Hodgson, V.V. Ter-Mikirtychev, H.J. Hoffman, W. Jordan, Diode-pumped, 220W ultra-thin slab Nd:YAG laser with near-diffraction limited beam quality, OSA TOPS vol. **68**, Advanced Solid State Lasers, eds.: M.E. Fermann, L.R. Marshall, p. 552, Washington: Optical Society of America 2002
3.177 K. Du, D. Li, H. Zhang, P. Shi, X. Wei, R. Diart, Electro-optically Q-switched Nd:YVO_4 slab laser with a high repetition rate and a short pulse width, Opt. Lett. **28**, 87, 2003

3.178 H. Zhang, P. Shi, D. Li, K. Du, Diode-end-pumped, electro-optically Q-switched Nd:YVO_4 slab laser and its second harmonic generation, Appl. Opt. **42**, 1681, 2003
3.179 M. Höfer, C. Schnitzler, G. Schmidt, R. Wester, R. Proprawe, Diode end pumped slab laser with variable reflectivity mirror, Internal paper, Fraunhofer Institut fur Lasertechnik, Aachen, FRG, 2000
3.180 Product brochure: Diode pumped laser INNOSLAB, EdgeWave GmbH, Steinbachstrasse 15, 52074 Aachen, Germany, 2001

Chapter 8 Resonators with Internal Optical Elements

3.181 R.C. Jones, A new calculus for the treatment of optical systems, J. Opt. Soc. Am. **32**, 486, 1942
3.182 H. Kogelnik, Imaging of optical modes-resonators with internal lenses, Bell Sys. Tech. J. **44**, 455, 1965
3.183 V. Evtuhov, A.E. Siegman, A twisted mode technique for obtaining axially uniform energy density in a laser cavity, Appl. Opt. **4**, 142, 1965
3.184 N. Kurauchi, W.K. Kahn, Rays and ray envelopes within stable optical resonators containing focusing media, Appl. Opt. **5**, 1023, 1966
3.185 C.J. Koester, S. Woodstock, E. Snitzer, Laser having a 90° polarization rotator between two rods to compensate for the effects of thermal gradients, US patent No. 3,484,714, December 1969
3.186 W. Koechner, D.K. Rice, Effect of birefringence on the performance of linearly polarized Nd:YAG lasers, IEEE. J. Quantum Electron. **QE-6**, 557, 1970
3.187 W.C. Scott, M. DeWitt, Birefringence compensation and TEM_{00} mode enhancement in a Nd:YAG laser, Appl. Phys. Lett. **18**, 3, 1971
3.188 W. Koechner, D.K. Rice, Birefringence of Nd:YAG laser rods as a function of growth direction, J. Opt. Soc. Am. **61**, 758, 1971
3.189 L. Casperson, Mode stability of lasers and periodic optical systems, IEEE J. Quantum Electron. **10**, 629, 1974
3.190 J. Junghans, M. Keller, H. Weber, Laser Resonators with polarizing elements - eigenstates and eigenvalues of polarization, Appl. Opt. **13**, 2793, 1974
3.191 V.G. Evdokimova, A.A. Mak, L.N. Soms, A.I. Shafarostov, Compensation of birefringence induced in laser systems by passive anisotropic elements, Sov. J. Quantum Electron. **5**, 1040, 1976
3.192 G.C. Dente, Polarization effects in resonators, Appl. Opt. **18**, 2911, 1979
3.193 G. Giuliani, Y.K. Park, R.L. Byer, Radial birefringent element and its application to laser design, Opt. Lett. **5**, 491, 1980
3.194 D.C. Hanna, C.G. Swayers, M.A. Yuratich, Telecopic resonators for large volume TEM_{00} mode operation, Opt. Quantum Electron. **13**, 493, 1981
3.195 D.J. Harter, J.C. Walling, Low-magnification unstable resonator used with ruby and alexandrite lasers, Opt. Lett. **11**, 706, 1986
3.196 J.M. Heritier, J. Henden, R. Aubert, Flashlamp pumped Nd:YAG lasers for scientific applications, Proceedings of the Society of Photo-Optical Instrumentation Engineers

vol. **609**, 167, 1986
3.197 J. Richards, Birefringence compensation in polarization coupled lasers, Appl. Opt. **26**, 2514, 1987
3.198 Q. Lü, S. Dong, H. Weber, Analysis of laser beam degradation caused by a birefringent Nd:YAG rod, Opt. Quantum Electron. **27**, 777, 1995
3.199 Q. Lü, N. Kugler, H. Weber, S. Dong, N. Müller, U. Wittrock, A novel approach for compensation of birefringence in cylindrical Nd:YAG rods, Opt. Quant. Electron. **28**, 57, 1996
3.200 K. Yasui, Efficient and stable operation of a high-brightness cw 500-W Nd:YAG rod laser, Appl. Opt. **35**(15), 2566, 1996
3.201 S. Konno, S. Fujikawa, K. Yasui, 80W cw TEM_{00} 1064nm beam generation by use of a laser-diode-side-pumped Nd:YAG rod laser, Appl. Phys. Lett. **70**, 2650, 1997
3.202 N. Kugler, S. Dong, Q.Lü, H. Weber, Investigation of the misalignment sensitivity of a birefringence-compensated two-rod Nd:YAG laser system, Appl. Opt. **36**, 9359, 1997
3.203 W.A. Clarkson, N.S. Felgate, D.C. Hanna, Simple technique for compensation of stress-induced birefringence in high power diode-end-pumped lasers, Conference on Lasers and Electro-Optics CLEO 98, Technical Digest, 102, Optical Society of America 1998
3.204 N. Hodgson, K. Griswold, W. Jordan, S. Knapp, A.A. Peirce, C.C. Pohalski, E. Cheng, J. Cole, D.R. Dudley, A.B. Petersen, W.L. Nighan jr., High power TEM_{00} mode operation of diode-pumped solid state lasers, Proceedings of the Society of Photo-Optical Instrumentation Engineers vol. **3611**, Laser Resonators II, 119, 1998
3.205 W.A. Clarkson, N.S. Felgate, D.C. Hanna, Simple method for the reducing depolarization loss resulting from thermally induced birefringence in solid-state lasers, Opt. Lett. **24**, 820, 1999
3.206 Y. Hirano, Y. Koyata, S. Yamamoto, K. Kasahara, T. Tajime, 208-W TEM00 operation of a diode-pumped Nd:YAG rod laser, Opt. Lett. **24**, 679, 1999
3.207 R, Hua, S. Wada, H. Tashiro, Principles and limitations of a quarter-wave plate for reducing the depolarization loss from thermally induced birefringence in Nd:YAG lasers, Opt. Commun. **175**, 189, 2000
3.208 S. Konno, T. Kojima, S. Fujikawa, K. Yasui, High-brightness 138W green laser based on an intravacity-frequency doubled diode-side-pumped Q-switched Nd:YAG laser, Opt. Lett. **25**, 105, 2000
3.209 I. Moshe, S. Jackel, Correction of birefringence and thermal lensing in nonreciprocal resonators by use of a dynamic imaging mirror, Appl. Opt. **39**, 4313, 2000
3.210 S. Konno, S. Fujikawa, K. Yasui, Efficient 153W cw TEM_{00} mode 1064nm beam generation by a laser-diode-pump Nd:YAG rod laser, OSA TOPS vol. **50**, Advanced Solid State Lasers, ed: C. Marshall, p. 11. Washington: Optical Society of America, 2001.
3.211 M. Schmid, Th. Graf, Sigle-rod Nd:YAG laser with efficient polarized output, OSA TOPS vol. **50**, Advanced Solid State Laser, ed. C. Marshall, p. 299. Washington: Optical Socienty of America, 2001

3.212 R. Hua, S. Wada, H. Tashiro, Versatile, compact, TEM_{00}-mode resonator for side-pumped single-rod solid-state lasers, Appl. Opt. **40**, 2468, 2001
3.213 M. Ostermeyer, G. Klemz, P. Kubina, R. Menzel, Quasi-continuous-wave birefringence compensated single- and dual-rod Nd:YAG lasers, Appl. Opt. **41**, 7573, 2002
3.214 B. Johnson, Spectra-Physics, Inc., Mountain View, CA, private communication, March 1999.

Part IV Open Resonators with Gain

Overview Literature:

4.1 A.E. Siegman: Lasers. Mill Valley: University Science Books 1986, pp. 457-490, 243-361
4.2 W. Koechner: Solid state laser engineering. Berlin Heidelberg New York London Paris Tokyo: Springer 1999, Chapters 2, 3, 7,8,10
4.3 P.W. Miloni, J.H. Eberly: Lasers. New York, Chichester, Brisbane, Toronto, Singapore: John Wiley & Sons 1988, pp. 283-364, 625-716

Chapters 9 and 10 Active Medium and Output Power

4.4 A.L. Schawlow, C.M. Townes, Infrared and optical masers, Phys. Rev. **112**, 1940, 1958
4.5 W.W. Rigrod, Gain saturation and output power of optical masers, J. Appl. Phys. **34**, 2602, 1963
4.6 L.M. Frantz, J.S. Nodvik, Theory of pulse propagation in a laser amplifier, J. Appl. Phys. **34**, 2346
4.7 A. Yariv, Energy and power considerations in injection and optically pumped lasers, Proc. IEEE, 1723, 1963
4.8 Y.A. Anan'ev, V.E. Sherstobitov, O.A. Shorokov, Calculation of the efficiency of a laser exhibiting large radiation losses, Sov. J. Quantum Electron. **1**, 65, 1971
4.9 Y.A. Anan'ev, L.V. Koval'chuk, V.P. Trusov, V.E. Sherstobitov, Method for calculating the efficiency of lasers with unstable resonators, Sov. J. Quantum Electron. **4**, 659, 1974
4.10 P.W. Miloni, Criteria for the thin-sheet approximation, Appl. Opt. **16**, 2794-2795, 1977
4.11 W. Rigrod, Homogeneously broadened cw lasers with uniform distributed loss, IEEE J. Quantum Electron. **14**, 377, 1978
4.12 D. Eimerl, Optical extraction characteristics of homogeneously broadened cw lasers with nonsaturating lasers, J. Appl. Phys. **51**, 3008, 1980
4.13 G.M. Schindler, Optimum output efficiency of homogeneously broadened lasers with constant loss, IEEE J Quantum Electron. **16**, 546, 1980

4.14 R.S. Galeev, S.I. Krasnov, Approximate method for calculation of output power of unstable telescopic resonators, Sov. J. Quantum Electron. **12**, 802, 1982
4.15 O. Svelto, Principles of Lasers, Plenum Press, 1982
4.16 Lui-teng-Lin, Analysis of energy extraction efficiency of unstable resonators, Final Report, Naval Research Laboratory, Washington DC, 1984
4.17 L.W. Casperson, Power characteristics of high magnification semiconductor lasers, Opt. Quantum Electron. **18**, 155, 1986
4.18 N. Hodgson, Optical resonators for high power lasers, Proceedings of the Society of Photo-Optical Instrumentation Engineers vol. **1021**, High power solid state lasers, 89, 1988
4.19 J. Eicher, N. Hodgson, Output power of slab and rod lasers, Proceedings of the Society of Photo-Optical Instrumentation Engineers vol. **1021**, High power solid state lasers, 147, 1988
4.20 J. Eicher, N. Hodgson, H. Weber, Output power and efficiencies of slab laser systems, J. Appl. Phys. **66**, 4608, 1989
4.21 J.M. Eggleston, L.M. Frantz, H. Injeyan, Derivation of the Frantz-Nodvik equation for zig-zag optical path, slab geometry laser amplifiers, IEEE J. Quantum Electron. **QE-25**, 1855, 1989
4.22 W. Demtröder, Laserspektroskopie, 2nd edition. Berlin, Heidelberg, New York, London, Paris, Tokyo, Hong Kong, Barcelona, Budapest: Springer Verlag 1991, pp. 41-65
4.23 T. Day, E.K. Gustafson, R.L. Byer, Sub-Hertz frequency stabilization of two diode laser-pumped Nd:YAG laser locked to a Fabry Perot interferometer, IEEE J. Quantum Electron. **28**(4), 1106, 1992

Chapter 11 Influence of Gain on Mode Structure and Loss

4.24 H. Statz, C.L. Tang, Problem of mode deformation in optical masers, J. Appl. Phys. **36**, 1816, 1965
4.25 H. Kogelnik, On the propagation of gaussian beams of light through lenslike media including those with a loss or gain variation, Appl. Opt. **4**, 1562, 1965
4.26 T. Li, J.G. Skinner, Oscillating modes in ruby lasers with nonuniform pumping energy distribution, J. Appl. Phys. **36**, 2595, 1965
4.27 A.G. Fox, T. Li, Effects of gain saturation on the oscillating modes of optical masers, IEEE J. Quantum Electron. **2**, 774, 1966
4.28 P.J. Warter, R.U. Martinelli, Some effects of nonuniform pumping on the mode structure of solid state lasers, J. Appl. Phys. **37**, 2103, 1966
4.29 A.G. Fox, T. Li, Effect of gain saturation on the oscillating modes of optical masers, IEEE J. Quantum Electron. **2**, 774, 1966
4.30 L. Casperson, A. Yariv, The Gaussian mode in optical resonators with a radial gain profile, Appl. Phys. Lett. **12**, 355, 1968
4.31 L. Casperson, A. Yariv, Gain and dispersion focusing in an high gain laser, Appl. Opt. **11**, 462, 1972

4.32 G.J. Ernst, W.J. Witteman, Mode structure of active resonators, IEEE J. Quantum Electron. **9**, 911, 1973
4.33 U. Ganiel, Y. Silberberg, Stability of optical resonators with an active medium, Appl. Opt. **14**, 306, 1975
4.34 E.A. Sziklas, A.E. Siegman, Mode calculations in unstable resonators with flowing saturable gain, Appl. Opt. **14**, 1874, 1975
4.35 G.T. Moore, R.J. McCarthy, Theory of modes in a loaded strip confocal unstable resonator, J. Opt. Soc. Am. **67**, 228, 1977
4.36 A. Hardy, Gaussian modes of resonators containing saturable gain medium, Appl. Phys. **19**, 3830, 1980
4.37 R. Hauck, F. Hollinger, H. Weber, Chaotic and Periodic Emission of high power solid state lasers, Opt. Commun. **47**, 2, 1983
4.38 F.D. Feick, J.R. Oldenettel, Gain effects on laser mode formation, J. Opt. Soc. Am. A **1**, 1097, 1984
4.39 N. Hodgson, Optical resonators for high power lasers, Proceedings of the Society of Photo-Optical Instrumentation Engineers vol. **1021**, High power solid state lasers, 89, 1988
4.40 W.P. Risk, Modeling of longitudinally pumped solid-state lasers exhibiting reabsorption losses, J. Opt. Soc. Am. B **5**, 1412, 1988
4.41 J. Frauchiger, P. Albers, H.P. Weber, Modeling of thermal lensing and higher order ring mode oscillation in end-pumped cw Nd:YAG lasers, IEEE J. Quantum Electron 28,1046, 1992
4.42 F. Salin, J. Squier, Gain guiding in solid state lasers, Opt. Lett. **17**, 2319, 1992
4.43 N. Hodgson, K. Griswold, W. Jordan, S. Knapp, A.A. Peirce, C.C. Pohalski, E. Cheng, J. Cole, D.R. Dudley, A.B. Petersen, W.L. Nighan jr., High power TEM_{00} mode operation of diode-pumped solid state lasers, Proceedings of the Society of Photo-Optical Instrumentation Engineers vol. **3611**, Laser Resonators II, 119, 1998
4.44 E.J. Grace, G.H.C. New, P.M.W. French, Simple ABCD matrix treatment for transversely varying saturable gain, Opt. Lett. **26**, 1776, 2001
4.45 O. Denchev, S. Kurtev, P. Petrov, Experimental investigation of saturable gain-guided modes, Appl. Opt. **41**, 1677, 2002

Chapter 12 Q-switched Laser Resonators

4.46 W.G. Wagner, B.A. Lengyel, Evolution of the giant pulse in a laser, J. Appl. Phys. **34**, 2040, 1963
4.47 C.C. Wang, Optical giant pulse from a Q-switched laser, Proc. IEEE **51**, 1767, 1963
4.48 A.A. Vuylsteke, Theory of laser generation switching, J. Appl. Phys. **34**, 1615, 1963
4.49 J.E. Midwinter, The theory of Q-switching applied to slow switching and pulse shaping for solid state lasers, Brit. J. Appl. Phys. **16**, 1125, 1965
4.50 J.E. Geusic, M.L. Hensel, R.G. Smith, A repetitively Q-switched, continuously pumped YAG:Nd laser, Appl. Phys. Lett. **6**, 175, 1965
4.51 R.B. Kay, G.S. Waldmann, Complete solutions to the rate equations describing Q-spoiled and PTM laser operation, J. Appl. Phys. **36**, 1319, 1965

4.52 A. Szabo, R.A. Stein, Theory of laser giant pulsing by a saturable absorber, J. Appl. Phys. **36**, 1562, 1965
4.53 R.G. Smith, M.F.Galvin, Operation of the continuously pumped, repetitively Q-switched YAlG:Nd laser, IEEE J. Quantum Electron. **QE-3**, 406, 1967
4.54 E.J. Woodbury, Five kilohertz repetition-rate pulse YAG:Nd laser, IEEE J. Quantum Electron. **QE-3**, 509, 1967
4.55 D.G. Carlson, Dynamics of a repetitively pump-pulsed Nd:YAG laser, J. Appl. Phys. **39**, 4369, 1968
4.56 I.W. McIntosh, Double etalon Q-switching of a continuously pumped Nd:YAG laser, Appl. Opt. **8**, 1991, 1969
4.57 R.B. Chesler, M.A. Karr, J.E. Geusic, An experimental and theoretical study of high repetition rate Q-switching in Nd:YAG lasers, Proceedings IEEE vol. **58**, 1899, 1970
4.58 J.E. Murray, S.E. Harris, Pulse lengthening via overcoupled internal second-harmonic generation, J. Appl. Phys. **41**, 609, 1970
4.59 R.B. Chesler, M.A. Karr J.E. Geusic, Repetitively Q-switched Nd:YAG-LiIO$_3$ 0.53-μ harmonic source, J. Appl. Phys. **41**, 4125, 1970
4.60 H.T. Powell, G.J. Wolga, Repetitive passive Q-switching of single-frequency lasers, IEEE J. Quantum Electron. **QE-7**, 213, 1971
4.61 G.D. Baldwin, Output power calculations for a continuously pumped Q-switched YAG:Nd^{3+} laser, IEEE J. Quantum Electron. **QE-7**, 220, 1971
4.62 M.K. Chun, J.T. Bischoff, Multipulsing behaviour of electrooptically Q-switched lasers, IEEE J. Quantum Electron. **QE-8**, 715, 1972
4.63 M.K. Chun, E.A. Teppo,Laser resonator: an electrooptically Q-switched Porro prism device, Appl. Opt. **15**, 1942, 1976
4.64 G.A. Badal'yan, V.A. Berenberg, B. A. Ermakov, Optimal operating regime for pulsed garnet lasers with a pulse repetition frequency around 1 kHz, Sov. J. Quantum Electron. **9**, 1134, 1979
4.65 W.E. Schmid, Pulse stretching in a Q-switched Nd:YAG laser, IEEE J. Quantum Electron. **QE-16**, 790, 1980
4.66 D.V. Ivanov et al., Chaos in a solid-state laser with periodically modulated losses, Phys. Lett. **89A**, 229, 1982
4.67 J. Richards, Unpolarized EQ Q-switched laser, Appl. Opt. **22**, 1306, 1983
4.68 D. Dangoisse et al., Laser chaotic attractors in crisis, Phys. Rev. Lett. **57**, 2657, 1986
4.69 D. Biswas et al., Experimental observation of oscillatory instabilities and chaos in a gain-modulated single-mode cw laser, Phys. Rev. **A35**, 456, 1987
4.70 L. Lindgren, L. Lundgren, Spike-mode oscillations in flashlamp pumped Nd lasers, J. Appl. Phys. **61**, 57, 1987
4.71 J.J. Degnan, Theory of the optimally coupled Q-switched laser, IEEE J. Quantum Electron. **QE-25**, 214, 1989
4.72 Y. Shimony, Z. Burshtein, Y. Kalisky, Cr^{4+}:YAG as passive Q-switch and Brewster plate in a pulsed Nd:YAG laser, IEEE J. Quantum Electron. **QE-31**, 1738, 1995
4.73 G. Xiao, M. Bass, A generalized model for passively Q-switched lasers including excited state absorption in the saturable absorber, IEEE J. Quantum Electron. **QE-**

33, 41, 1997

4.74 G. Xiao, M. Bass, Additional experimental confirmation of the predictions of a model to optimize passively Q-switched lasers, IEEE J. Quantum Electron. **QE-34**, 1142, 1998

4.75 A. Agnesi, S. Dell'Aqua, E. Piccinini, G. Reali, G. Piccinno, IEEE J. Quantum Electron. **QE-34**, 1480, 1998

4.76 Y.F.Chen, Y.P.Lan, S.C. Wang, Efficient high-power diode-end-pumped TEM_{00} Nd:YVO_4 laser with a planar cavity, Opt. Lett. **25**, 1016, 2000

4.77 N. Hodgson, D. Dudley, L. Gruber, W. Jordan, H. Hoffman, Diode end-pumped TEM_{00} Nd:YVO_4 laser with output power greater than 12W at 355nm, Conference on Lasers and Electro-Optics CLEO 2001, Technical Digest, p. 389. Washington: Optical Society of America 2001

4.78 R. Iffländer, Solid state lasers for material processing. Berlin, Heidelberg, New York, London, Paris, Tokyo: Springer 2001.

4.79 K.Du, D. Li, H. Zhang, P. Shi, X. Wei, R. Diart, Electro-optically Q-switched Nd:YVO_4 slab laser with a high repetition rate and a short pulse width, Opt. Lett. **28**, 87, 2003

Chapter 13 Resonators with Variable Internal Lenses

4.80 H. Kogelnik, Imaging of optical modes-resonators with internal lenses, Bell Syst. Tech. J. **44**, 455, 1965

4.81 C.M. Stickley, Laser brightness gain and mode control by compensation for thermal distortion, IEEE J. Quantum Electron. **2**, 511, 1966

4.82 L.M. Osterink, L.D. Foster, Thermal effects and transverse mode control in a Nd:YAG laser, Appl. Phys. Lett. **12**, 128, 1968

4.83 W. Koechner, Absorbed pump power, thermal profile and stress in a cw-pumped Nd:YAG laser rod, Appl. Opt. **9**, 1429, 1970

4.84 W. Koechner, Thermal lensing in a Nd:YAG laser rod, Appl. Opt. **9**, 2548, 1970

4.85 J.D. Foster, L.M. Osterink, Thermal effects in a Nd:YAG laser, J. Appl. Phys. **41**, 3656, 1970

4.86 H.K.V. Lotsch, W.C. Davis, The lensing effect of CO_2 Laser Plasma, Appl. Opt. **9**, 2725, 1970

4.87 F.A. Levine, TEM_{00} enhancement in cw Nd:YAG by thermal lensing compensation, IEEE J. Quantum Electron. **7**, 170, 1971

4.88 J. Steffen, J.P. Lörtschner, G. Herziger, Fundamental mode radiation with solid state lasers, IEEE J. Quantum Electron. **8**, 239, 1972

4.89 T.J. Gleason, J.S. Kruger, R.M. Curnutt, Thermally induced focusing in a Nd:YAG laser rod at low input powers, Appl. Opt. **12**, 2942, 1973

4.90 J.P. Lörtschner, J. Steffen, G. Herziger, Dynamic stable resonators: a design procedure, Opt. Quantum Electron. **7**, 505, 1975

4.91 P.H. Sarkies, A stable YAG resonator yielding a beam of very low divergence and high output energy, Opt. Commun. **31**, 189, 1979

4.92 R. Iffländer, H.P. Kortz, H. Weber, Beam divergence and refractive power of directly coated solid-state lasers, Opt. Comm. **29**, 223, 1979

4.93 H.P. Kortz, R. Iffländer, H. Weber, Stability and beam divergence of multimode lasers with internal variable lenses, Appl. Opt. **20**, 4124, 1981

4.94 A.J. Berry, D.C. Hanna, C.G. Swayers, High power single frequency operation of a Q-switched TEM_{00} mode Nd:YAG laser, Opt. Commun. **40**, 54, 1981

4.95 D.C. Hanna, C.G. Swayers, M.A. Yuratich, Telescopic resonators for large volume TEM_{00} mode operation, Opt. Quantum Electron. **13**, 493, 1981

4.96 J.E. Murray, Pulsed gain and thermal lensing in Nd:L_iYF, IEEE J. Quantum Electron. **19**, 288, 1983

4.97 V. Magni, Resonators for solid state lasers with large-volume fundamental mode and high alignment stability, Appl. Opt. **25**, 2039, 1986

4.98 S. DeSilvestri, P. Laporta, V. Magni, Misalignment sensitivity of solid-state laser resonators with thermal lensing, Opt. Commun. **59**, 43, 1986

4.99 R. Iffländer, H. Weber, Focusing of multimode laser beams with variable beam parameters, Optica Acta **33**, 1083, 1086

4.100 H. Weber, R.Iffländer, P. Seiler, High power Nd-lasers for industrial applications, Proceedings of the Society of Photo-Optical Instrumentation Engineers vol. **650**, 92, 1986

4.101 J.S. Uppal, J.C. Monga, F.D. Bhawalkar, Analysis of an unstable confocal ring laser with a thermally induced active medium, Appl. Opt. **25**, 1389, 1986

4.102 V. Magni, Multi-element stable resonators containing a variable lens, J. Opt. Soc. Am. A **4**, 1962, 1987

4.103 D. Metcalf, P. de Giovanni, J. Zachorowski, M. Leduc, Laser resonators containing self-focusing elements, Appl. Opt. **26**, 4508, 1987

4.104 N. Hodgson, H. Weber, High power solid state lasers with unstable resonators, Opt. Quantum Electron. **22**(39), 39, 1990

4.105 V. Magni, S. deSilvestri, L-J. Qian, O. Svelto, Rod-imaging supergaussian unstable resonator for high power solid state lasers, Opt. Commun. **94**, 87, 1992

4.106 N. Hodgson, C. Rahlff, H. Weber, Dependence of the refractive power of Nd:YAG on the intracavity-intensity, Opt. & Laser Techn. **25**(3), 179, 1993

4.107 N. Hodgson, G. Bostanjoglo, The near-concentric unstable resonator (NCUR) - an improved resonator design for high power solid state lasers, Opt. Commun. **99**, 75, 1993

4.108 N. Hodgson, H. Weber, Influence of spherical aberration of the active medium on the performance of Nd:YAG lasers, IEEE J. Quantum Electron. **29**, 2497, 1993

4.109 N. Hodgson, D.J. Golding, D. Eisel, Efficient high power operation at 1.44μm of Nd-doped crystals, Proceedings of the Society of Photo-Optical Instrumentation Engineers vol. **2206**, High power gas and solid state lasers, 426, 1994

4.110 R. Martinez-Herrero, P.M. Mejías, N. Hodgson, H. Weber, Beam-quality changes generated by thermally-induced spherical aberration in laser cavities, IEEE J. Quantum Electron. **31**, 2173, 1995

4.111 K. Yasui, Efficient and stable operation of a high-brightness cw 500-W Nd:YAG rod laser, Appl. Opt. **35**(15), 2566, 1996

4.112 D. Golla, M. Bode, S. Knoke, W. Schöne, A Tünnermann, 62-W cw TEM_{00} Nd:YAG laser side-pumped by fiber-coupled diode lasers, Opt. Lett. **21**, 210, 1996

4.113 D. Golla, M. Bode, S. Knoke, W. Schöne, F. von Alvensleben, A. Tünnermann, High power operation of Nd:YAG rod lasers pumped by fiber-coupled diode lasers, OSA TOPS vol. **1**, Advanced Solid State Lasers, p. 198. Washington: Optical Society of America, 1996.

4.114 S. Konno, S. Fujikawa, K. Yasui, 80W cw TEM_{00} 1064nm beam generation by use of a laser-diode-side-pumped Nd:YAG rod laser, Appl. Phys. Lett. **70**, 2650, 1997

4.115 N. Hodgson, K. Griswold, W. Jordan, S. Knapp, A.A. Peirce, C.C. Pohalski, E. Cheng, J. Cole, D.R. Dudley, A.B. Petersen, W.L. Nighan jr., High power TEM_{00} mode operation of diode-pumped solid state lasers, Proceedings of the Society of Photo-Optical Instrumentation Engineers vol. **3611**, Laser Resonators II, 119, 1998

4.116 Y. Hirano, Y. Koyata, S. Yamamoto, K. Kasahara, T. Tajime, 208-W TEM_{00} operation of a diode-pumped Nd:YAG rod laser, Opt. Lett. **24**, 679, 1999

4.117 M. Ostermeyer, R. Menzel, Single rod efficient Nd:YAG and Nd:YALO-lasers with average output powers of 46 and 47W in diffraction limited beams with $M^2<1.2$ and 100W with $M^2<3.7$, Opt. Commun. **160**, 251, 1999

4.118 R.J. Beach, E.C. Honea, S.B. Sutton, C.M. Bibeau, J.A. Skidmore, M.A. Emanuel, S.A. Payne, P.V. Avizonis, R.S. Monroe, D.G. Harris, High-average-power diode-pumped Yb:YAG lasers, Proceedings of the Society of Photo-Optical Instrumentation Engineers vol. **3889**, 246, 2000

4.119 E.C. Honea, R.J. Beach, S.C. Mitchell, J.A. Skidmore, M.A. Emanuel, S.B. Sutton, S.A. Payne, P.V. Avizonis, R.S. Mo nroe, D.G. Harris, High-power dual-rod Yb:YAG lasers, Opt. Lett. **25**, 805, 2000

4.120 S. Konno, S. Fujikawa, K. Yasui, Efficient 153W cw TEM_{00} mode 1064nm beam generation by a laser-diode-pumped Nd:YAG rod laser, OSA TOPS vol. **50**, Advanced Solid State Lasers, ed: C. Marshall, p. 11. Washington: Optical Society of America, 2001.

4.121 K. Furuta, S. Fujikawa, K. Yasui, Efficient high-beam-quality operation by use of diode-stacks-side-pumped quasi-cw Nd:YAG lasers, OSA TOPS vol. **50**, Advanced Solid State Lasers, ed: C. Marshall, p. 15. Washington: Optical Society of America, 2001.

4.122 D.Dudley, N. Hodgson, H. Hoffman, F. Kopper, Direct diode pumping of Vanadate lasers, Conference on Lasers and Electro-Optics CLEO 2002, Technical Digest, p. 176. Washington: Optical Society of America 2002

4.123 M. Ostermeyer, G. Klemz, P. Kubina, R. Menzel, Quasi-continuous-wave birefringence compensated single- and dual-rod Nd:YAG lasers, Appl. Opt. **41**, 7573, 2002

4.124 C. Kennedy, Helicoid modal analysis of laser oscillators with spherical aberration, Appl. Opt. **41**, 6991, 2002

4.125 A. Hicks, C. Wang, G. Wang., Advances in high power diode-pumped ultraviolet lasers, Proceedings of the Society of Photo-Optical Instrumentation Engineers vol. **5532**, Solid State Lasers, paper 17, 2004

4.126 W.A. Clarkson, D.C. Hanna, Resonators design considerations for efficient operation of solid-state lasers end-pumped by high-power diode-bars, in R. Kossowsky, M. Jelinek, J. Novak, eds., Optical Resonators - Science and Engineering, NATO ASI series, High Technology - Volume 45, 327-361, Kluwer Academic Publishing, 1998

Chapter 14 Resonators with Several Active Media

4.127 V.R. Kushnir, A.N. Nemkow, N.V. Shkunov, Influence of the resonator geometry on the output power of a laser with several active elements, Sov. J. Quantum Electron. **5**, 713, 1975

4.128 K.P. Driedger, R. Iffländer, H. Weber, Multirod resonators for high power solid state lasers with improved beam quality, IEEE. J. Quantum Electron. **24**, 665, 1988

4.129 J.M. Eggleston, Periodic resonators for average-power scaling of stable-resonator solid-state lasers, IEEE J. Quantum Electr. **24**, 1821, 1989

4.130 N. Hodgson, G. Bostanjoglo, H. Weber, Multirod unstable resonators for high power solid state lasers, Appl. Opt. **32**, 5902, 1993

4.131 W. Rath, T. Northemann, Industrial fast axial flow CO_2 laser series from 10 kW to 20 kW, Proceedings of the Society of Photo-Optical Instrumentation Engineers vol. **2206**, High power gas and solid state lasers, 185, 1994

4.132 V.V. Vasil'tsov, V.V. Zelenov, A.A. Lisin, D. Yu. Filimonov, Concept of technological high-power multichannel waveguide diffusion-cooled CO_2 lasers with electrodeless sound-frequency AC discharge, vol. **2206**, High power gas and solid state lasers, 206, 1994

Chapter 15 Misalignment Sensitivity of Output Power

4.133 R. Hauck, H.P. Kortz, H. Weber, Misalignment sensitivity of optical resonators, Appl. Opt. **19**, 598, 1980

4.134 J.L. Remo, Diffraction losses for symmetrically tilted plane reflectors in open resonators, Appl. Opt. **19**, 774, 1980

4.135 N.K. Berger, N.A. Deryugin, Y.N. Lukyanov, Y.E. Studenikin, Open misaligned spherical mirror resonators, Opt. Spectrosc. (USSR) **43**, 176, 1977

4.136 S. DeSilvestri, P. Laporta, V. Magni, Misalignment sensitivity of solid-state laser resonators with thermal lensing, Opt. Commun. **59**, 43, 1986

4.137 V. Magni, Resonators for solid state lasers with large-volume fundamental mode and high alignment stability, Appl. Opt. **25**, 2039, 1986

4.138 R. Hauck, N. Hodgson, H. Weber, Misalignment sensitivity of unstable resonators with spherical mirrors, J. Mod. Opt. **35**, 165, 1988

4.139 N. Hodgson, Optical resonators for high power lasers,Proceedings of the Society of Photo-Optical Instrumentation Engineers vol. **1021**, High power solid state lasers, 89, 1988

4.140 N. Hodgson, H. Weber, Misalignment sensitivity of stable resonators in multimode operation, J. Mod. Opt. **39**(9), 1873, 1992

Chapter 16 Resonators with Internal Nonlinear Elements

4.141 W. Koechner: Solid state laser engineering. Berlin, Heidelberg, New York, London, Paris, Tokyo: Springer 1999, Fifth edition, pp. 582-666

4.142 V.G. Dmitriev, G.G. Gurzadyan, D.N. Nikogosyan: Handbook of nonlinear optical crystals. Berlin, Heidelberg, New York, London, Paris, Tokyo: Springer 1991

4.143 J.A. Armstrong, N. Bloembergen, J. Ducuing, P.S. Pershan, Interactions between light waves in a nonlinear dielectric, Phys. Rev. **127**, 171, 1962

4.144 G.D. Boyd, D.A. Kleinman, Parametric interaction of focused Gaussian light beams, J. Appl. Phys. **39**, 3597, 1968

4.145 R.G. Smith, Theory of intracavity optical second-harmonic generation, IEEE J. Quantum Electron. **QE-6**, 215, 1970

4.146 J.M. Yarborough, J. Flak, C.B. Hitz, Enhancement of optical second harmonic generation by utilizing the dispersion of air, Appl. Phys. Lett.**18**, 70, 1971

4.147 D. Hon, High average power, efficient second harmonic generation. M.L. Stitch (editor): Laser handbook, volume 3. Amsterdam, New York, Oxford: North-Holland Publishing Company 1979

4.148 T. Baer, Large-amplitude fluctuations due to longitudinal mode coupling in diode-pumped intracavity-doubled Nd:YAG lasers, J. Opt. Soc. Am. B 3, 1175, 1986

4.149 D. Eimerl, High average power harmonic generation, IEEE J. Quantum Electron. **QE-23**, 575, 1987

4.150 M. Oak, S. Kubota, Stable intracavity doubling of orthogonal linearly polarized mode in diode-pumped Nd:YAG lasers, Opt. Lett. **13**, 805, 1988

4.151 J.T. Lin, C. Chen , Choosing a Nonlinear Crystal, Lasers & Optronics 11/87, 59, 1987

4.152 P.E. Perkins, T.S. Fahlen, 20-W average power KTP intracavity doubled Nd:YAG laser, J. Opt. Soc. Am. B **4**, 1066, 1987

4.153 G.E. James, E.M. Harrell II, C. Bracikowski, K. Wiesenfeld, R. Roy, Opt. Lett. **15**, 1141, 1990

4.154 C. Bracikowski, R. Roy, Chaos in a multimode solid-state laser system, Chaos **1**(1), 49-64, 1991

4.155 T.Y. Fan, Single-axial mode, intracavity doubled Nd:YAG laser, IEEE J. Quantum Electron. **27**, 2091, 1991

4.156 S.P. Velsko, M. Webb, L. Davis, C. Huang, Phase-matched harmonic generation in lithium triborate (LBO), IEEE J. Quantum Electron. **27**, 2182, 1991

4.157 H. Hemmati, J.R. Lesh, High repetition-rate Q-switched and intracavity doubled diode-pumped Nd:YAG laser, IEEE J. Quantum Electron. **28**, 1018, 1992

4.158 L.R. Marshall, A.D. Hays, Intracavity doubled mode-locked and cw diode-pumped lasers, IEEE J. Quantum Electron. **28**, 1158, 1992

4.159 D.W. Anthon, D.L. Sipes, T.J. Pier, M.R. Ressl, Intracavity doubling of cw diode-pumped Nd:YAG lasers with KTP, IEEE J. Quantum Electron. 28, 1148, 1992

4.160 H. Nagai, M. Kume, I. Ohta, H. Shimizu, M. Kazumura, Low-noise operation of a diode-pumped intracavity-doubled Nd:YAG laser using a Brewster plate, IEEE J. Quantum Electron. **28**, 1164, 1992

4.161 M.V. Ortiz, D.J. Kuizenga, J.H. Fair, High average power second harmonic generation with $KTiOPO_4$, OSA Proceedings on Advanced Solid State Lasers, vol. **18**. Washington: Optical Society of America 1992, pp. 302-304

4.162 K. Yasui, T. Kojina, Continuous wave 20W solid-state green laser, OSA Proceedings on Advanced Solid State Lasers, vol.**20**. Washington: Optical Society of America 1994, pp. 330-331

4.163 P.S. Durkin, S.G. Post, Compact, continuous wave 1.2W green, diode pumped solid-state laser, OSA Proceedings on Advanced Solid State Lasers, vol. **20**. Washington: Optical Society of America 1994, pp. 332-335

4.164 G.A. Rhines, R.A. Schwarz, P.F. Moulton, Diode laser pumped, continuous wave, intracavity-doubled Nd:YLF laser, OSA Proceedings on Advanced Solid State Lasers, vol. **20**. Washington: Optical Society of America 1994, pp. 336-339

4.165 J.-P. Meyn, G. Huber, Intracavity frequency doubling of a continuous-wave, diode-laser-pumped neodymium lanthanum scandium borate laser, Opt. Lett. **19**(18), 1436, 1994

4.166 L.Y. Lin, M. Ota, W. Wiechmann, S. Kubota, Longitudinally diode-pumped continuous-wave 3.5-W green laser, Opt. Lett. **19**(8), 189, 1994

4.167 F. Hanson, P. Poirier, Efficient intracavity frequency doubling of a high-repetition rate diode-pumped Nd:YAG laser, Opt. Lett.**19**(19), 1526, 1994

4.168 K. Suzuki, K. Shimomura, A. Eda, K. Muro, Low-noise diode-pumed intracavity-doubled laser with off-axially cut Nd:YVO_4, Opt. Lett. **19**(20), 1624, 1994

4.169 S. Helmfrid, K. Tatsuno, Stable single-mode operation of intracavity-doubled diode-pumped Nd:YVO_4 lasers, J. Opt. Soc. AM. B **11**, 436, 1994

4.170 F. Falcoz, F. Balembois, P. Georges, A. Brun, All-solid-state continuous-wave tunable blue light source by intracavity doubling of a diode-pumped Cr:LiSAF laser, Opt. Lett. **20**(11), 1274, 1995

4.171 T. Taira, T. Kobayashi, Intracavity frequency doubling and Q switching in diode laser pumped Nd:YVO_4 lasers, Appl. Opt. **34**(21), 4298, 1995

4.172 W. Wiechmann, L.Y. Lin, S. Kubota, Efficient 1 Watt single frequency cw green generation from an intracavity-doubled diode-pumped Nd:YVO_4 laser, OSA Proceedings on Advanced Solid State Lasers, vol. **21**. Washington: Optical Society of America 1995, pp. 96-97

4.173 D.G. Matthews, R.S. Conroy, B.D. Sinclair, N. MacKinnon, Blue microchip laser fabricated from Nd:YAG and $KnbO_3$, Opt. Lett. **21**(3), 198, 1996

4.174 W.L. Nighan Jr., J. Cole, Greater 6W of stable, 532nm, TEM_{00} output at 30% efficiency from an intracavity-doubled diode-pumped multiaxial mode Nd:YVO_4 laser, OSA Proceedings on Advanced Solid State Lasers vol. **22**. Washington: Optical Society of America 1996

4.175 K.I. Martin, W.A. Clarkson, D.C. Hanna, 3W of single-frequency output at 532nm by intracavity frequency doubling of a diode-bar-pumped Nd:YAG ring laser, Opt. Lett. **21**(12), 875, 1996

4.176 M.D. Selker, T.J. Johnson, G. Frangineas, J.L. Nightingale, D.K. Negus, > 8.5 Watts of single frequency 532nm light form a diode pumped intra-cavity doubled ring laser, Conference on Lasers and Electro-Optics CLEO 1996, Technical Digest, paper

CPD21. Washington, D.C. Optical Society of America 1996.
4.177 H, Nagai, M. Kume, A. Yoshikawa, K. Itoh, C. Hamaguchi, Periodic pulse oscillation in an intracavity-doubled Nd:YVO_4 laser, Appl. Opt. **35**, 5392, 1996
4.178 K.I. Martin, W.A. Clarkson, D.C. Hanna, Self-suppression of axial mode hopping by intracavity second-harmonic generation, Opt. Lett. **22**, 375, 1997
4.179 Y.F. Chen, T.M. Huang, C.L. Wang, L.J. Lee, Compact and efficient 3.2-W diode-pumped Nd:YVO_4/KTP green laser, Appl. Opt. **37**, 5727, 1998
4.180 Y.F. Chen, T.M. Huany, C.C. Wang, L. J. Lee, S.C. Wang, Theoretical and experimental studies of single-mode operation in diode pumped Nd:YVO_4/KTP green lasers: influence of KTP length, Opt. Commun. **152**, 319, 1998
4.181 Y.F. Chan, High-power diode-pumped Q-switched intracavity frequency doubled Nd:YVO_4 laser with a sandwich-type resonators, Opt. Lett. **24**, 1032, 1999
4.182 C. Czeranowsky, T. Kellner, G. Huber, All-solid-state continuous wave blue light laser source with1.3W output powerat 473nm, Conference on Lasers and Electro-Optics CLEO 1999, Technical Digest, p.3. Washington, D.C.:Optical Society of America 1999.
4.183 W.A. Clarkson, N.S. Felgate, D.C. Hanna, S.D. Butterworth, Frequency doubling of a diode-bar-pumped Nd:YAG laser at 946nm using non-critically phase-matched lithium triborate at high temperature, Conference on Lasers and Electro-Optics CLEO 1999, Technical Digest, p.4. Washington, D.C.:Optical Society of America 1999.
4.184 S. Konno, T. Kojima, S. Fujikawa, K. Yasui, High-brightness 138-W green laser based on an intracavity-frequency-doubled diode-side-pumped Q-switched Nd:YAG laser, Opt. Lett. **25**, 105, 2000
4.185 G. J. Friel, A.J. Kemp, T.K. Lake, B.D. Sinclair, Compact and efficient Nd:YVO_4 laser that generates a tunable single frequency green output, Appl. Opt. **39**, 4333, 2000
4.186 M. Schmidt, E. Heumann, C. Czeranowsky, G. Huber, Generation of 455nm radiation by intracavity doubling of a Nd:$LiLuF_4$ laser, Conference on Lasers and Electro-Optics CLEO 2001, Technical Digest, p.387. Washington, D.C.:Optical Society of America 2001.
4.187 S. Knoke, G. Hollemann, M. Nickel, H. Voelckel, Frequency doubled Nd:YVO_4 thin disk laser with 30% diode-to-green efficiency, Conference on Lasers and Electro-Optics CLEO 2001, Technical Digest, p.388. Washington, D.C.:Optical Society of America 2001.
4.188 N. Pavel, J. Saikawa, T. Taira, Intra-cavity doubling of a diode-pumped Nd:YAG laser passively Q-switched by Cr^{4+}:YAG saturable absorber, OSA TOPS vol. **50**, Advanced Solid State Lasers, ed.: C. Marshall, p. 246. Washington, D.C. Optical Society of America 2001.
4.189 A. Sennaroglu, Broady tunable continous-wave orange-red source based on intracavity-doubled Cr^{4+}:forsterite laser, Appl. Opt. **41**, 4356, 2002
4.190 C. Czeranowsky, E. Heumann, G. Huber, All-solid-state continous-wave frequency-doubled Nd:YAG-BiBO laser with 2.8W output power at 473nm, Opt. Lett. **28**, 432, 2003

4.191 R.S. Craxton, Theory of high efficiency third harmonic generation of high power Nd-glass laser radiation, Opt. Commun. **34**, 474, 1980

4.192 W. Seka, S.D. Jacobs, J.E. Rizzo, R. Boni, R.S. Craxton, Demonstration of high efficiency third harmonic conversion of high power Nd-glass laser radiation, Opt. Commun. **34**, 469, 1980

4.193 R.S. Craxton, High efficiency frequency tripling schemes for high-power Nd:glass lasers, IEEE J. Quantum Electron. **QE-17**, 1771, 1981

4.194 W.P. Risk, J.-C. Baumert, G.C. Bjorklund, F.M. Schellenberg, W. Lenth, Generation of blue light by intracavity frequency mixing of the laser and pump radiation of a miniature neodymium: yttrium aluminum garnet crystal, Appl. Phys. Lett. **54**, 789, 1989

4.195 B. Wu, N. Chen, C. Chen, D. Deng, Z. Xu, Highly efficient ultraviolet generation at 355nm in LiB_3O_5, Opt. Lett. **14**, 1080, 1989

4.196 J.L. Nightingale, Poynting vector walk-off compensation in type II phasematching, US patent no. 5,136,597, August 1992

4.197 S. Shichijyo, K. Yamada, K. Muro, Efficient intra-cavity sum-frequency generation of 490nm radiation by use of potassium niobate, Opt. Lett. **19**, 1022, 1994

4.198 A.J. Alfrey, Tripling of Diode-pumped Nd:YVO4 at high Q-switch repetition rates, Conference on Lasers and Electro-Optics CLEO 1996, Technical Digest, Postdeadline Paper CPD19, Washington: Optical Society of America 1996

4.199 J.W. Pieterse et al., Q-switched laser system providing UV light, US patent no. 5,835,513, November 1998

4.200 W.M. Grossman, J.D. Henrie, Third-harmonic generation with uncoated brewster-cut dispersive output facet, US patent no. 5,850,407, December 1998

4.201 D. Jacque, J. Capmany, F. Molero, J.G. Sole, Blue-light laser source by sum-frequency mixing in $Nd:YAl_3(BO_3)_4$, Appl. Phys. Lett. **73**, 3659, 1998

4.202 A.J. Alfrey, E. Sinofsky, High efficiency high repetition rate, intra-cavity tripled diode pumped solid state laser, US patent no. 6,002,695, December 1999

4.203 N. Hodgson, D. Dudley, L. Gruber, W. Jordan, H. Hoffman, Diode end-pumped TEM_{00} $Nd:YVO_4$ laser with output power greater than 12W at 355nm, Conference on Lasers and Electro-Optics CLEO 2001, Technical Digest, p. 389. Washington: Optical Society of America 2001

4.204 S. Konno, T. Kojima, S. Fujikawa, K. Yasui, High-average-power, high-repetition, diode-pumped third-harmonic Nd:YAG laser, Conference on Lasers and Electro-Optics CLEO 2001, Technical Digest, p. 391. Washington: Optical Society of America 2001

4.205 A.K. Mohamed, A. Mustellier, J.-P. Faleni, E. Rosenacher, Tunable ultraviolet intracavity tripled Ti:sapphire laser, Opt. Lett. **27**, 1457, 2002

4.206 Y.F. Chen, S.W. Tsai, S.C. Wang, Y.C. Huang, T.C. Lin, B.C. Wong, Efficient generation of continuous-wave yellow light by single-pass sum-frequency mixing of a diode-pumped $Nd:YVO_4$ dual-wavelength laser with periodically poled lithium niobate, Opt. Lett. **27**, 1809, 2002

4.207 D. Jacque, J.J. Romero, Y. Huang, Z.D. Luo, Tunable green laser source based on frequency mixing of pump and laser radiation from a $Nd:YVO_4$ crystal operating at

1342 nm with an intracavity KTP crystal, Appl. Opt. **41**, 6394, 2002
4.208 H. Caetano, T. Matsui, K. Sato, N. Ushiyama, M. Yoshimura, T. Mori, T. Sasaki, Efficient 355-nm generation in CsB_3O_5 crystal, Opt. Lett. **28**, 263, 2003
4.209 F. Xu, J. Liao, C.S. Guo, J.L. He, H.T. Wang, S.N. Zhu, Z. Wang, Y.Y. Zhu, N.B. Ming, Highly efficient direct third-harmonic generation based on control of the electro-optic effect in quasi-periodic optical super lattices, Opt. Lett. **28**, 429, 2003
4.210 Y. Yin, P. Chen, D. Donskoy, High power laser, US patent no. 6,584,134 B2, June 2003

4.211 D. Pohl, A new laser Q-switch-technique using stimulated Brillouin scattering, Phys. Lett. **24A**, 239, 1967
4.212 B. Y. Zel'dovich, V.I. Popovichev, V.V. Ragul'skii, F.S. Faizullov, Connection between the wave fronts of the reflected and the exciting light in stimulated Mandel'shtam-Brillouin scattering, JETP Lett. **15**, 109, 1972
4.213 J. Auyeung, 27 July 2004. Fekete, D.M. Pepper, A. Yariv, A theoretical and experimental investigation of the modes of optical resonators with phase-conjugate mirrors, IEEE J. Quantum Electron. **QE-15**, 1180, 1979
4.214 I.G. Zubarev, A.B. Mironov, S.I. Mikhailov, Single-mode pulse-periodic oscillator-amplifier system with wave-front reversal, Sov. J. Quantum Electron. **10**, 1179, 1980
4.215 V.I. Bedzrodnyi, F.I. Ibragimov, V.I. Kislenko, R.A. Petrenko, V.L. Strizhevskii, E.A. Tikhonov, Mechanism of laser Q switching by intracavity stimulated scattering, Sov. J. Quantum Electron. **10**, 382, 1980
4.216 P.A. Bélanger, A. Hardy, A.E. Siegman, Resonant modes of optical cavities with phase-conjugate mirrors: higher-order modes, Appl. Opt. **19**, 479, 1980
4.217 P.A. Bélanger, A. Hardy, A.E. Siegman, Resonant modes of optical cavities with phase-conjugate mirrors, Appl. Opt. **19**, 602, 1980
4.218 N.N. Ilichev, A.A. Malyutin, P.P Pashinin, Laser with diffraction-limited divergence and Q switching by stimulated Brillouin scattering, Sov. J. Quantum Electron. **12**, 1161, 1982
4.219 D.M. Pepper, Nonlinear optical phase conjugation, Opt. Eng. **21**, 155, 1982
4.220 A. Hardy, P.A. Bélanger, A.E. Siegman, Orthogonality properties of phase conjugate optical resonators, Appl. Opt. **21**, 1122, 1982
4.221 A. Hardy, S. Hochhauser, Higher-order modes of phase-conjugate resonators, Appl. Opt. **21**, 2330, 1982
4.222 P.A. Bélanger, Phase conjugation and optical resonators, Opt. Eng. **21**, 266, 1982
4.223 G. Giuliani, M.M. Denariez-Roberge, P.A. Bélanger, Transverse modes of a stimulated scattering phase-conjugate resonator, Appl. Opt. **21**, 3719, 1982
4.224 M. Slatkine, I.J. Bigio, B.J. Feldman, R.A. Fisher, Efficient phase conjugation of an ultraviolet XeF laser beam by stimulated Brillouin scattering, Opt. Lett. **7**, 108, 1982
4.225 M.C. Gower, KrF laser amplifier with phase-conjugate Brillouin retroreflectors, Opt. Lett. **7**, 423, 1982
4.226 P.A. Bélanger, C. Paré, M. Piché, Modes of phase-conjugate resonators with bounded mirrors, J. Opt. Soc. Am. **73**, 567, 1983

4.227 M. Cronin-Golomb, B. Fischer, G.O. White, A. Yariv, Passive phase conjugate mirror based on a self induced oscillation in an optical ring cavity, Appl. Phys. Lett. **42**, 919, 1983
4.228 V.E. Yashin, V.I. Kryzhanovskii, Apodization and spatial filtering of light beams in stimulated scattering, Opt. Spectrosc. (USSR) 55, 101, 1983
4.229 R.A. Fisher, Optical phase conjugation. New York:: Academic Press 1983
4.230 G.C. Valley, G.J. Dunning, Observation of optical chaos in a phase conjugate resonator, Opt. Lett. **9**, 513, 1984
4.231 E., Armandillo, Regenerative amplification in a XeCl excimer laser with a phase-conjugating Brillouin mirror, Opt. Commun. **49**, 198, 1984
4.232 D.M. Pepper, Nonlinear optical phase conjugation. M.L. Stitch, M. Bass (editors): Laser handbook, volume **4**. Amsterdam, New York, Oxford: North-Holland Publishing Company 1985
4.233 P. Yeh, Theory of phase-conjugate oscillators, J. Opt. Soc. Am. A **2**, 727, 1985
4.234 I.D. Carr, D.C. Hanna, Performance of a Nd:YAG oscillator/amplifier with phase-conjugation via stimulated Brillouin scattering, Appl. Phys. B**36**, 83, 1985
4.235 V.S. Arakelyan, G.E. Rylov, Laser with a wavefront-reversing mirror and Q switching by stimulated Brillouin backscattering, Sov. J. Quantum Electron. **15**, 433, 1985
4.236 K.V. Gratsianov, A.F. Kornev, V.V. Lyubimov, A.A. Mak, V.G. Pankov, A.I. Stepanov, Investigation of an amplifier with a composite active element and a stimulated Brillouin scattering mirror, Sov. J. Quantum Electron. **16**, 1544, 1986
4.237 H. Brusselbach, D.A. Rockwell, Spectral characterization of self-Q-switched phase-conjugate Nd:YAG laser, Conference on Laser and Electro-Optics 1986. Washington: Optical Society of America, 1986, p. 38
4.238 J.C. Valley, A review of stimulated Brillouin scattering excited with a broad-band pump laser, IEEE J. Quantum. Electron. **QE-22**, 704, 1986
4.239 I. Abramowitz, R.A. Fisher, eds., Phase conjugation, beam combining and diagnostics, Proc. SPIE **739**, 1987
4.240 M.J. Damzen, M.H.R. Hutchinson, W.A. Schroeder, Direct measurement of the acoustic decay times of hypersonic waves generated by SBS, IEEE J. Quantum Electron. **QE-23**, 328, 1987
4.241 M. Suggi, O. Sugihara, M. Ando, K. Sasaki, High locking efficiency XeCl ring amplifier injection locked by backward stimulated Brillouin scattering, J. Appl. Phys. **62**, 3480, 1987
4.242 P.W. Miloni, J.H. Eberly: Lasers. New York, Chichester, Brisbane, Toronto, Singapore: John Wiley & Sons 1988, pp. 625-716
4.243 D.A. Rockwell, A review of phase-conjugate solid-state lasers, IEEE J. Quantum Electron. **24**, 1124, 1988
4.244 P.P. Pashinin, E.J. Shkolvsky, Solid-state lasers with stimulated-Brillouin-scattering mirrors operating in the repetitive-pulse mode, J. Opt. Soc. Am. B **5**, 1957, 1988
4.245 M.J. Damzen, G.K.W. Wong, Enhancement of the phase-conjugate stimulated Brillouin scattering process using optical feedback, J. Mod. Opt. **35**, 483-490, 1988

4.246 W.A. Schroeder, M.J. Damzen, M.H.R. Hutchinson, Studies of a single-frequency stimulated-Brillouin-scattering phase-conjugate Nd:YAG laser oscillator, J. Opt. Soc. Am. B **6**, 171, 1989
4.247 M.D. Skeldon, R.W. Boyd, Transverse-mode structure of a phase-conjugate oscillator based on Brillouin enhanced four wave mixing, IEEE J. Quantum Electron. **QE-25**, 588, 1989
4.248 M.R. Osborne, W.A. Schroeder, M.J. Damzen, M.H.R. Hutchinson, Low-divergence operation of a long-pulse excimer laser using a SBS phase-conjugate cavity, Appl. Phys. B **48**, 351, 1989
4.249 G.K.W. Wong, M.J. Damzen, Investigations of optical feedback used to enhance stimulated scattering, IEEE J. Quantum Electron. **26**, 139, 1990
4.250 H. Meng, H.J. Eichler, Nd:YAG laser with phase-conjugating mirror based on stimulated Brillouin scattering in SF_6 gas, Opt. Lett. **16**, 569, 1991
4.251 A. Kummrow, H. Meng, Pressure dependence of stimulated Brillouin scattering in gases, Opt. Commun. **83**, 342, 1991
4.252 H.J. Eichler, R. Menzel, D. Schumann, 10-W single-rod Nd:YAG laser with stimulated Brillouin scattering Q-switching mirror, Appl. Opt. **31**(24), 5038, 1992
4.253 H.J. Eichler, R. König, R. Menzel, J.-J. Pätzold, J. Schwartz, SBS reflection of broad band XeCl excimer laser radiation: comparison of suitable liquids, J. Phys. D.:Appl. Phys. **25**, 1161, 1992
4.254 A.D. Case, P.J. Soan, M.J. Damzen, M.H.R. Hutchinson, Coaxial flash-lamp-pumped dye laser with a stimulated Brillouin scattering reflector, J. Opt. Soc. Am. B **9**, 374, 1992
4.255 R. Menzel, H.J. Eichler, Temporal and spatial reflectivity of focused beams in stimulated Brillouin scattering for phase conjugation, Phys. Rev. A **46**(11), 7139,1992
4.256 A. Kummrov, R. Menzel, D. Schumann, H.J. Eichler, Length tuning effect in SBS-lasers, Int. J. Nonlin. Opt. **2**(2), 261, 1993
4.257 R. Menzel, H.J. Eichler, Computation of stimulated Brioullin scattering (SBS) with focussed pump beams, Int. J. Nonlin. Opt. **2**(2), 255, 1993
4.258 S. Seidel, G. Phillipps, Pulse lengthening by intracavity stimulated Brillouin scattering in a Q-switched, phase-conjugated Nd:YAG laser oscillator, Appl. Opt. **32**(36), 7408, 1993
4.259 H.J. Eichler, A. Haase, R. Menzel, 125 Watt Nd-laser with SBS phase conjugation approaching the diffraction limit,Proceedings of the Society of Photo-Optical Instrumentation Engineers vol. **2206**, High power gas and solid state lasers, 456, 1994
4.260 P.J. Soan, M.J. Damzen, V. Aboites, M.H.R. Hutchinson, Long pulse self-starting stimulated Brillouin scattering resonator, Opt. Lett. **19**, 783, 1994
4.261 B. Barrientos, V. Aboites, M. Damzen, Temporal dynamics of an external-injection dye laser with a stimulated Brillouin scattering reflector, J. Mod. Opt. **26**, 97, 1995
4.262 M. Ostermeyer, A. Heuer, V. Watermann, R. Menzel, Single-rod Nd:laser with phase-conjugating SBS mirror and large-transversal-mode for average output powers above 20 Watts, Conference on Lasers and Electro-Optics CLEO 96,

CThN3, Optical Society of America, 1996
4.263 H.J. Eichler, A. Haase, O. Mehl, High-power Nd:YALO multirod amplifier system with near-diffraction-limited beam quality by SBS mirror, Conference on Lasers and Electro-Optics CLEO 96, CThN1, Optical Society of America, 1996
4.264 S. Seidel, G. Mann, High average power frequency conversion with phase conjugated laser systems, Proceedings of the Society of Photo-Optical Instrumentation Engineers vol. **2788**, High power lasers: gas and solid state lasers, 1996
4.265 H.J. Eichler, A. Haase, O. Mehl, Serial arrangement for high power solid-state laser amplifiers with phase conjugating mirror, Proceedings of the Society of Photo-Optical Instrumentation Engineers vol. **2788**, High power lasers : gas and solid state lasers, 1996
4.266 A. Heuer, J. Schultheiss, M. Ostermeyer, R. Menzel, Reflectivity and temporal response of phase conjugating SBS-mirrors for neodymium lasers, Physics Dept., University of Potsdam, Germany, (to be published)
4.267 M. Ostermeyer, K. Mittler, R. Menzel, Large TEM_{00} mode vol ume in laser resonators with phase conjugating SBS-mirror, Physics Dept., University of Potsdam, Germany, (to be published)
4.268 M. Ostermeyer, A. Heuer, R. Menzel, 27 Watt average output power with 1.2xDL beam quality from a single rod Nd:YAG laser with phase conjugating SBS-mirror, IEEE J. Quant. Electron **34(2)**, 372, 1998
4.269 B. Barrientos, V. Aboites, M. Damzen, Temporal dynamics of a ring dye laser with stimulated Brillouin scattering mirror, Appl. Opt. **35**, 5386, 1996
4.270 H.J. Eichler, A. Haase, R. Menzel, High beam quality by SBS phase conjugation of a single rod Nd-amplifier with up to 140W of average output power, Opt. Quant. Electron. **28**, 261, 1996
4.271 M. Ostermeyer, D. Lorenz, R. Menzel, Transverse , Nd:laser oscillators with phase-conjugating SBS mirrors for high average output powers and fundamental mode operation, Conference on Lasers and Electro-Optics CLEO 97, Technical Digest, 284 Optical Society of America, 1997
4.272 M. Ostermeyer, N. Hodgson, R. Menzel, High power, fundamental mode Nd:YALO laser using a phase-conjugate resonator based on SBS, Conference on Lasers and Electro-Optics CLEO 98, Technical Digest, 355 Optical Society of America, 1998
4.273 H.J. Eichler, A. Haase, O.Mehl, 500-W average output power MOPA system with high beam quality bv phase conjugation, Conference on Lasers and Electro-Optics CLEO 98, Technical Digest, 353, Optical Society of America, 1998
4.274 A. Heuer, N. Hodgson, R. Menzel, Efficient, low-threshold phase conjugation in a tapered optical fiber, Conference on Lasers and Electro-Optics CLEO 98, Technical Digest, 472, Optical Society of America, 1998

Part V Special Resonator Concepts

Overview Literature:

5.1 A.E. Siegman: Lasers. Mill Valley: University Sciences Books 1986, pp. 524-537, 891-922

5.2 D.R. Hall, P.E. Jackson (eds.): The physics and technology of laser resonators. Bristol, Philadelphia: Institute of Physics Publishing 1989

Chapter 17 Prism Resonators

5.3 G. Gould, S.F. Jacobs, P. Rabinowitz, T. Shultz, Crossed roof prism interferometer, Appl. Opt. **1**, 533, 1962

5.4 E.R. Peck, Polarization properties of corner reflectors and cavities, J. Opt. Soc. Am. **52**, 253,1962

5.5 S. Fujiwara, Optical properties of conic surfaces. I. Reflecting cone, J. Opt. Soc. Am. **52**, 287, 1962

5.6 L. Bergstein, W. Kahn, C. Shulman, A total-reflection solid state optical maser resonator, Proc. IRE **50**, 1833, 1962

5.7 M. Bertolotti, Matrix representation of geometrical properties of laser cavities, Il Nuovo Cimento **22**, 1242, 1963

5.8 L. Ronchi, Low-loss modes and resonances in a quasi 90° -roof mirror resonator, Appl. Opt. **12**, 93, 1973

5.9 Y.A. Anan'ev, Unstable prism resonators, Sov. J. Quantum Electron. **3**, 58, 1973

5.10 F. Pascaletti, L. Ronchi, Roof-mirror resonators, J. Opt. Soc. Am. **65**, 649, 1975

5.11 Y.A. Anan'ev, V.I. Kuprenyuk, V.V. Sergeev, V.E. Sherstobitov, Investigation of the properties of an unstable resonator using a dihedral corner reflector in a continuous-flow CO_2-laser, Sov. J. Quantum Electron. **7**, 822, 1977

5.12 M. Rioux, P.A. Bélanger, M. Cornier, High-order mode selection in a conical resonator, Appl. Opt. **16**, 1791, 1977

5.13 G. Zhou, L.W. Casperson, Modes of a laser resonator with a retro-reflecting roof mirror, Appl. Opt. **20**, 3542, 1981

5.14 D.K. Mansfield, K. Jones, A. Semet, L.C. Johnson, Properties of an optically pumped far-infrared rooftop resonator, Appl. Phys. Lett. **40**, 926, 1982

5.15 D.K. Mansfield, K. Jones, L.C. Johnson, A. Semet, Theory of the rooftop resonator: resonant frequencies and eigenpolarizations, Appl. Opt. **22**, 662, 1983

5.16 I.C. Kuo, T. Ko, Laser resonators of a mirror and corner cube reflector, Appl. Opt. **23**, 53, 1984

5.17 V.P. Trusov, B.N. Chumakov, Investigation of the properties of an unstable resonator with a dihedral corner reflector in a gas dynamic laser, Sov. J. Quantum Electon. **16**, 1257, 1986

5.18 Y.Z. Vimik, V.B. Gerasimov, A.L. Sivakov, Y.M. Treivish, Formation of fields in resonators with a composite mirror consisting of inverting elements, Sov. J. Quantum Electron. **17**, 1040, 1987

5.19 J.F. Lee, C.Y. Leung, Beam pointing direction changes in misaligned Porro prism resonator, Appl. Opt. **27**, 2701, 1988
5.20 B. Lü, B. Cai, Y. Liao, S. Xu, Z. Xin, Flowing air-water cooled slab Nd:glass laser, Proceedings of the Society of Photo-Optical Instrumentation Engineers SPIE **1021**, High power solid state lasers, 175, 1988
5.21 N. Hodgson, C. Rahlff, H. Weber, Wei Guang-hui, Properties of prism-resonators, Conference on Lasers and Electro-Optics 1991. Washington: Optical Society of America, 1991, pp. 438-441

Chapter 18 Fourier Transform Resonators

5.22 A.H. Paxton, T.C. Salvi, Unstable optical resonator with self-imaging aperture, Opt. Commun. **26**, 305, 1978
5.23 P.G. Gobbi, G.C. Reali, A novel unstable resonator configuration with a self filtering aperture, Opt. Commun. **52**, 195, 1984
5.24 E. Sklar, Fourier-transform ring laser, J. Opt. Soc. Am. A **1**, 537, 1984
5.25 P.G. Gobbi, S. Morosi, G.C. Reali, A.S. Zarkasi, A novel unstable resonator configuration with a self filtering aperture, Appl. Opt. **24**, 26, 1985
5.26 P.G. Gobbi, G.C. Reali, Mode analysis of a self filtering unstable resonator with a gaussian transmission aperture, Opt. Commun. **57**, 355, 1986
5.27 R. Barbini, A. Chigo, M. Giorgi, K.N. Iyer, A. Palucci, S. Ribezzo, Injection locked single mode high power low divergence TEA CO_2 laser using SFUR configuration, Opt. Commun. **60**, 239, 1986
5.28 V. Boffa, P. di Lazarro, G.P. Gallerano, G. Giordano, T. Hermsen, T. Letardi, C.E Zheng, Self-filtering unstable operation of XeCl excimer laser, IEEE J. Quantum Electr. **23**, 1241, 1987
5.29 L.H. Min, K. Vogler, Confocal positive branch-filtering unstable resonators for Nd:YAG-lasers, Opt. Commun. **74**, 79, 1989
5.30 A. Luches, V Nassisi, M.R. Perrone, Experimental characterization of a self-filtering unstable resonator applied to short pulse XeCl laser, Appl. Opt. **28**, 2047, 1989
5.31 P. di Lazarro, T. Hermsen, C. Zheng, A generalization of the self-filtering unstable resonator, IEEE J. Quantum Electron. **24**, 1543, 1988
5.32 P. di Lazarro, V. Nassisi, R. Perrone, Experimental study of a generalized self-filtering unstable resonator applied to an XeCl laser, IEEE J. Quantum Electron. **24**, 2284, 1988
5.33 A. Luches, V. Nassisi, M.R. Perrone, E. Radiotis, High mode volume self-filtering unstable resonator applied to a short pulse XeCl laser, Opt. Commun. **71**, 97, 1989
5.34 J.W. Chen, V. Nassisi, M.R. Perrone, Narrow-linewidth SFUR applied to a XeCl laser, Opt. Commun. **74**, 211, 1989
5.35 V. Kermene, A. Saviot, M. Vampouille, B. Colombeau, C. Froehly, Flattening of the spatial laser beam profile with low losses and minimal divergence, Opt. Lett. **17**(12), 859, 1992
5.36 N. Hodgson, B. Ozygus, F. Schabert, H. Weber, Asymmetric confocal resonator, Appl. Opt. **32**(18), 3190, 1993

5.37 F. Saviot, E. Mottay, M. Vampouille, B. Colombeau, Optical synthesis of a high-energy uniform and uniphase laser emission, Opt. Lett. **18**(24), 2117, 1993
5.38 V. Kermene, M. Vampouille, C. Froehly, B. Colombeau, Temporal laser beam construction under controlled mode filtering, Opt. Commun. **97**, 319, 1993
5.39 V. Couderc, O. Guy, A. Barthelemy, C. Froehle, F. Louradour, Self-optimized resonator for optical pumping of solid-state lasers, Opt. Lett. **19**(15), 1134, 1994
5.40 B. Colombeau, M. Vampouille, V. Kermene, A. Desfarges, C. Froehly, Spatial shaping of coherent waves inside a confocal resonator, Pure Appl. Opt. 3, 757, 1994

Chapter 19 Hybrid Resonators

5.41 A. Borghese, R. Canevari, V. Donati, L. Garifo, Unstable-stable resonators with toroidal mirrors, Appl. Opt. **15**, 3547, 1981
5.42 P.E. Dyer, D.J. James, Studies of a TEA CO_2 laser with a cylindrical mirror unstable resonator, Opt. Commun. **15**, 20, 1975
5.43 E.A. Phillips, J.P. Reilly, D.P. Northam, Off-axis unstable laser operation, Appl. Opt.**8**, 2241, 1976
5.44 G.W. Sutton, M.M. Weiner, S.A. Mani, Fraunhofer diffraction patterns from uniformly illuminated square output apertures with noncentered square obscurations, Appl. Opt. **15**, 2228, 1976
5.45 M.M. Weiner, Modes of empty off-axis unstable resonators with rectangular mirrors, Appl. Opt. **18**, 1828, 1979
5.46 O.L. Bourn, P.E. Dyer, A novel stable-unstable resonator for beam control of rare-gas halides lasers, Opt. Commun. **31**, 193, 1979
5.47 R. Simon, Laser cavities bounded by crossed cylindrical mirrors, J. Opt. Soc. Am. A**4**, 1953, 1987
5.48 N. Hodgson, T. Haase, H. Weber, Improved resonator design for rod lasers and slab lasers,Proceedings of the Society of Photo-Optical Instrumentation Engineers vol. **1277**, High power solid state lasers and applications, 70, 1990
5.49 K. Kuba, T. Yamamoto, S. Yagi, Improvement of slab-laser beam divergence by using an off-axis unstable-stable resonator, Opt. Lett. **15**, 121, 1990
5.50 N. Hodgson. T. Haase, Beam parameters, mode structure and diffraction losses of slab lasers with unstable resonators, Opt. Quantum Electron. **24**, 903, 1992
5.51 A. E. Siegman, Stable-unstable resonator design for a wide tuning-range free-e lectron laser, IEEE J. Quantum Electron. **28**, 1243, 1992
5.52 E. Snitzer, Cylindrical dielectric waveguide modes, J. Opt. Soc. Am. 51, 491, 1961
5.53 E.A.J. Marctili, R.A. Schmeltzer, Hollow metallic and dielectric waveguides for long distance optical transmission and lasers, Bell. Syst. Tech. J. 43, 1783, 1964
5.54 P.W. Smith, A waveguide gas laser, Appl. Phys. Lett. **19**, 132, 1971
5.55 A.N. Chester, R.L. Abrams, Mode losses in hollow-waveguide lasers, Appl. Phys. Lett. **21**, 576, 1972
5.56 R.L. Abrams, Coupling losses in hollow waveguide laser resonators, IEEE J. Quantum Electron. **QE-8**, 838, 1972

5.57 R.L. Abrams, W.B. Bridges, Characteristics of sealed-off waveguide CO_2 lasers, IEEE J. Quantum Electron. **QE-9**, 940, 1973
5.58 J.J. Degnan, D.R. Hall, Finite-aperture waveguide-laser resonators, IEEE J. Quantum Electron. **QE-9**, 901, 1973
5.59 J.J. Degnan, Waveguide laser mode patterns in the near and far field, Appl. Opt. **12**, 1026, 1973
5.60 R.L. Abrams and A. N. Chester, Resonator theory for hollow waveguide lasers, Appl. Opt. **13**, 2117, 1974
5.61 D. Marcuse, Theory of dielectric optical waveguides, in: Theory of dielectric optical waveguides, Y.H. Pau, ed. New York: Academic Press 1974
5.62 K.D. Laakmann, W.H. Steier, Waveguides: Characteristic modes of hollow rectangular dielectric waveguides, Appl. Opt. **15**, 1334, 1976
5.63 J.J. Degnan, The waveguide laser: a review, Appl. Phys. **11**, 1-33, 1976
5.64 D.M. Henderson, Waveguide lasers with intracavity electrooptic modulators: misalignment loss, Appl. Opt. **15**, 1066, 1976
5.65 H. Krammer, Field configurations and propagation constants of modes in hollow rectangular dielectric waveguides, IEEE J. Quantum Electron **QE-12**, 505, 1976
5.66 S. Avrillier and J. Verdonck, Coupling lasses in laser resonators containing a hollow rectangular dielectric waveguide, J. Appl. Phys. **48**, 4937, 1977
5.67 R.W. Abrams, Waveguide Gas Lasers, in Laser Handbook, vol.3. Amsterdam: North-Holland 1979, pp. 41-68
5.68 P.W. Smith, O.R. Wood II, P.J. Maloney, C.R. Adams, Transversely excited waveguide gas lasers, IEEE J. Quantum Electron. **QE-17**, 1166, 1981
5.69 K.D. Laakmann, Waveguide gas laser with transverse discharge excitation, European Patent Specification EP 0 003 280 B1, August 1981
5.70 J.L. Boulnois, G.O. Agrawal, Mode discrimination and coupling losses in rectangular-waveguide resonators with conventional and phase-conjugate mirrors, J. Opt. Soc. Am. **72**, 853, 1982
5.71 R. Gerlach, D. Wei, N.M. Amer, Coupling efficiency of waveguide laser resonators formed by flat mirrors: analysis and experiment, IEEE J. Quantum Electron. **QE-20**, 948, 1984
5.72 C.A. Hill, D.R. Hall, Coupling loss theory of single-mode waveguide resonators, Appl. Opt. **24**(9), 1283, 1985
5.73 C.A. Hill, D.R. Hall, Waveguide laser resonators with a tilted mirror, IEEE J. Quantum Electron. **QE-22**, 1078, 1986
5.74 D.R. Hall, C.A. Hill, Radiofrequency-Discharge-Excited CO_2 lasers, in Handbook of molecular lasers. New York, Basel: Marcel Dekker, Inc. 1987, pp. 165-198
5.75 S.J. Wilson, R.M. Jenkins, R.W.J. Devereuz, Hollow-core silica waveguides, IEEE J. Quantum Electron. **QE-23**, 52, 1987
5.76 C.A. Hill, Transverse modes of plane-mirror waveguide resonators, IEEE J. Quantum Electron. **24**, 1936, 1988
5.77 C.A. Hill, R.M. Jenkins, R.W.J. Devereuz, Transmission of linearly polarized infrared light through curved hollow dielectric waveguides, IEEE J. Quantum Electron. **24**, 618, 1988

5.78 J. Tulip, Carbon dioxide slab laser, United States Patent 4,719,639, January 1988
5.79 G. Allcock, A gas discharge structure for an RF excited gas laser, European Patent Application EP 0 283 161 A1, September 1988
5.80 H. Opower, High-power ribbon laser, European Patent Application EP 0 305 893 A2, March 1989
5.81 P.E. Jackson, H.J. Baker, D.R. Hall, CO_2 large-area discharge laser using an unstable-waveguide hybrid, Appl. Phys. Lett. **54**, 1950, 1989
5.82 C.J. Shackleton, K. M. Abramski, H.J. Baker, D.R. Hall, Lateral and transverse mode properties of CO_2 slab waveguides, Opt. Commun. **89**, 423, 1989
5.83 H. Gross, Calculation of Waveguide Lasers, Internal Report, Carl Zeiss, Oberkochen, Germany, January 1989
5.84 P.E. Jackson, D.R. Hall, C.A. Hill, Comparisons of waveguide folding geometries in a CO_2 z-fold laser, Appl. Opt. 28, 935, 1989
5.85 C.A. Hill, P.E. Jackson, D.R. Hall, Carbon dioxide waveguide lasers with folds and tilted mirrors, Appl. Opt. **29**(15), 2240, 1990
5.86 J. Nishima, K. Yoshizawa, Development of CO_2 laser excited by 2.45GHz microwave discharge, Proceedings of the Society of Photo-Optical Instrumentation Engineers vol. **1226**, High power gas lasers, 340, 1990
5.87 R. Nowack, H. Opower, U. Schaefer, K. Wessel, T. Hall, H. Krüger, H. Weber, High power CO_2 wavguide laser of the 1kW category, Proceedings of the Society of Photo-Optical Instrumentation Engineers vol. **1276**, CO_2 lasers and applications, 18, 1990
5.88 H. Opower, High power ribbon laser, United States Patent 4,939,738, July 1990
5.89 R. Alvarez-Chust, J.J. Jimenez-Lidon, A simplified and more general model for waveguide laser resonators, IEEE Photon. Tech. Lett. **2**(1), 24, 1990
5.90 K. Janulewicz, P. Szczepanski, Approximate analytical method of gain saturation analyisis of hollow waveguide lasers, Appl. Opt. **30**(27), 3818, 1991
5.91 J. Nishimae, K. Yoshizawa, M .Taki, Gas laser device, United States Patent 5,048,048, September 1991
5.92 R. Nowack, H. Opower, K. Wessel, H. Krüger, W. Haas, N. Menzel, Diffusion-cooled compact CO_2 higher power lasers, Laser und Optoelektronik **3/91**, 68. Stuttgart:AT-Fachverlag, 1991
5.93 M. Khelkal, F. Herlemont, Determination of effective optical constants of infrared CO_2 waveguide laser materials, Appl. Opt. **31**(21), 4175, 1992
5.94 R.M. Jenkins, R.J. Devereux, Effect of field regeneration on the TEM_{00} transmission characteristics of a circular-section waveguide, Appl. Opt. **31**(24), 5086, 1992
5.95 E.F. Yelden, H.J.J. Seguin, C.E. Capjack, S.K. Nikimb, H. Reshef, Multichannel laser resonators - an experimental study, Opt. Quantum Electron. **24**, 889, 1992
5.96 J.L. Hobart, J.M. Yarborough, J. Dallarosa, P. Gardner, RF excited CO_2 slab waveguide laser, United States Patent 5,123,028, June 1992
5.97 W. S. Mefferd, RF excited CO_2 slab waveguide laser, United States Patent 5,131,003, July 1992
5.98 J. Dallarosa, P. Gardner, RF excited CO_2 slab waveguide laser, United States Patent 5,131,004, July 1992

5.99 J.M. Yarborough, J.L. Hobart, J. Dallarosa, RF excited CO_2 slab waveguide laser, United States Patent 5,140,606, August 1992

5.100 W.S. Mefferd, RF excited CO_2 slab waveguide laser, United States Patent 5,155,739, October 1992

5.101 A.D. Colley, H.J. Baker, D.R. Hall, Planar waveguide, 1-kW cw, carbon dioxide laser excited by a single transverse rf discharge, Appl. Phys. Lett. **61**, 136, 1992

5.102 M. Khelkhal, F. Herlemont, RF excitation of a flowing gas CO_2 waveguide laser, IEEE J. Quantum Electron. **QE-29**, 818, 1993

5.103 C.L. Petersen, D. Eisel, J.J. Brzezinski, H. Gross, Mode and wavelength selectivity in slab-geometry CO_2 lasers, Proceedings of the Society of Photo-Optical Instrumentation Engineers vol. **2206**, High power gas and solid state lasers, 91, 1994

5.104 D.Eisel, H. Gross, C.L. Petersen, Slab-waveguide CO_2 laser, United States Patent 5,412,681, May 1995

5.105 A. Lapucci, A. Labate, F. Rossetti, S. Mascalchi, Hybrid stable-unstable resonators for diffusion-cooled CO_2 slab lasers, Appl. Opt. **35**(18), 3185, 1996

5.106 Rofin Sinar: slab waveguide CO_2 laser DC 015, 1996

5.107 K.M. Abramski, E.F. Plinski, P.A. Duda, J.S. Witkowski, Z. Borkowicz, Slab waveguide CO_2 laser with double Talbot filtration, Conference on lasers and Electro-Optics Europe 1996, CThC6, 1996

5.108 E.F. Plinski, K.M. Abramski, J.S. Wittkowski, Optical resonators for slab-waveguide lasers, in R. Kossowsky, M. Jelinek, J. Novak, eds., Optical Resonators - Science and Engineering, NATO ASI series, High Technology-volume 45,267-279, Kluwer Academic Publishing, 1998

5.109 A. Faulstich, H.J. Baker, D.R. Hall, Face-pumping of thin solid state, slab lasers with laser diodes, Opt. Lett. **21**, 594, 1996

5.110 C.T.A. Brown, C.L. Bonner, T.J. Warburton, D.P. Shepherd, A.C. Tropper, D.C. Hanna, H.E. Meissner, Thermally bonded planar waveguide lasers, Appl. Phys. Lett. **71**, 1139, 1997

5.111 A. Rameiz, C. Borel, B. Chambaz, B. Ferrand, D.P. Shepherd, T.J. Warburton, D.C. Hanna, A.C. Tropper, An efficient, diode-pumped, 2 μm Tm:YAG waveguide laser, Opt. Commun, **142**, 239, 1997

5.112 H.J. Baker, A.A. Chesworth, D. Pelaez-Millas, D.R. Hall, Power scaling of thin Nd:glass and Nd:YAG slab laser face-pumped by laser diodes, OSA TOPS vol. **19**, Advanced Solid State Lasers, 407, 1998

5.113 C.L. Bonner, T. Bhutta, D.P. Shepherd, A.C. Tropper, Double-clad structures and proximity coupling for diode-bar-pumped planar waveguide lasers, IEEE. J. Quant. Electron. **36**, 236, 2000

5.114 U. Griebner, J. Huschke, R. Grunwald, H. Schoennagel, G. Erbert, Cladding-pumped Yb:YAG planar waveguide laser, OSA TOPS vol **34**, Advanced Solid State Lasers, 431, 2000

5.115 R. Borghi, Analytical solution for the eigenmodes of closed waveguide resonators with small curvature mirrors, IEEE. J. Quant. Electron. **36**, 363, 2000

5.116 H.E. Meissner, O.R. Meissner, Laser-pumped compound waveguide lasers and amplifiers, United States Patent 6,160,824, December 2000

5.117 Q. Cao, H.J. Baker, D.R. Hall, Transverse mode propagation and gain coefficients in a planar waveguide CO_2 laser amplifier, IEEE J. Quant. Electron. **37**, 376, 2001
5.118 J.R. Lee, G.J. Friel, H.J. Baker, G.J. Hilton, D.R. Hall, A Nd:YAG planar waveguide laser operating at 121W output power with face-pumping by diode bars and its use as a power amplifier, OSA TOPS vol. **50**, Advanced Solid State Lasers, 36, Optical Society of America 2001
5.119 H.J. Baker, A.A. Chesworth, D.Pelaez Millas, D.R. Hall, A planar waveguide Nd:YAG laser with a hybrid waveguide-unstable resonator, Opt. Commun. **191**, 125, 2001
5.120 R.J. Beach, S.C. Mitchell, H.E. Meissner, O.R. Meissner, W.F. Krupke, J.M. McMahon, W.J. Bennett, Continuous-wave and passively Q-switched cladding-pumped planar waveguide lasers, Opt. Lett. **26**, 881, 2001
5.121a J.I. Mackenzie, C. Li, D.P. Shepherd, H.E. Meissner, End-pumped double-clad waveguide laser, Conference on Lasers and Electro-Optics CLEO 2001, Technical Digest, 582, Optical Society of America, 2001
5.121b J.I. Mackenzie, S.C. Mitchell, R.J. Beach, H.E. Meissner, D.P. Shephard, A 15W diode-side-pumped Tm:YAG waveguide laser at 2 um, Electron. Lett. **37**, 898, 2001.
5.122 J.I. Mackenzie, C. Li, D.P. Shepherd, H.E. Meissner, S.C. Mitchell, Longitudinally diode-pumped Nd:YAG double-clad planar waveguide laser, Opt. Lett. **26**, 698, 2001
5.123a H.J. Baker, J.R. Lee, D.R. Hall, Planar waveguide solid-state lasers, Conference on Lasers and Electro-Optics CLEO 2002, Technical Digest, 262, Optical Society of America, 2002
5.123b I.T. McKinnie, J.E. Koroshetz, W.S. Pelouch, D.D. Smith, J.R. Unternahrer, S.W. Henderson, Self-imaging waveguide Nd:YAG laser with 58% slope efficiency, Conference on Lasers and Electro-Optics CLEO 2002, Technical Digest, 262, Optical Society of America, 2002
5.124 J.I. Mackenzie, D.P. Shepherd, End-pumped, passively Q-switched Yb:YAG double-clad waveguide laser, Opt. Lett. **27**, 2161, 2002
5.125a J.I. Mackenzie, C. Li, D.P. Shepherd, R.J. Beach, S.C. Mitchell, Modeling of high-power continuous-wave Tm:YAG side-pumped double-clad waveguide lasers, IEEE J. Quant. Electron. **38**, 222, 2002
5.125b N. Hodgson, V.V. Ter-Mikirtychev, H.J. Hoffman, W. Jordan, Diode-pumped, 220W ultra-thin slab Nd:YAG laser with near-diffraction limited beam quality, OSA TOPS vol. **68**, Advanced Solid State Lasers, eds.: M.E. Fermann, L.R. Marshall, p. 552, Washington: Optical Society of America 2002
5.125c K. Sueda, Y. Tsujioka, H. Takahashi, S. Kawato, T. Kobayashi, High-power and high-efficiency LD pumped Yb:YAG micro-thickness slab laser. Advanced Solid State Photonics 2004, paper MB15, Santa Fe 2004, Optical Society of America 2004

Chapter 20 Resonators for Annular Gain Media

5.126 J.R. Pierce, Theory and Design of Electron Beams. New York: Van Nostrand: 1954, pp. 194-197

5.127 D. Herriot, H. Kogelnik, R. Kompfner, Off-axis paths in spherical mirror interferometers, Appl. Opt. **3**, 523, 1964

5.128 H. Kogelnik, T. Bridges, A nonresonant multipass CO_2 laser amplifier, IEEE J. Quantum Electron. **QE-3**, 95, 1967

5.129 R.G. Greenler, Multiple reflection of light between nonparallel reflecting surfaces, J. Opt. Soc. Am. **57**, 1062, 1973

5.130 D. Milam, H. Schlossberg, Emission characteristics of a tube-shaped laser oscillator, J. Appl. Phys. **44**, 2297, 1973

5.131 P. Burlamacchi, R. Partesi, High-efficiency coaxial waveguide dyelaser with internal excitation, Appl. Phys. Lett. **23**, 475, 1973

5.132 L.W. Casperson, Cylindrical laser resonators, J. Opt. Soc. Am. **63**, 25, 1973

5.133 L.W. Casperson, M. Shabbir Shekhani, Mode properties of annular gain lasers, Appl.Opt. **14**, 2653, 1975

5.134 R.A. Chodzko, S.B. Mason, E.F. Cross, Annular converging wave cavity, Appl. Opt. **9**, 2137, 1976

5.135 R.J. Freiberg, D.W. Fradin, P.P. Chenausky, Split-mode unstable resonator, Appl. Opt. **16**, 1192, 1977

5.136 A.H. Paxton, J.H. Erkkila, Annular converging wave resonator: new insights, Opt. Lett. **1**, 166-168, 1977

5.137 P.B. Mumola, H.J. Robertson, G.N. Steinberg, J.L. Kreuzer, A.W. McCullough, Unstable resonators for annular gain volume lasers, Appl. Opt. **17**, 936, 1978

5.138 J.W. Ogland, Mirror systems for uniform beam transformation in high-power annular lasers, Appl. Opt. **17**, 2917, 1978

5.139 D. Fink, Polarization effects of axicons, Appl. Opt. **18**, 581, 1979

5.140 W.R. Trutna, R.L. Byer, Multiple-pass Raman gain cell, Appl. Opt. **19**, 301, 1980

5.141 R.A. Chodzko, S.B. Mason, E.B. Turner, W.W. Plummer,Jr., Annular (HSURIA) resonators: some experimental studies including polarization effects, Appl. Opt. **19**, 778, 1980

5.142 W.P. Latham, Jr., Polarization effects in a half-symmetric unstable resonator with a coated rear cone, Appl. Opt. **19**, 1222, 1980

5.143 J.K. Guha, J.L. Martin, R.A. Mickish, E.E. Pape, Performance of a coated cone in an annular resonator, Appl. Opt. **20**, 3089, 1981

5.144 J.K. Guha, D. Kohler, R. Mickish, J. Martin, E. Pape, P. Briggs, C. Greninger, Performance of an annular resonator with a polarizer in the annular leg, Appl. Opt. **20**, 4135, 1981

5.145 L.W. Casperson, P.M. Schienert, Multipass resonators for annular gain lasers, Opt. Quantum Electron. **13**, 193, 1981

5.146 T.R. Ferguson, M.E. Smithers, Toric unstable resonators, Appl. Opt. **33**, 2122, 1984

5.147 A.H. Paxton, Propagation of high-order azimuthal Fourier terms of the amplitude distribution of a light beam: a useful feature, J., Opt. Soc. Am. A **1**,319, 1984

5.148 S. Marchetti, Multipass systems with mirrors of different radii, Optics and Laser Technology **18**(5), Oct. 1986
5.149 J.G. Xin, D.R. Hall, Multipass coaxial radiofrequency discharge CO_2 laser, Opt. Commun. **58**, 420, 1986
5.150 V.A. Seguin, H.J.J. Seguin, C.E. Capjack, S.K. Nikumb, H. Reshef, Multiple pass unstable resonator for an annular gain CO_2 laser, Appl. Opt. **25**, 3825, 1986
5.151 H. Schülke, G. Herziger, R. Wester, Multipass resonators for laser systems, Proceedings of the Society of Photo-Optical Instrumentation Engineers vol. **801**, High Power Lasers: Sources, laser-material interactions, high excitations, and fast dynamics, 45, 1987
5.152 J.G. Xin, D.R. Hall, Compact, multipass, single transverse mode CO_2 laser, Appl. Phys. Lett. **51**, 469, 1987
5.153 Y. Takada, H. Saito, T. Fujioka, Eigenmode of an annular resonator, IEEE J. Quantum Electron. **24**, 11, 1988

5.154 J.G. Xin, A. Duncan, D.R. Hall, Analysis of hyperboloidal ray envelopes in Herriot cells and their use in laser resonators, Appl. Opt. **28**, 4576, 1989
5.155 J.K. Jabczyński, A diffraction-free resonator, Opt. Commun. **77**, 292, 1990
5.156 A. Duncan, J.G. Xin, D.R. Hall, Herriot cell for large-area gas discharge lasers, Proceedings of the Society of Photo-Optical Instrumentation Engineers vol. **1224**, 312, 1990
5.157 U. Wittrock, H. Weber, B., Eppich, Inside-pumped Nd:YAG tube laser, Opt. Lett. **16**, 1092, 1991
5.158 N. Hodgson, Q. Lü, S. Dong, B. Eppich, U. Wittrock, High power solid state lasers in rod-, slab-, and tube geometry, Laser und Optoelektronik **23(3)**, 82, 1991
5.159 U. Habich, A. Bauer, P. Loosen, H.-D. Plum, Resonators for coaxial slow-flow CO_2 lasers, Proceedings of the Society of Photo-Optical Instrumentation Engineers vol. **1397**, Eighth International Symposium on Gas Flow and Chemical Lasers, 383, 1991
5.160 M. Morin, P.-A. Bélanger, Diffractive analysis of annular resonators, Appl. Opt. **31**, 1942, 1992
5.161 U. Wittrock, High power rod slab and tube lasers, NATO ASI, International School of Quantum Electronics, Elba, Italy, September 1992.
5.162 U. Wittrock, B. Eppich, O. Holst, Internally-pumped Nd:YAG tube laser with 10% efficiency and 1.8kW output power, Conference on Laser and Electo-Optics 1993, CWI 7. Washington: Optical Society of America, 1993
5.163 D. Ehrlichmann, U. Habich, H.-D. Plum, Azimuthal mode discrimination of annular resonators, Appl. Opt. **32**(33), 6582, 1993
5.164 D. Ehrlichmann, U. Habich, H.-D. Plum, Diffusion-cooled CO_2 laser with coaxial high frequency excitation and internal axicon, J. Phys. D: Appl. Phys. **26**, 183, 1993
5.165 U. Habich, H.-D. Plum, D. Ehrlichmann, P. Loosen, Tilted annular resonator, Proceedings of the Society of Photo-Optical Instrumentation Engineers vol. **2095**, 46, 1993
5.166 D. Ehrlichmann, U. Habich, H.-D. Plum, High-power CO_2 laser with coaxial waveguide and diffusion cooling, IEEE J. Quantum Electron. **29**(7), 2211, 1993

5.167 D. Ehrlichmann, U. Habich, H.-D. Plum, P. Loosen, Stable-unstable resonators for annular gain media, Proceedings of the Society of Photo-Optical Instrumentation Engineers vol. **2206**, High power gas and solid state lasers, 54, 1994
5.168 A. Lapucci, F. Rossetti, P. Burlamacchi, Beam properties of an R.F.-discharge annular CO_2 laser, Opt. Commun. **111**, 290, 1994
5.169 D. Ehrlichmann, U. Habich, H.-D. Plum, Ring resonator for lasers with annular gain media, Appl. Opt. **33**(30), 6919, 1994
5.170 D. Ehrlichmann, U. Habich, H.-D. Plum, P. Loosen, G. Herziger, Azimuthally unstable resonators for high-power CO_2 lasers with annular gain media, IEEE J. Quantum Electron. **30**(6), 1441, 1994
5.171 R.C. Wade, Optical resonators with annular gain media, in R. Kossowsky,M. Jelinek, J. Novak, eds., Optical Resonators - Science and Engineering, NATO ASI series, High Technology-volume 45, 211-224, Kluwer Academic Publishing, 1998
5.172 N. Hodgson, Beam Quality and Efficiency of Annular Gain Lasers, in R. Kossowsky,M. Jelinek, J. Novak, eds., Optical Resonators - Science and Engineering, NATO ASI series, High Technology-volume 45, 163-184, Kluwer Academic Publishing, 1998

Chapter 21 Ring Resonators

5.173 C.L. Tang, H. Statz, G. deMars, Jr., Spectral output and spiking behavior of solid-state lasers, J. Appl. Phys. **34**, 2289, 1963
5.174 M. Hercher, M. Young, C.B. Smoyer, Traveling-wave ruby laser with a passive optical oscillator, J. Appl. Phys. **36**, 3351, 1965
5.175 Yu. A. Anan'ev, N.A. Sventsitskaya, V.E. Sherstobitov, Properties of a laser with an unstable resonator, Sov. Phys. JETP **28**, 69, 1969
5.176 F.P. Schäfer, H. Müller, Tunable dye ring-laser, Opt. Commun. **2**, 407, 1971
5.177 A.R. Clobes, M.J. Brienza, Single-frequency traveling-wave Nd:YAG lasers, Appl. Phys. Lett. **21**, 265, 1972
5.178 H.W. Kogelnik, E.P. Ippen, A. Dienes, C.V. Shank, Astigmatically compensated cavities for cw dye lasers, IEEE J. Quantum Electron. **QE-8**, 373, 1972
5.179 J.M. Green, J.P. Hohimer, F.K. Tittel, Traveling-wave operation of a tunable cw dye laser, Opt. Commun. **7**, 349, 1973
5.180 P. dal Pozzo, R. Polloni, O. Svelto, F. Zaraga, An unstable ring resonator, IEEE J. Quantum Electron. **9**, 1061, 1973
5.181 R.J. Freiberg, P.P. Chenausky, C.J. Buczek, Unidirectional unstable ring lasers, Appl. Opt. **12**, 1140, 1973
5.182 R.J. Freiberg, P.P. Chenausky, C.J. Buczek, Asymmetric unstable traveling-wave resonators, IEEE J. Quantum Electron. **QE-10**, 279, 1974
5.183 P.del Pozzo, R. Pollini, O. Svelto, F. Zaraga, A doubly-confocal unstable ring resonator, Opt. Commun. **11**, 115, 1974
5.184 G. Marowsky, K. Kaufman, Influence of spatial hole burning on the output power of a cw dye ring laser, IEEE J. Quantum Electron. **QE-12**, 207, 1976
5.185 A.H. Paxton, T.C. Salvi, Unstable optical resonator with self-imaging aperture, Opt. Commun. **26**, 305, 1978

5.186 S.M. Jarrett, J.F. Young, High-efficiency single-frequency cw ring dye laser, Opt. Lett. **4**, 176, 1979
5.187 E.F. Ichshenko, E.F. Reshetin, Sensitivity to misalignment of an optical ring resonator with a focusing element, Opt. and Spectrosc. **46**, 202, 1979
5.188 Yu. A. Anan'ev, V.I. Kuprenyuk, V.E. Sherstobitov, Properties of unstable resonators with field rotation. I. Theoretical principles, Sov. J. Quantum Electron. **9**, 1105, 1979
5.189 T.F. Johnston, Jr., W. Proffitt, Design and performance of a broad-band optical diode to enforce one-direction traveling-wave operation of a ring laser, IEEE J. Quantum Electron. **QE-16**, 483, 1980
5.190 F.R. Faxvog, Modes of unidirectional ring laser, Opt. Lett. **5**, 285, 1980
5.191 O. Teschke, S. R. Teixeira, Unstable ring resonator nitrogen pumped dye laser, Opt. Commun. **32**, 287, 1980
5.192 K.E. Oughstun, P.A. Slaymaker, K.A. Bush, Intracavity spatial filtering in unstable ring resonator geometries, Part I - Passive cavity mode theory, IEEE J. Quantum Electron. **19**, 1558, 1983
5.193 E. Sklar, Fourier-transform ring laser, J. Opt. Soc. Am. A **1**, 537, 1984
5.194 E.M. Wright, D.P. O'Brein, W.J. Firth, Reciprocity and orthogonality relations for ring resonators, IEEE J. Quantum Electron. **20**, 1307, 1984
5.195 H.A. Haus, H. Statz, I.W. Smith, Frequency locking of modes in a ring laser, IEEE J. Quantum Electron. **QE-21**, 78, 1985
5.196 T.J. Kane, R.L. Byer, Monolithic, unidirectional single-mode Nd:YAG ring laser, Opt. Lett. **10**, 65, 1985
5.197 G.L. Lippi, J.R. Tr edicce, N.B. Abraham, F.T. Arecchi, Deterministic mode alternation, gaint pulses and chaos in a bidirectional CO_2 ring laser, Opt. Commun. **53**, 129, 1985
5.198 V.N. Smirnov, G.A. Strokovskii, Transverse mode formation in a ring laser with a 1-D diaphragm, Opt. Spectrosc. (USSR) **60**, 652, 1986
5.199 J.S. Uppal, J.C. Monga, D.D. Bhawalkar, Performance of a general asymmetric unstable Nd:glass ring laser, Appl. Opt. **25**, 97, 1986
5.200 J.S. Uppal, J. C. Monga, D.D. Bhalwalkar, Analysis of an unstable confocal ring laser with a thermally induced active medium, Appl. Opt. **25**(9), 389, 1986
5.201 P. Ru, L.M. Narducci, J.R. Tredicce, D.K. Bandy, L.A. Lugiato, The Gauss-Laguerre modes of a ring resonator, Opt. Commun. **63**(5), 310, 1987
5.202 T.J. Kane, A.C. Nilsson, R.L. Byer, Frequency stability and offset locking of a laser-diode-pumped Nd:YAG monolithic nonplanar ring oscillatior, Opt. Lett. **12**, 175, 1987
5.203 A.H. Paxton, Unstable ring resonator with an intracavity prism expander, IEEE J. Quantum Electron. **QE-23**, 241, 1987
5.204 T.J. Kane, E.A. P. Cheng, Fast frequency tuning and phase locking of a diode pumped Nd:YAG ring laser, Opt. Lett. **13**, 970, 1988
5.205 K.R. Calahan, C.M. Clayton, A.H. Paxton, Unstable ring resonator with a compact output beam: description and experimental evaluation, Appl. Opt. **27**(13), 2694, 1988

5.206 I. Schütz, S. Wiegand, R. Wallenstein, Diode-pumped solid state lasers, Laser und Optoelektronik **20**(3), 39-45, 1988
5.207 A.C. Nilsson, T. Day, A.D. Farinas, E.K. Gustafson, R.L. Byer, Narrow linewidth operation of diode-laser-pumped nonplanar ring oscillators, Springer Verlag Topics in Physics: Proceedings of the Fourth Symposium on Frequency and Metrology, Ancona, Italy, 1988
5.208 T. Day, A.C. Nilsson, M.M. Fejer, A.D. Farinas, E.K. Gustafson, C.D. Nabors, R.L. Byer, 30-Hz linewidth, diode-laser-pumped, Nd:GGG nonplanar ring oscillators by active frequency stabilization, Electron. Lett. **25**, 810, 1989
5.209 A.C. Nilsson, E.K. Gustafson, R.L. Byer, Eigenpolarization theory of monolithic nonplanar ring oscillators, J. Quantum Electron. **25**, 767, 1989
5.210 P. Fritschel, A. Jeffries, T. Kane, Frequency fluctuations of a diode-pumped Nd:YAG ring laser, Opt. Lett. **14**, 993, 1989
5.211 T. A. King, Ring laser resonators, in D.R. Hall, P.E. Jackson (eds.): Physics and technology of laser resonators. London: Adam Hilger 1989, pp. 62-79
5.212 F. Bretenaker, A. le Foch, J.P. Taché, Theoretical and experimental study of elliptical Gaussian-mode size dynamics in ring lasers, Phys. Rev. A **41**, 3792, 1990
5.213 T. Day, E.K. Gustafson, R.L. Byer, Active frequency stabilization of a 1.062-μm, Nd:GGG, diode-laser-pumped non-planar ring oscillator to less than 3 Hz of relative linewidth, Opt. Lett. **15**, 221, 1990
5.214 W.R. Trutna Jr., D.K. Donald, Two-piece, piezo-electrically tuned, single-mode Nd:YAG ring laser, Opt. Lett. **15**, 369, 1990
5.215 E.A.P. Cheng, T.J. Kane, High-power single-mode diode-pumped Nd:YAG laser using a monolithic nonplanar ring resonator, Opt. Lett. **16**, 478, 1991
5.216 T. Day, E.K. Gustafson, R.L. Byer, Sub-hertz relative frequency stabilization of two diode-laser-pumped Nd:YAG lasers locked to a Fabry-Perot interferometer, IEEE J. Quantum Electron. **28**, 1106, 1992
5.217 J. Harrison, A. Finch, J.H. Flint, P.F. Moulton, Broad-band rapid tuning of a single-frequency diode-pumped neodymium laser, IEEE J. Quantum Electron. **28**(4), 1123, 1992
5.218 K.I. Martin, W.A. Clarkson, D.C. Hanna, 3W of single frequency output at 532 nm by intracavity doubling of a diode-bar-pumped Nd:YAG ring laser, Opt. Lett. **21**(12), 875, 1996

Chapter 22 Single Mode Resonators

5.219 J.W. Evans, The birefringent filter, J. Opt. Soc. Am. **39**, 229, 1949
5.220 E. Evtuhov, A.E. Siegman, A 'twisted-mode' technique for obtaining axially uniform energy density in a laser cavity, Appl. Opt. **4**, 142, 1965
5.221 P. Zory, Single frequency operation of argon ion lasers, IEEE J. Quantum Electron. **QE-3**, 390, 1967
5.222 V.P. Belayev, V.A. Burmakin, A.N. Evtyunin, F.A. Korolyov, V.V. Lebedeva, A.I. Odintzov, High power single frequency argon ion laser, IEEE J. Quantum Electron. **QE-5**, 589, 1969
5.223 M. Francon, S. Mullik, Polarization Interferometers. Chichester: Wiley 1971

5.224 P.W. Smith, Mode selection in lasers, Proc. IEEE **60**, 422, 1972

5.225 H.W. Schröder, H. Dux, H. Welling, Single mode operation of cw dye lasers, Appl. Phys. **1**, 347, 1973

5.226 A.E. Siegman, An antiresonant ring interferometer for coupled laser cavities, laser output coupling, mode locking and cavity dumping, IEEE J. Quant. Electron. **QE-9**, 247, 1973

5.227 R.L. Barger, M.S. Sorem, J.L. Hall, Frequency stabilization of a cw dye laser, Appl. Phys. Lett. **22**, 583, 1973

5.228 A.L. Bloom, Modes of a laser resonator containing tilted birefringent plates, J. Opt. Soc. Am. **64**, 447, 1974

5.229 R. Beigang, G. Litfin, H. Welling, Frequency behavior and linewidth of a cw single mode color center laser, Opt. Commun. **22**, 269, 1977

5.230 M. Pinard, N. Leduc, G. Trenec, C.G. Aminoff, F. Laloe, Efficient single-mode operation of a standing wave dye laser, Appl. Phys. **19**, 399, 1978

5.231 K. Liu, M.G. Littman, Novel geometry for single-mode scanning of tunable lasers, Opt. Lett. **6**, 117, 1981

5.232 M.W. Fleming, A. Mooradian, Spectral characterisitics of external-cavity controlled semiconductor lasers, IEEE J. Quantum Electron. **QE-17**, 44, 1981

5.233 L. Goldberg, H.F. Taylor, A. Dandridge, J.F. Weller, R.O. Miles, Spectral characteristics of semiconductor lasers with optical feedback, IEEE J. Quantum Electron. **QE-18**, 555, 1982

5.234 J. H. Osmundsen, N. Gade, Influence of optical feedback on laser frequency spectrum and threshold conditions, IEEE J. Quantum Electron. **QE-19**, 465, 1983

5.235 W.T. Tsang, N.A. Olsson, R.A. Logan, High-speed direct single-frequency modulation with a large tuning rate and frequency excurstion in cleaved-coupled-cavity semiconductor lasers, Appl. Phys. Lett. **42**, 650, 1983

5.236 R. Wyatt, W.J. Devlin, 10 kHz linewidth 1.5μm InGaAsP external cavity laser with 55 nm tuning range, Electron. Lett. **19**, 110, 1983

5.237 M.G. Littman, Single-mode pulsed tunable dye laser, Appl. Opt. **23**, 4465, 1984

5.238 M.J. Adams, J. Buus, Two segment cavity theory for mode selection in semiconductor lasers, IEEE J. Quantum Electron. **QE-20**, 99, 1984

5.239 T.J. Kane, R.L. Byer, Monolithic, unidirectional single-mode Nd:YAG ring laser, Opt. Lett. **10**, 65, 1985

5.240 A.E. Siegman: Lasers. Mill Valley: University Science Books 1986, pp. 524-538

5.241 T.J. Kane, A.C. Nilsson, R.L. Byer, Frequency stability and offset locking of a laser-diode-pumped Nd:YAG monolithic nonplanar ring oscillator, Opt. Lett. **12**, 175, 1987

5.242 B. Dahmani, L. Hollberg, R. Drullinger, Frequency stabilization of semiconductor lasers by resonant optical feedback, Opt. Lett. **12**, 876-878, 1987

5.243 W. Fuhrmann, W. Demtröder, A widely tunable single-mode GaAs diode laser with external cavity, Appl. Phys. B **49**, 29, 1988

5.244 Ch. Salomon, D. Hills, J.L. Hall, Laser stabilization at the millihertz level, J. Opt. Soc. Am. B **5**, 1576, 1988

5.245 M. Houssin, M. Jardino, B. Gely, M. Desaintfuscien, Design performance of a few-kilohertz-linewidth dye laser stabilized by reflection in an optical resonator, Opt. Lett.**13**, 823, 1988
5.246 K.M. Abramski, D.R. Hall, Frequency stabilisation of lasers, in D.R. Hall, P.E. Jackson (eds.): The physics and technology of laser resonators. London: Adam Hilger 1989, pp. 117-131
5.247 J. Zayhovski, A. Mooradian, Frequency-modulated Nd:YAG microchip lasers, Opt. Lett. **14**, 618, 1989
5.248 T. Day, A.C. Nilsson, M.M. Fejer, A.D. Farinas, E.K. Gustafson, C.D. Nabors, R.L. Byer, 30-Hz linewidth, diode-laser-pumped, Nd:GGG nonplanar ring oscillators by active frequency stabilization, Electron. Lett. **25**, 810, 1989
5.249 R. Kallenbach, G. Zimmermann, D.H. McIntyre, T.W. Hänsch, R.G. DeVoe, A blue dye laser with sub-kilohertz stability, Opt. Commun. **70**, 56, 1989
5.250 D. Shoemaker, A. Brillet, C.N. Man, O. Crégut, G. Kerr, Frequency-stabilized laser-diode pumped Nd:YAG laser, Opt. Lett. **14**, 609, 1989
5.251 P. Fritschel, A. Jeffries, T. Kane, Frequency fluctuations of a diode-pumped Nd:YAG ring laser, Opt. Lett. **14**, 993, 1989
5.252 G.J. Kintz, T. Baer, Single-frequency operation in solid-state laser materials with short absorption depths, IEEE J. Quantum Electron. **26**, 1457, 1990
5.253 T. Day, E.K. Gustafson, R.L. Byer, Active frequency stabilization of a 1.062-μm, Nd:GGG, diode-laser-pumped nonplanar ring oscillator to less than 3 Hz of relative linewidth, Opt. Lett. **15**, 221, 1990
5.254 A. Hemmerich, D.H. McIntyre, D. Schropp Jr., D. Meschede, T.W. Hänsch, Optically stabilized narrow linewidth semiconductor laser for high resolution spectroscopy, Opt. Commun. **75**, 118, 1990
5.255 E.A.P. Cheng, T.J. Kane, High-power single-mode diode-pumped Nd:YAG laser using a monolithic nonplanar ring resonator, Opt. Lett. **16**, 478, 1991
5.256 T.Taira, A. Mukai, Y. Nozawa, T. Kobayashi, Single-mode oscillation of laser-diode-pumped Nd:YVO_4 microchip laser, Opt. Lett. **16**, 1955, 1991
5.257 F. Zhou, A.I. Ferguson, Frequency stabilization of a diode-laser-pumped microchip Nd:YAG laser at 1.3μm, Opt. Lett. **16**, 79, 1991
5.258 H. Nagai, M. Kume, A. Yoshikawa, K. Itoh, Low-noise operation (-140dB/Hz)in close-coupled Nd:YVO4 second-harmonic lasers pumped by single-mode laser diodes, Appl. Opt. **32**, 6610, 1991
5.259 D.T. Cassidy, D.M. Bruce, B.F. Ventrudo, Short-external-cavity module for enhanced single-mode-tuning of InGaAsP and AlGaAs semiconductor diode lasers, Rev. Sci. Instrum. **62**, 2385, 1991
5.260 W. Demtröder, Laserspektroskopie. Berlin, Heidelberg, New York, London, Paris Tokyo: Springer Verlag, pp. 176-191, 1991
5.261 T. Day, E.K. Gustafson, R.L. Byer, Sub-hertz relative frequency stabilization of two diode-laser-pumped Nd:YAG lasers locked to a Fabry-Perot interferometer, IEEE J. Quantum Electron. **28**, 1106, 1992
5.262 M. de Labachelerie, C. Lattrasse, P. Kemssu, P. Cerez, The frequency control of laser diodes, J. Phys. III France **2**, 1557, 1992

5.263 P.A. Ruprecht, J.R. Branderberg, Enhancing diode laser tuning with a short external cavity, Opt. Commun. **93**, 82, 1992
5.264 J. Harrison, A. Finch, J.H. Flint, P.F. Moulton, Broad-band rapid tuning of a single-frequency diode-pumped neodymium laser, IEEE J. Quantum Electron. **28**(4), 1123, 1992
5.265 N. Uehara, K. Ueda, 193-mHz beat linewidth of frequency-stabilized laser-diode-pumped Nd:YAG ring lasers, Opt. Lett. **18**, 505, 1993
5.266 N.M. Sampas, E.K. Gustafson, R.L. Byer, Long-term stability of two diode-laser-pumped nonplanar ring lasers independently stabilized to two Fabry-Perot interferometers, Opt. Lett. **18**, 947, 1993
5.267 M. Zhu, J.L. Hall, Stabilization of optical phase/frequency of a laser system: application to a commercial dye laser with external stabilizer, J. Opt. Soc. Am. B **10**, 802, 1993
5.268 N. Uehara, K. Ueda, Ultrahigh-frequency stabilization of a diode-pumped Nd:YAG laser with a high-power-acceptance photodetector, Opt. Lett. **19**, 728, 1994
5.269 S. Taccheo, S. Longhi, L. Pallaro, P. Laporta, C. Svelto, E. Bava, Frequency stabilization to a molecular line of a diode-pumped Er-Yb laser at 1533-nm wavelength, opt. Lett. **20**, 2420, 1995
5.270 L. Viana, S.S Vianna, M. Oriá, J.W.R. Tabosa, Diode laser mode selection using a long external cavity, Appl. Opt. **35**, 368, 1996
5.271 K.I. Martin, W.A. Clarkson, D.C. Hanna, 3W of single frequency output at 532 nm by intracavity doubling of a diode-bar-pumped Nd:YAG ring laser, Opt. Lett. **21**(12), 875, 1996
5.272 M. Hyodo, T. Carty, K. Sakai, Near shot-noise-level relative frequency stabilization of a laser-diode-pumped Nd:YVO_4 microchip laser, Appl. Opt. **24**(24), 4749, 1996
5.273 Coherent Laser Group, 8.5W of single frequency 532nm light from a diode pumped intra-cavity doubled ring laser, Optics and Photonics News 7(9), 51, 1996
5.274 A. Giesen, U. Brauch, I. Johannsen, M. Karszewski, C. Stewen, A. Voss, High-power near diffraction-limited and single frequency operation of Yb:YAG thin disc laser, OSA TOPS vol. **1**, Advanced Solid State Lasers, eds.: S.A. Payne, C. R. Pollock, p. 11. Washington, Optical Society of America, 1996
5.275 A. Giesen, U. Brauch, I. Johannsen, M. Karszewski, U. Schiegg, C. Stewen, A. Voss, Advanced tunability and high-power TEM00-operation of the Yb:YAG thin disc laser, OSA TOPS vol. **11**, Advanced Solid State Lasers, eds.: C.R. Pollock, W.R. Bosenberg, p. 280. Washington, Optical Society of America, 1997
5.276 S. Haroche, D. Kleppner, Cavity Quantum Electrodynamics, Physics Today **42**, 24, 1989
5.277 V.E. Weisskopf, E. Wigner, Berechnung der natürlichen Linienbreite auf Grund der Diracschen Lichttheorie, Z. f. Physik **63**, 54, 1930
5.278 M. Sargent III, M.O. Scully, W.E. Lamb jr. Laserphysics, Addison-Wesley 1974
5.279 M.O. Scully, M.S.Zubairy, Quantum Optics, Cambridge University Press 1997
5.280 E.M. Purcell, Spontaneous emission probabilities at radio frequencies, Phys. Rev. **69**, 681, 1946

5.281 I.M. Gérard, B. Sermage, B. Gayral, B. Legrand, E. Costard, V. Thierry-Mieg, Enhanced spontaneous emission by quantum boxes in a monolithic optical microcavity Phys. Rev. Lett. **81**, 1110, 1998
5.282 P. Michler, C. Becher, Photonen auf Bestellung, Physikal. Blätter **57**, 55, 2001
5.283 M. Bayer, Kontrolle von Licht in Mikroresonatoren, Physikal. Blätter **57**, 75, 2001
5.284 M. Bayer, T.L. Reinecke, F. Weidner, A. Larionov, A. McDonald, A. Fordel, Inhibition and enhancement of the spontaneous emission of quantum dots in structured microresonators, Phys. Rev. Lett. **86**, 3168, 2001
5.285 Y. Louer, M.D. Plimmer, P. Juncar, M.E. Himbert, F. Balembois, P. Georges, Nd:YLF laser at 1.3 μm for calcium atom optical clocks and precision spectroscopy of hydrogenic systems, Appl. Opt. **42**, 4867, 2003

Part VI Measurement Techniques

Chapter 23 Measurement of Laser Head Parameters

6.1 D. Findlay, R.A. Clay, The measurement of internal losses in 4-level lasers, Phys. Lett. **20**, 277, 1966
6.2 W.Koechner, Analytical model of a cw Nd:YAG laser, Laser Focus, April 1970
6.3 D.C. Burnham, Simple measurement of thermal lensing effects in laser rods, Appl. Opt. **9**, 1727, 1970
6.4 J. Fresquet, H. Irla, J. Roig, Measurement of the gain and refraction effects in a solid state laser, Appl. Opt. **18**, 175, 1979
6.5 N. Hodgson, H. Weber, Measurement of extraction efficiency and excitation efficiency of lasers, J. Mod. Opt. **35**, 807, 1988
6.6 K. Fuhrmann, N. Hodgson, F. Hollinger, H. Weber, Effective cross section of the Nd:YAG 1.0641μm transition, J. Appl. Phys. **62**, 4041, 1987
6.7 N. Hodgson, C. Rahlff, H. Weber, Dependence of the refractive power of Nd:YAG on the intracavity intensity, Opt. Las. Techn. **25**, 179, 1993
6.8 A. Sennaroglu, Determination of the stimulated-emission cross section in an end-pumped solid state laser from laser-induced pump saturation data, Opt. Lett. **26**, 500, 2000

Chapter 24 Measurement of Laser Beam Parameters

6.9 P. V. Avizonis, T. T. Doss, R. Heinlich, Rev. Sci. Instrument **38**, 331, 1967
6.10 G. Nemes, M. Nemes, I. E. Teodorescu, Phase space treatment of optical beams, Proceedings of the 3rd International School of Coherent Optics, Bucharest, September 1982
6.11 J.R. Fienup, Appl. Opt. **21**, 2758, 1982

6.12 M.V. Klein, T.E. Furtak: Optik. Berlin Heidelberg New York London Paris Tokyo: Springer 1988, pp. 457-486

6.13 W. A. E. Goethals, Geometrical optics of laser beams, in D.R. Hall, P.E. Jackson (eds.): The physics and technology of laser resonators. London: Adam Hilger 1989, pp. 143-153

6.14 M.W Sasnett, Propagation of multimode laser beams - the M^2 factor, in D.R. Hall, P.E. Jackson (eds.): The physics and technology of laser resonators. London: Adam Hilger 1989, pp. 132-142

6.15 G. Nemes, A phase space approach to beam pulse representations and measurements, NATO ASI, International School of Quantum Electronics, Erice, Italy, May 1990.

6.16 G. Herziger, E. Beyer, R. Kramer, P. Loosen, F. Rühl, Diagnostic system for measurement of the focus diameter of high-power CO_2 lasers, GCL Proceedings Jerusalem. Berlin, Heidelberg, New York, London, Paris, Tokyo: Springer-Verlag 1986

6.17 N. Hodgson, T. Haase, R. Kostka, H. Weber, Determination of laser beam parameters with the phase space beam analyser, Opt. Quantum Electron. **24**(9), 927, 1992

6.18 A. Cutolo, F. Ferreri, T. Iserna, R. Pierri, L. Zeni, Measurements of the waist and the power distribution across the transverse modes of a laser beam, Opt. Quantum Electron. **24**(9), 963, 1992

6.19 N. Reng, B. Eppich, Definition and measurements of high-power laser beam parameters, Opt. Quantum Electron. **24**(9), 973, 1992

6.20 D. Wright, P. Greve, J. Fleischer, L. Austin, Laser beam width, divergence and beam propagation factor - an international standardization approach, Opt. Quantum Electron. **24**(9), 993, 1992

6.21 A. Caprara, G.C. Reali, Time varying M^2 in Q-switched lasers, Opt. Quantum Electron. **24**(9), 1001, 1992

6.22 K.M. Du, G. Herziger, P. Loosen, F. Rühl, Measurement of the mode coherence coefficient, Opt. Quantum Electron. **24**(9), 1119, 1992

6.23 D. Wright, Beamwidths of a diffracted laser using four proposed methods, Opt. Quantum Electron. **24**(9), 993, 1992

6.24 Document ISO/TC 172/SC 9/WG 1 N 55, ISO/CD 11 145, Optics and optical instruments - Lasers and laser related equipment - Terminoloy, symbols and units of measure for the specification and testing of lasers and laser ensembles, Nov. 26, 1993

6.25 Document ISO/TC 172/SC 9/WG 1 N 56, ISO/CD 11 146, Optics and optical instruments - Lasers and laser related equipment - Test methods for laser beam parameters: Beam width, divergence angle and beam propagation factor, Nov. 26, 1993

6.26 B. Eppich, R. Kostka, N. Reng, H. Weber, High power laser beam characterisation, Proceedings of the Society of Photo-Optical Instrumentation Engineers vol. **2206**, High power gas and solid state lasers, 469, 1994

6.27 G. Nemes, A.E. Siegman, Measurement of all ten second-order moments of an stigmatic beam by use of rotating simple astigmatic (anamorphic) optics, J. Opt. Soc. Am. A **11**(8), 2257, 1994

6.28 R.D. Jones, T.R. Scott, Error propagation in laser beam spatial parameters, Opt. Quantum Electron. **26**, 25, 1995

6.29 R.D. Jones, T.R. Scott, Characterization of a clipped Gaussian beam, Proceedings of the Society of Photo-Optical Instrumentation Engineers vol. **2375**, Beam Control, Diagnostics, Standards, and Propagation, 360, 1995

6.30 T.F. Johnston, J.M. Fleischer, Calibration standard for laser beam profilers: method for absolute accuracy measurement with a Fresnel diffraction test pattern, Appl. Opt. **35**(10), 1719, 1996

6.31 J.W. Goodman, Introduction to Fourier Optics, New York: McGraw-Hill International Editions, 1996

6.32 T.F. Johnston, Jr., Beam propagation (M^2) measurement made as easy as it gets: the four-cuts method, Appl. Opt. **37**(21), 4840, 1998

6.33 Document EN ISO 11146: 1999E, Lasers and laser related equipment - Test methods for laser beam parameters: Beam widths, divergence angle and beam propagation factor (ISO 11146:1999), June 1999

6.34 D. Paganin, K.A. Nugent, Phys. Rev. Lett. **80**, 2586, 1998

6.35 J.M. Geary, Photonics Spectra **3/1999**, 135, 1999

6.36 B. Eppich, Definition, meaning and measurement of coherence parameters, Proceedings of the Society of Photo-Optical Instrumentation Engineers vol. **4270**, 71, 2001

6.37 B. Schäfer, K. Mann, Investigation of the propagation characteristics of excimer lasers using a Hartmann-Shack sensor, Rev. Scient. Instr. **71**, 2663, 2000

6.38 B. Schäfer, K. Mann, Determination of beam parameters and coherence properties of laser radiation by use of an extended Hartmann-Shack wave-front sensor, Appl. Opt. **41**, 2809, 2002

6.39 C. Gao, BIT-Beijing, private communication, 2002

6.40 K. Mann, Laser-Laboratorium Göttingen e.V., private communication, 2004

Some figures and data were taken from the following sources:

S.1 J. Becker, Herstellung von streuarmen und hochbelastbaren Beschichtungen für optische Anwendungen mit der Ionenstrahlsputtertechnik, PhD dissertation, Physics Department, University of Kaiserslautern, Germany, 1989

S.2 Melles Griot: Optics Guide **5**, page 4-18

S.3 G. Ripper, G. Herziger, Feinwerktechnik & Messtechnik **92**, Werkstoffbearbeitung mit Laserstrahlung, 301, 1984

S.4 F. Hollinger, Quasiperiodizität und Chaos bei der Emission von Nd-YAG Hochleistungslasern im transversalen Multimode-Betrieb, PhD dissertation, Physics Department, University of Kaiserslautern, Germany, 1987

S.5 B. Ozygus, Hochleistungs-Laser mit konfokalem Resonator, diploma thesis, Optical Insitute, Technical University Berlin, 1989

S.6 B.Ozygus, Verluste und Modenstrukturen optischer Resonatoren, internal report, Optical Insitute, Technical University Berlin, Germany 1988

S.7 T. Haase, Optische Resonatoren für Slablaser, diploma thesis, Optical Institute, Technical University Berlin, Germany 1991

S.8 R. Flieger, Instabile Resonatoren mit asphärischen Spiegeln, diploma thesis, Department of Mathematics and Sciences, Technical University Aachen, Germany 1991

S.9 G. Bostanjoglo, Nah- und Fernfeldverteilungen instabiler optischer Laserresonatoren mit apodisierten Reflektoren, PhD dissertation, Optical Insitute, Technical University Berlin, Germany 1995

S.10 C. Rahlff, Polarisationseffekte bei Hochleistungslaser, diploma thesis, Optical Institute, Technical University Berlin, Germany 1990

S.11 J. Eicher, Theoretische und experimentelle Ermittlung der Einzelwirkungsgrade von Festkörper-Lasern, PhD dissertation, Physics Department, University of Kaiserslautern, Germany 1990

S.12 H. Rademacher, Stabile Resonatoren mit Gradientenspiegel, diploma thesis, Optical Institute, Technical University Berlin, Germany 1997

S.13 N. Kugler, Der Einfluß der zirkularen Spannungsdoppelbrechung auf Strahlqualität und Leistung bei Festkörperlasern, diploma thesis, Optical Institute, Technical University Berlin, Germany 1995

S.14 NEC Nd:YAG Laser at Laser- und Medizin-Technologie gGmbH, Berlin, Germany

S.15 G. Bostanjoglo, Multirod solid state lasers with stable resonators using variable reflectivity mirrors, internal report, Laser und Medizin-Technogie gGmbH Berlin, Germany 1996

S.16 M. Kumkar, Theoretische und experimentelle Untersuchungen von Nd:YAG-Lasern und deren Skalierung in den kW-Bereich durch Multistab-Systeme, PhD dissertation, Optical Institute, Technical University Berlin, Germany 1994

S.17 S. Seidel, G. Mann, High average power second harmonic generation using a periodically Q-switched Nd:YAG laser, internal report, Laser- und Medizin-Technogie gGmbH Berlin, Germany 1996

S.18 Coherent Laser Group, 8.5W of single frequency 532nm light from a diode pumped intra-cavity doubled ring laser, Optics and Photonics News 7(9), 51, 1996

S.19 R. Menzel, Department of Physics, University of Potsdam, Germany, private communication 1996

S.20 D. Schulze, Eigenschaften phasenkonjugierender SBS-Zellen für Festkörperlaser, diploma thesis, Optical Institute, Technical University Berlin, Germany 1992

S.21 Courtesy of Uwe Zoske, Institut für Strahlwerkzeuge, University Stuttgart, Stuttgart Germany 1990

S.22 C.L. Petersen, Carl Zeiss, Inc., Humphrey Instruments, San Leandro, CA, private communication

S.23 Courtesy of U. Habich, Fraunhofer-Institut für Lasertechnik, Aachen, Germany

S.24 B. Eppich, Resonatoren für Rohr-Laser, diploma thesis, Optical Insitute, Technical University Berlin, Germany 1991

S.25 H. Schülke, Untersuchungen zur Strahlqualität von hochfrequenzangeregten CO_2-Hochleistungslasern, PhD dissertation, Department of Mechanical Engineering, Technical University Aachen, Germany 1987

5.26 W. Lobsiger, Optische Verzögerungsleitungen in Nd:YAG-Laser-Kavitäten zur Simulation langer Resonatoren, PhD dissertation, Phil. Nat. Fakultät, University Berne, Switzerland, 1975

S.27 K. Fuhrmann, Quantitative Untersuchung zum Einschwingverhalten frequenz-stabilisierter Nd-YAG Laser, diploma thesis, Department of Physics, University of Kaiserslautern, Germany 1986

S.28 K. Mann, Theoretische und experimentelle Untersuchungen zur Temperaturab-hängigkeit laserrelevanter Parameter in Alexandrit and Nd:YAG Lasern, Internal Report, Festkörper-Laser-Institut Berlin GmbH, Berlin, Germany 1989

S.29 B. Wedel, Strahlqualität und Wirkungsgrade von Mulitrod-Lasersystemen, diploma thesis, Optical Institute, Technical University Berlin, Germany 1990

S.30 P. Friedrich, Auswertung von fotografischen Aufnahmen mittels Mikrodensitometer zur Bestimmung der Divergenz von Laserstrahlen großer Energie, Thesis, Akademie der Wissenschaften, Berlin-Adlershof, German Democratic Republic 1985

S.31 D. Lorenz, Frequenzkonversion bei hohen mittleren Leistungen, diploma thesis, Optical Institute, Technical University Berlin, Germany 1994

S.32 K. Murdoch, D. Fortin, private communication, Coherent, Inc., Santa Clara, CA, June 2004

S.33 G. Wang, A. Hicks, private communication, Coherent, Inc. Santa Clara, CA, June 2004

S.34 References for temperature dependence of the thermal conductivity:
for YAG, :LuAG,YbAG: G.A. Slack, D.W. Oliver, Phys. Rev. **B4**, 592, 1971
for SiO2, MgO, quartz and Sapphire: Y.S. Touloukian, R.W. Powell, C.Y. Ho, P.G. Klemens, Thermophysical Porperties of Matter, IFI/Plenum, New York 1970

S.35 References for the temperature dependence of dn/dT:
for YAG: R. Wynne, J.L. Daneu, T.Y. Fan, Appl. Opt. **38**, 3282, 1999
for Al_2O_3: A.C. DeFranzo, B.G. Pazol, Appl. Opt. **32**, 2224, 1993

S.36 E. Garmire, Sources, modulators, and detectors for fiber-optic communication systems, in Fiber Optics Handbook, eds. M. Bass, E.W. Van Stryland. New York, Chicago, San Francisco, Lisbon, London, Madrid, Milan, New Delhi, San Juan, Seoul, Singapore, Sydney, Toronto: McGraw-Hill 2002

Figures without a reference in the caption represent unpublished results of the authors.

Index

Springer Series in
OPTICAL SCIENCES

Volume 1

1 **Solid-State Laser Engineering**
By W. Koechner, 5th revised and updated ed. 1999, 472 figs., 55 tabs., XII, 746 pages

Published titles since volume 80

80 **Optical Properties of Photonic Crystals**
By K. Sakoda, 2nd ed., 2004, 107 figs., 29 tabs., XIV, 255 pages

81 **Photonic Analog-to-Digital Conversion**
By B.L. Shoop, 2001, 259 figs., 11 tabs., XIV, 330 pages

82 **Spatial Solitons**
By S. Trillo, W.E. Torruellas (Eds), 2001, 194 figs., 7 tabs., XX, 454 pages

83 **Nonimaging Fresnel Lenses**
Design and Performance of Solar Concentrators
By R. Leutz, A. Suzuki, 2001, 139 figs., 44 tabs., XII, 272 pages

84 **Nano-Optics**
By S. Kawata, M. Ohtsu, M. Irie (Eds.), 2002, 258 figs., 2 tabs., XVI, 321 pages

85 **Sensing with Terahertz Radiation**
By D. Mittleman (Ed.), 2003, 207 figs., 14 tabs., XVI, 337 pages

86 **Progress in Nano-Electro-Optics I**
Basics and Theory of Near-Field Optics
By M. Ohtsu (Ed.), 2003, 118 figs., XIV, 161 pages

87 **Optical Imaging and Microscopy**
Techniques and Advanced Systems
By P. Török, F.-J. Kao (Eds.), 2003, 260 figs., XVII, 395 pages

88 **Optical Interference Coatings**
By N. Kaiser, H.K. Pulker (Eds.), 2003, 203 figs., 50 tabs., XVI, 504 pages

89 **Progress in Nano-Electro-Optics II**
Novel Devices and Atom Manipulation
By M. Ohtsu (Ed.), 2003, 115 figs., XIII, 188 pages

90/1 **Raman Amplifiers for Telecommunications 1**
Physical Principles
By M.N. Islam (Ed.), 2004, 488 figs., XXVIII, 328 pages

90/2 **Raman Amplifiers for Telecommunications 2**
Sub-Systems and Systems
By M.N. Islam (Ed.), 2004, 278 figs., XXVIII, 420 pages

91 **Optical Super Resolution**
By Z. Zalevsky, D. Mendlovic, 2004, 164 figs., XVIII, 232 pages

92 **UV-Visible Reflection Spectroscopy of Liquids**
By J.A. Räty, K.-E. Peiponen, T. Asakura, 2004, 131 figs., XII, 219 pages

93 **Fundamentals of Semiconductor Lasers**
By T. Numai, 2004, 166 figs., XII, 264 pages

Springer Series in
OPTICAL SCIENCES

94 **Photonic Crystals**
Physics, Fabrication and Applications
By K. Inoue, K. Ohtaka (Eds.), 2004, 209 figs., XV, 320 pages

95 **Ultrafast Optics IV**
Selected Contributions to the 4th International Conference
on Ultrafast Optics, Vienna, Austria
By F. Krausz, G. Korn, P. Corkum, I.A. Walmsley (Eds.), 2004, 281 figs., XIV, 506 pages

96 **Progress in Nano-Electro Optics III**
Industrial Applications and Dynamics of the Nano-Optical System
By M. Ohtsu (Ed.), 2004, 186 figs., 8 tabs., XIV, 224 pages

97 **Microoptics**
From Technology to Applications
By J. Jahns, K.-H. Brenner, 2004, 303 figs., XI, 335 pages

98 **X-Ray Optics**
High-Energy-Resolution Applications
By Y. Shvyd'ko, 2004, 181 figs., XIV, 404 pages

99 **Few-Cycle Photonics and Optical Scanning Tunneling Microscopy**
Route to Femtosecond Ångstrom Technology
By M. Yamashita, H. Shigekawa, R. Morita (Eds.) 2005, 241 figs., XX, 393 pages

100 **Quantum Interference and Coherence**
Theory and Experiments
By Z. Ficek and S. Swain, 2005, 178 figs., approx. 432 pages

101 **Polarization Optics in Telecommunications**
By J. Damask, 2005, 110 figs, XVI, 528 pages

102 **Lidar**
Range-Resolved Optical Remote Sensing of the Atmosphere
By C. Weitkamp (Ed.), 161 figs., approx. 416 pages

103 **Optical Fiber Fusion Splicing**
By A. D. Yablon, 2005, 100 figs., approx. 300 pages

104 **Optoelectronics of Molecules and Polymers**
By A. Moliton, 2005, 200 figs., approx. 460 pages

105 **Solid-State Random Lasers**
By M. Noginov, 2005, 149 figs., approx. 380 pages

106 **Coherent Sources of XUV Radiation**
Soft X-Ray Lasers and High-Order Harmonic Generation
By P. Jaeglé, 2005, 150 figs., approx. 264 pages

107 **Optical Frequency-Modulated Continuous-Wave (FMCW) Interferometry**
By J. Zheng, 2005, 137 figs., approx. 250 pages

CPSIA information can be obtained at www.ICGtesting.com
Printed in the USA
LVOW070452281212

313554LV00013B/246/P

9 780387 400785